MULTIPARTICLE DYNAMICS

To learn more about the AIP Conference Proceedings, including the Conference Proceedings Series, please visit the webpage
http://proceedings.aip.org/proceedings

MULTIPARTICLE DYNAMICS

XXXV International Symposium on Multiparticle Dynamics

Kroměříž, Czech Republic 9 – 15 August 2005

and

Workshop on Particle Correlations and Femtoscopy

Kroměříž, Czech Republic 15 – 17 August 2005

EDITORS

V. Šimák

Czech Technical University, Prague, Czech Republic

M. Šumbera

Nuclear Physics Institute ASCR, Řež/Prague, Czech Republic

Š. Todorova

Tufts University, Medford, Massachusetts

B. Tomášik

Niels Bohr Institute, Copenhagen, Denmark

SPONSORING ORGANIZATIONS

Faculty of Nuclear Sciences and Physical Engineering, Czech Technical University, Prague
Faculty of Mathematics and Physics, Charles University, Prague
Institute of Physics, Academy of Sciences of the Czech Republic
Nuclear Physics Institute, Academy of Sciences of the Czech Republic
Silesian University, Opava, Czech Republic

Melville, New York, 2006

AIP CONFERENCE PROCEEDINGS ■ VOLUME 828

Editors:

V. Šimák
Czech Technical University
Faculty of Nuclear Sciences and
Physical Engineering
Břehová 7
11519 Prague 1
Czech Republic

E-mail: simak@fzu.cz

Š. Todorova
Tufts University/CERN PH
CH-1211 Geneve 23
Switzerland

E-mail: sarka.todorova@cern.ch

M. Šumbera
Nuclear Physics Institute ASCR
250 58 Řež/Prague
Czech Republic

E-mail: sumbera@ujf.cas.cz

B. Tomášik
Niels Bohr Institute
Blegdamsvej 17
2100 Copenhagen O
Denmark

E-mail: boris.tomasik@nbi.dk

L.C. Catalog Card No. 2006923323
ISBN 0-7354-0320-1
ISSN 0094-243X
Printed in the United States of America

CONTENTS

XXXV INTERNATIONAL SYMPOSIUM ON MULTIPARTICLE DYNAMICS

SOFT INTERACTIONS

Chairpersons: V. Šimák, R. Peschanski, L. Šándor, and K. Fiałkowski

FLUCTUATIONS AND CORRELATIONS

Chairpersons: E. De Wolf and G. Gustafson

JET PHYSICS

Chairpersons: W. Kittel, L. S. Liu, J. G. Cramer, Y. Wu

PARTICLE PROPAGATION IN DENSE MATTER

Chairpersons: T. Csörgő and M. Šumbera

ASTROPARTICLE PHYSICS

Chairpersons: B. Shephard and G. Kozlov

HEAVY FLAVORS AND IDENTIFIED PARTICLES

Chairpersons: G. Kozlov, J. Rafelski, and W. Metzger

SMALL X-PHYSICS AND DIFFRACTION

Chairpersons: A. Valkárová, N. Schmitz, and Y. Hama

SPECIAL SESSION

Chairperson: Y. Hama

WORKSHOP ON PARTICLE CORRELATIONS AND FEMTOSCOPY

FEMTOSCOPY IN HEAVY ION COLLISIONS

Chairperson: J. Cramer

HOW CORRELATIONS REFLECT DYNAMICS

Chairperson: Y. Hama

RAPIDITY DEPENDENCE, CONSTRAINTS FROM FLOW, v_2 ...

Chairperson: S. Manly

SOURCE IMAGING
Chairperson: R. Lacey

BEYOND THE GAUSSIAN APPROXIMATION
Chairperson: H. Eggers

COMPARISONS OF DIFFERENT COLLIDING SYSTEMS INCLUDING COLLISIONS OF ELEMENTARY PARTICLES
Chairperson: Š. Todorova

FEMTOSCOPY WITH PENETRATING PROBES
Chairperson: T. Csörgő

MULTIPLE FSI INTERACTIONS
Chairperson: T. Csörgő

NON-IDENTICAL PARTICLE CORRELATIONS
Chairperson: R.Lednický

OTHER NEW RESULTS
Chairperson: S. Pratt

CORRELATIONS FROM EVENT GENERATORS
Chairperson: J. Pluta

CONDENSATION AND SQUEEZED STATES

Chairperson: R. Glauber

PREFACE

The International Symposium on Multiparticle Dynamics (ISMD) is a conference series held regularly every year in various countries. The 35th ISMD was held from August 9 to August 15, 2005 in Kroměříž, Moravia, Czech Republic. It was jointly organized by the Faculty of Nuclear Sciences and Physical Engineering of the Czech Technical University (CTU), the Faculty of Mathematics and Physics of the Charles University (CU), by the Nuclear Physics Institute and the Institute of Physics of the Academy of Sciences of the Czech Republic (ASCR), and by the Silesian University. The venue of the conference was the Secondary Vocation and Apprentice Training Centre in Kroměříž.

The conference reviewed and updated recent theoretical and experimental understanding of multiparticle production in high energy collisions. Altogether, 77 talks covering main results from all areas of multiparticle dynamics were delivered. About a half of the talks were devoted to collisions of ultra-relativistic nuclei with fresh results from RHIC. In addition to this some intriguing results from HERA and Tevatron colliders were presented. New theoretical results and study of future experiments were discussed too. Social part of the conference included excursion to the town of Olomouc as well as visit to the Archbishop Palace, Gardens and vine cellar in Kroměříž.

The ISMD conference was followed by the Workshop on Particle Correlations and Femtoscopy (WPCF) which was held in the same premises from August 15 to August 19, 2005. The WPCF was the first in newly established series of regular workshops aiming at critical and thorough analysis of the latest results on particle interferometry in high energy heavy ion collisions. Focus on this rather narrow subject was stimulated by wealth of new data coming steadily from RHIC and SPS experiments.

We were proud to welcome on both of these conferences a very distinguished participant: Professor Roy Glauber, the 2005 Nobel physics laureate.

Scientific programme of ISMD was put together in collaboration of I. Dremin and N. Xu (Soft interactions), E. de Wolf and R. Lednický (Fluctuations and Correlations), J. Chýla and Ch. Royon (Jet physics), C. Gagliardi and J. Rak (Particle Propagation in Dense Matter), J. Pinfold (Astroparticle Physics), J. Rafelski and K. Šafařík (Heavy Flavors and Identified Particles), H. Jung and V. Khoze (Small x-Physics and Diffraction). The WPCF scientific programme was organised by T. Csörgő, R. Lednický, M. Šumbera, and B. Tomášik, who were assisted by J.-e Alam, H. Appelshäuser, M. Asakawa, D. Brown, B. Buschbeck, J. Cramer, H. Eggers, Y. Hama, S. Hegyi, T. Humanic, W. Kittel, R. Lacey, M. Lisa, S. Manly, S. Padula, S. Panitkin, D. Peressounko, J. Pluta, S. Pratt, Y. Sinyukov, Š. Todorova, U. Wiedemann, C.-Y. Wong.

We would like to thank P. Hajný, director of the Secondary Vocation and Apprentice Training Centre, for excellent services provided to our conferences.

For their help with preparation and running the conference we thank our colleagues M. Bysterský, L. Fiala, Z. Hubáček, M. Lokajíček, M. Pachr, A. Valkárová, J. Švec, I. Zborovský. The support of the Physics Department of the Silesian University in Opava, in particular of students J. Juran, K. Mrázová and M.Vojik, was of great help to us.

V. Šimák and M. Šumbera
on behalf of local organizing committee

XXXV International Symposium on Multiparticle Dynamics

International Advisory Committee

A.Białas (Cracow)
T.Csörgő (Budapest)
E.De Wolf (Antwerp)
I.Dremin (Moscow)
K.Fiałkowski (Cracow)
G.Gustafson (Lund)
W.Kittel (Nijmegen)
L.S.Liu (Wuhan)
J.Morfin (Fermilab)
W.Ochs (Munich)
L.Šándor (Košice)
N.Schmitz (Munich)
V.Šimák (Prague)
A.Sissakian (Dubna)
C.I.Tan (Brown)
N.Xu (LBNL)
Y.F.Wu (Wuhan)

Local Organizing Committee

V.Šimák (CTU Prague), Chair
P.Lichard (Silesian U.Opava)
M.Lokajíček (IP ASCR Prague)
M.Pachr (CTU Prague)
M.Šumbera (NPI ASCR Řež/Prague)
Š.Todorova (Tufts U.)
A.Valkárová (Charles U.Prague)
I.Zborovský (NPI ASCR Řež/Prague)

Faculty of Nuclear Sciences and
Physical Engineering
Czech Technical University, Prague

Faculty of Mathematics and Physics
Charles University, Prague

Institute of Physics
Academy of Sciences
of the Czech Republic, Prague

Nuclear Physics Institute
Academy of Sciences
of the Czech Republic, Řež/Prague

Silesian University, Opava

Workshop on Particle Correlations and Femtoscopy

International Advisory Committee

H.Appelshäusser (Frankfurt U.)
M.Baker (BNL)
J.Cramer (UW Seattle)
T.Csörgő (KFKI Budapest)
B.Erazmus (Subatech Nantes)
Y.Hama (Sao Paulo U.)
T.Humanic (Ohio State U.)
W.Kittel (HEFIN Nijmegen)
R.Lacey (SUNY)
R.Lednický (IP ASCR Prague, JINR Dubna)
M.Lisa (Ohio State U.)
L.S.Liu (Wuhan U.)
S.Padula (Sao Paulo U.)
J.Pluta (Warsaw Technical U.)
S.Pratt (MSU East Lansing)
K.Šafařík (CERN)
P.Seyboth (MPI Munich)
Yu.Sinyukov (BITP Kiev)
T.Sugitate (Hiroshima U.)

Local Organizing Committee

M.Šumbera (NPI ASCR Řež/Prague), Chair
M.Bysterský (NPI ASCR Řež/Prague)
P.Lichard (Silesian U.Opava)
M.Pachr (CTU Prague)
B.Tomášik (NBI Copenhagen)
I.Zborovský (NPI ASCR Řež/Prague)

Faculty of Nuclear Sciences and
Physical Engineering
Czech Technical University, Prague

Nuclear Physics Institute
Academy of Sciences
of the Czech Republic, Řež/Prague

Silesian University, Opava

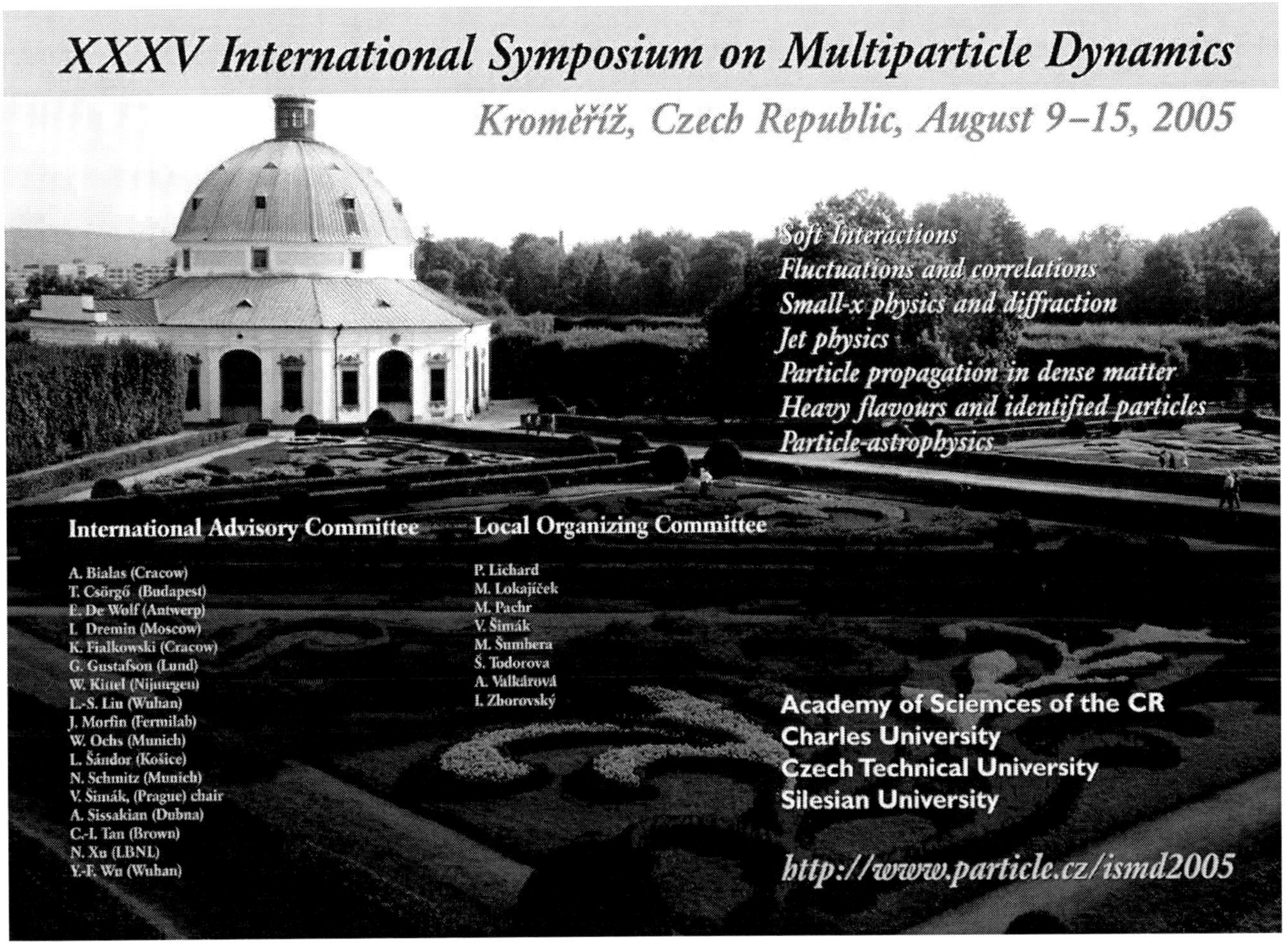

XXXV International Symposium on Multiparticle Dynamics

Kroměříž, Czech Republic, August 9–15, 2005

Soft Interactions
Fluctuations and correlations
Small-x physics and diffraction
Jet physics
Particle propagation in dense matter
Heavy flavours and identified particles
Particle-astrophysics

International Advisory Committee

A. Bialas (Cracow)
T. Csörgő (Budapest)
E. De Wolf (Antwerp)
I. Dremin (Moscow)
K. Fialkowski (Cracow)
G. Gustafson (Lund)
W. Kittel (Nijmegen)
L.-S. Liu (Wuhan)
J. Morfin (Fermilab)
W. Ochs (Munich)
L. Šándor (Košice)
N. Schmitz (Munich)
V. Šimák, (Prague) chair
A. Sissakian (Dubna)
C.-I. Tan (Brown)
N. Xu (LBNL)
Y.-F. Wu (Wuhan)

Local Organizing Committee

P. Lichard
M. Lokajíček
M. Pachr
V. Šimák
M. Šumbera
Š. Todorova
A. Valkárová
I. Zborovský

Academy of Sciemces of the CR
Charles University
Czech Technical University
Silesian University

http://www.particle.cz/ismd2005

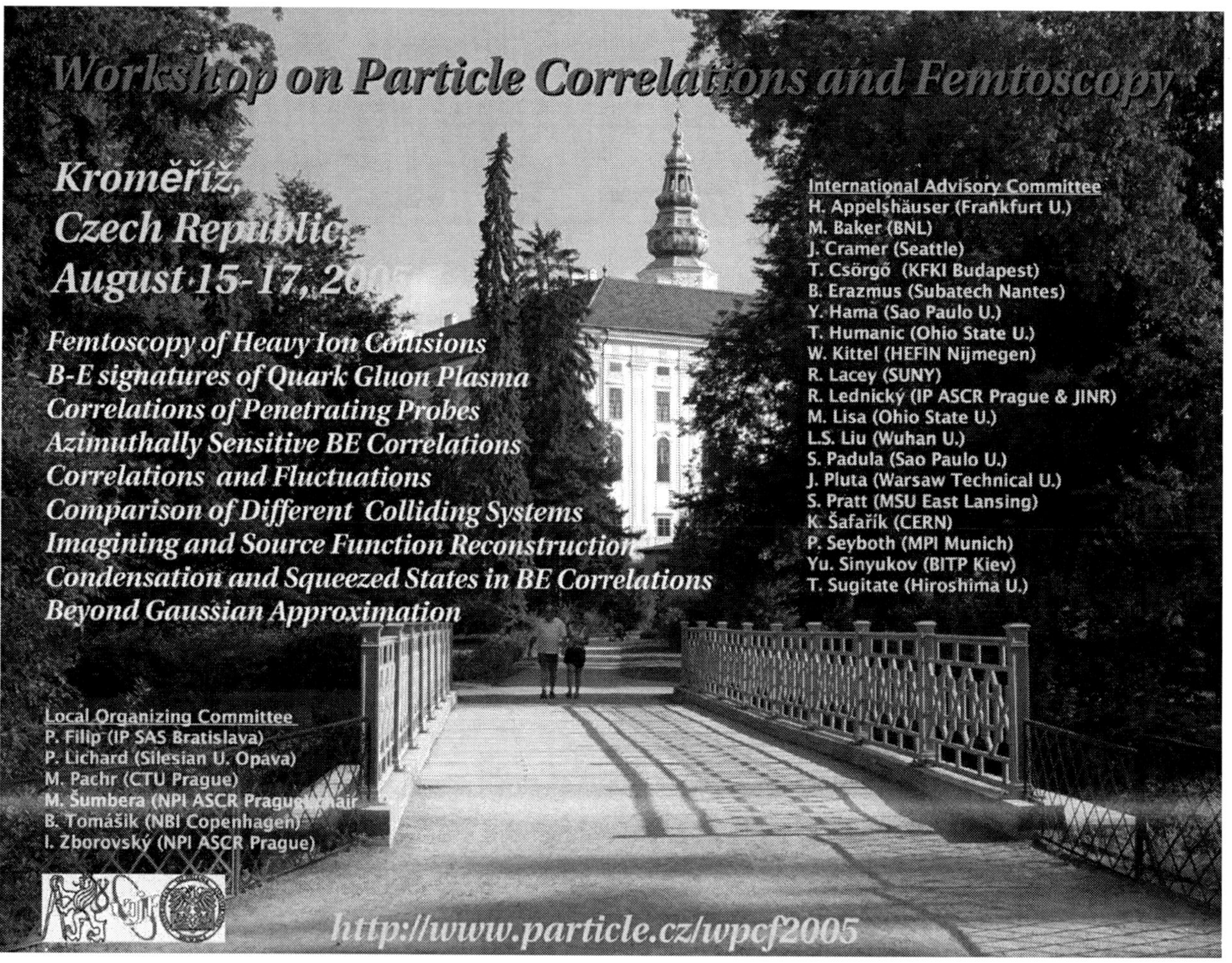

Workshop on Particle Correlations and Femtoscopy
Kroměříž,
Czech Republic,
August 15-17, 2
Femtoscopy of Heavy Ion Collisions
B-E signatures of Quark Gluon Plasma
Correlations of Penetrating Probes
Azimuthally Sensitive BE Correlations
Correlations and Fluctuations
Comparison of Different Colliding Systems
Imagining and Source Function Reconstructio
Condensation and Squeezed States in BE Correlations
Beyond Gaussian Approximation
International Advisory Committee
H. Appelshäuser (Frankfurt U.)
M. Baker (BNL)
J. Cramer (Seattle)
T. Csörgő (KFKI Budapest)
B. Erazmus (Subatech Nantes)
Y. Hama (Sao Paulo U.)
T. Humanic (Ohio State U.)
W. Kittel (HEFIN Nijmegen)
R. Lacey (SUNY)
R. Lednický (IP ASCR Prague & JINR)
M. Lisa (Ohio State U.)
L.S. Liu (Wuhan U.)
S. Padula (Sao Paulo U.)
J. Pluta (Warsaw Technical U.)
S. Pratt (MSU East Lansing)
K. Šafařík (CERN)
P. Seyboth (MPI Munich)
Yu. Sinyukov (BITP Kiev)
T. Sugitate (Hiroshima U.)
Local Organizing Committee
P. Filip (IP SAS Bratislava)
P. Lichard (Silesian U. Opava)
M. Pachr (CTU Prague)
M. Šumbera (NPI ASCR Pragu
B. Tomášik (NBI Copenhagen)
I. Zborovský (NPI ASCR Prague)
http://www.particle.cz/wpcf2005

International Symposia on Multiparticle Dynamics

I	Paris	France	1970
II	Helsinki	Finland	1971
III	Zakopane	Poland	1972
IV	Pavia	Italy	1973
V	Eisenach/Leipzig	German Democratic Republic	1974
VI	Oxford	United Kingdom	1975
VII	Tutzing/Munich	Federal Republic of Germany	1976
VIII	Kayersberg/Strasbourg	France	1977
IX	Tabor/Prague	Czechoslovakia	1978
X	Goa/Tifr	India	1979
XI	Bruges	Belgium	1980
XII	Notre Dame	USA	1981
XIII	Volendam/Amsterdam	Holland	1982
XIV	Lake Tahoe	USA	1983
XV	Lund	Sweden	1984
XVI	Kiryat Anavim	Israel	1985
XVII	Seewinkel/Vienna	Austria	1986
XVIII	Tashkent	USSR	1987
XIX	Arles	France	1988
XX	Dortmund	Federal Republic of Germany	1989
XXI	Wuhan	China	1990
XXII	Santiago de Compostela	Spain	1991
XXIII	Aspen	USA	1992
XXIV	Vietri sul Mare	Italy	1994
XXV	Stará Lesná	Slovakia	1995
XXVI	Faro	Portugal	1996
XXVII	Frascati	Italy	1997
XXVIII	Delphi	Greece	1998
XXIX	Rhode Island	USA	1999
XXX	Tihany	Hungary	2000
XXXI	Datong	China	2001
XXXII	Alushta	Ukraine	2002
XXXIII	Kraków	Poland	2003
XXXIV	Sonoma	USA	2004
XXXV	Kroměříž	Czech Republic	2005

XXXV INTERNATIONAL SYMPOSIUM ON MULTIPARTICLE DYNAMICS

SOFT INTERACTIONS

Chairpersons: V. Šimák, R. Peschanski, L. Šándor, and K. Fiałkowski

New PHOBOS results on event-by-event fluctuations

B.Alver [*], B.B.Back[†], M.D.Baker[**], M.Ballintijn[*], D.S.Barton[**], R.R.Betts[‡], A.A.Bickley[§], R.Bindel[§], A.Budzanowski[¶], W.Busza[*], A.Carroll[**], Z.Chai[**], V.Chetluru[‡], M.P.Decowski[*], E.García[‡], T.Gburek[¶], N.George[**], K.Gulbrandsen[*], S.Gushue[**], C.Halliwell[‡], J.Hamblen[||], G.A.Heintzelman[**], C.Henderson[*], I.Harnarine[‡], D.J.Hofman[‡], R.S.Hollis[‡], R.Hołyński[¶], B.Holzman[**], A.Iordanova[‡], E.Johnson[||], J.L.Kane[*], N.Khan[||], W.Kucewicz[‡], P.Kulinich[*], C.M.Kuo[††], W.Li[*], W.T.Lin[††], C.Loizides[*], S.Manly[||], A.C.Mignerey[§], R.Nouicer[**], A.Olszewski[¶], R.Pak[**], I.C.Park[||], C.Reed[*], L.P.Remsberg[**], M.Reuter[‡], E.Richardson[§], C.Roland[*], G.Roland[*], L.Rosenberg[*], J.Sagerer[‡], P.Sarin[*], P.Sawicki[¶], I.Sedykh[**], W.Skulski[||], C.E.Smith[‡], M.A.Stankiewicz[**], P.Steinberg[**], G.S.F.Stephans[*], A.Sukhanov[**], A.Szostak[**], J.-L.Tang[††], M.B.Tonjes[§], A.Trzupek[¶], C.Vale[*], G.J.van Nieuwenhuizen[*], S.S.Vaurynovich[*], R.Verdier[*], G.I.Veres[*], P.Walters[||], E.Wenger[*], D.Willhelm[**], F.L.H.Wolfs[||], B.Wosiek[¶], K.Woźniak[¶], A.H.Wuosmaa[†], S.Wyngaardt[**] and B.Wysłouch[*]

[*] *Massachusetts Institute of Technology, Cambridge, MA 02139-4307, USA*
[†] *Argonne National Laboratory, Argonne, IL 60439-4843, USA*
[**] *Brookhaven National Laboratory, Upton, NY 11973-5000, USA*
[‡] *University of Illinois at Chicago, Chicago, IL 60607-7059, USA*
[§] *University of Maryland, College Park, MD 20742, USA*
[¶] *Institute of Nuclear Physics PAN, Kraków, Poland*
[||] *University of Rochester, Rochester, NY 14627, USA*
[††] *National Central University, Chung-Li, Taiwan*

Abstract. We present new results from the PHOBOS experiment at RHIC on event-by-event fluctuations of particle multiplicities and angular distributions in nucleus-nucleus collisions at RHIC. Our data for Au+Au collisions at $\sqrt{s_{NN}} = 200$ GeV show that at a level of 10^{-4} or less, no rare, large-amplitude fluctuations in the total multiplicity distributions or the shape of the pseudorapidity distributions are observed. We however find significant short-range multiplicity correlations in these data, that can be described as particle production in clusters. In Cu+Cu collisions, we observe large final-state azimuthal anisotropies v_2. A common scaling behavior for Cu+Cu and Au+Au for these anisotropies emerges when fluctuations in the initial state geometry are taken into account.

Keywords: event-by-event fluctuations, collective flow
PACS: 25.75.-q,25.75.Gz,25.75.Ld

CP828, *Multiparticle Dynamics*
edited by V. Šimák, M. Šumbera, Š. Todorova, and B. Tomášik

OVERVIEW

PHOBOS is one of four experiments at the Relativistic Heavy Ion Collider (RHIC) at Brookhaven National Laboratory, studying collisions of heavy nuclei at energies up to $\sqrt{s_{NN}} = 200$ GeV. Our goal is to study the properties of strongly interacting matter at extreme temperature and density, where ab-initio numerical QCD calculations predict a phase transition to a system dominated by quark and gluon degrees of freedom. Our findings from the first four years of these studies have been summarized in [1]. There we argued that in Au+Au collisions, a dense, highly interacting system is created, with energy densities far in excess of the critical values for the QCD phase transition.

In this paper, we discuss three distinct studies of event-by-event fluctuations and particle correlations that were performed for Au+Au and Cu+Cu collisions at RHIC. These include a search for large-amplitude fluctuations in particle production that could be expected in the vicinity of the QCD phase transition [2, 3] and a measurement of multiplicity correlations that shows that final state particles are not produced independently. Finally, we argue that data on final state azimuthal anisotropies in Cu+Cu collisions at $\sqrt{s_{NN}} = 62$ and 200 GeV can be understood if fluctuations in the initial state geometry are considered. All three studies rely on the large pseudo-rapidity and azimuthal coverage of the PHOBOS multiplicity detector, which covers 2-π in azimuth over most of $|\eta| < 5.4$. Details of the experimental setup can be found in [4]. The analyses discussed here use the high statistics 200 GeV Au+Au dataset collected in the 2004 RHIC run (250M events) and the 62 and 200 GeV Cu+Cu datasets from the 2005 run, with 110 and 400 million events, respectively.

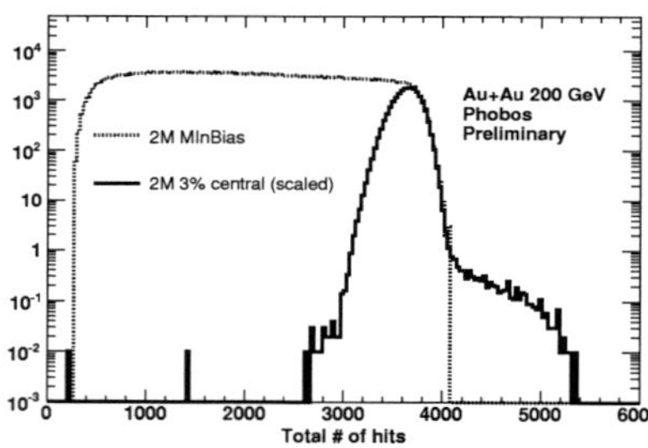

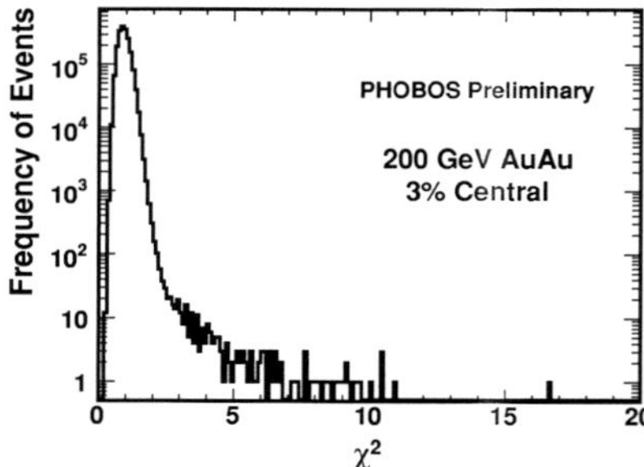

FIGURE 1. Left: Total number of hits per event observed in the multiplicity detector. Right: Reduced χ^2 distribution for event-by-event comparison of the charged hadron $dN/d\eta$ distributions in central Au+Au collisions at 200 GeV to ensemble average.

EVENT-BY-EVENT FLUCTUATIONS IN PHOBOS

Large amplitude multiplicity fluctuations

The high statistics Au+Au dataset, in combination with the large PHOBOS acceptance, allows us to perform a search for events that in a statistically quantifiable way

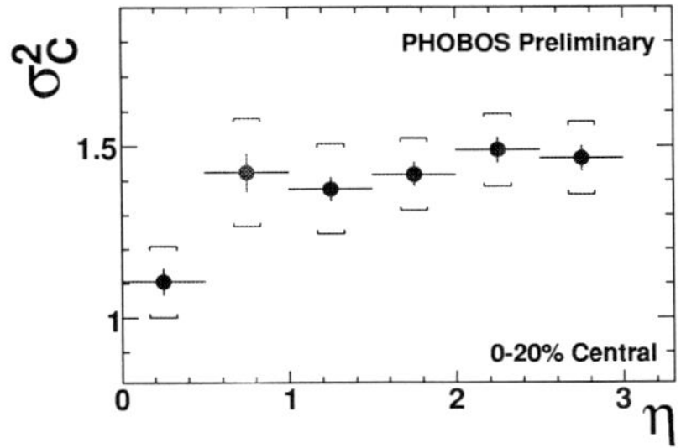

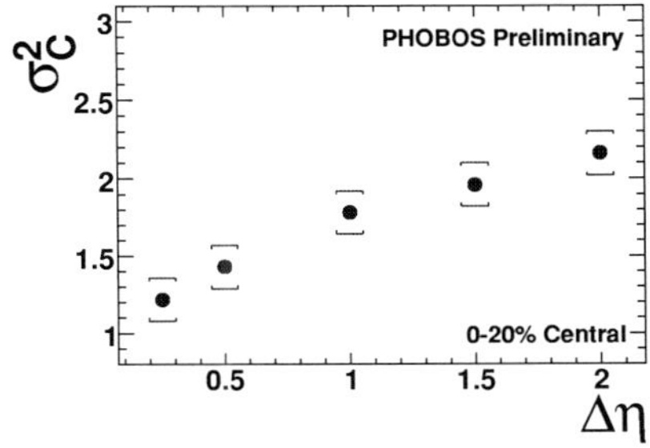

FIGURE 2. Dependence of the fluctuation measure σ_C^2 on the position η (left) and on the width $\Delta\eta$ (right) of the forward and backward multiplicity bins for the 20% most central Au+Au collisions at 200 GeV. Systematic uncertainties (90% C.L.) are shown as brackets.

differ from average Au+Au events. Possible mechanisms for the occurrence of such events might be the formation of droplets due to supercooling [3] or the formation of disoriented chiral condensates [2]. While the likelihood of such scenarios is unclear, an unbiased search for "unusual" events clearly is an important part of the RHIC physics program. The details of our strategy for identifying rare events is described in [5]. The results of these measurements for the most central two million events of our Au+Au dataset are shown in Fig. 1.

The left plot shows the distribution of the total number of hits in the PHOBOS multiplicity detector, which is closely correlated with the total charged particle multiplicity. In this distribution, a tail of high multiplicity events is visible, that contains about 10^{-4} of all central events. For the right hand plot, we determined the shape of the average uncorrected $dN/d\eta$ distribution for events in fine bins of vertex position. Similarly, the variance around the average shape in bins of η is extracted from the data. Using the average shape and variance obtained from the data, we then calculate the χ^2 of each individual event relative to the ensemble average. The resulting χ^2 distribution is shown in right-hand plot in Fig. 1. The distribution consists of a central core close to a $\chi^2/DoF \approx 1$ and again a tail out to large χ^2. For both event-by-event measures, about 0.01% of all events are found in a region where for purely statistical fluctuations no entries would be expected. Further studies of these unusual events have shown that their rate is linearly related to the instantaneous luminosity at which the collisions were recorded. At present, we therefore have no evidence for the existence of unusual physical fluctuations. Further studies are underway to set quantitative limits on various physical scenarious.

Forward-backward multiplicity correlations

Further examination of Fig. 1 shows that the width of the χ^2 distribution around unity is significantly larger than expected for a purely Poissonian production and detection of charged particles. I.e. particles appear to be produced in a correlated fashion, rather than one-by-one. This can be studied quantitatively using correlations of multiplicities in

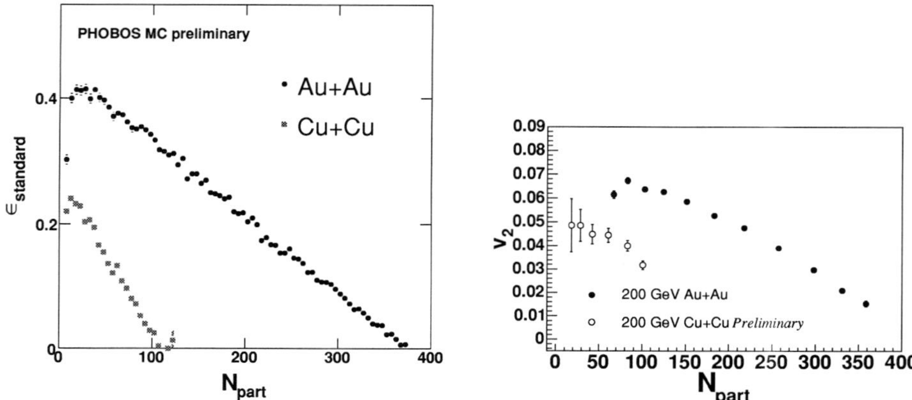

FIGURE 3. Left: Average eccentricity ε_{std}, of the collision zone in Cu+Cu (light symbols) and Au+Au (dark symbols) as a function of N_{part} for $\sqrt{s_{NN}}$ =200 GeV from PHOBOS Glauber Monte Carlo calculations. Right: Average elliptic flow coefficient v_2 measured near mid-rapidity, as a function of N_{part} for Cu+Cu collisions (open symbols) and Au+Au collisions (filled symbols) at $\sqrt{s_{NN}}$ =200 GeV. Only statistical errors are shown.

non-overlapping bins of pseudo-rapidity. We have performed such studies for symmetric pairs of pseudo-rapidity bins centered between $0.25 < \eta < 2.75$, varying the bin width, $\Delta\eta$, from 0.5 to 2.0. To quantify the relative fluctuations between the multiplicities N_F in the forward ($\eta > 0$) and N_B in the backward ($\eta < 0$) bins, we define the event-wise observable $C = (N_F - N_B)/\sqrt{N_F + N_B}$. This variable has the useful property that its variance σ_C^2 is one for independent particle emission, even when averaged over events from centrality bins of finite widths. After correction for detector and acceptance effects, which are described in [6, 7], σ_C^2 can be used to study short range correlations in particle production. If particles are produced as clusters which decay with a rapidity width smaller than the typical bin width chosen in the analysis, then σ_C^2 in the absence of other correlations will directly correspond to the cluster size k, i.e. the multiplicity of decay products from each cluster. In Fig. 2, we show the dependence of σ_C^2 on the position (left plot) and width (right plot) of the bins used in our analysis for central Au+Au events. The main result is that σ_C^2 is much larger than unity, in particular for larger $\Delta\eta$, indicating that particles in Au+Au collisions are not produced independently, but rather in clusters. The results are reminiscent of those obtained in similar analyses for $p+\bar{p}$ collisions [8]. The similarity of the results in A+A and $p+\bar{p}$ collisions, as well as the weak energy dependence seen in $p+\bar{p}$, could indicate that the cluster formation is a phenomenon related to common features at hadronization. This will be further tested by future studies of multiplicity correlations in Au+Au and Cu+Cu collisions as a function of collision energy. For further discussion and a comparison of the present results with event generators, see [7].

System size dependence of elliptic flow

The final correlation measurement discussed in this paper concerns elliptic flow, and the connection between the initial state conditions and the observed final state anisotropy. It has been argued in the past [9], that the observed flow coefficient v_2 for Au+Au closely follows the initial state eccentricity as a function of centrality. This can be seen by comparing the dependence of the Au+Au initial state eccentricity on the number of participants, N_{part}, from a Glauber calculation shown in the left plot of Fig. 3 and the Au+Au v_2 coefficient at mid-rapidity shown in the right plot of the same figure (see also [10]).

However, doubt on this crucial connection is cast by comparing the Cu+Cu calculation and data shown in the same plots. Whereas the average eccentricity for Cu+Cu tends to zero for the most central collisions, the corresponding v_2 values for Cu+Cu only drop to a value of $v_2 \approx 0.03$ even for the most central collisions. This would lead to the paradoxical conclusion that for the same N_{part} and therefore the same initial area density of produced particles, the Cu+Cu system is much more effective in translating an initial eccentricity into a final state anisotropy than the Au+Au system. Alternatively, large non-flow effects mimicking a dynamically generated anisotropy could be postulated for the Cu+Cu system.

However, a possible explanation unifying the observations in Au+Au and Cu+Cu can be found by examining the underlying definition of the initial state eccentricity. Commonly, this eccentricity, called ε_{std} in the following, is defined as the average eccentricity of the distribution of participating nucleons, relative to the known reaction plane, obtained for a certain centrality class in a Glauber calculation. This definition suffers from two potential problems: It averages out the fluctuations from event-to-event in the actual participant distributions. Finite number fluctuations will lead to an eccentric nucleon distribution even for collisions with impact parameter $b = 0$. With the standard definition, these fluctuations will be averaged to zero for central events. Furthermore, the minor axis for the actual event-by-event participant distribution will in general not coincide with the impact parameter vector. The eccentricity calculated relative to the reaction plane will therefore underestimate the true eccentricity of the nucleon distribution. To study these deficiencies, we have defined an alternative measure of the eccentricty in each centrality bin, where we calculate the eccentricity for each Glauber event relative to the principal axes of the actual participant distribution. By construction, this *participant eccentricity* , ε_{part}, will always be positive and will therefore average to a finite value even for the most central events. In addition, the smaller number of colliding nucleons, makes the difference between ε_{std} and ε_{part} particularly important for the Cu+Cu system relative to Au+Au. The result of a Glauber calculation for ε_{part} for Cu+Cu and Au+Au as a function of N_{part} can be seen in Fig. 4 (left plot). As expected, ε_{part} remains finite even for the most central Cu+Cu collisions.

Using ε_{part}, we can now attempt to identify a common scaling behaviour of Cu+Cu and Au+Au collisions over a large range of collision energies and centralities. This is shown in the right side of Fig. 4, plotting the ratio of $\langle v_2 \rangle / \langle \varepsilon_{part} \rangle$ versus the area density of produced particles [11, 12]. The data appear to exhibit a common scaling behavior over a large range in collision energy, suggesting that the efficiency for translating the initial state eccentricity estimated using ε_{part} into a final state anisotropy v_2 appears to

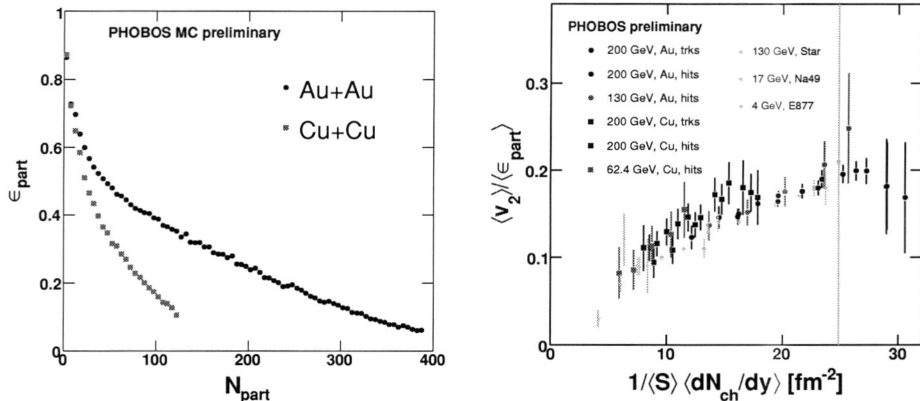

FIGURE 4. Left: Participant eccentricity ε_{part} of the collision zone in Cu+Cu (light symbols) and Au+Au (dark symbols) as a function of N_{part} for $\sqrt{s_{NN}} = 200$ GeV from Glauber calculations. Right: Ratio of v_2 coefficient to average participant eccentricity as a function of area density $1/\langle S \rangle \langle dN/dy \rangle$ for Au+Au collisions at $\sqrt{s_{NN}} = 130$ and 200 GeV and Cu+Cu collisions at $\sqrt{s_{NN}} = 62.4$ and 200 GeV, in comparison to lower energy data from NA49 and E877.

only depend on the initial area density achieved in the collision. Clearly, it is a fascinating question for future experiments whether this curve saturates at higher densities or continues to rise.

ACKNOWLEDGMENTS

This work was partially supported by U.S. DOE grants DE-AC02-98CH10886, DE-FG02-93ER40802, DE-FC02-94ER40818, DE-FG02-94ER40865, DE-FG02-99ER41099, and W-31-109-ENG-38, by U.S. NSF grants 9603486, 0072204, and 0245011, by Polish KBN grant 1-P03B-062-27(2004-2007), by NSC of Taiwan Contract NSC 89-2112-M-008-024, and by Hungarian OTKA grant (F 049823).

REFERENCES

1. B. B. Back *et al.*, [PHOBOS Collaboration], *Nucl. Phys. A* **757**, 28-101 (2005).
2. K. Rajagopal and F. Wilczek, *Nucl. Phys. B* **399**, 395-425 (1993).
3. I. N. Mishustin, *Phys. Rev. Lett.* **82**, 4779-4782 (1999).
4. B. B. Back *et al.* [PHOBOS Collaboration], *Nucl. Instrum. Meth. A* **499**, 603-623 (2003).
5. G. Stephans *et al.* [PHOBOS collaboration], Proceedings Quark Matter 2005.
6. Z. Chai *et al.* [PHOBOS Collaboration], arXiv:nucl-ex/0509027.
7. P. Steinberg *et al.* [PHOBOS collaboration], arXiv:nucl-ex/0510036.
8. K. Alpgard *et al.* [UA5 Collaboration], *Phys. Lett. B* **123**, 361 (1983).
9. K. H. Ackermann *et al.* [STAR Collaboration], *Phys. Rev. Lett.* **86**, 402-407 (2001).
10. S. Manly *et al.* [PHOBOS collaboration], arXiv:nucl-ex/0510031.
11. S. A. Voloshin and A. M. Poskanzer, *Phys. Lett. B* **474**, 27-32 (2000).
12. H. Heiselberg and A. M. Levy, *Phys. Rev. C* **59**, 2716-2727 (1999).

Similarity of Initial States in A+A and p+p Collisions in Constituent Quarks Framework

Rachid NOUICER

Chemistry Department, Brookhaven National Laboratory, Upton, NY 11973-5000, USA

Abstract. The multiparticle production results from A+A and p($\bar{p}$)+p collisions have been compared based on the number of nucleon participants and the number of constituent quark (parton) participants. In both normalizations, we observe that the charged particle densities in Au+Au and Cu+Cu collisions are similar for both $\sqrt{s_{NN}}$ = 62.4 and 200 GeV. This implies that in symmetric nucleus-nucleus collisions the charged particle density does not depend on the size of the two colliding nuclei but only on the collision energy. In the nucleon participants framework, the particle density at mid-rapidity as well as in the limiting fragmentation region from A+A collisions are higher than those of p($\bar{p}$)+p collisions at the same energy. Also the integrated total charged particle in A+A collisions as a function of number nucleon participants is higher than p($\bar{p}$)+p collisions at the same energy indicating that there is no smooth transition between peripheral A+A and nucleon-nucleon collisions. However, when the comparison is made in the constituent quarks framework, A+A and p($\bar{p}$)+p collisions exhibit a striking degree of agreement. The observations presented in this paper imply that the number of constituent quark pairs participating in the collision controls the particle production. One may therefore conjecture that the initial states A+A and p+p collisions are similar when the partonic considerations are used in normalization. Another interesting result is that there is an overall factorization of $dN_{ch}/d\eta$ shapes as a function of collision centrality between Au+Au and Cu+Cu collisions at the same energy, $\sqrt{s_{NN}}$ = 200 GeV.

Keywords: relativistic heavy ions collisions, nucleon participants, constituent quark participants
PACS: 25.75.-q, 25.75.Dw, 12.38.Mh, 24.10.Jv

1. INTRODUCTION

Since the first collisions were delivered at the Relativistic Heavy Ion Collider (RHIC), the experiments have obtained extensive results on multiparticle production in both nucleus-nucleus (A+A) and nucleon-nucleon (p ($\bar{p}$)+p) collisions at the same energies [1]. However, several aspects of the comparison between A+A and p($\bar{p}$)+p collisions are not well understood. For example it has been found that the particle density per participant pair, $N_{n-part}/2$, in A+A collisions is substantially higher than in p($\bar{p}$)+p collisions at $\sqrt{s_{NN}}$ = 200 GeV [2]. It has also been observed that the integrated total charged particle production, per participant nucleon pair, as a function of N_{n-part} is essentially constant and is higher than for p($\bar{p}$)+p collisions at the same energy [3] indicating that there is no smooth transition between the two systems. These comparisons are, however, based on scaling with the number of nucleon participants. In the following, I will show that the A+A and p($\bar{p}$)+p collisions have similar initial states if the results are scaled instead by the number of constituent quark participants, N_{q-part}. Within this framework, the similarity between A+A and p($\bar{p}$)+p collisions will be explored through global observables, which reflect the initial state of the system.

CP828, *Multiparticle Dynamics*
edited by V. Šimák, M. Šumbera, Š. Todorova, and B. Tomášik

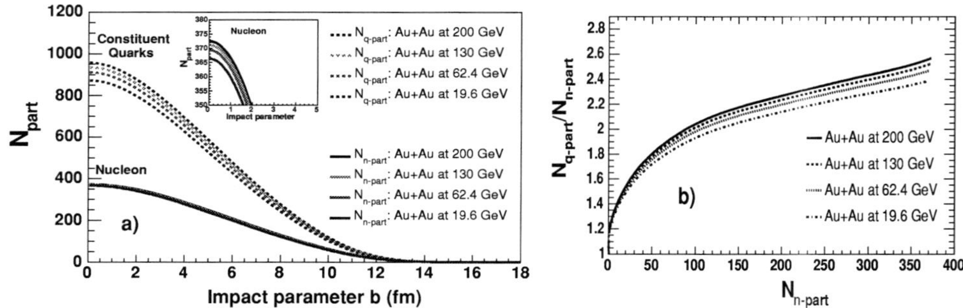

FIGURE 1. Panel a): number of nucleon participants (denoted N_{n-part}: solid curves) and number of constituent quark participants (denoted N_{q-part}: dashed curves) as a function of the impact parameter of Au+Au collisions at $\sqrt{s_{NN}}$ = 19.6, 62.4, 130 and 200 GeV. The inset figure represent a zoom on the solid curves. Panel b): ratio of N_{q-part}/N_{n-part} as a function of the number of nucleon participants.

2. CALCULATION OF THE NUMBER OF PARTICIPANTS

The constituent quark (parton) model has been introduced in Refs. [4]. The present work is a continuation of the study started by Ref. [5] which extends to global observables, namely the comparison of particle density and limiting fragmentation scaling in A+A and p($\bar{\text{p}}$)+p collisions.

The number of nucleon participants, N_{n-part} and the number of constituent quark participants, N_{q-part} are estimated using the nuclear overlap model in a manner similar to that used in Ref. [5]. The nuclear density profile is thus assumed to have a Woods-Saxon form,

$$n_A(r) = \frac{n_0}{1+exp[(r-R)/d]}, \tag{1}$$

where $n_0 = 0.17\ fm^{-3}$, $R = (1.12\,A^{1/3} - 0.86\,A^{-1/3})$ fm and d = 0.53 fm.

The number of nucleon participants, N_{n-part}, for nucleus-nucleus (A+B) collisions is calculated using the relation,

$$N_{n-part}|_{AB} = \int d^2sT_A(\vec{s})\{1-\left[1-\frac{\sigma_{NN}^{inel}T_B(\vec{s}-\vec{b})}{B}\right]^B\} + \int d^2sT_B(\vec{s}-\vec{b})\{1-\left[1-\frac{\sigma_{NN}^{inel}T_A(\vec{s})}{A}\right]^A\} \tag{2}$$

where $T(b) = \int_{-\infty}^{+\infty} dz n_A(\sqrt{b^2+z^2})$, is the thickness function. A and B are the mass number of the two colliding nuclei and the σ_{NN}^{inel} is the inelastic nucleon-nucleon cross section.

The number of constituent quark participants, N_{q-part}, is calculated in a similar manner by taking into account the following changes related to the physical realities: 1) the density is three times that of nucleon density with $n_0^q = 3n_0 = 0.51\ fm^{-3}$; 2) the cross sections $\sigma_{qq} = \sigma_{NN}^{inel}/9$; 3) the mass numbers of the colliding nuclei are three times their values, keeping the size of the nuclei same as in the case of N_{n-part}. For p($\bar{\text{p}}$)+p collisions the same procedure has been used to calculate the number of constituent quark

TABLE 1. Inelastic cross section for nucleon-nucleon collisions (σ_{NN}^{inel}) as function of colliding energy. My calculation for σ_{NN}^{inel} adopt similar manner as in Ref. [7]

$\sqrt{s_{NN}}$	19.6	53	56	62.4	130	200	540	630	900	1800
σ_{NN}^{inel}	31.5	35.0	35.3	36.0	39.3	42.0	48.0	48.6	51.0	56.0

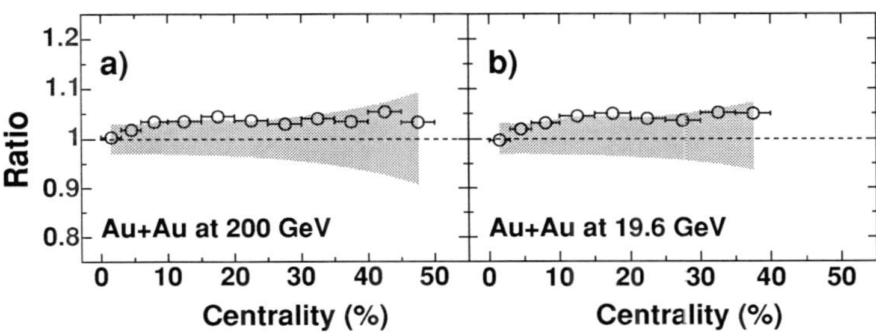

FIGURE 2. Quantitative evaluation of the model calculations expressed as the ratio of the average number of nucleon participants of the PHOBOS Glauber calculations [8] to the present work as a function of centrality. Panel a) and b) correspond to Au+Au collisions at $\sqrt{s_{NN}}$ = 200 and 19.6 GeV, respectively. The gray bands correspond to the systematic errors on the $\langle N_{n-part} \rangle$ of PHOBOS Glauber calculations.

participants by using $A = 3$ and $B = 3$ and considering nucleons as hard spheres of uniform radii of 0.8 fm [6].

Figure 1a) shows the number of nucleon (solid curves) and constituent quark (dashed curves) participants for Au+Au collisions as a function of the impact parameter. Figure 1b) presents the ratio of N_{q-part}/N_{n-part} as a function of N_{n-part} for Au+Au collisions at RHIC energies. The ratio shows that the correlation between N_{n-part} and N_{q-part} is not linear and that it depends slightly on the colliding energy. The inelastic nucleon-nucleon cross sections, σ_{NN}^{inel}, used in the present work are listed in TABLE 1.

Figure 2 presents the ratio of the N_{n-part} from the PHOBOS Glauber calculation based on HIJING [8] to the present calculation and shows good agreement (within systematic errors).

3. CHARGED PARTICLE DENSITIES

Figure 3 shows the primary charged particle density for central collisions at mid-rapidity divided by the number of participant nucleon pairs ($N_{n-part}/2$) and participant constituent quark pairs ($N_{q-part}/2$) as solid and open symbols, respectively. The plotted data are for Au+Au collisions at AGS, Pb+Pb collisions at the CERN SPS [9, 10, 11] and for Au+Au and Cu+Cu collisions at RHIC [12]. Also shown for comparison are results from p($\bar{p}$)+p collisions data [13, 14].

The particle density per nucleon participant pair for A+A collisions (solid points) shows an approximately logarithmic rise with $\sqrt{s_{NN}}$ over the full range of collision energies. The comparison of the particle density per nucleon of Au+Au to Cu+Cu collisions at the same energies, $\sqrt{s_{NN}}$ = 62.4 and 200 GeV indicates that in the symmetric nucleus-

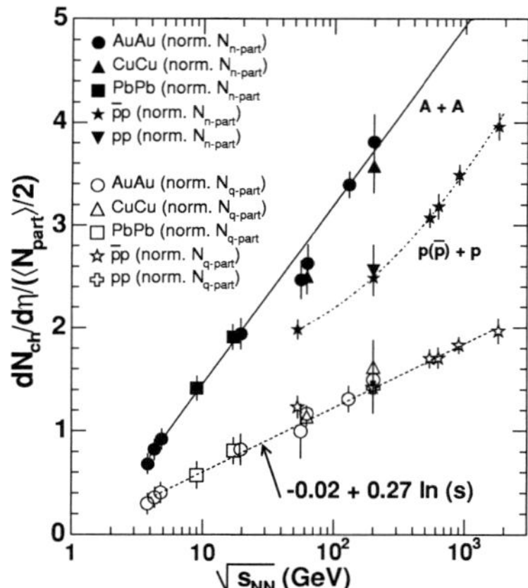

FIGURE 3. Particle density per constituent quark pair (open symbols) and particle density per nucleon participant pair (solid points) produced in central (6%) nucleus-nucleus (A+A) collisions as function of collision energy at AGS, SPS [9, 10, 11] and RHIC [12] and in $p(\bar{p})+p$ collisions at ISR energies [13] and $p+p$ at 200 GeV at RHIC [14]. The errors bars correspond to the systematic errors. The solid line represents a linear fit through solid points for A+A data, $f_{AA} = -0.287 + 0.757\ln(\sqrt{s})$. The dashed dotted line corresponds to the fit through solid points of $p(\bar{p})+p$ collisions, $f_{pp} = 2.25 - 0.41\ln(\sqrt{s}) + 0.09\ln^2(\sqrt{s})$. The dashed line corresponds to a linear fit trough the open symbols in the constituent quarks framework, $f_{p(\bar{p})p/AA} = -0.02 + 0.27\ln(\sqrt{s})$.

nucleus collisions the density per nucleon participant does not depend on the size of the two colliding nuclei but only on the collision energy. This means for Si+Si collisions at $\sqrt{s_{NN}}$ = 200 GeV, the particle density per nucleon participant will be similar to Au+Au collisions at the same energy. Figure 3 shows (solid points) that the charged particle multiplicity per participant nucleon pair in A+A collisions is higher compared to $p(\bar{p})$+p collisions at the same energy.

In contrast, we observe that the multiplicity per constituent quark participant pair (open symbols in figure 3) is similar for nucleus-nucleus collisions and nucleon-nucleon collisions at the same energy. It thus appears that using partonic participants accounts for the observed multiplicity in both A+A and $p(\bar{p})$+p collisions. One may therefore conjecture that the initial states in A+A and $p(\bar{p})$+p collisions are similar.

4. EXTENDED LONGITUDINAL SCALING

In general, the charged particle production in the limiting fragmentation region is thought to be distinct from that at mid-rapidity, although there is no obvious evidence for two separate regions at any of the RHIC energies. This observation is made based on the $dN_{ch}/d\eta$ distributions of charged particle presented in Ref. [15].

Figure 4 shows a comparison of the most central (6%) Au+Au and Cu+Cu collisions at several RHIC energies compared to $p(\bar{p})$+p ((inelastic and non single diffractive (NSD)) collisions at 200 GeV. When normalized to $N_{n-part}/2$, (figure 4a), we observe that the multiplicity in the limiting fragmentation region in A+A collisions is higher

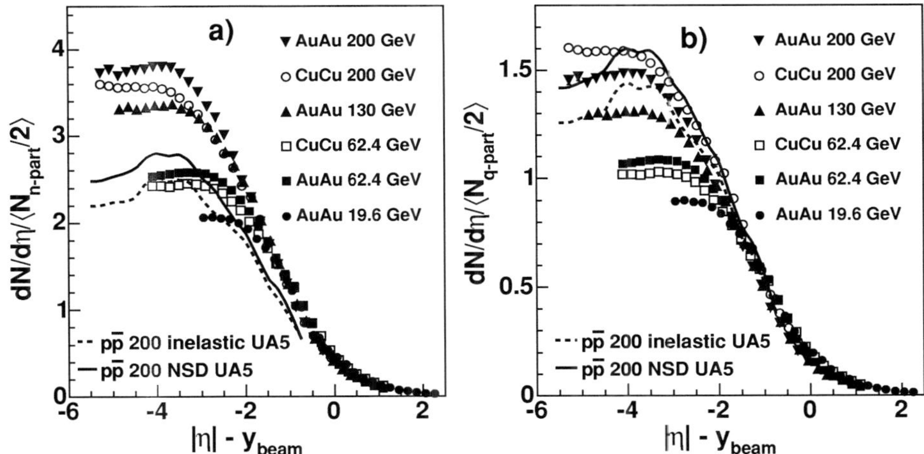

FIGURE 4. Pseudorapidity distributions of charged particle for Au+Au, Cu+Cu [12] collisions at RHIC energies compared to p($\bar{\text{p}}$)+p collisions at 200 GeV [13]. The distributions have been shifted to $\eta - y_{\text{beam}}$ in order to study the fragmentation regions in one of the nucleus rest frame. Panels a) and b) correspond to $dN_{ch}/d\eta$ distributions scaled to the number of nucleon participants and to the number of constituent quark participants, respectively. For clarity the systematic errors have been removed.

than for p($\bar{\text{p}}$)+p collisions. If, however, the comparison is carried out for multiplicities normalized to $N_{q-part}/2$ (figure 4 b), A+A and p($\bar{\text{p}}$)+p collisions exhibit a striking degree of agreement. Again, this observation implies that the number of constituent quark pairs participating in the collision controls the particle production.

5. SYSTEM SIZE INDEPENDENCE OF PSEUDORAPIDITY SHAPES

For 0-6% central collisions we find that the multiplicity shapes are essentially identical for Au+Au and Cu+Cu collisions and differ only by a constant factor,

$$R_{Au}^{Cu}(0-6\%) = \frac{dN/d\eta(Cu+Cu:0-6\%)}{dN/d\eta(Au+Au:0-6\%)}. \quad (3)$$

Figure 5 illustrates the fact that the same ratio between Au+Au and Cu+Cu collisions pseudorapidity distributions holds for all centrality bins (see figure 5) such that the shapes of the $dN_{ch}/d\eta$ distributions for the same centrality bin are similar for the two systems. The small difference at mid-rapidiy can be related to the difference of the mean P_T of charged particle in Cu+Cu and Au+Au collisions but it falls well within the systematic errors so this difference is not significant. It thus appears that the $dN_{ch}/d\eta$ shapes are independent of the overall size of the colliding nuclei, at least between the Cu+Cu and Au+Au systems studied here.

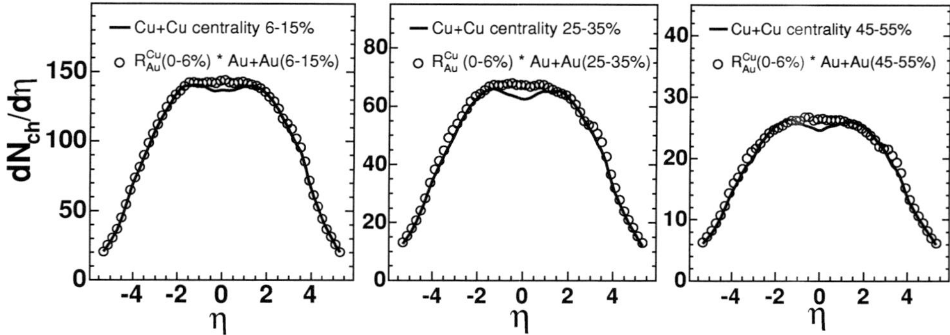

FIGURE 5. Comparison of $dN_{ch}/d\eta$ distributions of Cu+Cu to Au+Au collisions [12] at the same collision centrality and energy, 200 GeV, presented for different centrality bins. The $dN_{ch}/d\eta$ distributions of Au+Au collisions have been multiplied by factor which corresponds to the ratio of the measured $dN_{ch}/d\eta$ distributions of Cu+Cu central (6%) collisions to the measured $dN_{ch}/d\eta$ of Au+Au central (6%) collisions. For clarity, the systematic errors are not shown.

6. SUMMARY

I have shown that the charged particle multiplicity, both at mid-rapidity and in the limiting fragmentation region scale with the number of constituent quark participants, both in nucleus-nucleus systems of different sizes and in nucleon-nucleon collisions. This observation implies that both the overall the particle production and the distribution in pseudorapidity is controlled by at the participant quark level. In addition, I have shown that shapes of the charged particle multiplicity in Au+Au and Cu+Cu collisions at the same energies are very similar and that they differ only by a overall factor, even at different centralities given by the fraction of the overall cross section.

ACKNOWLEDGMENTS

I thank B. B. Back and M. D. Baker for discussions and a careful reading of the manuscript. This work was supported by U.S. DOE Grant No. DE-AC02-98CH10886.

REFERENCES

1. I. Arsene et al. *Nucl. Phys.*, **A757** 1 (2005); B. B. Back et al. *Nucl. Phys.*, **A757** 28 (2005).
2. B. B. Back et al. *Phys. Rev. Lett.*, **88** 22302 (2002).
3. B. B. Back et al. *Phys. Rev. C (Rapid Comm.) in press* (2005), e-Print nucl-ex/0409021.
4. A. Bialas, W. Czyz, and L. Lesniak *Phys. Rev.*, **D25** 2328 (1982) and references therin
5. S. Eremin and S. Voloshin *Phys. Rev.*, **C67** 064905 (2003).
6. C. Y. Wong, *World Scientific*, 161 (1994).
7. Bhaskar De and S. Bhattacharyya *Phys. Rev.*, **C71** 024903 (2005).
8. B. B. Back et al. *Phys. Rev.*, **C70** 021902(R) (2004).
9. L. Ahle et al., *Phys. Lett.*, **B476** 1 (2000); L. Ahle et al., *Phys. Lett.*, **B490** 53 (2000).
10. J. Bachler et al., *Nucl. Phys.*, **A661** 45 (1999).
11. C. Blume et al., *Proceeding of QM* , (2001).
12. G. Roland (for PHOBOS Collaboration) *Proceeding QM*, (2005); e-Print nucl-ex/0510042.
13. F. Abe et al., *Phys. Rev.*, **D41** 2330 (1990).
14. S. J. Snaders (for BRAHMS Collaboration), *Seminar presented at BNL* (2004).
15. B. B. Back et al. *Phys. Rev. Lett.*, **91**, 052303 (2003).

Radial and Elliptic Flow in High Energetic Nuclear Collisions

Xianglei Zhu[*,†], Hannah Petersen[*] and Marcus Bleicher[*]

[*]*Institut für Theoretische Physik, Johann Wolfgang Goethe-Universität, Max-von-Laue-Str. 1, D-60438 Frankfurt am Main, Germany*
[†]*Frankfurt Institute for Advanced Studies (FIAS), Max-von-Laue-Str. 1, D-60438 Frankfurt am Main, Germany*

Abstract.

Keywords: Relativistic heavy ion collisions, Collective flow, Phase transitions
PACS: 25.75.-q, 25.75.Nq, 25.75.Ld, 25.75.Dw, 24.10.Lx

INTRODUCTION

Lattice QCD (lQCD) calculations at finite chemical potential indicate a rapid increase of the thermodynamic pressure P with temperature above the critical temperature T_c for a cross-over (phase transition) from hadron gas to a Quark-Gluon Plasma (QGP)[1]. Numerous theoretical estimates suggest that this transition can be expected around 30 AGeV beam energy. Indeed, over the last several years' experimental studies of heavy-ion collisions in the $20-160$ AGeV energy regime, have revealed that many observables do show non-monotonous structures around $E_{\rm lab} = 30$ AGeV. It has been observed that there is, a sharp maximum in the K^+/π^+ ratio [2, 3], a step in the transverse momentum excitation function [3, 4] and an apparent softest point in the equation of state [5] all simultaneously at the lower SPS energy. The radial and elliptic flow (v_2) of the particles produced in a relativistic heavy ion collision are also intimately connected to the pressure and its gradients in the early stage of the reaction. Therefore these observables should also be sensitive to changes in the equation of state. For the present investigation, we study the excitation function, centrality and transverse momentum dependence of v_2 within the Ultra-relativistic quantum molecular dynamics (UrQMD) approach.

The UrQMD model is a relativistic transport model that employs hadronic and string degrees of freedom [22, 23]. It takes into account the formation and multiple scattering of ingoing and newly produced hadrons. It describes dynamically the generation of pressure in the hadronic/valence quark compression and expansion phase. Until now, only hadrons, valence quarks and valence di-quarks and their interactions are treated explicitly in this model. Gluonic degrees of freedom are not treated explicitly, but are implicitly present in strings. The UrQMD model reproduces the nucleon-nucleon, meson-nucleon and meson-meson cross section data in a wide kinematic range. And it allows for a systematic study of the change in the dynamics from elementary collisions to proton-nucleus and nucleus-nucleus reactions in a unique way without change in

CP828, *Multiparticle Dynamics*
edited by V. Šimák, M. Šumbera, Š. Todorova, and B. Tomášik

parameters.

The RHIC measurements on the identified particle v_2 and R_{AA} have shown the so-called number of constituent quark (NCQ) scaling behavior[12]. This scaling seems to come directly to the conclusion that the quark-gluon plasma phase has been created, and all the hadrons are created at the hadronization of the QGP by the recombination or coalescence of the partons[6, 7, 8]. Recently, the multi-strange particles have also been found to have as large v_2 as the light hadrons[9]. According to the observation on the radial flow of multi-strange particles, they should freeze-out right after the hadronization[9]. Therefore, their v_2 is usually regarded as a good measure of the partonic v_2. Also, the v_2 of multi-strange particles shows the same NCQ scaling behavior as that of light hadrons. However, one has to study whether this scaling is the unique behavior of the recombination/coalescence models or if it can also be obtained within other approaches not relying on the deconfinement condition.

While v_2 is one of the most important observables in heavy ion physics, its accurate and unambiguous experimental measurement is not trivial. Usually, the v_2 is measured with the reaction plane method [17], or the two-particle correlation method. In general, these two-particle methods are affected by the so called non-flow effects, which are the particle correlations not related to the reaction plane. In order to decrease the contribution of the non-flow effects to the flow measurement, the many-particle cumulant method was proposed[18]. However, as indicated in Refs. [19, 20, 21], the v_2 from many-particle cumulants might be affected by the event-by-event v_2 fluctuations. As shown in [21], the observed difference[12] between $v_2\{2\}$ and $v_2\{4\}$ (v_2 from 2 and 4-particle cumulants) can also be explained by definite amount of v_2 fluctuations. Thus, it is necessary to test the cumulant method with a transport model to disentangle the effects of non-flow effects and v_2 fluctuations on the cumulant method.

FLOW EXCITATION FUNCTIONS

Let us start with the energy dependence of the transverse momentum and the elliptic flow. As shown in Fig. 1, the $\langle p_T \rangle$ of the pions shows a monotonic rise with increasing energy, while the elliptic flow v_2 has a non-monotonic behavior. For low energies, v_2 increases with energy. However, it reaches a maximum around 30 AGeV beam energy and then it drops down. To show this qualitative change more clearly we plot the excitation function of v_2 divided by $\langle p_t \rangle$ in Fig. 1 (right). At low energies, both data and model yield a linear increase of $v_2/\langle p_t \rangle$. However, in the data (full diamonds) one can observe that after the first rise the value of this ratio stays constant with increasing energy. The present non-equilibrium study suggests that the initial increase of the scaled elliptic flow up to SPS energies might be due to viscosity effects (decreasing mean-free-path) in the hadronic gas. After this first increase the curve calculated by the UrQMD model levels off at 60% the experimental value.

Therefore, it is possible that above the energy range about $E_{lab} = 30$ AGeV partonic interactions have to be taken into account to describe the data as suggested in [13]. How can we analyze this question in the model, since there are no partonic degrees of freedom explicitly incorporated? In the current model exists a formation time for hadrons produced in the string fragmentation. The leading hadrons of the fragmenting

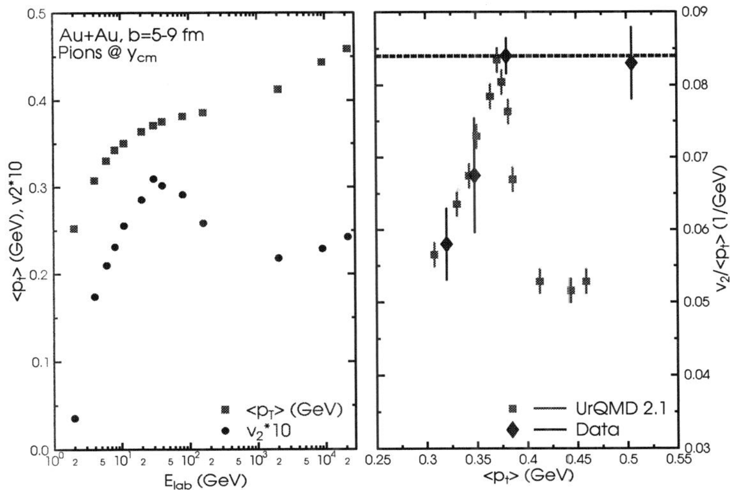

FIGURE 1. Left: Calculated excitation functions of mean transverse momentum and elliptic flow of pions at midrapidity for central Au+Au (Pb+Pb) reactions. Right: Excitation function of $v_2/\langle p_T \rangle$ of Pions in mid-central collisions from UrQMD (squares). Negatively charged particle data (diamonds) for v_2 are taken from Ref. [25] and $\langle p_T \rangle$ from Refs.[26, 27, 28, 29].

strings contain the valence quarks of the original excited hadron. These (di-)quark string ends are allowed to interact during their formation time with a reduced cross section defined by the additive quark model. These string ends that interact with a reduced cross section are dubbed as "unformed hadrons". Because most of the unformed hadrons do not interact with others during their formation time, the effective pressure is only related to the energy density of the formed hadrons according to the hadron gas EoS. We have calculated the energy density during heavy ion collisions at different beam energies (2 AGeV - 21.3 ATeV). From this, we extract the time corresponding to the maximum value of the energy density.

Fig. 2 shows the percentage of the energy density at that time of the reaction that is deposited in the "unformed hadrons". The fraction of "unformed hadrons" starts at zero for low energies and then rises fast to almost 100 %. Note that this fraction reaches 90% already around 30 AGeV beam energy, similar to the energy region where a phase transition is expected. As one can see, the energy density of the formed hadrons is much smaller than the total value, therefore the effective pressure of the formed hadrons in the model should also be too small to generate enough v_2 from this energy on. Thus, this finding supports the interpretation of the need for initial pressure from pre-QGP matter already at low SPS energies.

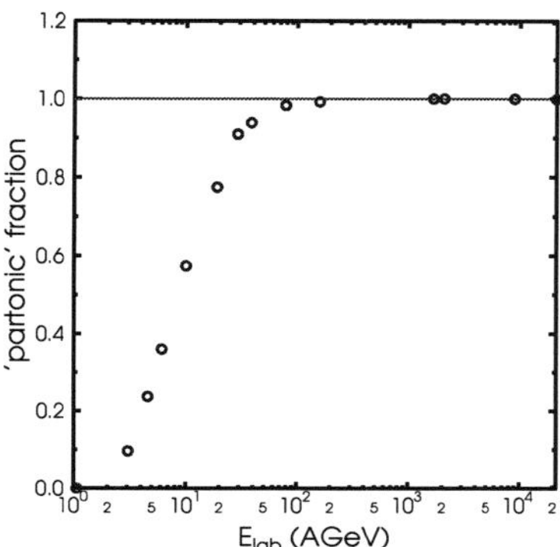

FIGURE 2. Percentage of the energy density in unformed hadrons as a function of the collision-energy for central Au+Au (Pb+Pb) reactions.

V_2 OF HADRONS AND NCQ-SCALING

The UrQMD result on the p_T dependence of the hadron v_2 is shown in Fig.3(A). When $p_T \leq 1.8$GeV/c, there is the same mass-ordering of the hadron v_2 as that shown in the data[12] and also by the hydrodynamical calculations[14]. When $p_T > 1.8$GeV/c, the v_2 of baryons is larger than those of mesons, which is also consistent with the RHIC data[12] but in contradiction to the hydro-model predictions. There no clear sign of thermalization in UrQMD final states, therefore, the qualitative agreement between the UrQMD results and the data at low and intermediate p_T probably indicates non-equilibrium dynamics in the real situation. When the p_T increases further, the v_2 of hadrons decreases fast. While in the RHIC data, at large p_T, the v_2 seems to show a behavior of saturation. As we will see in the next section, this RHIC v_2 measurements on the identified hadron might suffer from the non-flow effects. Thus, there might be a substantial method dependence in the v_2 analysis at large p_T in the current data. We stress that the multi-strange baryons in the present model do also show a similar v_2 as the ordinary baryons, although there is no assumption of deconfinement and QGP in the UrQMD model.

From Fig.3(A), one can also see a clear difference between the meson v_2 and baryon v_2. In Fig.3(B), the NCQ scaled hadron v_2 is shown. All the scaled v_2 including those of multi-strange baryons roughly fall in one line, except that at $p_T < 0.5$GeV/c. Deviation of pion v_2 from the scaling is expected and is might be caused by resonance decays[10]. We want to point out that the approximate NCQ-scaling also exists in another hadronic transport model, the RQMD model[11]. Therefore the constituent quark scaling is not a model-dependent feature. In fact, the observed NCQ-scaling in the UrQMD and RQMD model demonstrates that the NCQ scaling is not only a feature of quark coalescence models nor a unique signal of deconfinement.

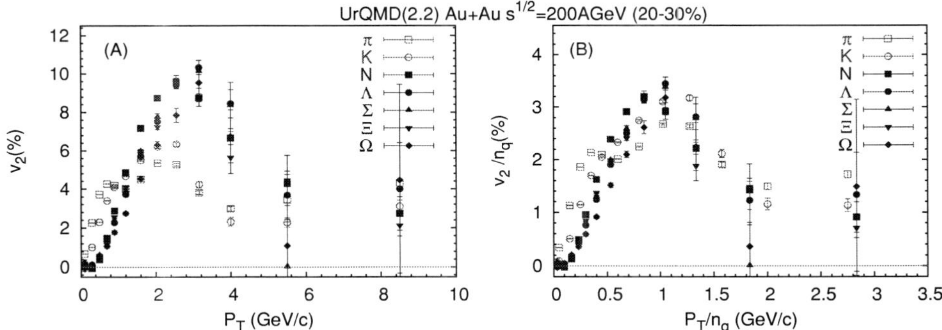

FIGURE 3. (A) UrQMD(2.2) results on the p_T dependence of hadron v_2, from semi-central Au+Au collisions (20-30%) at $\sqrt{s_{NN}}$ = 200 GeV; (B) The number of constituent quark scaled hadron v_2. The n_q referes to the number of constituent quarks in the hadron. At low $p_T/N_q \leq 0.5$ GeV/c, π does not follow the scaling perhaps caused by the resonance decay[10].

THE TEST OF THE CUMULANT METHOD

Let us turn to the test of the cumulant method in the UrQMD model. For the details, please refer to Ref. [24].

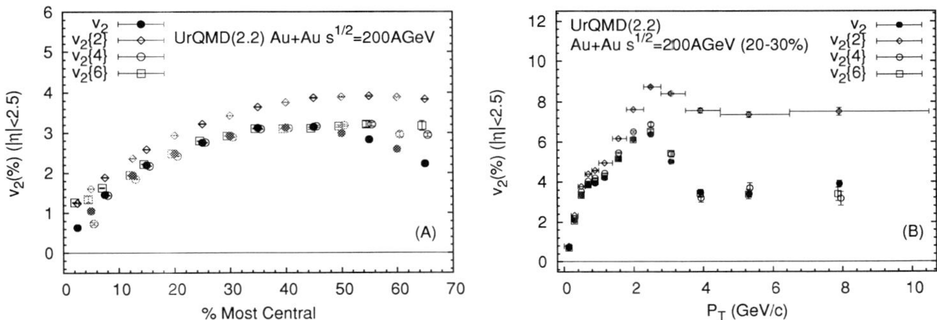

FIGURE 4. (A) The integral v_2 results($v_2\{2\}$,$v_2\{4\}$ and $v_2\{6\}$) from the cumulant method are compared to the exact v_2 in different centrality bins. The grey points are the corresponding results from the enlarged centrality bins which merge two of the original bins. (B) $v_2(p_T)$ in the semi-central collisions: results from the cumulant method are compared to the exact v_2

Firstly, let us show the calculated integral v_2 results with the cumulant method as a function of centrality in Fig.4(A). For mid-central collisions ($\sigma/\sigma_{tot} \sim 10-50\%$), the elliptic flow parameters extracted from two-particle cumulant ($v_2\{2\}$) deviates rather strongly from the exact v_2 as obtained from the known reaction plane. However, four particle cumulant ($v_2\{4\}$) and six particle cumulants ($v_2\{6\}$) show almost no difference and both agree well with the theoretically expected v_2. While, according to Ref. [21], the exact v_2 should be in the middle of $v_2\{2\}$ and $v_2\{4\}$ if their difference is mainly from the v_2 fluctuations. Therefore, we can conclude that for semi-central to semi-peripheral centralities the contribution of the v_2 fluctuations to the cumulant results is almost negligible and the difference between$v_2\{2\}$ and $v_2\{4\}$ or $v_2\{6\}$, is mainly due to non-flow effects in the UrQMD model.

However, from Fig.4(A), we have also seen that both $v_2\{4\}$ and $v_2\{6\}$ do not agree with the exact v_2 in the most central and the very peripheral bins. This means at central and very peripheral collisions, the v_2 fluctuations indeed play an important role as indicated in [21]. The deviations of $v_2\{n\}$ from the exact v_2 in the most central and very peripheral bins are qualitatively consistent with the expections within a simplified Monte-Carlo Glauber treatment [21].

Then, we show the p_T dependence of v_2 of particles within $|\eta| < 2.5$ in a semi-central (20%-30%) centrality bin in Fig.4(B). At large transverse momenta (p_T), non-flow contributions are expected to be large and might influence the results obtained by the cumulant method. As we can see in Fig.4(B), $v_2\{2\}$ is always larger than exact v_2. Especially towards large p_T, $v_2\{2\}$ stays roughly constant, while exact v_2 decreases when $p_T > 2.5$GeV/c. The saturation of $v_2\{2\}$ is consistent with STAR's $v_2\{2\}$ results [16]. This strong deviations point towards substantial contributions from non-flow effects in the two-particle cumulant method. The higher order cumulants do a much better job in reproducing the exact v_2. Here, the difference between $v_2\{4\}$ or $v_2\{6\}$ and the exact v_2 is much smaller especially at large p_T.

CONCLUSION

The v_2 excitation function from UrQMD shows a non-monotonic behavior around 30 AGeV beam energy. The lack of pressure in UrQMD at higher beam energies might be connected to the omission of partonic degrees of freedom in the present model. The constituent quark (NCQ) scaling of the hadron v_2 including multi-strange baryons is qualitatively reproduced by the UrQMD model. This demonstrates that NCQ scaling is not a unique feature of recombination and coalescence models. Finally, the cumulant method is tested with the UrQMD model. We found that the 4 and 6 particle cumulant methods work well for the mid-central 200AGeV Au+Au collisions, while the 2 particle cumulant method is always affected by non-flow effects. However, for the most central and very peripheral collisions, the cumulant method - even for higher order cumulants - is affected by the large v_2 fluctuations.

ACKNOWLEDGMENTS

We are grateful to the Center for the Scientific Computing (CSC) at Frankfurt for the computing resources. This work was supported by GSI and BMBF.

REFERENCES

1. Z. Fodor and S. D. Katz, JHEP **0203**, 014 (2002); Z. Fodor, S. D. Katz, and K. K. Szabo, Phys. Lett. B **568**, 73 (2003).
2. S. V. Afanasiev *et al.* [The NA49 Collaboration], Phys. Rev. C **66** (2002) 054902 [arXiv:nucl-ex/0205002].
3. M. Gazdzicki *et al.* [NA49 Collaboration], J. Phys. G **30** (2004) S701 [arXiv:nucl-ex/0403023].
4. C. Blume, J. Phys. G: Nucl. Part. Phys. 31, S57 (2005)
5. M. Bleicher, hep-ph/0509314.
6. R. J. Fries, nucl-th/0410085 and references therein.
7. R. Hwa and C.B. Yang, Phys. Rev. **C70**, 024904(2004) and reference therein.
8. D. Molnar and S. Voloshin, Phys. Rev. Lett. **91**, 092301(2003).
9. J. Adams, *et al.*, (STAR Collaboration), nucl-ex/0504022
10. X. Dong, *et al.*, *Phys. Lett.* B **597**, 328(2004).
11. Y. Lu, *et al.*, in preparation.
12. J. Adams *et al.* (STAR Collaboration), nucl-ex/0409033.
13. E. L. Bratkovskaya *et al.*, Phys. Rev. C **69** (2004) 054907 [arXiv:nucl-th/0402026].
14. P.F. Kolb, P. Huovinen, U.W. Heinz and H. Heiselberg, *Phys. Lett.* **B500**, 232 (2001); P. Huovinen, P.F. Kolb, U.W. Heinz, P.V. Ruuskanen and S.A. Voloshin, *Phys. Lett.* **B503**, 58 (2001)
15. M. Gyulassy, I. Vitev and X.-N. Wang, *Phys. Rev. Lett.* **86**, 2537 (2001).
16. J. Adams *et al.* (STAR Collaboration), *Phys. Rev. Lett.* **93**, 252301 (2004)
17. A.M. Poskanzer and S.A. Voloshin, *Phys. Rev.* C **58**, 1671 (1998).
18. N. Borghini, P.M. Dinh and J.-Y. Ollitrault, *Phys. Rev.* C **63**, 054906 (2001); N. Borghini, P.M. Dinh and J.-Y. Ollitrault, *Phys. Rev.* C **64**, 054901 (2001); N. Borghini, P.M. Dinh and J.-Y. Ollitrault, nucl-ex/0110016.
19. C. Adler *et al.* (STAR Collaboration), *Phys. Rev.* C **66**, 034904 (2002).
20. S.A. Voloshin, *Nucl. Phys.* A. **715**, 379 (2003).
21. M. Miller and R. Snellings, nucl-ex/0312008.
22. S.A. Bass *et al.*, *Prog. Part. Nucl. Phys.* **41**, 225 (1998); M. Bleicher *et al.*, *J. Phys. G: Nucl. Part. Phys.* **25**, 1859 (1999).
23. E.L. Bratkovskaya *et al.*, *Phys. Rev.* C **69**, 054907 (2004)
24. X. Zhu, M. Bleicher and H. Stöcker, nucl-th/0509081.
25. C. Alt *et al.* [NA49 Collaboration], *Phys. Rev.* C **68** (2003) 034903 [arXiv:nucl-ex/0303001].
26. Y. Akiba *et al.* [E802 Collaboration], *Nucl. Phys.* A **610**, 139C (1996).
27. R. Bramm, Diploma thesis, Institut für Kernphysik, 2002
28. H. Appelshauser *et al.* [NA49 Collaboration], *Phys. Rev. Lett.* **82**, 2471 (1999) [arXiv:nucl-ex/9810014].
29. C. Adler *et al.* [STAR Collaboration], *Phys. Rev. Lett.* **87**, 112303 (2001) [arXiv:nucl-ex/0106004].

Open charm production at RHIC

Xin Dong

Department of Modern Physics, University of Science and Technology of China - USTC, 96 Jinzhai Road, Hefei, Anhui 230026, China

Abstract. Recent experimental measurements on open charm production in proton-proton, proton(deuteron)-nucleus and nucleus-nucleus collisions at RHIC are reviewed. A comparison with theoretical prediction is made. Some unsettled issues call for precise measurements on directly reconstructed open charm hadrons.

Keywords: Heavy flavor, charm, collectivity, thermalization, cross section, R_{AA}, v_2
PACS: 25.75.Dw, 13.20.Fc, 13.25.Ft, 24.85.+p

INTRODUCTION

The ongoing four experiments at the Relativistic Heavy Ion Collider (RHIC) are designed to search for and measure the quark-gluon plasma (QGP), a new state of matter composed of deconfined, locally thermalized quarks and gluons. The equilibrated matter is expected to be described by the equation of state (EoS) with partonic degrees of freedom. The physics results from the first three-year runs at RHIC demonstrate that the partonic pressure gradient has been developed during the system evolution in heavy ion collisions. This bas been illustrated in the "white papers" from four experiments [1]. To determine the partonic EoS, the next task is to test the local and early thermalization hypothesis *experimentally*. Heavy quark (c, b) is an ideal probe in this direction.

In heavy ion collisions, the theoretical calculation shows that charm quarks are mostly created through initial gluon-gluon fusions [3]. Unlike light quark, heavy quark mass is dominated by its current quark mass - the mass originating from the coupling with the electroweak Higgs field [4]. Therefore heavy quark is an ideal penetrating probe to the rescatterings and thermalization at the early stage of heavy ion collisions. The measurement on charm quark production in proton-proton $(p+p)$ collisions not only provides a necessary reference for heavy ion collisions, but also enables us to test the pQCD calculations of both total and differential cross sections.

Charm quarks radiative energy loss in vacuum is characterized by the "dead-cone" effect [5]. Theoretical calculations predict that the suppression of the nuclear modification factor (R_{AA}) for charm quarks in central nucleus-nucleus (A + A) collisions is smaller than that of light quarks [6, 7, 8]. Most of these predictions were made based on the radiative energy loss mechanism and the medium properties to our knowledge (gluon density *etc.*). The interaction between charm quark with medium can also be reflected by the charm quark elliptic flow (v_2). The coalescence approach in a thermalized medium shows that charm hadrons may obtain a finite v_2 even if charm quarks have a zero v_2 [9]. The charm quark collectivity has been studied in an AMPT transport model, and the result shows that a large charm quark interacting cross section is needed to pro-

CP828, *Multiparticle Dynamics*
edited by V. Šimák, M. Šumbera, Š. Todorova, and B. Tomášik

duce the magnitude of v_2 comparable with that of light quarks [10]. Measurements of the charm quark collectivity will tell us the degree to which charm quarks interact with other partons, and then provide us with pivotal information on the early thermalization of light flavors.

CHARM PRODUCTION IN ELEMENTARY COLLISIONS

The first reconstruction of open charm hadrons through their hadronic decays was reported by the STAR Collaboration in Quark Matter 2004 [11] and was recently published in Ref. [12]: $D^0 \to K^-\pi^+$ (B.R.=3.83%,$p_T < 3$ GeV/c), $D^{*+} \to D^0\pi^+$ (B.R.=68%, $1 < p_T/(\text{GeV}/c) < 6$) and their charge conjugates. Event mixing technique was used to construct the combinatorial background in the invariant mass spectrum. STAR and PHENIX also reported the charm production cross section results from the measurements of non-photonic electrons mostly from heavy flavor decays in d + Au and $p+p$ collisions [11, 12, 13, 14] .

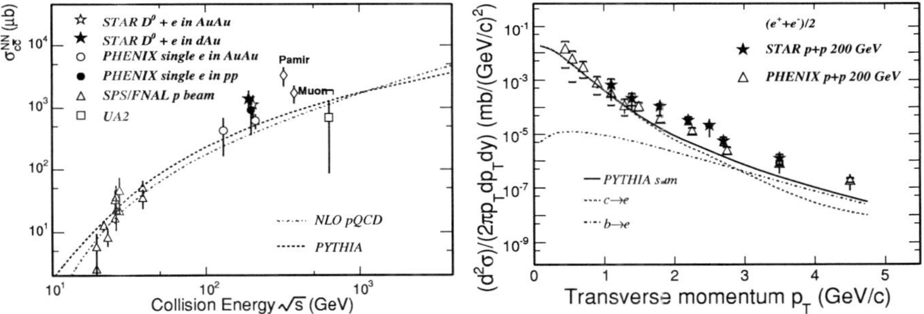

FIGURE 1. **Left:** Total $c\bar{c}$ cross section per nucleon-nucleon collision vs. the collision energy. The low energy data points are selected from fixed target experiments [15, 16]. The diamonds are taken from two cosmic ray measurements [17]. The dashed and dot-dashed lines are taken from [12]. The results from Au + Au system will be discussed in the next section. **Right:** Non-photonic electron spectrum in $p+p$ collisions compared with a PYTHIA model LO calculation with the parameter setting inspired from Au + Au 130 GeV data [18] with $\sigma_{c\bar{c}} = 658\ \mu$b.

Available data of charm cross section from various energies are shown in the left plot of Fig. 1. The results in Au + Au collisions at RHIC will be discussed in the next section. The dot-dashed curve depicts a typical next-to-leading (NLO) pQCD calculation result where parameters are optimized to fit the low energy data [19]. The dashed curve is a PYTHIA (version 6.152) calculation with the parton distribution function CTEQ5M1. Both the NLO pQCD calculation and the PYTHIA prediction give a total cross section of $300-450\ \mu$b, which is $2-3$ times lower than the experimental data. Data points from cosmic ray experiments also indicate a large cross section at $\sqrt{s} \sim 300$ GeV [17]. The discrepancy at RHIC may be attributed to higher order processes in the charm production at RHIC energy.

Except an overall normalization scale, PHENIX and STAR measured consistent electron spectral shape within errors. Fig. 1 right plot shows the measured non-photonic electron spectra from PHENIX and STAR in $p+p$ collisions. On that plot also shows a

PYTHIA calculation with parameters which describes the non-photonic electron spectrum well in Au + Au 130 GeV. The measured spectrum is clearly harder than the overall contribution from charm and bottom decays in the PYTHIA calculation. An important issue in obtaining the charm production in electron approach is how to determine the bottom contamination. A recent NLO pQCD investigation at RHIC energy shows the crossing point between the electron spectra from charm decay and bottom decay may vary in a broad p_T range ($\sim 3-10$ GeV/c) [20]. This will bring large uncertainties in the electron data from bottom decays. To establish a good reference for studying charm production in heavy ion collisions, precise measurements on reconstructed charm hadron with large p_T coverage are necessary.

OPEN CHARM PRODUCTION IN HEAVY ION COLLISIONS

Charm yields in heavy ion collisions are expected to be scaled by N_{bin} since most charm quark pairs are created in the initial hard processes. A recent publication from PHENIX Collaboration [21] reported the centrality dependence of non-photonic electron spectra in Au + Au collisions at $\sqrt{s_{NN}}$ = 200 GeV. The electron spectra in all centralities show approximate N_{bin} scaling with respect to the $p+p$ collisions. The charm total cross section in minimum bias Au + Au collisions is compatible with the that in $p+p$ collisions.

STAR also reported preliminary results on reconstructed D^0 signal in minimum bias Au + Au collisions using the same method as that in d + Au collisions [22]. They also reported centrality dependence of non-photonic electron measurements in Au + Au collisions [22, 23]. By combining D^0 and non-photonic electrons, the extracted total charm production cross section per nucleon-nucleon collision in minimum bias Au + Au collisions is also consistent with STAR d + Au results and PHENIX $p+p$ results.

The total cross section measurement is an important reference for charmonium production whose enhancement or suppression in central Au + Au collisions is thought to be a robust signal of QGP, so precise charm measurements in various centralities in Au + Au collisions are needed.

Fig. 2 shows recent results on the nuclear modification factor R_{AA} for non-photonic electrons in central Au + Au collisions from PHENIX [24] and STAR [22, 23]. Both experiments give a consistent, and *surprising* fact: the suppression factor of non-photonic electrons is $\sim 0.2-0.3$, which is almost at the same level as that of charged hadrons in similar p_T range. Two recent pQCD estimations in the radiative energy loss scenario are also shown in that figure [25, 26]. These approaches try to fix transport parameter (dN_g/dy or $\hat{q}$) boundaries by fitting to the R_{AA} for light hadrons and the boundaries obtained are $1000 < dN_g/dy < 3500$, and $4 < \hat{q}/(\mathrm{GeV}^2/\mathrm{fm}) < 14$ respectively. Fig. 2 shows the upper limit to which energetic partons lose the largest fraction of energies in the medium due to gluon bremsstrahlung from these two approaches. The comparison with data illustrates the suppression of electrons from charm decays can reach as low as that of light hadrons. However, if the bottom contribution is included according to pQCD calculations, the final electron R_{AA} will jump to $\sim 0.4-0.5$. This is a significant discrepancy compared to the data at $4 < p_T/(\mathrm{GeV}/c) < 7$. Hence, if the data on Fig. 2 are confirmed to be corret, this will bring at least two open issues: (i) if current radiative energy loss mechanism persists, there is no much room for bottom's contribution in the non-photonic electron spectrum up to $p_T \sim 7$ GeV/c. (ii) if bottom's contribution is as what is given by pQCD calculations (the crossing point is $\sim 3-5$ GeV/c), there must be other energy loss effects besides radiative gluons. These are challenges to theorists.

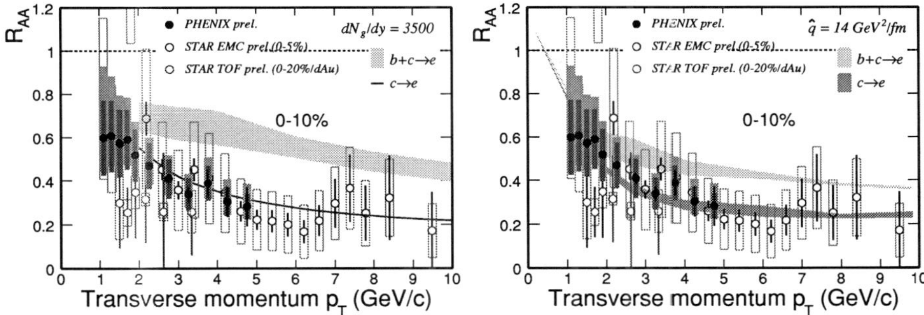

FIGURE 2. Recent measurements on non-photonic electron R_{AA} in central Au + Au collisions from PHENIX (top 0-10%) and STAR (top 0-5% from EMC and top 0-20% using d+Au as the reference from TOF) experiments compared with theoretical predictions (top 0-10%) from [25] and [26].

In several recent publications, some authors argue that because in momentum coverage $\gamma v \sim 1$, heavy quark is not ultrarelativistic, elastic collisional energy loss may play an important role when charm quarks traverse the medium [27, 28]. They computed the R_{AA} in hydrodynamic transport scenario with the transport property parameter such as diffusion coefficient, which gives the strong suppression as observed. The approach sheds light on the solution to the present discrepancy. To decouple two above issues, one should precisely measure reconstructed open charm hadrons instead of electrons. Certainly, a precise reference in $p+p$ collisions on reconstructed open charm hadrons is needed.

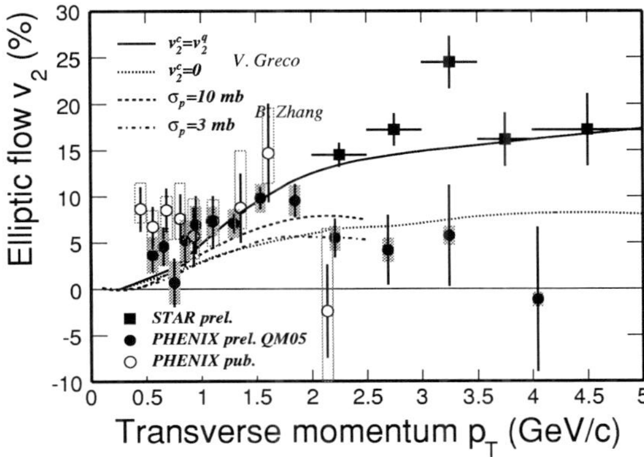

FIGURE 3. Recent measurements of non-photonic electron v_2 in minimum bias Au + Au collisions from PHENIX [24, 29] and STAR [30] experiments compared with theoretical predictions in Ref. [9, 10].

Fig. 3 shows recent results of non-photonic electron v_2 in minimum bias Au + Au collisions from PHENIX [24, 29] and STAR experiments [30]. At $p_T > 2$ GeV/c, two experiments have a big discrepancy there, although systematic errors from STAR are claimed to be $\sim 20-30\%$ in that region. In terms of the magnitude, the result shows non-photonic electron v_2 is comparable to other hadrons v_2 in $p_T < 2$ GeV/c. Two model predictions are also shown on the plot. Since at $p_T > 2$ GeV/c, there is inconsistency between two experiments and the bottom's contribution becomes important and uncertain, let us focus on the comparison at $p_T < 2$ GeV/c. In the coalescence approach through the thermalized system [9], the data points are consistent with that charm quark v_2 is the same as light quark v_2. Compared with the transport model calculations [10], the result favors charm quark has a large rescattering cross section, which indicates charm quark will have a finite v_2. Although it is hard to extract v_2 of charm hadrons from electron v_2, the data points suggest there can be a non-zero v_2 for charm quark. If the non-zero charm quark collectivity is confirmed, this strongly indicates light flavor thermalization at RHIC, as I argued in the beginning.

From the measurements of R_{AA} and v_2 for other particles, the radiative energy loss can contribute part of v_2, but not much. In the recent work discussed above [27], the authors attribute a large fraction of energy loss of heavy quarks to elastic collisions when $\gamma v \sim 1$. Then during the implementation of hydrodynamics, the flow of underlying medium will influence the heavy quark spectrum and heavy quark will pick up some flow. In this case, R_{AA} and v_2 are quite correlated, which means charm quark R_{AA} must be strong suppressed if large v_2 is observed. So the combination of measurements on both charm quark spectrum and v_2 is essential, especially at low p_T region. The above arguments are based on the hydro assumptions. Certainly, in the momentum region where hydrodynamic models fail, coalescence models or other hadronization schemes have been proved to be at work for light hardrons.

To test the medium response to heavy quarks, low p_T part is quite relevant. However, the charm decayed electron spectrum cannot disentangle different shapes in this p_T region due to smearing of the decay kinematics [31]. This effect can be also reflected on the R_{AA} of non-photonic electrons [16]. This indicates we need precise reconstructed open charm hadron spectrum measurement at low p_T , because the electron spectrum in this p_T region does not help to disentangle charm quark thermal property.

CONCLUSION AND OUTLOOK

The heavy flavor programme has started at RHIC extensively. Heavy flavor collectivity is expected to be an ideal probe to test the light flavor thermalization. Plenty of new and surprising results on open charm production at RHIC have been reported in recent conferences and publications. However, most of present measurements use the electrons from charm decays which brings large uncertainties. Due to the bottom contamination and the decay smearing, the electron approach can be only a placeholder. Precise measurements of spectrum and v_2 of reconstructed charm hadrons in a wide p_T range are needed. So current sub-detector upgrade proposals in pipe for PHENIX and STAR detectors are very important [32, 33, 34]. With the upgraded inner tracking detector, PHENIX and STAR can reconstruct secondary vertices of open charm decays with much lower background. We look forward to more exciting physical results from RHIC with upgraded detectors in the future.

ACKNOWLEDGMENTS

I would like to thank the conference organizers for inviting me to present the talk. I appreciate many constructive discussions with M. Djordjevic, L. Grandchamp, M. Gyulassy, H. Huang, J. Raufeisen, H.-G. Ritter, K. Schweda, P. Sorensen, A. Tai, R. Vogt, Q. Wang, X.-N. Wang, N. Xu, Z. Xu and H. Zhang. This work is partially supported by the National Natural Science Foundation of China under the Grant No. 10475071.

REFERENCES

1. I. Arsene et al. (BRAHMS Collaboration), *Nucl. Phys.* **A 757** (2005) 1; B.B. Back et al. (PHOBOS Collaboration), *Nucl. Phys.* **A 757** (2005) 28; J. Adams et al. (STAR Collaboration), *Nucl. Phys.* **A 757** (2005) 102; S.S. Adler et al. (PHENIX Collabration), *Nucl. Phys.* **A 757** (2005) 184.
2. X. Dong et al., *Phys. Lett.* **B 597** (2004) 328.
3. Z. Lin and M. Gyulassy, *Phys. Rev. Lett.* **77** (1996) 1222.
4. B. Müller, nucl-th/0404015.
5. Y. Dokshitzer and D. Kharzeev, *Phys. Lett.* **B 519** (2001) 199.
6. M. Djordjevic, M. Gyulassy and S. Wicks, *Phys. Rev. Lett.* **94** (2005) 112301.
7. B.W. Zhang, E.K. Wang, X-N. Wang, *Phys. Rev. Lett.* **93** (2004) 072301.
8. N. Armesto et al., *Phys. Rev.* **D 71** (2005) 054027.
9. V. Greco, C.M. Ko and R. Rapp, *Phys. Lett.* **B 595** (2004) 202.
10. B. Zhang, L.W. Chen, and C.M. Ko, nucl-th/0502056.
11. L. Ruan et al. (STAR Collaboration), *J. Phys.* **G 30** (2004) S1197; A. Tai et al. (STAR Collaboration), *J. Phys.* **G 30** (2004) S809; A.A.P. Suaide et al. (STAR Collaboration), *J. Phys.* **G 30** (2004) S1179.
12. J. Adams et al. (STAR Collaboration), *Phys. Rev. Lett.* **94** (2005) 062301.
13. Y. Kwon et al. (PHENIX Colaboration), *QM2005 proceedings.*
14. S.S. Adler et al. (PHENIX Collabration), nucl-ex/0508034.
15. S.P.K. Tavernier, *Rep. Prog. Phys.* **50** (1987) 1439, and references therein.
16. X. Dong, nucl-ex/0509011. Ph.D. Thesis, University of Science and Technology of China, 2005.
17. I.V. Rakobolskaya et al., *Nucl. Phys.* **B 112** (2003) 353c.
18. K. Adcox et al. (PHENIX Collaboration), *Phys. Rev. Lett.* **88** (2002) 192302.
19. R. Vogt, *Int. J. Mod. Phys.* **E 12** (2003) 211.
20. M. Cacciari, P. Nason and R. Vogt, hep-ph/0502203.
21. S.S. Adler et al. (PHENIX Collabration), *Phys. Rev. Lett.* **94** (2005) 082301.
22. H. Zhang et al. (STAR Collaboration), *QM2005 proceedings.*
23. J. Bielcik et al. (STAR Collaboration), *QM2005 proceedings.*
24. S.A. Butsyk et al. (PHENIX Collaboration), *QM2005 proceedings.*
25. M. Djordjevic et al., nucl-th/0507019.
26. N. Armesto, *QM2005 proceedings.*
27. G.D. Moore and D. Teaney, *Phys. Rev.* **C 71** (2005) 064904.
28. H. van Hees, V. Greco and R. Rapp, nucl-th/0508055.
29. S.S. Adler et al. (PHENIX Collaboration), *Phys. Rev.* **C 72** (2005) 024901.
30. F. Laue et al. (STAR Collaboration), *QM2005 proceedings.*
31. S. Batsouli et al., *Phys. Lett.* **B 557** (2003) 26.
32. STAR TOF Collaboration, STAR TOF Proposal, 2004.
33. STAR HFT Collaboration, STAR HFT Proposal, 2005.
34. A. Taketani et al. (PHENIX Vertex Group), *QM2005 proceedings.*

Soft and Hard Jets in QCD

I.M. Dremin

Lebedev Physical Institute, Moscow 119991, Russia
e-mail: dremin@lpi.ru

Abstract. Multiplicity of sets of soft jets with energies ranging in some interval is determined. The possible role of collective effects is discussed.

Keywords: gluon, jet, multiplicity
PACS: 12.39.Bx

The phenomenon of jet emission is well known in QCD and firmly established in experiment. The two-jet events in e^+e^--annihilation provide us with unique possibility to measure jets with fixed energy equal to that of colliding particles. In all other cases we have to deal with sets of jets with different energies. In three-jet events and in any high-p_t process the produced jets are somehow distributed in their energies. E.g., energies of gluon jets in three-jet process change from some lower limit determined by the requirement to separate this process to any value defined by experimental choice. Therefore, one has to deal with sets of jets with energies ranging in some interval.

Multiplicity of jets at a given energy has been calculated in QCD, and results agree quite well with experiment (see the reviews in [1, 2, 3]). This can be done for sets of jets as well [4]. The proper weights for jets with different energies in the set are provided by the QCD equations. The collective effects due to strings pulled apart during the separation of jets and color screening are somehow accounted in QCD. They lead, e.g., to different suppression of multiplicities between $q\bar{q}$ and qg pairs of jets. By measuring the multiplicity of sets of soft jets, this effect can become more pronounced.

To simplify the presentation, I consider gluodynamics (see also [5]). If the probability to create n particles[1] in a jet is denoted as P_n, the generating function G is defined as

$$G(z,y) = \sum_{n=0}^{\infty} P_n(y)(1+z)^n, \tag{1}$$

where z is an auxiliary variable, $y = \ln(p\Theta/Q_0) = \ln(2Q/Q_0)$ is the evolution parameter, defining the energy scale, p is the initial momentum, Θ is the angle of the divergence of the jet (jet opening angle), assumed here to be fixed, Q is the jet virtuality, $Q_0 = \text{const}$.

The gluodynamics equation for the generating function is written as

$$dG/dy = \int_0^1 dxK(x)\gamma_0^2[G(y+\ln x)G(y+\ln(1-x)) - G(y)], \tag{2}$$

[1] In what follows, we adopt the local parton-hadron duality hypothesis with no difference between the notions of particles and partons up to some irrelevant factor.

CP828, *Multiparticle Dynamics*
edited by V. Šimák, M. Šumbera, Š. Todorova, and B. Tomášik

where

$$\gamma_0^2 = \frac{6\alpha_S}{\pi}, \tag{3}$$

α_S is the coupling strength and the kernel $K(x)$ is

$$K(x) = \frac{1}{x} - (1-x)[2-x(1-x)]. \tag{4}$$

The equation for mean multiplicities follows from eq. (2):

$$\langle n(y)\rangle' = \int_0^1 dx\gamma_0^2 K(x)(\langle n(y+\ln x)\rangle + \langle n(y+\ln(1-x))\rangle - \langle n(y)\rangle). \tag{5}$$

As follows from eq. (5), the first two terms in the brackets correspond to mean multiplicities of two subjets, and their sum is larger than the third term denoting the mean multiplicity of the initial jet. Therefore, the integrand is positive. This does not contradict to the statement that for a given event the total multiplicity is a sum of multiplicities in the two subjets because the averages in eq. (5) are done at different energies.

The scaling property of the fixed coupling QCD [5] allows to look for the solution of the equation (5) with

$$\langle n\rangle \propto \exp(\gamma y) \quad (\gamma = const) \tag{6}$$

The anomalous dimension γ is determined from (5) as

$$\gamma = \gamma_0^2 \int_0^1 dx K(x)(x^\gamma + (1-x)^\gamma - 1) = \gamma_0^2 M_1(1,\gamma), \tag{7}$$

where

$$M_1(z,\gamma) = \int_0^z dx K(x)(x^\gamma + (1-x)^\gamma - 1). \tag{8}$$

For small enough γ and γ_0 one gets

$$\gamma \approx \gamma_0(1 - 0.458\gamma_0 + 0.213\gamma_0^2). \tag{9}$$

Now, according to the above discussion we define soft jets as those with sum of energies of belonging to them particles less than some x_0E. First, consider x_0=const and small. Then we should choose the upper limit of integration in eq. (5) equal to x_0.

One gets from (5) the mean multiplicity of a set of soft jets $\langle n_s\rangle$:

$$\frac{\langle n_s\rangle}{\langle n\rangle} = \frac{M_1(x_0,\gamma)}{M_1(1,\gamma)}. \tag{10}$$

For small x_0 it is

$$\frac{\langle n_s\rangle}{\langle n\rangle} \approx \frac{\gamma_0^2}{\gamma^2} x_0^\gamma N_1(x_0,\gamma), \tag{11}$$

$$N_1(x_0,\gamma) = 1 - \gamma^2 x_0^{1-\gamma} - \frac{2\gamma}{1+\gamma}x_0 + \frac{\gamma^2(3+\gamma)}{4}x_0^{2-\gamma} + \frac{3\gamma}{2+\gamma}x_0^2 - \frac{\gamma^2(2+\gamma)}{3}x_0^{3-\gamma}. \tag{12}$$

Thus we have found the energy dependence of mean multiplicity of particles in a set of subjets with low energies $E_s \leq x_0 E$. As expected for constant x_0, it is the same as the energy dependence of the total multiplicity with a different factor in front of it. Namely this dependence should be checked first in experimental data. Imposed on one another, these figures should coincide up to a normalization factor (10). This would confirm universality of gluons in jets.

Quite interesting is the non-trivial dependence of the normalization factor in eq. (10) on the parameter x_0, which does not coincide simply with x_0^γ. It reflects the structure of QCD kernel $K(x)$. The main dependence on the cut-off parameter x_0 is given for $x_0 \ll 1$ by the factor x_0^γ with the same power as in dependence of total multiplicity on energy. This corresponds to subjets with the largest energy of the set. However, with increase of x_0, this dependence is modified according to eqs (10)-(12). The negative corrections become more important in eq. (12). They are induced by subjets with energies lower than $x_0 E$. The decrease of the normalization factor corresponds to diminishing role of very low energy jets at higher initial energies. This should be also checked in experiment.

If plotted as a function of the maximum energy in a set of jets ε_m, the mean multiplicity is

$$\langle n_s \rangle \propto \varepsilon_m^\gamma \; [1 - \gamma^2 \left(\frac{\varepsilon_m}{E}\right)^{1-\gamma} - \frac{2\gamma}{1+\gamma}\left(\frac{\varepsilon_m}{E}\right) + \frac{\gamma^2(3+\gamma)}{4}\left(\frac{\varepsilon_m}{E}\right)^{2-\gamma} + \frac{3\gamma}{2+\gamma}\left(\frac{\varepsilon_m}{E}\right)^2 - \frac{\gamma^2(2+\gamma)}{3}\left(\frac{\varepsilon_m}{E}\right)^{3-\gamma}]. \tag{13}$$

It reminds eq. (6) with the correction factor in the brackets.

This is the consequence of the scaling property of the fixed coupling QCD which results in the jets selfsimilarity.

In principle, other definitions of soft jets are possible with $x_0 = x_0(E)$. Then one should solve the equation

$$\frac{d\langle n_s \rangle}{dE} = E^{\gamma-1}\gamma_0^2 M_1(x_0(E), \gamma), \tag{14}$$

which follows from eq. (5). For example, one can choose the jets with energies less than some fixed constant independent of the initial energy. This would imply ε_m =const or $x_0(E) \propto 1/E$, and the exact integration of eq. (14) is necessary. However, for qualitative estimates, eqs (11)-(13) can be used. They show that the ratio of average multiplicities (11) tends to a constant at high energies corresponding to the multiplicity at the upper limit. At lower energies, it slightly increases with energy due to increasing role of jets with energies closest to their upper limit.

It is well known that for running coupling the power dependence $s^{\gamma/2}$ is replaced by $\exp(c\sqrt{\ln s})$. The qualitative statement about the similar energy behaviour of mean multiplicities in soft and inclusive processes should be valid also.

The above results can be confronted to experimental data if soft jets are separated in 3-jet events or in high-p_t hadronic collisions. However, in our treatment we did not consider the common experimental cut-off which must be also taken into account. This is the low-energy cut-off imposed on a soft jet for it to be observable. It requires the soft jet not to be extremely soft. Otherwise the third jet is not separated and the whole event is considered as a 2-jet one. Thus the share of energy must be larger than some x_1, and the integration in eq. (8) should be from x_1 to x_0. For $x_1 \leq x_0 \ll 1$ one gets

$$\frac{\langle n_s \rangle}{\langle n \rangle} = \frac{\gamma_0^2}{\gamma^2}[\nu(x_0) - \nu(x_1)], \tag{15}$$

where the function $\nu(x)$ is easily guessed from eqs (11), (12). At $x_1 \ll x_0 \ll 1$ eq. (11) is restored.

The values for hard jets are obtained by subtracting these results from values for the total process.

I concentrated here on mean multiplicity but similar calculations have been done [4] for higher moments of multiplicity distributions, and they can be compared with experiment.

The main problem in comparison is related to the jet definition, i.e., to treatment of particles belonging to the regions between the jets. They are more influenced by the collective effects which appear even in e^+e^- annihilation due to strings pulled between jets and their mutual color screening. The experimental verification of the normalization factor in (10) (for small x_0 it is the behavior of N_1 (11)) becomes important because it would show that these collective effects are properly accounted in QCD in a wide energy interval.

Another interesting experimental aspect of collective effects is the energy distribution among jets in three-jet events. If boldly treated, the QCD diagram of the process implies that one of the quark jets remains untouched and must have the same energy as in two-jet events. Another (anti)quark jet emits a gluon, and they share this energy. Therefore, the energy distribution should have the two-bump structure with one bump at the primary energy and another one at smaller values (near half of it if the energy is shared equally between quarks and gluons). This statement is correct if collective effects due to string tension play minor role. In principle, string tension can lead to some collective energy flow from one jet to another one and change the shape of the energy distribution among three produced jets. Thus one can determine the role of collective effects by measuring this distribution.

In conclusion, the experimentally measured values of mean multiplicities of particles belonging to a set of soft jets can be compared with the obtained above theoretical predictions at different values of this share of energy. For a constant share, this dependence is the same as for the average total multiplicity but with non-trivial x_0-dependence of the factor in front of it. Some predictions are obtained for energy dependent cut-offs. The collective effects due to string tension are discussed. These results can be confronted to experiment.

ACKNOWLEDGMENTS

This work has been supported in part by the RFBR grants 03-02-16134, 04-02-16445, NSH-1936.2003.2.

REFERENCES

1. I.M. Dremin, UFN 164 (1994) 785; *Physics-Uspekhi* **37** (1994) 715.
2. I.M. Dremin and J.W. Gary, *Phys. Rep.* **349** (2001) 301.
3. V.A.Khoze and W. Ochs, *Int. J. Mod. Phys.* **A 12** (1997) 2949.
4. I.M. Dremin, *JETP Lett.* **81** (2005) 391.
5. I.M. Dremin and R.C. Hwa, *Phys. Lett.* **B 324** (1994) 477; *Phys. Rev.* **D 49** (1994) 5805.

Multihadron production features in different reactions

Edward K.G. Sarkisyan[*,†] and Alexander S. Sakharov[**,‡]

[*]*EP Division, Department of Physics, CERN, CH-1211 Geneva 23, Switzerland*
[†]*Department of Physics, the University of Manchester, Manchester M13 9PL, UK*
[**]*TH Division, Department of Physics, CERN, CH-1211 Geneva 23, Switzerland*
[‡]*Swiss Institute of Technology, ETH-Zürich, 8093 Zürich, Switzerland*

Abstract. We consider multihadron production processes in different types of collisions in the framework of the picture based on dissipating energy of participants and their types. In particular, the similarities of such bulk observables like the charged particle mean multiplicity and the pseudorapidity density at midrapidity measured in nucleus-nucleus, (anti)proton-proton and electron-positron interactions are analysed. Within the description proposed a good agreement with the measurements in a wide range of nuclear collision energies from AGS to RHIC is obtained. The predictions up to the LHC energies are made and compared to experimental extrapolations.

Keywords: multihadron production, nuclear collisions, participants, constituent quark
PACS: PACS numbers: 25.75.-q, 24.85.+p, 13.85.-t, 24.10.Nz, 13.66.Bc

1. High densities and temperatures of nuclear matter reached at RHIC provide us with an exceptional opportunity to investigate the matter at extreme conditions. Bulk observables such as multiplicity and particle densities (spectra) being sensitive to the dynamics of strong interactions, are of fundamental interest. Recent measurements at RHIC revealed striking evidences in the hadron production process including similarity in such basic observables like the mean multiplicity and the midrapidity density measured in complex ultra-relativistic nucleus-nucleus (AA) collisions *vs.* those obtained in relatively "elementary" e^+e^- interactions at the same centre-of-mass (c.m.) energy when number of participants ("wounded" nucleons [1] in AA collisions) are taken into account [2, 3]. The observation is shown to be independent of the c.m. energy per nucleon $\sqrt{s_{NN}} = 19.6$ GeV to 200 GeV. Assuming similar mechanisms of hadron production in both types of interactions which then depends only on the amount of energy transformed into particles produced, one would expect the same value of the observables to be obtained in hadron-hadron collisions at close c.m. energies. However, this is not the case: comparing measurements in hadronic data [4, 5] to the findings at RHIC, one obtains [2, 6, 7] quite lower values in hadron-hadron collisions. In the meantime, the RHIC dAu data at $\sqrt{s_{NN}} = 200$ GeV unambiguously point to the values of the mean multiplicity from $\bar{p}p$ data [2]. Moreover, recent CuCu RHIC data show no changes in the values of the bulk variables compared to those from AuAu collisions when properly normalised to the number of participants [8, 9].

The observations made earlier [2] and the recent ones [9] can be understood in the framework of a description proposed recently by us [10] and considered here. This description is based on a picture when the whole process of a collision is interpreted as the expansion and break-up into particles of an initial state, in which the whole available

CP828, *Multiparticle Dynamics*
edited by V. Šimák, M. Šumbera, Š. Todorova, and B. Tomášik

energy is assumed to be concentrated in a small Lorentz-contracted volume. There are no any restrictions due to the conservation of quantum numbers besides energy and momentum constraints allowing therefore to link the amount of energy deposited in the collision zone and features of bulk variables in different reactions. This description resembles the Landau hydrodynamical approach to multiparticle production [11] which has been found to give good description of the mean multiplicity AA, pp, e^+e^-, νp data [12, 13] as well as of pseudorapidity distributions at RHIC [6].

As soon as a collision of two Lorentz-contracted particles leads to the full thermalization of the system before extension, one can assume that the production of secondaries is defined by the fraction of participants energy deposited in the volume of the system at the collision moment. This implies that there is a difference between results of collisions of structureless particles like electron and composite particles like proton, the latter considered to be built of constituents. Indeed, in composite particle collisions not all the constituents deposit their energy when they form the Lorentz-contracted volume of the thermalized initial state. As a result, the leading particles [14], formed out of those constituents which are not trapped in the interaction volume, carry away a part of energy. Meantime, colliding structureless particles are ultimately stopped as a whole in the initial state of the thermalized collision zone depositing their total energy in the Lorentz-contracted volume and this energy is wholly available for production of secondaries.

We consider a single nucleon as a superposition of three constituent quarks due to the additive quark picture [15]. In this picture, most often only one quark from each nucleon contributes to the interaction with other quarks being spectators. Thus, the initial thermalized state is pumped in only by the energy of the interacting single quark pair and, so, only 1/3 of the entire nucleon energy is available for production of secondaries. Therefore, one expects that the resulting bulk variables like the multiplicity and rapidity distributions should show identical features in $\bar{p}p$ collisions at the c.m. energy $\sqrt{s_{pp}}$ and in e^+e^- interactions at the c.m. energy $\sqrt{s_{ee}} \simeq \sqrt{s_{pp}}/3$. Note that for the mean multiplicity, a similar behaviour was found in the beginning of LEP activity [16].

In AA collisions, more than one quark per nucleon interacts due to the large size of nucleus and the long path of interactions inside the nucleus. In central AA collisions, a contribution of constituent quarks rather than participating nucleons seem to determine the properties of produced particle distributions [17]. In headon collisions, the density of matter is almost saturated, so that all three constituent quarks from each nucleon may participate nearly simultaneously in collision depositing their energy coherently into the thermalized zone. Therefore, in the headon AA interactions at $\sqrt{s_{NN}}$ the bulk variables are expected to have the values similar to those from pp collisions at $\sqrt{s_{pp}} \simeq 3\sqrt{s_{NN}}$. This makes the most central collisions of nuclei akin to e^+e^- collisions at $\sqrt{s_{ee}} \simeq \sqrt{s_{NN}}$ in sense of the resulting bulk variables.

2. According to our consideration, in Fig. 1, we compare the c.m. energy dependence of the mean multiplicity in AA and e^+e^- interactions to that in pp/$\bar{p}$p collisions at $\sqrt{s_{NN}} = \sqrt{s_{ee}} = \sqrt{s_{pp}}/3$ from a few GeV to 200 GeV. For $\sqrt{s_{ee}} > M_{Z^0}$, we give the multiplicities averaged [31] from the recent LEP data at $\sqrt{s_{ee}} = 130$ GeV and 200 GeV: $23.35 \pm 0.20 \pm 0.10$ and $27.62 \pm 0.11 \pm 0.16$. Figure shows also the mean multiplicity fit to pp/$\bar{p}$p data [20] and the 3NLO pQCD [33] ALEPH fit to e^+e^- data [27].

From Fig. 1 one sees that the pp/$\bar{p}$p data are very close to the e^+e^- data at $\sqrt{s_{ee}} =$

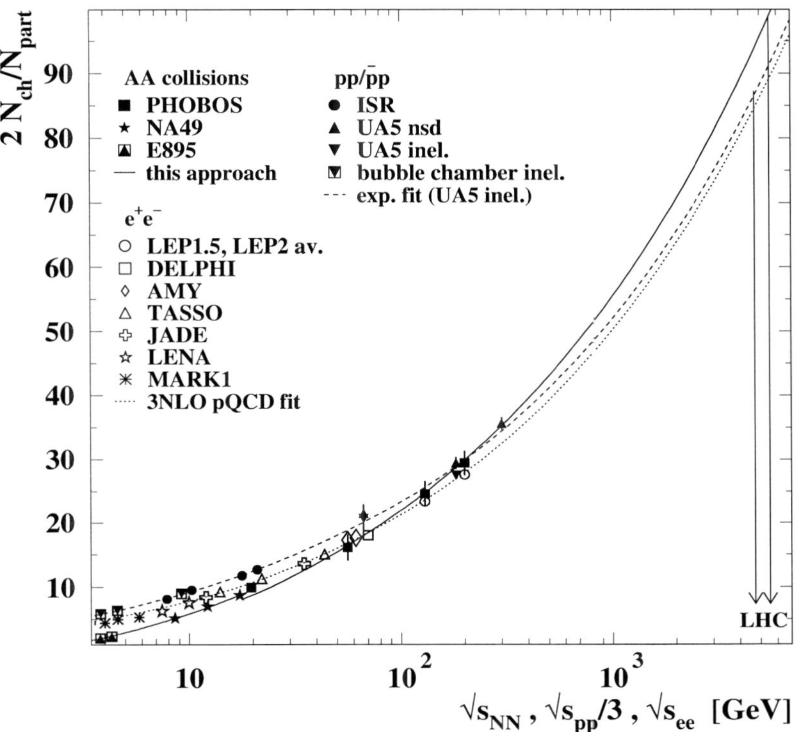

FIGURE 1. The charged particle mean multiplicity N_{ch} per participant pair ($N_{part}/2$) as a function of the c.m. energy. The solid and combined symbols show the multiplicity values from: most central heavy-ion (AA) collisions *vs.* c.m. energy per nucleon, $\sqrt{s_{NN}}$, measured by PHOBOS [2] (■), NA49 [18] (★), and E895 [19] (◭) (see also [2]); p̄p collisions, by UA5 (▲ for non-single diffractive, ▼ for inelastic events) at $\sqrt{s_{pp}} = 546$ GeV [20] and $\sqrt{s_{pp}} = 200$ and 900 GeV [21]; pp collisions (at lower $\sqrt{s_{pp}}$) from CERN-ISR (•) [22] and bubble chamber experiments [23, 24] (⧩) (the latter compiled and analysed in [25]). The inelastic UA5 data at $\sqrt{s_{pp}} = 200$ GeV is due to the extrapolation in [2]. The open symbols show the e^+e^- measurements: the high-energy LEP mean multiplicities (○) averaged here from the data at LEP1.5 $\sqrt{s_{ee}} = 130$ GeV [26, 27] and LEP2 $\sqrt{s_{ee}} = 200$ GeV [27, 28], and the lower-energy data by DELPHI [29] (□), TASSO (△), AMY (◇), JADE (+), LENA (⋆), and MARK1 (✳) experiments. (See refs. in [30, 31, 32] for e^+e^- and pp/p̄p data). The solid line shows the calculations from Eq. (2) based on our approach and using the corresponding fits (see text). The dashed and dotted lines show the fit [20] to the pp/p̄p data and the 3NLO perturbative QCD [33] ALEPH fit [27] to e^+e^- data. The arrows show the LHC expectations.

$\sqrt{s_{pp}}/3$. This nearness decreases the already small deficit in the e^+e^- data as the energy increases. The deviation can be attributed to the inelasticity factor, or leading particle effect [14] in pp/p̄p collisions, which is known to decrease with the c.m. energy. Then, at lower $\sqrt{s_{pp}}$, some fraction of the energy of spectators contributes more into the formation of the initial state as the spectators pass by. This leads to the excess of the mean multiplicity in pp/p̄p data compared to the e^+e^- data as it is seen in Fig. 1. Comparing further the average multiplicities from pp/p̄p collisions to those from AA ones, one finds that the data points are amazingly close to each other when the AA data are confronted the hadronic data at $\sqrt{s_{pp}} = 3\sqrt{s_{NN}}$. The inclusion of the tripling energy factor indeed allows to describe such a fundamental variable as the mean multiplicity *simultaneously* in e^+e^-, pp/$\bar{p}$p and central AA collisions for all energies. This shows that the multiparticle production process in headon AA collisions is derived by the energy deposited in the Lorentz-contracted volume by a single pair of effectively structureless nucleons similar to that in e^+e^- annihilation and of quark-pair interactions in pp/$\bar{p}$p collisions. Note that an examination of Fig. 1 reveals that not a factor 1/2 is needed to rescale $\sqrt{s_{pp}}$ to match the AA or e^+e^- data as earlier was assumed for the mean multiplicity while recognised to unreasonably shift the e^+e^- data on the pseudorapidity density at midrapidity when compared to the AA measurements [2]. This discrepancy finds its explanation in our consideration, within which the data on *both* the mean multiplicity and the midrapidity density (*vide infra*) are self-consistently matched for different reactions. Let us gain recall a factor 1/3 obtained earlier in [16] for $\sqrt{s_{pp}}$ for the pp mean multiplicity data relative to those from e^+e^- data, similar to our finding.

Fig. 1 shows that the mean multiplicities in different reactions are close starting from the SPS $\sqrt{s_{NN}}$, and become particularly close at $\sqrt{s_{NN}} \gtrsim 50$ GeV. However, at lower energies, the AA data are slightly below the e^+e^- and hadronic data and the nuclear data increase faster with energy than the pp and e^+e^- data do. On the other hand, as the c.m. energy increases above a few tens GeV, the AA data start to overshoot the e^+e^- data and reach the mean multiplicity values from $\bar{p}$p interactions. From this one concludes on two different energy regions of the multiparticle production in AA reactions. The observations made can be understood in terms of the overlap zone and energy deposition by participants [10]. Due to this, one would expect the differences to be more pronounced in midrapidity densities as discussed below.

3. In Fig. 2, we compare the pseudorapidity densities per participant pair at midrapidity as a function of $\sqrt{s_{NN}}$ from headon AA collisions at RHIC, CERN SPS and AGS to those of pp/$\bar{p}$p data from CERN and Fermilab plotted *vs.* $\sqrt{s_{pp}}/3$. Again one can see that up to the existing $\sqrt{s_{NN}}$ the data from hadronic and nuclear experiments are close to each other being consistent with our interpretation. The measurements from the two types of collisions coincide at $8 < \sqrt{s_{NN}} < 20$ GeV and are of the magnitude of the spread of AA data points at 200 GeV. However, above and below the 8-20 GeV region, there are visible differences in the midrapidity η-density values from AA *vs.* pp data. These indicate that, in contrast to the mean multiplicity which is a more global observable, the midrapidity density depends on some additional factor. As the densities are measured in the very central η-region, where the participants longitudinal velocities are zeroed, it is natural to assume that this factor is related to the size of the Lorentz-contracted volume of the initial thermalized system determined by participants.

To take into account the corresponding correction, let us consider our picture in the framework of the Landau model which is close to our description. Then, one finds for the ratio of the normalised charged particle rapidity density $\rho(y) = (2/N_{part})dN_{ch}/dy$ at the midrapidity value $y = 0$ in AA reaction, ρ_{NN}, to the density ρ_{pp} in pp/$\bar{p}$p interaction,

$$\rho_{NN}(0)/\rho_{pp}(0) = 2N_{ch}(L_{pp}/L_{NN})^{1/2} / \left(N_{part} N_{ch}^{pp}\right). \quad (1)$$

Here, N_{ch} (N_{ch}^{pp}) is the multiplicity in AA (pp/$\bar{p}$p) collision, $L = \ln[\sqrt{s}/(2m)]$, and m is the participant mass, *e.g.* the proton mass m_p in AA reaction. According to our interpretation, we compare in the ratio (1) $\rho_{NN}(0)$ to $\rho_{pp}(0)$ at $\sqrt{s_{NN}} = \sqrt{s_{pp}}/3$ and

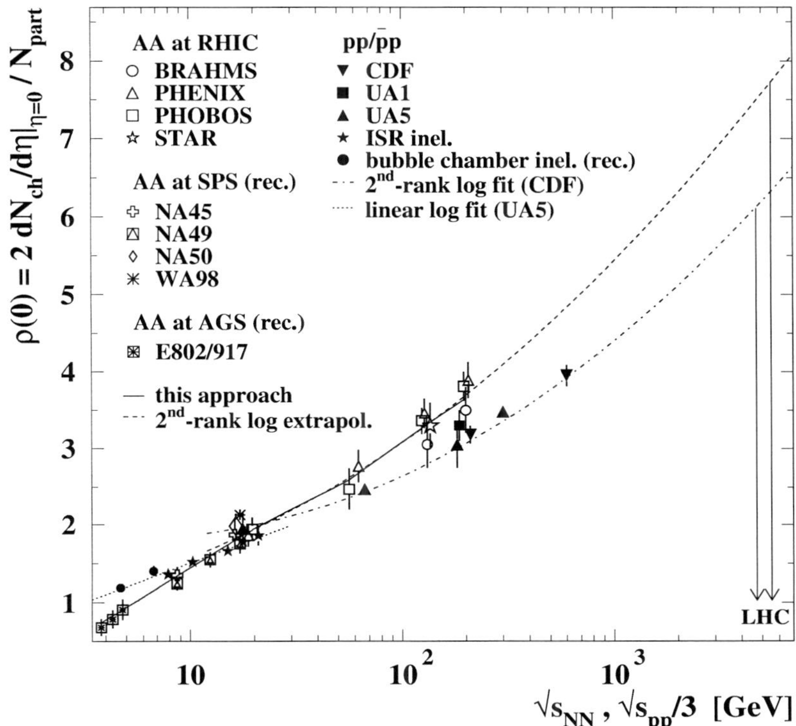

FIGURE 2. Pseudorapidity density $\rho(0)$ of charged particles per participant pair ($N_{\rm part}/2$) at midrapidity as a function of the c.m. energy of collision. The open and combined symbols show the pseudorapidity density values *vs.* c.m. energy per nucleon, $\sqrt{s_{\rm NN}}$, measured in the headon AA collisions by BRAHMS [6] (○), PHENIX [7] (△), PHOBOS [2] (□), and STAR [34] (☆), and the density values recalculated [7] from the measurements taken by CERES/NA45 [35] (+), NA49 [36] (⊠), NA50 [37] (◇) WA98 [38] (✳), E802, and E917 [39] (⊞). The nuclear data at $\sqrt{s_{\rm NN}}$ around 20 GeV and the RHIC data at $\sqrt{s_{\rm NN}} = 130$ GeV and 200 GeV are given spread horizontally for clarity. The solid symbols show the pseudorapidity density values *vs.* c.m. energy $\sqrt{s_{\rm pp}}/3$ as measured in non-single diffractive p̄p collisions by UA1 [40] (■), UA5 [4, 20] (▲), CDF [5] (▼), and from inelastic pp data from ISR [22] (★), and bubble chamber [24, 41] (●) experiments (the latter as recalculated in [4]). The solid line connects the predictions from Eq. (2). The dashed line gives the fit to the calculations using the 2nd order log-polynomial fit function analogous to that used [5] in p̄p data. The fit function from [5] is shown by the dashed-dotted line. The dotted line shows the linear log approximation of UA5 to inelastic events [4]. The arrows show the LHC expectations. Note that e^+e^- data at $\sqrt{s_{\rm ee}} = 14$ GeV to 200 GeV (not shown) follows the heavy-ion data [2].

consider a constituent quark of mass $\frac{1}{3}m_{\rm p}$ as a participant in pp/p̄p collisions and a proton as an effectively structureless participant in headon AA collisions. Then, Eq. (1) reads:

$$\rho_{\rm NN}(0) = 2N_{\rm ch}\,\rho_{\rm pp}(0)\,\sqrt{1-4\ln 3/\ln(4m_{\rm p}^2/s_{\rm NN})}\,/\,\big(N_{\rm part}N_{\rm ch}^{\rm pp}\big)\,. \qquad (2)$$

Using the fact that the transformation factor from y to η does not influence the above ratio and substituting the multiplicity values from Fig. 1 and of $\rho_{\rm pp}(0)$ from Fig. 2 into Eq. (2), one obtains the values of $\rho_{\rm NN}(0)$, displayed in Fig. 2 by solid line. One can see that the correction made provides good agreement between the calculated $\rho_{\rm NN}(0)$ values and the data. Eq. (2) shows the importance of the correction for the participant type to be introduced as argued above. One can see that our calculations account also for different types of rise of AA data below and above SPS region. Note that the same two regions recently have been indicated by PHENIX [7] from the ratio of the midrapidity *transverse energy* density to the pseudorapidity density. From these findings, one can expect the midrapidity transverse energy densities in pp/p̄p and headon AA collisions to be similar due to the description proposed here. Also, the SPS transition region properties discussed by NA49 [18], can be treated without any additional assumptions.

4. To estimate $\rho_{NN}(0)$ for $\sqrt{s_{NN}} > 200$ GeV, we extrapolated the values of Eq. (2) utilizing the function found [5] to fit well the p̄p data. The predictions for $\rho_{NN}(0)$ and the fit for p̄p data are shown in Fig. 2 by dashed and dashed-dotted lines, respectively.

The obtained $\rho_{NN}(0)$ show faster rise with $\sqrt{s_{NN}}$ than $\rho_{pp}(0)$. Our calculations, sharing the behaviour at SPS–RHIC energies with that up to the LHC ones, give $\rho_{NN}(0) \approx 7.7$ for LHC. From the CDF fit [5] and assuming it covers LHC energies, one finds $\rho_{pp}(0) \approx 6.1$. Our $\rho_{NN}(0)$ value for LHC is consistent with that of ≈ 6.1 given in the PHENIX extrapolation [7] within 1-2 particle error acceptable in the calculations we made. Our result is in a good agreement with the best ATLAS Monte Carlo tune [42]. Noticing that $\sqrt{s_{NN}}$ is near to $\sqrt{s_{pp}}/3$ at LHC, the close values of $\rho_{NN}(0)$ and $\rho_{pp}(0)$, predicted for LHC by us and estimated independently in [7, 42], demonstrates experimentally grounded description and predictive ability of our interpretation.

Solving Eq. (2) for $N_{ch}/(0.5N_{part})$ we predict the AA mean multiplicity energy dependence at $\sqrt{s_{NN}} > 200$ GeV. In this calculations, we use the fits of $\rho_{pp}(0)$ [5] and N_{ch}^{pp} [4] and our approximation for $\rho_{NN}(0)$, all shown in Figs. 1 and 2. From the resulted curve for $N_{ch}/(0.5N_{part})$ given in Fig. 1, one finds that the value obtained for LHC is just about 10% above the $N_{ch}^{pp}(\sqrt{s_{pp}})$ fit [20] prediction for LHC and about 3.3 times larger the AA RHIC data at $\sqrt{s_{NN}} = 200$ GeV. Again, this number is comparable with the estimate made by [7] and points out to no evidence for change to another regime as the $\sqrt{s_{NN}}$ increases by about two magnitudes from the top SPS energy. Nevertheless, one can see that the data obtained at the highest RHIC energy give a hint to some border-like behaviour of the mean multiplicity where the pp/p̄p data saturate the nuclear data, and another transition energy region is possible to be found (as at low energies). This makes AA experiments at $\sqrt{s_{NN}} > 200$ GeV of particular interest.

5. At the end, let us dwell on the following.

From our description, the mean multiplicity in *nucleon*-nucleus collisions is predicted to be of the same values as that in pp/p̄p data, and, moreover, almost no centrality dependence is expected for such type of interactions [10]. These predictions are well confirmed by various data from hadron-nucleus collisions at $\sqrt{s_{NN}} \approx 10$–20 GeV to recent RHIC dAu data at 200 GeV [2]. The same seems to be correct also for the pseudorapidity density at midrapidity, which is already supported to be a trend [2]. These findings remind about similar conclusions made about two decades ago [12].

The recent observation [8, 9] made at RHIC for multihadron data from CuCu collisions not to change compared to same c.m. energy AuAu data when scaled for the same participant numbers is also understood due to our description as already mentioned. Indeed, for the same number of participants, no difference in the bulk variables is expected as one moves from one type of (identical) colliding nuclei to another one at the same c.m. energy as soon as the same energy is deposited into the thermalization zone. Note that the proper definition of participants and, thus, of the energy available for particle production, as we discuss here, allows scaling within the constituent quark picture to be applied [8, 17, 43] to model the multihadron data at RHIC for different observables.

One of us (EKGS) is grateful to Organizers for invitation and partial financial support.

REFERENCES

1. A. Białas *et al.*, Nucl. Phys. B **111** (1976) 461; A. Białas, W. Czyż, Acta Phys. Pol. B **36** (2005) 905.
2. B.B. Back *et al.* (PHOBOS), Nucl. Phys. A **757** (2005) 28, and refs. therein.
3. P. Steinberg, J. Phys. G **30** (2004) S683; W. Busza, Acta Phys. Pol. B **35** (2004) 2873.
4. G.J. Alner *et al.* (UA5), Z. Phys. C **33** (1986) 1.
5. F. Abe *et al.* (CDF), Phys. Rev. D **41** (1990) 2330.
6. I. Arsene *et al.* (BRAHMS), Nucl. Phys. A **757** (2005) 1, and refs. therein.
7. K. Adcox *et al.* (PHENIX), Nucl. Phys. A **757** (2005) 184, and refs. therein.
8. R. Noucier, talk at Int. Symp. Multiparticle Dynamics 2005: these Proceedings.
9. G. Roland (for the PHOBOS Collab.), M. Konno (for the PHENIX Collab.): Quark Matter 2005.
10. E.K.G. Sarkisyan and A.S. Sakharov, hep-ph/0410324.
11. L.D. Landau, Izv. Akad. Nauk: Ser. Fiz. **17** (1953) 51.
12. E.L. Feinberg, Proc. Int. Conf. Elementary Particle Physics (Smolenice, 1985), p. 81; Relativistic Heavy Ion Physics (World Scientific, 1991), p. 341.
13. P.A. Carruthers, LA-UR-81-2221 ; P. Steinberg, nucl-ex/0405022.
14. M. Basile *et al.*, Nuovo Cim. A **66** (1981) 129, **73** (1983) 329.
15. V.V. Anisovich *et al.*, Quark Model and High Energy Collisions (World Scientific, 2005).
16. P.V. Chliapnikov and V.A. Uvarov, Phys. Lett. B **251** (1990) 192; for review, see W. Kittel and E.A. De Wolf, Soft Multihadron Dynamics (World Scientific, 2005).
17. S. Eremin and S. Voloshin, Phys. Rev. C **67** (2003) 064905.
18. S.V. Afanasiev *et al.* (NA49), Phys. Rev. C **66** (2002) 054902.
19. J.L. Klay (for the E895 Collab.), PhD Thesis (U.C. Davis, 2001), see [2].
20. G.J. Alner *et al.* (UA5), Phys. Rep. **154** (1987) 247.
21. R.E. Ansorge *et al.* (UA5), Z. Phys. C **43** (1989) 357.
22. W. Thomé *et al.*, Nucl. Phys. B **129** (1977) 365.
23. V.V.Ammosov *et al.*,Phys. Lett. B **42** (1972) 519; C.Bromberg *et al.*, Phys. Rev. Lett. **31** (1974) 254.
24. W.M. Morse *et al.*, Phys. Rev. D **15** (1977) 66.
25. E. De Wolf, J.J. Dumont, F. Verbeure, Nucl. Phys. B **87** (1975) 325.
26. G. Alexander *et al.* (OPAL), Z. Phys. C **72** (1996) 191.
27. A. Heister *et al.* (ALEPH), Eur. Phys. J. C **35** (2004) 457; P. Abreau *et al.* (DELPHI), Phys. Lett. B **372** (1996) 172; P. Achard *et al.* (L3), Phys. Rep. **399** (2004) 71.
28. G. Abbiendi *et al.* (OPAL), Eur. Phys. J. C **37** (2004) 25.
29. P. Abreau *et al.* (DELPHI), Z. Phys. C **70** (1996) 179.
30. P.D. Acton *et al.* (OPAL), Z. Phys. C **53** (1992) 539.
31. Particle Data Group, S. Eidelman *et al.*, Phys. Lett. B **592** (2004) 1, and refs. therein.
32. see `http://www.cern.ch/biebel/www/RPP04/`.
33. I.M. Dremin, J.W. Gary, Phys. Rep. **349** (2001) 301, and refs. therein.
34. C. Adler *et al.* (STAR), Nucl. Phys. A **757** (2005) 102, and refs. therein.
35. F. Ceretto (for the CERES/NA45 Collab., G. Agakichiev *et al.*), Nucl. Phys. A **638** (1998) 467c.
36. F. Siklér (for the NA49 Collab., J. Bächler *et al.*), Nucl. Phys. A **661** (1999) 45c.
37. M.C. Abreau *et al.* (NA50), Phys. Lett. B **530** (2002) 43.
38. M.M. Aggarwal *et al.* (WA98), Eur. Phys. J. C **18** (2001) 651.
39. L.Ahle *et al.*(E802), Phys.Rev. **C59** (1999)2173; B.Back *et al.*(E917), Phys.Rev.Lett. **86** (2001) 1970.
40. G. Arnison *et al.* (UA1), Phys. Lett. B **123** (1984) 108.
41. J. Whitmore *et al.*, Phys. Rep. **10C** (1974) 273.
42. A.M. Moraes, talk given at ATLAS Physics Week (CERN, Nov. 2004).
43. P.K. Netrakanti, B. Mohanty, Phys. Rev. C **70** (2004) 027901; B. De, S. Bhattacharyya, Phys. Rev. C **71** (2005) 024903.

Multiplicity Difference between Heavy and Light Quark Jets Revisited

Fabrizio Fabbri

INFN e Dipartimento di Fisica dell'Università, viale Berti Pichat 6/2, 40127 Bologna, Italy

Abstract. A peculiar prediction of perturbative QCD, obtained within the Local Parton Hadron Duality (LPHD) framework, is that the multiplicity difference $\delta_{Q\ell}$ between heavy and light quark jets produced in e^+e^- annihilation is energy independent. In the Modified Leading Logarithmic Approximation (MLLA) the corresponding constant is derived in terms of a few experimentally measurable quantities. While the energy independence of $\delta_{Q\ell}$ has been succesfully verified experimentally for b-quarks up to the highest LEP2 energy, its numerical prediction ($\delta_{b\ell}^{MLLA} = 5.5 \pm 0.8$) overestimates the experimental results. The work presented in this talk, done in collaboration with Yuri L. Dokshitzer, Valery A. Khoze and Wolfgang Ochs, shows that in the light of new experimental results and the improvement in the understanding of the experimental data, this prediction needs indeed a revision. We now find $\delta_{b\ell} = 4.4 \pm 0.4$, in better agreement with experiment, and we shaw that the remaining difference can be attributed largely to next-to-MLLA contributions, an important subset of which are identified and evaluated. The situation with charmed quarks is also reviewed.

Keywords: Multiplicity, Heavy Quarks, QCD, MLLA, e+e- Collisions
PACS: 13.66.Bc, 14.65.Fy, 14.65.Dw, 12.38.-t

INTRODUCTION

Multiple hadron production in hard processes is derived from the QCD parton cascade processes which are dominated by gluon bremsstrahlung. An essential difference in the structure of the energetic heavy and light quark jets ($\ell \equiv q = u,d,s$) results from the dynamical restriction on the phase space of primary gluon radiation in the heavy quark case: the gluon radiation off an energetic heavy quark Q with mass M and energy $E_Q \gg M$ is suppressed inside the forward angular cone with an opening angle $\Theta_0 = M/E_Q$, the so-called dead cone phenomenon [1, 2].

According to the concept of "Local Parton Hadron Duality" (LPHD), a direct consequence of this phenomenon should be observable in e^+e^- annihilation. In fact, the difference between the multiplicities of light hadrons 'accompanying' the heavy quark production in heavy-quark initiated events (excluding, then, decay products of Q-flavoured hadrons), $N_{Q\bar{Q}}(W)$, and the multiplicity of jets originated by light quarks at the same c.m.s. energy, $N_{q\bar{q}}(W)$ is predicted by QCD to be energy independent [3, 4]

$$N_{q\bar{q}}(W) - N_{Q\bar{Q}}(W) = const(W), \tag{1}$$

the corresponding constant being different for c- and b-quarks. At $W = 2E_Q \gg M \gg \Lambda_{QCD}$ the companion multiplicity $N_{Q\bar{Q}}(W)$ can be related to the particle yield in the

CP828, *Multiparticle Dynamics*
edited by V. Šimák, M. Šumbera, Š. Todorova, and B. Tomášik

light-quark events $e^+e^- \to q\bar{q}$ $(q = u, d, s)$ as [3, 4]

$$N_{q\bar{q}}(W) \; - N_{Q\bar{Q}}(W) \; = \; N_{q\bar{q}}(\sqrt{e}M) \cdot [1 + O(\alpha_s(M))] \, , \tag{2}$$

where we expressed their difference in terms of the light-quark event multiplicity at reduced (W independent) c.m.s. energy $W_0 = \sqrt{e}M$, $e = \exp(1)$.

In heavy quark e^+e^- annihilation events at c.m.s. energy W, the total mean charged multiplicity, $N^{ch}_{e^+e^- \to Q\bar{Q}}$, and the mean light charged hadron multiplicity accompanying the heavy quark production, $N^{ch}_{Q\bar{Q}}(W)$, are related according to the equation

$$N^{ch}_{e^+e^- \to Q\bar{Q}}(W) \; = \; N^{ch}_{Q\bar{Q}}(W) + n^{dc}_{Q} \, , \tag{3}$$

where n^{dc}_{Q} stands for the constant mean charged decay multiplicity of the two leading heavy hadrons. Therefore, within the MLLA framework the difference $\delta_{Q\ell}$ between the total multiplicity in heavy- and light-quark initiated events at the same c.m.s. energy is predicted to be

$$\delta^{MLLA}_{Q\ell} \; = N^{ch}_{e^+e^- \to Q\bar{Q}} - N^{ch}_{q\bar{q}} \; = \; n^{dc}_{Q} - N_{q\bar{q}}(\sqrt{e}M) \, , \tag{4}$$

which depends only on the heavy quark mass M, and remains W-independent. The main goal here is to explain why the previous numerical value of the MLLA prediction [3], in particular that for the b-quark $\delta^{MLLA}_{b\ell} = 5.5 \pm 0.8$, needs a revision. The situation with charmed quarks is considered as well. Finally, the size of an important subset of next-to-MLLA contributions which was evaluated in this work, is also discussed.

The limited amount of space available does not allow me to discuss here some important details, in particular those related to the identification and evaluation of dominant next-to-MLLA contributions. A complete and detailed presentation can be found in [5].

THEORETICAL PREDICTIONS CONFRONTED WITH EXPERIMENT

An updated compilation of $\delta_{b\ell}$ measurements in e^+e^- annihilation is shown in Fig. 1.

The figure is taken from [6] with the addition of the result from the VENUS experiment at $\sqrt{s} = 58$ GeV [7] as well as the preliminary result from DELPHI at $\sqrt{s} = 206$ GeV [8]. The dash-dotted line shown in Fig. 1 corresponds to the weighted average among all published results, $\langle \delta_{b\ell} \rangle = 3.12 \pm 0.14$, assuming uncorrelated measurements. Within the experimental uncertainties most data points are consistent with the original MLLA expectation [3], shown in Fig. 1 as a shaded area. However, the precise results from the OPAL, SLD, DELPHI and VENUS experiments, which dominate the weighted average value $\langle \delta_{b\ell} \rangle$, are definitely lower.

It should be emphasized that the numerical evaluation of the MLLA expression (4) relative to the b-quark

$$\delta^{MLLA}_{b\ell} = n^{dc}_{b} - N^{ch}_{q\bar{q}}(\sqrt{e}M_b) \tag{5}$$

relies strongly on experimentally measured quantities. In the original analysis [3] the light-quark mean charged multiplicity at $W = \sqrt{e}M_b = 8$ GeV, $N_{q\bar{q}}(8 \text{ GeV}) = 5.5 \pm 0.7$,

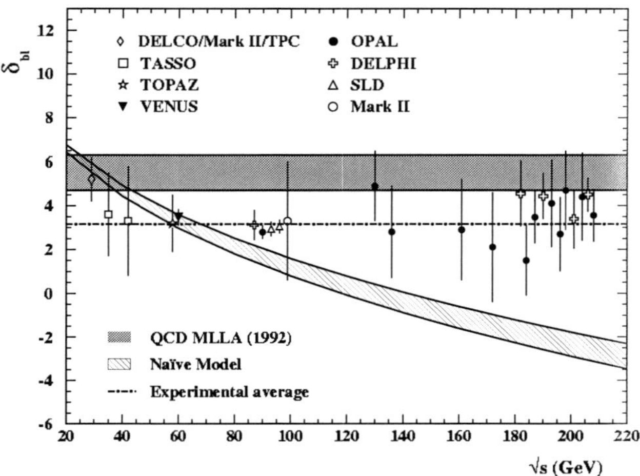

FIGURE 1. Compilation of the experimental measurements of $\delta_{b\ell}$. The 1992 MLLA expectation $\delta_{b\ell}^{MLLA} = 5.5 \pm 0.8$ (shaded area), and the weighted average $\langle \delta_{b\ell} \rangle = 3.12 \pm 0.14$ among all published results (dash-dotted line), are also shown.

was estimated by interpolation of experimental data at nearby energies, while $n_b^{dc} = 11.0 \pm 0.2$ was taken. This brought to the prediction $\delta_{b\ell}^{MLLA} = 5.5 \pm 0.8$.

Some new relevant experimental results became available since the analysis presented in [3], and we have investigated their impact on this prediction. We have found that there is practically no effect on the size of the first term in (5), i.e. the mean heavy hadron charged decay multiplicity n_b^{dc}. On the other end, the size of the second term, namely the light-quark mean multiplicity $N_{q\bar{q}}^{ch}$ evaluated at $\sqrt{s} = 8$ GeV, turned out to be sensibly affected.

As far as n_b^{dc} is concerned, we used the updated value $n_b^{dc} = 11.10 \pm 0.18$ coming from the most recent results of the LEP, SLD and CDF experiments on B-hadron production [9, 10]. This number has not practically changed with respect to the one used in the original MLLA analysis.

In order to evaluate the light-quark mean multiplicity $N_{q\bar{q}}^{ch}(8\text{ GeV})$, one should interpolate existing data on mean charged multiplicity and subtract the c-quark contamination, since at this energy also c-quark initiated events are produced, and they have a higher mean multiplicity compared to that produced by light quarks. With respect to the earlier analysis, we improved the data fitting procedure by considering in our study all available data in the energy range 1.4 - 91 GeV, rather than restricting to a very limited energy range as in [3]. All multiplicities measured above 10.5 GeV were corrected for the bias produced by the b-quark. We tested the consistency and the stability of the fit by varying the fitted energy intervals and by using different parameterisations. All fits gave very good χ^2 and we estimated the total mean charged hadron multiplicity as $N_{had}^{ch}(8\text{ GeV}) = 7.1 \pm 0.3$ (see [5] for a detailed discussion).

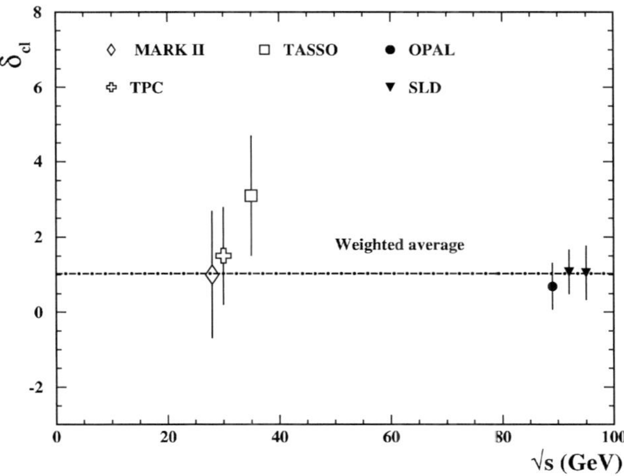

FIGURE 2. Compilation of the direct experimental measurements of $\delta_{c\ell}$.

In order to subtract the contribution from the c-quarks, we carefully studied the literature about the experimental results on the measurement of the multiplicity difference between the q- and c-quarks

$$\delta_{c\ell} = N_c^{ch} - N_{q\bar{q}}^{ch}. \tag{6}$$

At the time of the analysis of Ref. [3] only the results from MARK II, TPC and TASSO were available. These results are affected by large uncertainties, and we also noticed in the literature some inconsistencies in the evaluation of $\delta_{c\ell}$, which we corrected for. Much more precise results from OPAL [11] and SLD [12, 13] are now available, and in the present analysis the value of $\delta_{c\ell}$ used for the correction was reevaluated (see [5] for details). The experimental results from 29 GeV to 91 GeV are shown in Fig. 2, where one can see that the data on $\delta_{c\ell}$ also show a remarkable energy independence. A weighted average yields

$$\langle \delta_{c\ell} \rangle = 1.0 \pm 0.4. \tag{7}$$

This value is about a factor two smaller than that used in [3], and is more precise.

We finally corrected N_{had}^{ch} (8 GeV) for the effect of the 40% admixture of $c\bar{c}$ events using the new value as in (7), and find for the light quarks

$$N_{q\bar{q}}^{ch}(8\text{ GeV}) = 6.7 \pm 0.34. \tag{8}$$

We arrive then at the revised MLLA expectation for the multiplicity difference

$$\delta_{b\ell}^{MLLA} = n_b^{dc} - N_{q\bar{q}}^{ch}(\sqrt{e}M_b) = 4.4 \pm 0.4 \tag{9}$$

which is ~ 1.0 unit lower than the result reported in [3] and has half of its uncertainty. The comparison of the MLLA result (9) with the available experimental data on $\delta_{b\ell}$ in e^+e^- annihilation is shown in Fig. 3; here we included also the reevaluated results of

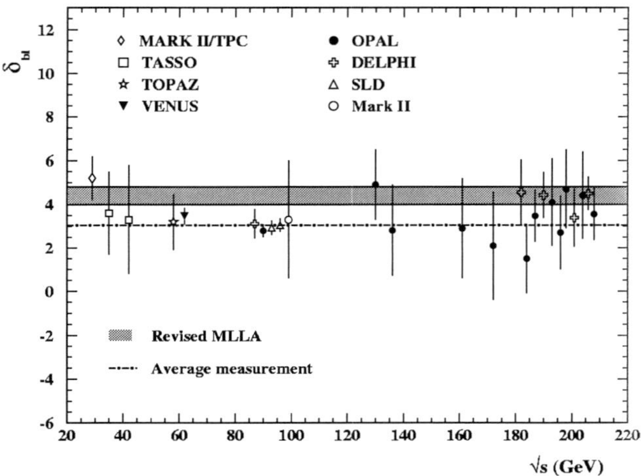

FIGURE 3. Experimental measurements of $\delta_{b\ell}$ plotted as a function of the c.m.s. energy, $\sqrt{s}$; data below 90 GeV reevaluated (see [5]). The revised MLLA expectation, $\delta_{b\ell}^{MLLA} = 4.4 \pm 0.4$, is indicated by the shaded area.

DELCO, MARK II, TPC, TASSO, TOPAZ and VENUS (see [5] for details). The new experimental average is given by $\delta_{b\ell}^{exp} = 3.14 \pm 0.14$.

Qualitatively, the previous conclusion remains valid: the experimental mean value is lower than the absolute value of the MLLA prediction. Quantitatively, however, the agreement between the data and the theory definitely improves.

Finally, we repeated the analysis for the c-quark data, restricting the multiplicity fits to the energy range 1.4 - 10.45 GeV. The predicted value at $W_0^c \equiv \sqrt{e} M_c = 2.7$ GeV is found to be $N_{q\bar{q}}^{ch}(2\ 7 \text{ GeV}) = 3.7 \pm 0.3$. Using the c-quark decay multiplicity $n_c^{dk} = 5.2 \pm 0.3$ we obtain the MLLA expectation for the charged particle multiplicity difference in the c-quark case

$$\delta_{c\ell}^{MLLA} = 1.5 \pm 0.4. \tag{10}$$

which is basically the same value as predicted in [3], $\delta_{c\ell}^{MLLA} = 1.7 \pm 0.5$. The result (10) is consistent with the new more precise experimental average given in (7). As in case of $\delta_{b\ell}$ the theoretical MLLA result lies now above the experimental value which is expected due to the presence of the higher order effects.

ESTIMATE OF THE NEXT-TO-MLLA TERMS OF THE ORDER OF $\alpha_S(M)N(M)$

Next-to-MLLA correction terms are copious and it is hard to collect them all. There are, however, some specific contributions that look *enhanced* and can be listed and estimated. These are contributions containing an additional (semi-dimensional) factor π^2. In particular, to predict the event multiplicity at the $\alpha_s N$ level, one has to take into consideration large angle two soft gluon systems ("dipole" contributions). Another correction of similar nature comes from the $1-z$ rescaling of the argument of the dead cone subtraction. It turns out to be numerically larger than the "dipole" contribution. Both these problems are discussed in detail in [5], where we show how a π^2 enhanced correction emerges. This correction to the companion multiplicity difference has been calculated in the present work:

$$N_{q\bar{q}}(W) - N_{Q\bar{Q}}(W) = N_{q\bar{q}}(\sqrt{e}M) \cdot \left\{ 1 + \frac{N_c \alpha_s(M)}{2\pi} \left[\frac{\pi^2}{24} + \left(\frac{\pi^2}{3} - \frac{5}{4} \right) \right] \right\}, \tag{11}$$

where M stands for the heavy-quark mass. The first term in the square bracket comes from the dipole corrections at large emission angles whereas the second one, coming from the $(1-z)$ rescaling, improves the description of the small angle emission from the heavy quark.

The result (11) is not claimed to be complete at this order but it includes the contributions considered to be dominant and shows the size of the next-to-MLLA terms. Remarkably, both corrections work in the same direction increasing the difference between the light and heavy quark companion multiplicities.

To gain insight at the quantitative level we estimate the size of the two large "π^2-contributions" shown in (11). From the 1-loop formula with $\Lambda = 250$ MeV and $n_f = 3$, we evaluate the strong coupling at scale M_b, $\alpha_s(M_b) = 0.23$, and finally obtain $\delta_{b\ell} \approx 2.6 \pm 0.4$. We conclude that the MLLA prediction is already close to the experimental data, and the remaining difference is of the order of the expected next-to-MLLA contributions.

CONCLUSIONS

As compared to the previous analysis, the updated MLLA prediction for the absolute value of the multiplicity difference between heavy- and light-quark initiated events in e^+e^- annihilation presented in this talk comes closer to the experimental data. The remaining difference is of the order of the next-to-MLLA corrections, the dominant part of which has been identified and evaluated in this work.

ACKNOWLEDGMENTS

I would like to thank the organizers for inviting me to attend this interesting conference, and for providing a friendly and constructive atmosphere.

REFERENCES

1. Yu.L. Dokshitzer, V.A. Khoze and S.I. Troyan, in: Proceedings of the *6th Int. Conf. on Physics in Collisions*, edited by M. Derrick, World Scientific, Singapore, 1987, p. 417.
2. Yu.L. Dokshitzer, V.A. Khoze and S.I. Troyan, *J. Phys.*, **G17**, 1481–1602 (1991).

3. B.A. Schumm , Yu.L. Dokshitzer, V.A. Khoze and D.S. Koetke, *Phys. Rev. Lett.*, **69**, 3025–3028 (1992).
4. V.A. Khoze and W. Ochs, *Int. J. Mod. Phys.*, **A12**, 2949–3120 (1997);
V.A. Khoze W. Ochs and J. Wosiek, In: Shifman, M. (ed.): At the frontier of particle physics, Vol. 2, pp. 1101–1194 (arXiv:hep-ph/0009298).
5. Yu.L. Dokshitzer, F. Fabbri, V.A. Khoze and W. Ochs, to appear in *Eur. Phys. J.*, **C**, (2005) (arXiv:hep-ph/0508074).
6. OPAL Collaboration: G. Abbiendi et al., *Phys. Lett.*, **B550**, 33–46 (2002), and references therein.
7. VENUS Collaboration: K. Okabe et al., *Phys. Lett.*, **B423**, 407–418 (1998).
8. P. Abreu and A. De Angelis, *DELPHI NOTE 2002-052 CONF 586*, June 2002.
9. ALEPH, CDF, DELPHI, L3, OPAL and SLD Collaborations: "Combined results on B-hadron production rates and decay properties", D. Abbaneo et al., *CERN-EP-2001-05* (arXiv:hep-ex/0112028), SLAC-PUB-9500.
10. OPAL Collaboration: R. Akers et al., *Z. Phys.*, **C61**, 209–222 (1994).
11. OPAL Collaboration: R. Akers et al., *Phys. Lett.*, **B352**, 176–186 (1995).
12. SLD Collaboration: K. Abe et al., *Phys. Lett.*, **B386**, 475–485 (1996).
13. SLD Collaboration: K. Abe et al., *Phys. Rev.*, **D69**:072003 (2004).

Cronin effect at RHIC

Ming Shao [1](for the STAR Collaboration)

University of Science and Technology of China, Hefei, Anhui 230026, China

Abstract. The Cronin effect is studied with identified particle spectra from 200GeV d+Au collision at RHIC. The nuclear modification factor R_{dAu} of baryons (proton etc.) rise faster than those of mesons (pions, kaons and phi etc.) and show a baryon/meson scaling behavior similar to that found at Au+Au collisions at same energy. The particle-species dependence of the Cronin effect is observed to be significantly smaller than that at lower energies. The eta asymmetry is also reported and shows contrary trends to prediction based on initial multiple parton scattering model. These measurements indicate that the final state effect plays an important role in the Cronin effect.

Keywords: Cronin effect, nuclear modification factor, pseudo-rapidity asymmetry, STAR
PACS: 25.75.Dw, 25.75.-q, 13.85.Ni

1. INTRODUCTION AND MOTIVATION

The Cronin effect [1] in high energy p+A and A+A collisions has gained renewed interest since many novel physical phenomena were discovered on the Relativistic Heavy Ion Collider (RHIC), especially the large suppression of hadron production at high p_T [3]. As an important baseline, p+A (d+A) collisions were proposed to clarify the role of initial/final state effect in A+A collisions. Thus Cronin effect, as one of the two main nuclear effect in p+A (d+A) collisions, must be firmly understood in order to reliably interpret the physics (partonic energy loss effect, etc.) in A+A collisions.

Cronin effect was first discovered in 1970s [1], featuring an enhancement of hadron production at intermediate p_T range in p+A relative to pp, when scaled by number of binary collisions. Partonic scattering at the initial impact [4] has been interpreted as main source of this effect. Models based on this picture can reproduce the Cronin effect of inclusive hadron production in d+Au collisions, as well as the suppression of charged hadron in Au+Au collisions at RHIC when coupling with partonic energy loss (jet quenching) model [5]. However, based on initial multiple parton scattering and independent fragmentation, one would expect same Cronin effect for different particle type (see, $p(\bar{p})$ and pions). This is not the case since the first discovery of Cronin effect, where experimentally the Cronin effect for $p(\bar{p})$ is larger than that for pions. On the other hand, some recent works have tried to interpret the Cronin effect as a final state effect, such as recombination [6] and coherent multiple scattering [7]. The recombination model has successfully reproduce the particle type dependence of hadron production at intermediate p_T in Au+Au collisions [8].

[1] Supported, in part, by the National Natural Science Foundation of China, under granted No. 10375062 and 10275060.

CP828, *Multiparticle Dynamics*
edited by V. Šimák, M. Šumbera, Š. Todorova, and B. Tomášik

The particle production in d+Au collisions at different rapidities can reflect the dynamics of nuclear and Bjorken-x dependence of the Cronin effect (and shadowing). The calculation based on initial multiple parton scattering is also employed to study the pseudo-rapidity dependence of particle production. A unique rapidity asymmetry of particle production in d+Au collisions is predicted [9], where the backward-to-forward (Au-side to d-side) particle ratio is greater than unity at low p_T , goes below unity at intermediate p_T, and approaches unity again at high p_T. Some other models based on gluon saturation [10] or parton recombination [11] predict a backward-to-forward particle ratio that is opposite to the predictions based on incoherent multiple partonic scattering.

Due to these complications, detailed measurements of particle production - the centrality dependence, pseudo-rapidity and p_T distribution - in d+Au collisions are needed for understanding the Cronin effect, and therefore underlying physics in Au+Au collisions. In this proceeding, identified hadron spectra for pion and proton, the pseudo-rapidity asymmetry of inclusive charged hadron, as well as some preliminary results from ϕ meson production from 200GeV d+Au collisions at RHIC, will be presented. Comparison to various model predictions will be made and the physics implication on Cronin effect will be discussed.

2. EXPERIMENT SETUP

The detector used for these studies was the Solenoidal Tracker at RHIC (STAR) [12]. The main tracking device is the Time Projection Chamber (TPC) which provides momentum information and particle identification for charged particles up to $p_T \sim 1.1$ GeV/c by measuring their ionization energy loss (dE/dx) [13]. Detailed descriptions of the TPC and d+Au run conditions have been presented in Ref [18].

A prototype time-of-flight detector (TOFr) based on multi-gap resistive plate chambers (MRPC) [14] was installed in STAR in the d+Au run. It extends particle identification up to $p_T \sim 3$ GeV/c for p and $\bar{p}$. By combining the particle identification capability of dE/dx from TPC and time-of-flight from TOFr, we are able to extend pion identification to ~ 3 GeV/c [15]. More information about the TOF system can be found in [16].

Charged particle multiplicity within $-3.8 < \eta < -2.8$ was measured by the Forward TPC [18] in the Au beam direction and served as the basis for d+Au centrality tagging scheme, as described in [5]. A separate centrality tag, which requires that a single neutron impinge on the Zero Degree Calorimeter [17] in the deuteron beam direction (ZDC-d), was also used.

3. RESULTS AND DISCUSSIONS

The pion and proton are identified by TOF up to $p_T = 3$ GeV/c in 200GeV d+Au collisions. The p_T distribution of particle yields for pion and proton are shown in Fig. 1 [19]. The errors shown in the plot are statistical (predominately smaller than the size of data symbols). The bin-to-bin systematic errors are $\sim 8\%$, while the normalization

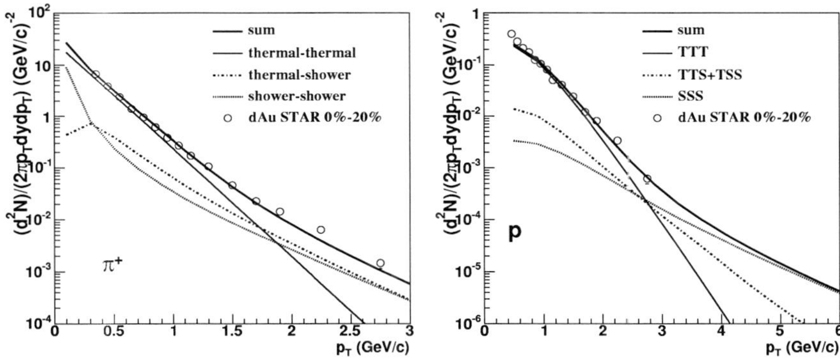

FIGURE 1. The invariant yields for π^+ (left) and proton (right) at $0\% - 20\%$ d+Au collisions as a function of p_T. The open circles are data points. The curves are the calculation results from recombination model. Sum represents the total contribution from recombination model. Thermal-thermal (or TTT) represents the soft contribution. The thermal-shower (or TTS+TSS) represents the contribution from the interplay between soft and hard components. The shower-shower (or SSS) represents the hard contribution.

uncertainty is around 10%. Using the method described in [11], the pion and proton spectra in d+Au collisions from the recombination model calculation are also plotted in the figure, shown as the curves on top of the data points. These plots show that the recombination model can reproduce the spectra of both pion and proton in minimum-bias (and centrality selected, not shown here) d+Au collisions.

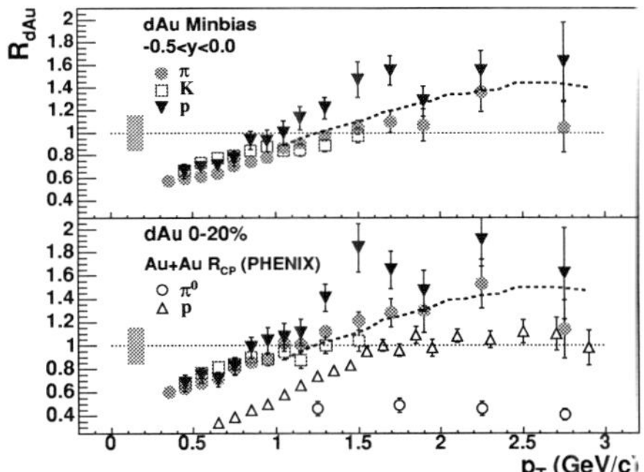

FIGURE 2. The identified particle $R_{dAu}(p_T)$ for minimum-bias and top 20% d+Au collisions. The filled triangles are for $p+\bar{p}$, the filled circles are for $\pi^+ + \pi^-$ and the open squares are for $K^+ + K^-$. Dashed lines are R_{dAu} of inclusive charged hadrons from [5]. The open triangles and open circles are R_{CP} of $p+\bar{p}$ and π^0 in Au+Au collisions measured by PHENIX [20]. Errors are statistical. The gray band represents the normalization uncertainty of 16%.

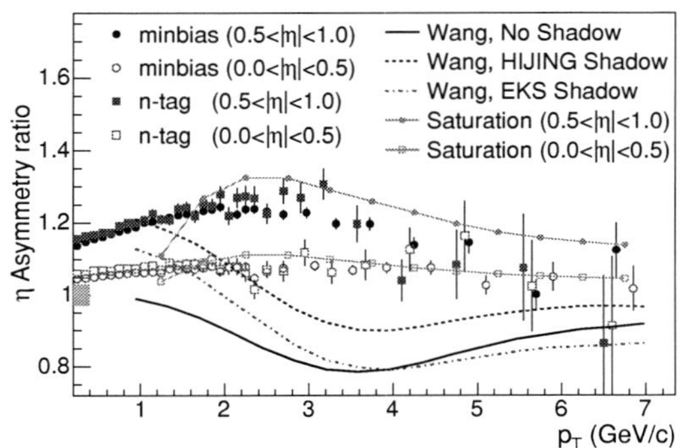

FIGURE 3. The ratio of charged hadron spectra in the backward rapidity to forward rapidity region for minbias and ZDC-d neutron-tagged events in 200GeV d+Au collisions. Calculations based on initial multiple partonic scattering [9] ($y = -1/y = 1$) for minbias events are also shown for cases with no shadowing (solid curve), HIJING shadowing (dashed curve) [23], and EKS shadowing (dot-dashed curve) [24]. Calculations in a gluon saturation model [10] for minbias events are shown for $0.5 < |\eta| < 1.0$ (filled circles with solid line) and for $0.0 < |\eta| < 0.5$ (open squares with solid line)

Nuclear modification factor in d+Au collisions are measured through comparison to the p+p spectrum, defined as

$$R_{dAu}(p_T) = \frac{d^2N^{dAu}/(2\pi p_T dp_T dy)}{T_{dAu}d^2\sigma^{pp}_{inel}/(2\pi p_T dp_T dy)}, \tag{1}$$

where $T_{dAu} =< N_{bin} > /\sigma^{pp}_{inel}$ describes the nuclear geometry. The results of $\pi^+ + \pi^-$, $K^+ + K^-$ and $p^+ + \bar{p}$ for minbias and central (20%) d+Au collisions are plotted in Fig. 2 [21], showing characteristic Cronin effect [1][22] in particle production with R_{dAu} less than unity at low p_T and above unity at $p_T \geq 1.0$ GeV/c. In both cases, the R_{dAu} of protons rise faster than R_{dAu} of pions and kaons. It is notable that the R_{dAu} of protons and anti-protons are greater than unity in both central and minimum-bias d+Au collisions while the proton and antiproton production follows binary scaling in all centralities in Au+Au collisions [20].

Fig. 3 [25] shows the p_T dependence of the asymmetry for minimum bias and ZDC-d neutron-tagged events. The asymmetry is obtained by taking ratios of inclusive backward (Au-side) to forward (d-side) p_T spectra. The ratio was taken between the $-1.0 < \eta < -0.5$ and $0.5 < \eta < 1.0$ as well as $-0.5 < \eta < 0.0$ and $0.0 < \eta < 0.5$ regions. The overall systematic uncertainty (indicated by the band) is less than 3%. The ratio taken within $|\eta| < 0.5$ is nearly constant in p_T, while at higher pseudo-rapidity the ratio slowly increases with p_T up to about p_T = 2.5 GeV/c, maximized at approximately 1.25, and then approaches unity beyond p_T = 5 GeV/c, indicating the absence of nuclear effects at high p_T . For the ZDC-d neutron-tagged events, the ratio exhibits nearly the same p_T dependence as minimum bias events.

Also shown in Fig. 3 are the calculation of the asymmetry in the incoherent multiple partonic scattering framework [9] with various nuclear shadowing parameterizations [23][24], as well as minimum bias gluon saturation results [10]. The measurements

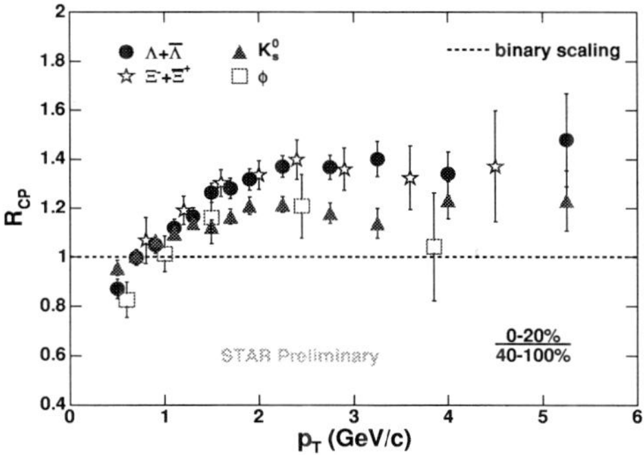

FIGURE 4. The ratio of central($0 \sim 20\%$) collisions over peripheral($40 \sim 100\%$) collisions (R_{CP}) normalized by N_{bin} for ϕ, Λ, Ξ and K_s^0 in d+Au collisions at 200GeV d+Au collisions.

show an opposite trend at intermediate p_T to the theoretical calculations in [9] and thus suggest that multiple scattering of partons in the initial state alone cannot reproduce the observed pseudo-rapidity asymmetry. The gluon saturation prediction exhibits a stronger p_T dependence of pseudo-rapidity asymmetry than actually observed, over-predicting the magnitude of the asymmetry at high pseudo-rapidities. However, it does show the same trend of increasing asymmetry with increasing centrality. The idea of recombination [11], as a final state effect, can also give calculations qualitatively consistent with the measurements of the pseudo-rapidity asymmetry as a function of p_T (not shown in this article). More measurements and quantitative comparisons to these models will be interesting and important to understand the particle production at intermediate p_T.

The parton coalescence/recombination model has been successfully applied in describing the scaling behavior of particle production at intermediate p_T in Au+Au collisions, based on meson/baryon type. It was suggested this final state effect can be pronounced even in d+Au collisions [11], a system considered to be predominated by initial state effect. ϕ meson is a good candidate to test whether particle species dependence still exists in d+Au system, since it has a mass similar to that of the Λ baryon. Preliminary R_{CP} measurements of ϕ meson are shown in Fig. 4 [26], as a function of p_T. Also shown for comparison are R_{CP} for Λ, Ξ and K_s^0. The R_{CP} of baryons are larger than those of mesons at intermediate p_T, which imply the particle production at this p_T region is divided by the particle's types. R_{dAu} measurements [26], which is not shown here, give similar implication. This particle species dependence seems to indicate significant final state effect at d+Au collisions.

4. CONCLUSIONS

In summary, Cronin effect is studied at 200GeV d+Au collisions at RHIC. Pion and proton are identified up to $p_T = 3$ GeV/c. Their spectra are reported and can be reproduced by the recombination model. R_{dAu} of protons rises faster than that of pions and kaons. Ratios of backward-to-forward pseudo-rapidity transverse momentum distributions are above unity for p_T below 5 GeV/c. The initial multiple scattering of partons alone cannot reproduce the observed pseudo-rapidity asymmetry, while the latest calculations in gluon saturation model and parton recombination model stand in qualitative agreement with the measurements. Preliminary results of the nuclear modification factors for ϕ meson in 200 GeV d+Au collisions show that the R_{CP} and R_{dAu} are divided into two groups in intermediate pt range, where the R_{CP}/R_{dAu} of baryons (Λ, Ξ and proton) are larger than that of mesons (ϕ, K_s^0, pion and kaon). This is similar to that observed in Au+Au collisions, where the recombination model has been successfully applied. Based on all these measurements, we can we conclude that the Cronin effect in $\sqrt{s_{NN}}$ = 200 GeV d+Au collisions is not the initial state effect only, and the final state effect plays an important role. Understanding of the Cronin effect should involve both effects simultaneously.

REFERENCES

1. J.W. Cronin et al.,*Phys. Rev. Lett.* **31**, 1426 (1973); J.W. Cronin et al., *Phys. Rev.* **D 11**, 3105 (1975).
2. D. Antreasyan et al., *Phys. Rev.* **D 19**, 764 (1979); P.B. Straub et al., *Phys. Rev. Lett.* **68**, 452 (1992).
3. STAR Collaboration, J. Adams et al., *Phys. Rev. Lett.* **91**, 172302 (2003); PHENIX Collaboration, S.S. Adler et al., *Phys. Rev. Lett.* **91**, 072301 (2003).
4. M. Lev and B. Petersson, *Z. Phys.* **C 21**, 155 (1983); T. Ochiai et al., *Prog. Theor. Phys.* **75**, 288 (1986); Y. Zhang, G. Fai, G. Papp, G.G. Barnafoldi, P. Levai, *Phys. Rev.* **C 65**, 034903 (2002); I. Vitev, M. Gyulassy, *Phys. Rev. Lett.* **89**, 252301 (2002).
5. STAR Collaboration, J. Adams et al., *Phys. Rev. Lett.* **91**, 072304 (2003).
6. R. C. Hwa and C. B. Yang, *Phys. Rev.* **C 67**, 064902 (2003); R. J. Fries et al., *Phys. Rev. Lett.* **90**, 202303 (2003); V. Greco, C. M. Ko and P. Levai, *Phys. Rev. Lett.* **90**, 202302 (2003).
7. J. W. Qiu and I. Vitev, *Phys. Rev. Lett.* **93**, 262301 (2004); J. W. Qiu and I. Vitev, hep-ph/0405068.
8. STAR Collaboration, J. Adams et al., *Phys. Rev. Lett.* **92**, 052302 (2004);
9. X.N. Wang, *Phys. Lett.***B 565**, 116 (2003).
10. D. Kharzeev, Y.V. Kovchegov and K. Tuchin, hep-ph/0405045;
11. R.C. Hwa and C.B. Yang, *Phys. Rev. Lett.* **93**, 082302 (2004); R.C. Hwa and C.B. Yang, *Phys. Rev.* **C 70**, 037901 (2004); R.C. Hwa and C.B. Yang, *Phys. Rev.* **C 71** , 024902 (2005).
12. K.H. Ackermann et al., *Nucl. Instr. Meth.* **A 499**, 624 (2003).
13. M. Anderson et al., *Nucl. Instr. Meth.* **A 499**, 659 (2003).
14. B. Bonner et al., *Nucl. Instr. Meth.* **A 508**, 181 (2003); M. Shao et al., *Nucl. Instr. Meth.* **A 492**, 344 (2002).
15. STAR Collaboration, M. Shao et al., *J. Phys.* **G 31**, S85-S92 (2005), Contribution to the proceedings of the Hot Quarks 2004; M. Shao et al., nucl-ex/0505026.
16. J. Wu et al., *Nucl. Instr. Meth.* **A 538**, 243 (2005); W.J. Llope et al., *Nucl. Instr. Meth.* **A 522**, 252 (2004).
17. C. Adler et al., *Nucl. Instrum. Meth.* **A 461**, 337 (2001).
18. K.H. Ackermann et al., *Nucl. Instrum. Meth.***A 499**, 713 (2003).
19. L. Ruan, *Ph.D. thesis*, University of Science and Technology of China, nucl-ex/0503018.
20. PHENIX Collaboration, K. Adcox et al., *Phys. Lett.* **B 561**, 82 (2003); PHENIX Collaboration, S.S. Adler et al., *Phys. Rev. Lett.* **91**, 172301 (2003).
21. STAR Collaboration, J. Adams et al., *Phys. Lett.* **B 616**, 9 (2005).
22. A. Accardi, *Contribution to the CERN Yellow report on Hard Probes in Heavy Ion Collisions at the LHC*, hep-ph/0212148; X.N. Wang, *Phys. Rev.* **C 61**, 064910 (2000).
23. S.Y. Li and X.N. Wang, *Phys. Lett.* **B 527**, 85 (2002).
24. K.J. Eskola, V.J. Kolhinen and C.A. Salgado, *Eur. Phys. J.* **C 9**, 61 (1999).
25. STAR Collaboration, J. Adams et al., *Phys. Rev.* **C 70**, 064907 (2004).
26. STAR Collaboration, X. Cai et al., *Quark Matter 2005*, August 4-9, 2005, Budapest, Hungary.

Particle multiplicities and fluctuations in 200 GeV Au-Au collisions

Giorgio Torrieri*, Sangyong Jeon*,† and Johann Rafelski**

*Department of Physics, McGill University, Montreal, QC H3A-2T8, Canada
†RIKEN-BNL Research Center, Upton NY, 11973, USA additional visiting address
**Department of Physics, University of Arizona, Tucson AZ 85721, USA

Abstract. We use the statistical hadronization model (SHM) to describe hadron multiplicity yields and fluctuations. We consider 200 GeV Au-Au collisions, and show that both event averaged yields of stable particles and resonances, and event-by-event fluctuation of the K/π ratio can be described within the SHM using the same set of thermal parameters, provided that the phase space occupancy parameter value is significantly above chemical equilibrium, and the freeze-out temperature is $\sim$ 140 MeV. We present predictions that allow to test the consistency of our results.

The statistical hadronization model (SHM) [1, 2, 3] has been extensively applied to the study of soft particle production in hadronic systems. When it includes the full spectrum of hadronic resonances [4], the SHM, with judiciously chosen (fitted) parameters, quantitatively describes the abundances of all hadrons produced in heavy ion collisions at all considered reaction energies [5].

The ability of the SHM to describe not just averages, but event-by-event multiplicity fluctuations has not been widely investigated. Event-by-event fluctuations have attracted theoretical [6, 7, 8, 9] and experimental interest, both as a consistency check for existing models [6, 7] and as a way to search for new physics [8].

The objective of this work is to determine weather the SHM can describe both yields and fluctuations with the same parameters. We obtain a good fit to 200 GeV RHIC experimental data including both yields and fluctuations measurements, discuss the results in the context of the bulk properties of the matter created at RHIC, and present predictions allowing further tests of the model.

The statistical hadronization model assumes that final states are produced in proportion to their phase space size. The first and second cumulants of this probability distribution give, respectively, the average value over all events its event-by-event fluctuation.

In this work we use the Grand Canonical (GC) ensemble, as implemented in open-source software [10], to calculate fluctuations and yields. We motivate this choice by the fact that the considered RHIC experiments observe the mid-rapidity slice of the system, comprising roughly 1/8 of the total hadron multiplicity. The further assumption of (approximate) boost invariance at mid-rapidity allows to image this rapidity slice into a domain in configuration space. The matter content in this space domain is expected to be in contact and exchanging energy and conserved quantum numbers with the unobserved regions. This than creates the GC system we consider on an event-by-event basis.

If the freeze-out temperature throughout the system is the same, a rather simple model can be used to obtain both yields and fluctuations [11, 12]. However, one should note

CP828, *Multiparticle Dynamics*
edited by V. Šimák, M. Šumbera, Š. Todorova, and B. Tomášik

that even if practically all produced particles originate in statistical model processes, their fluctuations could comprise novel creation mechanisms, related to their formation dynamics. However, such novel mechanisms are not present in all observables, and the observables we consider here seem to follow non-equilibrium SHM calculations.

The final state particle yield can then be computed as a function of the particle properties, resonance decay tree, freeze-out temperature and fugacities (technical details are found in our recent report [13]). While the temperature controls the particle yield dependence on the mass, the fugacity λ describes both the yield of conserved quantities (such as baryon number, charge and strangeness) across all particles, and the absolute yields which depend on the degree of chemical equilibration. It is common to introduce chemical potentials associated with each conservation law, $\mu = T \ln \lambda$, while the fugacities associated with the chemical nonequilibrium condition are called γ.

Detailed balance requires that the particle fugacity be given by the conserved charge fugacity to the power of the particle's 'charge', generalized to contain all conserved quantities (electrical charge, baryon number etc.) . Thus, for a particle with $q, (\overline{q})$ light (anti)quarks, $s, (\overline{s})$ strange (anti)quarks and isospin I_3 we have

$$\lambda_i^{\mathrm{eq}} = \lambda_q^{q-\overline{q}} \lambda_s^{s-\overline{s}} \lambda_{I_3}^{I_3} \tag{1}$$

However, the condition of chemical equilibrium might no longer hold when the fireball is rapidly expanding. Thus, chemical parameters acquire a *kinetic* (time-dependent) component parametrized in terms of phase space occupancies,

$$\lambda_i = \lambda_i^{\mathrm{eq}} \gamma_q^{q+\overline{q}} \gamma_s^{s+\overline{s}}, \tag{2}$$

where $\gamma_q = 1, \gamma_s = 1$ at equilibrium. Even in chemical nonequilibrium the particle fugacity λ_i is the parameter controlling the particle yield, and the first and second cumulants can be calculated from the partition function in the usual way [14, 15].

If the expanding system undergoes a fast phase transition from a Quark Gluon Plasma (QGP) to a hadron gas (HG), chemical non-equilibrium [16] and super-cooling [17] are expected to arise given the requirement that entropy has to increase while the transition occurs from a high to low entropy density phase. The virtue of a hadronization temperature near 140 MeV, and an over-saturated phase space ($\gamma_q \sim 1.5, \gamma_s \sim 2$), is a match of both the energy and entropy density between the QGP and HG phases.

Fits to experimental data at both SPS and RHIC energies support these values of $\gamma_{q,s}$. Moreover, best fit $\gamma_{q,s} > 1$ arises for a critical reaction energy [5] (corresponding to the energy of the K/π "horn" [18]) and system size [19], as expected from the interpretation of γ_q as a manifestation of a phase transition. However, even though the fits performed in [5] strongly favor the chemical non-equilibrium, they do not rule out equilibrium models. The equilibrium model remains marginally compatible with data giving a less convincing but statistically still non-absurd validity.This happens since in a fit considering only particle yields, the chemical non-equilibrium phase space occupancies γ_s and γ_q correlate with freeze-out temperature [5], making a distinction between a higher temperature equilibrated freeze-out ($\gamma_q = 1, \gamma_s \leq 1$) scenario and a supercooled scenario where $\gamma_{q,s} > 1$ difficult. When full chemical nonequilibrium is allowed for, $\gamma_q \simeq 1.6$ and $T = 140$ MeV is found. The best fit freeze-out temperature when full chemical equilibrium is assumed varies between studies, ranging from $T =$

155 MeV in latest SHARE based studies [5], $T = 165$ MeV for those carried out 2 years ago by STAR experimental group [20], to $T = 177$ MeV offered in the initial RHIC data exploration in which strange particles were not yet fully incorporated [21].

The study presented in [13] makes it clear that the dependence of the fluctuation

$$\sigma_X^2 = \frac{\langle X^2 \rangle - \langle X \rangle^2}{\langle X \rangle} \tag{3}$$

on T and γ_q is different, allowing us to independently determine these two variables. A higher temperature decreases the fluctuations with respect to the Poisson value $\sigma = 1$, expected for a Boltzmann distribution, since it introduces greater particle correlations arising from increased resonance decay contribution. Increasing γ_q rapidly increases fluctuations of quantities related to pions, due to the fact that at $\gamma_q > 1$ λ_π rapidly approaches $e^{m_\pi/T}$, giving fluctuations an extra increase compared to yields [12, 13].

By virtue of the implied physical picture, equilibrium models generally assume a long time span between chemical (particle production) and thermal (particle scattering) freeze-out, which would alter considerably the multiplicity of directly detectable resonances. In the chemical non-equilibrium supercooled freeze-out picture, however, it is natural to assume that particle scattering after emission is negligible [16] and thus one can in most cases assume that the thermal and chemical freeze-out temperatures are the same. Hence, a reliable way to probe the re-interaction period would be instrumental for our understanding of how the fireball produced in heavy ion collisions breaks up.

We have recently shown [12] that a comparison of fluctuations to directly detected resonances probes the interval between chemical and thermal freeze-out. Consider, for example, σ_{K^+/π^-}. The numerator and the denominator terms in this ratio are linked by a large correlation term due to the $K^{*0}(892) \to K^+\pi^-$ decay. This correlation probes the $K^{*0}(892)$ abundance at the initial *chemical* freeze-out, since subsequent rescattering of $K^* \to K\pi$ decay products *or* on-shell K^* regeneration from in-medium $K^+\pi^-$ pairs does not alter the final abundance of π^+ and a K^-. On the other hand, a direct measurement of the $K^{*0}(892)/K^-$ ratio through invariant mass reconstruction measures the $K^{*0}(892)$ abundance at *thermal* freeze-out, after all rescattering/regeneration ceased. Hence, comparing the K^+/π^- fluctuation to the $K^{*0}(892)/K^-$ ratio provides a gauge for the effect of the hadronic reinteraction period on particle abundances. A strong constraint arises in a model where chemical and thermal freeze-out coincide, as both observables must be described with the same set of statistical parameters determined by global yields of other particles. In this way one can argue that the K^+/π^- fluctuation and the $K^{*0}(892)/K^-$ relative yield measured by invariant mass method offer a decisive test of the chemical non-equilibrium sudden hadronization reaction picture.

While they are phenomenologically powerful, fluctuation measurements are also vulnerable to systematic effects which need to be carefully considered. Volume and centrality fluctuations, difficult to describe in a model-independent way, can be taken care of by considering event by event fluctuations of particle ratios, where the fluctuation in volume cancels out, event by event, to first order. This leaves, however, possibly large effects due to limited experimental acceptance. These can usually be subtracted by considering fluctuations measured within fake events, created using mixed event techniques [23]. Such "static" fluctuation, with, by definition, no correlations between particles, can be described by a purely Poisson term, where fluctuation is governed by particle yields:

$$\sigma_{stat}^2 = \frac{1}{\langle N_1 \rangle} + \frac{1}{\langle N_2 \rangle} \tag{4}$$

It can also be seen that mixed-events, made from tracks measured in a given detector, also contain the effect due to that detector's acceptance.

Subtracting σ_{stat} from the total fluctuation leaves the "dynamical" fluctuation term:

$$\sigma^{dyn} = \sqrt{\sigma^2 - \sigma_{stat}^2}, \tag{5}$$

which comprises the physically interesting effects such as resonance decay correlations, Bose-Einstein correlations, and eventual new dynamics. Provided certain assumptions for the detector response function hold (see appendix A of [23]), this dynamical fluctuation is a "robust", detector-independent observable.

When using measured σ^{dyn}s in fits to experimental data, the data-sample should include σ^{dyn}, particle yields (which are needed to determine the Poisson contribution to σ_{stat} as per Eq. (4)) and particle ratios. We have performed a fit incorporating all STAR ratios given in Ref. [20], with the exception of the Δ^{++}/p, which the STAR collaboration has since begun to reevaluate, and we also for the present ignore the $\overline{\Omega}/\Omega > 1$, which cannot be fitted with the SHM [13, 24]. On the other hand, we have included in the fit procedure the preliminary value of $\sigma_{K/\pi}^{dyn}$ measured by STAR [22], as well as the published yield for ϕ [25] and π^- [26]. We assume [26, 27] full feed-down correction for $K_{S,L} \rightarrow \pi^{\pm}$ and $\Lambda \rightarrow \pi$ weak decays, and no correction for $\Lambda \rightarrow p$.

The fit parameters include the overall normalization, the freeze-out temperature T, λ_{q,s,I_3} and $\gamma_{q,s}$. We also require, by implementing them as "data-points", strangeness, electrical charge and baryon number conservation: $\langle s - \bar{s} \rangle = 0$ $\langle Q \rangle / \langle B \rangle = (\langle Q \rangle / \langle B \rangle)_{Au} = 0.4$. The fit's statistical significance (P_{true}) profile is shown in the left panel of Fig. 1. To obtain a profile for P_{true} a fit was performed for each fixed parameter point on the abscissa, all fit parameters except the one shown on the abscissa were varied. It can be seen that the fit tightly constrains γ_q well above the equilibrium value, accompanied by $T \sim 140$ MeV, in good agreement of the prediction of the supercooled hadronization scenario.

The right panel of Fig. 1 shows the sensitivity of $\sigma_{K/\pi}^{dyn}$ to γ_q and temperature, and explains why the correlation between T and γ_q disappears when fluctuations are taken into account. As can be seen, a fit assuming the chemical equilibrium ($\gamma_q = 1, T = 155$ MeV [5]). misses $\sigma_{K/\pi}^{dyn}$ by many standard deviations. On the contrary the chemical nonequilibrium fit seems to be right where this fluctuation is measured. Introducing exact conservation for strangeness within the observed window (canonical ensemble) would decrease the theoretical σ_K [28], thereby increasing the chemical equilibrium theory to experiment discrepancy. It is only through $\gamma_q > 1$ that $\sigma_{K/\pi}^{dyn}$ increases to the point where it becomes compatible with the experimental value.

An eye assessment of the fit's goodness is provided by the left panel of Fig. 2. As can be seen, the fit gives an adequate description of all particle yields, including the

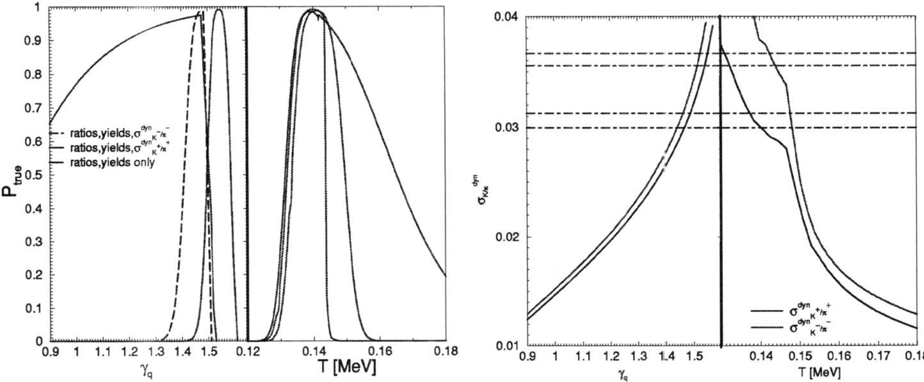

FIGURE 1. Left: statistical significance (P_{true}) profiles for freeze-out Temperature T and light quark phase space occupancy γ_q obtained in the fit shown in Fig. 2. Right: Sensitivity of $\sigma^{dyn}_{K/\pi}$ to freeze-out temperature and γ_q for the fit in Fig. 2. At each point in the abscissa, a fit is performed , with only the particle yield data-points used, varying all fit parameters except the one on the abscissa. $\sigma^{dyn}_{K/\pi}$ is evaluated using the best fit parameters. The dot-dashed lines refer to the experimental limits [22] for the σ_{K^+/π^+} (blue, lower values) and σ_{K^-/π^-} (red, higher values). See [13] for further details.

resonance $(K^{*0}(892)+\overline{K^{*0}}(892))/K^-$ and $\Lambda(1520)/\Lambda$. It can also adequately describe the event-by-event fluctuations of K^+/π^+ *or* K^-/π^-.

$K^\pm/\pi^\pm$ fluctuations do not directly test the simultaneous freeze-out hypothesis, since $K^-\pi^-$ and $K^+\pi^+$ are not correlated by resonances. To test sudden freeze-out, we have used the best fit parameters to predict the yields of several resonances subject to current experimental investigation (ρ^0,$f_0(950)$,Δ^{++},$\Sigma^{*+}(1385)$,$\Xi^*(1530)$) as well as the dynamical fluctuation of the ratio of their decay products (π^+/π^-,$p/\pi^\pm$,$\Lambda/\pi^\pm$,$\Xi/\pi^\pm$). The result is shown in Fig 2, right panel.

Note the significant difference between ratios such as p/π^+ and p/π^-, while the fluctuations of p/π^+ and $\overline{p}/\pi^-$ are substantially identical. This systematics, which repeats itself when the Λ/π ratio is considered, is due to the correlations provided by the leading resonance decay ($\Delta \to p\pi$, $\Sigma^*(1385) \to \Lambda\pi$). Thus, the combined measurement of the resonance and the ratio of decay products yields a very powerful constraint on the simultaneous freeze-out model considered here, and it will be interesting to see to what extent will the model agree with data. In particular, the difference between $\sigma^{dyn}_{K^+/\pi^+}$ and $\sigma^{dyn}_{K^-/\pi^-}$ is intriguing, since the isospin chemical potential required to reproduce it ($\lambda_{I3} \sim 0.96$) is excluded through ratios such as the π^+/π^-. It remains to be seen weather this result is due to experimental systematics not analyzed within the preliminary measurement, or weather additional theoretical insights are needed to describe it.

In conclusion, we have shown that the SHM can describe both the yields and event-by-event fluctuations measured so far in RHIC 200 GeV Au-Au collisions, provided that phase space is saturated above equilibrium and the system is super-cooled with respect to the phase transition temperature. We have justified this scenario in the context of a fast

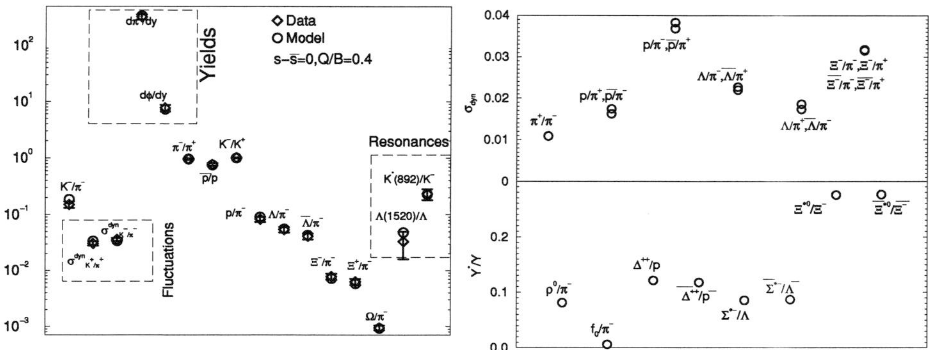

FIGURE 2. Left: Fit of preliminary 200 GeV data, including the $K^{\pm}/\pi^{\pm}$ fluctuations and the $K^*(892)$ and $\Lambda^*(1520)$ resonance Right: Predictions of resonances and event-by-event fluctuations of their decay product ratios with the best fit parameters.

phase transition from a high-entropy phase, and argued that the simultaneous description of yields and fluctuations is consistent with an explosive freeze-out, where interactions after hadronization are negligible. We have predicted experimental observables suitable for testing this model further, and eagerly await more published data to determine to what extent can the SHM account for both yields and fluctuations in light and strange hadrons produced in heavy ion collisions.

ACKNOWLEDGMENTS

Work supported in part by grants from the U.S. Department of Energy (J.R. by DE-FG02-04ER41318) the Natural Sciences and Engineering research council of Canada, the Fonds Nature et Technologies of Quebec. G. T. thanks the Tomlinson foundation for support. S.J. thanks RIKEN BNL Center and U.S. Department of Energy [DE-AC02-98CH10886] for providing facilities essential for the completion of this work.

REFERENCES

1. E. Fermi, *Prog. Theor. Phys.* **5**, 570 (1950).
2. I. Pomeranchuk, *Proc. USSR Academy of Sciences* (in Russian) **43**, 889 (1951).
3. L. D. Landau, *Izv. Akad. Nauk Ser. Fiz.* **17** (1953) 51.
4. R. Hagedorn, *Suppl. Nuovo Cimento* **2**, 147 (1965).
5. J. Letessier and J. Rafelski, arXiv:nucl-th/0504028.
6. S. Jeon, V. Koch, K. Redlich and X. N. Wang, *Nucl. Phys. A* **697**, 546 (2002)
7. S. Jeon and V. Koch, *Phys. Rev. Lett.* **83**, 5435 (1999)
8. S. Jeon and V. Koch, *Phys. Rev. Lett.* **85**, 2076 (2000)
9. C. Pruneau, S. Gavin and S. Voloshin, *Phys. Rev. C* **66**, 044904 (2002) [arXiv:nucl-ex/0204011].
10. G. Torrieri, *et al.*, Comm. in Computer physics in press arXiv:nucl-th/0404083
11. J. Cleymans, K. Redlich, *Phys. Rev. C* **60**, 054908 (1999).
12. G. Torrieri, S. Jeon, J. Rafelski Submitted to PRL [arXiv:nucl-th/0503026].
13. G. Torrieri, S. Jeon and J. Rafelski, arXiv:nucl-th/0509067.

14. J. Letessier and J. Rafelski, Cambridge Monogr. *Part. Phys. Nucl. Phys. Cosmol.* **18**, 1 (2002),
15. Material usually presented in physical chemistry textbooks, see for example: *http://www2.mcdaniel.edu/Chemistry/ch307.notes/Chemical%20Equilibrium.html*
16. J. Rafelski and J. Letessier, *Phys. Rev. Lett.* **85**, 4695 (2000).
17. T. Csorgo and L. Csernai, arXiv:hep-ph/9312330
18. D. Flierl *et al.* [NA49 Collaboration], arXiv:nucl-ex/0410041.
19. J. Rafelski, J. Letessier and G. Torrieri, *Phys. Rev. C* **72**, 024905 (2005) [arXiv:nucl-th/0412072].
20. O. Barannikova [STAR Collaboration], arXiv:nucl-ex/0403014 and references therein.
21. P. Braun-Munzinger, D. Magestro, K. Redlich and J. Stachel, *Phys. Lett. B* **518**, 41 (2001).
22. S. Das [STAR Collaboration], arXiv:nucl-ex/0503023.
23. C. Pruneau, S. Gavin and S. Voloshin, *Phys. Rev. C* **66**, 044904 (2002) [arXiv:nucl-ex/0204011].
24. M. Bleicher *et al.*, *Phys. Rev. Lett.* **88**, 202501 (2002) [arXiv:hep-ph/0111187].
25. J. Adams *et al.* [STAR Collaboration], *Phys. Lett. B* in press, [arXiv:nucl-ex/0406003].
26. J. Adams *et al.* [STAR Collaboration], *Phys. Rev. Lett.* **92**, 112301 (2004) [arXiv:nucl-ex/0310004].
27. O. Barannikova [STAR Collaboration], private communication
28. V. V. Begun, M. Gazdzicki, M. I. Gorenstein and O. S. Zozulya, *Phys. Rev. C* **70**, 034901 (2004) [*Phys. Rev. C* 72, 014902 (2005)] [arXiv:nucl-th/0404056].

Strange particle production at HERA

Leszek Zawiejski
for the ZEUS and the H1 Collaborations

*Institute of Nuclear Physics Polish Academy of Sciences
Cracow, Poland*

Abstract. The selected results on the strange particle production in deep inelastic *ep* scattering from ZEUS and H1 experiments at HERA are reviewed.

Keywords: Strange particles, Hadronization, Bose-Einstein correlations, Pentaquarks
PACS: 12.39.Mk, 13.60.Hb, 13.60.-r, 13.90.+i, 14.20.Jn

INTRODUCTION

The studies of the strange particles production in lepton-hadron scattering may help in understanding of the features of the transition from partons to hadrons. The observed hadrons can result from the fragmentation of the struck quark or the proton remnant. Recently, the large statistics of deep inelastic scattering (DIS) events collected by ZEUS and H1 experiments at HERA (for ZEUS it corresponds to 120 pb^{-1} integrated luminosity) allow to investigate many aspects of the strange baryons and mesons production mechanism. They include the differential cross sections for baryon and meson production, the ratio baryon to meson, baryon-antibaryon asymmetry and the strange quark fragmentation universality. The results have been compared with the popular hadronization Monte Carlo models [1, 2, 3] for the different values of the strangeness supression factor λ_s. [1] In other analyses also Bose-Einstein correlations for the charged and neutral kaon pairs and search for the strange pentaquarks have been performed. These new results bring new informations on the hadronization process.

INCLUSIVE PRODUCTON OF Λ, $\bar{\Lambda}$ AND K_S^0

The differential cross sections with respect to various variables of the strange kaons and baryons production have been measured by ZEUS in deep inelastic scattering (DIS) [4] and compared with Monte Carlo simulation [5].

In Fig. 1a the differential spectra for baryons $\Lambda+\bar{\Lambda}$ and mesons K_S^0, baryon-antibaryon asymmetry $(\Lambda - \bar{\Lambda})/(\Lambda + \bar{\Lambda})$ and baryon to meson ratio $\Lambda+\bar{\Lambda}/K_S^0$ as the function of the kinematic variable Q^2 is shown. The cross sections for baryons and mesons production are reasonable described by Monte Carlo simulations. [2] No asymmetry Λ - $\bar{\Lambda}$ was found

[1] It descibes the production probability of s quark relative to u or d quark: λ_s=P(s)/P(u); P(d)=P(u).
[2] ARIADNE MC, version 4.08 [1] interfaced to HERACLES via DJANGOH [5].

CP828, *Multiparticle Dynamics*
edited by V. Šimák, M. Šumbera, Š. Todorova, and B. Tomášik

though it can be expected in the case of the baryon number transport from initial proton to baryon system. The baryon to meson ratio can be descibed by Monte Carlo except the small Q^2 region where MC tends to underestimate the data. The similar behaviour was observed when ploted those distributions, as the function of the kinematic variable x. In this case the baryon to meson ratio was underestimated by MC in the lower x region. Figure 1b shows the same distributions as a function of pseudorapidity η in

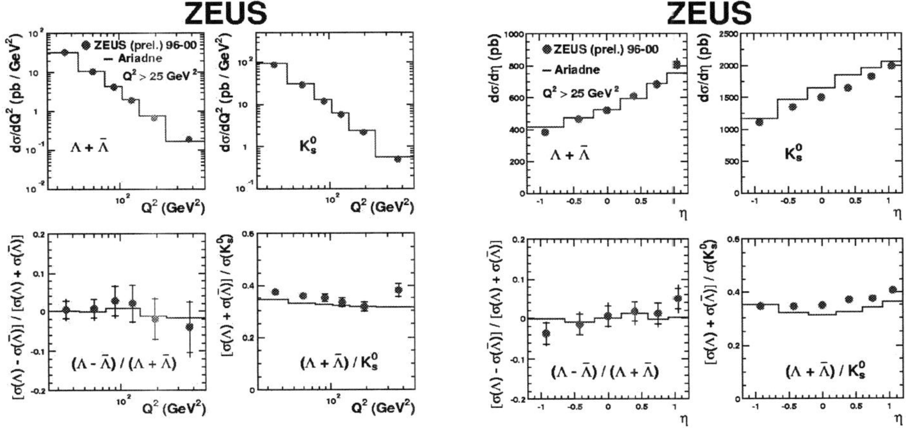

FIGURE 1. Cross sections and ratios compared with ARIADNE: (a) as a function of Q^2 and (b) as a function of η.

the LAB frame. For the most of values of η the MC distribution for K_S^0 overstimates the data. Also there is disagreement in ratio of the baryon to meson, particularly in $\eta > 0$ region related to proton fragmentation. To look closer at this case an additional studies was done in the Breit frame which allows on the separation of the proton remnant fragmentation region from this coming from the scattered quark. Using the so called scaled momentum variable $x_p = 2p/Q$ as fragmentation variable, where p is the particle momentum measured in the Breit frame, the x_p distributions for K_S^0 and Λ were measured and compared with the results of the two different Monte Carlo: ARIADNE + JETSET [1, 3] and HERWIG [2]. In the case of the JETSET simulation, two different values of strangeness supression factor λ_s were used: 0.2 and 0.3. It was found that HERWIG can not describe the data for both the proton and the current fragmentation regions. The ARIADNE simulation can descibe the data if the value $\lambda_s = 0.2$ was used for current region and $\lambda_s = 0.3$ for the proton hemisphere. The result of the production ratio of Λ to K_S^0 in the Breit frame leads to the similar conclusion. The default value of $\lambda_s = 0.3$ is used in JETSET simulation which reasonable describes the LEP data. In contrary to this the DIS results suggest rather smaller value of $\lambda_s \approx 0.2$.

BOSE-EINSTEIN CORRELATIONS (BEC)

The space-time characteristics of the kaons emission region were studied by ZEUS [6] throught measurements of the Bose-Einstein correlation function $R(Q_{12})$ for the charged $K^{\pm}K^{\pm}$ and neutral $K_S^0K_S^0$ pairs. Introducing a double ratio method to remove other than BEC corrrelations with so-called mixed-event sample P_{mix}, contained pairs of kaons from different events, the two-particle correlation function $R(Q_{12})$ was defined:

$$R(Q_{12}) = \frac{P(Q_{12})^{data}}{P_{mix}(Q_{12})^{data}} / \frac{P(Q_{12})^{MC,noBEC}}{P_{mix}(Q_{12})^{MC,noBEC}} \tag{1}$$

The invariant variable Q_{12}, was define as 4-momenta difference of the two particles: $Q_{12} = \sqrt{-(p_1 - p_2)^2}$. The standard Goldhaber parametrisation of $R(Q_{12})$ for the case of a Gausian shape of emission source was used:

$$R(Q_{12}) = \alpha(1 + \lambda e^{-Q_{12}^2 r^2})(1 + \delta Q_{12}), \tag{2}$$

where r descibes the radius of the emitting source and λ - a strength of the effect. The α and δ parameters describe the overal normalization and background. The result of the

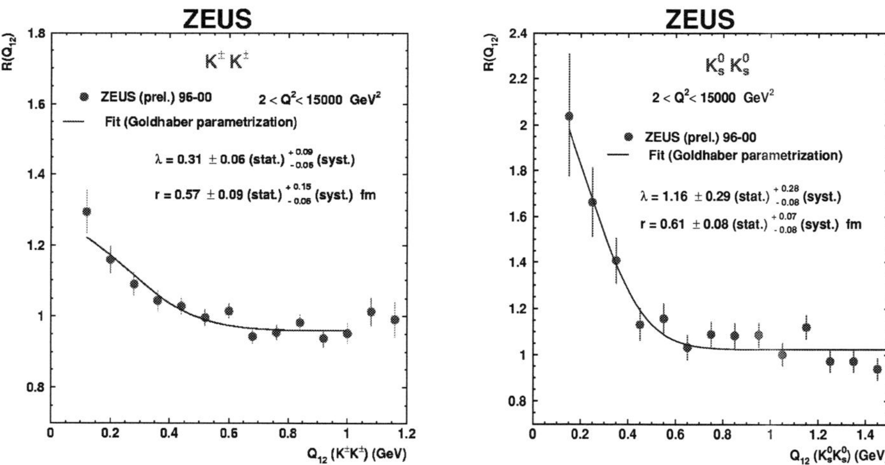

FIGURE 2. Bose-Einstein correlation function: (a) for $K^{\pm}K^{\pm}$ pairs and (b) for $K_S^0K_S^0$ pairs.

fit (2) to charged and neutral kaon pairs is shown in Fig. 2 together with the extracted values for radius r and λ. The value of the radius for charged kaons is consistent with that for charged pions in DIS [7] and those obtained for neutral kaons (Fig. 2b). However the λ values are different for charged and neutral kaons: it is significantly larger for neutral ones. This is mainly effect of the influence of a $f_0(980)$ resonance in low Q_{12} region, which is not well descibed in the ZEUS Monte Carlo. In extended analysis it was checked that small contribution of f_0 can significantly decrease λ with very small changes in radius. Figure 3a shows comparison of the values of the r and λ measured in

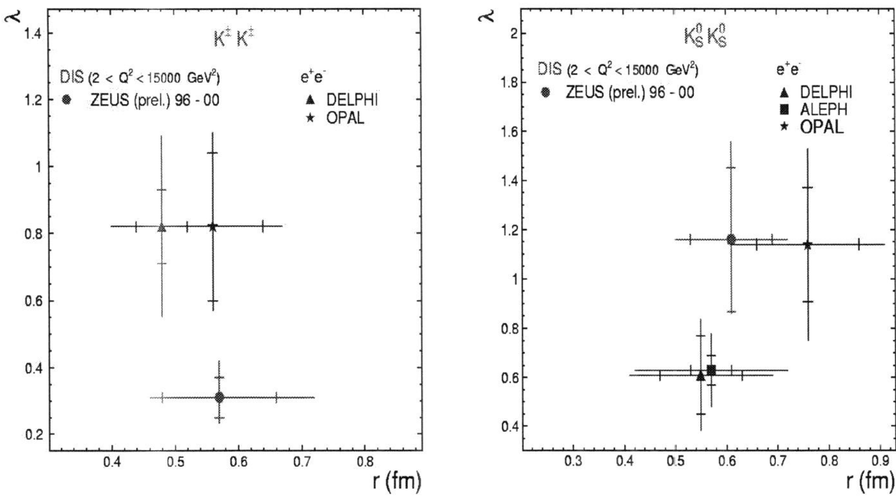

FIGURE 3. DIS and e^+e^- (LEP) results for r and λ: (a) for charged kaons and (b) for neutral kaons.

DIS with LEP results [8, 9] for charged kaons. The reasonable agreement for radius was found. The smaller value of λ can be related to fact that DIS data populate mostly proton fragmentation region where the production of resonances in the final state is large. This can lead to significant increase of the non-prompt kaons coming from resonances decay and to the lower value for λ. For neutral kaons comparison with LEP results is shown in Fig. 3b. The value of λ is similar to OPAL [10] and larger in comparison with DELPHI [8] and ALEPH [11]. But in both experiments the final results were corrected for the influence of $f_0(980)$ resonance. This was not done for OPAL.

Previously LEP results for BEC and FDC [3] indicate that the radius can depends on the hadron mass. Figure 4 shows such LEP results together with the recent ZEUS DIS results. The DIS points are for pions and kaons and LEP for pions, kaons, protons and Λ's. The r(m) dependence was compared with predictions coming from theoretical calculations using Heisenberg uncertainty relations and QCD approach together with the virial theorem [12, 13]. Both theoretical calculations well describe the data. [4] The string (LUND) model [12, 15] however does not predict such dependence. More results for heavier particles in DIS measurements (proton, Λ) will help to to clarify such situation.

---

[3] The Fermi-Dirac correlations were measured for protons and Λ ($\bar{\Lambda}$).

[4] The other theoretical description of the mass dependence of the source radius, can be found in [14].

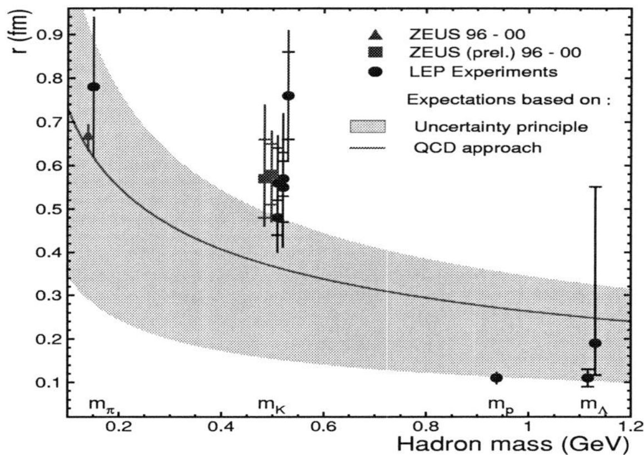

FIGURE 4. The dependence of the radius on the hadron mass. Data and the theoretical predictions.

SEARCH FOR PENTAQUARKS

The existence of baryon states with more than classical 3 quarks is not forbiden by QCD. The specific prediction given by Diakonov et al [16] for mass and width of the exotic five quarks baryon (pentaquark) Θ^+ $(uudd\bar{s})$ inspired many exeperimental groups to look for it in different processes. Recently search for such strange pentaquarks in DIS and photoproduction were reported by ZEUS and H1 collaborations [17, 18]. Figure 5a shows invariants mass distribution of $K^0_S p\bar{p}$ for one of possible decay channel of Θ^+. The clear peak near 1522 MeV and width close to 8 MeV was found at the level of 4.6 σ in DIS data. For photoproduction the signal is absent. In further analysis it was found that photoproduction is not the proper place to look for such state as the large multiplicity creates the significant combinatorial background which lead to small signal to background ratio. The properties of Θ^+ production was studied looking for such signal in two separate hemispheres: one with significant contributions from proton fragments $(\eta > 0)$ and other one dominated by pure fragmentation $(\eta < 0)$. It was found that Θ^+ is mostly produced in forward rapidity hemisphere. In Fig. 5b this observation is illustrated. From other side the similar type of analysis done for well reconstructed $\Lambda(1520)$ decaying into $K^{\pm}\bar{p}(p)$, gives the similar number of reconstructed Λ's in both η regions. The result obtained for Θ^+ suggests that in its production the diquark fragmentation mechanism was involved [19] which is not a case for Λ.

In contrary to ZEUS the H1 experiment [18] has not found the Θ^+ signal in the similar effective mass distribution even supplying the similar selection for low momentum protons (see Fig. 6a). But the upper limit of Θ^+ cross section obtained in the Q^2 range :$20 < Q^2 < 100$ GeV^2 (see Fig. 6b) : $\sigma(\Theta^+) \sim (80 - 115)$ pb does not contradict the ZEUS observation ($\sim$ 120 pb for $Q^2 > 20$ GeV^2).

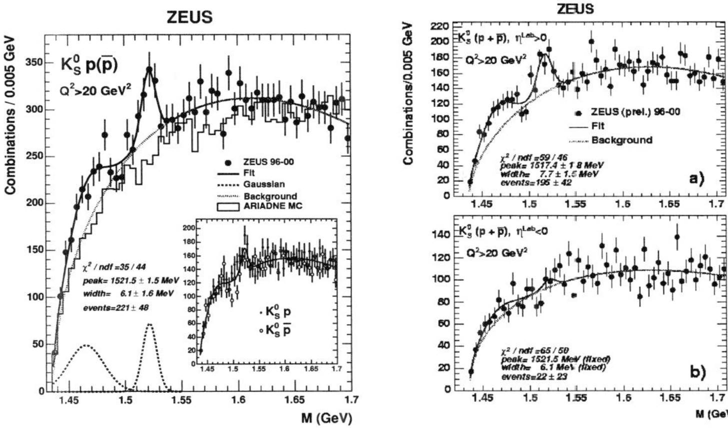

FIGURE 5. Peak in invariant-mass spectrum for $K_S^0\, p\, (\bar{p})$ decay channel: (a) in the whole phase space and (b) in the proton fragmentation and scattered quark fragmentation hemispheres.

The ZEUS also searches for heavy strange pentaquark decaying to $\Xi\pi$ [20]. The possible existence of such state was reported previously by NA49 experiment [21] at 1862 MeV with width < 18 MeV. No signal was found in the mass region suggested by NA49 experiment.

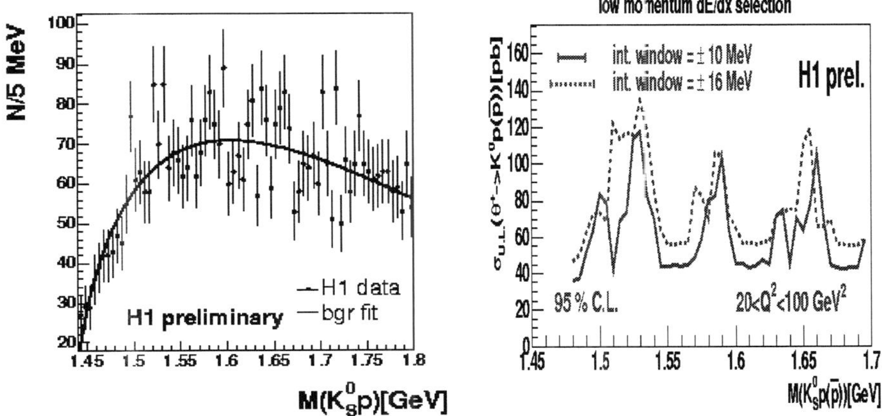

FIGURE 6. H1 results in the case of the low momentum proton selection: (a): Invariant-mass spectrum for $K_S^0 p\, (\bar{p})$ and (b): Upper limits of the cross section.

ACKNOWLEDGMENTS

I would like to thank Krystyna Olkiewicz, Anna Galas and Sergei Chekanov for their help and useful discussions and suggestions.

REFERENCES

1. L. Lönnblad, Comp. Phys. Comm., **71**, 15 (1992).
2. G. Marchesini, B.R. Weber, Nucl. Phys. **B 238**, 1 (1984); G. Marchesini et al., Comp. Phys. Comm. **67**, 465 (1992).
3. T. Sjöstrand, Comp. Phys. Comm. **82**, 74 (1994).
4. ZEUS Coll., A. Cottrell, "Polarization and asymmetries in neutral strange particle production", in Proc. DIS 2005 (Madison, April 27 - May 1, 2005, USA); ZEUS Coll., S. Chekanov et al., "Neutral strange particle production in deep inelastic scattering at HERA", Paper submitted to the XXII Inter. Symp. on Lepton-Photon Inter. at High Energy, LP2005, Uppsala, June 30 - July 5, 2005, Sweden.
5. H. Spiesberger, Event Generation for *ep* Interactions at HERA Including Radiative Proceses, 1998,URL; http://www.desy.de/~hspiesb/djangoh.html.
6. ZEUS Coll., A. Galas "Neutral and charged kaon Bose-Einstein correlations in DIS", in Proc. DIS 2005 (Madison, April 27 - May 1, 2005, USA); ZEUS Coll., S. Chekanov et al., "Bose-Einstein correlations of neutral and charged kaons in deep inelastic scattering at HERA", Paper submitted to the XXII Inter. Symp. on Lepton-Photon Inter. at High Energy, LP2005, Uppsala, June 30 - July 5, 2005, Sweden.
7. H1 Coll., C. Adloff et al., Z. Phys. **C 75**, 437 (1997);
ZEUS Coll., S. Chekanov et al., Phys. Lett. **B 583**, 231 (2004).
8. DELPHI Coll., P. Abreu et al., Phys. Lett. **B 379**, 330 (1996).
9. OPAL Coll., G. Abbiendi et al., Eur. Phys. **C 21**, 23 (2001).
10. OPAL Coll., R. Akers et al., Z. Phys. **C 67**, 389 (1995).
11. ALEPH Coll., S. Schael et al., Phys. Lett.. **B 611**, 66 (2005).
12. G. Alexander, Rep. Prog. Phys. **66**, 481 (2003); G. Alexander, Acta Phys. Polon. **B** 35, 69 (2004).
13. G. Alexander et al., Phys. Lett. **B 452**, 159 (1999).
14. A. Bialas et al., Phys. Rev. **D 62**, 114007; A. Bialas et al., Acta Phys. Polon. **B 32**, 2901 (2001).
15. B. Andersson et al., Phys. Rep. **97**, 31 (1983).
16. D. Diakonov, V. Petrov, M.V. Polyakov, Z. Phys.**A359**, 305 (1997).
17. ZEUS Coll., S. Chekanov et al., Phys. Lett. **B 591**, 7 (2004).
18. H1 Coll., A. Aktas et al., "Search for the pentaquark Θ^+ decaying to $K_S^0 p$ with the H1 detector at HERA", Paper submitted to the XXII Inter. Symp. on Lepton-Photon Interactions at High Energy, LP2005, Uppsala, June 30 - July 5, 2005, Sweden.
19. S. Chekanov, hep-ph/0502098.
20. ZEUS Coll., S. Chekanov et al., Phys. Lett. **B 610**, 212 (2005).
21. NA49 Coll., C. Alt et al., Phys. Rev. Lett. **92**, 042003 (2004).

Heavy-Flavor Collectivity – Light-Flavor Thermalization at RHIC

K. Schweda [1]

Lawrence Berkeley National Laboratory, One Cyclotron Rd MS70R0319, Berkeley, CA 94720, USA

Abstract. Flow measurements of multi-strange baryons from Au+Au collisions at RHIC energies demonstrate that collectivity develops before hadronization, among partons. To pin down the partonic EOS of matter produced at RHIC, the status of thermalization in such collisions has to be addressed. We propose to measure collective flow of heavy-flavor quarks, e.g. charm quarks, as an indicator of thermalization of light flavors (u,d,s). The completion of the time of flight barrel and the proposed upgrade with a μVertex detector for heavy-flavor identification in STAR are well suited for achieving these goals.

Keywords: Ultra-relativistic nuclear reactions, quark-gluon plasma, collectivity, thermalization, heavy-flavor quarks
PACS: 25.75.Dw, 25.75.Ld

INTRODUCTION

Quantum Chromo–Dynamics (QCD) is the theory of strong interactions. Lattice calculations of QCD predict that at a critical temperature of $T_c \simeq 170$ MeV a phase transition of ordinary nuclear matter to a deconfined state of quarks and gluons occurs [1]. Quarks and gluons are not confined in hadrons any more; they become asymptotically free. Under the same conditions, chiral symmetry is approximately restored and quark masses are reduced from their large effective values in hadronic matter to their small bare masses.

In ultra–relativistic nuclear collisions, a system with a temperature larger than the critical temperature T_c is expected to be created. The development of collectivity at the partonic level (among quarks and gluons) and the degree of thermalization are closely related to the equation of state of partonic matter: Re-scattering among constituents and the density profile lead to the development of collective flow. In case of sufficient re-scattering, the system might be able to reach local thermal equilibrium.

In this paper, we show by means of flow measurements of multi-strange baryons that at RHIC energies collectivity develops before hadronization, among partons. We further suggest to measure heavy-flavor (c,b) collective flow to probe thermalization of light quarks.

---

[1] Present address: University of Heidelberg, Physikalisches Institut, Philosophenweg 12, 69120 Heidelberg, Germany.

CP828, *Multiparticle Dynamics*
edited by V. Šimák, M. Šumbera, Š. Todorova, and B. Tomášik

MULTI-STRANGE HADRON FLOW - PARTONIC COLLECTIVITY

In ultra-relativistic nuclear collisions, measured final-state transverse–momentum spectra can be fit within a hydrodynamically motivated approach, with a kinetic freeze–out temperature T_{fo} and a mean collective flow velocity $\langle\beta_T\rangle$ as the relevant parameters [2]. Figure 1 shows results of those fits from Au+Au collisions at $\sqrt{s_{\mathrm{NN}}}$= 200 GeV at RHIC in the T_{fo}-$\langle\beta_T\rangle$ plane. Dashed and solid lines represent 1-σ and 2-σ contours, respectively. As the collisions become more and more central, the bulk of the system dominated by the yields of π, K, p appears to be cooler and develops stronger collective flow, representing a strongly interacting system expansion. At the most central collisions, the tem-

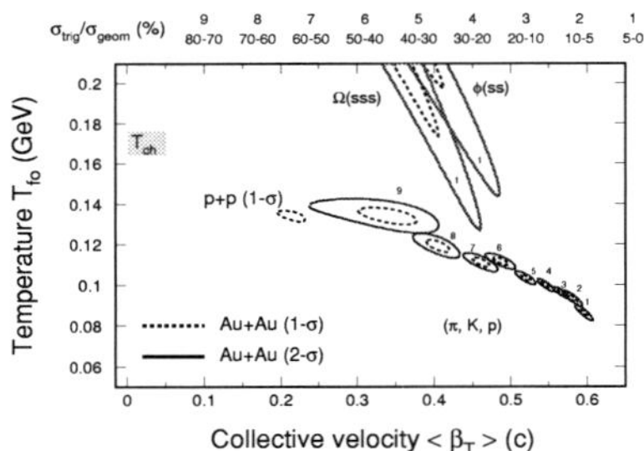

FIGURE 1. 1–σ (dashed lines) and 2–σ (solid lines) contours for the transverse radial flow velocity $\langle\beta_T\rangle$ and the kinetic freeze-out temperature parameter T_{fo} derived from hydrodynamically motivated fits to particle spectra. The results for π, K and p, are numbered from 1 (most central) to 9 (most peripheral) Au+Au collisions and p+p collisions [3]. Results for the multi-strange hadrons ϕ and Ω are shown in the top of for most central Au+Au collisions only. The numbers on the top give the fraction of total hadronic cross section for centrality bins 1-9. This figure has been taken from [4].

perature parameter and the velocity are $T_{\mathrm{fo}} \sim 100$ MeV and $\langle\beta_T\rangle \sim 0.6$(c), respectively. On the other hand, for the same collision centrality, the multi-strange hadrons ϕ and Ω freeze-out at a higher temperature $T_{\mathrm{fo}} \sim 180$MeV, close to the point at which chemical freeze-out occurs [5]. A similar behavior was also observed in Au+Au collisions at $\sqrt{s_{\mathrm{NN}}}$= 130 GeV [6].

Multi-strange hadrons might have smaller hadronic cross sections [7] and therefore decouple from the fireball early, perhaps right at the point of hadronization. This would explain the low $\langle\beta_T\rangle$ and higher temperature parameter. Most importantly, the finite value of $\langle\beta_T\rangle$ therefore must be cumulated prior to hadronization - via partonic interactions.

Elliptic flow, due to its self-quenching nature, is an early stage signal [8]. In non-central nuclear collisions, the initial overlap zone between the colliding nuclei is spatially deformed. If the matter produced in the reaction zone re–scatters efficiently, this spatial anisotropy is transferred into momentum space and the initial, locally isotropic, momentum distribution develops anisotropies. This anisotropy in momentum space is quantified by the second Fourier coefficient v_2, the elliptic flow parameter. Results on elliptic flow measurements at RHIC are shown in Fig. 2, (a) for π, K^0_S, p and $\Lambda+\overline{\Lambda}$ [9, 10],

(b) double-strange $\Xi^- + \overline{\Xi}^+$ [11] and (c) triple-strange $\Omega^- + \overline{\Omega}^+$ [11]. The elliptic flow parameter increases with p_T and then saturates at larger p_T. In the lower p_T region, a mass ordering is observed with lighter particles exhibiting larger elliptic flow parameters. The shaded bands show results from hydrodynamical calculations for π (upper edge) to Ω (lower edge), assuming zero mean free path length and therefore infinitely fast re–scattering which leads to instantaneous local thermal equilibrium distributions. These calculations qualitatively describe the experimental results in the lower p_T region, especially the observed mass ordering. As can be seen in Fig. 2, even the multi-strange baryons $\Xi^- + \overline{\Xi}^+$ (b) and $\Omega^- + \overline{\Omega}^+$ (c) do significantly flow. This suggests that collective flow of multi-strange baryons $\Xi^- + \overline{\Xi}^+$ and $\Omega^- + \overline{\Omega}^+$ indeed develops before hadronization - among partons.

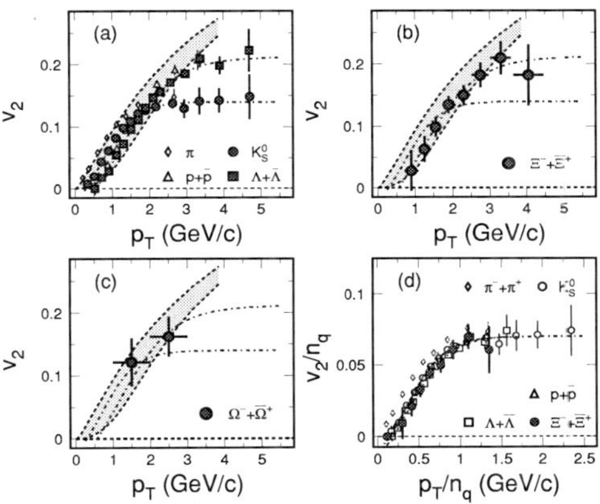

FIGURE 2. Results on elliptic flow measurements at RHIC for (a) π, K^0_S, p and $\Lambda + \overline{\Lambda}$ [9, 10], (b) double-strange $\Xi^- + \overline{\Xi}^+$ [11] and (c) triple-strange $\Omega^- + \overline{\Omega}^+$ [11]. Values for v_2 versus p_T both scaled by the number n of constituent quarks are shown in panel (d). The shaded bands show results from hydrodynamical calculations for π (upper edge) to Ω (lower edge). The dash-dotted lines are results from empirical fit-functions [12] for baryons (upper) and mesons (lower).

In the intermediate p_T region (2-6 GeV/c), the calculations overshoot the data. At these momenta, the mean free path length is relatively large leading to deviations from hydrodynamic behavior. In this region, the saturation level depends on particle type: Baryons saturate at larger values than mesons. The dash-dotted lines are results from empirical fit-functions [12]. This particle type dependence is accounted for in quark coalescence models [13]. In these hadronization models, hadrons are dominantly formed by coalescing massive constituent quarks from a partonic system with the intrinsic assumption of collective flow among these partons. These models predict a universal scaling of the observed elliptic flow v_2 and the hadron transverse momentum p_T with the number of constituent quarks n (meson, $n = 2$, baryon, $n = 3$). The accordingly n-scaled values for v_2 versus p_T are shown in Fig. 2(d). Above a parton momentum $p_T/n >$ 0.7 GeV/c, the predicted universal scaling holds within experimental uncertainties. An exception might be π which can be attributed to the contribution of feeddown from resonances with π in the decay channel [12]. The successful prediction of quark coalescence models further supports the idea that collectivity develops at the partonic stage at RHIC. The important question and maybe the final step to a QGP discovery at RHIC is the status of thermalization of light quarks.

HEAVY-FLAVOR COLLECTIVITY AS A PROBE OF LIGHT-FLAVOR THERMALIZATION

Heavy-flavor quarks are special probes because of their heavy mass. If chiral symmetry is restored in a QGP, light quarks obtain their small current masses. On the other hand, heavy quarks get almost all their mass from their coupling to the Higgs field [14]. Thus, heavy quarks stay heavy - even in a QGP. The observation of heavy-quark collective flow indicates multiple interactions among partons. This would suggest that light quarks are thermalized.

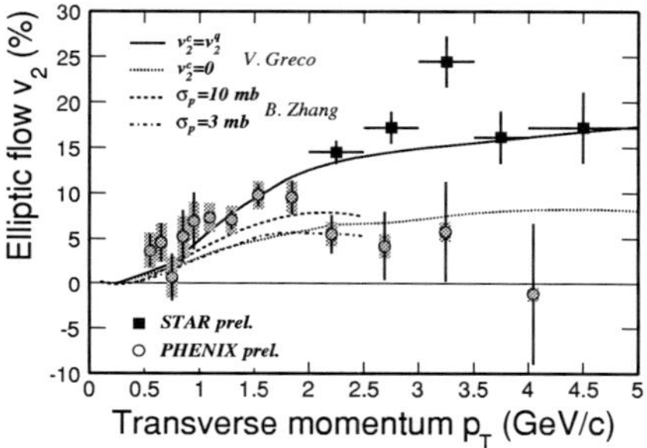

FIGURE 3. Results of elliptic flow measurements on electrons from heavy-flavor semileptonic decays [17] as a function of p_T from STAR (closed squares) and PHENIX (open circles). The curves show results from a quark coalescence model [18] assuming identical flow of heavy and light quarks (solid) and no heavy quark flow (dotted) and microscopic calculations using different partonic cross sections [19] of 10 mb (dashed) and 3 mb (dash-dotted).

First results on heavy-flavor production at RHIC have been reported from observing electrons stemming from the decay of heavy-flavor quarks [15, 16]. Recent results of elliptic flow measurements on electrons from heavy-flavor semileptonic decays [15, 17] are shown in Fig. 3 as a function of p_T from STAR (closed squares) and PHENIX (open circles). The electron momentum range p_T=0.5-2.0 GeV/c corresponds to heavy-flavor hadron p_T=1.0-4.0 GeV/c. In this region, the values of v_2 are significantly different from zero. The curves show results from calculations within a quark coalescence model [18] assuming identical flow of heavy and light quarks (solid) and no heavy quark flow (dotted) and microscopic calculations using different partonic cross sections [19] of 10 mb (dashed) and 3 mb (dash-dotted). Both models support the idea of heavy-flavor collectivity at RHIC, while the unexpectedly large cross section needed to describe the experimental data comes as a surprise. It is possible, as argued by several theorists, that elliptic flow and energy loss of heavy quarks are correlated [20, 21, 22, 23]. It is therefore very interesting to study both elliptic flow and nuclear modification factors.

However, due to the decay kinematics, important information on heavy-flavor dynamics is smeared out [24]. It seems that we do not fully understand the underlying mechanism of heavy-flavor interaction with the dense medium. At higher p_T, therefore, it is also important to measure distributions from directly reconstructed D-mesons in order to isolate the bottom contributions in collisions at RHIC.

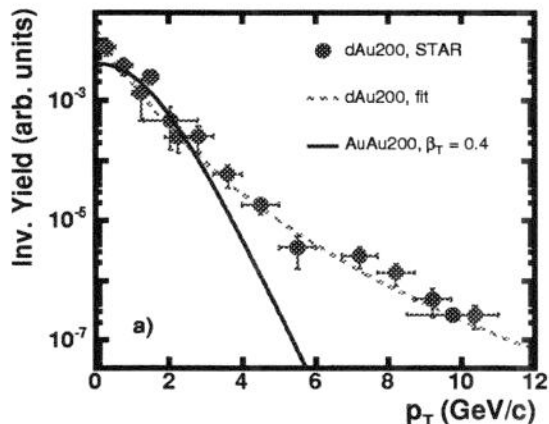

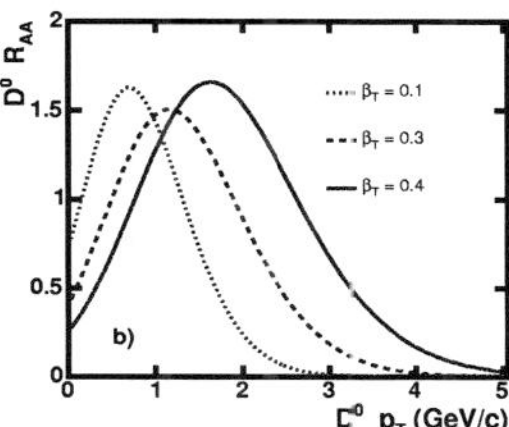

FIGURE 4. (a) The measured invariant yield of D-mesons from direct reconstruction through the invariant mass of decay-daughter candidates in d+Au collisions at $\sqrt{s_{NN}}$=200 GeV as a function of p_T [25, 26]. The dashed line shows the fit-result of a pQCD inspired power-law function. A prediction from hydro-dynamically inspired model calculations is shown by the solid line. (b) The modification of the D^0 spectrum as a function of transverse momentum for three different flow velocities.

The measured invariant yield of D-mesons from direct reconstruction through the invariant mass of decay-daughter candidates in d+Au collisions at $\sqrt{s_{NN}}$=200 GeV is shown in Fig. 4(a) as a function of p_T [25, 26]. The spectrum steeply falls with increasing momentum, followed by a long tail at high p_T. The dashed line shows the fit-result of a pQCD inspired power-law function, describing the experimental data over the whole momentum range. A prediction from hydro-dynamically inspired model calculations is shown by the solid line. In these calculations, a kinetic freeze-out temperature T_{fo} = 160 MeV and an average flow velocity $\langle\beta_T\rangle$= 0.4 (in units of speed of light) was assumed. Both curves are normalized to the same yield in the momentum range p_T= 0 - 14 GeV/c. The presence of collective flow modifies the spectrum, its shape changes from concave (no-flow in d+Au collisions) to convex (flow in Au+Au collisions).

The modification of the D^0 spectrum is further quantified by taking the ratio R_{AA} of the spectra expected from Au+Au collisions (flow) relative to d+Au collisions (non-flow). Figure 4(b) shows the modification of the D^0 spectrum as a function of average transverse momentum for three different flow velocities with $\langle\beta_T\rangle$= 0.1 (dotted), 0.2 (dashed) and 0.4 (solid). The modification is in the order of 30-50% with the maximum moving to larger momentum with increasing flow velocity.

Due to the large multiplicities of π,K,p and the rather small production cross section for charm-hadrons, the combinatorial background in the invariant mass distribution is roughly 1000 times larger than the signal [27]. Extending particle identification by time of flight information will improve the statistical significance by a factor of five. This large combinatorial background leads to systematic uncertainties of extracted charm-hadron yields in the order of 30%. On the other hand, elliptic flow modulates particle yields with respect to the reaction plane in the order of 10%. To overcome these large systematic uncertainties and make precise heavy-flavor elliptic flow measurements feasible, we propose to upgrade STAR with μ-vertex capabilities to identify heavy-flavor hadrons through their displaced decay vertex [28].

SUMMARY

Elliptic flow measurements have demonstrated that partonic collectivity, collective flow of partons, develops in 200 GeV Au+Au collisions at RHIC. To pin down the partonic EOS of matter produced at RHIC, the status of thermalization in such collisions has to be addressed. Since the masses of heavy-flavor quarks, e.g. charm quarks, are much larger than the maximum possible excitation of the system created in the collision, heavy-flavor collective motion could be used to indicate the thermalization of light flavors (u,d,s). The completion of the time of flight barrel and the proposed upgrade with a μVertex detector for heavy-flavor identification in STAR are well suited for achieving these goals.

ACKNOWLEDGMENTS

Discussions with Drs. X.Dong, J. Gonzalez, Y. Lu, H.G. Ritter, P. Sorensen, L. Ruan, N. Xu, Z. Xu and H. Zhang are gratefully acknowledged.

REFERENCES

1. F. Karsch, Nucl. Phys. A698 (2002) 199c.
2. E. Schnedermann, J. Sollfrank and U. Heinz, Phys. Rev. C48 (1993) 2462.
3. J. Adams, *et al.*, Phys. Rev. Lett. 92 (2004) 112301.
4. K.Schweda and N.Xu, Acta Phys. Hung. A22, (2005) 103.
5. P. Braun-Munzinger *et al.*, Phys. Lett. B344 (1995) 43; P. Braun-Munzinger, I. Heppe, and J. Stachel, Phys. Lett. B465 (1999) 15.
6. J. Adams *et al.*, Phys. Rev. Lett. 92 (2004) 182301.
7. H. van Hecke, H. Sorge, and N. Xu, Phys. Rev. Lett. 81 (1998) 5764.
8. H. Sorge, Phys. Rev. Lett. 82 (1999) 2048.
9. S.S. Adler *et al.*, Phys. Rev. Lett. 91 (2003) 182301.
10. J. Adams *et al.*, Phys. Rev. Lett. 92 (2004) 052302.
11. J. Adams *et al.*, Phys. Rev. Lett. 95 (2005) 122301.
12. X. Dong *et al.*, Phys.Lett. B597 (2004) 328.
13. For an overview see R. Fries, J. Phys. G30 (2004) 853.
14. B. Müller, hep-ph/0410115.
15. X. Dong, *these proceedings*.
16. K. Adcox *et al.*, Phys. Rev. Lett. 88 (2002) 192303;
J. Adams *et al.*, Phys. Rev. Lett. 94 (2005) 062301.
17. X. Dong, *Proceedings of the Quark Matter 2005 Conference*, to appear in Nucl. Phys. A.
18. V. Greco *et al.*, Phys.Lett. B595 (2004) 202.
19. B. Zhang, L.-W. Chen, and C-M. Ko, Phys.Rev. C72 (2005) 024906.
20. M. Djordjevic *et al.*, nucl-th/0507019; Magdalena Djordjevic, Miklos Gyulassy, and Simon Wicks, Phys. Rev. Lett. 94 (2005) 112301.
21. N. Armesto *et al.*, Phys. Rev. D71 (2005) 054027; hep-ph/0501225
22. H. van Hees, V. Greco, and R. Rapp, nucl-th/0508055.
23. G.D. Moore and D. Teaney, Phys. Rev. C71 (2005) 064904.
24. S. Batsouli *et al.*, Phys. Lett. B557 (2003) 26.
25. A. Tai, *these proceedings*.
26. A. Tai, J. Phys. G30 (2004) S809.
27. H. Zhang, *Proceedings of the Quark Matter 2005 Conference*, to appear in Nucl. Phys. A.
28. K. Schweda, *Proceedings of the Quark Matter 2005 Conference*, to appear in Nucl. Phys. A.

Bose-Einstein correlations from "within"

O.V. Utyuzh*, G. Wilk* and Z.Włodarczyk†

*The Andrzej Sołtan Institute for Nuclear Studies; Hoża 69; 00-681 Warsaw, Poland
†Institute of Physics, Świętokrzyska Academy, Świętokrzyska 15; 25-406 Kielce, Poland

Abstract. We describe an attempt to model numerically Bose-Einstein correlations (BEC) from "within", i.e., by using them as the most fundamental ingredient of some Monte Carlo event generator (MC) rather than considering them as a kind of (more or less important, depending on the actual situation) "afterburner", which inevitably changes original physical content of the MC code used to model multiparticle production process.

Keywords: Bose-Einstein correlations; Statistical models; Fluctuations
PACS: 25.75.Gz 12.40.Ee 03.65.-w 05.30.Jp

Introduction. The problem of BEC is so well known that we shall skip introductory remarks (referring in that matter to other presentations at this conference and to [1] for more information) and instead we shall start right away with our subject concerning *the proper numerical modelling of BEC*, which we call *BEC from within*. Although phenomenon of BEC is with us from the very beginning of the systematic investigation of multiparticle production processes, its modelling is still virtually nonexisting. With the exception of attempts presented in [2, 3] all other approaches are tacitly assuming that on the whole BEC constitute only a small effect and it is therefore justify to add it in some way to the already known outputs of the MC event generators widely used to model results of high energy collisions (in the form of the so called *afterburner*) [1]. There are two types of such afterburners:

(*a*) those modifying accordingly energy-momenta of identical secondaries (and correcting afterwards the whole sample for energy-momentum conservation) - they apply to each event separately;

(*b*) those selecting events which already have (due to some fluctuation present in any MC code) right energy-momenta of identical secondaries and counting them (by introducing some weights) many times (see [1] for references) - in this case energy-momentum balance is left intact but instead the particle spectra provided by MC code are distorted (albeit in most cases only slightly); they apply only to all events.

Example of modifications introduced by afterburner. It must be stressed that modifications that such afterburners introduce to *physical background* behind a given MC code were never investigated. Tacit assumption made is that they are small and therefore irrel-

[1] Methods presented in [2] were never used in practice, only [3] applied their approach to e^+e^- annihilation data.

CP828, *Multiparticle Dynamics*
edited by V. Šimák, M. Šumbera, Š. Todorova, and B. Tomášik

evant [2]. The problem is that with the MC codes in use at present it is practically impossible to estimate the nature and strength of such changes. In can be done only by using some simple scheme of cascading, for example simple cascade model for hadronization developed by us some time ago [4]. This model *per se* leads to no BEC (cf. left panel of Fig. 1). To obtain effect of BEC one has to add to it some simple afterburner, for example the one proposed by us in [5]. In it one preserves the whole space-time and energy-momentum structure of each event but changes the charge assignment to secondaries in such way that clusters of identically charged particles occur (the original number of $(+)$, $(-)$ and neutrals remains the same). This automatically leads to BEC [3]. The right panel of Fig. 1 shows what are the changes introduced by this afterburner in the original cascade: *the multicharged vertices* occur now (with total charge in the whole event remaining conserved). It means that, *in principle*, one could obtain BEC directly from MC (i.e., without any afterburner) allowing in it for the appearance of such vertices (according to some prescribed scheme - for example as some multicharged clusters, such possibility has been already mentioned in [6])). However, it would be then extremely difficult to run such cascade till the very end without producing spurious multicharged particles not observed in nature. Our afterburner can be then considered as a kind of shortcut realization of such possibility with well defined physical consequences. We argue therefore that each afterburner changes original MC code it is attached to in a similar fashion but this statement cannot be at the moment substantiated[4].

BEC from "within". The above observation was one of our motivations to look for the MC scheme, which would be build around BEC rather than starting from some single-particle observables. This idea has been for the first time used in [3] where particles were selected from a grand-canonical-like distribution with temperature T and chemical potential μ chosen in such way as to describe rapidity and multiparticle distributions[5] and particles were then distributed into some rapidity cells of given width, each cell containing only particles of the same charge; the method was then (quite successfully) used to describe e^+e^- annihilation data. Introduction of these cells was the real origin of the BEC observed. Actually the idea that BEC demands that particles of the same charge are emitted from some cells (named *elementary emitting cells* - EEC) was

---

[2] Actually the frequent practice is to use the original MC data to obtain single particle spectra and to calculate the corresponding BEC *afterwards* by means of some afterburner.

[3] Notice that such reassignment of charges results in the effective bunching of momenta very much similar to that assumed in the method (a) mentioned above [1].

[4] Notice that the problem, which is clearly visible in the CAS model, is not at all straightforward in other approaches. However, at least in the string-type models of hadronization, one can imagine that it could proceed through the formation of charged (instead of neutral) color dipoles, i.e., by allowing formation of multi(like)charged systems of opposite signs out of vacuum when breaking the string. Because only a tiny fraction of such processes seems to be enough in getting BEC in the case of CAS model, it would probably be quite acceptable modification. It is worth to mention at this point that there is also another possibility in such models, namely when strings are nearby in the phase space one can imagine that production of given charge with one string enhances emission of the same charge from the string nearby - in this case one would have a kind of *stimulated emission* discussed already in [7].

[5] Actually, in [3], which was using information theory approach based on Shannon entropy, T and μ are, in fact, two lagrange multipliers obtained from energy conservation and charge conservation constraints.

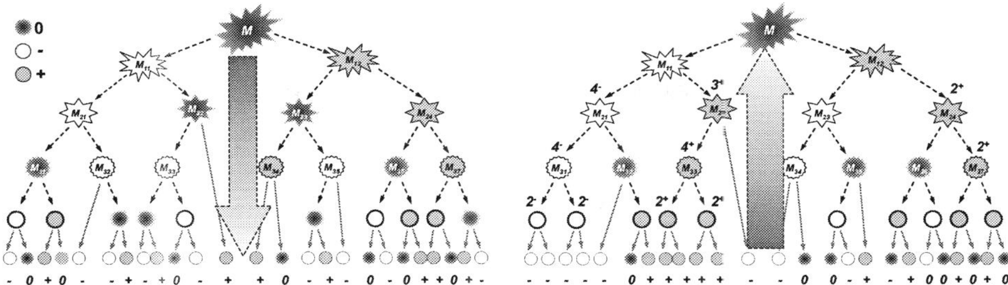

FIGURE 1. Example of charge flows in MC code using simple cascade model for hadronization [4]: left panel - no effect of BEC observed; right panel - after applying afterburner described in [5] (based on new assignement of charges to the produced particles) one has BEC present at the cost of appearance of multicharged vertices.

proposed earlier in [8] and consist also cornerstone of our present proposal, which can be regarded as generalization of that presented in [3]. This is nothing else but attempt to numerically realize the bunching of particles as quantum statistical effect used already in connection with BEC long time ago [7] and is done by dividing all available energy among particles (taken here as being pions) in such way that a number n_{cell} of EEC's, each containing particles of the same charge, is formed, with multiplicity of particles in each EEC followinig Bose-Einstein (or geometrical) distribution[6]. When energy distribution from which particles are selected is thermal-like then $P(n_{cell})$ is Poissonian and the total multiplicity distribution is of Pòlya-Aeppli type [9], closely resembling Negative Binomial distribution obtained in the so called *clan model* [10] (which differs only by the fact that particles in clans are distributed according to logarithmic distribution, not geometrical one [1]). Particles in a given EEC can have energies spread around the energy E_1 of the first particle defining this EEC with some width σ. With such energy spreading allowed one gets quite reasonable results for $C_2(Q_{inv})$ distributions (see [1] for details)[7].

Here we would like to present extension to this algorithm, which in addition to bunching accounts also for the symmetrization of the two-particle wave function (not used before) and allows to obtain in addition to $C_2(Q_{inv})$ also $C_2(Q_{x,y,z})$, i.e., in a sense it is 3-dimensional extension of our algorithm. This extension is based on the observation that symmetrization correlates the energy-momenta of particles with their space-time locations. The bunching of particles considered before was done only in the energy-momentum space and left us with a number of EEC's, each with a number of particles with well defined energies, E_i (and momenta $p_i = (E_i^2 - m^2)^{1/2}$). Each EEC is build

[6] To this end to the first particle selected in a given EEC one adds (up to first failure after which new EEC is selected) another particles with probability $P = P_0 \cdot e^{-E/T}$, where $P_0 \in (0,1)$ is constant, E is energy of the first particle and T parameter (corresponding to temperature in thermal models). Such form of P ensures the characteristic Bose-Einstein form of energy distribution.

[7] Actually, as was shown in [11] (see also [12]), this spreading is crucial to obtain the proper shape of C_2 function.

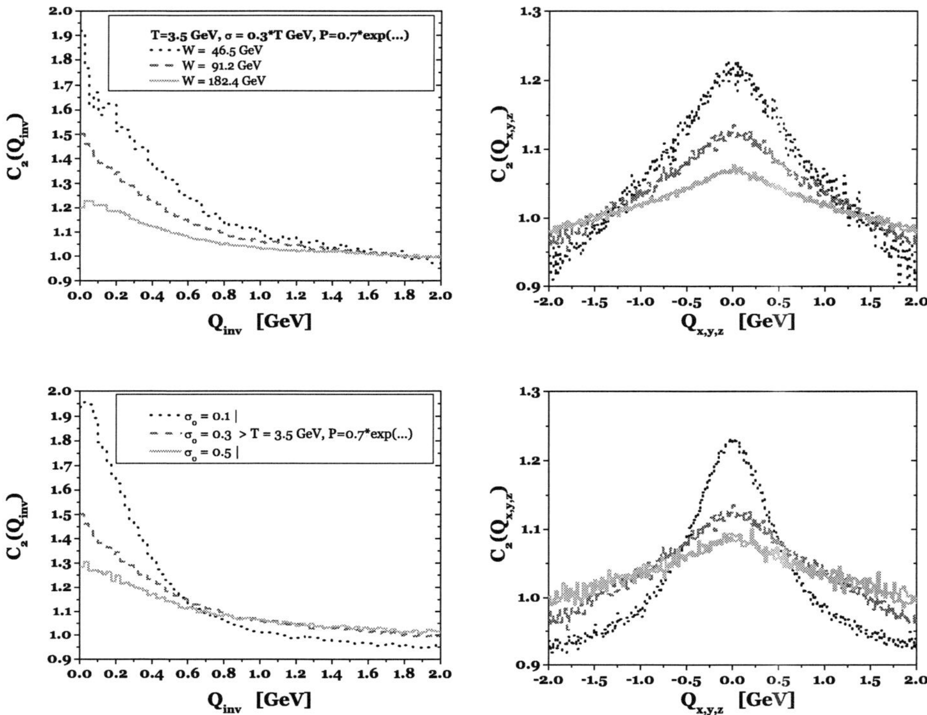

FIGURE 2. Example of results obtained with 3-dimensional version of our model. Upper panels show results for different mass of the hadronizing source. Lower panel demonstrates dependence on the allowed energy spread in EEC.

around some "seed" particle, which is taken as particle $i = 1$. This was enough to get $C_2(Q_{inv})$ (see [1]), but not $C_2(Q_{x,y,z})$ involving components of p_i, $p_{i(x,y,z)}$. To assign them one has first choose some space-time positions for particles in a given EEC taking them from some distribution function $\rho(r,t)$. Actually in what follows we shall use only *static source approximation*, i.e., hadronization is instantaneously and therefore $\rho(r,t) = \rho(r)$. Now $p_{i(x,y,z)}$ have to be correlated with the corresponding space positions, $r_i = (x_i, y_i, z_i)$, in the way emerging from the symmetrization of the wave functions resulting (for the plane wave approximation) in the famous $1 + \cos(\delta p \cdot \delta r)$ expression. Technically this is achieved by accepting only such momenta $p_{i(x,y,z)}$, which for given r_i lead to $\cos(\delta p \cdot \delta r) \leq 2 \cdot Rand - 1$ where *Rand* is random number uniformly chosen from interval $(0, 1)$.

In Fig. 2 we present examples of our new results (the elder results can be found in [1]) obtained with full, 3-dimensional version of our model for $\rho(r)$ being sphere of radius $R = 1$ fm and assuming that all $p_{i(x,y,z)}$ are spherically symmetric. As one can see now in addition to $C_2(Q_{inv})$ one has also corresponding to it $p_{i(x,y,z)}$ $C_2(Q_{x,y,z})$.

So far we are assuming direct pion production. However, the inclusion of Coulomb and other final state interactions in our approach is straightforward - one must simply change $\cos(\cdots)$ term arising from plane wave approximation used to form obtained by some distorted wave function. One can also easily include resonances and allow for finite life time of the emitting source. Finally, so far, for simplicity reason, only two particle symmetrization effects have been accounted for. Namely, in a given EEC all particles are symmetrized with the particle number 1 being its seed, they are not symmetrized between themselves. This seems to be justified because majority of our EEC's contain only $1-2$ or 3 particles. But to fully account for multiparticle effects one should simply add other terms in addition to the $\cos(\ldots)$ used above. This, however, would result in dramatic increase of the calculational time[8]. Nevertheless the effect of including at least terms when symmetrization between, say, particles 2 and 3 are added to the already present symmetrization between 1 and 2 and 1 and 3, must be carefully investigated before any final conclusion is to be reached.

Summary. To summarize, we are proposing numerical scheme of modelling quantum statistical phenomenon represented by BEC occurring in all hadronization processes. Distinctive features of our scheme not present in other propositions are:

- identical particles are emitted from EEC's and only these particles are subjected to BEC;
- inside each EEC particles are distributed according to the geometrical (or Bose-Einstein) distribution;
- altogether they show characteristic Bose-Einstein form of distribution of energy.

As result we obtain a kind of *quantum clan model* with Negative-Binomial-like multiplicity distributions and characteristic shape of $C_2(Q_{inv})$ function [1] (notice that we automatically include in this way BEC to all orders given by the maximal number of particles in a given EEC). To get also $C_2(Q_{x,y,z})$ one has to use some additional space-time information and the characteristic $1+\cos(\delta r\cdot\delta p)$ correlations between space-time and energy-momenta induced by the symmetrization of the respective wave functions. So far this is only a case study, we cannot yet offer any attempt to compare it with experimental data. On the other hand our approach offers new understanding of the way in which BEC are entering hadronization process.

We shall close with remark that there are attempts in the literature to model numerically BE condensation [13] (or to use notion of BE condensation in other branches of science as well [14]) using ideas of bunching of some quantities in the respective phase spaces.

---

[8] In fact, this converges to the proposition presented long time ago in [2], the only hope is that in our case symmetrization is performed within given EEC and therefore the number of terms of $\cos(\ldots)$ type involved is rather limited, whereas in [2] the whole source had to be symmetrized at once resulting with number of terms growing like $n!$ where n is observed multiplicity.

ACKNOWLEDGMENTS

OU is grateful for support and for the warm hospitality extended to him by organizers of the ISMD2005. Partial support of the Polish State Committee for Scientific Research (KBN) (grant 621/E-78/SPUB/CERN/P-03/DZ4/99 (GW)) is acknowledged.

REFERENCES

1. O. V. Utyuzh, G. Wilk, Z. Wlodarczyk, "Quantum Clan Model description of Bose Einstein Correlations", hep-ph/0503046, to be published in *Acta Phys. Hung. A - Heavy Ion Phys.* (2005); cf. also " Numerical modelling of quantum statistics in high-energy physics", hep-ph/0410398.
2. W. Zajc, *Phys. Rev.* **D 53**, 3396 (1987); R.L. Ray, *Phys. Rev.* **C 57**, 2523 (1998); J. G. Cramer, "Event Simulation of High-Order Bose-Einstein and Coulomb Correlations", Univ. of Washington preprint (1996, unpublished).
3. T. Osada, M. Maruyama and F. Takagi, *Phys. Rev.* **D 59**, 014024 (1999).
4. O. V. Utyuzh, G. Wilk and Z. Włodarczyk, *Phys. Rev.* **D 61**, 034007 (1999).
5. O. V. Utyuzh, G. Wilk and Z. Włodarczyk, *Phys. Lett.* **B 522**, 273 (2001) and *Acta Phys. Polon.* **B 33**, 2681 (2002).
6. B. Buschbeck and H. C. Eggers, *Nucl. Phys.(Proc. Suppl.)* **B 92**, 235 (2001).
7. E. E. Purcell, *Nature* **178**, 1449 (1956); A. Giovannini and H. B. Nielsen, "Stimulated emission model for multiplicity fluctuations", The Niels Bohr Institute preprint, NBI-HE-73-17 (unpublished) and "Stimulated emission effect on multiplicity distribution" in: *Proc. of the IV Int. Symp. on Multip. Hadrodynamics*, Pavia 1973, Eds. F. Duimio, A. Giovannini and S. Ratii, p. 538.
8. M. Biyajima, N. Suzuki, G. Wilk and Z. Włodarczyk, *Phys. Lett.* **B 386**, 297 (1996).
9. J. Finkelstein, *Phys. Rev.* **D 37**, 2446 (1988); Ding-wei Huang, *Phys. Rev.* **D 58**, 017501 (1998).
10. A. Giovannini and L. Van Hove, *Z. Phys.* **C 30**, 381 (1986); P. Carruthers and C. S. Shih, *Int. J. Mod. Phys.* **A 2**, 1447 (1986).
11. G. A. Kozlov, O. V. Utyuzh and G. Wilk, *Phys. Rev.* **C 68**, 024901 (2003) and *Ukr. J. Phys.* **48**, 1313 (2003).
12. K. Zalewski, *Lecture Notes in Physics* **539**, 291 (2000).
13. R. Kutner and M. Regulski, *Comp. Phys. Com.* **121-122**, 586 (1999) and R. Kutner, K. W. Kehr, W. Renz and R. Przeniosło , *J.Phys.* **A 28**, 923 (1995).
14. A. E. Ezhov and A. Yu. Khrennikov, *Phys. Rev.* **E 71**, 016138 (2005); K. Stalinas, "Bose-Einstein condensation in dissipative systems far from thermal equilibrium", cond-mat/0001436, and "Bose-Einstein condensation in classical systems", cond-mat/0001347.

Gluon Dominance Model

Kokoulina E.S.

JINR, Dubna, Moscow region,
141980, Russia
GSTU, LPR, October av. 48, Gomel
246746, Belarus

Abstract. A new way to study the multiplicity production in high energy processes is proposed. It is based on the multiplicity distribution description by the schemes taking into account the formation of quark-gluon system and hadronization. This investigations revealed an active role of gluons and the universal mechanism of their hadronization.

Keywords: multiplicity distributions, hadronization
PACS: 13.65+i, 13.75.-n, 13.85.-t, 13.85.Rm

INTRODUCTION

At present there is neither single consistent theory nor convincing model which could explain all the results obtained at RHIC and SPS [1]. It is evident that in these experiments the local thermal equilibrium is reached and thermalized matter is produced. Thermal statistical models based on this approach explain yields of different hadrons at high energy [1]. At the same time these models describe well the particle yields in e^+e^- and pp interactions where, as generally accepted, thermalization is impossible [2]. So the intensive study and insight of these processes remain to be very importance.

To investigate multiparticle production (MP) at high energy in these processes, a two stage model was proposed [3, 4]. It is based on the use of QCD and the phenomenological scheme of hadronization. The model describes well multiplicity distributions (MD) and their moments in e^+e^- annihilation, pp and $p\bar{p}$ interactions and gives complementary information to a better understanding of MP of relativistic heavy ion collisions.

This model confirms the fragmentation mechanism of hadronization in e^+e^- annihilation and the transition to recombination mechanism in hadron and nucleus interactions. It explains the shoulder structure in MD at higher energies and the behavior of f_2 in $p\bar{p}$ annihilation at few tens GeV/c by including intermediate quark topologies. The mechanism of the soft photons (SP) production as a sign of hadronization and estimates of their emission region size is proposed [5].

The e^+e^- annihilation is the most suitable process to study MP. It can be realized through the formation of virtual γ or Z^0–boson which then decays into two quarks: $e^+e^- \to (Z^0/\gamma) \to q\bar{q}$. The e^+e^-–reaction is simple for analysis, as the produced state is pure $q\bar{q}$. To describe the process of parton fission (quarks and gluons) at big virtuality, pQCD may be applied. This stage can be named as the stage of cascade. When partons get small virtuality, pQCD cannot be applied. Therefore phenomenological models are used to describe hadronization in this case.

The probabilistic nature of parton fission in QCD has been established in [6].

CP828, *Multiparticle Dynamics*
edited by V. Šimák, M. Šumbera, Š. Todorova, and B. Tomášik

E^+E^- - ANNIHILATION

A.Giovannini [7] had proposed to describe quark and gluon jets as markovian branching processes with three elementary contributions: gluon fission, quark bremsstrahlung and quark pair production. He constructed a system of differential equations for generating functions (GF) and obtained solutions of MD for quark jet

$$P_m^{(q)}(Y) = \frac{k_p(k_p+1)\dots(k_p+m-1)}{m!}\left(\frac{\overline{m}}{\overline{m}+k_p}\right)^m\left(\frac{k_p}{k_p+\overline{m}}\right)^{k_p}. \quad (1)$$

These MD are known as negative binomial distribution (NBD). The GF for them is $Q^{(q)}(z,Y) = \sum_{m=0}^{\infty} z^m P_m(Y) = [1+\overline{m}/k_p(1-z)]^{-k_p}$, where $\overline{m}$ is the mean gluon multiplicity, Y – QCD evolution variable and k_p - NBD parameter.

In [3] MD (1) was taken for the cascade stage. For the hadronization a sub narrow binomial distribution (BD) was added

$$P_n^H = C_{N_p}^n \left(\frac{\overline{n}_p^h}{N_p}\right)^n \left(1-\frac{\overline{n}_p^h}{N_p}\right)^{N_p-n}, \quad Q_p^H(z) = \left[1+\frac{\overline{n}_p^h}{N_p}(z-1)\right]^{N_p}, \quad (2)$$

where $C_{N_p}^n$ - binomial coefficient, $\overline{n}_p^h$ and N_p $(p=q,g)$ have a sense of mean multiplicity and maximum number of secondary hadrons formed from parton (p) at its passing of the second stage.

We have chosen BD from the analysis of experimental data on e^+e^-- annihilation. Second correlation moments f_2 were negative at low energies (less than 9 GeV). The choice of such distributions was the only one that could describe the experiment. We suppose that the hypothesis of soft decoloration is right. Therefore the hadronization stage is added to the parton stage by means of a factorization. We introduce parameter $\alpha = N_g/N_q$ to distinguish the hadrons, produced from quark or gluon. MD of hadrons in e^+e^- - annihilation can be written as follows ($N_q = N$, $\overline{n}^h = \overline{n}_q^h$):

$$P_n(s) = \sum_{m=0}^{M_g} P_m^{(q)} C_{(2+\alpha m)N}^n \left(\frac{\overline{n}^h}{N}\right)^n \left(1-\frac{\overline{n}^h}{N}\right)^{(2+\alpha m)N-n}. \quad (3)$$

The expression (3) describes well the experimental data [8] from 14 to 189 GeV [9] (e.g. Fig. 1). The mean gluon multiplicity $\overline{m}$ has a tendency to rise, but slower than the logarithmic one. Parameter k_p was been related with temperature in [10]. The fact that $\alpha < 1$ is the evidence that hadronization of a gluon is softer than a quark. It is surprising that gluon parameters of hadronization (N_g, $\overline{n}_g^h$) remain constant without considerable deviations in spite of the indirect finding: $N_g \sim 3-4$ and $\overline{n}_g^h \sim 1$ (e.g. Fig. 2). Therefore we can draw a conclusion about the universality of gluon hadronization in e^+e^- annihilation in the sufficient wide energy region.

It was shown [11] that the ratio of factorial cumulative moments over factorial moments changes the sign as a function of the order. The calculation of this ratio by using (3) was done in [3]. It has been obtained that the period of oscillations is equal to 2 before Z^0-region and increases at higher energies.

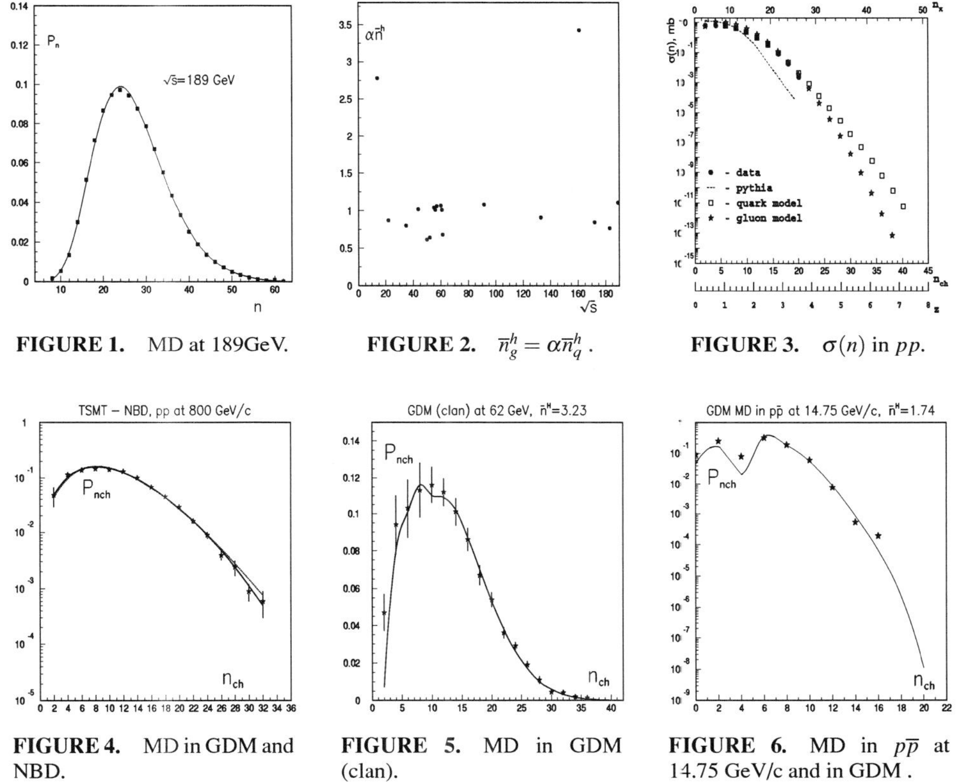

FIGURE 1. MD at 189GeV.

FIGURE 2. $\bar{n}_g^h = \alpha \bar{n}_q^h$.

FIGURE 3. $\sigma(n)$ in pp.

FIGURE 4. MD in GDM and NBD.

FIGURE 5. MD in GDM (clan).

FIGURE 6. MD in $p\bar{p}$ at 14.75 GeV/c and in GDM .

PP INTERACTIONS

The study of MP in pp interactions is implemented in the framework of the project "Thermalization" [12]. This project is aimed at studying the collective behavior of secondary particles and advancement to the high multiplicity region (HMR) beyond available data [13] in pp interactions at 70 GeV/c. The calculation by the MC PHYTHIA code has shown that the standard generator predicts a value of the cross section which is in a reasonably good agreement with data at small multiplicity $n_{ch} < 10$, but it underestimates the value $\sigma(n_{ch})$ by two orders of the magnitude at $n_{ch} = 20$. The existing models are very much sensitive in this region [14], also (Fig. 3).

We suppose that after the inelastic collision the part of the energy of the initial impact protons is transformed to the inside energy. Several quarks and gluons become free and form quark-gluon system (QGS). Partons which can produce hadrons are named the active ones. Two schemes were proposed [4]. In the first scheme the parton fission inside the QGS is taken into account (the scheme with a branch). If we are not interested in what is going inside QGS, we come to the scheme without a branch (TSMT).

At the beginning of research we took a model with active quarks and gluons. Parameters of that model had values which differed very much from parameters obtained in e^+e^-- annihilation, especially for hadronization. It was one of the main reasons to refuse the scheme with active quarks. So, we reserve quarks remained inside of the leading particles. All of the newly born hadrons were formed by active gluons. That is why we began to name this model – the gluon dominance model (GDM).

The Poisson distribution was chosen as the simplest MD for active gluons which appeared for the first time after the collision. The number of these gluons fulfils the role of the impact parameter for nucleus. To describe MD in the gluon cascade (fission), Farry distribution was used with GF $Q_k^B(z) = z^k/\overline{m}^k \left[1 - z\left(1 - \frac{1}{\overline{m}}\right)\right]^{-k}$, where k – the number of initial gluons in QGS, $\overline{m}$ – the mean multiplicity of them in the end of the branch. On the second stage some of active gluons can leave QGS ("evaporate") and transform to real hadrons. Our BD (2) for hadrons on the stage of hadronization is as follows

$$P_n^H(m) = C_{\delta m N_g}^{n-2} \left(\frac{\overline{n}_g^h}{N_g}\right)^{n-2} \left(1 - \frac{\overline{n}_g^h}{N_g}\right)^{\delta m N_g - (n-2)}. \tag{4}$$

In (4) parameter δ is the ratio of evaporated gluons leaving QGS, to all active gluons. From the comparison with data [13] we have obtained that a maximum possible number of hadrons from a single gluon looks very much like the number of partons in the glob of cold QGP of L.Van Hove [15], the branch processes are weak. The fraction of released gluons is equal to $\delta = 0.47 \pm 0.01$ (the same as in [16]). A part of active gluons does not convert into hadrons. They stay in QGS and can become sources of soft photons (SP).

In the scheme without a branch we consider that evaporated gluons have Poisson MD, too. Using the idea of the convolution of two stages $P_n(s) = \sum_m P_m^P(s) P_n^H(m)$ as well as the BD for hadronization, we obtain MD in pp-collisions. The comparison GDM with the data [13] (Fig. 3) gives values of parameters $N_g = 4.24 \pm 0.13$, $\overline{n}_g^h = 1.63 \pm 0.12$ and it is in agreement with the values obtained in e^+e^- annihilation [9]. We are restricted in sum to $m = 6$. Our estimation of the maximal possible observable number of charged particles is ≤ 26.

We have also got MD for neutral mesons and total multiplicity by using mean multiplicities of π^0-mesons [17] and active gluons, expected approximate equality of probabilities of the formation charged and neutral particles from single gluon at hadronization and the above-mentioned idea of the convolution [5]. The maximum observable number of them is estimated as 16, total - as 42, the parameter $\overline{n}_g^h = 1.036 \pm 0.041$ (h = π^0).

The analysis of the mean multiplicity of π^0– mesons versus the number of charged particles n_{ch}, allows to set limitations to the number of neutral mesons at given n_{ch} and indicates the absence of AntiCentauro events (Centauro may be in HMR). The obtained estimations of probabilities for charged and neutral hadrons production from gluon while its passing through hadronization permits to get "the charged hadron/pion" ratio [18] in pp interactions. At 69 GeV/c this ratio is equal to 1.19 ± 0.25.

GDM describes well MD in the region of $100 - 800$ GeV/c (Fig. 4 and table 1 in [5]). The number of active gluons, their mean multiplicity, N_g and $\overline{n}_g^h$ increase slowly. A growth of $\overline{n}_g^h$ in pp interactions indicates a possible change mechanism of hadronization of gluons in comparison with e^+e^- annihilation. It is considered that in the last case the partons transform to hadrons by the fragmentation mechanism at the absence of the thermal medium. Our MD analysis gives $\overline{n}_g^h \sim 1$ for this hadronization [9]. The recombination is specific for the hadron and nucleus processes. In this situation a lot of quark pairs from gluons appear almost simultaneously and recombine to various hadrons [19]. The value $\overline{n}_g^h$ becomes bigger $\sim 2-3$, that indicates this transition. The recombination mechanism provides justification for applying the statistical model to describe ratios of hadron yields and the explanation of the collective flow of quarks [19].

In [7] MP is described by means of a clan mechanism and emphasizes the gluon nature of clan. Our GDM allows to give a concrete content for the clan. The clan model uses the logarithmic distribution (LD) in a single clan. LD are similar to our BD.

At the SPS energy the shoulder structure appears in MD [20]. As it was marked in the branch scheme, the gluon fission is strengthened at higher energies. The independent evaporation of gluon sources of hadrons may be realized as single gluons as groups from two and more fission gluons. Following [7] we name such groups - clans. GDM with two kinds of clans [5] describes data [21] very well (Fig. 4). Moreover "the charged hadron/pion" ratio at 62 GeV is equal to ~ 1.6 and agrees with Au-Au peripheral interactions at 200 GeV and with pp interactions at 53 GeV [18].

The specific feature of GDM is the dominance of active gluons in MP. We expect the emergence of many of them in nucleus collisions at RHIC and the formation of a new kind of matter (QGP) at high energy. Our QGS can be a candidate for this.

Experiments [22] have shown that the measured cross sections of SP are several times larger than the expected ones from QED. The phenomenological glob model explains the SP excess [15]. We consider that at a certain moment the QGS or excited new hadrons may set in an almost equilibrium state during a short period of time. That is why, to describe massless photons, we have used the black body emission spectrum [23]. At 70 GeV/c an inelastic cross section is equal to $\sim 40mb$, the SP formation cross section is about $4mb$ [12] and since $\sigma_\gamma \simeq n_\gamma(T) \cdot \sigma_{in}$ then $n_\gamma \approx 0.1$. If n_γ and temperature $T(p)$ (p–momentum) are known, then we can estimate the emission region size L of SP. The obtained values $L \sim 4-6fm$ [5, 24].

In conclusion one can show how GDM may explain experimental $p\bar{p}$ annihilation data at tens GeV/c [25]. The differences between $p\bar{p}$ and pp inelastic topological cross sections ($\Delta\sigma_n(p\overline{p}-pp) = \sigma_n(p\overline{p}) - \sigma_n(pp)$) point out the contribution of different mechanisms in MP. The negative values of f_2 indicate the predominance of the hadronization stage in MP. According to MGD, the active gluons are a basic source of secondary hadrons.

There are three valent $q\overline{q}$-pairs at the initial stage of annihilation . They can turn to the "leading" mesons which consist of: a) valent quarks or b) valent and sea quarks [24]. In the case a) three "leading" neutral pions (the "0"–topology) or two charged and one neutral "leading" mesons ("2"–topology) may form. In b) case the "4"- and "6"-topologies are realized. The neutron and antineutron formation is possible, too.

The active gluons emerge together with the formation of intermediate topology. At this region hadronization dominates since f_2 is negative. In the simple case for m active gluons GF $Q_m(z) = [1+\overline{n}^h/N(z-1)]^{mN}$, and $f_2 = -m(\overline{n}^h)^2/N$. If m grows while increasing the energy of the colliding particles, then f_2 will decrease almost linearly from m that agrees with the data [25].

According to GDM [5] and taking into account an intermediate charged topology ("0", "2" and "4") with active gluons, GF $Q(z)$ for final annihilation MD ($\Delta\sigma_n(p\overline{p} - pp)/\sum_n \Delta\sigma_n(p\overline{p} - pp)$) may be written as the convolution gluon and hadron components

$$Q(z) = c_0 \sum_m^{M_0} P_m^G [Q^H(z)]^m + c_2 z^2 \sum_m^{M_2} P_m^G [Q^H(z)]^m + c_4 z^4 \sum_m^{M_4} P_m^G [Q^H(z)]^m.$$

The parameters of c_0, c_2 and c_4 are parts of intermediate topologies. GDM describes data (Fig. 6) with the ratio $c_0 : c_2 : c_4 = 15 : 40 : 0.05$ and $M_0 \sim M_2 \sim 1-2, \quad M_4 \sim 4$ [5]. The carried out investigations allow one to understand deeper the MP nature.

I thank the Organizing committee for a wonderful possibility to take part in ISMD2005.

REFERENCES

1. RHIC collaboration white paper: http//www.phenix.bnl.gov/WWW/infor/comment.
2. D. Kharseev, *Nucl.Phys.* **A 715**, 35–44 (2003).
3. V.I. Kuvshinov, E.S. Kokoulina, *Acta Phys. Pol.* **B13**, 553–558 (1982).
4. E.S. Kokoulina and V.A. Nikitin, *Int. School-Seminar APMP*, Gomel, Belarus, 221–236 (2003); E.S. Kokoulina, *Acta Phys.Polon.* **B35**, 295–302 (2004).
5. E.S. Kokoulina and V.A. Nikitin, *ISHEPP*, 2004, hep-ph/0502224; P.F. Ermolov et al., *ISHEPP*, 2004, hep-ph/0503254.
6. K. Konishi, A. Ukawa, G. Veneciano, *Nucl.Phys.* **B 157**, 45–107 (1979).
7. A. Giovannini, *Nucl. Phys.* **B 161**, 429-448 (1979); A. Giovannini and R. Ugocioni, hep-ph/0405251.
8. W. Braunschweig et al. *Z. Physik* **C 45**, 193– 208 (1989); P. Abreu P. et al. *Z.Phys.* **C 52**, 271–281 (1991); G. Alexander et al. *Z.Physik.* **C 72**, 191–206 (1996); K. Acketstaff et al. *Z.Physik.* **C 75**, 193 (1997); G. Abbiendi et al. *Eur.Phys.J.* **C16** 185–210 (2000).
9. E.S. Kokoulina, *XI Ann.Sem. NPCS*, Minsk, Belarus, pp.15 (2002); E.S. Kokoulina, *ISMD 2002, Multiparticle dynamics*, pp. 340–343, Alushta, Ukraine.
10. E.S. Kokoulina and V.I. Kuvshinov, *Vesti AN BSSR*. Ser.Phys.-Math.Sc.(Rus. ed)3, 49–54 (1988).
11. K. Abe et al. *Phys. Lett.* **B 371**, 149–156 (1996); I.M. Dremin, *Phys.Lett.* **B 341**, 95–98 (1994).
12. P.F. Ermolov et al. *Yad. Phys.* **67**, 108–114 (2004).
13. V.V. Babintsev et al. *IFVE-76-25*, Protvino (1976); V.V. Ammosov et al., *Phys. Let.* **42B**, 519–521 (1972).
14. O.G. Chikilev and P.V. Chliapnikov, *Yad. Phys.*,**55**, 820–825 (1991).
15. P. Lichard and L. Van Hove, *Phys. Let.* **B245**, 605–608 (1990).
16. A.H. Mueller, *Nucl.Phys.* **B715**, 20–34 (2003).
17. V.S. Murzin and L.I. Sarycheva. "Secondary multiplicities" in *Interactions of high energy hadrons*. Nauka, Moscow, 1983, pp.194-207; V.G. Grishin, *Phys. El.Part. and At.Yad.*, **10**, 608– 656 (1979).
18. K. Adcox et. al, *Nucl.Phys.* **A757**:184-283(2005); B. Alper et al, *Nucl. Phys.* **B100**, 237–290 (1975).
19. R.C. Hwa and C.B. Yang, *Phys. Rev.*, **C67**, 034902 (2003); R.Hwa, *Acta Phys.Polon.*, **C67**:227-234, (2005).
20. G.I. Alner et al. *Z. Phys.*, **C33**, 1–6 (1986); R.E. Ansorge et al. *Z.Phys.* **C43**, 357–399 (1989).
21. A. Breakstoune et al., *Phys.Rev.*, **D30**, 528–535 (1984).
22. P.V. Chliapnikov et al., *Phys. Let.*, **141B**, 276–280 (1984); S. Banerjee et al., *Phys.Let.*,**B305**,182–186 (1983); J. Antos et al., *Z.Phys.*, **C59**, 547–554 (1993).
23. Yu.B. Rumer and M.Sh. Ryvkin. “The equilibrium radiation. The photon gas.” in *Thermodynamics, statistical physics and kinetics*. "Nauka", Moscow, 1977, pp. 229–233.
24. M.K.Volkov et al., *Particles and Nuclei, Letters*, **1**, 16–23 (2004).
25. J.G. Rushbrooke and B.R. Webber, *Phys. Rep.* **C44**, 1–92 (1978); E. Klempt et al., *Phys. Rep.* **413**, 197-317 (2005).

Longitudinal, Azimuthal and Multiplicity Dependences of Mean Transverse Momentum and Transverse Momentum Correlations in π^+p and K$^+$p Collisions at 250GeV/c

Huang Yanping* and Wu Yuanfang*

*Institute of Particle Physics, Hua-Zhong Normal University, Wuhan 430079, China

Abstract. Rapidity, azimuthal and multiplicity dependence of inclusive mean transverse momentum and transverse momentum correlation is studied in π^+p and K$^+$p collisions at 250 GeV/c. It is found for the first time that rapidity dependence of two-particle transverse momentum correlations is different from those of inclusive mean transverse momentum but both have similar multiplicity dependence. In particular, the transverse momentum correlation is boost-invariant. The strong azimuthal dependence of transverse momentum correlation comes from the constraint of energy-momentum conservation. The results are compared with those from the PYTHIA Monte Carlo generator. The similarities and differences with those from current heavy ion experiments are discussed.

Keywords: transverse momentum, correlation, longitudinal, azimuthal, multiplicity
PACS: 13.85.Hd, 25.75.Gz

INTRODUCTION

Since transverse momenta of the final state particles are produced after the collision and carry the information of system expansion, their fluctuations are considered to be a good probe for the formation of quark-gluon plasma (QGP) and a trace of local thermal equilibrium [1]. All current heavy ion experiments report substantial dynamical fluctuations of transverse momentum. Preliminary PHENIX and STAR data on Au+Au collisions show that p_t fluctuations increase as centrality increases [2, 3]. However, it is unclear how these fluctuations differ form those observed in elementary collisions, where no QGP is expected.

In general, it is believed that transverse expansion extends over a wide rapidity range and correlates with longitudinal expansion [4]. How the transverse expansion relates to the longitudinal one is unexplored even for the elementary collisions. Detailed investigation is important for a correct evolution picture and for understanding those phenomena in current heavy ion experiments. This investigation is possible only for an experiment with full acceptance, such as NA22 experiment. The NA22 experiment setup and data selection will be presented in the second section.

It is argued that the inclusive mean transverse momentum is sensitive to the expansion velocity and the transverse momentum correlation measures the collective transverse expansion velocity [4]. In this paper, we will consider the rapidity, azimuthal and multiplicity dependence of the mean transverse momentum and the two-particle transverse momentum correlation. The measure methods of the two-particle transverse momentum

CP828, *Multiparticle Dynamics*
edited by V. Šimák, M. Šumbera, Š. Todorova, and B. Tomášik

correlation are described in the third section. The results and discussions are presented in the forth section. And the fifth section presents a summary of this work.

EXPERIMENTAL SETUP AND DATA SELECTION

The NA22 experiment has been performed at CERN in the European Hybrid Spectrometer(EHS), equipped with the Rapid Cycling Bubble Chamber (RCBC) as an active vertex detector, which has excellent momentum resolution and full 4π acceptance, and exposed to a 250 GeV/c tagged positive meson enriched beam.

The data in this analysis include π^+p and K$^+$p collisions at 250 GeV/c. Events are accepted when measured and reconstructed charge multiplicity are consistent, charge balance is satisfied, no electron (positron) is detected. In particular, possible contamination from secondary interactions is suppressed by a visual scan; γ conversions near the primary vertex are removed by electron identification. Since no statistically significant differences are seen between the results for π^+ and K$^+$ induced reactions, the two data samples are combined for the purpose of this analysis. A total of 44 524 non-single-diffractive events is obtained after all necessary selections, as described in detail in [5].

MEASURES OF TWO-PARTICLE TRANSVERSE MOMENTUM CORRELATION

Since the first measure for transverse momentum fluctuation Φ_{p_t} was suggested [6], a number of other ones have been recommended to extract the genuine dynamical fluctuations such as $\sigma^2_{dynamic}$, Σ_{p_t}, F_{p_t}, and $\Delta\sigma^2_{p_t,n}$. Among those measures, the common and essential part is the event two-particle transverse momentum correlation $\sum_{i=1}^{n_{\rm ch}}\sum_{j=1,i\neq j}^{n_{\rm ch}} p_{ti}p_{tj}$. Hence, in the following, we will focus only on this most simple correlation.

There are two normalization schemes for this correlation. One is to normalize by the number of correlation pairs $n_{\rm ch}(n_{\rm ch}-1)$ of an event in considered window and the corresponding average of event mean transverse momentum $\langle\bar{p}_{\rm t}\rangle$, i.e.,

$$R\left(p_{ti},p_{tj}\right)=\left\langle\frac{\sum_{i=1}^{n_{\rm ch}}\sum_{j=1,i\neq j}^{n_{\rm ch}} p_{ti}p_{tj}}{n_{\rm ch}(n_{\rm ch}-1)}\right\rangle\frac{1}{\langle\bar{p}_t\rangle^2}\,, \tag{1}$$

where $n_{\rm ch}$ is the number of charged particles and $p_{\rm ti}$ is the transverse momentum of the ith particle. The bar is the average over all particles in an event and $\langle\cdots\rangle$ is the average over all event. The correlation R defined in this way is the *event mean two-particle transverse momentum correlation*. It is one unit more than the measure $\langle\Delta p_{\rm ti}\Delta p_{\rm tj}\rangle=\langle\frac{(p_{\rm ti}-\langle\bar{p}_{\rm t}\rangle)(p_{\rm tj}-\langle\bar{p}_{\rm t}\rangle)}{n_{\rm ch}(n_{\rm ch}-1)}\rangle/\langle p_{\rm t}\rangle^2$ used by the STAR Collaboration [8].

The other scheme is to normalize by the average number of correlation pairs $\langle n_{\rm ch}(n_{\rm ch}-1)\rangle$ and the inclusive mean transverse momentum $\langle p_{\rm t}\rangle$, i.e.,

$$R'\left(p_{ti},p_{tj}\right)=\frac{\left\langle \sum\limits_{i=1}^{n_{\rm ch}}\sum\limits_{j=1,i\neq j}^{n_{\rm ch}} p_{ti}p_{tj}\right\rangle}{\left\langle n_{\rm ch}\left(n_{\rm ch}-1\right)\right\rangle\left\langle p_{t}\right\rangle^{2}}\,. \tag{2}$$

This normalization is strongly recommended by S. Voloshin, et al. [7, 4] as it is directly related to the well defined two-particle correlation and is unlike the so-called "bad" ratio-like observable, such as R. The similarly normalization $\langle \Delta p_{\rm ti}\Delta p_{\rm tj}\rangle = \langle \sum\limits_{i=1}^{n_{\rm ch}}\sum\limits_{j=1,i\neq j}^{n_{\rm ch}}(p_{\rm ti}-\langle p_{\rm t}\rangle)(p_{\rm tj}-\langle p_{\rm t}\rangle)\rangle/[\langle n_{\rm ch}(n_{\rm ch}-1)\rangle\langle p_{\rm t}\rangle^{2}]$ is not directly related to the corresponding two-particle correlation R' since it contains an extra $p_{\rm t}$-$n_{\rm ch}$ correlation.

It is clear that a good measure should be sensitive to the underlying dynamics and less dependent on the effects from detector acceptance(rapidity, azimuthal and $p_{\rm t}$ region considered). Due to the advantage of NA22 experiment data described above, it provides the best test of the robustness of these two measures — R and R'.

RESULTS AND DISCUSSIONS

In Fig. 1, $\langle p_{\rm t}\rangle$, R and R' are presented for different central rapidity windows with $|y|<Y_{\rm c}$ (upper row), as well as for a unit rapidity bin at different positions (lower row). The full circles and triangles are the data and the corresponding open ones are the results of PYTHIA 5.720. From Fig. 1(a) and 1(c), we can see that the value of $\langle p_t\rangle$ is small in the central, target and projectile regions but relatively large in the two mid-rapidity regions

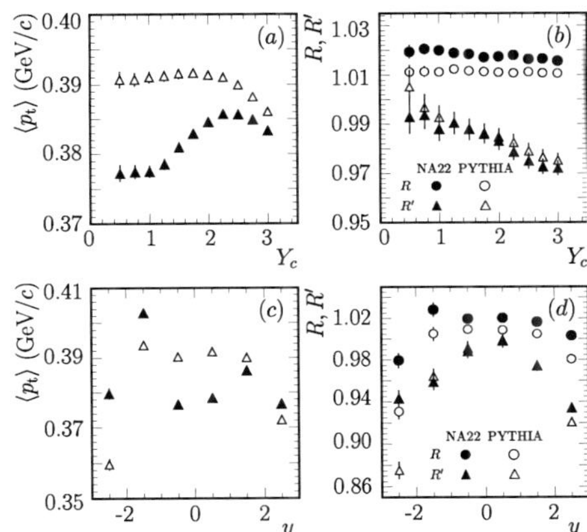

FIGURE 1. The mean transverse momentum $\langle p_{\rm t}\rangle$ (a) and (c), and the two-particle transverse momentum correlation R and R', (b) and (d), in different central rapidity windows with $|y|<Y_{\rm c}$(upper row), as well as in a unit rapidity window at different positions(lower row).

[1,2] and [-1,-2]. This result is consistent with the famous sea-gull effect [9]. On the other hand, for Au-Au collisions at 200GeV [2], it keeps decreasing with increasing size of central rapidity window. This shows that the sea-gull effect is smeared in those nuclear collisions, but remnant effect exsits.

The results from PYTHIA are flat in a rather wide central rapidity region $[-2,2]$ and decrease in target and projectile regions. Though the difference between data and PYTHIA is large in the central rapidity region, they tend to be close in the full rapidity region. But PYTHIA gives lower estimates in both target and projectile regions, cf. Fig. 1(c).

The event two-particle transverse momentum correlation measured by R as shown in Fig. 1(b) and (d) is independent of the size and position of rapidity window, except that in the left-most and right-most bins $[-3,-2]$ and $[2,3]$, the value of R is lower than that in other bins. It is caused by the unidentified protons contribution in $[-3,-2]$, where the rapidity distribution is not completely symmetric to the region $[2,3]$. These results agree with those of the most central Au-Au collisions at 200GeV but are different from those of peripheral collisions, where the dependence on the size of rapidity window is observed [8].

It can be seen from the same figures that R' decreases with increasing rapidity window and depends on its position, in a way similar to the rapidity density distribution [10]. This shows that R' still contains the acceptance effects which is proportional to rapidity density distribution. So R is well normalized and more robust than R'. The rapidity dependence of the mean transverse momentum correlation agree with the sea-gull effect, suggesting that $\langle p_t \rangle$ depends both on the size and position of the rapidity window. The rapidity independence of the correlation R demonstrates that the transverse momentum correlations are longitudinal momentum independent or boost invariant, similar to the boost invariance of the charge balance function found recently [10].

The azimuthal dependence of $\langle p_{\mathrm{t}} \rangle$ and R are shown in Fig. 2(a) and (b), respectively, where ϕ_{cut} is the size of the azimuthal window and $\phi < \phi_{\mathrm{cut}}$. It can be seen from Fig. 2(a) that $\langle p_{\mathrm{t}} \rangle$ is independent of the azimuthal size within experimental error bars. On the other hand, R increases with increasing of the azimuthal size.

This dependence can be understood by the constraints of energy and momentum conservation. The constraint of energy conservation suppresses large transverse momentum particles in the same side of the ϕ plane, so the correlation decreases when $\phi_{\mathrm{cut}} < \pi$. The

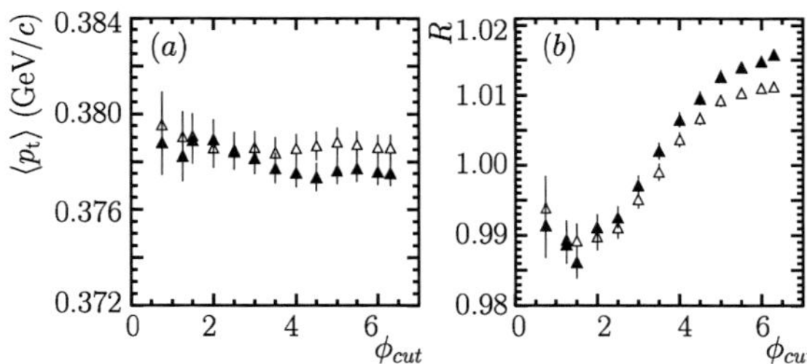

FIGURE 2. The mean transverse momentum $\langle p_{\mathrm{t}} \rangle$ (a) and two-particle transverse momentum correlation R (b) for different azimuthal angular cuts.

TABLE 1. Comparison of the two different experiments

	NA22	NA49
Rapidity cut	$1.1 < y < 2.6$	
Transverse momentum cut	$0.005 < p_t < 1.5$ (GeV/c)	
Azimuthal cut	$\phi_{cut} = 2\pi$	$\phi_{cut} < 2\pi$
colliding nucleus	$\pi^+ + p$ $k^+ + p$	p − p
Φ_{p_t}(MeV/c)	10.91 ± 1.54	2.2 ± 0.3

increase of the correlation at $\pi < \phi_{\rm cut} < 2\pi$ originates from back-to-back correlation due to transverse momentum conservation in the full ϕ plane.

This result shows that the transverse momentum correlation is indeed influenced by the size of the azimuthal window and explains why the transverse momentum fluctuation, measured by Φ_{p_t}, from NA22 [11] is larger than that from NA49, as pointed out in [12]. The results are listed in Table 1. The value of $2.2 \pm 0.3 MeV/c$ measured by the NA49 experiment for pp collision at $158 GeV/c$ is considerably less than the value $10.91 \pm 1.54 MeV/c$ measured by the NA22 experiment. So, although $\langle p_t \rangle$ is uniformly distributed in the ϕ plane, the two-particle transverse momentum correlations vary with the size of the azimuthal window, due to the constraint of energy and momentum conservation. PYTHIA reproduce these dependence.

The multiplicity dependence of $\langle p_t \rangle$ and the correlation R is presented in Fig. 3, respectively. Three multiplicity intervals have been taken into account, $1 < n_{ch} < 6$, $6 \leq n_{ch} \leq 8$ and $n_{ch} > 8$. Both $\langle p_{\rm t} \rangle$ and R decrease with the increase of multiplicity. The latter is consistent with the centrality dependence of $\langle \Delta p_{\rm ti} \Delta p_{\rm tj} \rangle$ given by STAR and HIJING for nuclear collisions at different colliding energies, though the results from HIJING are much lower than the data of STAR [8]. So the smaller is the event-multiplicity, the stronger is the event mean two-particle transverse momentum correlation.

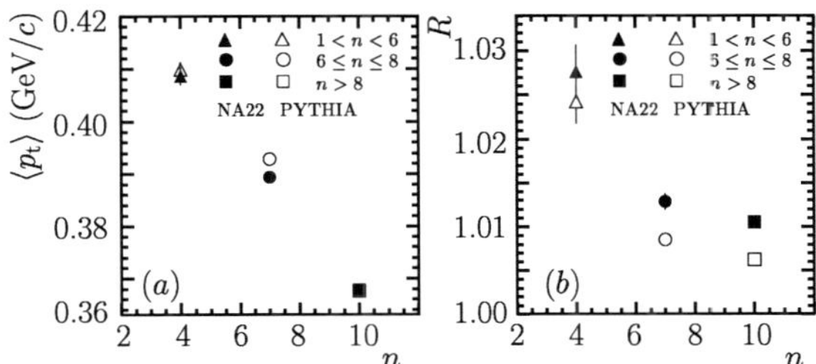

FIGURE 3. The multiplicity dependencies of mean transverse momentum $\langle p_{\rm t} \rangle$ (a) and two-particle transverse momentum correlation R (b).

SUMMARY

Those results can be summarized as follows:

1. In agreement with the well-known sea-gull effects, $\langle p_t \rangle$ depends both on the size and position of the rapidity bin. The event mean correlation R is independent of the size and position of the rapidity bin, contrary to the correlation R'. This reveals for the first time that the transverse momentum correlation is longitudinal momentum independent. PYTHIA can roughly describe the rapidity dependence of R but not that of the $\langle p_t \rangle$.

2. $\langle p_t \rangle$ is uniformly distributed in the ϕ plane, but due to the constraint of energy-momentum conservation, R strongly depends on the size of the azimuthal window.

3. $\langle p_t \rangle$ and R decrease with increasing multiplicity. This is consistent with the centrality dependence of $\langle \Delta p_{ti} \Delta p_{tj} \rangle$, as reported by STAR at different colliding energies range from 20 to 200 GeV [8], while $\langle p_t \rangle$ increases with increasing multiplicity when colliding energy is above ISR [13].

ACKNOWLEDGMENTS

Yanping Huang is gratefull for the tireless guide of Prof. Liu Lianshou in preparing the presentation of the work and for the financial support form the organizer of the conference. This work is also supported in part by the National Natural Science Foundation of China with project No.10375025 and No.10475030 and the Ministry of Education of China with project No. CFKSTIP-704035.

REFERENCES

1. L. Van Hove, *Z. Phys.* **C 21**, 93 (1984); J. Kapusta and A. Vischer, *Phys. Rev.* **C 52**, 2725 (1995); L. Stodolsky, *Phys. Rev. Lett.* **75**, 1044 (1995); S. Gavin, *Phys. Rev. Lett.* **92**, 162301 (2004).
2. STAR Coll., J. Adams, et al., *Phys. Rev. Lett.* **90**, 172301 (2003); R. L. Ray, *Nucl. Phys.* **A 715**, 45(2003); STAR Coll., C. A. Pruneau, nucl-ex/0401016.
3. PHENIX Coll., S. S. Adler et al., *Phys. Rev. Lett.***93**, 092301 (2004); PHENIX Coll., K. Adcox et al., *Phys. Rev.* **C 66**, 024901 (2002), PHENIX Coll., J. Nystrand, *Nucl. Phys.* **A 715**, 603 (2003).
4. S. A. Voloshin, *Nucl. Phys.* **A 749**, 287-290 (2005); nucl-ex/0505003.
5. NA22 Coll., M. Adamus et al., *Z. Phys.* **C 32**, 476 (1986); NA22 Coll., M.R. Atayan et al., *Eur. Phys. J.* **C 21**, 271 (2001).
6. M. Gaździcki and St. Mrówczyński, *Z. Phys.* **C 54**, 127 (1992); M. Gaździcki, A. Leonidov and G. Roland, *Eur. Phys. J.* **C 6**, 365 (1999).
7. S. A. Voloshin, V. Koch and H. G. Ritter, *Phys. Rev.* **C 60**, 024901 (1999); C. Pruneau, S. Gavin and S. Voloshin, *Phys. Rev.* **C 66**, 044904 (2002).
8. STAR Coll., G. D. Westfall, *J. Phys.* **G 30** S1389 (2004); J. Adams, et al., nucl-ex/0504031.
9. J. Pernegr, V. Šimák and M. Votruba, *Nuovo Cim.* **17**, 129 (1960); M.Bardadin et al., Proc. Sienna Conf. on Elem. Part. (1963), p.628; NA22 Coll., N. M. Agababyan, *Phys. Lett.* **B 320**, 411 (1994).
10. NA22 Coll., M. R. Atayan, et al., hep-ex/0506027.
11. NA22 Coll., M. R. Atayan, et al., *Phys. Rev. lett.* **89**, 12 (2002)
12. NA49 Coll., T. Anticic et al., *Phys. Rev.* **C 70**, 034902 (2004).
13. NA22 Coll., V. V. Aivazyan, et al., *Phys. Lett.* **B 209**, 103(1998); X. N. Wang and R. C. Hwa, *Phys. Rev.* **D 39**, 187(1989).

FLUCTUATIONS AND CORRELATIONS

Chairpersons: E. De Wolf and G. Gustafson

Study of order parameters through fluctuation measurements by the PHENIX detector at RHIC

Kensuke Homma and the PHENIX collaboration

Physical Science, Graduate School of Science, Hiroshima University, Kagamiyama 1-3-1, Higashi-hiroshima, 739-8526 Japan.

Abstract. Fluctuations in high energy heavy ion collisions can be sensitive observables to critical phenomena in the QCD phase transition. In general the phase transition can be observed as an appearance of a discontinuity or a divergence on quantities related with an order parameter around a critical temperature. The correlation length related with the order parameter to discuss the 2nd order phase transition can be extracted from the multiplicity fluctuation or the density correlation of produced charged particles. An experimental study of the fluctuation has been performed by the PHENIX detector through direct fits to the negative binomial distribution. We will present the result in Au+Au collisions at $\sqrt{s_{NN}} = 200$ in Relativistic Heavy Ion Collider(RHIC).

Keywords: QCD, fluctuation, density correlation, correlation length, 2nd order phase transition

INTRODUCTION

RHIC experiments have probed the state of strongly interacting dense medium with properties consistent with partonic pictures. However, the information on the phase transition has not been quantified yet. Is the transition from the partonic matter to the hadronic matter the first order or second order transition? Are there interesting critical phenomena such as multicritical points? In this letter, we will focus whether the transition is the second order or not based on the spatial size dependence of the multiplicity or density fluctuations with dominantly low p_t charged particles where instant fluctuations such as high p_t jets are expected to be negligible.

In order to connect the density fluctuations with the 2nd order phase transition in general form, Landau's theory[1][2] is briefly reviewed here. The theory is supposed to be qualitatively correct in a limit where fluctuations of an order parameter ϕ itself are small compared to the mean fluctuations of the system with temperature. Therefore, it is valid up to the vicinity of a critical temperature T_c but not at T_c. It describes the relation between a free energy density g and an order parameter ϕ for a given external field h as a function of the system temperature T with the volume V.

$$g(T,\phi,h) = g_0(T) + \frac{1}{2}A(\nabla\phi)^2 + \frac{1}{2}a(T)\phi^2 + \frac{1}{4}b\phi^4 - h\phi, \tag{1}$$

where terms with odd powers are neglected due to the symmetry of the order parameter and $b > 0$ is required to discuss the 2nd order phase transition. Since the order parameter should vanish above a critical temperature T_c, it is natural for the coefficient $a(T)$ to be expressed as $a(T) = a_0(T - T_c)$, while b is usually assumed to be constant in the vicinity of T_c. In the absence of the external field and no spatial fluctuation due to it, the

CP828, *Multiparticle Dynamics*
edited by V. Šimák, M. Šumbera, Š. Todorova, and B. Tomášik

equilibrium value of ϕ should correspond to the value which minimizes the free energy. Hence

$$\left(\frac{\partial g}{\partial \phi}\right)_{\phi_0} = a(T)\phi_0 + b\phi_0^3 = 0. \tag{2}$$

From this, two solutions are obtained

$$\phi_0^2 = 0 \quad \text{for} \quad T \geq T_c \tag{3}$$

$$\phi_0^2 = -\frac{a(T)}{b} = -\frac{a_0(T-T_c)}{b} \neq 0 \quad \text{for} \quad T < T_c. \tag{4}$$

In the presence of the external field and the spatial fluctuation due to it, similarly ϕ is obtained by minimizing the free energy

$$\left(\frac{\partial g}{\partial \phi}\right) = -A\nabla^2\phi(x) + a\phi(x) + b\phi(x)^3 - h(x) = 0. \tag{5}$$

It is instructive to consider a spatially periodic external field with a spatial position x and a spatial frequency k and the effect on ϕ such as

$$h(x) = \delta h_k e^{ikx} \tag{6}$$

$$\phi(x) = \phi_0 + \delta\phi_k e^{ikx}. \tag{7}$$

Then Eq.(5) becomes

$$(a(T)\phi_0 + b\phi_0^3) + (Ak^2 + a + 3b\phi_0^2)\delta\phi_k e^{ikx} - \delta h_k e^{ikx} = 0. \tag{8}$$

Susceptibility defined as $\chi_k = \delta\phi_k/\delta h_k$ is obtained for each temperature region

$$\chi_{+k} = \frac{1}{Ak^2 + a(T)} = \frac{A^{-1}}{k^2 + a(T)/A} = \frac{A^{-1}}{k^2 + \xi_+^{-2}} \quad \text{for} \quad T > T_c \tag{9}$$

$$\chi_{-k} = \frac{1}{Ak^2 - 2a(T)} = \frac{A^{-1}}{k^2 - 2a(T)/A} = \frac{A^{-1}}{k^2 + \xi_-^{-2}} \quad \text{for} \quad T < T_c, \tag{10}$$

where ξ is named as a correlation length and defined as

$$\xi_+ \equiv \sqrt{\frac{A}{a_0|T - Tc|}} \tag{11}$$

$$\xi_- \equiv \sqrt{\frac{A}{2a_0|T - Tc|}}. \tag{12}$$

It is known that the Fluctuation-Dissipation theorem describes the direct connection between spatial correlations of order parameters and the susceptibility with the following relation[3][4]

$$< (\phi(x) - <\phi>) >< (\phi(x') - <\phi>) > = k_B T \chi(x - x') \tag{13}$$

where x and x' are different spatial point. A Fourier transformation of the relation becomes

$$< (\phi_k(x) - < \phi_k >) >< (\phi_{k'}(x') - < \phi_{k'} >) >= \frac{k_B T}{V} \chi_k. \tag{14}$$

Given the susceptibility defined in Eq.(9) or (10), an inverse Fourier transformation gives a function shape for the correlation function

$$< (\phi(x) - < \phi >) >< (\phi(x') - < \phi >) >\sim k_B T \exp(-|x - x'|/\xi), \tag{15}$$

where the one dimensional spatial variable x is assumed. In more general dimensions, the function shape usually contains a power of the distance and the exponential dumping factor[5]. One important note here is that **the observation of the divergence of ξ is a direct evidence of the 2nd order phase transition whatever the detailed mechanism of the phase transition is** , since the above derivations of ξ through the spatial correlation of the order parameter is quite general.

In this study we choose the rapidity particle density as an order parameter. The single and two particle rapidity density operator are defined as

$$\rho_1(y_1) = \sum_i \delta(y_1 - y_i) \tag{16}$$

$$\rho_2(y_1, y_2) = \sum_i \sum_{j;j \neq i} \delta(y_1 - y_i)\delta(y_2 - y_j), \tag{17}$$

where one should bear in mind $i \neq j$ must be satisfied. A general correlation function $G(y_1 - y_2)$ is expressed as

$$G(y_1 - y_2) =< (\rho_1(y_1) - < \rho_1(y_1) >)(\rho_1(y_2) - < \rho_1(y_2) >) > \tag{18}$$

$$=< \rho_1(y_1)\rho_1(y_2) > - < \rho_1(y_1) >< \rho_1(y_2) > \tag{19}$$

$$\equiv \frac{1}{2}(< \rho_1(y_1) > + < \rho_1(y_2) >)\delta(y_1 - y_2) + C_2(y_1, y_2), \tag{20}$$

where the last equation is based on the consideration that the density operator in the second equation does not exclude the case of $i = j$[3][6]. This treatment allows one to discuss a self correlation at $y_1 = y_2$. C_2 in Eq.(20) is the usual two particle correlation function, which is defined as

$$C_2(y_1, y_2) =< \rho_2(y_1, y_2) > - < \rho_1(y_1) >< \rho_1(y_2) > . \tag{21}$$

Based on the one dimensional function form in Eq.(15) and the discussion above, the general correlation function can be parametrized to perform the fit to experimental data as follows;

$$G(y_1 - y_2) = ae^{-\delta y/\xi} + b. \tag{22}$$

The mathematical connection between the second rank normalized factorial moment F_2 and the two particle correlation function is expressed as[7]

$$F_2 - 1 = \frac{\int^{\delta y} G(y_1 - y_2) dy_1 dy_2}{\int^{\delta y} < \rho_1(y_1) >< \rho_1(y_2) > dy_1 dy_2} \tag{23}$$

where δy is defined as $|y_1 - y_2|$ and the original $C_2(y_1, y_2)$ is replaced by $G(y_1 - y_2)$ here.

In the following study, we use an indirect parameter k of the Negative Binomial Distribution(NBD) instead of F_2 itself, which is defined as

$$P_{k,\mu}(n) = \frac{\Gamma(n+k)}{\Gamma(n-1)\Gamma(k)} \left(\frac{\mu/k}{1+\mu/k}\right) \frac{1}{1+\mu/k}, \tag{24}$$

where μ corresponds to the mean value and the relation with F_2 is expressed as[8]

$$k^{-1} = F_2 - 1. \tag{25}$$

This is because NBD can provide us an approximate probability distribution which enables us to estimate how inefficient or dead areas of the detector system bias the k parameter and to obtain the true value of k based on the estimation, while factorial moment itself does not give us any specific models on the distribution function which resulted the observed factorial moment. With the pseudo rapidity η instead of the rapidity y, the final relation between the NBD k parameter and the pseudo rapidity interval size $\delta\eta$ for the parametrization given in Eq.(22) is expressed as

$$k^{-1}(\delta\eta) = F_2 - 1 = \frac{2a\xi^2(\delta\eta/\xi - 1 + e^{-\delta\eta/\xi})}{\delta\eta^2} + \frac{b}{2}. \tag{26}$$

ANALYSIS AND RESULT

Au+Au collision events taken by the PHENIX detector[9] at $\sqrt{S_{NN}} = 200$GeV have been analyzed. The events have been triggered by the Beam-Beam Counter(BBC) in $|\eta| = 3.0 - 3.9$ and Zero-Degree Calorimeters in $|\eta| > 6$. The charged particles were reconstructed by a drift chamber and two multi-wire chambers with pad readouts without a magnetic field in order to enhance the low p_t particles. Each charged tracks were required to point to the primary collision vertex within 5cm which were reconstructed by BBC. We have checked the detector stability rigorously for a run range we have analyzed. In addition, the dead areas in the tracking system have been identified and the effects on the NBD parameters are carefully studied and corrected for the final result.

Fig.1 shows the charged particle multiplicity distributions in each pseudo rapidity interval from 1/8 to 8/8 of the full rapidity coverage of $|\eta| < 0.35$ of the tracking system with top 10% events in the collision centrality. In order to plot the distribution in all rapidity intervals, the distributions were provided as a function of the number of tracks n normalized to the mean multiplicity $< n >$ obtained by the NBD fit. The solid curves were determined by performing the NBD fit. As seen from the figure, it is good enough to assume NBD as a baseline multiplicity distribution to obtain the integrated correlation function through the k parameter in Eq.(26).

Fig.2 shows the k parameter as a function of the rapidity interval $\delta\eta$ for all centrality classes as indicated inside the figure. The solid line indicates the fit result with Eq.(26). The fits are remarkably well and the parametrization is actually reasonable.

Fig.3 shows the two particle correlation length ξ and b as a function of the number of participants. The number of participants N_{part} was obtained from the centrality classes

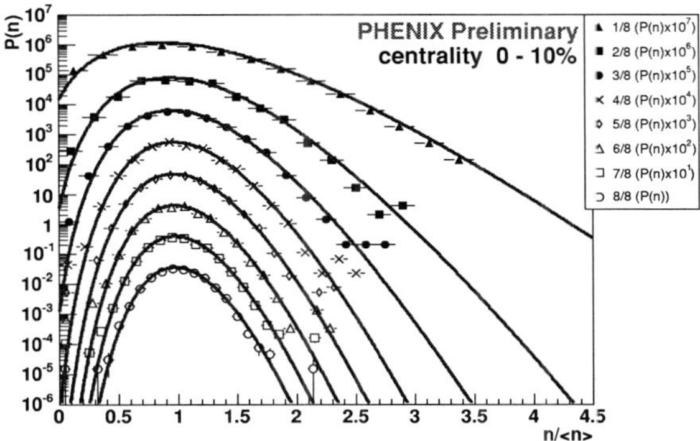

FIGURE 1. Multiplicity distributions in each $\delta\eta$ indicated inside the figure measured in 0-10% centrality bin in Au+Au collisions at $\sqrt{S_{NN}} = 200$GeV. The horizontal axis is normalized by the mean multiplicities. The vertical axis is scaled by the factors indicated inside the figure.

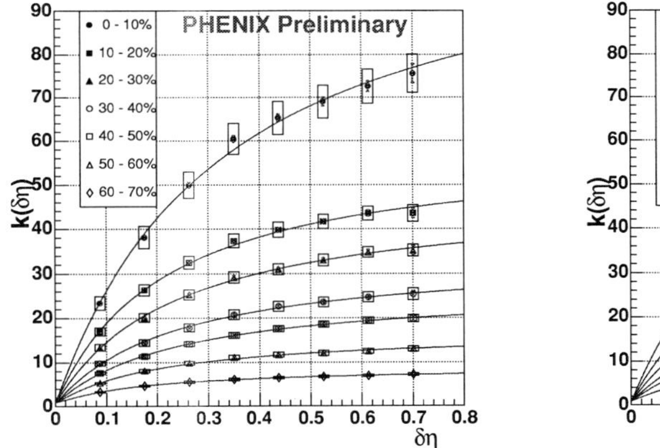

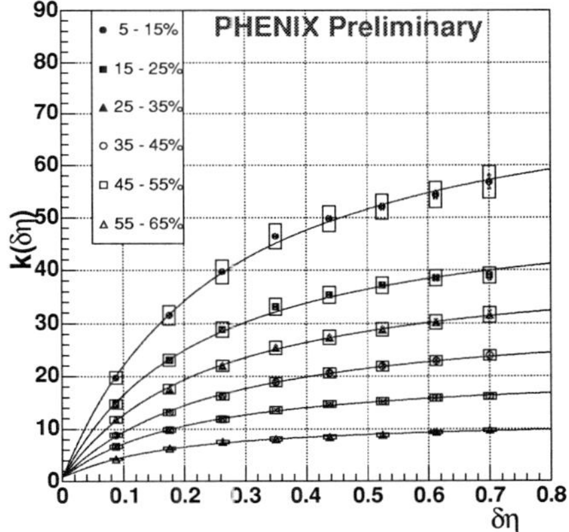

FIGURE 2. Corrected NBD k parameters as a function of pseudo rapidity interval sizes measured in from 0% to 70%(left) and from 5% to 65% centralities with the 10% binning. The error bars indicate statistical errors and boxes indicate systematic errors dominated by ambiguities on the correction factors due to the effect of dead areas of the detector system.

based on the Glauber model which is explained in [10] in detail. If one assumes an ideal gas and the energy density is proportional to the multiplicity ($\propto N_{part}$), N_{part} can be assumed to be proportional to T^3 where T is the system temperature. The one exponent fit with $\xi = N_{part}^{\alpha}$ gives us $\alpha = -0.72 \pm 0.03$. There is no significant divergent behavior in the correlation length. With Eq.(20), (22) and (23), one can expect that the parameter b should decrease as the mean rapidity density increases. Actually the N_{part} dependence of b is qualitatively understandable.

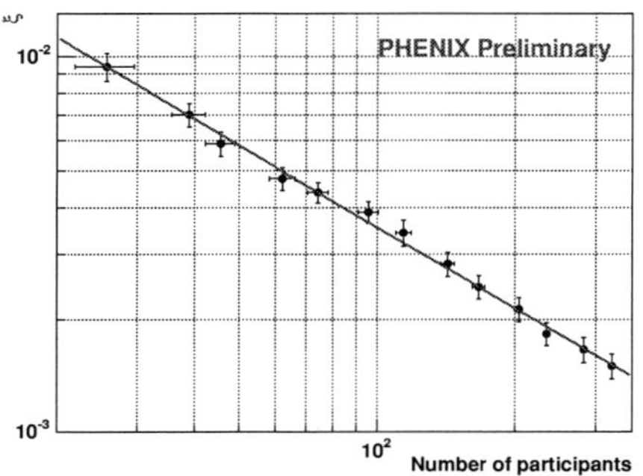

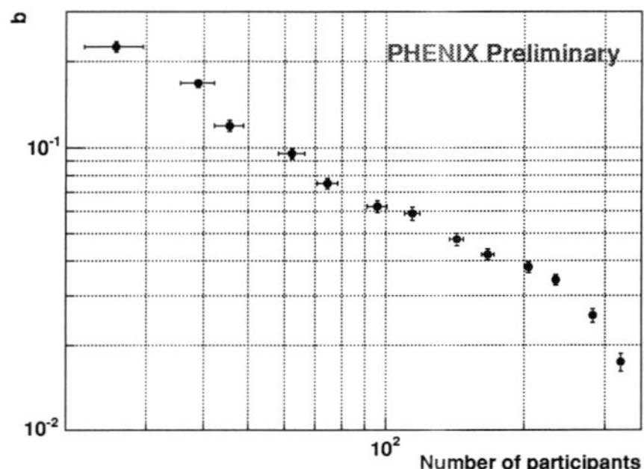

FIGURE 3. Extracted parameters ξ(left) and b(right) in Eq.(26) as a function of the number of participants. The solid line in the left figure is the result of the one slope fit in logarithmic scales.

SUMMARY AND PROSPECTS

The multiplicity distributions measured in Au+Au collisions at $\sqrt{S_{NN}} = 200\text{GeV}$ well agree with the negative binomial distribution. The correlation length has been extracted based on the function by relating pseudo rapidity density correlations and the susceptibility through the Fluctuation-Dissipation theorem. The function can fit k vs. $\delta\eta$ in all centralities remarkably well. The correlation length indicates a power law behavior as a function of the number of participants. There is no significant divergent behavior in the correlation length at the level of the present systematic errors. In the present dynamic range of $\delta\eta$, we have performed the fit based on Eq.(26) by fixing a at 1.0, because large parameter correlations between a and ξ were seen. We plan to extend the fit range to much narrower $\delta\eta$ regions. More steady results on the correlation length will be published in near future.

REFERENCES

1. L. D. Landau, *Phys. Zurn. Sowjetunion* **11**, 546 (1937).
2. V. L. Ginzburg, and L. Landau, *Sov. Phys. - JETP* **20**, 1064 (1950).
3. E. M. Lifshitz, and L. D. Landau, *Statistical Physics 3rd edition*, Nauka, Moscow, 1976.
4. W. Gebhardt, and U. Krey, *Phasenübergänge und kritische Phänomene*, Friedr. Vieweg & Sohn Verlagsgesellschaft, Braunschweig/Wiesbaden, 1980.
5. M. E. Fisher, *Physica* **28**, 172 (1962).
6. H. E. Stanley, *Introduction to phase transitions and critical phenomena*, CLARENDON PRESS, OXFORD, 1971.
7. P. Carruthers, and I. Sarcevic, *Phys. Rev. Lett.* **63**, 1562 (1989).
8. W. A. Zajc, *Phys. Lett.* **B175**, 219 (1986).
9. K. Adcox, et al., *Nucl. Instrum. Meth.* **A499**, 469 (2003).
10. K. Adcox, et al., *Phys. Rev. Lett.* **86**, 3500 (2001).

Chiral symmetry restoration, pion opacity, and the RHIC HBT puzzle

John G. Cramer* and Gerald A. Miller†

*Department of Physics, Box 351560, University of Washington, Seattle, WA 98195-1560, USA, email=cramer@phys.washington.edu
†Department of Physics, Box 351560, University of Washington, Seattle, WA 98195-1560, USA, email=miller@phys.washington.edu

Abstract. We present a relativistic quantum wave-mechanical optical model treatment of the opacity and refractive effects of pions emerging from a hot dense expanding source. The Klein-Gordon equation with Bjorken-cylinder geometry is solved for distorted waves representing the emerging pions. These waves are combined with a hydro-inspired pion source function to predict HBT radii and pion spectra. The model, when using a very deep potential well, has produced excellent HBT radii and spectrum fits to RHIC $\sqrt{s_{NN}} = 200$ GeV Au+Au data from STAR, reproducing the observed small source sizes, $R_O/R_S \approx 1$, the low-momentum behavior of the radii, and the pion spectrum in shape and magnitude. The calculations represent a new tool for investigating the presence and characteristics of chiral symmetry restoration in heavy ion collisions.

Keywords: Chiral symmetry, HBT

INTRODUCTION

Many of the "signals" from analysis of Au+Au collisions at RHIC suggest that a quark gluon plasma (QGP) has been created in the initial stages of the collision. A major problem with such an interpretation has been that a QGP scenario would require a large source that has expanded for a long time before freeze out and has a long duration for emission of pions. On the other hand, analysis of RHIC data using HBT interferometry has been interpreted as indicating a relatively small unexpanded source with a very short pion emission duration. In the literature, this conundrum has been called the "RHIC HBT Puzzle".

In an effort to understand the origins of this problem, we have undertaken a new approach to RHIC physics that required the application of quantum wave mechanics and the nuclear optical model to the medium produced by the colliding systems. We formulated a new relativistic quantum mechanical description of the collision medium that included collective flow as well as absorption and refraction in a complex potential. We solved the Klein-Gordon wave equations for pions in the medium and calculated overlap integrals with these wave functions to obtain predictions of pion spectra and HBT radii.

CP828, *Multiparticle Dynamics*
edited by V. Šimák, M. Šumbera, Š. Todorova, and B. Tomášik

FORMALISM

Extensive details of the formalism used in our model have been presented in a long paper recently submitted to Physical Review C[1], and the formal approach will only be summarized here. Briefly, we treat the dynamics of the observed pions from their point of initial emission (not from freezeout). We separate the emitted particles into "channels", and explicitly threat only the channel including pions that participate in Bose-Einstein symmetrization leading to the observed HBT correlation "bump".

We apply the nuclear optical model to the pions in that channel as they traverse the hot dense medium of the collision fireball and emerge into the vacuum. We deal with other channels of the problem (e.g., halo pions, pions from long-lived resonances, reaction-channel absorption of pions, ...) through the use of an imaginary potential that removes pions from the channel of interest. We solve the Klein-Gordon wave equation in a partial-wave expansion and numerically calculate the wave functions of the pions in the channel of interest. We do not explicitly use a freezeout hypersurface, but rather allow the optical potential to describe the interactions of the emitted pions with the medium. Figure 1 shows 3-dimensional plots of calculated wave functions and compares them with wave functions with no real potential and with wave functions based on the eikonal approximation.

FIGURE 1. (Color online) Wave functions. The figures show the absolute square of the calculated wave functions times the density $\rho(b)$. For each (row) value of K_T $(.125, 1.0, 3.0\,\mathrm{fm}^{-1})$, the full calculation, the calculation including only the imaginary part of the optical potential and the eikonal approximation are shown horizontally. The out direction is parallel to the lower right edge of each picture.

We employ a "hydro-inspired" multi-dimensional Gaussian emission function $S_0(\vec{x},\vec{p})$ to describe the probability of pion emission as a function of position and momentum in the medium. We then combine this with optical model wave functions to obtain $S(\vec{x},\vec{p})$, a distorted-wave emission function (DWEF) that is used to calculate the pion correlation function and spectrum. The optical potential used is not explicitly time dependent, but it acts only over a relatively short time interval limited by the emission function.

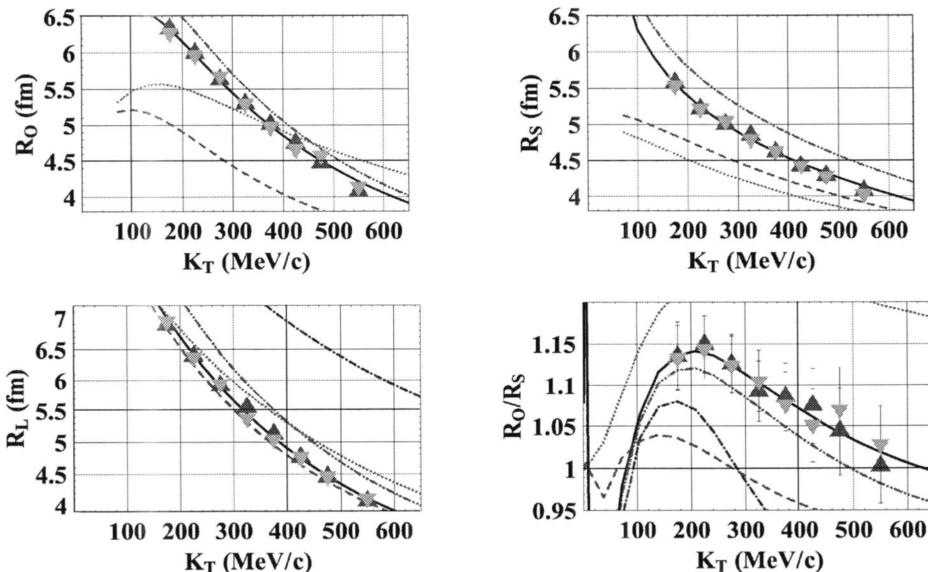

FIGURE 2. (Color online) HBT Radii R_s,R_o,, R_l and the ratio R_o/R_s; Data: $\nabla \Rightarrow \pi^+\pi^+$; $\triangle \Rightarrow \pi^-\pi^-$. Curves: solid $\Rightarrow$ full calculation; dotted $\Rightarrow \eta_f = 0$ (no flow); dashed $\Rightarrow$ Re[U]=0 (no refraction); dot-dashed) $\Rightarrow$ U=0 (no optical potential), double-dot-dashed $\Rightarrow$ substituting Boltzmann for Bose-Einstein thermal distribution.

APPLICATION

Numerical calculations applying this formalism were placed under the control of a Marquardt-Levenberg chi-squared minimization program that varied up to 12 model parameters to obtain the best fit to STAR $\sqrt{s_{NN}}$=200 GeV Au+Au pion spectrum[3] and HBT radii[4]. We note that we calculate the HBT radii by explicitly evaluating the correlation function near its half-maximum point and calculating the Gaussian radius that would give this value. We find that this method gives stable results even in the region of low average momentum K, while the widely-used second-moment method is unreliable in that region.

Because the pion channel that is calculated does includes only those pions that participate in producing the correlation "bump", we must correct the measured spectrum data by multiplying by values from a linear fit to the experimental values of $\sqrt{\lambda}$, where λ is the HBT "dilution" parameter. We use these corrected spectrum values in the fits, and then correct the predicted spectrum by multiplication by $1/\sqrt{\lambda}$. The dotted line in Fig. 3 shows the "raw" uncorrected prediction.

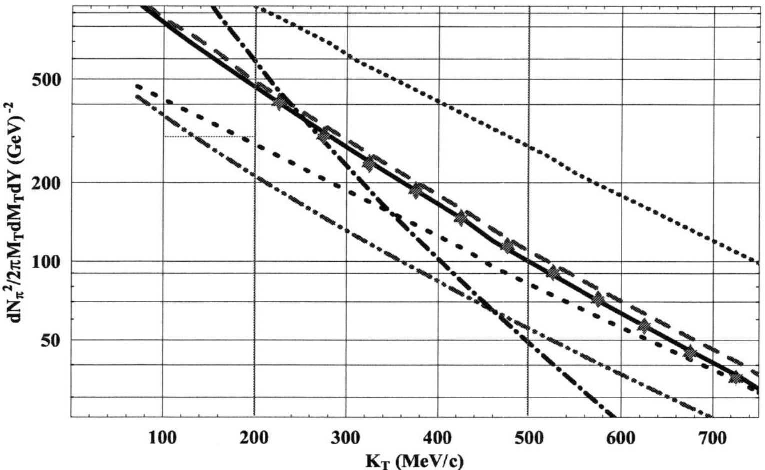

FIGURE 3. (Color online) Pion transverse momentum spectrum. Data: $\nabla \Rightarrow \pi^+$; $\triangle \Rightarrow \pi^-$. Curves: solid $\Rightarrow$ full calculation; dotted $\Rightarrow \eta_f = 0$ (no flow); dashed $\Rightarrow$ Re[U]=0 (no refraction); dot-dashed) $\Rightarrow U$=0 (no optical potential), double-dot-dashed $\Rightarrow$ substituting Boltzmann for Bose-Einstein thermal distribution. The short dashed line shows the "raw" uncorrected spectrum prediction.

Our initial expectation was that the imaginary part of the optical potential would be important for simulating pion absorption, while the real potential with its refractive effects was included mainly for formal reasons. To our surprise, when the fitting began the real potential grew deeper and deeper as the fit improved, until it was essentially as deep as the pion mass. This result suggested to us that the pion must be losing mass in the hot dense medium of the collision because chiral symmetry had been partially restored in the medium. Therefore, we gave the optical potential the momentum dependence that is consistent with chiral symmetry restoration. The result of this inclusion was impressive. Good fits to the STAR data, giving a chi-squared of about 2.7 per data point and 5.1 per degree of freedom, were obtained. These results have been published in Physical Review Letters[2]. Some of these fits are shown in Fig. 2.

NON-GAUSSIAN CORRELATION FUNCTIONS

It is well known that the identical pion correlation functions observed experimentally are only approximated by a Gaussian distribution, as seen from deviations from a Gaussian in the half-maximum region and an exponential falloff in the tail region. This leads to systematic uncertainties in the determination of HBT radii. To investigate this phenomenon, we have used our DWEF code to investigate the deviation of the calculated correlation functions from a Gaussian shape. Figure 4 shows a linear plot of one correlation function and a logarithmic plot of another, both calculated at an average pair momentum of 100 MeV/c.

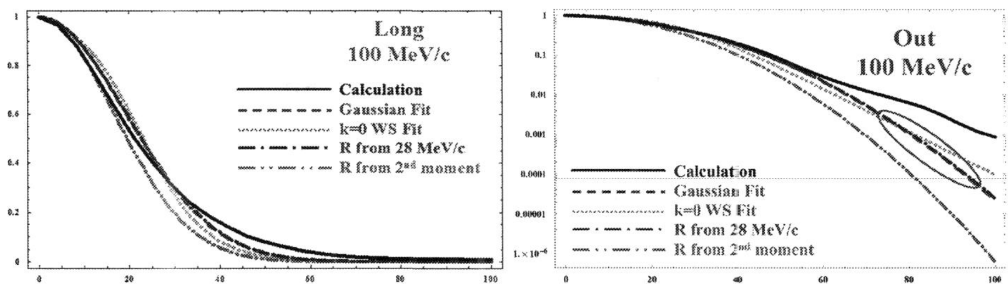

FIGURE 4. (Color online) Correlation Functions: A linear plot (left) of a calculated correlation function in the "Long" direction with several fits, and a logarithmic plot (right) of a calculated correlation function in the "Out" direction with several fits described in the text.

We have fitted such calculated correlation functions in several ways: (1) one-parameter fit with a Gaussian (red dashed); (2) one-parameter fit with a Woods-Saxon distribution that has the diffuseness set to give the function a kurtosis of k=0 (green dotted); (3) deduce a Gaussian radius from the correlation function at q=28 MeV/c, i.e., approximately at the half-maximum point (blue dot-dash); and (4) deduce a Gaussian radius from the 2nd moment (or curvature) of the correlation function near q=0 (magenta dot-dot-dash). As can be seen from the plots, methods (1) and (3) agree well with each other and are closest to the calculated correlation function, while the 2nd moment method (4) fails by giving a radius that is too large. The k=0 Woods-Saxon distribution is further from the calculated function than (1) and (3), and while it does go to an exponential is the tail region (circled region) it does not reproduce the falloff of the calculated correlation function. Our conclusion is that the method of choice is the determination of the HBT radius using method (3), as we have done in Fig. 2 above.

CONCLUSION

These results indicate that roughly half of the pions are absorbed while traversing the medium, with those surviving emitted primarily from a localized "bright ring" near the outer surface of the medium. Moreover, the emerging pions must regain their mass by expending a sizable fraction of their kinetic energy in climbing out of the very deep well made by the real potential. When these effect are properly taken into account, the pion source size and emission duration are consistent with a QGP scenario. Further, in most lattice gauge studies of heated and compressed nuclear matter the chiral phase transition and the transition to a quark-gluon plasma occur under about the same conditions. Our inferred observation of a chiral phase transition at RHIC is therefore consistent with the presence of a quark-gluon plasma transition in RHIC collisions.

In closing, we note that this work may represent the first direct observation of a chiral phase transition in a multiparticle system. Other experimental support of chiral symmetry restoration comes from the structure of highly excited states of the nucleon. Therefore, we have developed a new tool for relativistic heavy ion physics, which we plan to use for investigating the onset and properties of chiral symmetry restoration as a function of energy and centrality in relativistic heavy ion collisions, using data at the very wide range of energies and systems that has already been provided by experiments at the AGS, SPS, and RHIC.

REFERENCES

1. Gerald A. Miller and John G. Cramer, *Phys. Rev.* **C** (Submitted for publication, 2005); nucl-th/0507004
2. John G. Cramer, Gerald A. Miller, Jackson M. S. Wu, and Jin-Hee Yoon, *Phys. Rev. Lett.* **94**, 102302 (2005).
3. J. Adams, *et al.*, *Phys. Rev. Lett.* **92**, 112301 (2004).
4. J. Adams, *et al.*, *Phys. Rev.* **C71** 044906 (2005).

Energy and rapidity dependence of the electric charge correlations at 20-158GeV beam energies at the CERN SPS (NA49)

P. Christakoglou, A. Petridis, M. Vassiliou for the NA49 collaboration

Physics Department - University of Athens - 15771 - Athens, Greece

Abstract.
Electric charge correlations are studied with the Balance Function method for central Pb + Pb collisions at the CERN - SPS. The results on centrality selected Pb + Pb interactions at 40 and 158 AGeV are presented for the first time for two different rapidity intervals:

The mid-rapidity region where a decrease of the width with increasing centrality of the collision is observed and the forward rapidity region where this effect vanishes. This could suggest a delayed hadronization scenario.

In addition, the results from a first attempt to study the energy dependence of the Balance Function throughout the whole SPS energy range, will also be presented. The suitably scaled decrease of the width is approximately constant for the intermediate energies (30 to 80 AGeV) and gets stronger for the highest SPS and RHIC energies. On the other hand, both URQMD and HSD simulation results show no dependence on the collision energy.

Keywords: Balance Function,correlations
PACS: <25.75.Gz>

1. INTRODUCTION

The study of correlations and fluctuations is expected to provide additional information on the reaction mechanism of high energy nuclear collisions [1, 2, 3, 4]. In particular, many event-by-event signatures have already been analyzed [5]. Another important measure of correlations, the Balance Function (BF), was introduced by Bass, Danielewicz and Pratt [6]. It measures the correlation of the oppositely charged particles produced during a heavy ion collision and its width can be related to the time of hadronization. The BF is derived from the charge correlation function that was used to study the hadronization of jets in p+p collisions at the ISR [7] and $e^- + e^+$ annihilations at PETRA [8]. The first results on the BF were obtained for Au+Au collisions by the STAR collaboration at RHIC [9] and for Pb+Pb collisions by the NA49 collaboration at the CERN-SPS [10].

The motivation for studying the Balance Function comes from the idea that hadrons are produced locally as oppositely charged particle pairs. Particles of such a pair are separated in rapidity due to the initial momentum difference and secondary interactions with other particles. Particles of a pair that were created earlier are separated further in rapidity because of the expected large initial momentum difference and the long lasting rescattering phase. On the other hand, oppositely charged particle pairs that were created later are correlated within a smaller interval Δy of the relative rapidity. Our aim is to measure the degree of this separation of the balancing charges and to find possible

CP828, *Multiparticle Dynamics*
edited by V. Šimák, M. Šumbera, Š. Todorova, and B. Tomášik

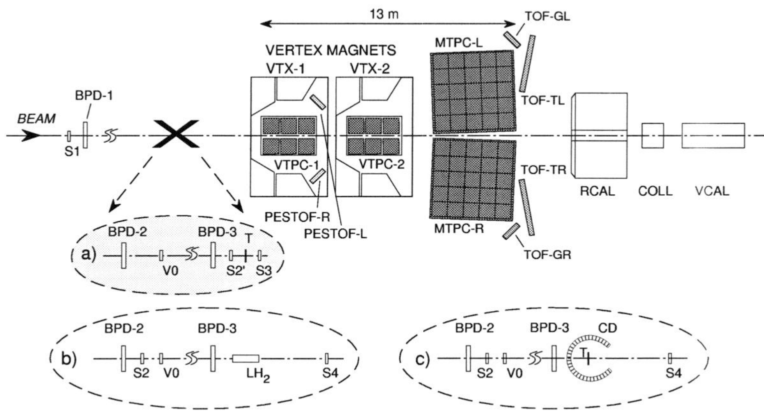

FIGURE 1. The NA49 experimental setup.

indications for delayed hadronization.

In order to examine the pseudo-rapidity (η) correlation of charged particles the BF is defined as a difference of the correlation function of oppositely charged particles and the correlation function of like-charge particles normalized to the total number of particles. The definition of the BF reads [6]:

$$B(\Delta\eta) = \frac{1}{2}\Big[\frac{N_{+-}(\Delta\eta) - N_{--}(\Delta\eta)}{N_-} + \frac{N_{-+}(\Delta\eta) - N_{++}(\Delta\eta)}{N_+}\Big]. \tag{1}$$

The most interesting property of the BF is its width. Early stage hadronization is expected to result in a broad BF, while late stage hadronization leads to a narrower distribution [6]. The width of the BF can be characterized by the weighted average $\langle\Delta\eta\rangle$:

$$\langle\Delta\eta\rangle = \sum_{i=0}^{k}(B_i \cdot \Delta\eta_i) / \sum_{i=0}^{k} B_i, \tag{2}$$

where i is the bin number of the BF histogram.

In the following sections we will firstly describe in brief the NA49 experimental setup. The next two sections will be dedicated to the new NA49 experimental results on the rapidity study and the energy scan that was performed using the Balance Function. We will conclude with the summary.

2. EXPERIMENTAL SETUP

The NA49 detector [11] is a wide acceptance hadron spectrometer for the study of hadron production in collisions of hadrons or heavy ions at the CERN SPS. The main components are four large - volume Time Projection Chambers (TPCs) (Fig. 1) which are capable of detecting 80% of some 1500 charged particles created in a central Pb+Pb collision at 158A GeV.

The targets are C (561 mg/cm^2), Si (1170 mg/cm^2) disks and a Pb (224 mg/cm^2) foil for ion collisions and a liquid hydrogen cylinder (length 20 cm) for hadron interactions. They are positioned about 80 cm upstream from VTPC-1.

The centrality of a collision is selected (on-line for central Pb+Pb, Si+Si and C+C and off-line for minimum bias Pb+Pb, Si+Si and C+C interactions) by a trigger using information from a downstream calorimeter (VCAL), which measures the energy E_0 of the projectile spectator nucleons.

3. RAPIDITY DEPENDENCE

In order to investigate the properties of hadronization in heavy ion collisions, as proposed by the BF methodology [6], we studied the system size and centrality dependence of the width of the BF in two SPS energies ($\sqrt{s} = 17.2$ GeV and $\sqrt{s} = 8.8$ GeV) and in two different pseudo-rapidity intervals. The defined pseudo-rapidity regions for the highest energy are $2.6 \leq \eta \leq 3.9$ named as mid-rapidity region and $4.0 \leq \eta \leq 5.4$ as forward rapidity region. The corresponding pseudo-rapidity regions for the second energy are $1.8 \leq \eta \leq 3.2$ (mid-rapidity region) and $3.3 \leq \eta \leq 4.7$ (forward rapidity region).

Fig. 2 shows the width of the BF distributions for real, HIJING [12] and shuffled data as a function of the mean number of wounded nucleons, for the two pseudo-rapidity regions analyzed (mid-rapidity region-left plot, forward rapidity region-right plot) for the $\sqrt{s_{NN}} = 17.2$ GeV case. There is an apparent centrality dependence for real data in the mid-rapidity region while this dependence vanishes in the forward region. The same effect one may observe for $\sqrt{s_{NN}} = 8.8$ GeV in fig. 3.

One can also point out that for the smaller system, such as p+p up to Si+Si, there is no significant difference in the actual values of the widths between the two rapidity regions for both energies. The difference in the absolute values appears if one studies the more central Pb+Pb collisions for both cases.

4. ENERGY DEPENDENCE

Finally, an attempt was made to study the energy dependence of the Balance Function was studied within the NA49, where the most central Pb+Pb events were analyzed throughout the whole available SPS energy range. These data samples passed once again through the shuffling mechanism [9, 10] so that we could have an estimate of the highest value of the width for each energy. The pseudo-rapidity interval analyzed for each energy was limited to the same range (1.4 units) and was located around mid-rapidity. In order to quantify the decrease of the width for the different energies, we

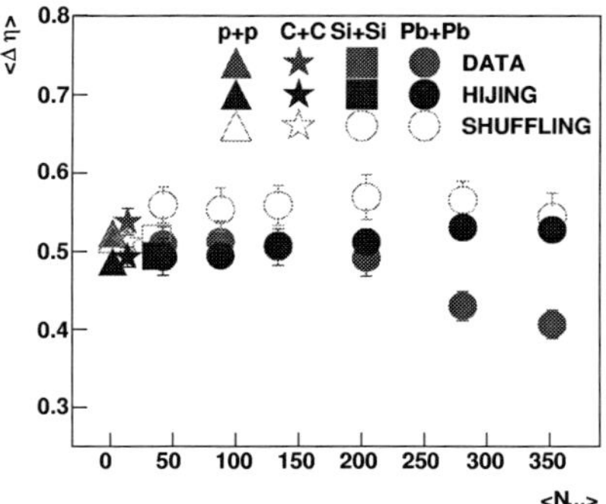

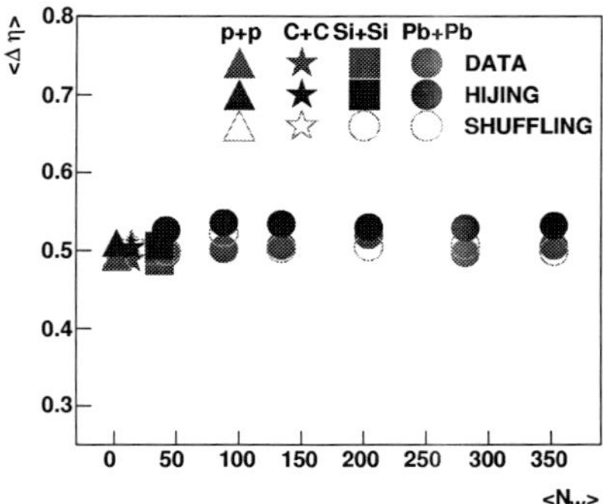

FIGURE 2. The system size and centrality study for the measured width of the Balance Function for charged particles at $\sqrt{s_{NN}} = 17.2$ GeV as a function of the mean number of wounded nucleons for the two different rapidity intervals analyzed: the mid-rapidity region (left plot) and the forward rapidity region (right plot).

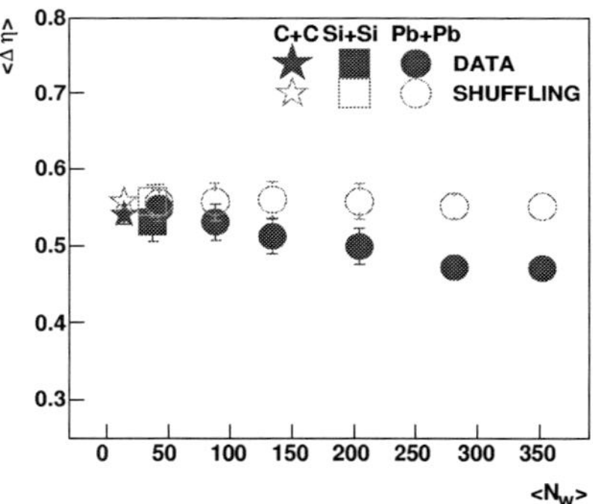

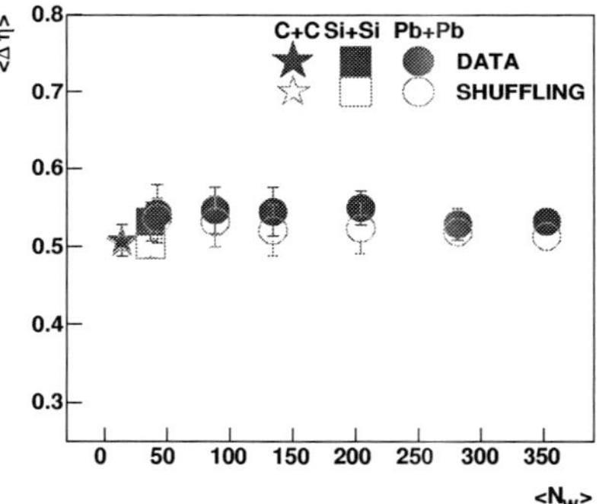

FIGURE 3. The system size and centrality study for the measured width of the Balance Function for charged particles at $\sqrt{s_{NN}} = 8.8$ GeV as a function of the mean number of wounded nucleons for the two different rapidity intervals analyzed: the mid-rapidity region (left plot) and the forward rapidity region (right plot).

introduced the normalized parameter W which is defined by the following equation:

$$W = \frac{100 \cdot (\langle \Delta\eta \rangle_{shuffled} - \langle \Delta\eta \rangle_{data})}{\langle \Delta\eta \rangle_{shuffled}}. \tag{3}$$

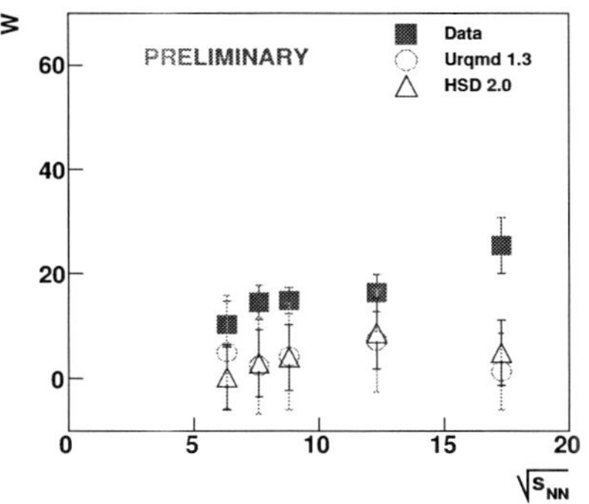

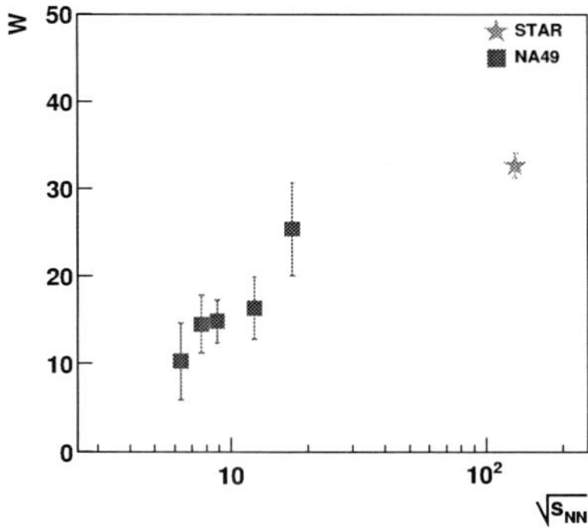

FIGURE 4. The dependence of the normalized parameter W on the $\sqrt{s_{NN}}$ for central Pb+Pb collisions in the SPS energy range after applying the NA49 acceptance filter (left plot) and for central Au+Au collisions at RHIC (right plot).

In other words this parameter indicates how many standard deviations is the measured percentage away from the estimated shuffled value.

The left plot of fig. 4 shows the dependence of the normalized W parameter on the $\sqrt{s_{NN}}$. As far as the data is concerned we notice a first indication of an energy dependence. This plot indicates an energy dependence at the level of $(40.4 \pm 19.3)\%$ for our phase-space detector acceptance.

In addition, two models were used in order to further investigate this energy dependence. The Ultra-relativistic Quantum Molecular Dynamics model (UrQMD) [13] and the Hadron-String Dynamics (HSD) transport approach [14] are two microscopic models used to simulate (ultra)relativistic heavy ion collisions in the energy range from Bevalac and SIS up to AGS, SPS and RHIC. The points from the corresponding analysis of both UrQMD and HSD generated events throughout the whole SPS energy range can also be seen in fig. 4, where one can notice no sign of energy dependence of the W parameter.

The right plot of fig. 4 shows the dependence of the normalized W parameter on the $\sqrt{s_{NN}}$ for the whole SPS energy range as well as for Au+Au collisions at $\sqrt{s_{NN}} = 130$ GeV. The corresponding RHIC point tends to be even higher than the last SPS one, indicating an additional rise of the W parameter towards RHIC leaving an open question for the LHC energies.

5. SUMMARY

In this paper, we presented the latest results of the BF obtained by the NA49 collaboration. We have analyzed data coming from different systems and centrality classes for two different energies in two totally different rapidity regions. We have seen that although there is a system size and centrality dependence of the width of the BF when analyzed in the mid-rapidity region, this effect is not apparent in the forward region.

As far as the energy scan is concerned, data show a first indication of an energy dependence which is not apparent in the two models used (HSD and UrQMD). The normalized W parameter indicates a rise from the lowest SPS energy to the highest SPS and RHIC energies.

ACKNOWLEDGMENTS

This work was supported by the University of Athens/Special account for research grants, the US Department of EnergyGrant DE-FG03-97ER41020/A000, the Bundesministerium fur Bildung und Forschung, Germany, the Polish State Committee for Scientific Research (2 P03B 130 23, SPB/CERN/P-03/Dz 446/2002-2004, 2 P03B 04123), the Hungarian Scientific Research Foundation (T032648, T032293, T043514), the Hungarian National Science Foundation, OTKA, (F034707), the Polish-German Foundation, and the Korea Research Foundation Grant (KRF-2003-070-C00015).

REFERENCES

1. See for recent results: Proc. of Quark Matter 2004, *J. Phys. G* **30**, (2004).
2. T. Alber et al. (*NA35 Collaboration*), *Z. Phys. C* **64**, 195 (1994).
   S. Afanasiev et al.(*NA49 Collaboration*), *Phys. Rev. C* **66**, 054902 (2002).
   R. Stock, *Nucl. Phys. A* **661**, 282c (1999).
3. J. Collins, M. Perry, *Phys. Rev. Lett.* **34**, 1353 (1975).
   P. Braun-Munzinger, J. Stachel, Christof Wetterich, *Phys. Lett. B* **596**, 61-69 (2004).
4. M. Gazdzicki, M. Gorenstein, *Acta Phys. Polon.* **B30**, 2705 (1999)
   M. Gazdzicki (*NA49 Collaboration*) Proc. of Quark Matter 2004, *J. Phys. G* **30**, S701 (2004).
5. S. Jeon and V. Koch, *Phys. Rev. Lett.* **85**, 2076 (2000).
   H. Heiselberg and A. D. Jackson, *Phys. Rev. C* **C63**, 064904 (2001).
   E. V. Shuryak and M. A. Stephanov, *Phys. Rev. C* **C63**, 064903 (2001).
   S. V. Afanasiev et al. (NA49 Collaboration), *Phys. Rev. Lett.* **86**, 1965 (2001).
   G. Georgopoulos, P. Christakoglou, A. Petridis and M. Vassiliou, *Proc. of "Correlations and Fluctuations 2002"*.
   T. Anticic et al. (NA49 Collaboration), *arXiv:hep-ex/0311009*.
6. S. A. Bass, P. Danielewicz and S. Pratt, *Phys. Rev. Lett.* **85**, 2689 (2000).
7. D. Drijard et al, *Nucl. Phys.* **B155**, 269 (1979)
   D. Drijard et al, *Nucl. Phys. B* **166**, 233 (1980)
   I.V. Ajinenko et al., *Nucl. Phys. C* **43**, 37 (1989).
8. R. Brandelik et al., *Phys. Lett. B* **100**, 357 (1981)
   M. Althoff et al., *Z. Phys. C* **17**, 5 (1983)
   H. Aihara et al., *Phys. Rev. Lett.* **53**, 2199 (1984)
   H. Aihara et al., *Phys. Rev. Lett.* **57**, 3140 (1986)
   P.D. Acton et al., *Phys. Lett. B* **305**, 415 (1993).
9. J. Adams et al., (STAR Collaboration) *Phys. Rev. Lett.* **90**, 172301 (2003).
10. C. Alt et al., (NA49 Collaboration), *Phys. Rev. C* **71**, 034903 (2005).
    P. Christakoglou et al., (NA49 Collaboration), *Nucl. Phys. A* **749**, 279-282 (2005).
11. S. Afanasiev et al. (NA49 Collab.) *Nucl. Instrum. Meth.* **A430**, 210 (1999).
12. Xin-Nian Wang, Miklos Gyulassy, *Phys. Rev. C* **71, 1859-8996 (1999).**
13. M. Bleicher et al., *Nucl. Part. Phys.* **25**, 1859-189 (1999).
14. W. Ehehalt, W. Cassing, *Nucl. Phys. A* **602**, 449-486 (1996).

Boost Invariance and Multiplicity Dependence of the Charge Balance Function in π^+p and K$^+$p collisions at 22 GeV

Li Na* and Wu Yuanfang*

*Institute of Particle Physics, Hua-Zhong Normal University, Wuhan 430070, China

Abstract.
Boost invariance and multiplicity dependence of the charge balance function are studied in π^+p and K$^+$p collisions at 22 GeV. Charge balance, as well as charge fluctuations, are found to be boost invariant over the whole rapidity region, but both depend on the size of the rapidity window. It is also found that the balance function becomes narrower with increasing multiplicity, consistent with the narrowing of the balance function when centrality and/or system size increase, as observed in current relativistic heavy ion experiments.

Keywords: Boost invariance, Balance function, Multiplicity dependence
PACS: 13.85.Hd, 25.75.Gz

INTRODUCTION

Charge balance and charge flow are measures of rapidity correlations between oppositely charged particles and have been used to study hadronization in hadron-hadron [1] as well as in lepton-hadron [2] and e^+e^- [3] collisions. In the form of a charge balance function (BF) [4], they have recently gained new interest in the field of relativistic heavy ion collisions. A narrowing of the balance function is suggested as a signature of delayed hadronization, of the formation of a Quark-Gluon Plasma (QGP) during the early stage of a collision. The integral of the balance function is related to the event-by-event charge fluctuations [5], which are expected to be suppressed in a QGP [6].

So far, two heavy ion experiments [7, 8] have measured the balance function at various centralities and for different colliding nuclei. A narrowing of the balance function is indeed observed with increasing centrality of the collision and with increasing size of the colliding nuclei. The measured charge fluctuations, on the other hand, are consistent with those expected for a hadronic gas [9, 10, 11].

Before drawing any conclusions from the observed narrowing of the balance function, it is necessary to know how the BF behaves in hadron-hadron collisions, where no QGP is expected, and how the limited detector acceptance influences its width. Since current heavy ion experiments cover only a limited rapidity region, the measured BF's do not correspond to that for the full rapidity region. Whether the results from different heavy ion experiments are comparable or not depends on the influence of the acceptance.

The balance function in finite rapidity window is defined as

$$B(\delta y|Y_{\mathrm{w}}) = \frac{1}{2}\left[\frac{\langle n_{+-}(\delta y)\rangle - \langle n_{++}(\delta y)\rangle}{\langle n_{+}\rangle} + \frac{\langle n_{-+}(\delta y)\rangle - \langle n_{--}(\delta y)\rangle}{\langle n_{-}\rangle}\right]. \qquad (1)$$

CP828, *Multiparticle Dynamics*
edited by V. Šimák, M. Šumbera, Š. Todorova, and B. Tomášik

Here, $n_{+-}(\delta y)$,$n_{++}(\delta y)$, $n_{--}(\delta y)$ are the numbers of pairs of opposite- and like-charged particles satisfying the criteria that they fall into the rapidity window Y_{w} and that their relative rapidity equals δy; n_+ and n_- are the numbers of positively and negatively charged particles, respectively, in the interval Y_{w}.

Based on the assumption of longitudinal boost invariance [5], Jeon and Pratt proposed a relation between the balance function in a rapidity window $B(\delta y|Y_{\mathrm{w}})$ and in the full rapidity range $B(\delta y|\infty)$ [5],

$$B(\delta y|Y_{\mathrm{w}}) = B(\delta y|\infty)\left(1 - \frac{\delta y}{Y_{\mathrm{w}}}\right). \tag{2}$$

Conventionally, boost invariance refers to particle density being independent of rapidity, as originally assumed in [12]. While this may be correct in a very restricted region at mid-rapidity for the rapidity density itself, boost invariance of the balance function only required that the charge correlation between final state particles be the same in any longitudinally-Lorentz-transformed frame. Whether the BF is boost invariant over the whole rapidity region or only in the central region, cannot be simply deduced from the corresponding shape of the rapidity density distribution. This important issue has not yet been investigated in either its theoretical or experimental aspects.

DATA SAMPLE

Boost invariance and multiplicity dependence of the charge balance function is studied on π^+p and K$^+$p data at 250 GeV/c ($\sqrt{s} = 22$ GeV) of the NA22 experiment. This experiment was equipped with a rapid cycling bubble chamber as an active vertex detector, had excellent momentum resolution and 4π acceptance. The latter feature allows, for the first time, to study the properties of the balance function in full phase space.

Since no statistically significant differences are seen between the results for π^+ and K$^+$ induced reactions, the two data samples are combined for the purpose of this analysis. A total of 44 524 non-single-diffractive events is obtained after all necessary selections, as described in detail in [13]. In particular, possible contamination from secondary interactions is suppressed by a visual scan and the requirement that overall charge balance be satisfied within the whole event; γ conversions near the primary vertex are removed by electron identification.

MAIN RESULTS

Boost-invariance of Balance Function

In Fig. 1, the balance function is shown for five cms rapidity windows of width $Y_{\mathrm{w}} = 3$, located at different positons, $[-3,0]$ (open stars), $[-2,1]$ (open crossed), $[-1,2]$ (open circles), $[0,3]$ (open triangles). In this and following figures, errors are smaller than the size of the symbols. The five functions coincide within the experimental errors, except that a few points in $[-3,0]$ are somewhat lower than the others. This is caused by very

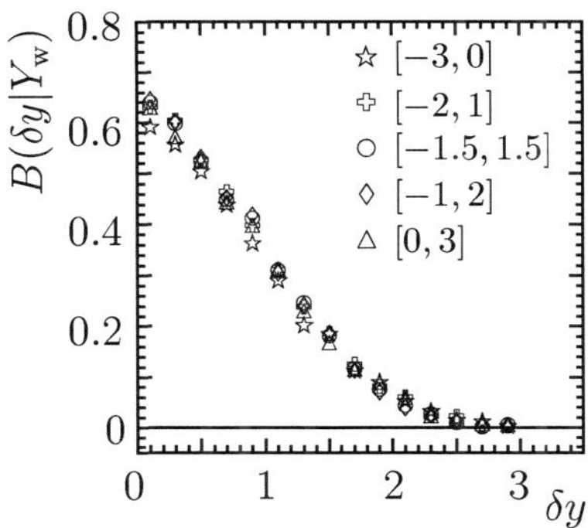

FIGURE 1. The balance function for five different positions of a rapidity window of size $Y_w = 3$: $[-3,0]$ (open stars), $[-2,1]$ (open crossed), $[-1,2]$ (open circles), $[0,3]$ (open triangles).

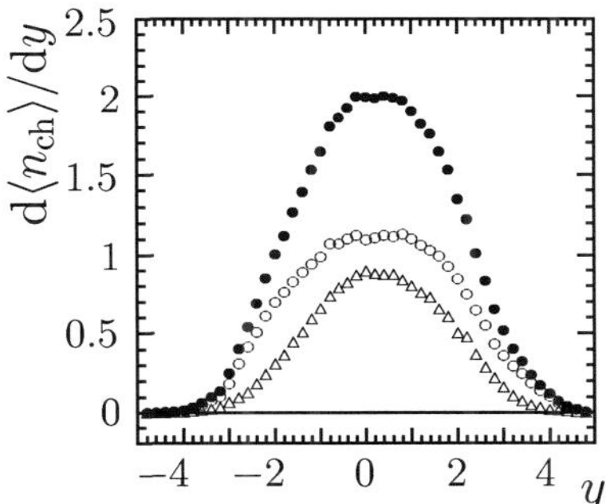

FIGURE 2. Center-of-mass rapidity distribution of positively (open circles), negatively (open triangle), and all charged (solid circles) particles.

low multiplicities in the rapidity region $[-3,-2]$, where unidentified protons contribute and where the rapidity distribution is not completely symmetric to the region $[2,3]$. The figure demonstrates that, despite a strong rapidity dependence of the particle density given in Fig. 2, the balance function is largely independent of the position of the rapidity window, i.e., the charge correlation is essentially the same in any longitudinally-Lorentz-transformed frame.

Balance Function in different widths of rapidity windows

Since boost invariance of the BF is found to be valid over the whole rapidity region, it is now interesting to verify if the BF in a limited rapidity window can be deduced from that in the full rapidity region by Eq.(2), and vice versa. In Fig. 3, the balance function, $B(\delta y|Y_w)$ (solid points), for four rapidity windows (central in Fig. 3a, non-central in Fig. 3b), is compared to $B(\delta y|\infty)(1-\frac{\delta y}{Y_w})$ (open points) obtained for the corresponding

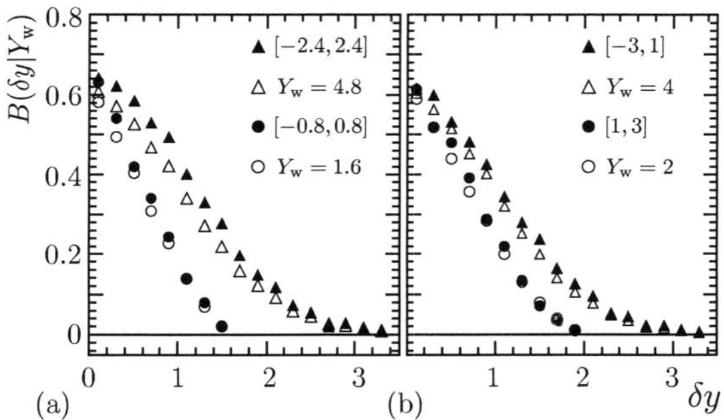

FIGURE 3. The balance functions $B(\delta y|Y_w)$ (solid symbols)(a) for two central rapidity windows, $[-2.4, 2.4]$ (triangles) and $[-0.8, 0.8]$ (circles) and (b) two asymmetric rapidity windows $[-3, 1]$ (triangles), and $[1, 3]$ (circles), compared with corresponding $B(\delta y|\infty) \cdot (1 - \frac{\delta y}{Y_w})$ (open symbols).

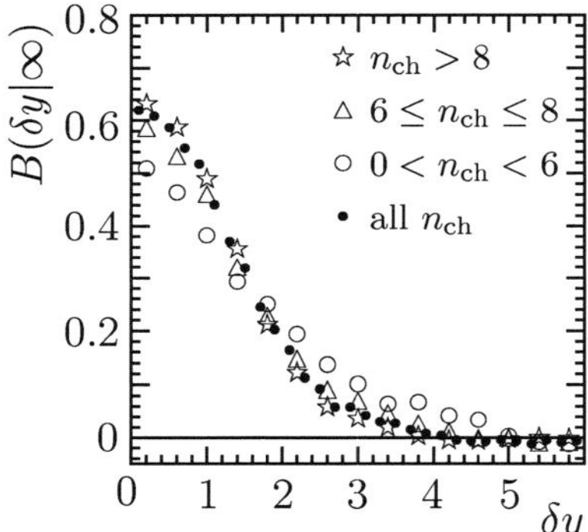

FIGURE 4. The balance function for all charged particles (full circles) and for three multiplicity intervals, $0 < n_{ch} < 6$ (open circles), $6 \leq n_{ch} \leq 8$ (open triangles), $n_{ch} > 8$ (open stars).

window from the BF in the full region. The data confirm that the relation Eq.(2) is indeed approximately satisfied, independently of size or position of the window. This result is especially useful for experiments with limited acceptance, in particular for the current heavy ion experiments.

Fig. 3 further illustrates that the BF becomes narrower with decreasing Y_w, in agreement with Eq.(2).

The Multiplicity dependence of Balance Function

In Fig. 4, the full-rapidity BF, $B(\delta y|\infty)$, is presented for all charged particles (full circles) and for the three multiplicity intervals, $0 < n_{ch} < 6$ (open circles), $6 \leq n_{ch} \leq 8$ (open triangles), $n_{ch} > 8$ (open stars). This is, at least qualitatively, consistent with the narrowing of the balance function with increasing centrality observed in current heavy

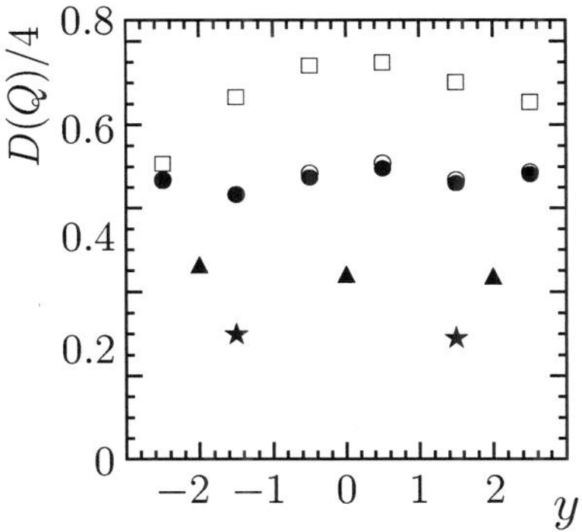

FIGURE 5. $D(Q)/4$ versus he position of a rapidity window of size $Y_w = 1.0$ (circles), 2.0 (triangles), 3.0 (stars). Open circles and open squares are $D(Q)/4$ under the same transverse momentum and azimuthal angle cuts as STAR ($p_t > 0.1$GeV/c) and PHENIX ($p_t > 0.2$GeV/c and $\Delta\phi = \pi/2$) with a rapidity window of size $Y_w = 1.0$.

ion experiments [7, 8]. So, before a narrowing of the BF with increasing centrality and increasing mass number of the colliding nuclei can be interpreted as due to the formation of a QGP, the multiplicity effect observed here should be properly accounted for.

Relation between Balance Function and charge fluctuation

Since the charge fluctuation $D(Q)$ is approximately related to the BF by

$$\frac{D(Q)}{4} = 1 - \int_0^{Y_w} B(\delta y|Y_w)\mathrm{d}\delta y + \mathscr{O}\left(\frac{\langle Q\rangle}{\langle n_{ch}\rangle}\right) \tag{3}$$

where $Q = n_+ - n_-$ and $n_{ch} = n_+ + n_-$, it is interesting to see how the charge fluctuation changes with position and size of the rapidity window. For this purpose, $D(Q)/4$ is presented in Fig. 5 for different positions and sizes of a rapidity window, $Y_w = 1.0$ (circles), 2.0 (triangles), 3.0 (stars). The results confirm that for a given window size its value is independent of the position of that window, in agreement with the boost invariance of the balance function. The data also show that $D(Q)$ is sensitive to the size of the observed window. So it is necessary to give the exact size of the rapidity region when the fluctuation is treated quantitatively.

As has been demonstrated in [14], $D(Q)$ also depends on the acceptance in transverse momentum and azimuthal angle. In Fig. 5, $D(Q)/4$ is also presented under the same transverse momentum and azimuthal angle cuts as used in STAR ($p_t > 0.1$GeV/c) and PHENIX ($p_t > 0.2$GeV/c and $\Delta\phi = \pi/2$) with a rapidity window of size $Y_w = 1.0$, by open circles and open squares, respectively. The transverse momentum cut used by STAR has little influence on the result, while the combined transverse momentum and azimuthal cut used by PHENIX destroys the boost invariance of $D(Q)$. These results show that a limited acceptance in transverse momentum and azimuthal angle can destroy the boost invariance of charge fluctuations. Furthermore, it has been verified that the effect is the larger the larger if the percentage of particles lost.

SUMMARY

The results of this paper can be summarized as follows:

1. In contrast to the strong dependence of the particle density on rapidity, the BF is invariant under a longitudinal boost over the whole rapidity region. This property allows to determine the BF in full rapidity, $B(\delta y|\infty)$, from a measurement with limited rapidity acceptance.
2. The balance function becomes narrower with decreasing size of the window. Therefore, only the full-rapidity BF can be used in comparing data from different experiments.
3. The balance function becomes narrower with increasing multiplicity, an effect also observed in heavy ion interactions when the centrality of the collision increases.
4. The charge fluctuations are boost invariant boost invariant but depend on the size of the rapidity window.

ACKNOWLEDGMENTS

Li Na is thankful for the great help given by Prof. Liu Lianshou for preparing the talk, and the partly financial support given by the organizers.This work is supported in part by the National Natural Science Foundation of China with project No.10375025 and No.10475030 and the Ministry of Education of China with project No.CFKSTIP-704035.

REFERENCES

1. D. Drijard *et al.*, Nucl. Phys. B **155**, 269(1979) and **166**, 233(1980); M. Barth *et al.*, Z. Phys. C **16**, 291(1983); D.H. Brick *et al.*, Z. Phys. C **31**, 59(1986); I.V. Ajinenko *et al.*, Z. Phys. C **43**, 37(1989).
2. P. Allen *et al.*, *Phys. Lett. B* **69**, 77(1982); M. Arneodo *et al.*, *Z. Phys. C* **36**, 527(1987).
3. R. Brandelik *et al.*, *Phys. Lett. B* **100**, 357(1981); M. Althoff *et al.*, *Z. Phys. C* **17**, 5(1983); H. Aihara *et al.*, *Phys. Rev. Lett.* **53**, 2199(1984) and **57**, 3140(1986).
4. S. A. Bass, P. Danielewicz, and S. Pratt, *Phys. Rev. Lett.* **85**, 2689(2000).
5. S. Jeon, and S. Pratt, *Phys. Rev. C* **65**, 044902(2002).
6. S. Jeon, and V. Koch, *Phys. Rev.Lett.* **85**, 2076(2000); M. Asakawa, U. Heinz and B. Müller, *Phys. Rev. Lett.* **85**, 2072(2000); V. Koch, M. Bleicher and S. Jeon, *Nucl. Phy. A* **698**, 261(2002).
7. J. Adams et al. (STAR Coll.), *Phys. Rev. Lett.* **90**, 172301(2003).
8. C. Alt et al. (NA49 Coll.), *Phys. Rev. C* **71**, 034903(2005).
9. J. T. Mitchell, *J. Phys. G* **30**, S819(2004).
10. J. Adams et al. (STAR Coll.), *Phys. Rev. C* **68**, 044905(2003); C. A. Pruneau, *Heavy Ion Phys.***21**, 261(2004).
11. K. Adcox et al. (PHENIX Coll.), *Phys. Rev. Lett.* **89**, 082301(2002); J. Nystrand, *Nucl. Phys. A***715**, 603(2003).
12. R.P. Feynman, *Phys. Rev. Lett.* **23**, 1415(1969).
13. M. Adamus et al. (NA22 Coll.), *Z. Phys. C* **32**, 476(1986); M. R. Atayan et al. (NA22 Coll.), *Eur. Phys. J. C***21**, 271(2001).
14. M. R. Atayan et al. (NA22 Coll.), *Phys. Rev. D* **71**, 012002(2005).

Moments of the phase-space density, coincidence probabilities, and entropies of a multiparticle system

A.Bialas

M.Smoluchowski Institute of Physics, Jagellonian University, Reymontc 4, 30-059 Krakow, Poland
e-mail:bialas@th.if.uj.edu.pl

Abstract. A method to estimate moments of the phase-space density from event-by-event fluctuations is reviewed and its accuracy analyzed. Relation of these measurements to the determination of the entropy of the system is discussed. This is a summary of the results obtained recently together with W.Czyz and K.Zalewski [1, 2].

Keywords: coincidence probability, entropy, phase-space density
PACS: 05.30.-d,05.40.-a,05.50.+q

MOMENTS OF PHASE-SPACE DENSITY AND COINCIDENCE PROBABILITIES

Consider the normalized phase-space distribution $W(X,K)$ of a system of M particles produced in a high-energy collision. Here $X = X_1,...,X_M$ and $K = K_1,...,K_M$ stand for two sets of vectors decribing, respectively, particle positions and momenta. The moments of phase-space *density* are defined by

$$< [D(X,K)]^l >= \int W(X,K)[MW(X,K)]^l dX dK = M^l < [W(X,K)]^l > \tag{1}$$

with $l = 1,2,\ldots$

As the first step we investigated the relation of the moments $< [W(X,K)]^l >$ to the *coincidence probabilities* in the system, defined as

$$C_{l+1} \equiv Tr[\rho]^{l+1}. \tag{2}$$

where ρ is the density matrix describing the system.

The name "coincidence probability" is justified by the observation that *in a diagonal representation* of ρ we have

$$C_{l+1} = \sum_i [p_i]^{l+1} \tag{3}$$

where p_i is the probability for the state i to occur and thus C_{l+1} is the probability to find $l+1$ identical states of the system. Note, however, that if the phase-space distribution

CP828, *Multiparticle Dynamics*
edited by V. Šimák, M. Šumbera, Š. Todorova, and B. Tomášik

has a non-trivial dependence on X, the density matrix in the momentum representation is certainly non-diagonal. Indeed, the density matrix is the Fourier transform [3]

$$\rho(p;p') = \int dX e^{i(p-p')X} W[X,(p+p')/2] \tag{4}$$

and is diagonal only if W does not depend on X. This is the origin of the complications which were the subject of our investigation [1].

We have studied the phase-space distribution $W(X,K)$ of the general form

$$W(X,K) = \frac{1}{(L_x L_y L_z)^M} G[(X-\hat{X})/L] F(K) \tag{5}$$

where the function $G(u)$ satisfies three conditions

$$\int du G(u) = 1\,; \quad \int du u G(u) = 0\,; \quad \int du u^2 G(u) = 1. \tag{6}$$

The first condition defines $F(K)$ as the normalized momentum distribution; the second one defines $\hat{X}$ as the average value of X; the third one defines L^2 as the average of $(X-\hat{X})^2$.

Using (2) it can be shown [2] that the coincidence probabilities C_{l+1} are related to the moments of $W(X,K)$. In the simplest case of Gaussian distributions one obtains

$$C_{l+1} = (2\pi)^{3Ml} < [W(X,K)]^l > [(1-\delta_x^{(l)})(1-\delta_y^{(l)})(1-\delta_z^{(l)})]^M \tag{7}$$

with

$$\delta_{x,y,z}^{(1)} = 0\,; \quad \delta_{x,y,z}^{(l)} = \frac{a_l}{L_{x,y,z}^2 < K_{x,y,z}^2 >} \geq 0\,; \; l \geq 2. \tag{8}$$

where a_l are constants.

Thus C_2 is simply identical to the average phase-space density, whereas for higher l's the correction vanishes in the limit of very large size of the system. This last result can be readily understood: the correction reflects the non-diagonality of the density matrix in the momentum representation. In the limit of very large $L's$, corresponding to a very large volume of the system, the density matrix approaches a diagonal form and thus correction must disappear.

EXPERIMENTAL ESTIMATE OF MOMENTS OF PHASE-SPACE DENSITY AND OF COINCIDENCE PROBABILITIES

It was suggested [4]-[6] that the coincidence probabilities defined by (2) can be estimated from measurements of the coincidences between *events* produced in high-energy collisions. This estimate, and the factors determining its accuracy were considered in [1], the investigation I am going to report in this section.

The *experimental* coincidence probabilities C_{l+1}^{exp} are defined [4] as

$$C_l^{exp} = \frac{N_l}{N_{tot}} \tag{9}$$

where N_l denotes the number of l-plets of identical events found in the investigated sample. $N_{tot} = N(N-1)...(N-l+1)$ denote the total number of l-plets of events in the sample (N is the total number of events). Thus the measurement of C_l^{exp} reduces to the count of the number of coincidences between the observed events.

One sees from (9) that C_l^{exp} measures coincidences between real events observed in experiment. On the other hand, as seen from (3), C_l describes coincidences between the states of the system in a diagonal representation. Therefore these objects are not identical.

Since the observed events -apart from the number and quantum numbers of the produced particles- are also characterized by particle momenta, to obtain a nontrivial result it is necessary to discretize the momentum distribution. Of course the obtained value of C_l^{exp} depends on the adopted discretization and, consequently, the procedure does not yield an unique answer [4, 5].

In the investigation I am reporting here [1] we have looked for an "optimal" discretization, i.e. a procedure which brings the measurement of C_l^{exp} as close as possible to C_l. Using again $W(X,K)$ in the form given by (5), we found a formula for the size of the binning in momentum space which approximately reproduces the required result. Here I describe only the simplest case, most interesting from the practical point of view.

Let the momentum space be split into bins of size $\Delta_x, \Delta_y, \Delta_z$. Then the optimal size is given by the condition

$$\omega \equiv \Delta_x \Delta_y \Delta_z = \frac{(2\pi g_l)^3}{L_x L_y L_z}. \tag{10}$$

Note that the product $L_x L_y L_z$ is proportional to the volume of the system in configuration space. Here $g_l = \int [G(u)]^l du$ depends weakly on l (thus for each l the discretization should be sligthly different).

Using this discretization we have

$$(2\pi)^{3Ml} < [W(X,K)]^l >= \Phi_l \, C_{l+1}^{exp} \tag{11}$$

where

$$\Phi_l = \frac{\sum_{bins} < [F(K)]^{l+1} >_{bin}}{\sum_{bins} [< F(K) >_{bin}]^{l+1}} \tag{12}$$

and the average inside a bin is defined in the standard way:

$$< [F(K)]^l >_{bin} = \frac{1}{\omega^M} \int_{\omega^M} [F(K)]^l d^{3M} K. \tag{13}$$

The "correction factor" Φ_l tends to 1 if the bins are very small, i.e., in the limit of very large volume of the system. For small systems one may estimate it from the meausured single-particle particle distribution and correlation functions[1].

Eq.(11) allows to estimate $< [W(X,K)]^l >$ and thus, through (7), also C_{l+1}.

RENYI AND SHANNON ENTROPIES

The coincidence probabilities C_l are often represented in terms of Renyi entropies H_l:

$$H_l \equiv \frac{1}{1-l} \log C_l. \tag{14}$$

They are related to the Shannon entropy S of the system by the limiting procedure

$$S \equiv Tr[\rho \log \rho] = \lim_{l \to 1} H_l. \tag{15}$$

Since our method allows to estimate C_l (and thus also H_l) only for integer values, $l = 2,3,4...$, to obtain information on S it is necessary to perform extrapolation which may introduce large uncertainties [7]. However, since $\delta_l \geq 0$, $\Phi \geq 1$ and because of the mathematical inequality $S \geq H_l$ [8], the method provides a way to determine the lower limit on S, a quantity of great interest [9].

MULTIPLICITY DISTRIBUTION

Until now we have discussed the case of fixed multiplicity. It is, however, not difficult to see what is the effect of multiplicity distribution. For the moments of phase-space density we have

$$< [D(X,K)]^l > = \sum_M P(M) < [D_M(X,K)]^l > \tag{16}$$

where $P(M)$ is the multiplicity distribution and $< [D_M(X,K)]^l >$ is the moment of phase-space density at fixed multiplicity.

For the coincidence probabilities one obtains

$$C_l = \sum_M [P(M)]^l C_l(M) \tag{17}$$

which shows that at large l only multiplicities close to the most probable one contribute effectively to C_l. This formula is automatically realized when one measures C_l^{exp} summing over all multiplicities.

---

[1] Φ_l depends only on momentum distribution.

DISCUSSION

The accuracy of the method depends on correct estimate of Φ_l. It is clear that when the bins are very small, $\Phi_l \approx 1$. Thus the accuracy increases with increasing volume of the system. Large volume brings also $(2\pi)^{3Ml} < [W(X,K)]^l >$ close to C_{l+1}. It follows that, as a way to estimate the entropies of the system, the coincidence method is particularly suitable for large systems, as encountered, e.g., in heavy ion collisions.

Although the accuracy increases when bins are very small, the applicability of the method is reduced because it is more difficult to find coincidences, unless the statistics is indeed enormous. This difficulty may be solved by considering a limited part of phase-space and thus to measure local rather than global entropy of the system.

In conclusion, we have investigated the accuracy of the determination of moments of phase-space density and of Renyi entropies from measured coincidence probabilities of events observed in high-energy collisions. The optimal discretization, as well as the accuracy of the procedure were shown to depend on the volume of the system. The proposed method does not demand any assumptions about the thermodynamic properties of the system, in particular it does not assume the thermodynamic equilibrium. It thus may be of particular interest for testing the standard assumptions of the models of quark-gluon plasma. Moreover, it can serve as a quantitative measure of the deviation from the equilibrium. Further discussion may be found in [1, 2].

ACKNOWLEDGMENTS

This report was prepared with the active participation of Wieslaw Czyz and Kacper Zalewski. Discussions with Robi Peschanski and Jacek Wosiek are also acknowledged. Thanks are due to Vlada Simak for the kind hospitality at Kromeriz.

REFERENCES

1. A.Bialas, W.Czyz and K.Zalewski, hep-ph/0506233; *Acta Phys. Pol.* **B36** (2005) 3109, [hep-ph/0508289].
2. A.Bialas, W.Czyz and K.Zalewski, in preparation.
3. See, e.g., Hillery et al., *Phys.Rep.* **106** (1984) 121.
4. A.Bialas and W.Czyz, *Phys. Rev.* **D61** (2000) 074021.
5. A.Bialas and W.Czyz, *Acta Phys. Pol.* **B31** (2000) 687.
6. A.Bialas, W. Czyz and J.Wosiek, *Acta Phys. Pol.* **B30** (1999) 107 .
7. K.Zyczkowski, *Open Sys. and Information Dyn.* **10** (2003) 297.
8. See, e.g., C.Beck and F.Schloegl, *Thermodynamics of chaotic systems*, Cambridge University Press, Cambridge, 1993.
9. See e.g., B.Muller and K.Rajagopal, hep-ph/0502174.

Entropy Analysis in π^+p and K$^+$p Collisions at $\sqrt{s} = 22$ GeV

LI Zhiming

(for the NA22 Collaboration)

Institute of Particle Physics, Huazhong Normal University, Wuhan 430079, China; Radboud University/NIKHEF, NL-6525 ED Nijmegen, The Netherlands.

Abstract. We report on an entropy analysis using Ma's coincidence method on π^+p and K$^+$p collisions at $\sqrt{s} = 22$ GeV. A scaling law and additivity properties of Rényi entropies and their charged-particle multiplicity dependence are investigated. The results are compared with those from the PYTHIA Monte Carlo model.

Keywords: entropy coincidence method scaling additivity

PACS: 13.85.-t, 13.85.Hd, 05.10.-a

INTRODUCTION

The assumption of thermodynamic equilibrium is commonly used when discussing a multiparticle final system, especially in central heavy ion collisions. There are many different ways or methods to test the equilibrium of the system. One of the most promising ones seems to be[1]

$$\left.\frac{\partial S(E,n,V)}{\partial E}\right|_{n,V} = \frac{1}{T}. \tag{1}$$

A test of this equation requires measurement of the temperature T, the number of particles n, the energy E, and most importantly the entropy S of the system. To obtain T, we can measure the slope of the transverse momentum distribution, n is a direct measurement in the experiment, for E we can measure the energies of all particles created in the collision. The real difficulty is the measurement of entropy. Here, we use the coincidence probability method proposed by Ma[2] and adapted by Białas and Czyż[3, 4].

Entropy was introduced into multiparticle dynamics by Šimák et al[5]. Recently, it has been suggested that the measurement of entropy could provide a novel tool for the investigation of local properties[6, 7, 8, 9]. In heavy ion collisions this can be used in the search for the formation of a quark gluon plasma (QGP) [1]. It also has proved effective for the study of systems of particles created in multiplicative branching processes [9] as present in high-energy QCD jet fragmentation. Finally, entropy measurements offer an additional tool to analyze event-to-event fluctuations and particle correlations [4, 7].

Before applying it to heavy ion collisions, we test the sensitivity of this method by applying it to the more basic hadron-hadron collisions, where no equilibrium is expected. In this note, we present the results of a study of entropy in π^+p and K$^+$p collisions at

CP828, *Multiparticle Dynamics*
edited by V. Šimák, M. Šumbera, Š. Todorova, and B. Tomášik

250 GeV/c incident momentum ($\sqrt{s} = 22$ GeV) collected by the NA22 experiment. We investigate the dependence on discretization of the system, the effects of particle correlations by testing scaling and additivity properties of the entropy measures, as well as the multiplicity dependence of the Rényi entropies in rapidity space. We also compare the results of the NA22 data to those of the Monte Carlo event generator PYTHIA 5.7 [10] used for inelastic, non-single-diffractive pion-proton collisions with Bose-Einstein Correlations (BEC), as well as to a simple random production model.

PROCEDURE AND METHOD

The original definition of the standard Shannon entropy in thermal statistics is:

$$S = -\sum_j p_j \ln p_j. \tag{2}$$

Here, $p_j = n_j/N$ denotes the probability to obtain a specific configuration of the system, where n_j is the number of events in such a configuration and N is the total number of events. The sum runs over all possible configurations.

Following Ma's method [2] and as explained in [4], the first step in the entropy measurement is to determine coincidence probabilities. For every event, a certain phase space region is divided into M bins of equal size. An event is then characterized by the number of particles, m_i, in each bin, i.e., by a set of integer numbers $s \equiv \{m_i\}$, where $i = 1, ...M$.

After counting how many times, n_s, the set s appears in the whole event sample, one can determine the sums

$$N_k = \sum_s n_s(n_s - 1)...(n_s - k + 1). \tag{3}$$

This gives the total numbers of observed coincidences of k configurations.

The coincidence probability of k configurations is then given by

$$C_k = \frac{N_k}{N(N-1)...(N-k+1)}, \tag{4}$$

where N, as above, is the total number of events in the sample. Only states with $n_s \geq k$ contribute to C_k.

From the coincidence probabilities, one calculates the Rényi entropies defined as

$$H_k \equiv -\frac{\ln C_k}{k-1}. \tag{5}$$

The Shannon entropy S is formally equal to the limit of H_k as $k \to 1$. It can easily be proved that for large enough N it reduces to the form defined in Eq. (2).

One of most attractive features of Ma's coincidence method compared to the standard method of using Eq. (2) directly is that the statistical error decreases very fast with increasing number of configurations. Moreover, it has been demonstrated [6] that Ma's method yields more stable results than the standard method.

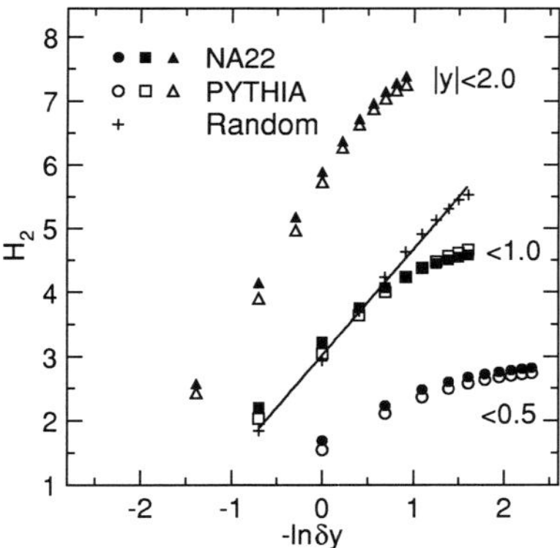

FIGURE 1. The dependence on the dividing bin sizes of the second Rényi entropy calculated in central rapidity windows.

For a system close to thermal equilibrium, and if the phase space subdivision is sufficiently fine-grained, the scaling relation

$$H_k(lM) = H_k(M) + \ln l \Rightarrow S(lM) = S(M) + \ln l \tag{6}$$

holds. Here, M and lM are the numbers of bins in two different discretizations. If scaling holds, one should observe a linear relation if $H_k(M)$ is plotted as a function of $-\ln \delta y$, where $\delta y = \Delta y / M$ is the bin size.

Another feature is additivity: the entropies measured in a region R which is the union of two non-overlapping and independent regions R_1 and R_2 satisfy

$$H_k(R) = H_k(R_1) + H_k(R_2) \Rightarrow S(R) = S(R_1) + S(R_2). \tag{7}$$

Furthermore, it is suggested [7] that the dependence of the Rényi entropies on particle multiplicity n carries important information on the produced system. A Rényi entropy proportional to n is indicative of an equilibrated system with no strong long-range correlations. On the other hand, for non-equilibrated systems or a superposition of subsystems with different properties, proportionality to $\ln n$ is expected.

We will test the above scaling and additivity properties and check this multiplicity dependence of the Rényi entropies on the NA22 data.

RESULTS

We first test scaling in the central rapidity region by using Rényi entropy H_2 in Fig. 1. The solid circles, squares and triangles represent the NA22 data for three different central rapidity windows $|y| < 0.5, 1.0, 2.0$, respectively. In all windows, H_2 is flattening with increasing $-\ln \delta y$. This means that the scaling law does not hold, i.e. that there is no thermal equilibrium. This implies that there are strong particle correlations in the system under investigation. The corresponding open symbols represent the results from the PYTHIA model. PYTHIA has the same trend as the data, but some discrepancies exist.

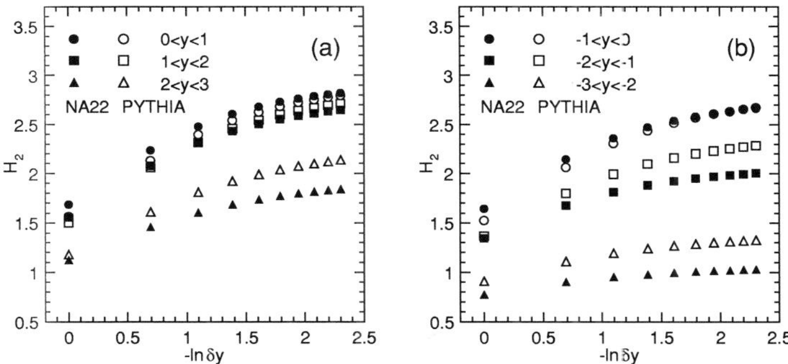

FIGURE 2. The dependence on the bin size of the second Rényi entropy calculated in (a) forward and (b) backward hemispheres in different non-central rapidity windows.

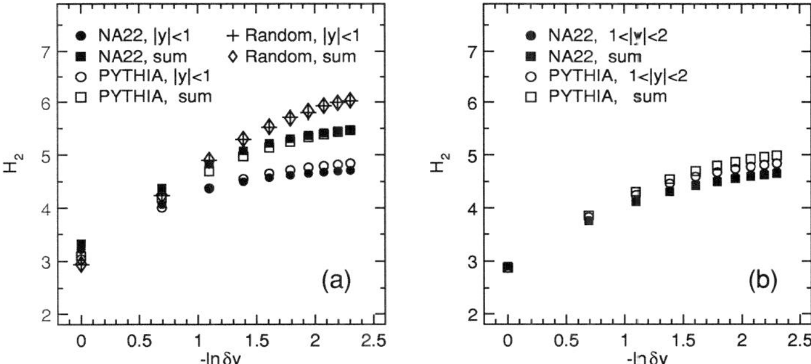

FIGURE 3. The second Rényi entropy as a function of $-\ln \delta y$ measured in two (a) adjacent and (b) separated intervals.

The cross symbols in Fig. 1 show the results from a random model generated as: take the total multiplicity distribution as observed in the NA22 data in the rapidity widow $|y| < 1.0$, but the multiplicity fluctuations in each bin for every event as Poisson with the mean value as in the data. In this model, there are no correlations between particles in every event, so it should give a straight-line relationship, as is indeed observed in the figure.

In Fig. 2 we show H_2 in various non-central rapidity windows of fixed size $\Delta y = 1$. Due to the asymmetry of the proton and meson fragmentation regions in the NA22 experiment, we plot forward and backward hemispheres separately in Fig. 2 (a) and (b), respectively. We see that, again, the H_2 values are flattening with increasing $-\ln \delta y$, so that no linear relationship is found. The corresponding open symbols for the PYTHIA results have the same trend, but overestimate the data in the peripheral regions. This is presumably due to too weak correlations in the model for these regions.

In order to test additivity, we calculate H_2 in two ways: one is a direct measurement in the rapidity window $|y| < 1$ (solid circles in Fig. 3 (a)), the other the sum of the values obtained from the two adjacent sub-windows $-1 < y < 0$ and $0 < y < 1$ (solid squares).

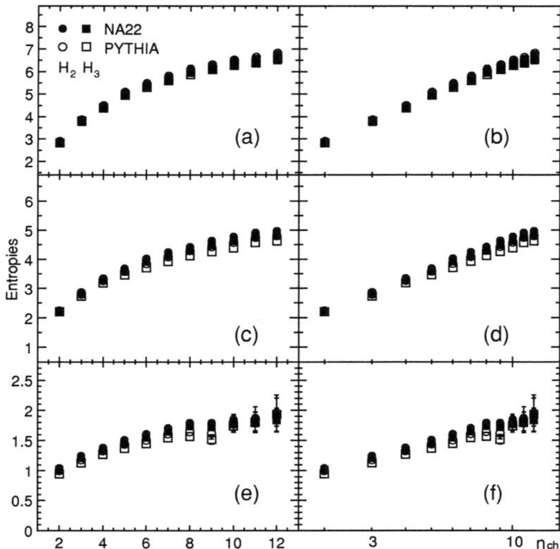

FIGURE 4. The dependence of the second and third Rényi entropies on charged-particle multiplicity in linear (left) and logarithmic (right) scale, with (a)(b) $M = 6$-fold division in the rapidity window $|y| < 3$, (c)(d) $M = 4$ in $|y| < 2$ and (e)(f) $M = 2$ in $|y| < 1$.

A clear difference is observed between these two results. The H_2 obtained from the sum is larger than that calculated directly, and the difference increases with increasing $-\ln \delta y$. So, additivity is not observed here. The reason is that strong correlations exist between the particles belonging to these two adjacent rapidity windows. The crosses and diamonds in the same figure represent the results from the random model. We find that additivity is valid for this model, which confirms the role of correlations in the data.

Fig. 3 (b) shows the results in the non-adjacent regions separated by two rapidity units, $-2 < y < -1$ and $1 < y < 2$. In this case, the difference between the direct measurement and the sum is very small, since correlations between particles belonging to widely separated rapidity intervals are small in our data. The open symbols for PYTHIA still somewhat overestimate the data.

In Fig. 4, we show the charged-particle multiplicity n_{ch} dependence of H_2 (solid circles) and H_3(solid squares) for three different rapidity windows. The left column is in linear scale, the right column in logarithmic scale. We see that a logarithmic rather than a linear relation is observed. This confirms that there is no thermal equilibrium in the system. The open symbols are the results from the PYTHIA model. They agree quite well with the data here.

SUMMARY

We have analyzed the entropy properties of multiparticle production in π^+p and K$^+$p collisions at 250 GeV/c. By using Rényi entropies, we find that neither the scaling nor the additivity properties are valid and that the multiplicity dependence is logarithmic rather than linear. These observations confirm the expectation that no thermal equilibrium is reached in hadron-hadron collisions at $\sqrt{s} = 22$ GeV.

The PYTHIA Monte Carlo model, in general, agrees with the data. However, significant deviations exist. In particular, the model overestimates the values in the peripheral rapidity regions, presumably due to too weak correlations. This shows that Rényi entropies provide a sensitive measure of multiparticle correlations.

RHIC has already collected data on gold-gold and pp collisions at 200GeV. It would be interesting to investigate the entropy properties on this high-temperature, high-density system, which may create the long expected QGP. Our results presented here should provide a valuable guide to the interpretation of future results from high energy heavy ion collisions.

ACKNOWLEDGMENTS

We thank A. Białas, K. Fiałkowski, W. Kittel, W.J. Metzger, and R. Wit for constructive discussions and remarks. This work is also supported in part by the National Natural Science Foundation of China under Grant 10475030 and the Ministry of Education of China and by the Royal Dutch Academy of Sciences under the Project numbers 01CDP017, 02CDP011 and 02CDP032.

REFERENCES

1. A. Białas, W. Czyż, and J. Wosiek, *Acta Phys. Pol. B* **30**, 107–117 (1999).
2. S. K. Ma, and M. Payne, *Phys. Rev. B* **24**, 3984–3990 (1981).
3. A. Białas, and W. Czyż, *Phys. Rev. D* **61**, 074021-1–074021-7 (2000).
4. A. Białas, and W. Czyż, *Acta Phys. Pol. B* **31**, 687–692 (2000).
5. V. Šimák, M. Šumbera, and I. Zborovský, *Phys. Lett. B* **206**, 159–162 (1988); M. Pachr, V. Šimák, M. Šumbera, and I. Zborovský, *Mod. Phys. Lett. A* **7**, 2333–2339 (1992);
6. K. Fiałkowski, and R .Wit, *Phys. Rev. D* **62**, 114016-1–114016-4 (2000).
7. A. Białas, and W. Czyż, *Acta Phys. Pol. B* **31**, 2803–2817 (2000).
8. A. Białas, and W. Czyż, *Acta Phys. Pol. B* **32**, 2793–2800 (2001).
9. A. Białas, W. Czyż, and A. Ostruszka, *Acta Phys. Pol. B* **34**, 69–85 (2003).
10. T. Sjöstrand, *Comp. Phys. Comm.* **82**, 74–89 (1994).

On the Measure of Dynamical Event-by-Event Transverse Momentum Fluctuations

Fu Jinghua

Department of Engineering Physics, Tsinghua University, Beijing 100084, P.R.China
Institute of Particle Physics, Huazhong Normal University, Wuhan 430079, P.R.China

Abstract. The fluctuations of theoretically defined event mean transverse momentum can be expressed in terms of inclusive multi-particle distributions and is experimentally measurable. It is equivalent to the dynamical fluctuations of the experimentally defined event average transverse momentum when multiplicity is fixed. They are generally different from each other when multiplicity varies.

Keywords: Relativistic Heavy Ion Collisions, Event-by-Event Transverse Momentum Fluctuations, Dynamical Fluctuations, Statistical Fluctuations
PACS: 13.85.Hd, 25.75.-q, 24.60.Ky, 24.10.Lx

INTRODUCTION

In the conventional investigation of high energy multi-particle production, all the events in a collision process are taken as a whole and the distributions, fluctuations and correlations inside the sample are studied without distinguishing the individual events. This kind of study is usually referred to as an inclusive one. Motivated by the high multiplicity events produced in central relativistic heavy ion collisions and theoretical perspective[1], people started to carry on the study event-by-event, with the aim of exploring the possible existence of new physics which might be neglected by the conventional investigation. One of the difference in these two kinds of studies is that the inclusive distributions can readily be extracted from the experimental measurement by averaging over a large number of events, while obtaining single event dynamics is complicated due to the statistical fluctuations coming from the limited number of particles in a single event.

Event-by-event average transverse momentum fluctuations in relativistic heavy ion collisions have been proposed as a probe of critical temperature fluctuations at the QCD phase transitions[2, 3] and/or as a searching for the onset of thermalization[4, 5]. It is one of the earliest and mostly studied event-by-event fluctuation measurements in relativistic heavy ion physics. Though measurements of event-by-event transverse momentum fluctuations have been made at several SPS and RHIC experiments, there is no standard observable for it agreed upon in literature. Different variables are used in describing the magnitude of the fluctuations, such as Φ_{p_t}[4, 6, 7], σ^2_{dynam}[8], Σ_{p_t}[9], F_{p_T}[10] and $\Delta\sigma^2_{p_t,n}$[11]. In spite of various formalism, all the measurements start with the experimentally defined event average transverse momentum

$$M_{p_t} = \frac{1}{n}\sum_{i=1}^{n} p_{t\,i}, \tag{1}$$

CP828, *Multiparticle Dynamics*
edited by V. Šimák, M. Šumbera, Š. Todorova, and B. Tomášik

and measure its (variance) fluctuations $\sigma^2(M_{p_t}) = \langle M_{p_t}^2 \rangle - \langle M_{p_t} \rangle^2$, where n is the number of particles observed in an event. Collision dynamics dictates single event transverse momentum probability distributions $P_{\text{eve}}(p_t)$, while M_{p_t} is determined by the measured multiplicity distribution which generally does not give directly the probability distribution. Poisson (or Bernoulli) distribution smear out $P_{\text{eve}}(p_t)$, particularly for small n. Statistical fluctuations caused by finite number of particles are generally estimated either by direct calculations[8] where $\sigma^2(M_{p_t})_{\text{stat}} = \sigma^2(p_t)_{\text{incl}}/n$, assuming independent particle emission and constant multiplicity n, or by mixing-event method where $\sigma^2(M_{p_t})_{\text{stat}} = \sigma^2(M_{p_t})_{\text{mix}}$. The dynamical M_{p_t} fluctuations is gotten by subtracting $\sigma^2(M_{p_t})_{\text{stat}}$ from $\sigma^2(M_{p_t})$, i.e.

$$\sigma^2(M_{p_t})_{\text{dyn}} = \sigma^2(M_{p_t}) - \sigma^2(M_{p_t})_{\text{stat}}. \tag{2}$$

This problem could also be considered from the other point of view by starting with the theoretically defined event mean transverse momentum. Though the theoretically defined event mean transverse momentum itself is not measurable experimentally we will demonstrate that its fluctuations can be related to inclusive multi-particle distributions and thus is experimentally measurable. A comparison between these two types of fluctuation measurements will also be made both analytically and with HIJING monte carlo simulations.

ANALYTICAL CONSIDERATION

Theoretically, event mean transverse momentum $M_{p_t,\text{dyn}}$ can be expressed as

$$M_{p_t,\text{dyn}} = \int_\Delta p_t P_{\text{eve}}(p_t) dp_t \tag{3}$$

where $P_{\text{eve}}(p_t)$ is single particle p_t distribution of an event and the integral is taken over transverse momentum region Δ. By dividing Δ into M bins, the variance of $M_{p_t,\text{dyn}}$ in event-space can be expressed as

$$\begin{aligned}\sigma^2(M_{p_t,\text{dyn}}) &= \sigma^2\left(\int_\Delta p_t P_{\text{eve}}(p_t) dp_t\right) = \sigma^2\left(\sum_{m=1}^{M}(p_t)_m P_m\right) \\ &= \left\langle \sum_{m=1}^{M}(p_t)_m^2 \frac{n_m(n_m-1)}{\langle n\rangle^2}\right\rangle + \left\langle \sum_{m=1}^{M}\sum_{m'=1,m\neq m'}^{M}(p_t)_m(p_t)_{m'}\frac{n_m n_{m'}}{\langle n\rangle^2}\right\rangle \\ &\quad - \frac{\left\langle\sum_{m=1}^{M}(p_t)_m n_m\right\rangle^2}{\langle n\rangle^2},\end{aligned} \tag{4}$$

where P_m is the probability of a particle falling into the mth phase space cell δ_m, n_m is the number of particles measured in δ_m and $(p_t)_m$ is the p_t value in the mth bin. Factorial moment formula[12]

$$\left\langle \sum_{m=1}^{M}(p_t)_m P_m^q \right\rangle = \left\langle \sum_{m=1}^{M}(p_t)_m \frac{n_m(n_m-1)\cdots(n_m-q+1)}{\langle n\rangle^q}\right\rangle, \tag{5}$$

is used in getting the result of Eq.(4), which can be generalized to higher order straightforwardly[13].

To express $\sigma^2(M_{p_t,\mathrm{dyn}})$ into a physically more transparent form, we notice that the qth order factorial moment of the multiplicity distribution can be expressed as

$$\langle n(n-1)\cdots(n-q+1)\rangle_\Delta = \int_\Delta dp_{t1}\cdots\int_\Delta dp_{tq}\rho_q(p_{t1},\ldots,p_{tq}), \tag{6}$$

where $\rho_q(p_{t1},\ldots,p_{tq})$ is the inclusive q-particle distribution function. Then the variance of $M_{p_t,\mathrm{dyn}}$ can be expressed in terms of inclusive one- and two- particle distribution function ρ_1 and ρ_2 as

$$\begin{aligned}\sigma^2(M_{p_t,\mathrm{dyn}}) &= \frac{1}{\langle n\rangle^2}\int_\Delta\int_\Delta p_{t1}p_{t2}\rho_2(p_{t1},p_{t2})dp_{t1}dp_{t2} - \frac{1}{\langle n\rangle^2}\left[\int_\Delta p_t\rho_1(p_t)dp_t\right]^2 \\ &= \frac{1}{\langle n\rangle^2}\left\langle\sum_{i=1}^{n}\sum_{j=1,i\neq j}^{n} p_{ti}p_{tj}\right\rangle - \frac{1}{\langle n\rangle^2}\left\langle\sum_{i=1}^{n} p_{ti}\right\rangle^2,\end{aligned} \tag{7}$$

and the variance of the theoretically defined event mean transverse momentum $M_{p_t,\mathrm{dyn}}$ becomes directly measurable.

Now the question is whether the dynamical mean p_t fluctuations derived here are the same as those proposed previously. $\sigma^2(M_{p_t,\mathrm{dyn}})$ of Eq.(5) can also be written as

$$\sigma^2(M_{p_t,\mathrm{dyn}}) = \sigma^2\left(\sum_{i=1}^{n} p_{ti}/\langle n\rangle\right) - \sigma^2\left(\sum_{i=1}^{n} p_{ti}/\langle n\rangle\right)_{\mathrm{stat}} = \sigma^2\left(\sum_{i=1}^{n} p_{ti}/\langle n\rangle\right)_{\mathrm{dyn}}, \tag{8}$$

where $\sigma^2(\sum_{i=1}^{n} p_{t_i}/\langle n\rangle)_{\mathrm{stat}} = \langle\sum_{i=1}^{n} p_{t_i}{}^2\rangle/\langle n\rangle^2$, assuming independent particle emission. $\sum_{i=1}^{n} p_{t_i}$ is the event total transverse momentum which is very similar to event total transverse energy. Eq.(8) indicate $\sigma^2(M_{p_t,\mathrm{dyn}})$ measure the dynamical fluctuations of event total transverse momentum divided by the square of average sample multiplicities, while, from Eq.(2), $\sigma^2(M_{p_t})_{\mathrm{dyn}}$ measure the dynamical fluctuations of event average transverse momentum. They are identical only when event multiplicity is fixed. Actually Eq.(4) (or (7)) is derived for event sample with non-fixed event multiplicities where Poisson statistical particle distribution is assumed. For fixed multiplicity, Bernoulli statistics should be used[12] and the corresponding formula for $\sigma^2(M_{p_t,\mathrm{dyn}})$ is given by

$$\sigma^2(M_{p_t,\mathrm{dyn}})|_n = \frac{1}{n(n-1)}\sum_{i=1}^{n}\sum_{j=1,i\neq j}^{n}\langle(p_{ti}-\langle p_{ti}\rangle)(p_{tj}-\langle p_{tj}\rangle)\rangle = \frac{n}{n-1}\sigma^2(M_{p_t})_{\mathrm{dyn}}. \tag{9}$$

For fixed multiplicity n, $\sigma^2(M_{p_t,\mathrm{dyn}})$ is nothing but $\langle\Delta p_{t,i}\Delta p_{t,j}\rangle$, i.e. the two particle correlation measurement suggested by STAR[14]. When event multiplicity n varies, the fluctuation of $\sum_{i=1}^{n} p_{t_i}/\langle n\rangle$ and that of M_{p_t} would be different. $\sigma^2\left(\sum_{i=1}^{n} p_{t_i}/\langle n\rangle\right)$ explicitly depends on fluctuations in multiplicity, while multiplicity fluctuations in $\sigma^2\left(M_{p_t}\right)$ are suppressed.

HIJING SIMULATIONS

Next we use Hijing simulations to study the fluctuations measured by $\sigma^2(M_{p_t})_{\text{dyn}}$ and $\sigma^2(M_{p_t,\text{dyn}})$ respectively. Hijing-1.382 was used to generate 1,500,000 $\sqrt{s_{NN}} = 200$ GeV min-bias Au-Au collision events. The default Hijing setting is used with jet quenching on and decay of particles off. Generated charge particles with pseudo-rapidity $|\eta| < 1$ and transverse momentum 0.2 GeV/c$< p_t <$2 GeV/c are used in the analysis.

First, we consider fixed multiplicity event samples. Multiplicities are fixed at $n = 5, 10, 20, 50, 100, 200, 500$. Due to limited statistics, we did not go to higher multiplicity. Different fluctuation measurements mentioned above, $\sigma^2(M_{p_t,\text{dyn}})$, $\sigma^2(M_{p_t})$, $\sigma^2(M_{p_t})_{\text{stat}} = \sigma^2(p_t)_{\text{incl}}/n$ and $\sigma^2(M_{p_t})_{\text{dyn}} = \sigma^2(M_{p_t}) - \sigma^2(M_{p_t})_{\text{stat}}$ were calculated and plotted in Fig. 1. $\sigma^2(M_{p_t,\text{dyn}})$ was calculated in two ways, by using Eq.(4) (labelled as $\sigma^2(M_{p_t,\text{dyn}})_1$ in the figure) and by using Eq.(7) (labelled as $\sigma^2(M_{p_t,\text{dyn}})_2$ in the figure).

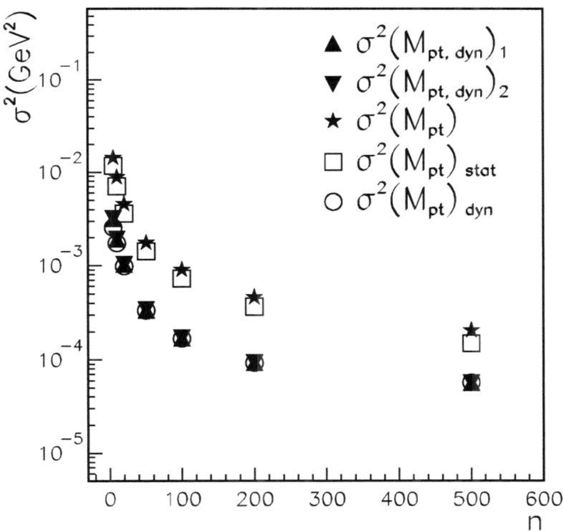

FIGURE 1. Event-by-event mean p_t fluctuation measurements $\sigma^2(M_{p_t,\text{dyn}})$, $\sigma^2(M_{p_t})$, $\sigma^2(M_{p_t})_{\text{stat}}$ and $\sigma^2(M_{p_t})_{\text{dyn}}$ at fixed multiplicity $n = 5, 10, 20, 50, 100, 200, 500$ for Au-Au collisions at $\sqrt{s_{NN}} = 200$GeV from Hijing-1.382.

The results indicate, for a given multiplicity, $\sigma^2(M_{p_t,\text{dyn}})$ (upward and downward full triangles) has similar result as $\sigma^2(M_{p_t})_{\text{dyn}}$ (open circles). The two types of measurement of dynamical event-by-event mean p_t fluctuations are essentially the same for fixed multiplicity event samples. They are slightly different from each other only in very low multiplicity and become consistent for high multiplicity event samples. However, $\sigma^2(M_{p_t})_{\text{dyn}}$ it is order of magnitude less than the corresponding statistical fluctuations $\sigma^2(M_{p_t})_{\text{stat}}$ (open squares). In measuring event average p_t fluctuations $\sigma^2(M_{p_t})$ (full stars), statistical contribution is dominant even for high multiplicity samples. Instead of subtracting two relatively large quantity to get an order of magnitude smaller quantity, it is better to measure $\sigma^2(M_{p_t,\mathrm{dyn}})$ directly.

Second, we study event samples with multiplicity n varies from event to event. For simplicity, number of participant N_{part} is taken to be fixed from peripheral to central collisions at $N_{\mathrm{part}} = 10, 50, 100, 150, 200, 250, 300$. The measured $\sigma^2(M_{p_t,\mathrm{dyn}})$, $\sigma^2(M_{p_t})$ $\sigma^2(M_{p_t})_{\mathrm{stat}}$ and $\sigma^2(M_{p_t})_{\mathrm{dyn}}$ are shown in Fig. 2.

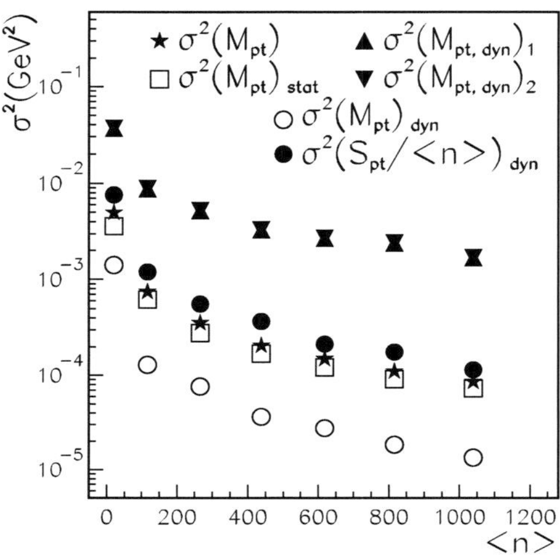

FIGURE 2. Event-by-event mean p_t fluctuation measurements $\sigma^2(M_{p_t,\mathrm{dyn}})$, $\sigma^2\left(S_{p_t}/\langle n\rangle\right)_{\mathrm{dyn}}$, $\sigma^2(M_{p_t})$, $\sigma^2(M_{p_t})_{\mathrm{stat}}$ and $\sigma^2(M_{p_t})_{\mathrm{dyn}}$ at number of participant $N_{\mathrm{part}} = 10, 50, 100, 150, 200, 250, 300$ for Au-Au collisions at $\sqrt{s_{NN}} = 200\mathrm{GeV}$ from Hijing-1.382.

The results indicate, in this case, fluctuations measured by $\sigma^2(M_{p_t,\mathrm{dyn}})$ is much larger than that of $\sigma^2(M_{p_t})_{\mathrm{dyn}}$ because of their different dependence on multiplicity fluctuations. $\sigma^2(M_{p_t,\mathrm{dyn}})$ measured by using Eq.(4) or Eq.(7) strongly depend on initial nuclear geometry fluctuations because of its dependence on multiplicity fluctuations. In experiments, where we do not have a good control of impact parameters, the best way to measure $\sigma^2(M_{p_t,\mathrm{dyn}})$ might still be the mixing events method by using Eq.(8). The results of $\sigma^2\left(\sum_{i=1}^{n} p_{ti}/\langle n\rangle\right)_{\mathrm{dyn}}$ are also plotted in Fig. 2 (foll dots, $S_{p_t} = \sum_{i=1}^{n} p_{ti}$). We can see $\sigma^2\left(\sum_{i=1}^{n} p_{ti}/\langle n\rangle\right)_{\mathrm{dyn}}$ is lower than $\sigma^2(M_{p_t,\mathrm{dyn}})_1$ (or $\sigma^2(M_{p_t,\mathrm{dyn}})_2$) for every N_{part} sample because the volume fluctuations is subtracted. However, it is still higher than $\sigma^2(M_{p_t})_{\mathrm{dyn}}$ and their dependence on average multiplicities are similar. It would be interesting to see how they behave in experiments.

SUMMARY

In summary, we studied the fluctuation of theoretically defined event mean transverse momentum $M_{p_t,\mathrm{dyn}}$ and find it can be expressed in terms of inclusive particle distributions. $\sigma^2(M_{p_t,\mathrm{dyn}})$ measure the dynamical fluctuations of event total transverse momentum divided by the square of average sample multiplicities. On the other hand, the previously proposed $\sigma^2(M_{p_t})_{\mathrm{dyn}}$ measure the dynamical fluctuations of experimentally defined event average transverse momentum. When multiplicity varies from event to event, the fluctuations measured by $\sigma^2(M_{p_t,\mathrm{dyn}})$ and $\sigma^2(M_{p_t})_{\mathrm{dyn}}$ could be different. Careful study is needed when we try to get temperature fluctuations from the measured dynamical M_{p_t} fluctuations. It would be interesting to measure both of them in experiments.

ACKNOWLEDGMENTS

This work was supported in part by the NSFC under Project No.10305004, by SRF for ROCS, SEM, by China Postdoctoral Science Foundation and by Cultivation Fund of the Key Scientific and Technical Innovation Project, Ministry of Education of China (NO. 704035)

REFERENCES

1. *Proceeding of the Bielefeld Workshop, 1982*, eidted by M. Jacob and H. Satz, World Scientific, Singapore, 1982 .
2. L. Stodolsky, *Phys.Rev.Lett.*, **75**, 1044 (1995) .
3. M. Stephanov, K. Rajagopal and E. Shuryak, *Phys.Rev.Lett.*, **81**, 4816 (1998) .
4. M. Gaździcki, S.Mrówczyński, *Z. Phys. C* **54**, 127 (1992) .
5. S. Gavin, *Phys. Rev. Lett.* **92**, 162301 (2004).
6. H. Appelshäuser,*et al.* (NA49 Collaboration), *Phys. Lett. B* **459**, 679 (1999) .
7. M. R. Atayan, *et al.* (NA22 Collaboration), *Phys. Rev. Lett.* **89**, 121802 (2002) .
8. S. A. Voloshin, V. Koch and H. G.Ritter, *Phys. Rev. C* **60**, 024901 (1999) .
9. D. Adamová, *et al.* (CERES Collaboration), *Nucl. Phys. A* **727**, 97 (2003) .
10. K. Adcox, *et al.* (PHENIX Collaboration), *Phys. Rev. C* **66**, 024901 (2002) .
11. J. Adams, *et al.* (STAR Collaboration), nucl-ex/0308033 .
12. A. Białas and R. Peschanski, Nucl. Phys. B**273**, 703 (1986) .
13. Fu Jinghua and Liu Lianshou, *Phys. Rev. C* **68**, 064904 (2003).
14. Gary D. Westfall (for the STAR collaboration), *l. Phys. G* **30**, S345 (2004) .

Multiplicity structure of inclusive and diffractive scattering at HERA.

Benoit Delcourt, On the behalf of H1 and ZEUS collaborations.

LAL,Orsay (France) and H1

Abstract.
We present here a review of the progresses of Zeus and H1 about multiplicities of charged particles in e-p scattering. It includes a comparison of H1 and Zeus results on multiplicities, and studies on multiplicities in ep vs e+e- and pp (Zeus), on dependence of multiplicities on various variables in DIS and diffractive DIS, and of KNO scaling for DIS and Diffractive DIS (DDIS).

Keywords: deep inelasting scattering,diffraction

A COMPARISON OF H1 AND ZEUS RESULTS ON MULTIPLICITIES.

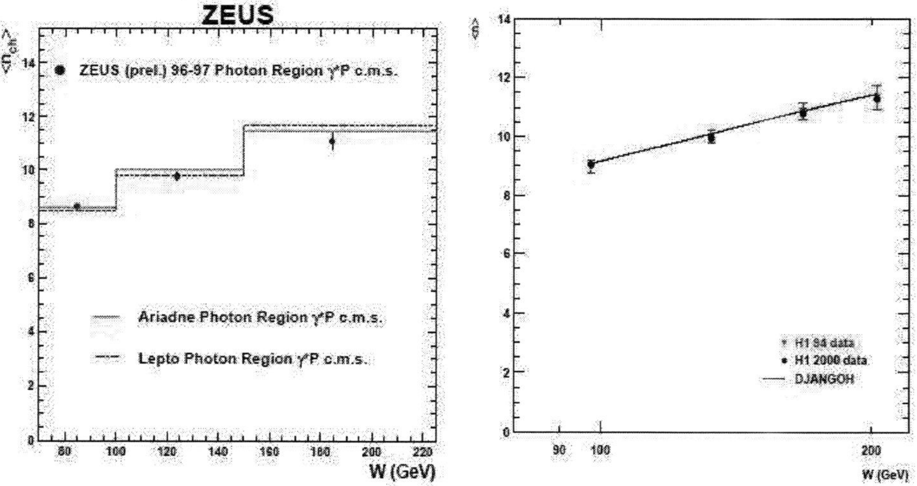

FIGURE 1. *Multiplicities in the photon region of the HCM frame for Zeus(left) and H1(right).*

Both experiments, [1],[2],[3] and [4], compute their multiplicities in the so called hadronic center of mass (HCM) frame, which includes the virtual photon and the final hadrons; the count of charged particles is limited to the so called 'photon region', that is the forward hemisphere of the photon, where, for example, the proton remnants are not included. W is the invariant mass of the hadronic system.

CP828, *Multiparticle Dynamics*
edited by V. Šimák, M. Šumbera, Š. Todorova, and B. Tomášik

The selection of events for both experiments include the following main cuts:
Energy of scattered electron $E_e < 12GeV$,
z of the vertex $|z_{vertex}| < 30cms$ for H1 and $|z_{vertex}| < 50cms$ for Zeus,
$Q^2 > 10GeV^2$ for H1, and $Q^2 > 25GeV^2$ for Zeus.
$0.05 < y < 0.65$ for H1 and $0.05 < y < 0.95$ for Zeus.
The tracks are included in the multiplicity calculation when:
their transverse momentum P_t is $< 100MeV/c$ for H1 and $< 150MeV/c$ for Zeus.
their polar angle θ is between 15 and 165 degrees for H1, 20 and 160 degrees for Zeus, corresponding to pseudorapidities $-2 < \eta < 2$ for H1 and $-1.75 < \eta < 1.75$ for Zeus.

These selections being generally more restrictive for Zeus that for H1, one would, may be, expect larger multiplicities for H1 than for Zeus; the figure 1 shows that, in contrary, there is agreement. As a conclusion, the multiplicities are not very different, but it would be useful that both experiments do the same cuts before doing this comparison.

A STUDY BY ZEUS ON MULTIPLICITIES IN EP VS E+E-.

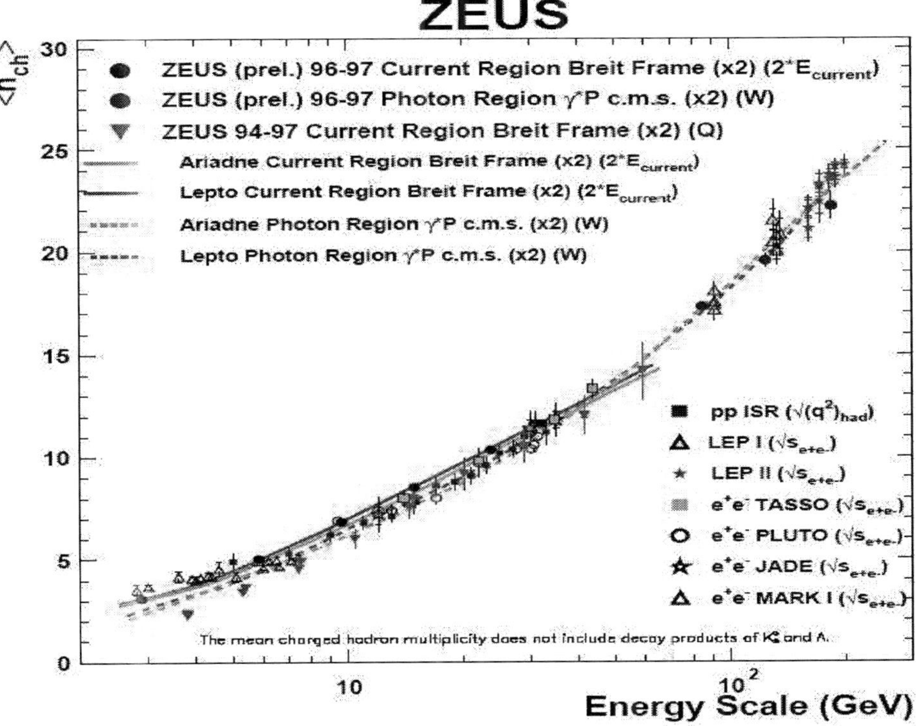

FIGURE 2. *Corrected multiplicities of e-p (Zeus), both in HCM and Breit frames, compared to e+e- and p-p.*

Zeus has made a comparison of multiplicities at HERA versus multiplicities in e+e- experiments. Two different frames were considered:

-i- the HCM frame, see above. This frame is a high energy frame, as it takes a good proportion of final hadrons.

-ii- the Breit frame, in which the virtual photon is on the z axis and the scattered quark is incoming with a momentum $Q/2$ and scattered in the backward direction with an opposite momentum $-Q/2$, where Q^2 is the usual momentum transfer of the beam electron. Here also, one considers only the 'current' region, that is the forward hemisphere of the photon. This frame is a low energy frame as it removes the main part of hadrons, which are in the other hemisphere. The problem with it is that it neglects gluons, whether they are in the initial state (the so called boson-gluon fusion) or in the final state (QCD Compton); to take account of this, the analysis of Zeus takes the energy scale from the total energy seen by the calorimeter in the current hemisphere of the Breit frame, rather than the energy in this frame calculated from the scattered electron.

In both cases, the multiplicity found in e-p collisions has to be multiplied by 2, as one looks only to particles in one hemisphere, versus 2 hemispheres for e+e-. The comparison is done in [1] and [3], with references in [3], and shown on fig.2. . It is seen that all the e+e- and the e-p data fall along the same curve, which can be fit using the Monte-Carlos Lepto 6.5 [5] and Ariadne 4.08[6]. The HCM frame of HERA data are compared to LEP data, whereas the Breit frame ones are compared to lower energy e+e- data. A comparison with p-p data is also made.

Question: e+e- data include weak decays of D's and B's, which do not enter the QCD framework. Is there a correction for this? Answer: Zeus data were corrected for decays of K_s^0's and Λ's but e+e- data included the decays of D's and B's. There might be a correction to consider.

Question: $p-p$ data are difficult to compare with e+e- and e-p. Answer: see [7].

DEPENDENCE OF MULTIPLICITIES UPON VARIOUS VARIABLES IN DIS AND DIFFRACTIVE DIS (DDIS).

Both experiments have searched for exotic behaviours of multiplicities in DIS. Within QCD, one expects generally, but not always, multiplicities to depend only upon the invariant mass of the considered system. Zeus has searched without success for a dependence upon Bjorken x and Q^2 in Deep inelastic scattering (DIS), see fig.3 left for the x dependence; the effective mass is $M_{eff} = \sqrt{(\Sigma E_i)^2 - (\Sigma \vec{P}_i)^2}$ where the index i runs over all signals in the calorimeter. H1 has also looked to this dependence, see fig.3 right for the Q^2 plots, and it has added the search in the diffractive DIS sector (DDIS).

In diffraction, there are two sequential reactions happening: first the proton emits a colorless object, the Pomeron, with an x named x_P, and then a quark from the inside of this Pomeron scatters with the virtual photon, with an x named $\beta = Q^2/(Q^2 + M_x^2)$, where M_x is the mass of the final hadronic system coming from the photon-Pomeron scattering (the X system). The emission of the Pomeron and its scattering with the virtual photon are factorisable, and, for instance, the Bjorken x is: $x = x_P.\beta$.

H1 has unsuccessfully searched for a possible dependence of the multiplicity of

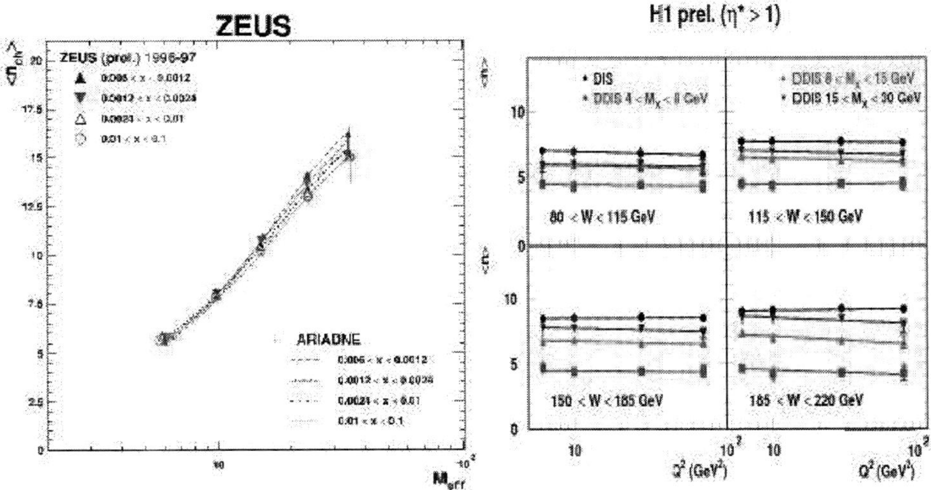

FIGURE 3. *Left: Zeus, multiplicity vs effective Mass for different $x_{bjorken}$. Right: H1, multiplicity in both DIS and DDIS as functiion of Q^2 for various W bins.*

the X system upon the β, fig.4 left.

Moreover, in diffraction, the Pomeron could be replaced in a part of the phase space by a meson, which is named Reggeon in this case. This complicates the simple picture of diffraction. The multiplicity is one of the parameters which has sensitivities different for the Pomeron and the Reggeon: the Pomeron is expected to have a larger gluonic content than the Reggeon, and, so, to produce more quarks, and finally more hadrons [8].

In fig.4 right , we show the multiplicities as functions of W, for different bins of M_x. On the same plots are shown the predictions from the Rapgap M.C.[9] for the cases of Pomeron and Reggeon. At low M_x, both components agree with data. At medium M_x's, $8 < M_x < 15GeV/c^2$, the Pomeron model seems better than the Reggeon. But at $M_x > 15GeV/c^2$, it is seen that the data agree better with the Reggeon model at low W, and with the Pomeron model at higher W; as $x_P = (Q^2 + M_x^2)/(Q^2 + W^2)$ if we neglect the proton mass, this means that the Pomeron model holds for lower x_P's, and the Reggeon at higher x_P's.

Question: is it normal that the M.C. curves increase with W at fixed M_x, as the multiplicities would only depend on M_x? Answer: the widths of bins of M_x are rather wide; also the H1 data are for a restricted part of the phase space: pseudorapidity in HCM frame $\eta^* > 1$, because of some losses of acceptance below 1.

KNO SCALING.

The invariance of the quantity $< n > P_n$ as a function of $n/ < n >$ for different $< n >$'s is known as the KNO scaling [10]. This property has been shown to be true in

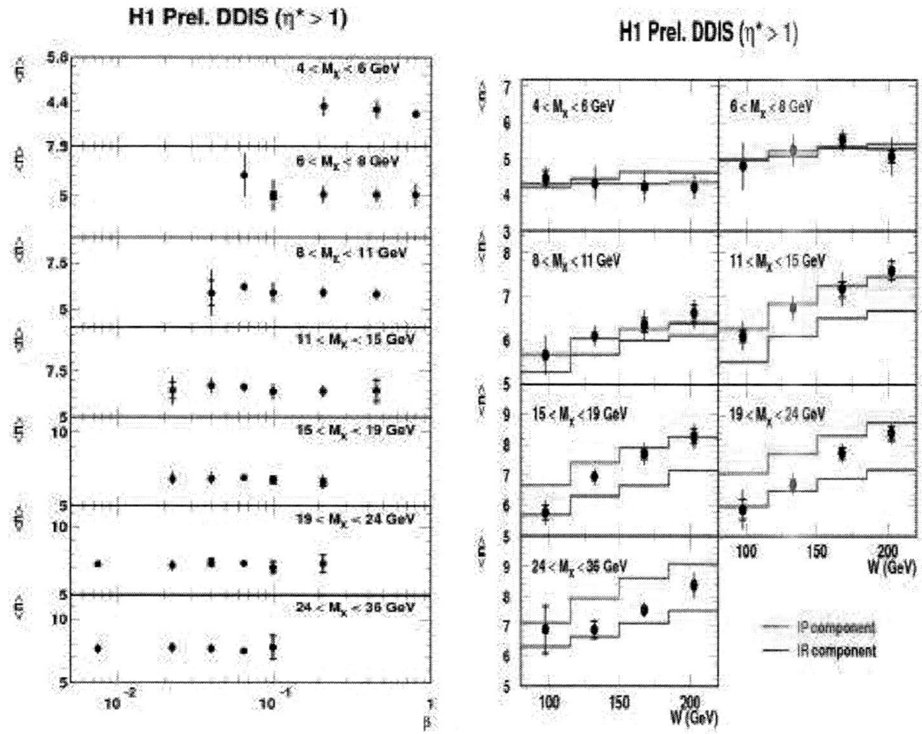

FIGURE 4. *Multiplicities for different M_x bins as functions of: left, β , right, W.*

different circumstances, but it is enough to say that the different QCD models ([5] and [6] for example) agree with it. Fig. 5 shows that it holds for DIS and DDIS; this fig.5 is for only negative particles, but it is also true for positive ones and for both signs.

Nevertheless, more precise verifications are possible; they include measurements of so called moments F_n. For instance, $F_2 = < n.(n-1) > / < n >^2$. These moments should be energy independent if KNO scaling holds. This has been already checked in the past by H1[4] for DIS, but this is also the case for a simple Poisson law, as they are then equal to 1. But, to disentangle KNO scaling from this Poisson law (which does not fulfill the KNO scaling), one should also check that $< n > /D$, where D is the dispersion of the multiplicity curve (or its root mean square) is independant of $< n >$, and so on energy.

CONCLUSION

Multiplicity measurements are going on in H1 and Zeus. The main achievements have been, for Zeus the comparison with e+e- data, even if the problem of decays of

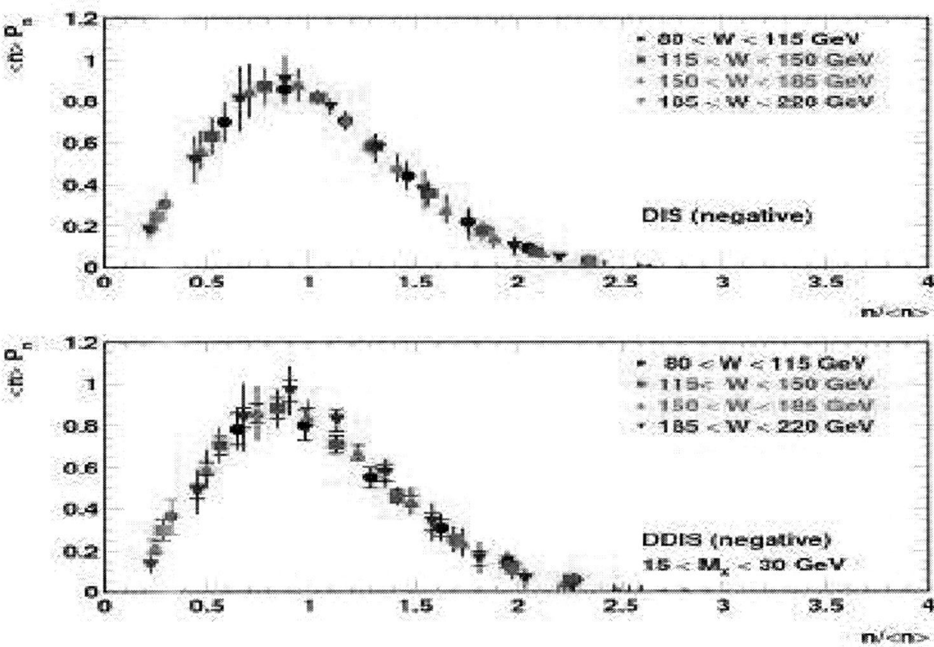

FIGURE 5. *KNO scaling for DIS(top) and DDIS (bottom).*

D's and B's should now be tackled, and for H1 the indication that the Pomeron does not account alone for diffractive DIS, and a Reggeon is also needed at higher x_P's.

REFERENCES

1. Proceedings of the DIS-2005 conference, 27 April to 1 May 2005, Madison (USA). Session Hadronic final states. Michelle Rosin: charged multiplicity distributions.
2. ibidem.Session Diffraction. Tinne Anthonis: The hadronic final state in diffraction.
3. XXII International Symposium on Lepto-Photon Interactions at High Energy. June 30-July 5, 2005, Uppsala, Sweden. Session QCD/HS. Contributed paper 283, from Zeus.
4. Multiplicities were given for H1 in Aid et all,Z.Phys.C 72, page 573 (1996). What is shown here is the result of a preliminary analysis using data of 2000, which agree with the former.
5. G.Ingelman,A.Edin and J.Rathsman, Comp.Phys.Comm. 101,page 108,(1997)
6. L.Loennblad, Comp.Phys.Comm. 71,page 15,(1992)
7. M.Basile et all, Lettere al Nuovo Cimento,41N, 9, page 293 (1984).
8. Hannes Jung, private communication.
9. H.Jung, Comp.Phys.Comm.,86,147,(1995)
10. Z.Koba, H.B.Nielsen, P.Olesen, Nucl.Pys.B40, page 317(1972).

High Order Multiplicity Moments

K. Fiałkowski* and R. Wit*

*M. Smoluchowski Institute of Physics
Jagellonian University
30-059 Kraków, ul.Reymonta 4, Poland

Abstract. We analyze the ratios of cumulants to factorial moments both for data and for the Monte Carlo generated events. For the PYTHIA generated events the moments are investigated both for the full and restricted range of phase-space and for the jets reconstructed from single particle momenta. The results cast doubts on the validity of extended local parton-hadron duality and suggest the possibility of more effective experimental investigations concerning the origin of the observed structure in the dependence of moments on their order.

Keywords: Multiplicity moments, parton - hadron duality
PACS: 12.90.+b, 13.66.Bc

INTRODUCTION

In this note we discuss the moments of multiplicity distributions proposed by the Pavia group [1]: $H_q = K_q/F_q$. Here the factorial moments are defined in the standard way

$$F_q = \sum_n \frac{n!}{(n-q)!} P(n),$$

and the cumulants may be calculated from a recursive formula

$$K_q = F_q - \sum_i \frac{(q-1)!}{(i-1)!(q-i)!} K_{q-i} F_i.$$

The results of the Pavia group for the new moments initiated vivid activity because of their apparent compatibility with the predictions of perturbative QCD at the NLLA level [2]. In both cases the dependence of moments on their order was found to be non-monotonical: after a minimum (with negative value) at $q = 5$, a hint of oscillations at higher q was seen. Later, however, some doubts appeared about the origin of the observed structure. In particular, the simple cut in the smooth multiplicity distribution (e.g. of the negative binomial type) was shown to produce similar effects [3]. In this note we analyze the behaviour of the moments calculated from the PYTHIA generator [4] for the electron-positron annihilation at LEP-I energy and compare them, where possible, with the experimental L3 results [5], [6]. We see in both cases a similar structure. It is unlikely to be due to the NNLO order QCD perturbative effects, as the generator does not contain explicitly such components. A more plausible explanation is to relate the structure more generally to the multicomponent character of the production process [7]. An example of such "multicomponent expansion" for Z decays may be a separation of

CP828, *Multiparticle Dynamics*
edited by V. Šimák, M. Šumbera, Š. Todorova, and B. Tomášik

two- and three-jet classes of events; we have checked that parametrizing these classes by negative binomials with the proper ratio of parameters and coefficients one reproduces qualitatively the pattern seen in data [8]. Thus any generator reproducing correctly the two- and three-jet components may be expected to reproduce also qualitatively the observed structure.

We include also some proposals for future investigations. An extended version of this communication will be published in Journal of Physics G.

DATA AND MONTE CARLO RESULTS FOR Z DECAYS

In this section we discuss the LEP-I data of L3 collaboration [5] published recently (and compared with JETSET [9] and HERWIG [10] Monte Carlo results). We remind these results and compare them with the moments calculated from PYTHIA. The comparison is only qualitative, as we do not use the L3 detector simulation program, which served to unfold the experimental multiplicity distribution [5]. Nevertheless, we will see that some surprising features of data and MC results are confirmed.

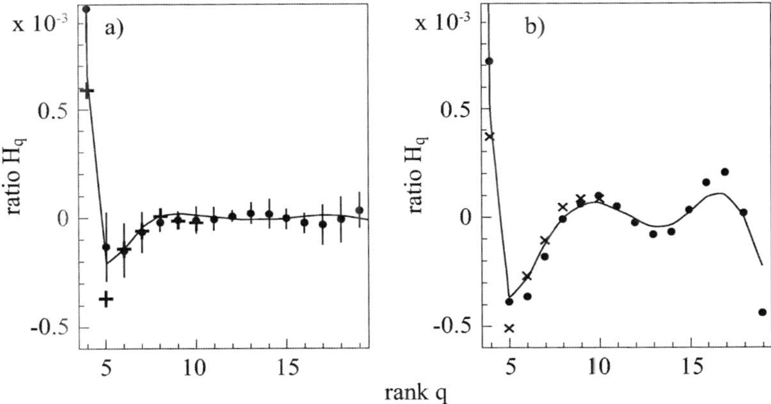

FIGURE 1. The L3 results [5] for H_q moments calculated from full (a) and truncated (b) multiplicity distribution. Black points represent the data, solid curve the events generated from JETSET. PYTHIA events result in values shown as crosses and x-s.

The values of H_q moments calculated from the unfolded multiplicity distributions are reproduced in Fig.1a for the orders $q = 4 \div 19$ together with the expectations of JETSET and PYTHIA. It is obvious that both JETSET and PYTHIA agree with data well. However, it is equally obvious that the errors shown cannot be interpreted as simple uncorrelated statistical errors. The JETSET results never deviate from data by more than one standard deviation (and for the majority of moments by more than half SD). This is statistically improbable and suggests that the errors are not simple independent statistical errors. The authors recognize this problem, and they do not use directly the experimental multiplicity distributions. Instead they use the multiplicity distributions truncated by removing about 0.035% of highest multiplicity events. With such a truncation the errors are visibly reduced and the clear structure appears both in data and MC results shown in Fig.1b: a minimum with negative values of H_q at $q = 5 \div 6$, and possible oscillations with maxima for q around $10 \div 11$ and $16 \div 17$.

In our opinion the lack of fluctuations above one standard deviation and the reduction of errors after the removal of some data suggest that the errors shown are misleading. Thus for our simulations using PYTHIA we do not calculate errors from standard statistical formulae (relating the errors of the q-th moment to some moments of order $2q$), but use instead the uncertainty evaluated from the spread of results for different samples of independently generated events. We use the statistics, for which the errors are comparable to the size of symbols in the figures: 4M events for each sample.

HIGHER MOMENTS IN RESTRICTED PARTS OF PHASE-SPACE

The data and simulations discussed above concerned the full phase-space multiplicity distributions. It is well known, however, that the multiplicity distributions in limited regions of phase-space depend significantly on the definition of the limits.

The LEP1 data have sufficient accuracy and statistics to investigate higher moments for various regions of phase-space. Thus we have performed calculations of such moments for the events generated with PYTHIA, defining phase-space regions by cuts in values of CM momenta.

We have generated samples of $4M$ events and calculated the moments for orders $2 \div 10$ for regions defined by simple inequality for the values of the CM three-momenta p: $p < \varepsilon n$, where $\varepsilon = 0.2 GeV$, and $n = 1 \div 10$.

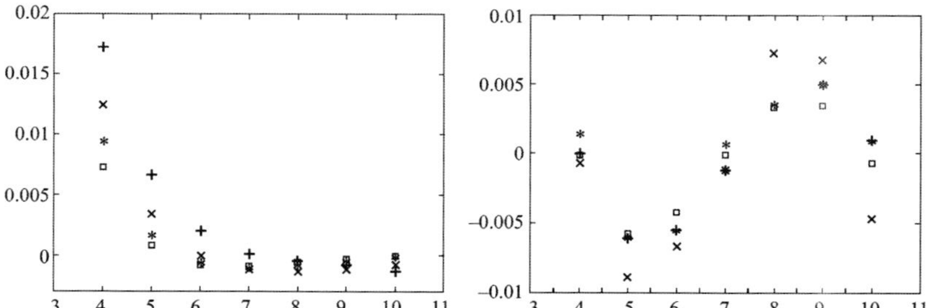

FIGURE 2. The H_q moments calculated from PYTHIA events in limited parts of phase-space (see text) for L3 values of the parameters without (a) and with (b) a cut removing highest multiplicities. Crosses, x-s, stars and squares represent the results for n = 4, 6, 8 and 10, respectively.

We show only the results for $L3$ parameters (the results for the default PYTHIA parameters are very similar). The minimum (with negative values) in the dependence of moments H_q on their order q occurs for small phase space regions very late (at q around 8) and shifts gradually to smaller values of q for increasing range of CM momenta. However, even for the widest range of momenta investigated, the position and shape of minimum differs significantly from that for full phase-space. The results are shown in Fig.2a; for transparency we show only points for n =4,6,8 and 10.

In the previous chapter we have seen that the minimum for full phase-space was strongly enhanced by introducing cuts in the multiplicity distributions. Thus we repeated our calculations cutting off the highest multiplicities contributing 0.001 to the full cross-section. The results, shown in Fig. 2b, are quite surprising: now the position of minimum is practically independent on the size of phase-space region selected. It will be very interesting to see if the data show the same patterns when investigated with- and without extra cuts in multiplicity.

HIGHER MOMENTS FOR HADRONS AND FOR JETS

As noted in the introduction, the existence of minimum (with negative values) in the q-dependence of moments H_q was related to the higher order corrections in perturbative QCD. It is obvious, however, that the perturbative QCD calculations yield the multiplicity distributions for gluons, and not for single hadrons. The identifications of higher moments for those two distributions means a rather bold extension of the assumption of parton-hadron duality, usually applied only to the average values.

To estimate the reliability of such an extension we applied to the generated events a standard PYTHIA clustering algorithm (PYCLUS) and investigated the multiplicity distributions of reconstructed jets (which may be expected to correspond to the gluon distribution more closely, than the distribution of single hadrons). The average number of jets depends strongly on the values of two parameters, which define the maximal phase-space distance between the particles added to the existing jet and the maximal CM momentum for the slowest particles in CM (which form a separate cluster/jet).

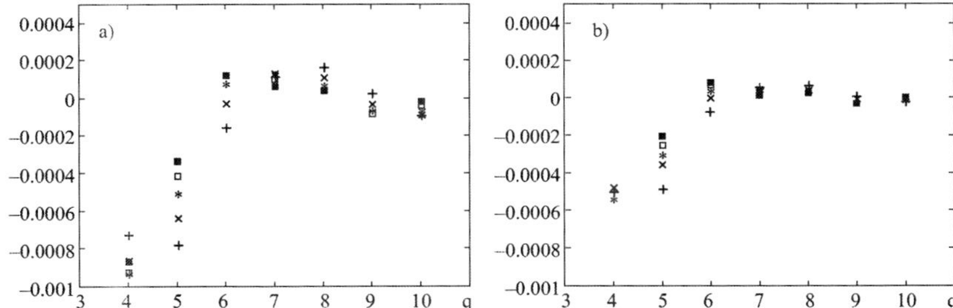

FIGURE 3. The H_q moments for jets reconstructed from hadrons for PYTHIA events generated with L3 parameters. In Fig. (a) only charged hadrons are used; in Fig. (b) all stable particles are counted. Crosses, x-s, stars, open and closed squares correspond to the increasing average number of particles in a jet (from about 1.1 to 1.5 in Fig. (a) and from 1.3 to 2 in Fig. (b).

We show in Fig.3 the ten lowest moments for the distributions of jets defined with those parameters spanning a range of 0.02 – 0.1 GeV for L3 parameters, with cut in the multiplicity. The values of moments for jets and for hadrons are very different, even for relatively small jets (when the multiplicity in both cases is similar). When we count not only charged particles, but all the stable particles, we observe also a clear difference between the values of higher moments for jets and hadrons.

An experimental investigation of the H_q moments for jets has been performed by the L3 collaboration [6]. Although the jet definition used (Durham algorithm) was different than the default algorithm from PYTHIA used by us, the results are very similar to those presented in Fig. 3a: with increasing "jet size" the pattern of moments changes significantly.

We regard this dependence as a suggestion that the extended local parton-hadron duality (ELPHD), in which one identifies the (charged) hadron multiplicity distribution with the parton distribution at small fixed virtuality, is not reliable. We should note, however, that the opposite conclusion was drawn from the same data [11]. The dependence of the H_q moments on their order q was investigated there for the parton jets in a MC model for varying jet resolution parameter Q_c. A good description of L3 data was found for jets as well as for hadrons. This was regarded as a support for ELPHD.

The agreement with the L3 data is very good indeed for $Q_c \geq 1GeV$. For small Q_c the agreement is reasonable, although not perfect. In particular, the ELPHD prediction for hadrons is practically identical to that for jets with $Q_c = 100MeV$, whereas in data there is a marked difference between the moments. This difference is similar to that seen for our PYTHIA events: a clear minimum for $q = 5$ seen for hadrons (Fig.1) moves to $q = 4$ already for smallest jets (Fig.3a). Thus there seems to be no clear "hadron limit" for parton MC, although the differences are not big.

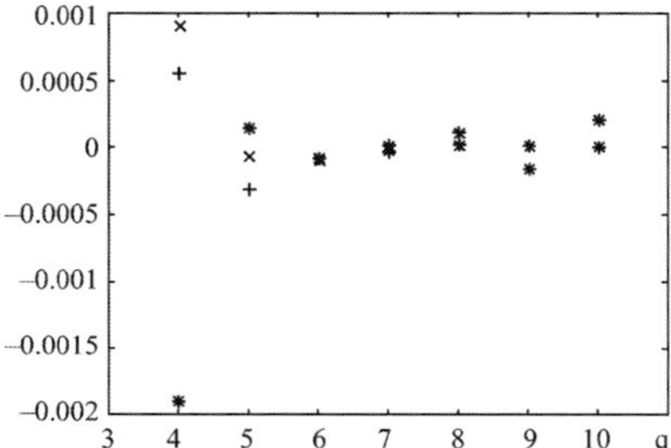

FIGURE 4. The H_q moments for PYTHIA events generated with L3 parameters. Crosses denote the results for charged hadrons, x-s are for all stable particles and stars for positive hadrons.

To verify further the possibility of describing hadron data by ELPHD we have compared the moments calculated for charged hadrons (pions, kaons, protons and antiprotons) with the moments calculated for all stable particles (thus adding photons, neutral kaons, neutrons and antineutrons). The results, shown in Fig.4 suggest that a relatively deep minimum at $q = 5$ is to a large extent due to the charge conservation effect. Indeed, the results for positive hadrons (shown also in Fig.4) do not exhibit a pattern seen for all charged hadrons. A deep minimum appears for $q = 4$, and the small oscillations with a very short period follow. In this case it is rather obvious that the data will confirm our findings, as the charge conservation is correctly built in PYTHIA and the charged-neutral correlations are known to be well described by this generator.

The choice of charged hadrons of both signs as a hadronic counterpart of partons (mostly gluons) in ELPHD seems in fact worse than other possible choices. The measured cross section for odd multiplicities of charged hadrons should be zero for an ideal detector, whereas the distribution of gluons should be smooth without any discrimination of odd multiplicity values. This suggests that the agreement of moments calculated for partons and charged hadrons is to a large extent accidental and is due mostly to the similar influence of cuts and the two-component mechanism (mentioned in section 2) in both distributions. In our opinion, one should rather try to find parameters, for which the parton MC would agree with the distribution of positive (or all) hadrons.

CONCLUSIONS AND OUTLOOK

Using the PYTHIA generator we have investigated the multiplicity distributions for hadrons coming from the Z decay and calculated the ratios of cumulants to factorial moments. The results for charged hadrons in full phase-space are compatible with the L3 data.

We find that the moments calculated from the multiplicity distributions in the limited part of phase-space differ significantly from that for the full phase-space: the minimum in the dependence of H_q on q shifts to higher values of q for smaller range of momenta. However, introducing an universal truncation on the multiplicity distributions for different ranges of momenta, we recover a stable position of the minimum. It will be interesting to see if this pattern of generated events will be confirmed in the real data.

We have also investigated the moments for the jets, defined as clusters of particles close in momentum space. We find that even for very small jets, containing in average only slightly more than one hadron, the pattern of moments is different from that for hadrons. Since the distributions of such jets may be expected to correspond more closely to the distributions of partons, one may wonder if it is meaningful to compare the moments measured for the observed hadrons to the moments calculated for partons. We feel that our results cast doubts on the possibility of such an extension of "local parton-hadron duality" beyond the average quantities (moments of order one). Again, performing more such investigations on the real data would be highly desirable.

ACKNOWLEDGMENTS

We are grateful to A Białas and A. Kotański for reading the manuscript and for helpful remarks. Fruitful critical remarks by W. Ochs and W. Metzger are gratefully acknowledged.

REFERENCES

1. I.M. Dremin *et al.*, *Phys. Lett.* **B336**, 119 (1994).
2. I.M. Dremin, *Phys. Lett.* **B313**, 209 (1993).
3. R. Ugoccioni, A. Giovannini and S. Lupia, *Phys. Lett.* **D59**, 094020 (1999).
4. T. Sjöstrand *et al.*, *Comp. Phys. Comm.* **135**, 238 (2001).
5. P. Achard *et al.*, L3 Collaboration, *Phys. Lett.* **B577**, 109 (2003).
6. D.J. Mangeol, Correlations in the Charged-Particle Multiplicity Distribution, Ph.D. thesis, e-print hep-ph/0110029.
7. I.M. Dremin, e-print hep-ph/0404092, proceedings of the Moriond meeting 2004.
8. A similar model for hadron-hadron collisions was first suggested by A. Giovannini and R. Ugoccioni, *Phys. Rev.* **B336**, 119 (1994).
9. T. Sjöstrand, *Comp. Phys. Comm.* **82**, 74 (1994).
10. G. Marchesini *et al.*, *Comp. Phys. Comm.* **67**, 465 (1992).
11. M.A. Buican, C. Förster and W. Ochs, *Eur. Phys. J.* **C31**, 57 (2003); W. Ochs, *Acta Phys. Pol.* **B35**, 429 (2004).

JET PHYSICS

Chairpersons: W. Kittel, L. S. Liu, J. G. Cramer, and Y. Wu

Top Quark Physics at Hadron Colliders

Jochen Cammin

University of Rochester, Department of Physics & Astronomy, Bausch & Lomb Hall, Rochester, NY 14627, USA

Abstract. This talk presents the latest results on top quark physics from the Fermilab Tevatron and gives prospects for top quark studies at the Large Hadron Collider.

Keywords: Top quark, Tevatron, Fermilab, LHC, Cern, D0, CDF, ATLAS, CMS
PACS: 14.65.Ha

INTRODUCTION

The top quark is of special interest to the experiments at the Fermilab Tevatron and the upcoming Large Hadron Collider (LHC) at CERN. It is the particle that we know least of, and it may play an important rôle in our understanding of electroweak symmetry breaking. Furthermore, due to its large mass, signs of new physics may appear first in the top quark sector.

In the Standard Model, the top quark decays almost exclusively via $t \rightarrow Wb$. The top quark decay channels are therefore characterized by the decay modes of the W. For $t\bar{t}$ production, the 'all hadronic' channel (both W decay into qq') has the largest branching fraction but suffers from large QCD backgrounds. The 'lepton plus jets' channel is usually considered the 'golden channel' for $t\bar{t}$ studies, since it still has a sizable branching fraction and the backgrounds are manageable. The cleanest final state is the 'dilepton channel', where both Ws $\rightarrow \ell\nu$, $\ell = e$ or μ, but the branching fraction is small.

TOP PHYSICS AT THE TEVATRON

As of August 2005, the Fermilab Tevatron has delivered more than 1 fb^{-1} of $p\bar{p}$ collision data at a center-of-mass energy of 1.96 TeV to the D0 and CDF detectors. Current analyses utilize up to 400 pb^{-1} of this data. The focus of top quark studies at the Tevatron consists of the $t\bar{t}$ cross section, mass and other properties of the top quark and search for single top production.

$t\bar{t}$ cross section

Pair production is the main production mode for top quarks at the Tevatron. The theoretical prediction for the $t\bar{t}$ cross section at $\sqrt{s} = 1.96$ TeV is about 7 pb [1]. D0 and CDF have measured the $t\bar{t}$ cross section in the lepton+jets, dilepton and all hadronic final states. The analyses can be grouped into two categories: topological analyses, where

CP828, *Multiparticle Dynamics*
edited by V. Šimák, M. Šumbera, Š. Todorova, and B. Tomášik

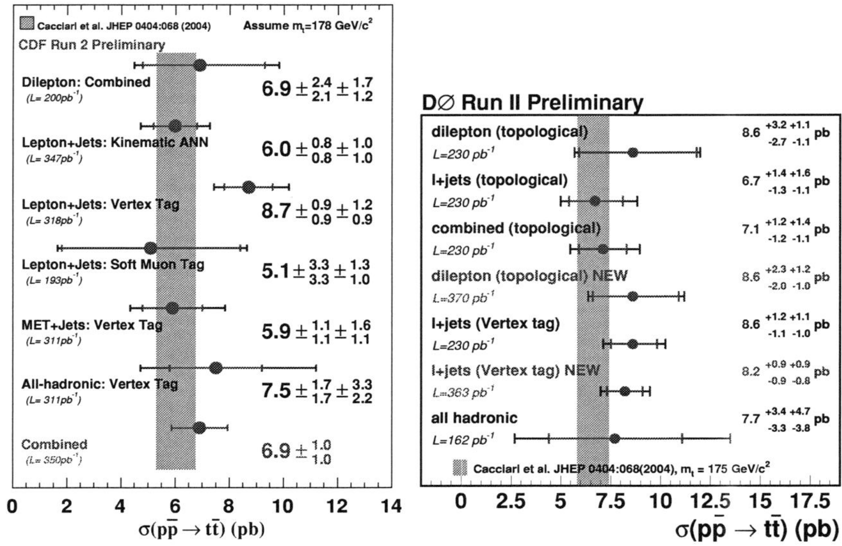

FIGURE 1. Summary of $t\bar{t}$ cross section measurements at CDF and D0.

quantities like kinematic variables are usually combined using multivariate techniques to discriminate the background from the $t\bar{t}$ signal, and b-tagged analyses, which make use of either the large lifetime of b hadrons to find displaced tracks or vertices or of semileptonic decays of the b quark in order to isolate the signal process. Once the background N_{bkg} is estimated and the selection efficiency, ε, and acceptance, A, are known, the $t\bar{t}$ production cross section is calculated by $\sigma_{t\bar{t}} = \left(N_{obs} - N_{bkg}\right) / \left(\varepsilon \cdot A \cdot \int \mathscr{L} \mathrm{d}t\right)$, where $\int \mathscr{L} \mathrm{d}t$ is the integrated luminosity, and N_{obs} is the number of selected events in data. N_{obs} is usually measured as a function of the number of jets in the event. Signal events are expected to accumulate in the four-jet bin in the lepton+jets final state and in the two-jet bin in the dilepton final state. A summary of all $t\bar{t}$ cross section measurements by CDF and D0 is give in Fig. 1 [2, 3].

Mass of the top quark

Knowing the mass of the top quark with good precision is of fundamental interest. For example, a small uncertainty on m_t helps to constrain predictions for the mass of the Higgs boson. In order to reach high precision on m_t, CDF and D0 have applied several advanced analysis techniques. The top quark mass has been measured in the lepton+jets and dilepton channels with and without b-tagging. B-tagging is useful to suppress the background as well as to reduce the combinatorics when assigning the reconstructed objects to the final states of the decays of the two top quarks. To give two examples, the following sections describe the template method by CDF and the matrix element method by D0.

The currently most precise CDF analysis uses the lepton+jets final state and b-tagging. The data set is divided into subsamples with 0, 1, and 2 tagged jets. The jet energy calibration is obtained from the W mass constraint. Templates for the top and W masses are obtained from Monte Carlo simulations, and then are fitted to the reconstructed mass distribution in data. The combination of final state objects that is most likely the correct combination is found by a χ^2 minimization using reconstructed W and top quark masses and four-vectors of the jets, the lepton and unclustered energy. A total likelihood to extract the best top quark mass is constructed from the four subsamples, where the likelihood for each subsample is the product of likelihoods for the reconstructed W and top masses, the observed events and expected background events. This method yields a preliminary top quark mass of $m_{\rm t} = (173.5^{+2.7}_{-2.6}(\text{stat.}) \pm 3.0(\text{JES} + \text{syst.}))$ GeV

The D0 matrix element method makes maximal use of the event information by calculating a probability for each event to be signal or background, based on the leading-order (LO) matrix element for signal and background processes. This approach led to the most precise single measurement of the top quark mass with Run I data [4]. The probability is given by $P(x;M_{top}) = \frac{1}{\sigma}\int d^n\sigma(y;M_{top})dq_1dq_2f(q_1)f(q_2)W(x,y)$, where $\sigma(y;M_{top})$ contains the LO matrix element, $f(q)$ describes the probability distribution of the initial partons having momentum q, and $W(x,y)$ is the probability of a parton variable y to be measured as variable x. This method, too, makes use of the intrinsic jet energy calibration from hadronic W decays and fits simultaneously the jet energy scale (JES) and the top quark mass. The total event probability is thus given by $P_{\rm evt}(x;m_{\rm t},JES) = f_{\rm top}\cdot P_{\rm sgn}(x;m_{\rm t},JES) + (1-f_{\rm top})\cdot P_{\rm bkg}(x;JES)$. Finally, a likelihood is maximized as a function of $m_{\rm t}$. The analysis requires exactly four jets and no b-tagging is used. 150 candidate events are selected in a data sample of 320 pb^{-1}, and the fraction of $t\bar{t}$ events obtained from a fit is 0.36. The preliminary result for the top quark mass in this study is $m_{\rm t} = (169.5 \pm 4.4(\text{stat.} + \text{JES})^{+1.7}_{-1.6}(\text{syst.}))$ GeV.

The combination of all measurements from Run I and Run II (Fig. 2), gives a preliminary world average of $m_{\rm t} = (172.7 \pm 1.7[\text{stat.}] \pm 2.4[\text{syst.}])$ GeV. The published Run I analyses contribute with 22.8% (CDF: 1.9%, D0: 20.9%) to the combined result, and the preliminary Run II analyses with 77.3% (CDF: 44%, D0: 33.3%). With this value of the top quark mass, electroweak fits yield new limits for the mass of the Standard Model Higgs boson: $m_{\rm H^0} = 91^{+45}_{-32}$ GeV and $m_{\rm H^0} < 186$ GeV (95% CL) [5].

Search for single top production

Electroweak single top production has not yet been seen observed at the Tevatron. The next-to-leading order predictions for the t-channel and s-channel processes are $\sigma_t = (1.98 \pm 0.25)$ pb and $\sigma_s = (0.88 \pm 0.11)$ pb. The general signature of single top events (with $W \to \ell\nu$) are a high p_T lepton, missing transverse energy and two jets. CDF uses a template method and a data set of 162 pb^{-1}, whereas the D0 analysis is based on 270 pb^{-1} and features a likelihood discriminant approach. Limits on the single top production cross section from CDF are $\sigma_{s+t} < 17.8$ pb, $\sigma_t < 10.1$ pb, $\sigma_s < 13.6$ pb. The D0 limits are $\sigma_t < 5.0$ pb and $\sigma_s < 6.4$ pb.

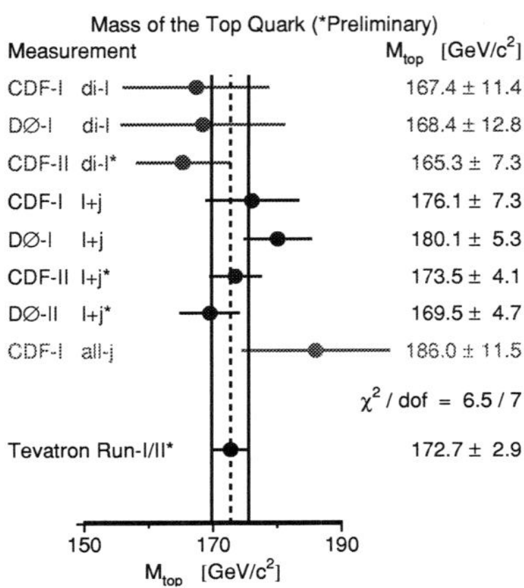

FIGURE 2. Measurements of the top quark mass at the Tevatron.

Other top quark related analyses

Measurements of the ratio **R=BR(t→Wb)/BR(t→Wq)** are a test of the Standard Model which predicts $R \approx 1$. Both CDF and D0 have measured this decay ratio. CDF yields $R = 1.12^{+0.27}_{-0.23}(\text{stat}+\text{syst})$ (with 162 pb^{-1} of data) and D0 yields $R = 1.03^{+0.19}_{-0.17}(\text{stat}+\text{syst})$ (with 230 pb^{-1} of data). This translates to the following limits: CDF: $R > 0.61$ @95% *CL*, D0: $R > 0.64$ @95% *CL* [7, 8].

D0 has searched for **narrow width heavy resonances** $X \rightarrow t\bar{t}$ in the lepton+jets final state using a lifetime tag to identify b jets using 370 pb^{-1} of data. The background consists of non-resonant Standard Model $t\bar{t}$ production, W+jets and multijet events with fake leptons. In technicolor models, a heavy resonance Z' is excluded at the 95% CL with $M_{Z'} < 680$ GeV for $\Gamma_{Z'} = 0.012 \cdot M_Z$ [6].

Both CDF and D0 have measured the **W helicity** to search for deviations from the Standard Model. The SM suppresses right-handed polarization states and predicts $f_0 = 0.703 \pm 0.012$ for the longitudinal polarization. The CDF measurement yields $f_+ < 0.18$ at 95% CL (Run I), $f_0 = 0.25^{+0.35}_{-0.21}$ $(194, 162\ \text{pb}^{-1})$ and the DØ result is $f_+ < 0.25$ at 95% CL (230 pb^{-1}) [9, 10].

TOP PHYSICS AT THE LHC

At the Large Hadron Collider, protons will be brought to collisions at a center-of-mass energy of 14 TeV. The luminosity will be $\mathscr{L} = 10^{33}$ cm^{-2}s^{-1} in the initial phase, with the design luminosity being about ten times larger. Such high energies and luminosities will make the LHC be a top quark factory, producing in the order of 10^7 top quark pairs per year. The large statistics allow for calibration of the detector, precision measurements and tests for new physics. For example, with an integrated luminosity of 10 fb^{-1}, the statistical error on the top mass is expected to be about 0.07 GeV. Top quark pairs will also be a major background to new physics processes, thus it is important to study top quarks in great detail.

Mass of the top quark

As for the Tevatron, the golden channel will be the lepton+jets final state, and b-tagging will be crucial to reduce physics background and combinatorics. Monte Carlo studies performed within ATLAS and CDF to measure m_t start by reconstructing the hadronic side of the $t\bar{t} \rightarrow \ell\nu + \text{jets}$ process. First, the pair of non b-tagged jets with m_{jj} closest to m_W is assigned to the hadronic W decay. Then a b-tagged jet is assigned to the reconstructed W, and the combination that gives the highest p_T jjb object is chosen. The top quark mass is then measured from the hadronic decay as m_{jjb}. The expected total uncertainty on m_t with this method is about 2 GeV [11]. Further improvements are expected if both top quark decays are reconstructed. But since the neutrino four-vector cannot be measured completely, this approach requires kinematically fitting the whole event, and hence excellent understanding of energy and momentum resolutions, in particular the missing E_T resolution is essential.

Apart from the conventional channels, there are also several alternatives to measure m_t. One can use high p_T $t\bar{t}$ events ($p_T > 200$ GeV) so that the event is clearly divided into two hemispheres. This topology helps to reduce the combinatorics. Jets from $t \rightarrow jjb$ will overlay due to the large boost, so the energy is reconstructed from calorimeter cells rather than from individual jets. After subtracting the underlying event contribution, one obtains a reconstructed top quark mass, which is independent of the reconstruction cone size. A systematic uncertainty of 1.7 GeV is estimated for this measurement [11].

Another approach is to use rare decays of B hadrons to $J/\Psi \rightarrow \mu\mu$ in the b jet associated to the leptonic decaying W, and semi-muonic decays of the other b quark. This gives a very clean signature, but the branching ratio is only 5.3×10^{-5}. About 1000 events per year can be expected at design luminosity. The top quark mass is strongly correlated with the invariant mass $M_{\ell J/\Psi}$ and can be extracted from a calibration curve. A total systematic uncertainty of less than 1 GeVon m_t can be expected in this channel [13].

Single top

At the LHC, the cross sections for the three main diagrams for single top production are Wg fusion: 244 pb, Wt: 60 pb and W*: 10 pb. Measuring single-top events allows for a precise determination of the W-t-b vertex and coupling strength. The main backgrounds to the process are $t\bar{t}$ production, Wbb and Wjj events. With 30 fb^{-1} it is expected to measure V_{tb} with a precision of 0.4% in Wg fusion, 1.4% in the Wt and 2.7% in the W* process. This is substantially better than the current theoretical uncertainties of 6% in Wg fusion, 6% in the Wt and 5% in the W* process [11].

Other properties of top quarks

Due to the short lifetime of top quarks, the **spin information** in $t\bar{t}$ events is not diluted by hadronization. Measuring the asymmetry $\mathscr{A}$ of finding top and anti-top in the same or opposite polarization tests the Standard Model properties of the top quark and probes the presence of non-Standard Model interactions. The asymmetry is defined by

$$\mathscr{A} = \frac{N(t_L\bar{t}_L + t_R\bar{t}_R) - N(t_L\bar{t}_R + t_R\bar{t}_L)}{N(t_L\bar{t}_L + t_R\bar{t}_R) + N(t_L\bar{t}_R + t_R\bar{t}_L)},$$

where the SM predictions for the processes $gg \to t\bar{t}$ and $qQ \to t\bar{t}$ are $\mathscr{A}(gg) = 0.431$, $\mathscr{A}(qQ) = -0.469$ so that, at the LHC, the expected total asymmetry is $\mathscr{A}(\text{LHC}) = 0.311$. Using dilepton final states and the angle between lepton and the $t\bar{t}$ rest frame as an observable, CMS estimates statistical and systematic uncertainties of $\Delta\mathscr{A}$(stat)=0.035, $\Delta\mathscr{A}$(syst)=0.028 with 30 fb^{-1} of data.

Flavor changing neutral currents would occur in decays such as $t \to qZ$, $t \to q\gamma$, and $t \to qg$ and are strongly suppressed in the Standard Model with a branching fraction of less than 10^{-10}. However, these decay fractions can be substantially enhanced in supersymmetric, multi-Higgs doublet or exotic quark models. ATLAS estimates a sensitivity to these branching fractions of 10^{-3}–10^{-4} with 100 fb^{-1} [11].

REFERENCES

1. Cacciari, M. and Frixione, S. and Mangano, M. L. and Nason, P. and Ridolfi, G., *The* $t\bar{t}$ *cross-section at 1.8 TeV and 1.96 TeV: A study of the systematics due to parton densities and scale dependence*, JHEP 0404:068 (2004).
2. CDF Top Group homepage, for latest updates see http://www-cdf.fnal.gov/physics/new/top/top.html.
3. D0 Top Group homepage, for latest updates see http://www-d0.fnal.gov/Run2Physics/top/top_public_web_pages/top_public.html.
4. D0 collaboration, *An Improved Measurement of the Top Quark Mass*, Nature 429, 638 2004.
5. LEP Electroweak Working Group http://lepewwg.web.cern.ch/LEPEWWG/plots/summer2005/.
6. D0 collaboration, *Search for a* $t\bar{t}$ *resonance in* $p\bar{p}$ *collisions at* $\sqrt{s} = 1.96$ TeV *in the lepton+jets final state*, D0 conference note 4880.
7. CDF Collaboration, *Measurement of B(t* $\to$ *W b)/B(t* $\to$ *W q) at the Collider Detector at Fermilab*, Phys. Rev. Lett. **95**, 102002 (2005).
8. D0 collaboration, *Simultaneous measurement of B(t* $\to$ *W b)/B(t* $\to$ *W q) at D0*, D0 conference note 4833.
9. CDF collaboration, *Measurement of the W boson polarization in top decay at CDF at* $\sqrt{s} = 1.8$ TeV, Phys. Rev. **D71** (2005), 031101.
10. D0 collaboration, *Search for right-handed W bosons in top quark decay*, Phys. Rev. D **72**, 011104(R) (2005).
11. ATLAS collaboration, *ATLAS: Detector and physics performance technical design report. Volume 1*, CERN-LHCC-99-14.
12. CDF and D0 collaborations, *Combination of CDF and D0 results on the top-quark mass*, hep-ex/0507091.
13. A. Kharchiliva, *Top Mass determination in leptonic final states with J/Ψ*, CMS-NOTE-1999-065.

Particle production and saturation at HERA

Cyrille Marquet

Service de physique théorique, CEA/Saclay, 91191 Gif-sur-Yvette cedex, France
URA 2306, unité de recherche associée au CNRS

Abstract. Perturbative QCD in the high-energy limit describes the evolution of scattering amplitudes with increasing energy towards and into the so-called saturation regime. Comparisons of the predictions with experimental data for a number of observables led to significant progress and understanding. We discuss the case of particle-production cross-sections measured at HERA and argue that these measurements have the potential to provide evidence for the saturation regime of QCD.

INTRODUCTION

In the Regge limit of perturbative QCD, *i.e.* when the centre-of-mass energy in a collision is much bigger than the fixed hard scale of the problem, parton densities inside the colliding particules grow with increasing energy, leading to the growth of scattering amplitudes. When the parton density becomes too large and the scattering amplitudes approach the unitarity limit, one enters in a regime called saturation [1].

The transition to the saturation regime is characterized by the so-called saturation scale which is an intrinsic hard scale of the problem. Contributions to the scattering amplitudes which are neglected as higher twist in the Bjorken limit of perturbative QCD become important in the saturation regime: leading-twist gluon distributions are no more sufficient to describe scattering at high energies.

A consistent approach is to express physical observables in terms of the leading terms in an expansion with respect to the inverse of the center-of-mass energy: this is called the eikonal approximation. This formalism is well-suited because, as the energy increases, density effects and non-linearities that lead to saturation and unitarization of the scattering amplitudes can be taken into account.

In the following, after introducing the eikonal approximation, we discuss the case of particle-production cross-sections measured at HERA. We concentrate on the phenomenology for diffractive observables: we review their model descriptions and investigate their potential to provide evidence for parton saturation.

HIGH-ENERGY SCATTERING AND SATURATION

Let us start with the eikonal approximation for quarks and gluons scattering at high energies. When a system of partons propagating at nearly the speed of light passes through a target and interacts with its gluon fields, the dominant couplings are eikonal: the partons have frozen transverse coordinates and the gluon fields of the target do not vary during the interaction. This is justified since the time of propagation through the

CP828, *Multiparticle Dynamics*
edited by V. Šimák, M. Šumbera, Š. Todorova, and B. Tomášik

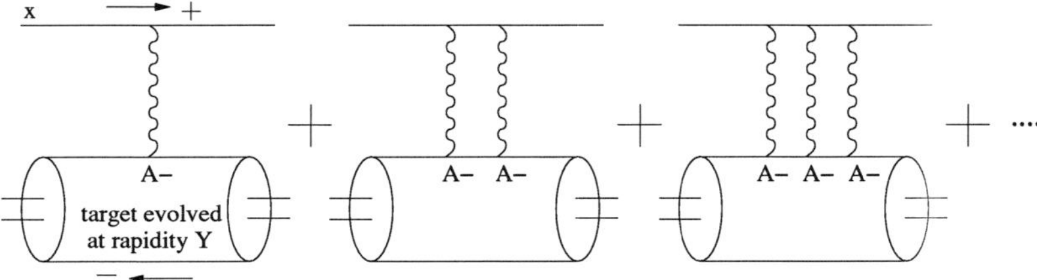

FIGURE 1. Eikonal propagation of a quark with transverse position x through a target evolved at rapidity Y. The eikonal phase $W_F(x)$, see formula (2), resums all number of gluon exchanges.

target is much shorter than the natural time scale on which the target fields vary. The effect of the interaction with the target is that the partonic components of the incident wavefunction pick up eikonal phases: if $|(\alpha,x)\rangle$ (resp. $|(a,x)\rangle$) is the wavefunction of an incoming quark of color index $\alpha\in[1,N_c]$ (resp. gluon of color index $a\in[1,N_c^2-1]$) and transverse position x (the irrelevant degrees of freedom like spins or polarizations are not explicitly mentioned), then the action of the $S-$matrix is (see for instance [2]):

$$S|(\alpha,x)\rangle\otimes|t\rangle=\sum_{\alpha'}[W_F(x)]_{\alpha\alpha'}|(\alpha',x)\rangle\otimes|t\rangle\ ,\quad S|(a,x)\rangle\otimes|t\rangle=\sum_b W_A^{ab}(x)|(b,x)\rangle\otimes|t\rangle\ , \tag{1}$$

where $|t\rangle$ denotes the initial state of the target. The phase shifts due to the interaction are the color matrices W_F and W_A, the eikonal Wilson lines in the fundamental and adjoint representations respectively, corresponding to propagating quarks and gluons. They are given by

$$W_{F,A}(x)=P\exp\{ig_s\int dz_+T^a_{F,A}A^a_-(x,z_+)\} \tag{2}$$

with A_- the gauge field of the target and $T^a_{F,A}$ the generators of $SU(N_c)$ in the fundamental (F) or adjoint (A) representations. We use the light-cone gauge $A_+=0$ and P denotes an ordering in the light-cone variable z_+ along which the incoming partons are propagating. As displayed in Fig.1, all number of gluons exchanges are included in (1), which shows the leading term of an expansion with respect to the inverse of the center-of-mass energy squared $s\sim e^Y$ where Y is the total rapidity.

For an incoming state $|\Psi_{in}\rangle$, the outgoing state $|\Psi_{out}\rangle=S|\Psi_{in}\rangle\otimes|t\rangle$ emerging from the eikonal interaction is obtained by the action of the $S-$matrix on the partonic components of $|\Psi_{in}\rangle$ as indicated by formula (1). The outgoing wavefunction $|\Psi_{out}\rangle$ is therefore a function of the Wilson lines (2). When calculating physical observables from $|\Psi_{out}\rangle$, one obtains objects that are target averages of traces of Wilson lines (the traces come from the color summation). For instance, the simplest of these objects is

$$T_{q\bar{q}}(x,x';Y)=1-\frac{1}{N_c}\left\langle\mathrm{Tr}\left(W_F^\dagger(x')W_F(x)\right)\right\rangle_Y\ , \tag{3}$$

namely the $q\bar{q}-$dipole scattering amplitude (x, x' : positions of the quark and antiquark) off a target evolved at rapidity Y. The target average has been denoted $\langle\ .\ \rangle_Y$ and contains the Y dependence. The amplitude (3) enters for instance in the DIS total cross-section (see next section). More generally, observables are functions of (3) or more complicated amplitudes. Let us introduce another one of them, which we shall need later:

$$T^{(2)}_{q\bar{q}}(x,x';y,y';Y) = 1 - \frac{1}{N_c^2}\left\langle \mathrm{Tr}\left(W_F^\dagger(x')W_F(x)\right)\mathrm{Tr}\left(W_F^\dagger(y')W_F(y)\right)\right\rangle_Y \quad . \tag{4}$$

This is the scattering amplitude for a set of two dipoles (x,x') and (y,y'). The amplitudes (3) and (4) take values between 0 (transparency) and the black-disk (saturation) limit 1.

To actually compute these amplitudes, one has to evaluate the averages $\langle\ .\ \rangle_Y$ which amounts to calculating averages of Wilson lines: $\langle f[A]\rangle_Y = \int \mathscr{D}A f[A]U_Y[A]$ where the target wavefunction $U_Y[A]$ represents the probability to find a given field configuration inside the target evolved at rapidity Y. The information contained in the target averages is mainly non-perturbative but the evolution towards higher rapidities $dU_Y[A]/dY$ can be computed perturbatively, at least in the leading-logarithmic approximation. Several equations have been established with different degrees of approximations, we shall not discuss them here and the reader can refer to [3] for more details. Let us only mention the Balitsky-Kovchegov saturation equation (BK) [4] which is a closed equation for $T_{q\bar{q}}$ obtained in a mean-field approximation. We shall refer to the BK equation later on when we link observables to the dipole amplitudes (3) and (4) and discuss phenomenology.

SATURATION PHENOMENOLOGY AT HERA

In deep inelastic scattering (DIS), a photon of virtuality Q^2 collides with a proton. In an appropriate "dipole" frame, the virtual photon undergoes the hadronic interaction via a fluctuation into a dipole (see Fig.2, left); the dipole then interacts with the target proton and one has the following factorization

$$\sigma_{DIS}(Q^2,Y) = \int d^2r\ \psi(|r|,Q^2)\ 2\int d^2b\ T_{q\bar{q}}\left(b-\frac{r}{2},b+\frac{r}{2};Y\right) \tag{5}$$

which relates the DIS total cross-section σ_{DIS} to the $q\bar{q}-$dipole amplitude $T_{q\bar{q}}$. The function $\psi(r,Q^2) = \int dz\ |\phi^\gamma(r,z;Q^2)|^2$ is obtained from the well-known wavefunction $\phi^\gamma(r,z;Q^2)$ which describes the splitting of the photon onto a dipole of transverse size r and with the antiquark carrying a fraction of photon longitudinal momentum z. Note that in this case, not all the information on $T_{q\bar{q}}$ is relevant as the impact parameter b is integrated out: only the cross-section $\sigma_{q\bar{q}}(r,Y) = 2\int d^2b\ T_{q\bar{q}}\left(b-\frac{r}{2},b+\frac{r}{2};Y\right)$ is needed.

Measurements of σ_{DIS} at HERA have had a great impact on saturation phenomenology, especially the discovery of geometric scaling [5]: the fact that $\sigma_{DIS}(Q^2,Y)$ is a function of the single variable $Q^2/Q_s^2(Y)$ with the saturation scale $Q_s^2(Y)\sim\exp(\lambda Y)$ and $\lambda\simeq 0.28$. Indeed this has a natural explanation in terms of traveling-wave solutions [6] of the BK equation $T_{q\bar{q}}(r,b=0,Y) = T_{q\bar{q}}(rQ_s(Y))$. However, this result is obtained neglecting the impact parameter dependence of $T_{q\bar{q}}$ and considering $\sigma_{q\bar{q}}(r,Y) = S_P\times T_{q\bar{q}}(r,b=0,Y)$ where S_P is the transverse area of the proton fitted to the data.

In order to understand better and study more consistently the impact parameter dependence of $T_{q\bar{q}}$, the authors of [7] have looked at diffractive vector-meson production (see Fig.2, center). In diffractive deep inelastic scattering, the proton gets out of the collision intact and there is a rapidity gap between that proton and the final state X. When

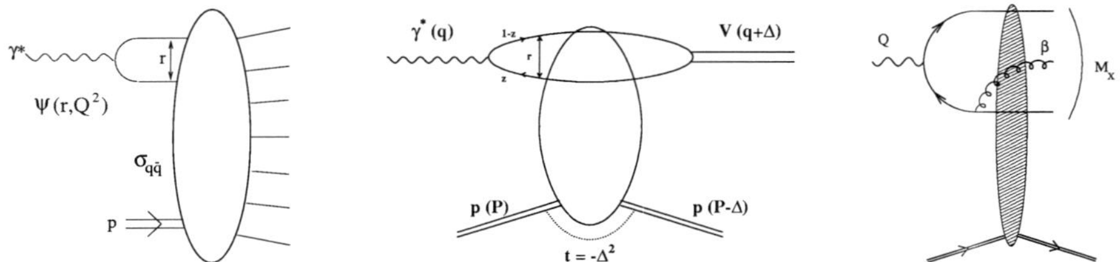

FIGURE 2. Three processes measured in virtual photon-proton collisions at HERA: DIS total cross-section (left), diffractive vector-meson production (center), and diffractive photon dissociation (right).

the final state is a vector meson, the momentum transfer Δ dependence of the cross-section is related to the impact parameter dependence of the dipole amplitude. Indeed the cross-section reads ($t=-\Delta^2$)

$$\frac{d\sigma}{dt}=\frac{1}{4\pi}\left|\int d^2r\ \Psi(|r|,Q^2,M_V^2)\ \int d^2b\ e^{ib.\Delta}\ T_{q\bar{q}}\Big(b-\frac{r}{2},b+\frac{r}{2};Y\Big)\right|^2 \tag{6}$$

where the function $\Psi(r,Q^2,M_V^2)=\int dz\ \phi^{\gamma}(r,z;Q^2)\phi^{V}(r,z;M_V^2)$ is obtained from both the photon wavefunction ϕ^{γ} and the final-state meson (whose mass has been denoted M_V) wavefunction ϕ^{V}. By analysing data on $\rho-$meson production at fixed Q^2 and $Y\simeq 7$, the authors extracted the dipole $S-$matrix $S_{q\bar{q}}(r,b;Y)=1-T_{q\bar{q}}\big(b-\frac{r}{2},b+\frac{r}{2};Y\big)$ as a function of b for a fixed size r_Q with $r_Q^2\sim 4/(Q^2+M_V^2)$. Three different sets of data at different Q^2 have been used. Their results are shown on the left plot of Fig.3; the shaded area on the left is an uncontrolled region due to the lack of large$-t$ data. The plot shows that the b dependence cannot be neglected and that $T_{q\bar{q}}(r\sim 1\ \text{GeV}^{-1},b\sim 0;Y\sim 7)\simeq 0.4$. This significant value of $T_{q\bar{q}}$ indicates that HERA could be entering the saturation regime.

As the importance of the impact parameter had been pointed out, a phenomenological model for the dipole amplitude $T_{q\bar{q}}$ with an impact parameter profile was proposed in [8]. With that saturation parametrization, the authors could well reproduce the data for diffractive J/Ψ production at HERA: the t spectrum (6) as well the the Q^2 and Y dependences of the total cross-section $\sigma_{J/\Psi}$. Their results are displayed on the center and right plots of Fig.3 where one can see the good agreement of the model with the data. A successful description of the same data using numerical simulations of the BK equation was also given in [9], confirming the compatibility of saturation predictions.

However in all the model descriptions of t spectra, the impact parameter dependence was introduced by hand as one had not extracted any information on the b depencence of $T_{q\bar{q}}$ from saturation equations. That moderated the impact of the results mentioned above. Interestingly, it was recently [10, 11] pointed out that important information can be obtained from the BK equation when looking at the Δ dependence of $\tilde{T}(r,\Delta;Y)=\int d^2b\ e^{ib.\Delta}\ T_{q\bar{q}}\big(b-\frac{r}{2},b+\frac{r}{2};Y\big)$. For instance, the geometric scaling property was extended: $\tilde{T}(r,\Delta;Y)=\tilde{T}(rQ_s(\Delta,Y))$ with $Q_s^2(\Delta,Y)\sim\Delta^2\exp(\lambda Y)$. Parametrizing the dipole amplitude with the momentum transfer instead of the impact parameter opens a new approach to analyse the data. An experimental confirmation of geometric scaling at non-zero momentum transfer would represent a significant success for saturation.

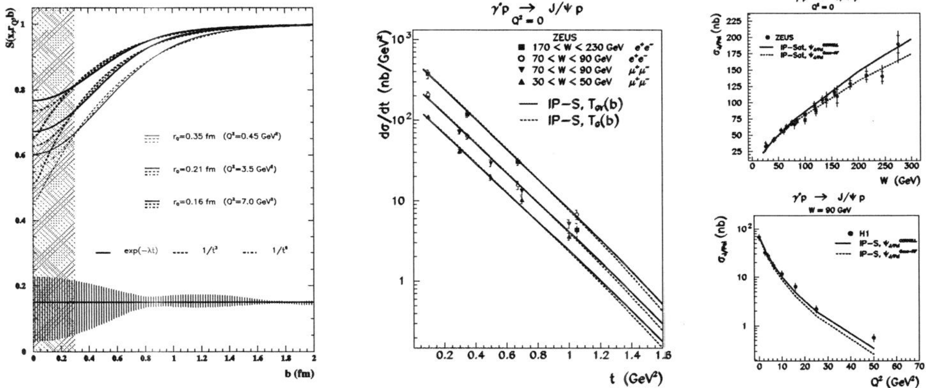

FIGURE 3. Left plot: the $q\bar{q}-$dipole $S-$matrix extracted from the diffractive $\rho-$meson production data in [7]; this shows the impact parameter dependence for three different dipole sizes r_Q and a rapidity $Y \simeq 7$. Center and right plots: diffractive J/Ψ production at HERA; $d\sigma/dt$ as a function of t (center), $\sigma_{J/\Psi}$ as a function of $W \sim \exp(Y/2)$ (top right) and Q^2 (bottom right); comparison with the model of [8] is shown.

Let us finally consider high-mass diffraction. If the final state diffractive mass M_X is much bigger than Q, the dominant configurations to the final state come from the $q\bar{q}g$ component of the photon wavefunction (see Fig.2, right) or from higher Fock states, i.e. from the photon dissociation. By contrast, if $M_X \ll Q$, the dominant configurations come from the $q\bar{q}$ component as it was the case for vector-meson production. Let us then consider the kinematical regime where $\beta \equiv Q^2/(Q^2+M_X^2) \ll 1$ and investigate the $q\bar{q}g$ component. The right plot of Fig.2 represents the diffractive production of a gluon with transverse momentum k and rapidity $\log(1/\beta)$ in the collision of the photon with the target proton. Provided k is a hard scale, the gluon momentum spectrum is given by [12]

$$\frac{M_X d\sigma}{d^2k dM_X} = \frac{\alpha_s N_c^2}{2\pi^2 C_F} \int d^2r\, \psi(|r|,Q^2) \int d^2b\, \mathbf{A}(k, r-\frac{b}{2}, r+\frac{b}{2}; \Delta\eta) \cdot \mathbf{A}^*(k, r-\frac{b}{2}, r+\frac{b}{2}; \Delta\eta) \tag{7}$$

where $\Delta\eta = Y - \log(1/\beta)$ is the rapidity gap. The two-dimensional vector $\mathbf{A}$ is given by

$$\mathbf{A}(k,x,x';\Delta\eta) = \int \frac{d^2z}{2\pi}\, e^{-ik.z} \left[\frac{z-x}{|z-x|^2} - \frac{z-x'}{|z-x'|^2}\right] \left(T_{q\bar{q}}^{(2)}(x,z;z,x';\Delta\eta) - T_{q\bar{q}}(x,x';\Delta\eta)\right). \tag{8}$$

Interestingly enough, independently of the form of the dipole amplitudes $T_{q\bar{q}}$ and $T_{q\bar{q}}^{(2)}$, the behavior of the observable $k^2\ d\sigma/d^2kdM_X$ as a function of the gluon transverse momentum k is the following [13]: it rises as k^2 for small values of k and falls as $1/k^2$ for large values of k. A maximum occurs for a value k_0 which is related to the inverse of the typical size for which the $T-$matrices approach one; in other words, the maximum k_0 reflects the scale at which unitarity sets in. If the energy is large enough so that the saturation scale Q_s is hard, unitarity will come as a consequence of parton saturation and $k_0 \sim Q_s$. If not the case, unitarity will be rather due to non-perturbative physics.

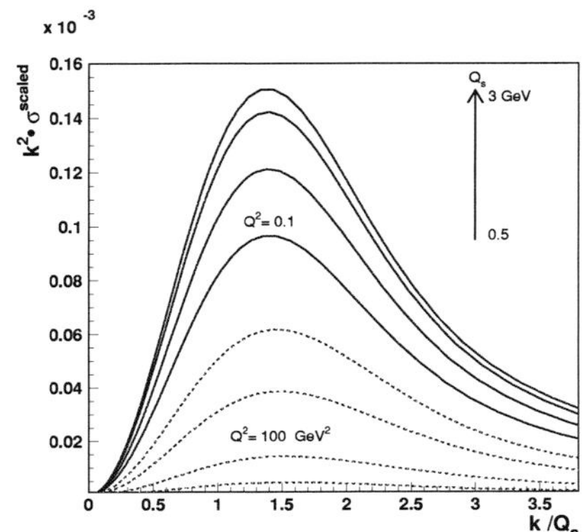

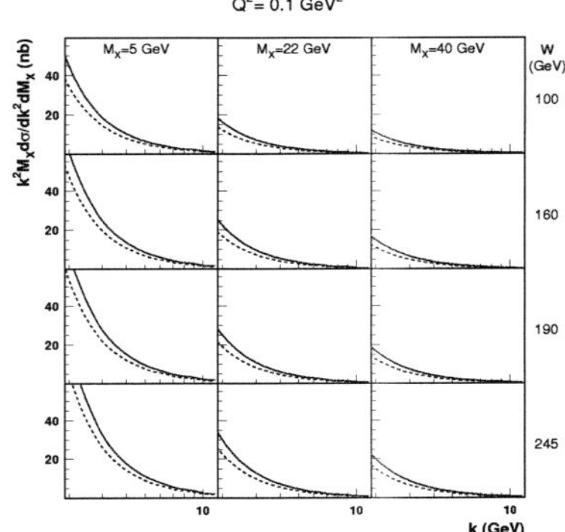

FIGURE 4. The diffractive gluon production cross-section $k^2 M_X d\sigma/dk^2 dM_X$. Left plot: as a function of the rescaled gluon transverse momentum k/Q_s for two extreme values of Q^2 equal to 0.1 and 100 GeV^2 and four values of the saturation scale $Q_s = 0.5, 1, 2, 3$ GeV. Right plot: as a function of the jet transverse momentum k and in the HERA energy range for $Q^2 = 0.1$ GeV^2 and different values of diffractive mass M_X and energy W; full lines: only the light quarks are included in ψ, dashed lines: charm is also included.

In the saturation case, the model of [13] for the dipole amplitudes allows to plot the whole k spectrum (7). This is shown on the left plot of Fig.4 and one can indeed see that the spectrum features a maximum peaked around $k_0 \simeq 1.4\ Q_s$ independently of Q^2 and Q_s. Measuring this cross-section at HERA would offer a unique opportunity to test if saturation plays a role in diffraction at the present energies. On the right plot of Fig.4, the cross-section is plotted in the HERA energy range for different values of M_X and total energy $W \sim e^{Y/2}$, corresponding to different values of Q_s. The saturation scale is the one extracted [14] from the F_2 data. As expected for realistic jet transverse momenta, $k > Q_s$ and the data would lie on the perturbative side of the bump. There is a big difference in the rise towards the bump between the lowest (top right) and highest (bottom left) Q_s bins. A confirmation of this behavior would certainly favor the saturation scenario.

REFERENCES

1. L. V. Gribov, E. M. Levin and M. G. Ryskin, *Phys. Rep.* **100** (1983) 1.
2. A. Hebecker, *Phys. Rept.* **331** (2000) 1; A. Kovner and U. Wiedemann, *Phys. Rev.* **D64** (2001) 114002.
3. L. McLerran, these proceedings; R. Peschanski, these proceedings.
4. I. Balitsky, *Nucl. Phys.* **B463** (1996) 99; Y.V. Kovchegov, *Phys. Rev.* **D60** (1999) 034008.
5. A. M. Stasto, K. Golec-Biernat and J. Kwiecinski, *Phys. Rev. Lett.* **86** (2001) 596.
6. S. Munier and R. Peschanski, *Phys. Rev. Lett.* **91** (2003) 232001; *Phys. Rev.* **D69** (2004) 034008.
7. S. Munier, A. M. Stasto and A. H. Mueller, *Nucl. Phys.* **B603** (2001) 427.
8. H. Kowalski and D. Teaney, *Phys. Rev.* **D68** (2003) 114005.
9. E. Gotsman, E. Levin, M. Lublinsky, U. Maor and E. Naftali, *Acta Phys. Polon.* **B34** (2003) 3255.
10. C. Marquet, R. Peschanski and G. Soyez, *Nucl. Phys.* **A756** (2005) 399.
11. C. Marquet and G. Soyez, *Nucl. Phys.* **A760** (2005) 208.
12. C. Marquet, *Nucl. Phys.* **B705** (2005) 319; *Nucl. Phys.* **A755** (2005) 603c.
13. K. Golec-Biernat and C. Marquet, *Phys. Rev.* **D71** (2005) 114005; hep-ph/0509034.
14. K. Golec-Biernat and M. Wüsthoff, *Phys. Rev.* **D59** (1999) 014017; *Phys. Rev.* **D60** (1999) 114023.

Jet Physics and the Underlying Event at the Tevatron

Rick Field
(for the CDF & D0 Collaborations)

Department of Physics, University of Florida, Gainesville, Florida, 32611, USA

Abstract: Tevatron Run 2 results on the inclusive jet cross section (MidPoint and K_T algorithm) and the b-jet and $b\bar{b}$-jet cross section (MidPoint algorithm) are presented and compared with theory. The CDF b-jet $\bar{b}$-jet $\Delta\phi$ distribution is compared with theory and with the D0 jet#1-jet#2 $\Delta\phi$ distribution. The understanding and modeling of the "underlying event" in Run 2 at the Tevatron is reviewed and new CDF results are presented.

Keywords: QCD, Jets, Hadron Collider.
PACS: 12.38.-t, 12.38.Bx, 12.38.Qk.

The study of proton-antiproton collisions in Run 2 at the Tevatron is teaching us a lot about how QCD works. Comparing data with theory will lead to improved QCD Monte-Carlo models and to more precise parton distribution functions. In Run 2 at the Tevatron we are studying the inclusive jet cross section using both the MidPoint cone algorithm and the K_T algorithm [1]. We are studying heavy flavor jets (*i.e.* b-jets) and jets produced in association with photons, W bosons, and Z bosons. We are studying jet-jet correlations, jet fragmentation (jet shapes, momentum distributions, two-particle correlations), and we are making good progress in understanding and modeling the "underlying event" in hard scattering processes. Here I will only be able to show a little bit of what we have learned.

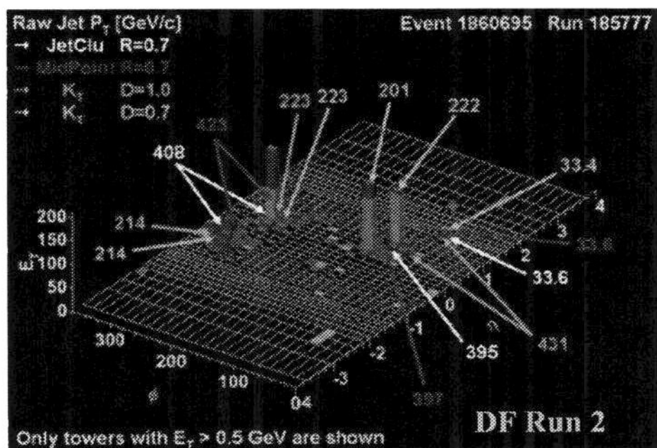

FIGURE 1. Shows the transverse energy of calorimeter towers with $E_T > 0.5$ GeV for an event in the CDF detector. The MidPoint algorithm combines the two clusters into one "jet" with $p_T = 423$ GeV/c while the K_T algorithm (D = 0.7) finds two "jets" with $p_T = 223$ GeV/c and 214 GeV/c.

Experimentally we measure "jets" at the detector (*i.e.* calorimeter) level by observing the energy in each calorimeter cell as illustrated in Fig. 1. Of course the

CP828, *Multiparticle Dynamics*
edited by V. Šimák, M. Šumbera, Š. Todorova, and B. Tomášik

"jet" cross section depends on ones choice of jet algorithm. Each jet algorithm is a different observable and comparing the results of different jet algorithm teaches us about QCD. Of course, what is measured in the calorimeter must be corrected for detector efficiency which is done by comparing the QCD Monte-Carlo models at the particle (*i.e.* generator level) with the result after detector simulation. I believe that experimenters should publish what they measure (*i.e.* observables at the particle level with the "underlying event"). However, to compare with NLO parton level calculations one must go one step further. The NLO parton level does not have fragmentation or an "underlying event" (*i.e.* beam-beam remnants, initial and final-state parton showers, multiple-parton interactions, resonance decays, etc.). There are three approaches for comparing data corrected to the particle level (*i.e.* hadron level) with parton level calculations. The first approach is to neglect the difference and to compare the hadron level data directly with the parton level calculation. Fig. 2 shows the inclusive jet cross section using the MidPoint algorithm (R = 0.7, f_{merge} = 0.5) for two rapidity bins as measured by D0. D0 compares the experimentally measured hadron level prediction directly with the NLO parton level theory curves and assumes that the parton level to hadron level corrections are small for jets above 50 GeV. The agreement between the parton-level theory prediction and the measured hadron-level is quite good over 10 decades!

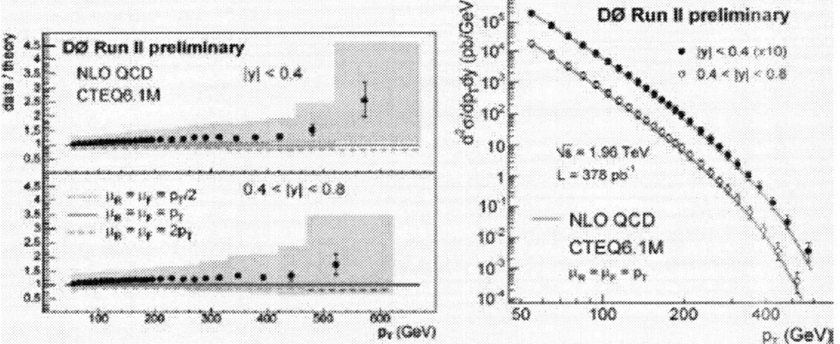

FIGURE 2. The D0 Run 2 inclusive jet cross section using the MidPoint algorithm (R = 0.7, f_{merge} = 0.50) compared with parton-level NLO QCD. The hadron-level data are compared directly with the parton-level NLO QCD.

Another approach for comparing what is measured at the particle level in the detector with the NLO parton level theory is to use the QCD Monte-Carlo models and try to extrapolate the data to the parton level. This requires removing the "underlying event" and correcting for fragmentation effects. Fig. 3 shows the inclusive jet cross section using the MidPoint algorithm (R = 0.7, f_{merge} = 0.75) in the central region as measured by CDF compared with the parton level NLO QCD prediction, where the data have been extrapolated (*i.e.* corrected) to the parton level. Fig. 3 shows that the hadron level to parton level correction factors are significant for P_T(jet) < 300 GeV/c (they come mostly from the "underlying event"). The agreement between the theory and data is very good.

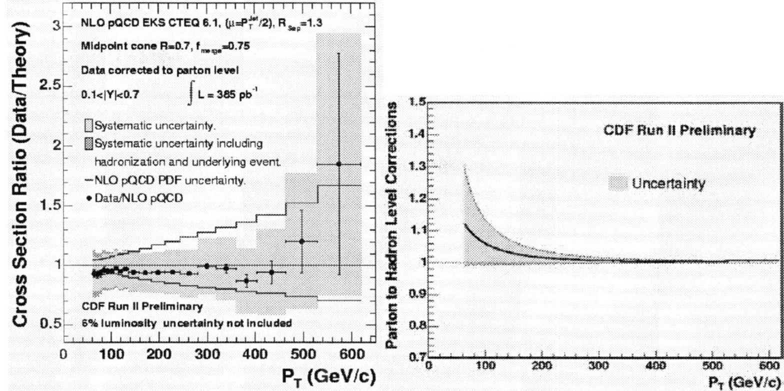

FIGURE 3. The CDF Run 2 inclusive jet cross section using the MidPoint algorithm (R = 0.7, f_{merge} = 0.75) compared with parton-level NLO QCD (*left*). The data have been extrapolated (*i.e.* corrected) to the parton level using the parton to hadron correction factor (*right*). The hadron-level data are multiplied by the reciprocal of this factor.

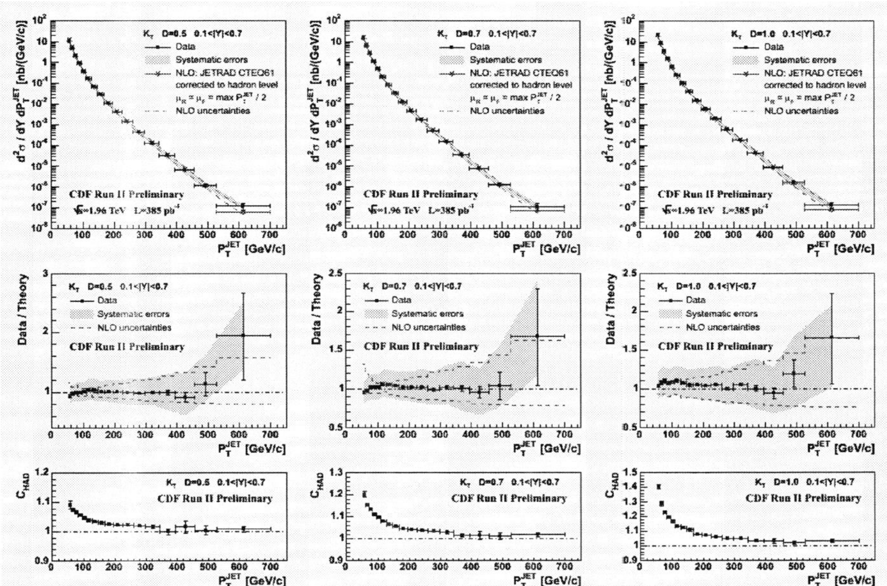

FIGURE 4. The CDF Run 2 inclusive jet cross section using the K_T algorithm with D = 0.5, 0.7, and 1.0. The data are at the particle level (with an "underlying event") and the NLO parton level (CTEQ61M) has been corrected for fragmentation effects and for the "underlying event" (with correction factors C_{HAD}).

A third approach for comparing what is measured at the particle level in the detector with the NLO parton level theory is to use the QCD Monte-Carlo models to correct the NLO parton level theory by adding in the effects of fragmentation and the "underlying event". I prefer this approach. It is much better to correct the theory to the hadron level (with an "underlying event") than it is to extrapolate a perfectly good experimental observable to something that is not observable (*i.e.* parton level). Fig. 4

shows the CDF Run 2 inclusive jet cross section using the K_T algorithm. Here the data are at the particle (*i.e.* hadron level) and the NLO parton level theory has been corrected to the particle level. As for the MidPoint algorithm, the parton level to hadron level corrections are significant for P_T(jet) < 300 GeV/c (coming mostly from the "underlying event"). The agreement between the theory and data is good. Most theorists prefer the K_T algorithm over cone algorithms, however, it must be demonstrate that the K_T algorithm will work in the collider environment where there is an "underlying event". Fig. 4 shown that the K_T algorithm works fine at the Tevatron The parton to hadron correction factors for the K_T algorithm are similar to the MidPoint algorithm correction factors.

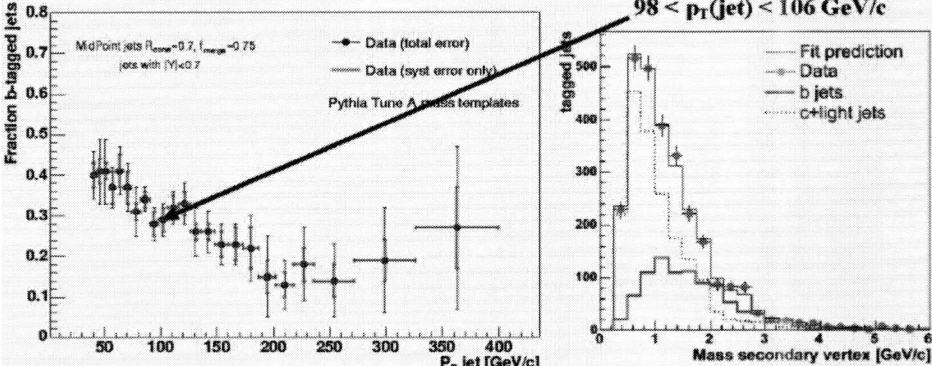

FIGURE 5. (*left*) Shows the fraction of b-tagged jets as a function of the jet p_T. (*right*) Shows the fit to the secondary vertex mass for the bin 98 < p_T(jet) < 106 GeV/c.

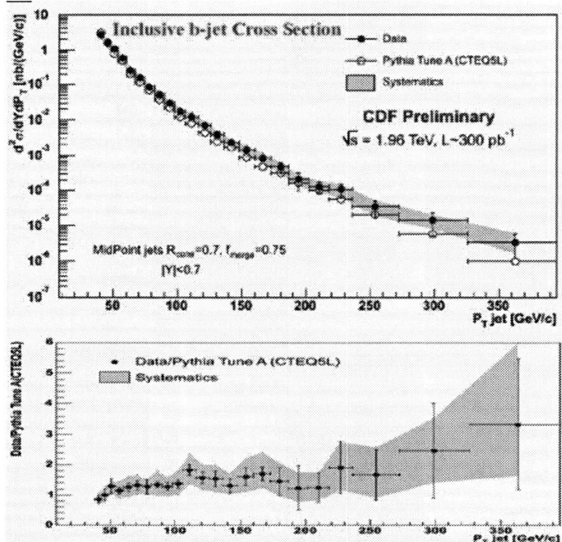

FIGURE 6. (*top*) Shows the CDF Run 2 b-jet inclusive cross section at 1.96 TeV compared with PYTHIA Tune A. (*bottom*) Shows the ratio data/theory for PYTHIA Tune A.

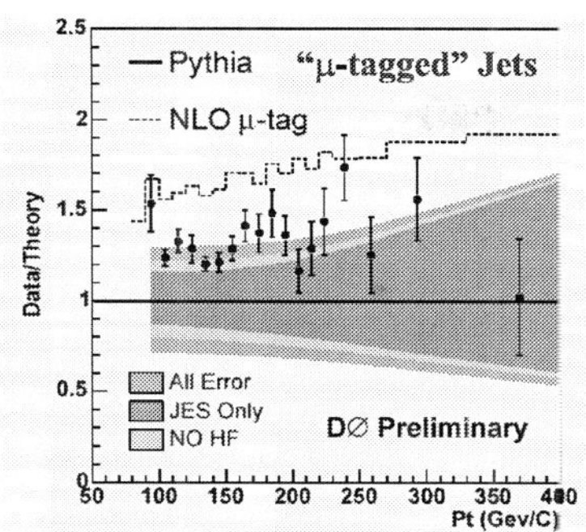

FIGURE 7. Shows the DO Run 2 "μ-tagged" jet inclusive cross section at 1.96 TeV compared with PYTHIA and NLO QCD.

At CDF b-jets are identified by studying the invariant mass of the charged particles emanating from the secondary vertex which is displaced slightly from the primary interaction vertex due to the long lifetime of the heavy b-quark. As shown in Fig. 5, the fraction of b-tagged jets is determined by fitting (on a bin-by-bin bases) the secondary vertex invariant mass to templates determined from PYTHIA Tune A [2-3]. Fig. 6 shows the resulting CDF b-jet inclusive cross section at 1.96 TeV compared with PYTHIA Tune A. The ratio of the data to PYTHIA Tune A is constant with a value of about 1.4. We expect that NLO corrections will account the factor of 1.4

At D0 they study heavy flavor jets by requiring a muon in a jet (*i.e.* inside R = 0.5). Searching for muons in jets enhances the heavy flavor content of the jet. Fig 7 shows the DO Run 2 "μ-tagged" jet inclusive cross section compared with PYTHIA Tune A and NLO QCD. The ratio of the data to PYTHIA is constant with a value of about 1.2, which is similar to the CDF result.

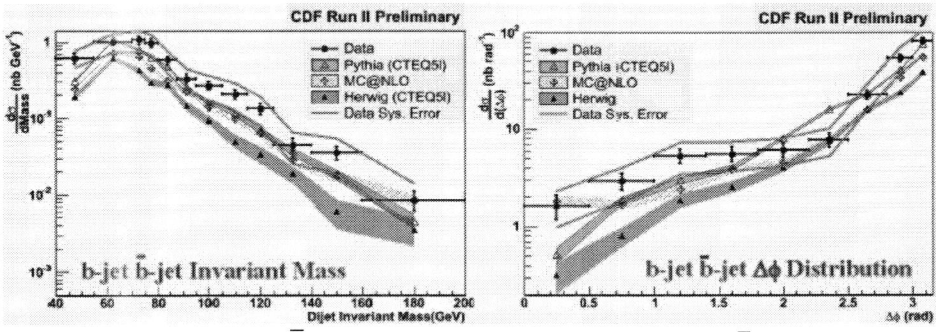

FIGURE 8. Shows the CDF Run 2 $b\bar{b}$ dijet invariant mass distribution (left) and the b -jet $\bar{b}$ -jet $\Delta\phi$ distribution (right) at 1.96 TeV compared with PYTHIA Tune A, HERWIG, and MC@NLO.

Fig. 8 shows the CDF Run 2 $b\bar{b}$ dijet invariant mass distribution at 1.96 TeV compared with PYTHIA Tune A, HERWIG [4], and MC@NLO [5] and Table 1 shows the integrated $b\bar{b}$ dijet cross section. PYTHIA Tune A fits the data better than HERWIG or MC@NLO. The is because PYTIA Tune A has been tuned to fit the "underlying event" at the Tevatron by adjusting the multiple-parton interactions.

HERWIG and MC@NLO (with uses HERWIG) do not include multiple-parton interactions do not have enough activity in the "underlying event". JIMMY [6] is a model of multiple parton interaction which can be combined with HERWIG (and MC@NLO) to enhance the "underlying event" thereby improving the agreement with data. When JIMMY is added to MC@NLO then agreement is improved. Both the inclusive jet cross section and the b-jet cross section depend sensitively on the "underlying event".

Table 1. The CDF Run 2 integrated $b\bar{b}$ dijet cross section (E_T(b-jet#1) > 30 GeV, E_T(b-jet#2) > 20 GeV, $|\eta$(b-jets)$| < 1.2$) at 1.96 TeV compared with PYTHIA Tune A, HERWIG, MC@NLO, and MC@NLO + JIMMY).

CDF (preliminary)	34.5 ± 1.8 ± 10.5 nb
PYTHIATuneA (CTEQ5L)	38.7 ± 0.6 nb
HERWIG (CTEQ5L)	21.5 ± 0.7 nb
MC@NLO	28.5 ± 0.6 nb
MC@NLO+ JIMMY	35.7 ± 2.0 nb

Fig. 8 also shows the b-jet $\bar{b}$-jet $\Delta\phi$ distribution compared with PYTHIA Tune A, HERWIG, and MC@NLO. PYTHIA Tune A and MC@NLO do a good job in describing the $b\bar{b}$ $\Delta\phi$ distribution. It is not an accident that PYTHIA Tune A roughly agrees with the data. I tuned the initial-state radiation in PYTHIA Tune A (*i.e.* PARP(67)) to agree with the CDF Run 1 $b\bar{b}$ $\Delta\phi$ distribution [7]. For MC@NLO the agreement is a prediction. For PYTHIA Tune A the agreement is a "tune", but it does show consistency between the CDF Run 1 analysis and the preliminary Run 2 results.

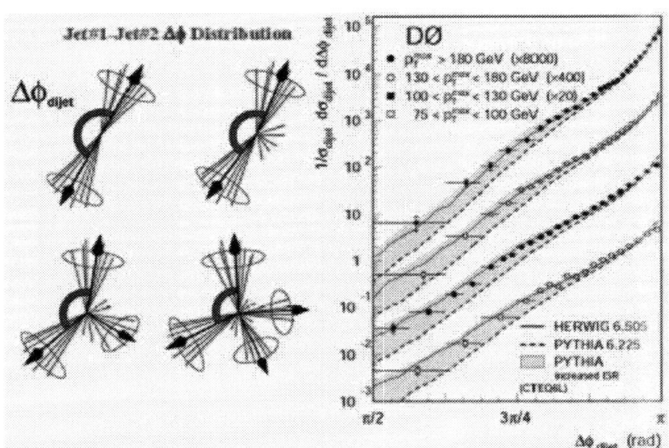

FIGURE 9. Shows the D0 Run 2 jet#1-jet#2 $\Delta\phi$ distribution at 1.96 TeV compared with PYTHIA (*default*) and PYTHIA Tune A (upper edge of the shaded regions). Jet#1 and jet#2 are the leading two jets (MidPoint algorithm, R = 0.7, f_{merge} = 0.5).

Fig. 9 shows the D0 Run 2 jet#1-jet#2 $\Delta\phi$ distribution at 1.96 TeV, where jet#1 and jet#2 are the leading two jets. Here again PYTHIA Tune A does a good job in describing the data.

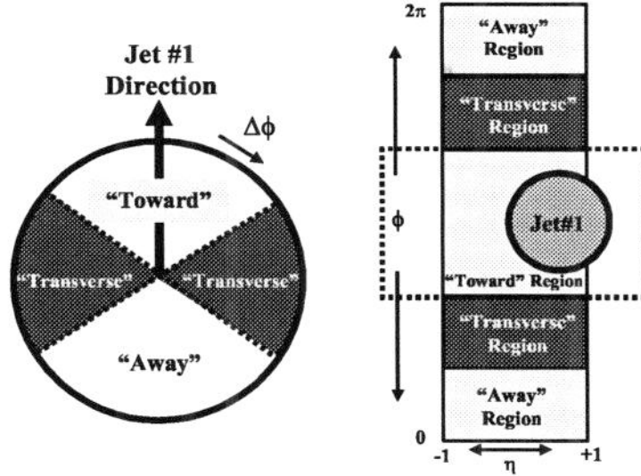

FIGURE 10. Illustration of correlations in azimuthal angle $\Delta\phi$ relative to the direction of the leading jet (MidPoint, R = 0.7, f_{merge} = 0.75) in the event, jet#1. The angle $\Delta\phi = \phi - \phi$jet#1 is the relative azimuthal angle between charged particles (or calorimeter towers) and the direction of jet#1. The "transverse" region is defined by $60^\circ < |\Delta\phi| < 120^\circ$ and $|\eta| < 1$. We examine charged particles in the range $p_T > 0.5$ GeV/c and $|\eta| < 1$ and calorimeter towers with $E_T > 0.1$ GeV and $|\eta| < 1$, but allow the leading jet to be in the region $|\eta(\text{jet\#1})| < 2$.

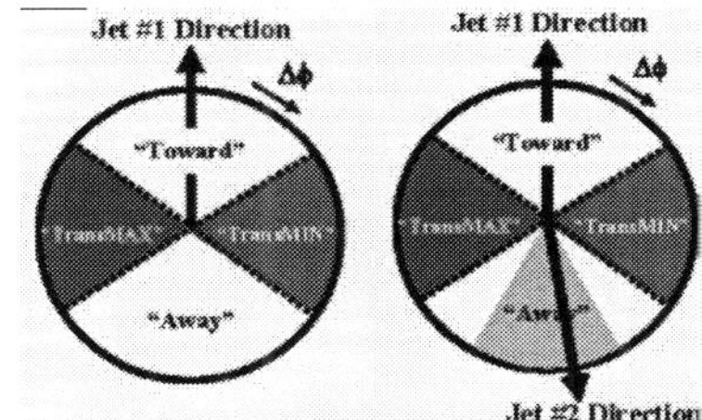

FIGURE 11. Illustration of correlations in azimuthal angle $\Delta\phi$ relative to the direction of the leading jet (highest P_T jet) in the event, jet#1 for "leading jet" events (*left*) and "back-to-back" events (*right*). Events in which there are no restrictions placed on the on the second highest P_T jet, jet#2, are referred to as "leading jet" events. Events with at least two jets where the leading two jets are nearly "back-to-back" ($\Delta\phi_{12} > 150^\circ$) with P_T(jet#2)/P_T(jet#1) > 0.8 and P_T(jet#3) < 15 GeV/c are referred to as "back-to-back" events. In both cases the angle $\Delta\phi = \phi - \phi$jet#1 is the relative azimuthal angle between charged particles (or calorimeter towers) and the direction of jet#1. On an event by event basis, we define "transMAX" ("transMIN") to be the maximum (minimum) of the two "transverse" regions, $60^\circ < \Delta\phi < 120^\circ$ and $60^\circ < -\Delta\phi < 120^\circ$. "TransMAX" and "transMIN" each have an area in η-ϕ space of $\Delta\eta\Delta\phi = 4\pi/6$. The overall "transverse" region defined in Fig. 10 includes both the "transMAX" and the "transMIN" region.

We have seen that both the inclusive jet cross section and the b-jet cross section depend sensitively on the "underlying event". At CDF we are working to understand and model the "underlying event" at the Tevatron. We use the topological structure of hadron-hadron collisions to study the "underlying event" [8-10]. The direction of the leading calorimeter jet is used to isolate regions of η-ϕ space that are sensitive to the "underlying event". As illustrated in Fig. 10, the direction of the leading jet, jet#1, is used to define correlations in the azimuthal angle, $\Delta\phi$. The angle $\Delta\phi = \phi - \phi_{jet\#1}$ is the relative azimuthal angle between a charged particle (or a calorimeter tower) and the direction of jet#1. The "transverse" region is perpendicular to the plane of the hard 2-to-2 scattering and is therefore very sensitive to the "underlying event". Furthermore, we consider two classes of events. We refer to events in which there are no restrictions placed on the second and third highest P_T jets (jet#2 and jet#3) as "leading jet" events. Events with at least two jets with $P_T > 15$ GeV where the leading two jets are nearly "back-to-back" ($|\Delta\phi_{12}| > 150^o$) with P_T(jet#2)/P_T(jet#1) > 0.8 and P_T(jet#3) < 15 GeV are referred to as "back-to-back" events. "Back-to-back" events are a subset of the "leading jet" events. The idea here is to suppress hard initial and final-state radiation thus increasing the sensitivity of the "transverse" region to the "beam-beam remnant" and the multiple parton scattering component of the "underlying event".

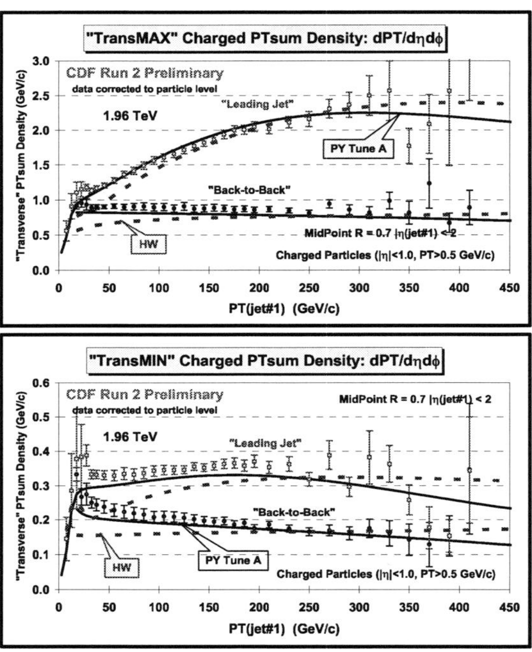

FIGURE 12. CDF Run 2 data at 1.96 TeV on scalar PTsum density of charged particles, dPTsum/dηdφ, with $p_T > 0.5$ GeV/c and $|\eta| < 1$ in the "transMAX" region (*top*) and the "transMIN" region (*bottom*) for "leading jet" and "back-to-back" events defined in Fig. 11 as a function of the leading jet P_T compared with PYTHIA Tune A and HERWIG. The data are corrected to the particle level (with errors that include both the statistical error and the systematic uncertainty) and compared with the theory at the particle level (*i.e.* generator level).

As illustrated in Fig. 11, we define a variety of MAX and MIN "transverse" regions which helps separate the "hard component" (initial and final-state radiation) from the "beam-beam remnant" component. MAX (MIN) refer to the "transverse" region containing the largest (smallest) scalar p_T sum of charged particles or the region containing the largest (smallest) scalar E_T sum of particles. Since we will be studying regions in η-φ space with different areas, we construct densities by dividing by the area. For example, the PTsum density, dPTsum/dηdφ, corresponds the amount of charged particle ($p_T > 0.5$ GeV/c) scalar p_T sum per unit η-φ, and the transverse energy density, $dE_T/d\eta d\phi$, corresponds the amount of scalar E_T sum of all particles per unit η-φ. One expects that "transMAX" will pick up the hardest initial or final-state radiation while both "transMAX" and "transMIN" should receive "beam-beam remnant" contributions. Hence one expects "transMIN" to be more sensitive to the "beam-beam remnant" component of the "underlying event", while the "transMAX" minus the "transMIN" (*i.e.* "transDIF") is very sensitive to initial and final-state radiation. This idea, was first suggested by Bryan Webber, and implemented by in a paper by Jon Pumplin [11]. Also, Valaria Tano studied this in her CDF Run 1 analysis of maximum and minimum transverse cones [12].

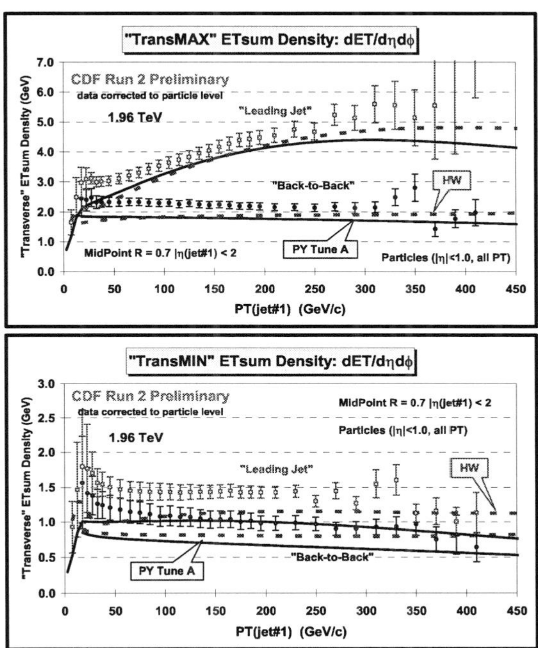

FIGURE 13. CDF Run 2 data at 1.96 TeV on the ETsum density, $dE_T/d\eta d\phi$, for particles with $|\eta| < 1$ in the "transMAX" region (*top*) and the "transMIN" region (*bottom*) for "leading jet" and "back-to-back" events defined in Fig. 11 as a function of the leading jet P_T compared with PYTHIA Tune A and HERWIG. The data are corrected to the particle level (with errors that include both the statistical error and the systematic uncertainty) and compared with the theory at the particle level (*i.e.* generator level).

Fig. 12 compares the data on the density of charged particles and the charged PTsum density in the "transverse" region corrected to the particle level ($p_T > 0.5$) for "leading jet" and "back-to-back" events with PYTHIA Tune A and HERWIG at the particle level. As expected, the "leading jet" and "back-to-back" events behave quite differently. For the "leading jet" case the "transMAX" densities rise with increasing P_T(jet#1), while for the "back-to-back" case they fall with increasing P_T(jet#1). The rise in the "leading jet" case is, of course, due to hard initial and final-state radiation, which has been suppressed in the "back-to-back" events. The "back-to-back" events allow for a more close look at the "beam-beam remnant" and multiple parton scattering component of the "underlying event" and PYTHIA Tune A (with multiple parton interactions) does a better job describing the data than HERWIG (without multiple parton interactions).

The "transMIN" densities are more sensitive to the "beam-beam remnant" and multiple parton interaction component of the "underlying event". The "back-to-back" data show a decrease in the "transMIN" densities with increasing P_T(jet#1) which is described fairly well by PYTHIA Tune A (with multiple parton interactions) but not by HERWIG (without multiple parton interactions). The decrease of the "transMIN" densities with increasing P_T(jet#1) for the "back-to-back" events is very interesting and might be due to a "saturation" of the multiple parton interactions at small impact parameter. Such an effect is included in PYTHIA Tune A but not in HERWIG (without multiple parton interactions).

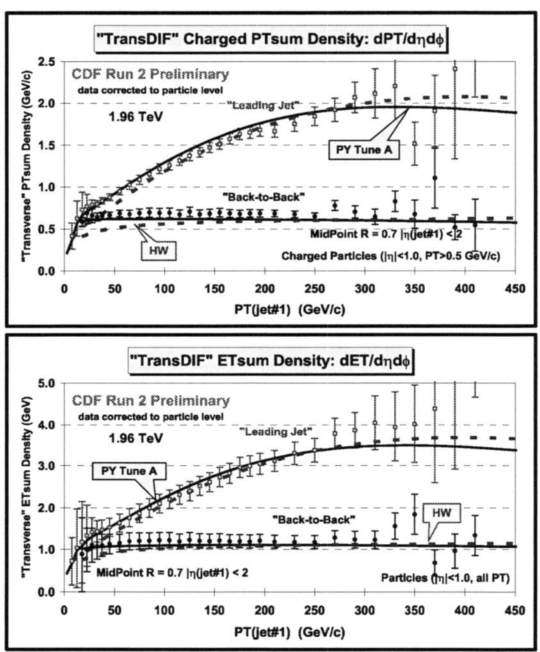

FIGURE 14. CDF Run 2 data at 1.96 TeV on the difference of the "transMAX" and "transMIN" region ("transDIF" = "transMAX" minus "transMIN") for "leading jet" and "back-to-back" events defined in Fig. 11 as a function of the leading jet P_T compared with PYTHIA Tune A and HERWIG. The data are corrected to the particle level (with errors that include both the statistical error and the systematic uncertainty) and compared with the theory at the particle level (*i.e.* generator level).

Fig. 13 shows the data corrected to the particle level for the ET_{sum} density, $dE_T/d\eta d\phi$, in the "transverse" region for "leading jet" and "back-to-back" events compared with PYTHIA Tune A and HERWIG at the particle level. The data on scalar ET_{sum} density has been corrected to correspond to all particles (all p_T, $|\eta| < 1$). Neither PYTHIA Tune A or HERWIG produce enough energy in the "transverse" region. HERWIG has more "soft" particles than PYTHIA Tune A does slightly better in describing the energy density in the "transMAX" and "transMIN" region.

Fig. 14 shows the difference of the "transMAX" and "transMIN" region ("transDIF" = "transMAX" minus "transMIN") for "leading jet" and "back-to-back" events compared with PYTHIA Tune A and HERWIG. "TransDIF" is more sensitive to the hard scattering component of the "underlying event" (*i.e.* initial and final state radiation).

CONCLUSIONS

We have measured the inclusive jet cross section at the Tevatron using the MidPoint algorithm with R = 0.7 and f_{merge} = 0.5 (D0), using the MidPoint algorithm with R = 0.7 and f_{merge} = 0.75 (CDF), and using the K_T algorithm(CDF). The data agree well with the NLO parton-level theory after the data is extrapolated to the parton-level or the parton-level theory is corrected for the "underlying event" and fragmentation. The K_T algorithm works fine at the Tevatron collider, which has positive implications for the LHC.

We have also measured the b-jet inclusive cross section and the $b\bar{b}$ dijet invariant mass distribution. There measurements at hadron colliders provide an important test of QCD. Past Tevatron measurements of b-quark production indicated a possible "excess" with respect to QCD predictions. However, the b-jet cross is in agreement expectations. The data are about a factor of 1.4 larger than the prediction of PYTHIA Tune A, however, this is to be expected since PYTHIA is a "leading log order" model. One cannot expect it correctly predict the precise amount of "flavor excitation" and "gluon splitting". We are working on the comparisons with MC@NLO [5].

CDF has done extensive studies of the "underlying event" at the Tevatron. PYTHIA Tune A (with multiple parton interactions) does a good job in describing the charged particles (p_T > 0.5 GeV/c, $|\eta|$ < 1) in the "underlying event" (*i.e.* "transverse" regions) for both "leading jet" and "back-to-back" events. HERWIG (without multiple parton interactions) does not have enough activity in the "underlying event" for P_T(jet#1) less than about 150 GeV. Both PYTHIA Tune A and HERWIG underestimate the energy density in the "transMAX" and "transMIN" regions. However, they both fit the "transDIF" energy density. This indicates that the excess energy density seen in the data probably arises from the "soft" component of the "underlying event" (*i.e.* beam-beam remnants and/or multiple parton interactions).

We see interesting dependence of the "underlying event" on the transverse momentum of the leading jet (*i.e.* the Q^2 of the hard scattering). For the "leading jet" case the "transMAX" densities rise with increasing P_T(jet#1), while for the "back-to-back" case they fall with increasing P_T(jet#1). The rise in the "leading jet" case is due to hard initial and final-state radiation, which has been suppressed in the "back-to-back" events. The "back-to-back" data show a decrease in the "transMIN" densities with increasing P_T(jet#1). The decrease of the "transMIN" densities with increasing P_T(jet#1) for the "back-to-back" events is very interesting and might be due to a "saturation" of the multiple parton interactions at small impact parameter. Such an effect is included in PYTHIA Tune A (with multiple parton interactions) but not in HERWIG (without multiple parton interactions). PYTHIA Tune A does predict this decrease, while HERWIG shows an increase (due to increasing initial and final state radiation).

REFERENCES

1. S. D. Ellis and D. E. Soper, Phys. Rev. **D48**, 3160 (1993).
2. *Min-Bias and the Underlying Event at the Tevatron and the LHC*, talk by R. Field at the Fermilab ME/MC Tuning Workshop, Fermilab, October 4, 2002. *Toward an Understanding of Hadron Collisions: From Feynman-Field until Now*, talk by R. Field at the Fermilab Joint Theoretical Experimental "Wine & Cheese" Seminar, Fermilab, October 4, 2002.
3. T. Sjostrand, Phys. Lett. **157B**, 321 (1985); M. Bengtsson, T. Sjostrand, and M. van Zijl, Z. Phys. **C32**, 67 (1986); T. Sjostrand and M. van Zijl, Phys. Rev. **D36**, 2019 (1987).
4. G. Marchesini and B. R. Webber, Nucl. Phys **B310**, 461 (1988); I. G. Knowles, Nucl. Phys. **B310**, 571 (1988); S. Catani, G. Marchesini, and B. R. Webber, Nucl. Phys. **B349**, 635 (1991).
5. *The MC and NLO 3.1 Event Generator*, Stefano Frixione and Bryan R. Webber, CAVENDISH-HEP-05-09, hep-ph/0506182 (2005). *Matching NLO QCD and Parton Showers in Heavy Flavor Production*, Stefano Frixione, Paolo Nason, and Bryan R. Webber, JHEP 0308:007 (2003).
6. *Multiparton Interactions in Photoproduction at HERA*, J.M. Butterworth, J.R. Forshaw, and M.H. Seymour, Z. Phys. **C7**, 637-646 (1996).
7. *Measurements of Bottom Anti-Botootm Azimuthal Production Correlations in Proton-Antiproton Collisions at 1.8 TeV*, CDF Collaboration (D. Acosta et al.), Phys. Rev. **D71**, 092001 (2005),
8. *Charged Jet Evolution and the Underlying Event in Proton-Antiproton Collisions at 1.8 TeV*, The CDF Collaboration (T. Affolder et al.), Phys. Rev. **D65**, 092002 (2002).
9. *The Underlying Event in Large Transverse Momentum Charged Jet and Z-boson Production at 1.8 TeV*, talk presented by Rick Field at DPF2000, Columbus, OH, August 11, 2000.
10. *A Comparison of the Underlying Event in Jet and Min-Bias Events*, talk presented by Joey Huston at DPF2000, Columbus, OH, August 11, 2000. *The Underlying Event in Jet and Minimum Bias Events at the Tevatron*, talk presented by Valeria Tano at ISMD2001, Datong, China, September 1-7, 2001.
11. *Hard Underlying Event Corrections to Inclusive Jet Cross-Sections*, Jon Pumplin, Phys. Rev. **D57**, 5787 (1998).
12. *The Underlying Event in Hard Interactions at the Tevatron Proton-Antiproton Collider*, CDF Collaboration (D. Acosta et al.), Phys. Rev. **D70**, 072002 (2004).

Forward jet production at HERA

Leif Jönsson

Physics Department, Lund University
Box 118
SE-221 00 Lund, Sweden
e-mail: leif.jonsson@hep.lu.se

Abstract. Measurements in the low Bjorken-x region of deep inelastic electron-proton scattering events, containing a jet of high transverse momentum close to the flight direction of the proton, offer high sensitivity to BFKL-like dynamics. The results of recent measurements at HERA by the H1 and ZEUS experiments are presented.

Keywords: ep-scattering, DIS, HERA H1, ZEUS, forward jet
PACS: 13.60.Hb,13.87.-a

INTRODUCTION

Deep inelastic scattering (DIS) of electrons against protons offers unique possibilities to study the partonic structure of matter, especially since the hardness of the interaction can be varied by selecting the energy and virtuality of the exchanged photon. HERA has extended the available region in the Bjorken scaling variable, x_{Bj}, down to values of $x_{Bj} \simeq 10^{-4}$, for values of the four momentum transfer squared, Q^2, larger than a few GeV^2, where perturbative calculations in QCD are expected to be valid. At these low x_{Bj} values, a parton in the proton can induce a QCD cascade, consisting of several subsequent parton emissions, before eventually an interaction with the virtual photon takes place. Measurement of the forward jet production cross section at small x_{Bj} is of special interest for studies of parton dynamics, since jets emitted close to the proton direction lie well away in rapidity from the photon end of the evolution ladder [1, 2] .

THEORETICAL CALCULATIONS AND QCD BASED MODELS

Fixed order QCD calculations have been performed up to NLO accuracy. The corresponding order in α_S is always related to the observable, in the sense that NLO inclusive jet production means $\mathcal{O}(\alpha_S^1)$, whereas for di-jet production it means $\mathcal{O}(\alpha_S^2)$. Expressed in general terms, LO is the lowest order in which the observable obtains a non-zero value, and NLO is the next higher order. The DISENT [3, 4] and DISASTER++ [5, 6] programs calculate LO(α_S) and NLO(α_S^2) calculations of di-jet production via direct photon interaction. Both programs use the subtraction method in the numerical integration to handle singularities due to soft and collinear radiation. The MEPJET [7] and JetViP [8, 9] programs instead use the phase space slicing method to avoid singularities in the cross section calculations of di-jet final states. The JetViP program also offers the possibility to account for contributions from interactions with the partons of the photon,

CP828, *Multiparticle Dynamics*
edited by V. Šimák, M. Šumbera, Š. Todorova, and B. Tomášik

so called resolved photon processes. Since a convolution of the pointlike term in the photon parton density function (pdf) with the NLO resolved matrix elements provides two gluons in the final state it has been regarded as an approximation to the NNLO cross section for direct photon interactions.

Even higher orders are approximated by parton cascades which are generated via evolution equations, where the probability to have splittings of the type $q \to qg$, $g \to q\overline{q}$ and $g \to gg$ are calculated. There are different types of evolution equations, taking different parts of the full calculation into account and evolving the density functions in different variables.

The RAPGAP [10] and LEPTO [11] Monte Carlo programs use LO matrix elements supplemented with initial and final state parton showers (MEPS) generated according to the DGLAP evolution scheme [12, 13, 14, 15, 16] for the description of DIS processes. RAPGAP also offers the possibility to include contributions from processes with resolved photons [17, 18]. The ARIADNE [19] program is based on the colour dipole model (CDM) [20, 21], which assumes gluon emissions to originate from independently radiating colour dipoles. The DJANGO program [22] provides an interface between HERACLES [23], which simulates QED-radiative effects, and any of the above programs.

The CASCADE Monte Carlo program [24, 25] is based on the CCFM formalism [26, 27, 28, 29]. Two different parametrizations (set-1 and set-2) [30] have been used to describe the pdf of the proton.

INCLUSIVE JET PRODUCTION

Single differential cross sections

The ZEUS experiment has studied inclusive jet production in three different phase space regions [31]. Comparisons are made to the predictions of fixed order QCD calculations and of QCD-based parton shower models in order to explore the effects of higher order α_S processes. The data sample corresponds to a luminosity of $38.7 \pm 0.6 pb^{-1}$.

The phase space regions are restricted by the application of various kinematic cuts. In the so called *global phase space region*, at least one jet with transverse energy larger than 6 GeV, as reconstructed in the laboratory frame by the longitudinally invariant k_t-algorithm [32, 33], should fall in the rapidity range from -1 to 3. The forward hemisphere corresponds to positive rapidities. With no further restrictions, QPM-events are expected to dominate strongly.

The more restrictive *BFKL phase space region* is defined by introducing some additional cuts. The vector, which represents the direction of motion of the total hadronic final state in the event, should point into the backward hemisphere. This condition typically favours low x processes, where the effects of BFKL-dynamics [34, 35, 36] are expected to appear. Further, the events should also contain a reconstructed jet in the forward hemisphere. This combination of requirements essentially excludes QPM-events and the sample is expected to be dominated by multijet events. Finally, events with E_T-ordered parton evolution (DGLAP) are suppressed by requiring E_T^2 of the forward going jet to be approximately equal to Q^2.

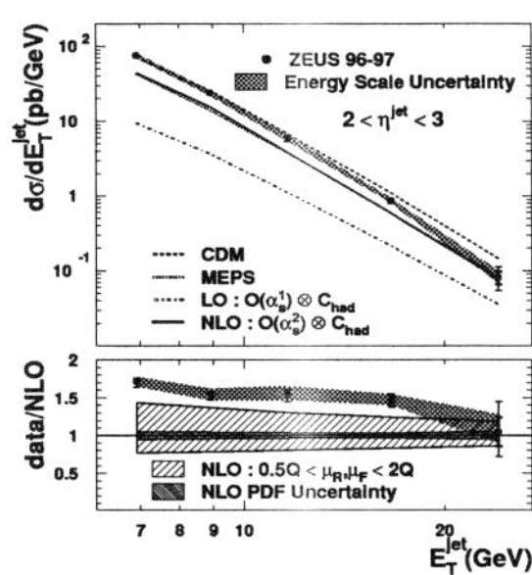

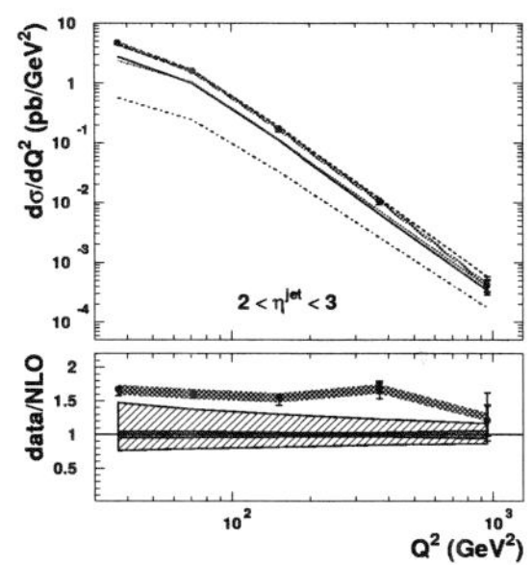

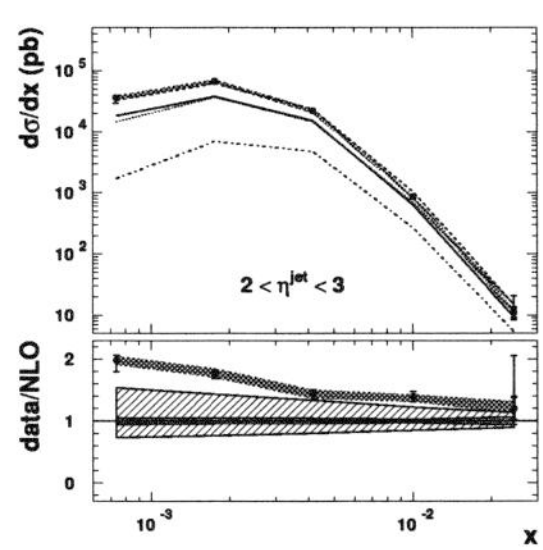

FIGURE 1. *Differential cross sections (dots) in the forward BFKL phase space region for inclusive jet production in neutral current DIS with $E_T^{jet} > 6$ GeV, $2 < \eta^{jet} < 3$, $Q^2 > 25$ GeV^2, $y > 0.04$, $\cos\gamma_h < 0$ and $0.5 < E_T^{jet}/Q^2 < 2$ as functions of (a) E_T^{jet}, (b) Q^2 and (c) x. The $\mathcal{O}(\alpha_S)$ (dot-dashed lines) and $\mathcal{O}(\alpha_S^2)$ (solid lines) QCD calculations are shown.*

The *forward BFKL phase space region* is defined by the same cuts as above except that the reconstructed jet should be in the extreme forward region ($2 < \eta < 3$).

LEPTO and ARIADNE are used to generate DIS events, which are subsequently detector simulated. In the event generations the CTEQ4M parametrization [37] of the proton parton density is used. Predictions from fixed order α_S calcualtions are obtained from the DISENT program, using five flavours, and the renormalization and factorization scales set to Q^2. Here the proton pdf's are given by CTEQ6 [38].

The measured cross sections are presented as a function of E_T^{jet}, η^{jet}, x_{Bj} and Q^2. The general observation in all phase space regions is that ARIADNE is in good agreement with data everywhere. The MEPS model fails to reproduce the rapidity distribution and the small x_{Bj} behaviour, and especially in the forward BFKL region it is in complete disagreement with data, which can be seen from Fig. 1. Fixed order calculations exhibit a behaviour, which is not too different to the MEPS model and thus also fails to describe the distributions in the forward BFKL region.

The observed excess of data over fixed order calculations and over the DGLAP based models can not be accomodated within the experimental and theoretical uncertainties. This indicates the importance of contributions from higher order processes and from non-E_T-ordered parton emission, of which the latter conclusion is supported by the good agreement with CDM. It should, however, be noted that no comparison with models including resolved photons has been performed.

The H1 experiment has measured the forward jet cross section using a data sample based on an integrated luminosity of 13.7 pb^{-1} [39]. The enlarged statistics allow to study more differential distributions than previously presented [40, 41], and to use new observables compared to other measurements [31].

DIS events are obtained by applying the cuts $E_{e'} > 10$ GeV, $156^o < \theta_e < 175^o$, $0.1 < y < 0.7$ and 5 $GeV^2 < Q^2 < 85$ GeV^2, where E_e' is the energy of the scattered electron, θ_e the polar angle, and y is the inelasticity of the exchanged photon. Jets are defined using the k_t-jet algorithm [32, 33] with combined calorimeter and track information [42] as input. The jet algoritm is applied in the Breit-frame and with the

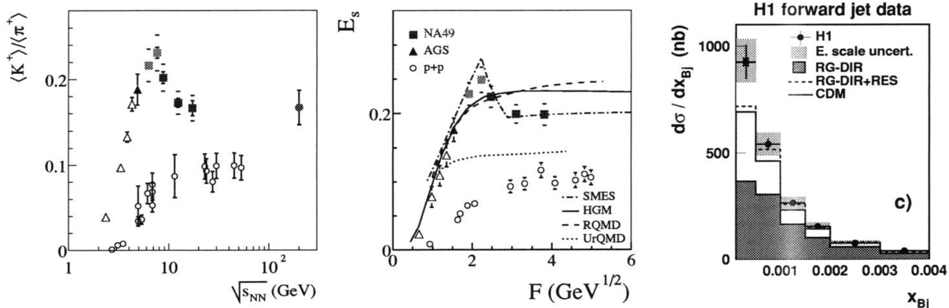

FIGURE 2. *The hadron level cross section for forward jet production as a function of xbj compared to NLO predictions from DISENT (a) and to QCD Monte Carlo models (b and c). The shaded band around the data points shows the error from the uncertainties in the calorimetric energy scales. The hatched band around the NLO calculations illustrates the theoretical uncertainties in the calculations, estimated as described in the text. The inner error bars show the statistical errors. The outer error bars represent the statistical errors added in quadrature to those systematic not already included in the error bands. The dashed line in (a) shows the LO contribution.*

p_t-recombination scheme. A forward jet is defined in the laboratory system as having $p_{t,jet} > 3.5$ GeV and being in the angular range $7^o < \theta_{jet} < 20^o$. The phase space for DGLAP evolution is suppressed by the additional requirement $0.5 < p_{t,jet}^2/Q^2 < 5$.

The forward jet measurements by H1 have been compared to the predictions of the RAPGAP [10], ARIADNE [19] and CASCADE [24] programs and to NLO calculations from DISENT [3, 4] and NLOJET++ [43]. The CTEQ6M [38] parametrization of the parton distributions in the proton has been used in the NLO calculations and in the event generation by RAPGAP and ARIADNE. The renormalisation scale in DISENT is given by the average p_t^2 of the di-jets from the matrix element, while the factorisation scale is given by the average p_t^2 of all forward jets in the selected sample. In NLOJET++ both scales are set to the average p_t^2 of the forward jet and the two hardest jets in the event. The scales in RAPGAP are p_t^2 for direct interactions and $Q^2 + p_t^2$ for resolved. In CASCADE the renormalization scale is $m^2 + p_t^2$, where m is the quark mass and p_t the transverse momentum of the di-jets from the matrix element. The factorization scale is $\hat{s} + Q_t^2$, where $\hat{s}$ is the invariant mass squared of the di-quark system and Q_t its transverse momentum.

Fig. 2a shows that the NLO di-jet calculations from DISENT are significantly larger than the LO contribution at small x_{Bj}. This reflects the fact that the contribution from forward jets in the LO scenario is suppressed by kinematics. The somewhat improved agreement at higher x_{Bj} can be understood from the fact that the range in the longitudinal momentum fraction which is available for higher order emissions decreases.

From Fig. 2b it is seen that the CCFM model (both set-1 and set-2) predicts a somewhat harder x_{Bj} distribution, which results in a comparatively poor description of the data.

Fig. 2c shows that the DGLAP model with direct photon interactions alone (RG-DIR) gives results similar to the NLO di-jet calculations and falls below the data, particularly in the low x_{Bj} region. The description of the data by the DGLAP model is significantly improved if contributions from resolved virtual photon interactions are included (RG-DIR+RES). However, there is still a discrepancy in the lowest x_{Bj}-bin, where a possible BFKL signal would be expected to show up most prominently. The CDM model, which gives emissions that are non-ordered in transverse momentum, shows a behaviour similar to the RG DIR+RES model.

Triple differential cross section

The triple differential cross section $d\sigma/dx_{Bj}dQ^2dp_{t,jet}^2$ is presented as a function of x_{Bj} for three regions in Q^2 and $p_{t,jet}^2$. Based on the ratio $r = p_t^2/Q^2$ three different kinematic regions can be defined. In the region $r \sim 1$ parton emissions ordered in p_t are suppressed and the data are best described by the DGLAP-resolved model and CDM. Direct photon interactions should dominate in the region $r < 1$ and this is also the region where the DGLAP-direct model comes closest to the data. Typical for resolved photon processes is that $r > 1$ and as expected the DGLAP-resolved model provides a good overall description of the data as does also CDM. As already observed in the single differential measurement the CCFM model predicts a somewhat harder x_{Bj} distribution than seen in the data, which is valid over the full kinematic range. The general trend of the NLO calculations from DISENT is that they agree better with data as x_{Bj}, Q^2 and $p_{t,jet}^2$ increase. The plots are not shown here.

Three jet production

By requiring reconstruction of the two hardest jets in the event in addition to the forward jet, different kinematic regions can be investigated by applying cuts on the jet momenta and their rapidity separation. All three jets are required to have transverse momenta greater than 6 GeV. The three jets are ordered in rapidity according to $\eta_{forw-jet} < \eta_{jet2} < \eta_{jet1}$ and the rapidity differences are defines as $\Delta\eta_1 = \eta_{jet2} - \eta_{jet1}$ and $\Delta\eta_2 = \eta_{forw-jet} - \eta_{jet2}$. From Fig. 3 it is seen that NLOJET++ gives good agreement with the data if the two additional hard jets are emitted in the central region ($\Delta\eta_2$ large). However, the more the additional hard jets are shifted to the forward region ($\Delta\eta_2$ small), the less well are the data described by NLOJET++. When both $\Delta\eta_1$ and $\Delta\eta_2$ are small, it is possible that one or both of the additional jets originate from gluon radiation close in rapidity space to the forward jet, which implies that the p_t-ordering by default is broken.

As can be seen from Fig. 4 CDM provides the best description of the data while the other models, including the DGLAP-resolved model, fail in most of the bins.

The conclusion is that additional breaking of the p_t ordering is needed compared to what is included in the resolved photon model.

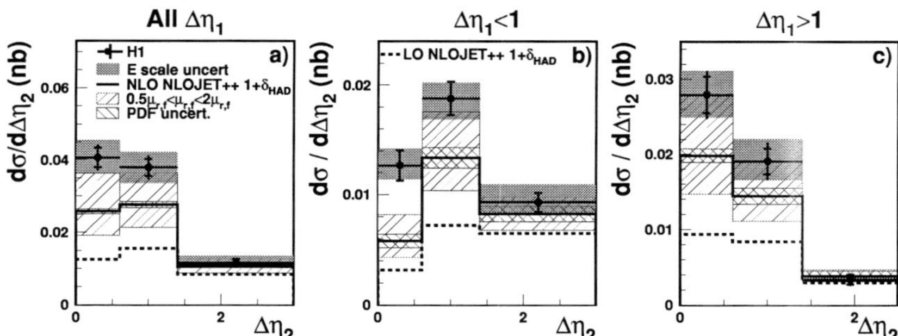

FIGURE 3. *The cross section for events with a reconstructed high transverse momentum di-jet system and a forward jet as a function of the rapidity separation between the forward jet and the most forward-going additional jet, $\Delta\eta_2$. Results are shown for the full sample and for two ranges of the separation between the two additional jets, $\Delta\eta_1 < 1$ and $\Delta\eta_1 > 1$. The data are compared to the LO and NLO predictions of three-jet final state calculations by NLOJET++ $(1+\delta_{HAD})$. The band around the data points illustrates the error due to the uncertainties in the calorimetric energy scales. The band around the NLO calculations illustrates the theoretical uncertainties in the calculations. The inner error bars show the statistical errors. The outer error bars represent the statistical errors added in quadrature to those systematic not already included in the error bands.*

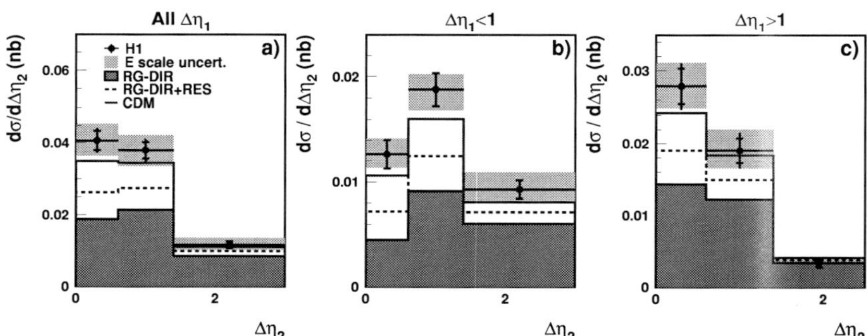

FIGURE 4. *The cross section for events with a reconstructed high transverse momentum di-jet system and a forward jet as a function of the rapidity separation between the forward jet and the most forward-going additional jet, $\Delta\eta_2$. Results are shown for the full sample and for two ranges of the separation between the two additional jets, $\Delta\eta_1 < 1$ and $\Delta\eta_1 > 1$. The data are compared to the predictions of RAPGAP DIR, RAPGAP DIR+RES and CDM. The band around the data points illustrates the error due to the uncertainties in the calorimetric energy scales. The inner error bars show the statistical errors. The outer error bars represent the statistical errors added in quadrature to those systematic not already included in the error bands.*

ACKNOWLEDGMENTS

I would like to thank the organizers for a pleasant conference.

REFERENCES

1. A. H. Mueller, *Nucl. Phys. Proc. Suppl.* **18C**, 125 (1991).
2. A. H. Mueller, *J. Phys. G* **17** (1991) 1443.
3. S. Catani and M. H. Seymour, *Phys. Lett. B* **378** (1996) 287 [hep-ph/9602277].
4. S. Catani and M. H. Seymour, *Nucl. Phys. B* **485** (1997) 291 [Erratum-ibid. B **510** (1997) 503] [hep-ph/9605323].
5. D. Graudenz, arXiv:hep-ph/9708362.
6. D. Graudenz, arXiv:hep-ph/9710244.
7. E. Mirkes and D. Zeppenfeld, *Phys. Rev. Lett.* **78**, 428 (1997) [arXiv:hep-ph/9609231].
8. G. Kramer and B. Potter, *Phys. Lett. B* **453**, 295 (1999) [arXiv:hep-ph/9901314].
9. B. Potter, *Comput. Phys. Commun.* **119**, 45 (1999) [arXiv:hep-ph/9806437].
10. H. Jung, *Comput. Phys. Commun.* **86** (1995) 147.
11. G. Ingelman, A. Edin and J. Rathsman, *Comput. Phys. Commun.* **101**, 108 (1997) [arXiv:hep-ph/9605286].
12. V. N. Gribov and L. N. Lipatov, *Sov. J. Nucl. Phys.* **15** (1972) 675 [Yad. Fiz. **15** (1972) 1218].
13. V. N. Gribov and L. N. Lipatov, *Sov. J. Nucl. Phys.* **15** (1972) 438 [Yad. Fiz. **15** (1972) 781].
14. L. N. Lipatov, *Sov. J. Nucl. Phys.* **20** (1975) 94 [Yad. Fiz. **20** (1974) 181].
15. G. Altarelli and G. Parisi, *Nucl. Phys. B* **126** (1977) 298.
16. Y. L. Dokshitzer, *Sov. Phys. JETP* **46** (1977) 641 [*Zh. Eksp. Teor. Fiz.* **73** (1977) 1216].
17. H. Jung, L. Jönsson and H. Küster, *"Resolved photon processes in DIS and small x dynamics,"* [hep-ph/9805396].
18. H. Jung, L. Jönsson and H. Küster, *Eur. Phys. J. C* **9** (1999) 383 [hep-ph/9903306].
19. L. Lönnblad, *Comput. Phys. Commun.* **71** (1992) 15.
20. B. Andersson, G. Gustafson, L. Lönnblad and U. Pettersson, *Z. Phys. C* **43** (1989) 625.
21. L. Lönnblad, *Z. Phys. C* **65** (1995) 285.
22. K. Charchula, G. A. Schuler and H. Spiesberger, *Comput. Phys. Commun.* **81** (1994) 381.
23. A. Kwiatkowski, H. Spiesberger and H. J. Möhring, *Comput. Phys. Commun.* **69** (1992) 155.
24. H. Jung and G. P. Salam, *Eur. Phys. J. C* **19** (2001) 351 [hep-ph/0012143].
25. H. Jung, *Comput. Phys. Commun.* **143** (2002) 100 [hep-ph/0109102].
26. M. Ciafaloni, *Nucl. Phys. B* **296** (1988) 49.
27. S. Catani, F. Fiorani and G. Marchesini, *Phys. Lett. B* **234** (1990) 339.
28. S. Catani, F. Fiorani and G. Marchesini, *Nucl. Phys. B* **336** (1990) 18.
29. G. Marchesini, *Nucl. Phys. B* **445** (1995) 49 [hep-ph/9412327].
30. M. Hansson and H. Jung, *"Status of CCFM: Un-integrated gluon densities"*, [hep-ph/0309009].
31. S. Chekanov *et al.* [ZEUS Collaboration], hep-ex/0502029.
32. S. Catani, Y. L. Dokshitzer, M. H. Seymour and B. R. Webber, *Nucl. Phys. B* **406** (1993) 187.
33. S. Catani, Y. L. Dokshitzer and B. R. Webber, *Phys. Lett. B* **285** (1992) 291.
34. E. A. Kuraev, L. N. Lipatov and V. S. Fadin, *Sov. Phys. JETP* **44** (1976) 443 [*Zh. Eksp. Teor. Fiz.* **71** (1976) 840].
35. E. A. Kuraev, L. N. Lipatov and V. S. Fadin, *Sov. Phys. JETP* **45** (1977) 199 [*Zh. Eksp. Teor. Fiz.* **72** (1977) 377].
36. I. I. Balitsky and L. N. Lipatov, *Sov. J. Nucl. Phys.* **28** (1978) 822 [Yad. Fiz. **28** (1978) 1597].
37. H. L. Lai *et al.*, *Phys. Rev. D* **55**, 1280 (1997) [arXiv:hep-ph/9606399].
38. J. Pumplin, D. R. Stump, J. Huston, H. L. Lai, P. Nadolsky and W. K. Tung, *JHEP* **0207**, 012 (2002) [arXiv:hep-ph/0201195].
39. A. Aktas *et al.* [H1 Collaboration], arXiv:hep-ex/0508055.
40. C. Adloff *et al.* [H1 Collaboration], *Nucl. Phys. B* **538** (1999) 3 [hep-ex/9809028].
41. J. Breitweg *et al.* [ZEUS Collaboration], *Eur. Phys. J. C* **6** (1999) 239 [hep-ex/9805016].
42. C. Adloff *et al.* [H1 Collaboration], *Z. Phys. C* **74** (1997) 221 [hep-ex/9702003].
43. Z. Nagy and Z. Trocsanyi, *Phys. Rev. Lett.* **87** (2001) 082001 [hep-ph/0104315].

Jet production in deep inelastic scattering at HERA

M.R. Sutton

University College London, Dept of Physics and Astronomy, Gower Street, London WC1E 6BT, Great Britain

Abstract.
A number of the most recent results from the wealth of precision HERA data on high transverse energy jet production in deep inelastic scattering are reviewed. These measurements are confronted with predictions from next-to-leading order (NLO) Quantum Chromodynamics and allow the extraction of the strong coupling constant, α_s, and have been used in QCD fits of the parton distribution functions in the proton.

Keywords: jets, deep inelastic scattering, HERA
PACS: 13.87.-a;

INTRODUCTION

The production of high transverse energy jets in the hadronic final state of deep inelastic scattering (DIS) at HERA provides a unique laboratory for the study of the perturbative QCD sub-process. The large kinematic region available at HERA allows the detailed study of the hadronic final state with a variation of the hard scale, provided by the virtuality, Q^2, of the exchanged boson, or the jet transverse energy, E_T, over many orders of magnitude.

At sufficiently high transverse energies, the relative contribution from long distance, non-perturbative effects such as hadronisation is expected to be small. The high precision data now available at these scales enable more meaningful comparison with next-to-leading order (NLO) parton-level calculations, allowing questions relevant to the description of the QCD hard subprocess, such as the possible size of higher-order corrections and the choice of appropriate scale, to be addressed.

In the region of the Q^2 and $x_{\rm Bj}$ kinematic plane covered by jet measurements at HERA, the parton distributions, particularly the quark densities, are constrained by fits [1, 2, 3] to inclusive charged- and neutral-current cross sections [4, 5, 6, 7]. Measurements of jet production in DIS, such as inclusive jet production in the Breit frame, or dijet production are already $\mathcal{O}(\alpha_s)$ processes at leading order and directly sensitive to the gluon distribution in the proton. Where the hadronic final state is sufficiently well described by the NLO predictions, these data can be used for extractions of the QCD coupling, α_s and as additional input to the QCD fits to the proton parton distribution functions to complement that from inclusive DIS data.

This paper presents several measurements from the ever increasing wealth of precision jet data in DIS from the HERA experiments and illustrates how it is helping to add to our understanding of perturbative QCD.

CP828, *Multiparticle Dynamics*
edited by V. Šimák, M. Šumbera, Š. Todorova, and B. Tomášik

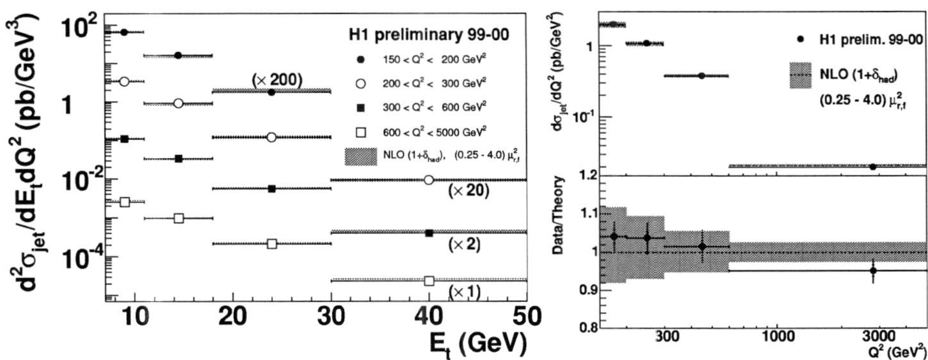

FIGURE 1. Inclusive jet production in the Breit frame from H1

JETS IN DEEP INELASTIC SCATTERING

Within the framework of perturbative QCD, the jet cross section in DIS can be factorised into a convolution of the short distance sub-process cross section, $\mathrm{d}\hat{\sigma}_a$, with the parton distribution functions, f_a, within the proton,

$$\mathrm{d}\sigma_{\mathrm{jet}} = \sum_{a=q,\bar{q},g} \int \mathrm{d}x\, f_a(x,\mu_F^2) \times \mathrm{d}\hat{\sigma}_a(x,\alpha_s(\mu_R^2),\mu_R^2,\mu_F^2) \times (1+\delta_{\mathrm{had}}).$$

For comparison with the data, which is corrected to the hadron level, a hadronisation correction is applied to the parton level calculation.

Inclusive jet production at high Q^2

For inclusive production of jets with high E_T in the Breit frame, the $\mathcal{O}(\alpha_s^0)$ single quark born term in the final state is suppressed since the quark in this frame has no transverse energy. This means that inclusive jet production in the Breit frame is an order $\mathcal{O}(\alpha_s)$ process since an additional parton must be radiated to balance the E_T of the observed jet, even if the jet from this additional parton is unobserved. This leads to better infrared behaviour of the cross section and smaller renormalisation scale uncertainties and a contribution from the gluon distribution in the proton already in the leading order diagrams.

Recent measurements of inclusive jet production in the Breit frame from the H1 [8] and ZEUS [9] collaborations are shown in Figures 1 and 2 for the ranges $150 < Q^2 < 5000$ GeV and $Q^2 > 125$ GeV respectively.

Also shown are the predictions from NLO QCD which describe both sets of data reasonably well over the whole range of E_T and Q^2. For this Q^2 range, the hadronisation uncertainty is typically less than 1% although the theory uncertainty is generally larger than the experimental uncertainty, particlularly at lower Q^2 where the statistical precision is high. The renormalisation scale uncertainty is typically below 5% so that the

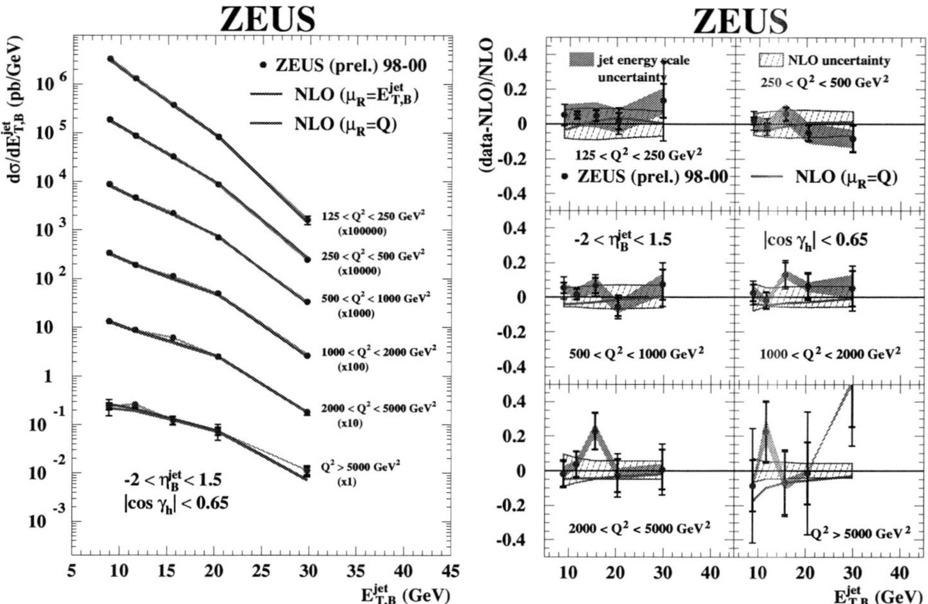

FIGURE 2. Inclusive jet production in the Breit frame from ZEUS

strong coupling can be extracted which yields the values

$$
\begin{aligned}
\alpha_s(M_Z) &= 0.1197 \pm 0.0016(\text{exp})^{+0.0046}_{-0.0048}(\text{th}), \\
\alpha_s(M_Z) &= 0.1196 \pm 0.0011(\text{stat})^{+0.0019}_{-0.0025}(\text{exp})^{+0.0029}_{-0.0017}(\text{th.})
\end{aligned}
$$

from the H1 and ZEUS data respectively.

Three jet production

The description of the QCD subprocess can also be studied for higher order processes, such as three jet production. Both ZEUS and H1 have measured the DIS cross section for three-jet production [10, 11] in the Breit frame. Figure 3 shows the ratio of the two-jet to three-jet cross section as a function of Q^2 for jets measured in the Breit frame by the H1 collaboration. In this ratio, the parton distribution uncertainty partially cancels, and although the data are well described by the calculation, as with the case of the inclusive jet data the theory uncertainty is still larger than the experimental precision at lower Q^2. The value of α_s extracted from the data for each Q^2 bin is also shown in the figure.

The overall values obtained by the H1 and ZEUS collaborations are

$$
\alpha_s(M_Z) = 0.1175 \pm 0.0017(\text{stat}) \pm 0.0050(\text{exp})^{+0.0054}_{-0.0068}(\text{th.})
$$

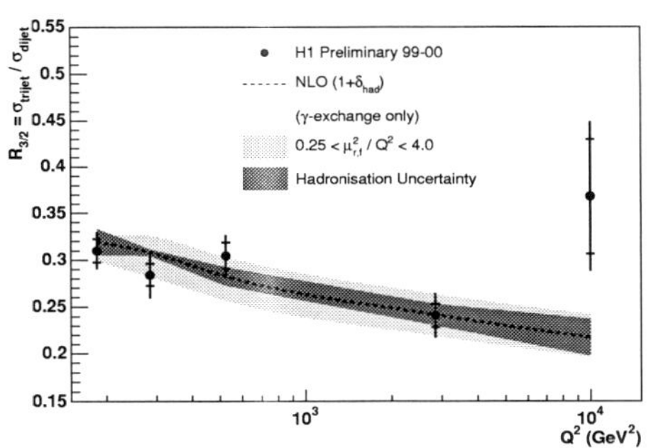

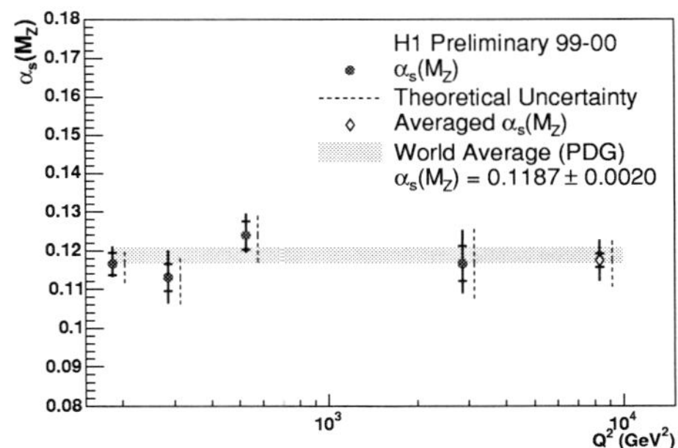

FIGURE 3. Three jet production in the Breit frame

$$\alpha_s(M_Z) \quad = \quad 0.1179 \pm 0.0013(\text{stat})^{+0.0028}_{-0.0046}(\text{exp})^{+0.0064}_{-0.0046}(\text{th})$$

respectively, again consistent with each other, although for this data, the theoretical uncertainties are larger, due to the additional constraints on the sub-leading jets. The extracted values are both consistently below that obtained for the inclusive data, although both sets of values are consistent within the relevant uncertainties.

USING JET DATA IN QCD FITS

In QCD fits to inclusive DIS data, the sensitivity to the gluon density typically enters only as a higher order correction to the Born term. In the case of jet data, the cross section has leading order contributions from the gluon distribution in the proton through the boson-gluon fusion diagram. Therefore including the jet data in the QCD fit should be able to better constrain the gluon distribution in the the proton than the use of inclusive DIS data alone.

Although NLO calculations of jet cross sections have been available for some time, the intergration over the phase space including the jet algorithm typically cannot be performed analytically so that their use in PDF fitting is prohibitive since the integration over the phase space needs to be performed anew for each point over the parameter space of the PDF set being fitted.

To use the complete NLO jet calculation rigourously in the QCD fit, the ZEUS collaboration have taken the approach of deconvoluting the PDF and α_s dependence of the cross section from the subprocess and storing the results in a grid, G, of the hard scale, Q, and the momentum fraction, ξ, of the parton from the proton entering the hard sub process,

$$d\sigma_i^{\text{jet}} = \sum_{n=1,2}\sum_{a=q,\bar{q},g}\sum_j\sum_k f_a(\xi_k, Q_j^2) \times \alpha_s^n(Q_j^2) \times G_{i,a,n,\xi_k,Q_j^2}.$$

This enables the NLO calculation to be performed only once, and the convolution with the PDF to be performed using a fast summation for any PDF set. This yields results consistent with the full calculation to better than 0.5%.

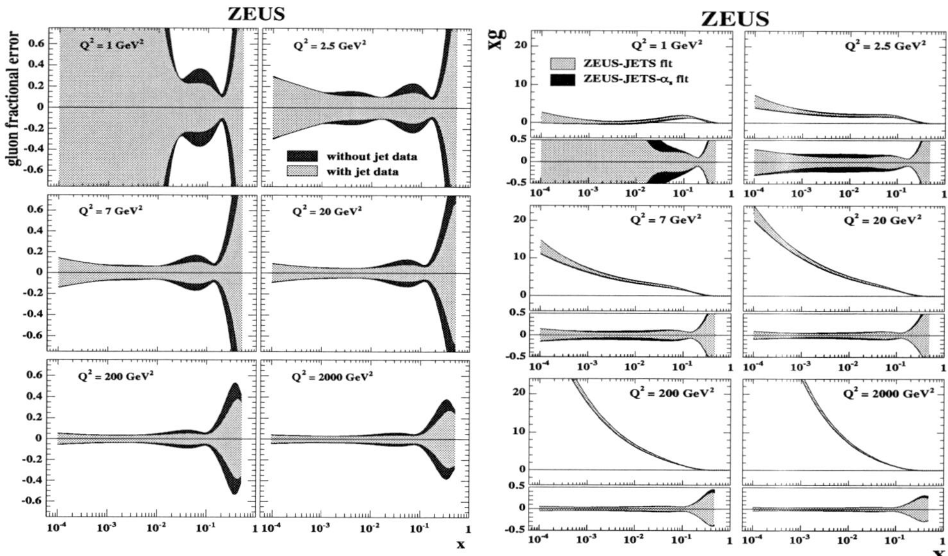

FIGURE 4. The gluon uncertainty a) for the fits with, and without jet data,, b) with and without α_s as a free parameter.

Using the inclusive neutral- and charged-current data to constrain the quark distributions, including the data from inclusive jet production in the Breit frame [12], and inclusive dijet data from photoproduction [13] is seen to more tightly constrain the gluon distribution for medium x than before.

This is illustrated in Figure 4(left) which shows the uncertainty on the gluon from the ZEUS-JETS fit [14] when including and excluding the jet data. A clear improvement of almost a factor of two is seen over the medium x region, with this improvement persisting to high x.

Since the Jet cross section is sensitive to α_s through the $\gamma q \to qg$ subprocess as well as the $\gamma g \to q\bar{q}$ subprocess, this partialy breaks the strong correlation of the gluon distribution with α_s seen in previous fits to inclusive DIS data alone. The result of this is illustrated in Fig. 4(right) which shows the gluon distribution uncertainty for fits both with α_s constrained, and with α_s as a free parameter in the fit, clearly illustrating that for the fit where the coupling is a free parameter, the gluon distribution, particularly at large Q^2, does not become unconstrained.

The resultant value of α_s fitted simultaneously with the parton distribution functions in the proton using both the inclusive DIS data and the jet data from DIS and photoproduction is

$$\alpha_s(M_Z) = 0.1183 \pm 0.0007(\text{uncorr}) \pm 0.0022(\text{corr}) \pm 0.0016(\text{norm}) \pm 0.0008(\text{model})$$

where the estimated uncertainty from terms beyond NLO ± 0.0050 [14]. This result, using HERA data alone, represents the first fully consistent use of jet data in a PDF fit and yields a value consistent with

$$\alpha_s(M_Z) = 0.1182 \pm 0.0027,$$

the Bethke world average [15]. The χ^2 of the fit as a function of α_s when including the jet data is much much more tightly than when using the inclusive charged- and neutral-current data alone.

SUMMARY AND OUTLOOK

The HERA collider continues to produce some of the most precise QCD jets data in the world. The first genuinely consistent simultaneaous extraction of the strong coupling and the parton distrubution functions within the proton has been performed and the corresponding value of α_s is competitive with the current world avereage. More HERA data are currently available, and the continued running of the HERA-II collider will provide even more, so allowing even tighter constraints to be made on α_s, the parton distribution functions within the proton and increasing our understanding of perturbative QCD.

ACKNOWLEDGMENTS

The author would like to thank PPARC for the support of an advanced fellowship to enable attendance at this conference.

REFERENCES

1. J. S. A.D.Martin, R. Roberts, and R.Thorne, *Eur. Phys. J.* **C23** p. 73 (2002).
2. J. S. A.D.Martin, R. Roberts, and R.Thorne, *Eur. Phys. J.* **C28** p. 455 (2002).
3. J.Pumplin et al., *JHEP* **0207** p. 12 (2002).
4. The ZEUS Collaboration, *Eur. Phys. J.* **C28 2**, 175–201 (2003).
5. The ZEUS Collaboration, *Eur. Phys. J.* **C21 3**, 443–471 (2001).
6. The H1 Collaboration, *Eur. Phys. J.* **C30** pp. 1–32 (2003).
7. The ZEUS Collaboration, *Eur. Phys. J.* **C32** pp. 1–16 (2003).
8. The H1 Collaboration, *Abs: 629, contributed paper to HEP 2005 International Europhysics Conference on High Energy Physics EPS (2005), Lisboa, Portugal* (2005).
9. The ZEUS Collaboration, *Abs: 375, contributed paper to HEP 2005 International Europhysics Conference on High Energy Physics EPS (2005), Lisboa, Portugal* (2005).
10. The ZEUS Collaboration, *DESY Report DESY-05-019, submitted to Eur. Phys. J.* (2005).
11. The H1 Collaboration, *Abs: 390, contributed paper to the XXXII International Symposium on Lepton-Proton Interactions at High Energy LP2005* (2005).
12. The ZEUS Collaboration, *Phys. Let.* **B547** pp. 164–180 (2002).
13. The ZEUS Collaboration, *Eur. Phys. J.* **C23** p. 615 (2002).
14. The ZEUS Collaboration, *Eur. Phys. J.* **B 8**, 195–201 (2005).
15. S. Bethke, *hep-ex/0407021* (2004).

Energy Flow and Leading Neutron Production at HERA

Wenbiao Yan

On behalf of the H1 and ZEUS Collaborations
DESY, Hamburg, Germany
E-mail: yanwb@mail.desy.de

Abstract. The azimuthal asymmetry of hadrons in the semi-inclusive process e+p → e+h+X is studied in deep inelastic scattering at HERA. A measurement of the dijet cross section with a leading neutron is also reported, the P_T^2 distribution of the leading neutron is investigated.

Keywords: azimuthal asymmetry, leading neutron production, HERA
PACS: 12.38.Qk, 13.85.Ni

INTRODUCTION

The semi-inclusive process e+p → e+h+X in deep inelastic scattering (DIS), where h is an observed hadron, provides an important test of pQCD in hadron production. An interesting variable is the distribution of the azimuthal angle ϕ of the hadron (measured in the hadronic centre-of-mass frame HCM), between the lepton plane, defined by the incoming and outgoing lepton, and the hadron plane, defined by the outgoing hadron and the virtual boson. The azimuthal asymmetry in the ϕ distribution is due to the non-zero transverse momentum of the hadron. The contribution from the intrinsic momentum of a quark in the proton is small at high momentum transfer Q^2 [1]. The contribution from higher order QCD processes such as the QCD Compton and boson-gluon fusion processes weakly depends on Q^2 and persists at high Q^2. In this paper, we use the energy flow method to investigate the azimuthal asymmetry, where the range of investigated phase space is increased with respect to previous publications [2]. We compare experimental data with Monte Carlo (MC) predictions incorporating leading order (LO) and parton showers and with a next-to-leading-order (NLO) pQCD calculations.

There is a significant fraction of events with a baryon in the hadronic final state carrying a large fraction of incoming proton energy in ep collisions. It is interesting to investigate the production mechanism for low transverse momentum baryons. In this paper, we study leading neutron production e+p → e+n+X. The leading neutron can result from fragmentation of the proton remnant or one pion exchange. In the one pion exchange model, the virtual photon interacts with a pion from the proton, i.e. $\sigma_{ep\to enX} = f_{\pi/p}(x_L,t) \times \sigma(e\pi \to e'X)$, where t is the square of the four-momentum of the exchanged pion. The outgoing neutron of energy E_B carries a fraction $x_L = E_B/E_p$ of the incoming proton energy. $f_{\pi/p}$ is the flux of virtual pions emitted by the proton. One feature of the exchange model is vertex factorization, which predicts that the cross section on baryon vertex variables will be independent of lepton vertex variables. We test vertex factorization by comparing events in photoproduction ($Q^2 \sim 0$ GeV^2) and

CP828, *Multiparticle Dynamics*
edited by V. Šimák, M. Šumbera, Š. Todorova, and B. Tomášik

DIS ($Q^2 \geq 1$ GeV^2) events. In this paper, we study the p_T^2 of the leading neutron, and report the dijet cross section in leading neutron production. We compare experimental results with Monte Carlo and pQCD calculations.

AZIMUTHAL ASYMMETRY IN ϕ DISTRIBUTION

The semi-inclusive DIS cross section $d\sigma/d\phi$ in the HCM frame is

$$\frac{d\sigma^{ep\to ehX}}{d\phi} = A + B\cos(\phi) + C\cos(2\phi) + D\sin(\phi) + E\sin(s\phi) \tag{1}$$

where the azimuthal asymmetry is denoted by parameters B, C, D and E. The parameters can be determined by first moments, i.e.

$$\langle\cos(\phi)\rangle = \frac{B}{2A} \quad \langle\cos(2\phi)\rangle = \frac{C}{2A} \quad \langle\sin(\phi)\rangle = \frac{D}{2A} \quad \langle\sin(2\phi)\rangle = \frac{E}{2A} \tag{2}$$

The mean values $\langle\cos(n\phi)\rangle$ and $\langle\sin(n\phi)\rangle$ (n=1,2) are measured by the ZEUS Collaboration [3] in neutral current DIS events. The kinematic region is defined by $100 < Q^2 < 8000$ GeV^2 and $0.01 < x < 0.1$. The measured azimuthal asymmetry in terms of $\langle\cos(\phi)\rangle$, $\langle\cos(2\phi)\rangle$, $\langle\sin(\phi)\rangle$ and $\langle\sin(2\phi)\rangle$ determined by the energy flow method are presented in figure 1 as a function of pseudorapidity η of the hadron with transverse momenta $P_T^{LAB} > 150$ MeV. The mean values $\langle\sin(\phi)\rangle$ and $\langle\sin(2\phi)\rangle$ are small and consistent with zero, this is consistent with predictions of LO MC programs LEPTO and ARIADNE. The measured $\langle\cos 2\phi\rangle$ is about zero for $\eta^{HCM} < -2.0$ and positive for higher η^{HCM}. LEPTO and ARIADNE provide a good description of the data. The mean value $\langle\cos\phi\rangle$ is negative for $\eta^{HCM} < -2.0$ and positive for higher η^{HCM}, while the LEPTO and ARIADNE predictions are always negative in the whole η^{HCM} range. The NLO predictions, made using DISENT [4] and corrected for hadronization effects and Z^0 exchange effects, describe the data points better than LEPTO and ARIADNE.

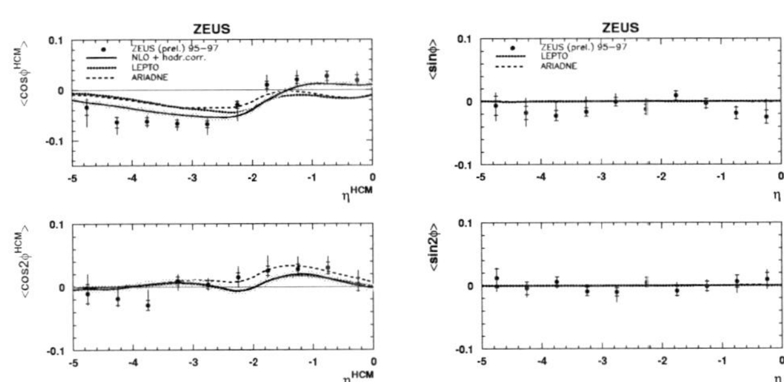

FIGURE 1. The measured $\langle\cos(n\phi)\rangle$ and $\langle\sin(n\phi)\rangle$ (n=1,2) as a function of η of hadrons. The NLO DISENT prediction and LO MC predictions by LEPTO and ARIADNE are also shown.

The azimuthal asymmetry is also analyzed as a function of the detected hadron's minimum transverse energy E_T^{HCM}(min). In this case, the lowest order contribution come

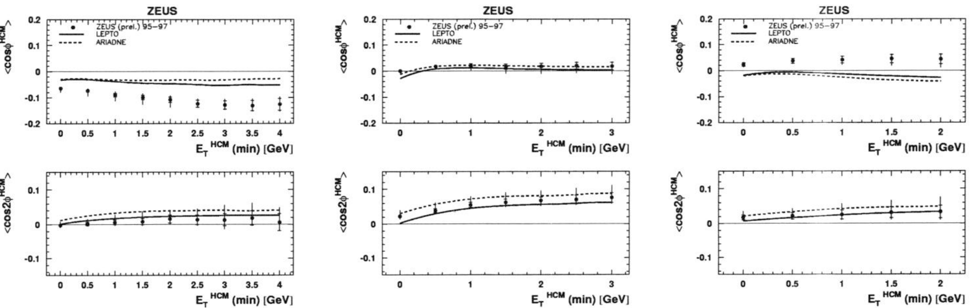

FIGURE 2. The measured $\langle \cos(n\phi) \rangle$ (n=1,2) as a function of hadron's minimum transverse energy E_T^{HCM}(min) in three regions of η^{HCM}: $-5.0 < \eta^{HCM} < -2.5$ (left), $-2.5 < \eta^{HCM} < -1.0$ and $-1.0 < \eta^{HCM} < 0.0$ (right). The lines are LEPTO (solid line) and ARIADNE (dashed line) predictions.

from QCDC and BGF processes, and the contributions from Born level are suppressed. The experimental results are shown in figure 2 in three regions of η^{HCM}: $-5.0 < \eta^{HCM} < -2.5$ (left), $-2.5 < \eta^{HCM} < -1.0$ and $-1.0 < \eta^{HCM} < 0.0$ (right). The same trends are seen as a function of E_T^{HCM}(min) as in the inclusive case.

P_T^2 DISTRIBUTION OF LEADING NEUTRON

The ZEUS Collaboration has measured the p_T^2 distribution of the leading neutron in photoproduction ($Q^2 < 0.02$ GeV^2) and DIS ($Q^2 > 2$ GeV^2) [5]. The p_T^2 distribution for different x_L bins in photoproduction and DIS, normalized to unity at $p_T^2 = 0$, is shown in figure 3 (left). The photoproduction distribution is clearly steeper in the range $0.6 < x_L < 0.9$. The line on each plot is a fit to a function of form $\exp(-bp_T^2)$, which gives a good description of experimental data. The fitted b in DIS as a function of x_L is in figure 3 (middle). Fitted b is consistent with zero below $x_L = 0.3$; and b rises linearly in the range $0.30 < x_L < 0.85$, and then decrease slightly at higher x_L. The pion flux $f_{\pi/p}$ is not an exponential in p_T^2, but, the different models of $f_{\pi/p}$ can be fitted with $\exp(-bp_T^2)$ at fixed x_L, where the resulting b is a model prediction. The figure 3 (middle) shows that there is no model which describes the data over the whole x_L range. As we know, one pion exchange is expected to dominate for $0.6 < x_L < 0.9$. Hence one can use the measured b to critical test models.

For the observable $\Delta b = b(\gamma p) - b(DIS)$, some systematic uncertainties are cancelled. The measured Δb is in figure 3 (right) as a function of x_L, where $b(\gamma p)$ is clearly larger than $b(DIS)$ in the range $0.6 < x_L < 0.9$ with Δb between 0.5 and 1.0 GeV^{-2}. The depletion of neutrons at large p_T^2 is qualitatively described by absorption models. According to absorption models [6], the p_T^2 distribution is given by the $n - \pi$ separation whose size, which is inversely proportional to p_T^2. Rescattering processes remove neutron from small $n - \pi$ system hence with large p_T^2. Thus rescattering results in a depletion of neutron at high p_T^2 in photoproduction relative to DIS.

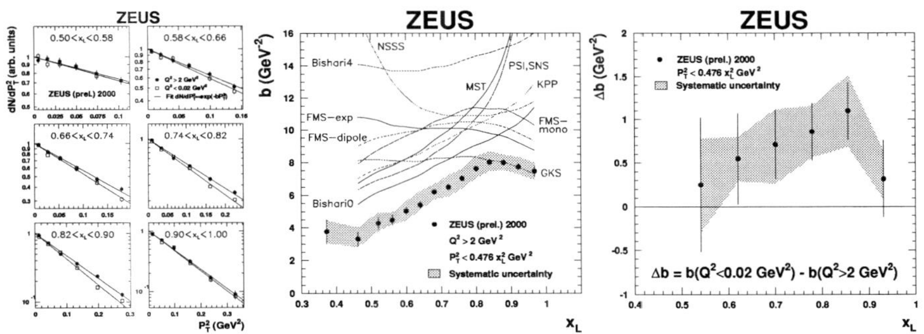

FIGURE 3. Left: p_T^2 distribution of the leading neutron for different x_L bins in photoproduction and DIS events. Middle: dependence of fitted b on x_L in DIS, Right: dependence of Δb on x_L.

DIJET CROSS SECTION WITH A LEADING NEUTRON

Leading neutron production is investigated further by requiring two jets with large transverse momenta e+p $\rightarrow$ e+n+jet+jet+X. This allows a more detailed comparison with MC predictions. We study the dependence of the cross section on jet related variables, such as $x_\gamma^{jet} = \sum_{i=1,2} E_{T,i}^{jet} \exp(-\eta_i^{jet})/(2yE_e)$ where E_T is the jet transverse energy, η^{jet} is the jet pseudorapidity, y is the inelasticity and E_e is the incoming electron energy.

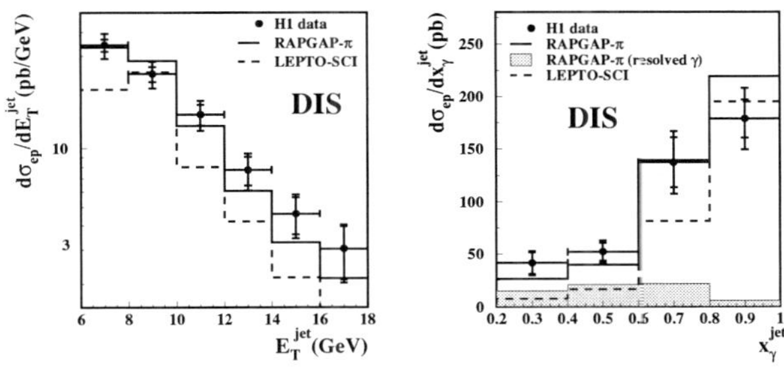

FIGURE 4. Dijet cross sections with a leading neutron as functions of E_T^{jet} and x_γ^{jet} in DIS.

The dijet cross sections with a leading neutron measured by the H1 Collaboration [8] as a function of E_T^{jet} and x_γ^{jet} are presented in figure 4 (DIS: $2 < Q^2 < 80$ GeV^2) and figure 5 (photoproduction: $Q^2 < 0.01$ GeV^2). The data are well described by the pion exchange model in DIS and photoproduction taking into account a 20% normalization uncertainty in the data. LEPTO with soft color interactions is too low to describe the dijet cross sections with a leading neutron, however it describes well leading neutron production in inclusive DIS [7]. The contribution from resolved photons is important at low x_γ^{jet} in DIS events. PYTHIA without multi-parton interactions (MI) and pion

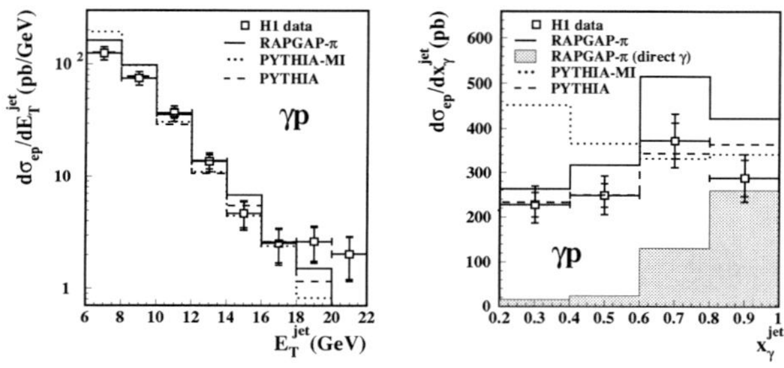

FIGURE 5. Dijet cross sections with a leading neutron as functions of E_T^{jet} and x_γ^{jet} in photoproduction.

exchange describes the data in photoproduction, but PYTHIA with MI is too high at low x_γ, while the MI is needed to describe inclusive dijet in the considered E_T^{jet} range.

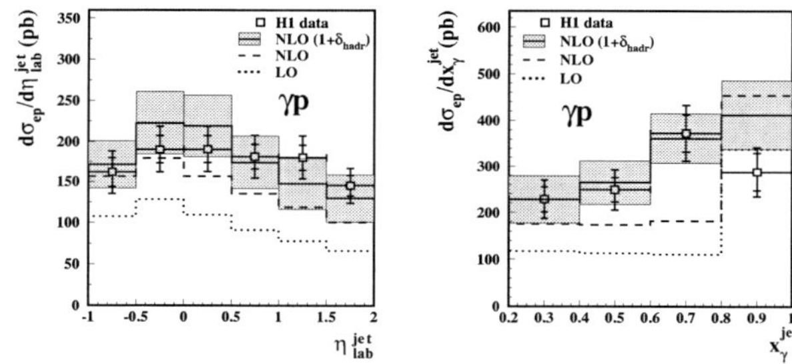

FIGURE 6. Dijet cross sections with a leading neutron as functions of η_{Lab}^{jet} and x_γ^{jet} in photoproduction.

NLO QCD predictions on the dijet cross sections with a leading neutron in photoproduction [9], corrected for hadronization effects, are shown in figure 6 with the measured cross section as functions of η_{Lab}^{jet} and x_γ^{jet}. There is good agreement between the NLO calculation and the measured data.

Figure 7 shows the ratio of the dijet cross section f_{ln} with and without the requirement of a leading neutron as functions of E_T^{jet} and x_γ^{jet} in photoproduction. If the hard interaction is independent of neutron production, f_{ln} should be independent of jet variables. f_{ln} is almost independent of E_T^{jet}, however f_{ln} shows a strong dependence on x_γ^{jet}, also on η^{jet} and x_p^{jet} (not shown). These dependences can only partly described by PYTHIA, which provides some estimates of possible phase space effects. If the leading neutron production is described by Rapgap-π and the inclusive dijet data by PYTHIA with multiparton interactions, this comparison suggests the dijet production with a leading neutron differs from inclusive dijet production.

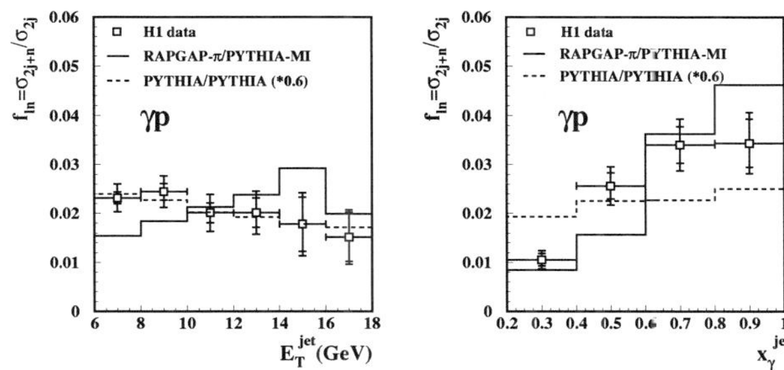

FIGURE 7. Ratio of dijet cross section with and without the requirement of a leading neutron, f_{ln}, as functions of E_T^{jet} and x_γ^{jet} in photoproduction.

SUMMARY

For the azimuthal asymmetry of hadrons in DIS, the NLO prediction shows better agreement with experimental data than LO MC programs, however some discrepancies are still visible. The form $\exp(-bp_T^2)$ describes well the p_T^2 distribution of leading neutron production in photoproduction and DIS. There is no model which describes the measured b over the whole range, and the measured b is different in photoproduction and in DIS. The dijet cross section with a leading neutron is described by pion exchange MC models. The NLO predictions describe the data in photoproduction. The dependence of f_{ln} on x_γ^{jet} and η^{jet} show the violation of factorization.

ACKNOWLEDGMENTS

I would like to thank my colleagues from H1 and ZEUS for their help in preparing this paper.

REFERENCES

1. R. N. Cahn, *Phys. Lett.*, **B78**, 269 (1978).
2. ZEUS Collaboration, J. Breitweg *et al.*, *Phys. Lett.*, **B481**, 199 (2000); ZEUS Collaboration, S. Chekanov *et al.*, *Phys. Lett.*, **B551**, 3 (2003).
3. ZEUS Collaboration, Paper 277, *XXII International Symposium on Lepton-Photon Interactions at High Energy*, June 30-July 5, 2005, Uppsala, Sweden.
4. S. Catani and M. H. Seymour, *Nucl. Phys.* **B485**, 291 (1997).
5. ZEUS Collaboration, Paper 296, *XXII International Symposium on Lepton-Photon Interactions at High Energy*, June 30-July 5, 2005, Uppsala, Sweden.
6. U. D'Alesio and H. J. Pirner, *Eur. Phys. J.*, **A7**, 109 (2000).
7. H1 Collaboration, C. Adloff *et al.*, *Eur. Phys. J.*, **C6**, 587 (1999).
8. H1 Collaboration, A. Aktas *et al.*, *Eur. Phys. J.*, **C41**, 273 (2005).
9. M. Klasen and G. Kramer, *Phys. Lett.*, **B508**, 259 (2001).

Jets in Photoproduction and at Low Q^2 at HERA

Kamil Sedlák on behalf of the H1 and ZEUS Collaborations

Department of Physics, Oxford University, Oxford, UK

Abstract. Recent H1 and ZEUS measurements of two and three jet cross sections in electron-proton interactions in the photoproduction region or at low photon virtuality, Q^2, are presented and compared to NLO QCD calculations. Phase space regions where data are not well reproduced by the NLO theory are pointed out. A recent ZEUS measurement of color dynamics sensitive to the underlying gauge group in QCD is presented.

Keywords: Jets; Photoproduction; DIS; Low Q2; HERA
PACS: 13.85.Hd

INTRODUCTION

The production of jets with high transverse momenta in electron-proton collisions is described within perturbative quantum chromodynamics (QCD) by the hard interaction of photons with partons inside the proton. Jet cross sections are successfully described by next-to-leading order (NLO) QCD calculations in most of the HERA kinematic range. However, regions of phase space have been observed for which NLO QCD predictions do not reproduce the data satisfactorily.

The measurement of jet production in the low Q^2 region, where Q^2 is the photon virtuality, is particularly suitable for the investigation of effects related to the photon structure. For $Q^2 \lesssim E_T^2$, where E_T is the transverse energy of the leading jet[1] (jets), the interaction can be conveniently described by the sum of two contributions. In the "direct photon" process the photon interacts as a whole with a parton from the proton, whereas in the "resolved photon" process it acts as a source of partons, which interact with partons from the proton. The parton distributions functions (PDF) of the photon are introduced in the latter case.

The dijet cross sections are often studied as a function of the variable x_γ defined as

$$x_\gamma = \frac{\sum\limits_{j=1,2} (E_j^* - p_{z,j}^*)}{\sum\limits_{\text{hadrons}} (E^* - p_z^*)}, \tag{1}$$

where the sum in the numerator runs over the two leading jets and the sum in the denominator includes the full hadronic final state. Neglecting the masses of the partons and beam particles, the variable x_γ represents an estimate of the fraction of the photon four-momentum carried by the parton involved in the hard scattering.

---

[1] E_T is assumed to be the hard scale of the process. In general, Q^2 should be smaller than the square of the renormalisation scale, μ_r^2, when introducing the resolved photon concept.

CP828, *Multiparticle Dynamics*
edited by V. Šimák, M. Šumbera, Š. Todorova, and B. Tomášik

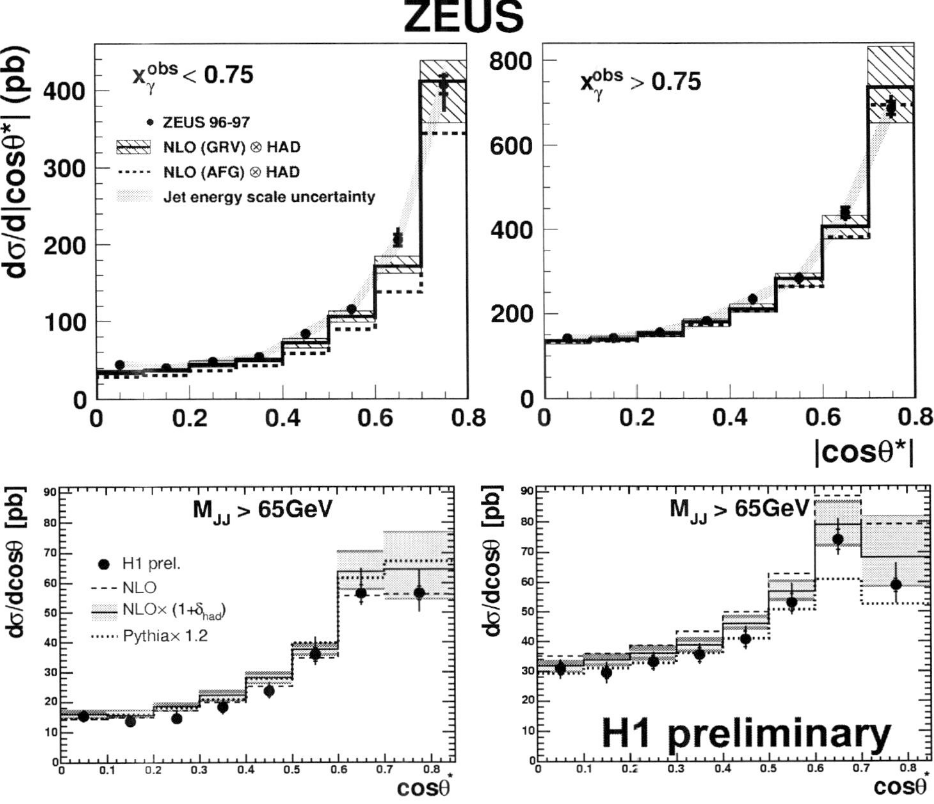

FIGURE 1. Cross section as a function of $|\cos\theta^*|$ measured by ZEUS (upper plots) and H1 (lower plots). The plots in the left column correspond to the region of $x_\gamma < 0.75$ ($x_\gamma < 0.8$), the plots in the right column to $x_\gamma > 0.75$ ($x_\gamma > 0.8$) for ZEUS (H1), respectively. Data are compared to the NLO predictions corrected for the hadronisation effects.

PHOTOPRODUCTION OF DIJETS

A published ZEUS measurement [1] as well as recent H1 preliminary results [2] of the measurement of dijet cross sections in the photoproduction region are compared with the NLO QCD predictions. As an example, dijet cross sections are shown in Fig. 1 as a function of $|\cos\theta^*|$, where θ^* denotes angle between the two leading jets in their centre-of-mass system. The phase space of the measurements is defined mainly by the cuts on E_T of the two leading jets ($E_{T\text{jet1}} > 14\,\text{GeV}$, $E_{T\text{jet2}} > 11\,\text{GeV}$ in the case of ZEUS and $E_{T\text{jet1}} > 25\,\text{GeV}$, $E_{T\text{jet2}} > 15\,\text{GeV}$ in the case of H1; see [1, 2] for further details about the event selection). The dijet cross section in Fig. 1 are shown for two different regions of x_γ. The data at low x_γ rise more rapidly at high $|\cos\theta^*|$ than those at high x_γ. This is consistent with a difference in the dominant propagators for direct (quark propagator) and resolved (gluon propagator) photon events[2].

Considering the theoretical and experimental uncertainties, the NLO predictions give a reasonable description of the data. This is also true for the photoproduction dijet cross sections plotted more differentially, e.g. as a function of E_T of the leading jet in different pseudorapidity regions [1, 2]. Due to the precision of these data, they are now used to constrain the mid- to high-x gluon distribution functions in the proton [3] as well as to constrain PDFs of the photon.

TRANSITION BETWEEN THE PHOTOPRODUCTION AND DIS

While dijet cross sections in the photoproduction region are quite well reproduced by the NLO QCD calculations, both H1 and ZEUS have reported significant discrepancies between the measurements and theory predictions of dijet cross section in the low Q^2 region, especially at low x_γ and for low E_T jets [4, 5]. The discrepancy is exemplified in Fig. 2, where dijet cross sections are shown triple differentially as a function of x_γ in different bins of Q^2 and E_T. In Fig. 2, the variable E_T denotes the transverse energies of the jets with the highest and the second highest transverse energy measured in the photon-proton centre-of-mass frame, so that each event contributes twice to the distributions, not necessarily in the same bin. The data are compared with the NLO QCD calculations: DISENT, JETVIP and NLOJET++. JETVIP denoted as "full" corresponds to the sum of the direct and resolved photon components, all the other predictions are just the direct photon contributions. NLOJET++ is run in two modes – either as the NLO QCD prediction for dijets (2 jet mode) or as the NLO QCD prediction for three jet production (3 jet mode). The latter can be compared to dijet cross section just in the region of $x_\gamma < 1$, where only events with at least three jets can contribute (see the discussion in [6]). Thus, NLOJET++ in the 3 jet mode calculates dijet cross section up to one order higher than the other NLO QCD programs mentioned above.

Fig. 2 illustrates that data are well reproduced by NLO QCD predictions at high x_γ, at high E_T or in the high Q^2 region. As we go to the low x_γ, low E_T and low Q^2 region, quite significant discrepancies are observed. The prediction of NLOJET++ in three jet mode and the prediction of JETVIP that includes also the resolved photon contributions get closer to the data, which indicates that higher order effects are probably missing in the theory predictions.

A similar conclusion has been drawn from the ZEUS measurement [5].

COLOR DYNAMICS IN PHOTOPRODUCTION

Recently, new ZEUS preliminary measurements of three jet cross sections have been presented [7] as a function of jet angular variables Θ_H, $\cos\alpha_{23}$ and $\cos\beta_{KSW}$, which are constructed from different combinations of planes and vectors defined by jets and

[2] More exactly, the resolved photon contributions are driven by both gluon and quark propagators, while the gluon propagator is absent in LO direct photon processes. The quark propagator is proportional to $(1-|\cos\theta^*|)^{-1}$ and the gluon propagator is proportional to $(1-|\cos\theta^*|)^{-2}$.

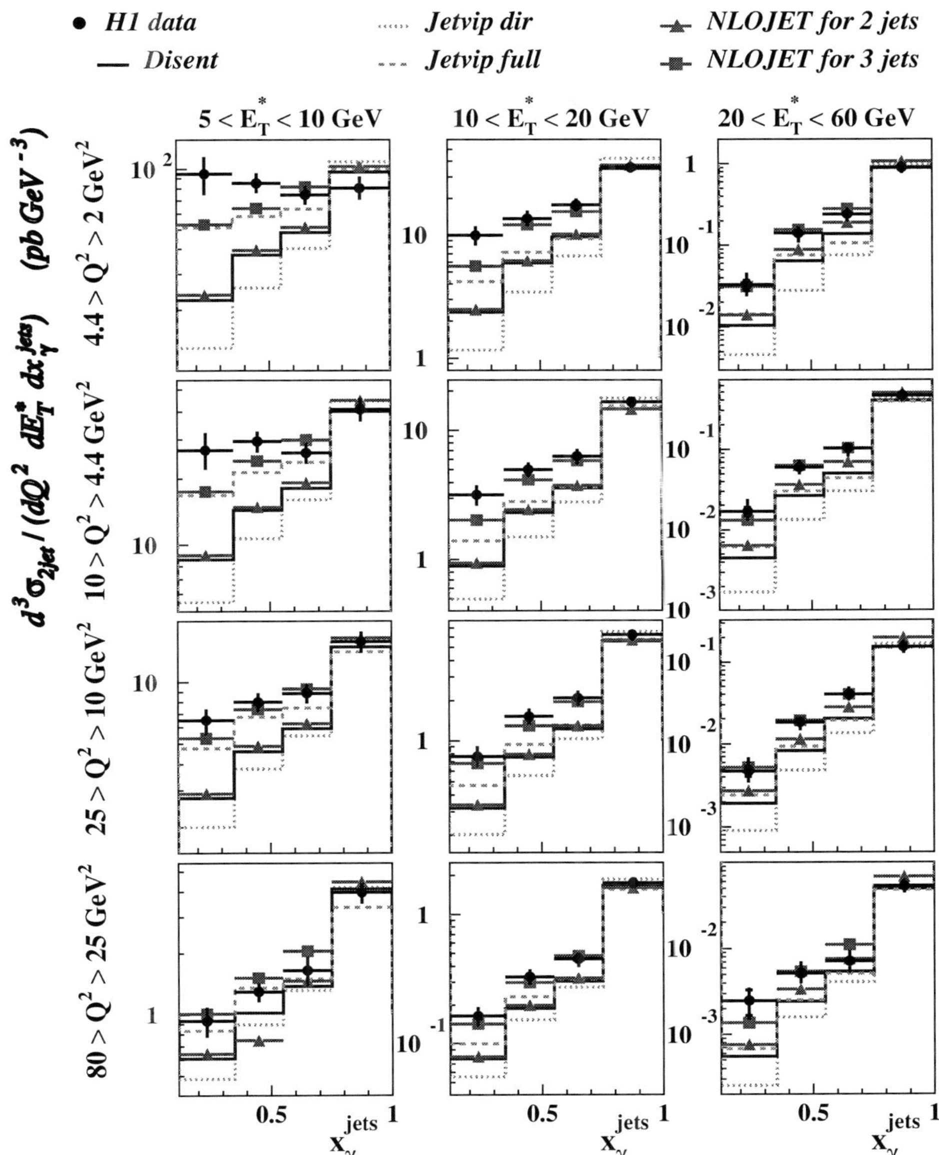

FIGURE 2. Triple differential dijet cross section $d^3\sigma_{2jet}/dQ^2 dE_T\, dx_\gamma$ with asymmetric E_T cuts (see [4] for details). The data are compared to NLO direct photon calculations using DISENT (full line) and JETVIP (dotted line), the sum of NLO direct and resolved photon contributions of JETVIP (dashed line) and the NLOJET++ predictions in 2-jet (triangles) and 3-jet (squares) mode. All calculations are corrected for hadronisation effects.

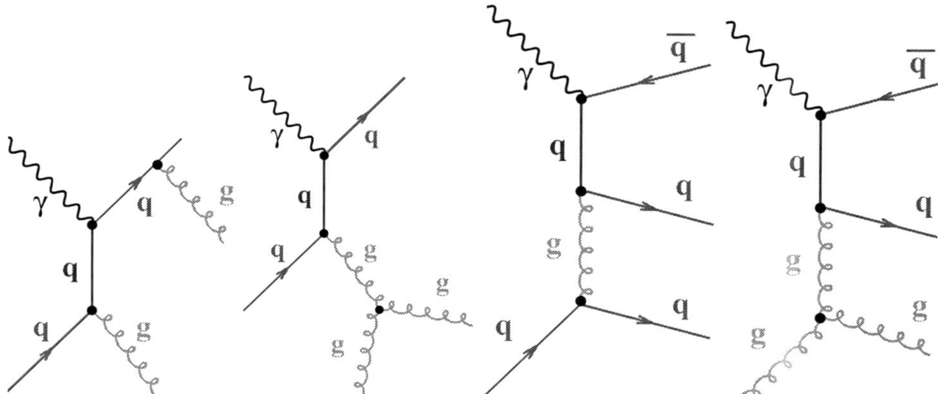

FIGURE 3. Examples of diagrams for the photoproduction of three jet events through direct photon processes in each color configuration.

the electron-proton beam. These variables are sensitive to different color configurations (i.e. to different combinations of color factors C_F, C_A and T_F) that correspond to four different types of diagrams for the photoproduction of three-jet events through direct photon processes (Fig. 3). The C_F, C_A and T_F color factors are a physical manifestation of the underlying group structure in QCD. They represent the relative strengths of the processes $q \rightarrow qg$, $g \rightarrow gg$ and $g \rightarrow q\overline{q}$ respectively. Due to different spins of quark and gluon, the color factors give rise to a specific pattern of angular correlations between the final state jets.

In Fig. 4, the color components have been combined in such a way as to reproduce the color structure of a theory based on the non-abelian group SU(N) for N=3 and in the limit of large N, the abelian group $U(1)^3$ and, as an extreme choice, a calculation with $C_F = 0, C_A = 3$ and $T_F = 1/2$. The data clearly disfavour the last example with $C_F = 0$ and a theory based on SU(N) in the limit of large N. On the other hand, no distinction between the SU(3) and $U(1)^3$ can be done from the presented angular distributions.

SUMMARY

Jet production in the region of $Q^2 < E_T^2$ in electron-proton collisions keeps providing us a precise and important information about the proton and photon structure that uniquely complements other measurements. On the other hand, the region of low x_γ, low E_T and low Q^2 (i.e. $Q^2 \sim 1\,\mathrm{GeV}^2$) is not satisfactorily described by the present theory. The measurement of angular correlations in three jet photoproduction is consistent with the admixture of color configurations as predicted by the SU(3) gauge group.

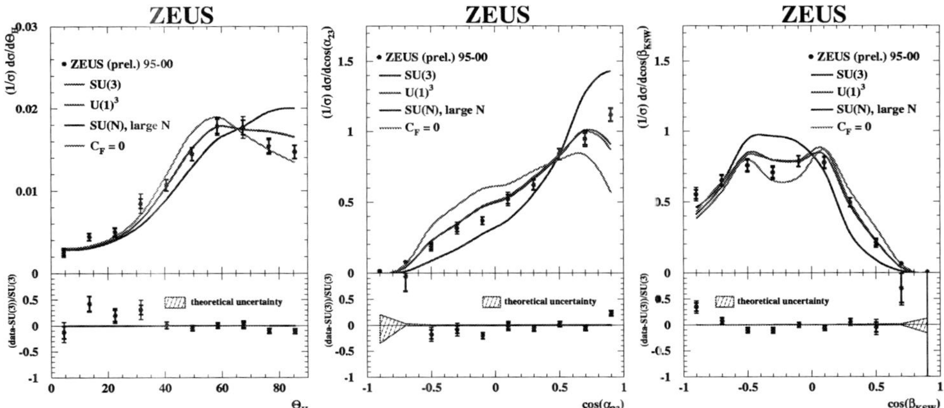

FIGURE 4. Normalised differential *ep* cross sections for three jet photoproduction integrated over $E_T > 14$GeV and $-1 < \eta < 2.5$ in the kinematic region defined by $Q^2 < 1\text{GeV}^2$, $0.2 < y < 0.85$ and $x_\gamma > 0.7$ as function of Θ_H, $\cos\alpha_{23}$ and $\cos\beta_{KSW}$. The data are compared to NLO QCD calculations for direct photon processes based on SU(3) (solid lines), $\text{U}(1)^3$ (dashed lines), SU(N) in the limit of large N (dot-dashed lines) and $C_F = 0$ (dotted line).

ACKNOWLEDGMENTS

This talk was supported by a Marie Curie Intra-European Fellowships within the 6th European Community Framework Programme.

REFERENCES

1. S. Chekanov et al., *Eur. Phys. J.* **C 23** (2002) 615.
2. I. Strauch, "Jets with high transverse momenta in photoproduction at HERA," DESY-THESIS-2004-047.
3. S. Chekanov et al., *Eur. Phys. J.* **C 42** (2005) 1.
4. A. Aktas et al., *Eur. Phys. J.* **C 37** (2004) 141.
5. S. Chekanov et al., *Eur. Phys. J.* **C 35** (2004) 487.
6. J. Chyla, J. Cvach, K. Sedlak and M. Tasevsky, *Eur. Phys. J.* **C 40** (2005) 469.
7. J. Terron, "Study of Color Dynamics in Photoproduction at HERA", proceedings of the 13th International Workshop on Deep Inelastic Scattering (DIS 2005), April 27 – May 1, 2005, Madison, Wisconsin, USA.

The Color Glass Condensate: An Intuitive Physical Description

Larry McLerran

Physics Department PO Box 5000, Brookhaven National Laboratory, Upton, NY 11973 USA

Abstract. I argue that the scattering of very high energy strongly interacting particles is controlled by a new, universal form of matter, the Color Glass Condensate. This matter is predicted by QCD and explains the saturation of gluon densites at small x. I motivate the existence of this matter and describe some of its properties.

WHAT IS THE COLOR GLASS CONDENSATE?

The original ideas for the Color Glass Condensate were derived from consideration of the results for the HERA data on the gluon distribution function shown in Fig. 1(a) [1].

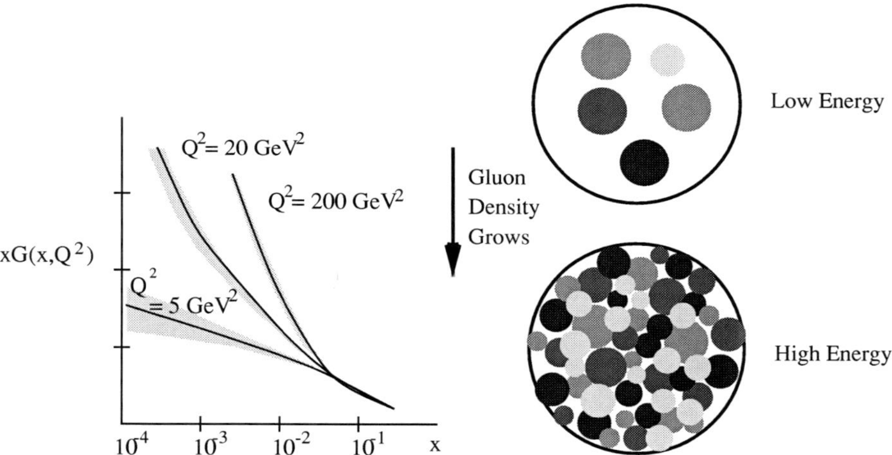

FIGURE 1. (a)The HERA data for the gluon distribution function as a function of x for various values of Q^2. (b) A physical picture of the low x gluon density inside a hadron as a function of energy.

The gluon density is rising rapidly as a function of decreasing x. This was expected in a variety of theoretical works,[2]-[4] and has the implication that the real physical transverse density of gluons must increase.[2]-[3],[5]. This follows because total cross sections rise slowly at high energies but the number of gluons is rising rapidly. This is shown in Fig. 1(b). This led to the conjecture that the density of gluons should become limited, that is, there is gluon saturation [2]-[3], [5].

CP828, *Multiparticle Dynamics*
edited by V. Šimák, M. Šumbera, Š. Todorova, and B. Tomášik

The low x gluons therefore are closely packed together. The strong interaction strength must become weak, $\alpha_S \ll 1$. Weakly coupled systems should be possible to understand from first principles in *QCD* [5]-[6].

This weakly coupled system is called a Color Glass Condensate for reasons we now enumerate [6]:

- **Color** The gluons which make up this matter are colored.
- **Glass** The gluons at small x are generated from gluons at larger values of x. In the infinite momentum frame, these larger momentum gluons travel very fast and their natural time scales are Lorentz time dilated. This time dilated scale is transferred to the low x degrees of freedom which therefore evolve very slowly compared to natural time scales. This is the property of a glass.
- **Condensate** The phase space density

$$\rho = \frac{1}{\pi R^2} \frac{dN}{dyd^2 p_T} \tag{1}$$

  is generated by a trade off between a negative mass-squared term linear in the density which generates the instability, $-\rho$ and an interaction term $\alpha_S \rho^2$ which stabilizes the system at a phase space density $\rho \sim 1/\alpha_S$. Because $\alpha_S << 1$, this means that the quantum mechanical states of the system associated with the condensate are multiply occupied. They are highly coherent, and share some properties of Bose condensates. The gluon occupation factor is very high, of order $1/\alpha_S$, but it is only slowly (logarithmically) increasing when further increasing the energy, or decreasing the transverse momentum. This provides saturation and cures the infrared problem of the traditional BFKL approach [7].

Implicit in this definition is a concept of fast gluons which act as sources for the colored fields at small x. These degrees of freedom are treated differently than the fast gluons which are taken to be sources. The slow ones are fields. There is an arbitrary X_0 which separates these degrees of freedom. This arbitrariness is cured by a renormalization group equation which requires that physics be independent of X_0. In fact this equation determines much of the structure of the resulting theory as its solution flows to a universal fixed point [6]-[9].

There is evidence which supports this picture. One piece is the observation of limiting fragmentation. This phenomena is that if particles collide at some fixed center of mass energy and the distribution of particles are measured as a function of their longitudinal momentum from the longitudinal momentum of one of the colliding particles, then these distributions do not change as one goes to higher energy, except for the new degrees of freedom that appear. This is true near zero longitudinal momentum in the center of mass frame because new degrees of freedom appear as the center of mass energy is increased. In the analogy with the CGC, the degrees of freedom, save the new ones added in at low longitudinal momentum, are the sources. The fields correspond to the new degrees of freedom. The sources are fixed in accord with limiting fragmentation. One generates an effective theory for the low longitudinal momentum degrees of freedom as fixed sources above some cutoff, and the fields generated by these sources below the cutoff. A recent measurement of limiting fragmentation comes from the Phobos experiment at RHIC shown in Fig. 2 [10].

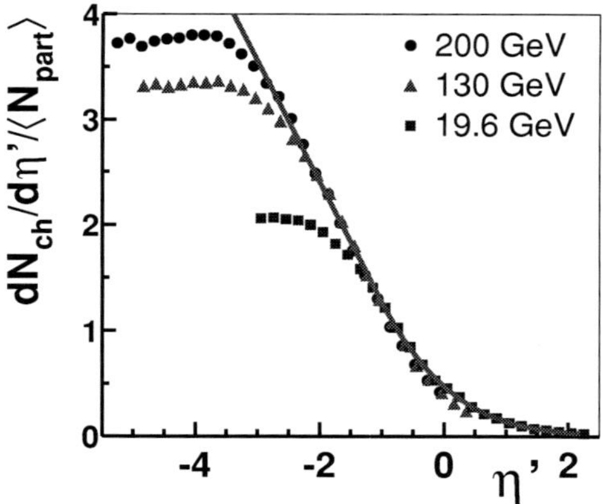

FIGURE 2. Limiting fragmentation and the RHIC data.

Of course the perfect scaling of the limiting fragmentation curves is only an approximation. As shown by Jalilian-Marian, the limiting fragmentation curves are given by the total quark, antiquark and gluon distribution functions of the fast particle measured at a momentum scale Q^2_{sat} appropriate for the particle that it collides with [11]. The saturation momentum Q_{sat} will play a crucial role in our later discussion. It is a momentum scale which is determined by the density of gluons in the CGC

$$\frac{1}{\pi R^2}\frac{dN}{dy} \sim \frac{1}{\alpha_S} Q^2_{sat} \tag{2}$$

The saturation momenta turns out to depend on the total beam energy because the longitudinal momentum scale of the target particle at fixed x of the projectile will depend upon the beam energy. It is nevertheless remarkable how small these violations appear to be.

The CGC may be defined mathematically by a path integral:

$$Z = \int_{X_0} [dA][dj] exp\left(iS[A,j] - \chi[j]\right) \tag{3}$$

What this means is that there is an effective theory defined below some cutoff in x at X_0, and that this effective theory is a gluon field in the presence of an external source j. This source arises from the quarks and gluons with $x \geq X_0$, and is a variable of integration. The fluctuations in j are controlled by the weight function $\chi[j]$. It is $\chi[j]$ which satisfies renormalization group equations which make the theory independent of X_0.[8]-[14],[6]. The equation for χ is called the JIMWLK equation. This equation reduces in appropriate limits to the BFKL and DGLAP evolution equations [4], [15]. The theory above is mathematically very similar to that of spin glasses.

There are a variety of kinematic regions where one can find solutions of the renormalization group equations which have different properties. There is a region where the gluon density is very high, and the physics is controlled by the CGC. This is when typical momenta are less than a saturation momenta which depends on x,

$$Q^2 \leq Q_{sat}^2(x) \tag{4}$$

The dependence of x has been evaluated by several authors, [2],[16]-[18], and in the energy range appropriate for current experiments has been determined by Triantafyllopoulos to be

$$Q_{sat}^2 \sim (x_0/x)^{\lambda}\ GeV^2 \tag{5}$$

where $\lambda \sim 0.3$. The value of x_0 is not determined from the renormalization group equations and must be found from experiment.

There is also a region of very high Q^2 at fixed x, where the density of gluons is small and perturbative QCD is reliable. It turns out there is a third region intermediate between high density and low where there are universal solutions to the renormalization group equations and scaling in terms of Q_{sat}^2.[17] In this region and in the region of the CGC, distribution functions are universal functions of only $Q^2/Q_{sat}^2(x)$. The extended scaling region is when

$$Q_{sat}^2 \leq Q^2 \leq Q_{sat}^4/\Lambda_{QCD}^2 \tag{6}$$

WHAT IS THE CGC GOOD FOR?

The CGC provides a unified description of deep inelastic structure functions, of deep inelastic diffraction and of hadron-hadron collisions at high energies. It is the high energy limit of QCD. As such, it has many test to pass before being accepted as a correct description. Over the last several years, there have been many qualitative and semi-quantitative successes of this description. It also provides an intuitively plausible and mathematically consistent description of such phenomena.

In the future years, there will be increasingly stringent tests arising at RHIC, LHC and potentially eRHIC. Theoretically, we are just beginning to understand the properties of this matter. New ideas concerning the structure of the underlying theory and the breadth of phenomena it describes are changing the way we think about high energy density matter.

ACKNOWLEDGEMENTS

I gratefully acknowledge conversations with Edmond Iancu, Kazu Itakura, Miklos Gyulassy, Dima Kharzeev Genya Levin, and Raju Venugopalan on the subject of this talk

This manuscript has been authorized under Contract No. DE-AC02-98CH0886 with the U. S. Department of Energy.

REFERENCES

1. J. Breitweg et. al. *Eur. Phys. J.* **67**, 609 (1999).
2. L. V. Gribov, E. M. Levin and M. G. Ryskin, *Phys. Rept.* **100**, 1 (1983).
3. A. H. Mueller and Jian-wei Qiu, *Nucl. Phys.* **B268**, 427 (1986); J.-P. Blaizot and A. H. Mueller, *Nucl. Phys.* **B289**, 847 (1987).
4. L.N. Lipatov, *Sov. J. Nucl. Phys.* **23** (1976), 338;
   E.A. Kuraev, L.N. Lipatov and V.S. Fadin, *Sov. Phys. JETP* **45** (1977), 199;
   Ya.Ya. Balitsky and L.N. Lipatov, *Sov. J. Nucl. Phys.* **28** (1978), 822.
5. L. D. McLerran and R. Venugopalan, *Phys. Rev.* **D49**, 2233(1994); 3352 (1994); **D50**, 2225 (1994).
6. E. Iancu, A. Leonidov and L. D. McLerran, *Nucl. Phys.* **A692**, 583 (2001); E. Ferreiro E. Iancu, A. Leonidov and L. D. McLerran, *Nucl. Phys.* bf A710,373 (2002).
7. E. Iancu and L. McLerran, *Phys.Lett.* **B510**, 145 (2001).
8. J. Jalilian-Marian, A. Kovner, L. McLerran and H. Weigert, *Phys. Rev.* **D55** (1997), 5414.
9. J. Jalilian-Marian, A. Kovner, A. Leonidov and H. Weigert, *Nucl. Phys.* **B504** (1997), 415; *Phys. Rev.* **D59** (1999), 014014.
10. B. Back et. al. *Phys. Rev. Lett.* **91**, 052303 (2003).
11. J. Jalilian-Marian, nucl-th/0212018
12. I. Balitsky, *Nucl. Phys.* **B463** (1996), 99.
13. Yu. V. Kovchegov, *Phys. Rev.* **D60** (1999), 034008; *ibid.* **D61** (2000), 074018.
14. A. H. Mueller, *Phys.Lett.* **B523**, 243 (2001)
15. V.N. Gribov and L.N. Lipatov, *Sov. Journ. Nucl. Phys.* **15** (1972), 438; G. Altarelli and G. Parisi, *Nucl. Phys.* **B126** (1977), 298; Yu. L. Dokshitzer, *Sov. Phys. JETP* **46** (1977), 641.
16. E. Levin and K. Tuchin, *Nucl. Phys.* **A691**, 779 (2001)
17. E. Iancu, K. Itakura and L. McLerran, *Nucl. Phys.* **A708**, 327 (2002).
18. A. H. Mueller and V. N. Triantafyllopoulos, *Nucl. Phys.* **B640**, 331 (2002). D. N. Triantafyllopoulos, *Nucl. Phys.* **B648**, 293 (2003).

Verification of Z-scaling in pp Collisions at RHIC

M.Tokarev* and I.Zborovský†

*JINR, 141980 Dubna, Moscow region, Russia
†NPI, 25068 Řež, Czech Republic

Abstract. New experimental data on inclusive spectra of identified particles produced in *pp* collisions at the RHIC are used to test *z*-scaling. Energy and multiplicity independence of the scaling function is established. The RHIC data confirm *z*-scaling observed at U70, ISR, SpS and Tevatron energies. The obtained results are of interest to search for new physics phenomena of particle production in high transverse momentum and high multiplicity region at RHIC, Tevatron and LHC.

Keywords: Proton-proton collisions, high energy, high multiplicity, scaling
PACS: 13.85.Hd, 13.85.Ni, 13.87.Fh

INTRODUCTION

Study of scaling regularities in high energy collisions is always subject of intense experimental and theoretical investigations [1]-[9]. Some scalings can reflect fundamental symmetries in Nature. Basic principles to study such symmetries at small scales are self-similarity, locality and fractality. New scaling (*z*-scaling) for description of high-p_T particle production in inclusive reactions was established in [10]. Properties of *z*-presentation of numerous experimental data confirm self-similarity, locality and fractality of hadron interactions at high energies. The Relativistic Heavy Ion Collider (RHIC) at the Brookhaven National Laboratory (BNL) gives wide possibilities to perform experimental measurements and test scaling regularities in a new physics domain.

We present results of analysis of new data on high-p_T particle spectra obtained at the RHIC. The data confirm *z*-scaling observed at the U70, ISR, SpS and Tevatron.

Z-SCALING

Search for an adequate, physically meaningful but still sufficiently simple form of the self-similarity parameter z plays a crucial role in our approach. For inclusive reactions we define the scaling variable

$$z = \frac{s_{\perp}^{1/2}}{W} \qquad (1)$$

as ratio of the minimal transverse kinetic energy $s_{\perp}^{1/2}$ of underlying constituent subprocess and relative number W of such configurations of the colliding system which can contribute to production of inclusive particle with the momentum p. The number of configurations is expressed via the multiplicity density $dN/d\eta|_0$ at pseudorapidity

CP828, *Multiparticle Dynamics*
edited by V. Šimák, M. Šumbera, Š. Todorova, and B. Tomášik

$\eta = 0$ and kinematical characteristics x_1, x_2 and y of the subprocess as follows

$$W = (dN/d\eta|_0)^c \cdot \Omega(x_1,x_2,y), \tag{2}$$

where

$$\Omega(x_1,x_2,y) = (1-x_1)^{\delta_1}(1-x_2)^{\delta_2}(1-y)^{\varepsilon}. \tag{3}$$

Here x_1 and x_2 are momentum fractions of the colliding objects (hadrons or nuclei). The y is momentum fraction of outgoing constituent from the subprocess carried by the inclusive particle. The δ_1, δ_2 and ε are anomalous fractal dimensions of the incoming and outgoing objects, respectively. The variable z has character of a fractal measure

$$z = z_0\Omega^{-1}. \tag{4}$$

Its divergent part Ω^{-1} describes resolution at which the collision of the constituents can be singled out of inclusive reaction. With increasing resolution the measure z tends to infinity. The x_1, x_2 and y are determined in a way to minimize the resolution $\Omega^{-1}(x_1,x_2,y)$ taking into account the energy-momentum conservation of the binary subprocess written in the form

$$(x_1P_1 + x_2P_2 - p/y)^2 = (x_1M_1 + x_2M_2 + m_2/y)^2. \tag{5}$$

Here P_1, P_2 and M_1, M_2 are 4-momenta and masses of the colliding objects. The p is 4-momentum of the inclusive particle. The parameter m_2 is minimal mass introduced to satisfy the internal conservation laws (for baryon number, isospin, strangeness,...).

The relative number W of the configurations which include the constituent subprocess is expressed via entropy of the rest of the colliding system as follows

$$S = \ln W. \tag{6}$$

Using equations (2) and (3), we get

$$S = c\ln\left[dN/d\eta|_0\right] + \ln\left[(1-x_1)^{\delta_1}(1-x_2)^{\delta_2}(1-y)^{\varepsilon}\right] \tag{7}$$

Exploiting analogy with the thermodynamical formula

$$S = c_V \ln T + R\ln V + const. \tag{8}$$

we can consider multiplicity density $dN/d\eta|_0$ as a quantity characterizing "temperature" of the colliding system and the parameter c as "heat capacity" of the medium. The second term in (7) is related to volume of the configurations in space of the momentum fractions which can contribute to production of the inclusive particle with the momentum p. Note that minimal resolution Ω^{-1} of the fractal measure z with respect to constituent subprocesses corresponds to maximal entropy S of the rest of the system.

In accordance with self-similarity principle we search for the scaling function

$$\psi(z) = \frac{1}{N\sigma_{in}}\frac{d\sigma}{dz} \tag{9}$$

depending on the single variable z. Here σ_{in} is the inelastic cross section of the inclusive reaction and N is particle multiplicity. The function $\psi(z)$ is expressed in terms of the

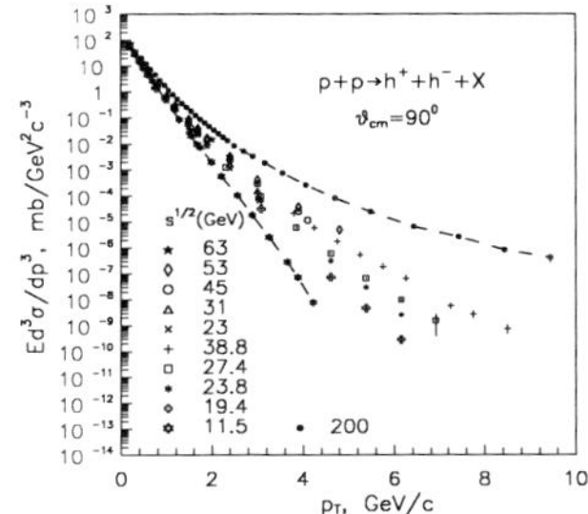

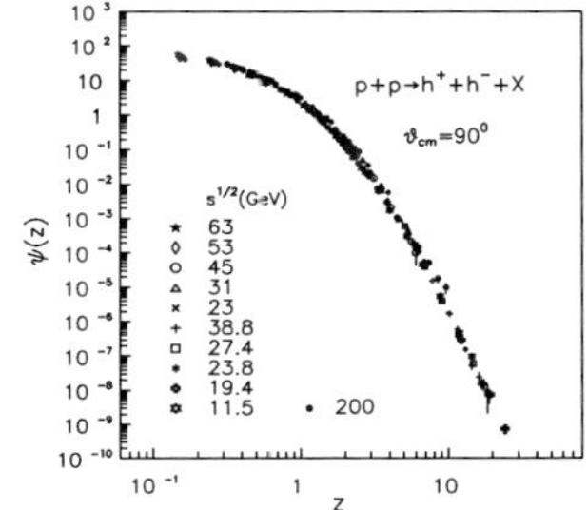

FIGURE 1. (a) Inclusive cross sections of charged hadrons produced in pp collisions at $\sqrt{s}=11.5-63$ and 200 GeV and $\theta_{cm}\simeq 90^0$ as a functions of the transverse momentum p_T. The experimental data are taken from [12]-[15] and [11]. (b) The corresponding scaling function.

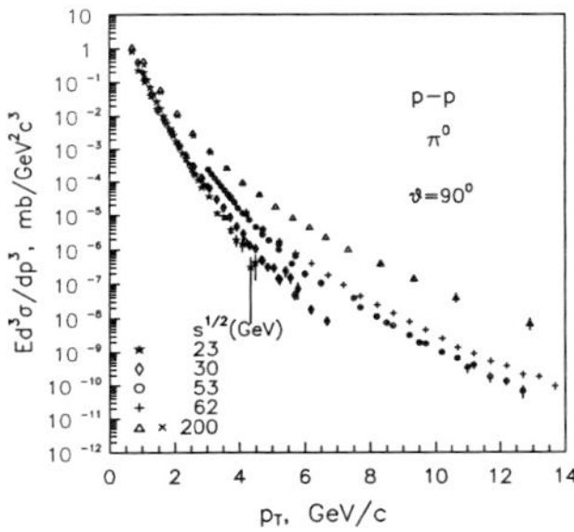

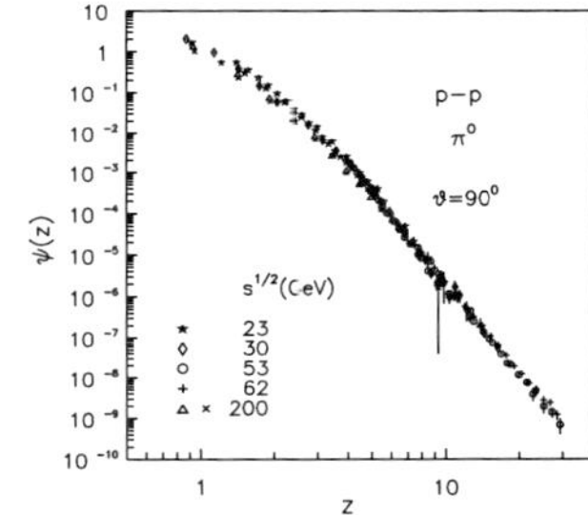

FIGURE 2. The dependence of the inclusive cross section of π^0-meson production on the transverse momentum p_T in pp collisions at $\sqrt{s}=30,53,62$ and 200 GeV and the angle θ_{cm} of 90^0. The experimental data are taken from [17, 18, 19, 20, 21] and [16]. (b) The corresponding scaling function.

experimentally measured inclusive invariant cross section $Ed^3\sigma/dp^3$ and multiplicity density $dN/d\eta$ as follows

$$\psi(z)=-\frac{\pi s}{(dN/d\eta)\sigma_{in}}J^{-1}E\frac{d^3\sigma}{dp^3}. \tag{10}$$

Here s is the center-of-mass collision energy squared. The Jacobian J of transformation to the variables (z,η) depends on the momentum p of the inclusive particle. The $\psi(z)$ has meaning of probability density to produce inclusive particle with the corresponding value of the variable z.

Z-SCALING AT RHIC

We analyze experimental data on minimum bias pp spectra of different hadrons $(h^{\pm},\pi^0,\pi^-,K_S^0)$ measured at the RHIC. Comparison of the RHIC data with data obtained at the U70, ISR, SpS and Tevatron is used to test z-scaling.

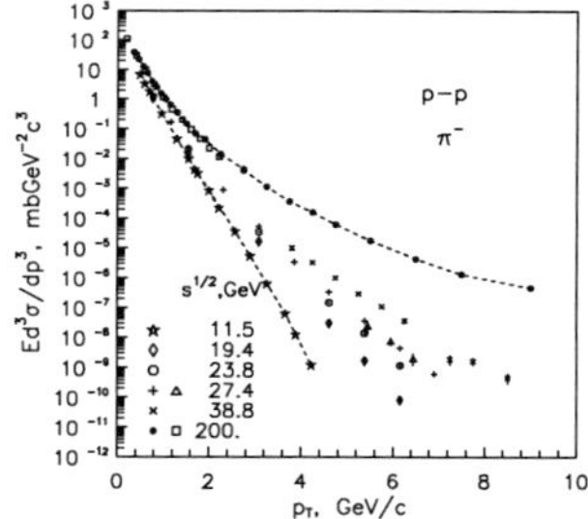
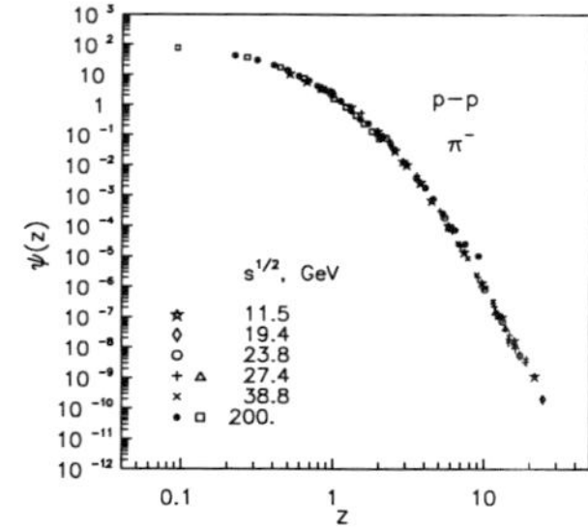

FIGURE 3. (a) Inclusive cross sections of charged hadrons produced in pp collisions at $\sqrt{s} = 11.5 - 38.8$ and 200 GeV and $\theta_{cm} \simeq 90^0$ as a functions of the transverse momentum p_T. The experimental data are taken from [12, 13, 14] and [11]. (b) The corresponding scaling function.

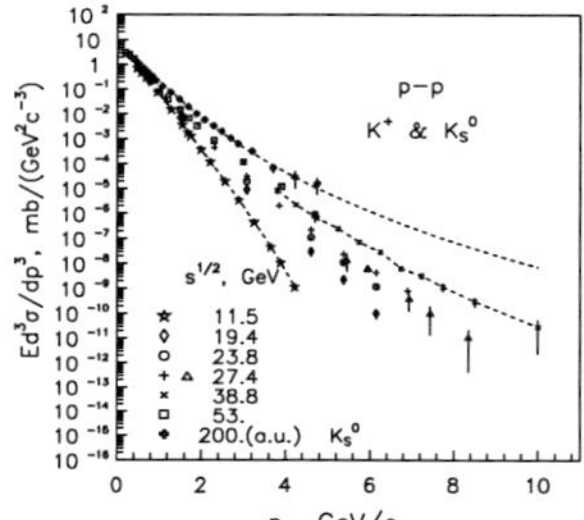
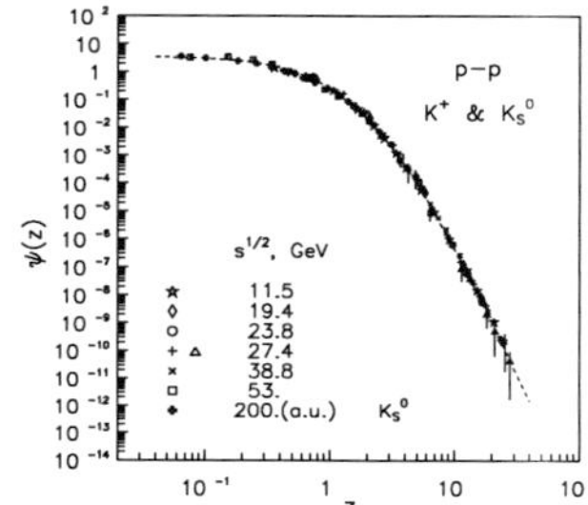

FIGURE 4. (a) The inclusive cross sections of K^+- and K_S^0-mesons produced in pp collisions in the central rapidity range as a function of the transverse momentum at $\sqrt{s} = 11.5 - 53$ GeV and 200 GeV. Experimental data are taken from [12, 13, 14, 15] and [23]. (b) The corresponding scaling function.

Energy independence of $\psi(z)$

The high-p_T spectra of charged hadrons produced in pp collisions at the energy $\sqrt{s} = 200$ GeV within $|\eta| < 0.5$ were measured by the STAR Collaboration [11]. The inclusive cross sections obtained at the U70 [12], Tevatron [13, 14], ISR [15] and RHIC are presented in Fig.1a. The spectra have strong energy dependence at high p_T. Fig.1b shows z-presentation of the same data. The scaling function demonstrates energy independence and power law, $\psi(z) \sim z^{-\beta}$, for $z > 4$. Results of analysis of the STAR data at $\sqrt{s} = 200$ GeV confirm z-scaling observed at lower energies.

The PHENIX Collaboration measured the inclusive spectrum of π^0-mesons produced in pp collisions at $\sqrt{s} = 200$ GeV for $|\eta| < 0.35$ and p_T up to 13 GeV/c [16]. The p_T- and z-presentations of data for π^0-meson spectra obtained at ISR [17, 18, 19, 20, 21] and RHIC are shown in Figs.2a and 2b, respectively. Energy dependence of the inclusive cross sections increases with p_T. The PHENIX data at $\sqrt{s} = 200$ GeV confirm energy independence and power law of the scaling function for π^0-mesons.

The STAR Collaboration measured the inclusive spectrum of π^--mesons produced in pp collisions at $\sqrt{s} = 200$ GeV for $|\eta| < 0.5$ up to $p_T = 9$ GeV/c [22]. The data

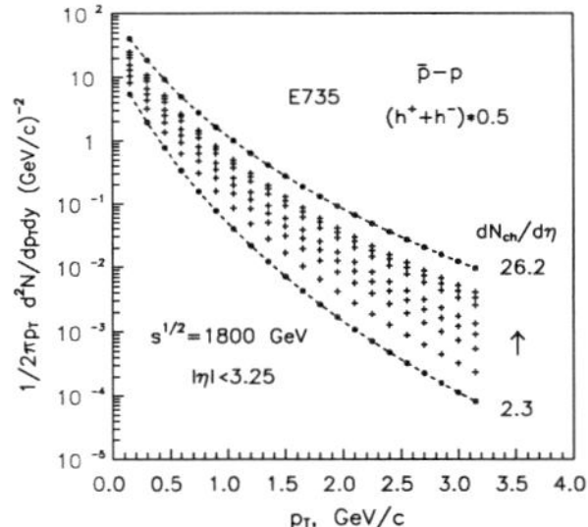

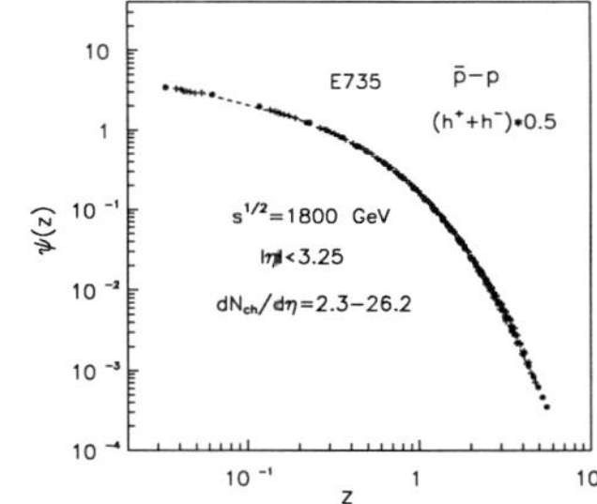

FIGURE 5. a) Multiplicity dependence of charged hadron spectra in $\bar{p}p$ collisions at $\sqrt{s} = 1800$ GeV. Experimental data are obtained by the E735 Collaboration [24]. (b) The corresponding scaling function.

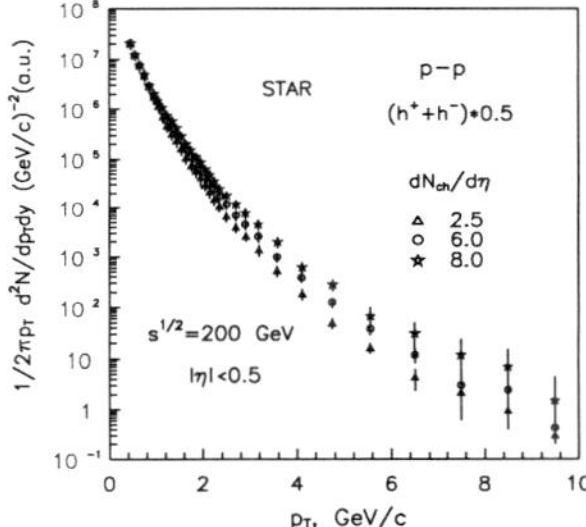

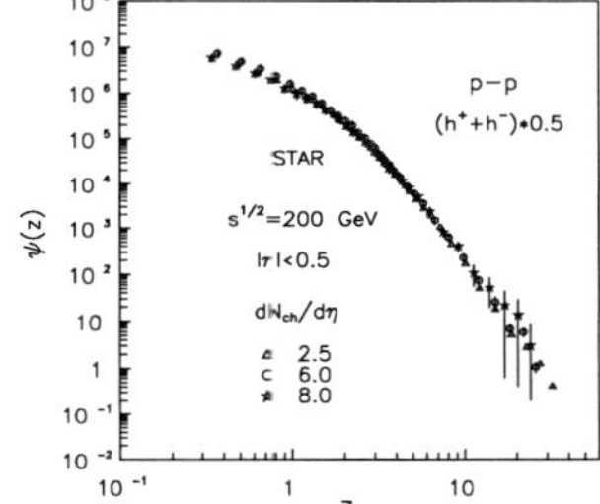

FIGURE 6. (a) Multiplicity dependence of charged hadron spectra in pp collisions at $\sqrt{s} = 200$ GeV. Experimental data are obtained by the STAR Collaboration [25]. (b) The corresponding scaling function.

obtained at the STAR, U70 [12] and Tevatron [13, 14] are shown in Fig.3a. The energy dependence of the p_T-spectra is in contrast with the scaling depicted in Fig.3b.

The p_T-spectra and the scaling function $\psi(z)$ for data on K^+ and K_S^0-mesons obtained at the U70 [12], Tevatron [13, 14], ISR [15] and RHIC [23] are presented in Figs. 4a and 4b, respectively. The shape of the scaling function for K_S^0 is found to be in good agreement with $\psi(z)$ for K^+-mesons. Using parametrization of $\psi(z)$, the dependence of inclusive spectrum of K_S^0-mesons on transverse momentum at $\sqrt{s} = 200$ GeV is plotted by the dashed line in Fig.4a.

Multiplicity independence of $\psi(z)$

We analyze data on charge hadron production in pp and $p\bar{p}$ collisions at different multiplicities and energies. The E735 Collaboration measured the multiplicity dependence of charged hadron spectra in $p\bar{p}$ collisions at $\sqrt{s} = 1800$ GeV for $dN_{ch}/d\eta = 2.3-26.2$, $|\eta| < 3.25$ and $p_T = 0.15-3$ GeV/c [24]. Strong dependence of the spectra on multiplicity is shown in Fig.5a. The z-presentation of the data is plotted in Fig.5b. Independence of the scaling function $\psi(z)$ on multiplicity was found for same value of $c = 0.25$.

The STAR Collaboration measured multiplicity dependence of inclusive spectra of charged hadrons produced in pp collisions at $\sqrt{s} = 200$ GeV for $|\eta| < 0.5$ [25]. Fig.6a demonstrates strong dependence of the spectra on multiplicity. The STAR data confirm multiplicity independence of the scaling function $\psi(z)$ established in $p\bar{p}$ collisions at higher energies. The value of the heat capacity $c = 0.25$ is found to be the same as for UA1[26], E735 and CDF [27] data.

Additional confirmation of the z-scaling was obtained at RHIC. The scaling manifests self-similarity and fractality in hadron interactions at high energies.

ACKNOWLEDGMENTS

The authors would like to thank Yu.Panebratsev for his support of this work. The investigations have been partially supported by the IRP AVOZ10480505 and by the Grant Agency of the Czech Republic under the contract No. 202/04/0793.

REFERENCES

1. R. P. Feynman, *Phys. Rev. Letters*, **23**, 1415–1417 (1969).
2. J. D. Bjorken, *Phys. Rev.*, **179**, 1547–1553 (1969).
3. P. Bosted *et al.*, *Phys. Rev. Letters*, **49**, 1380–1383 (1972).
4. J. Benecke *et al.*, *Phys. Rev.*, **188**, 2159–2169 (1969).
5. A. M. Baldin, *Sov. J. Part. Nucl.*, **8**, 429–477 (1977).
6. V. S. Stavinsky, *Sov. J. Part. Nucl.*, **10**, 949–995 (1979).
7. Z. Koba, H. B. Nielsen, and P. Olesen, *Nucl. Phys.*, **B40**, 317–334 (1972).
8. V. A. Matveev, R. M. Muradyan, and A. N. Tavkhelidze, *Sov. J. Part. Nucl.*, **2**, 5–32 (1971).
9. S. Brodsky, and G. Farrar, *Phys. Rev. Letters*, **31**, 1153–1156 (1973).
10. I. Zborovský, Yu. A. Panebratsev, M. V. Tokarev, and G. P. Škoro, *Phys. Rev.*, **D54**, 5548–5557 (1996); I. Zborovský, M. V. Tokarev, Yu. A. Panebratsev, and G. P. Škoro, *Phys. Rev.*, **C59**, 2227–2240 (1999); M. V. Tokarev, O. V. Rogachevski, and T. G. Dedovich, *J. Phys. G: Nucl. Part. Phys.*, **26**, 1671–1696 (2000); M. Tokarev, I. Zborovský, Yu. Panebratsev, and G. Skoro, *Int. J. Mod. Phys.*, **A16**, 1281–1301 (2001); I. Zborovský, and M. Tokarev, hep-ph/0506003.
11. J. Adams *et al.*, *Phys. Rev. Letters*, **91**, 172302 (2003).
12. V. V. Abramov *et al.*, *Sov. J. Nucl. Phys.*, **31**, 937–946 (1980); *Sov. J. Nucl. Phys.*, **41**, 700–710 (1985).
13. J. W. Cronin *et al.*, *Phys. Rev.*, **D11**, 3105–3123 (1975); D. Antreasyan *et al.*, *Phys. Rev.*, **D19**, 764–778 (1979).
14. D. Jaffe *et al.*, *Phys. Rev.*, **D40**, 2777–2795 (1989).
15. B. Alper *et al.*, *Nucl. Phys.*, **B87**, 19–40 (1975).
16. S. S. Adler *et al.*, *Phys. Rev. Letters*, **91**, 241803 (2003).
17. A. L. S. Angelis *et al.*, *Phys. Letters*, **B79**, 505–510 (1978).
18. C. Kourkoumelis *et al.*, *Phys. Lett.*, **B83**, 257–260 (1979).
19. C. Kourkoumelis *et al.*, *Z. Phys.*, **C5**, 95–104 (1980).
20. D. Lloyd Owen *et al.*, *Phys. Rev. Letters*, **45**, 89–93 (1980).
21. K. Eggert *et al.*, *Nucl. Phys.*, **B98**, 49–72 (1975).
22. O. Barannikova (STAR Collaboration), In: *Proceedings of the Quark Matter 2005, August 4–9, 2005, Budapest, Hungary*; http://qm2005.kfki.hu/
23. M. Heinz (STAR Collaboration), nucl-ex/0505025.
24. T. Alexopoulos *et al.*, *Phys. Letters*, **B336**, 599–604 (1994).
25. J. E. Gans, PhD Thesis, Yale University, USA (2004).
26. G. Arnison *et al.*, *Phys. Letters*, **B118**, 167–172 (1982).
27. D. Acosta *et al.*, *Phys. Rev.*, **D65**, 072005 (2002).

PARTICLE PROPAGATION IN DENSE MATTER

Chairpersons: T. Csörgő and M. Šumbera

Refractive Distortions of Two-Particle Correlations

Scott Pratt

Department of Physics and Astronomy, Michigan State University
East Lansing, Michigan 48824 email

Abstract. Using optical model calculations it has recently been shown that refractive phenomena from the collective mean field can significantly alter the sizes inferred from two-pion correlations. We demonstrate that such effects can be accounted for in classical calculations if mean field effects are included.

Keywords: Correlations, Femtoscopy, Heavy Ion Collisions
PACS: 25.70.Nq

The six-dimensional two-pion correlation function $C(\mathbf{P},\mathbf{Q})$, measured as a function of the total and relative momenta, has provided crucial insight into the space-time development of heavy-ion reactions [1]. Recently, the refractive effects of particles traversing the mean field were calculated by symmetrizing quantum waves in the frame work a time-independent complex optical potential [2, 3]. If the pions left a region of lower mass, it was found that the lensing effect of the potential was to distort the extracted source dimensions by a few tens of percent. The effect was analagous to lensing effects from Coulomb mean fields, which were shown to explain different apparent source sizes for positive and negative pions at AGS energies [4, 5].

The refractive corrections studied in Refs. [2, 3] allowed for a more physical interpretation of correlation data from RHIC. When ignoring the corrections it appears that the fireball grows to a transverse (to the beam) radius of ≈ 13 fm, expanding at a speed $\approx 0.7c$, and rapidly disintegrates at a time of ≈ 10 fm/c [6]. The puzzling aspect of this picture is that the fireball surface must expand at a speed of $0.7c$ from the initial collision to grow from its initial size of 6 fm to 13 fm in 10 fm/c, not allowing any time for the matter to accelerate. Correlation analyses typically provide three dimensions: $R_{\rm long}$, the longitudinal size defined parallel to the beam, $R_{\rm out}$, the dimension of the phase space packet perpendicular to the beam and outward along the direction of the pair's momentum, and $R_{\rm side}$, the sideward dimension which is perpendicular to both the pair momentum and the beam axis. The apparent sideward dimension in [2, 3] was shown to be increased by the refractive effects of the potential, which would suggest the true radius of the fireball was somewhat smaller than the 13 fm previously believed. Reducing the size by one or two fm would provide a more physically plausible picture of the reaction's evolution.

The calculations in Ref. [2, 3] involved solving for single-particle outgoing wave functions in the presence of complex optical potentials, then using the symmetrized product as a correlation weight. Quantum calculations such as these have some drawbacks. First, implementation becomes complictated if one were to account for time-varying poten-

CP828, *Multiparticle Dynamics*
edited by V. Šimák, M. Šumbera, Š. Todorova, and B. Tomášik

tials, or potentials without the spherical and boost symmetries assumed in [2, 3]. Secondly, finding expressions for the mean field, especially the imaginary part of the optical potential, would be quite involved. Finally, it is difficult to extend such calculations beyond the case of identical particles as one then needs to consider the quantum three-body problem. For these reasons, we wish to study whether the effects can be calculated by considering classical trajectories through the mean field, then using the asymptotic phase space density to generate correlations. Given that the inferred diameters of the RHIC fireballs are often near 25 fm, one might expect that classical considerations could be valid except at very low p_t.

For non-interacting identical particles, two-particle probabilities can be calculated in terms of the outgoing phase space density [1].

$$\begin{aligned} C(\mathbf{P},\mathbf{Q}) &= 1+\int d^3r\mathscr{S}_{\mathbf{P}}(\mathbf{r})\cos(\mathbf{Q}'\cdot\mathbf{r}), \\ \mathscr{S}_{\mathbf{P}}(\mathbf{r}) &= \frac{\int d^3r'_a d^3r'_b f(\mathbf{P}'/2,\mathbf{r}'_\mathbf{a},t')f(\mathbf{P}'/2,\mathbf{r}'_b,t')\delta(\mathbf{r}'_a-\mathbf{r}'_b-\mathbf{r})}{\int d^3r'_a d^3r'_b f(\mathbf{P}'/2,\mathbf{r}'_\mathbf{a},t')f(\mathbf{P}'/2,\mathbf{r}'_b,t')}, \end{aligned} \tag{1}$$

where the primes denote the positions measured in the pair frame, and $\mathbf{Q}'=\mathbf{p}'_a-\mathbf{p}'_b$ is the relative momentum in that frame. Since $\mathscr{S}$ describes the separations of particles with the same velocity, the expression is independent of t' once it satistifies the criteria of being beyond the point at which particles interact with the remainder of the system, including mean-field interactions.

An alternative approach is to treat interaction with the mean field separately, define the source function so that it describes the points at which particles had their last interaction other than through the mean field, then use the outgoing wave function, which is a solution to the equations of motion using the mean field, to describe the evolution from x to the asymptotic momentum state [2, 3, 4]. This involves replacing the phase factor used to describe the evolution from x to its asymptotic state $\mathbf{p}$,

$$e^{i(p_a-p_b)\cdot x}\rightarrow\phi^*(p_a,x)\phi(p_b,x). \tag{2}$$

The two-particle probability is then,

$$\begin{aligned} \frac{dN}{d^3p_a d^3p_b} &= \int d^4x\,\tilde{s}(p_a,x)|\phi(p_a,x)|^2\int d^4x\,\tilde{s}(p_b,x)|\phi(p_b,x)|^2 \\ &+ \left|\int d^4x\,\tilde{s}(P/2,x)\phi^*(p_a,x)\phi(p_b,x)\right|^2. \end{aligned} \tag{3}$$

The correlation function is the ratio of the two terms. Here, ϕ is the outgoing single-particle wave function describing evolution through the mean field. The source functions $\tilde{s}(p,x)$ now refer to the points where a particle had its last non-mean-field interaction.

Both approaches should be equivalent if the classical trajectories would correctly describe propagation through the mean field. We present an "apples-to-apples" comparison of the two approaches using the same complex optical potential,

$$U(r)=(U_R+iU_I)\frac{1}{e^{(r-R)/a}+1}, \tag{4}$$

where a is the diffuseness parameter. If $a = 0$, the potential is a step function. We solve for the distortion to the apparent sideward radii for particles which originate isotropically from points of radius r_0 from the center of the cylinder. For identical particles, radii can be determined from correlations using the expression [1],

$$\langle x_i x_j \rangle = \frac{1}{2} \frac{d^2 C(\mathbf{P}, \mathbf{Q})}{dQ_i dQ_j} \bigg|_{\mathbf{Q}=0}, \tag{5}$$

where the radii represent the dimensions of the asymptotic phase space density, not the dimensions of the emission points. Using Eq. (3) which gives the correlation function in terms of the distributions points for last collisions convoluted with outgoing wave functions, one can then apply Eq. (5) to write an expression for the variance of the sideward size,

$$R^2_{\text{side}} \equiv \langle y^2 \rangle = \frac{\int d^4x \tilde{s}(p,x) \frac{d}{dp_y} \phi^*(\mathbf{p},x) \frac{d}{dp_y} \phi(\mathbf{p},x)}{\int d^4x \tilde{s}(p,x) \phi^*(\mathbf{p},x) \phi(\mathbf{p},x)}, \tag{6}$$

where the asymptotic momentum $\mathbf{p}$ moves along the x axis and ϕ are asymptotic outgoing wave functions.

Calculations were performed for $\tilde{s}$ being independent of the direction of $\mathbf{p}$ with emission points being confined to a radius of r_0. The potential parameters were $U_R = m^2_{\text{med}} - m^2_{\text{vac}}$ with the in-medium mass $m_{\text{med}} = 50$ MeV/c^2 and $U_I = m_\pi \cdot \Gamma_0$ with $\Gamma_0 = 100$ MeV. This corresponds to a classical decay rate of $\Gamma = \Gamma_0 m_\pi / E$. The fact that the decay rate falls $\sim 1/E$ is characteristic of a scalar form for the optical potential. The lower panel of Fig. 1 shows the distortion of R_{side}, defined as the ratio of R_{side} with the potential to R_{side} with $U = 0$, as a function of p_t. The distance scales for the potential were $R = 10$ fm and $a = 3$ fm. Results are shown for three different values of r_0: 5 fm, 10 fm and 15 fm. The distortions rise at low p_t with stronger distortions for small r_0. Also displayed in Fig. 1 are escape probabilities. Quantum mechanically, the probability that a particle escapes the fireball is

$$P_{\text{escape}} = \frac{\int d^4x \tilde{s}(p,x) |\phi(\mathbf{p},\mathbf{r})|^2}{\int d^4x \tilde{s}(p,x)}. \tag{7}$$

Since low p_t particles spend more time in the fireball, and since the decay rate falls as $1/E$, escape rates are small at low p_t.

The corresponding classical calculations were performed by solving for the trajectories of particles through the mean field and are shown alongside the corresponding quantum calculations in Fig. 1. The apparent sideward source sizes from the calculations used for Fig. 1 agree well with one another, differing by only a percent or two for $p_t > 100$ MeV/c. Even at $p_t = 40$ MeV/c, the calculations agree to better than 10%. As explained above, the agreement should be expected to worsen for smaller a. Repeating the calculations for $a = 0.5$ fm, discrepancies increased to the level of a few percent for $p_t > 100$ MeV/c, and notably higher for $p_t \sim 50$ MeV/c. However, it is hard to physically motivate such a sudden change in densities.

One can also understand the connection between classical and eikonal approaches such as those discussed in Refs. [7, 8, 9]. Since eikonal phases appear in interferences

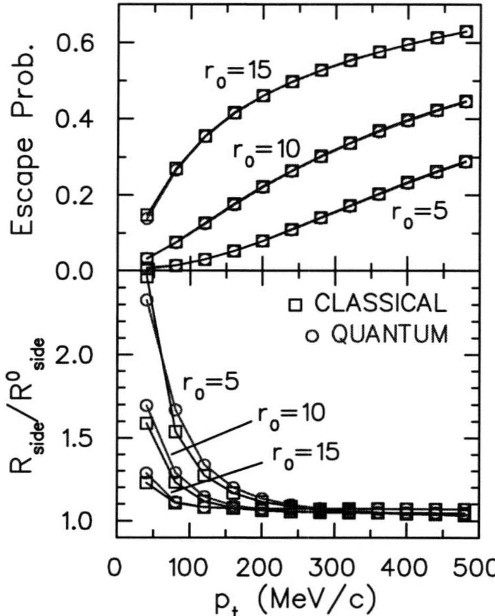

FIGURE 1. The ratio of the apparent sideward dimension to the dimension without mean field is shown as a function of p_t for the optical potential described in the text. Distortions are calculated for both quantum calculations (circles) and classical trajectory calculations (squares). The source function describing the final collision points was confined to an intial radius r_0=5,10 or 15 fm. The upper panel shows the probability that such particles escape without being absorbed due to the imaginary part of the optical potential. The two approaches agree within a few percent.

as $\delta(p_1,x) - \delta(p_2,x)$, and since the derivatives of phase shifts can be identified as the spatial offset incurred by slowing down or speeding up in a classical path through the potential, an connection between eikonal and classical approaches ensues for $p_1 \sim p_2$ [10].

In order to better illustrate the physics of the refractive distortion we consider a simple example where pions are emitted from the surface of a cylinder of radius R, where the in-medium mass of the pion at the surface is $m_{\rm med}$ and the asymptotic vacuum mass is $m_{\rm vac}$. We assume the mass returns to its vacuum value exponentially,

$$m^2(r) = m^2_{\rm vac} + (m^2_{\rm med} - m^2_{\rm vac})e^{-(r-R)/a}. \tag{8}$$

From time-reversal arguments a thermal source will emit particles of given momentum with the same trajectories as those that describe absorption. The asymptotic trajectories of particles with momentum $p_x = p_t$, $p_y = 0$, that intersect with the cylinder have a uniform distribution of impact parameters up to a maximum b as illustrated in Fig. 2. The effect of an attractive mean field is to stretch the width of the phase space cloud by a factor $b_{\rm max}/R$. The sideward dimension measured in two-particle correlations is stretched by this factor.

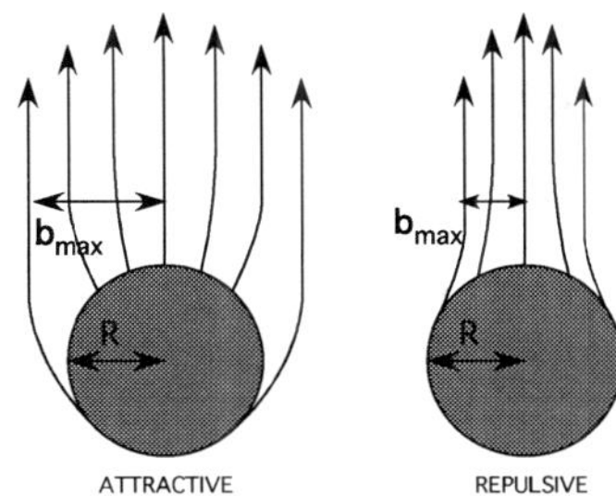

FIGURE 2. Trajectories that lead to the same asymptotic momentum are altered by an attractive mean field (left). The phase space distribution for particles of this given momentum are widened in coordinate space by a factor b_{max}/R. Repulsive mean fields (right) reduce the size of the region of the outgoing particles.

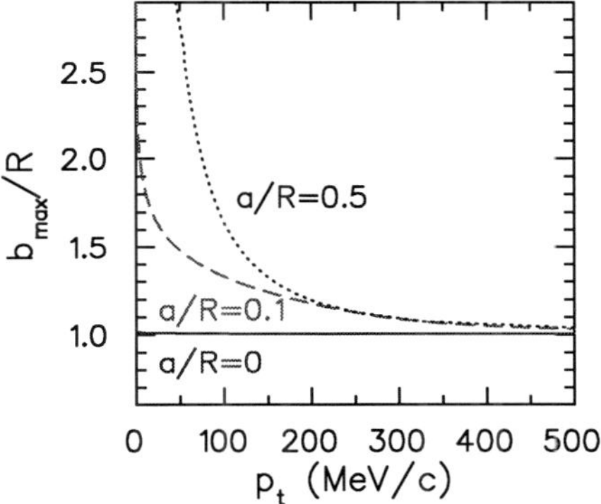

FIGURE 3. Assuming a static cylindrical source, the distortion to the sideward dimension due to the mean field is shown for attractive scalar fields. The field lowers the pion mass to 50 MeV/c^2 and falls off exponentially outside the emitting radius R with a distance scale a. The distortion is stronger for larger ranges a and at lower p_t.

The maximum impact parameter for capture can be found by combining conservation of energy and angular momentum, $L = bp_t$,

$$\begin{aligned} p_t^2 &= p_r^2 + V_{\text{eff}}(r) \\ V_{\text{eff}}(r) &= \frac{b^2 p_t^2}{r^2} + m^2(r) - m_{\text{vac}}^2. \end{aligned} \tag{9}$$

One can analytically solve for the maximum impact parameter $b_{\max}$ from which one can overcome the potential barrier with a give p_t. The high p_t behavior is independent of a, and in the case where $a > R/2$, the result is independent of a for all p_t.

Figure 3 shows the ratio $b_{\max}/R$ as a function of p_t for several values of a given the case of a lighter in-medium pion mass, $m_{\text{med}} = 50$ MeV/c^2. For $a = 0$, the mean field has no effect as the capture cross section does not extend beyond the cylinder. For the saturating value, $a = R/2$, the cross section is significantly enhanced. For $a = R/10$, the enhancement is identical to the saturating value at large p_t and is somewhat reduced at low p_t. The enhancement of the apparent sideward size is characteristic of an attractive mean field. This is consistent with Liouvilles theorem, which states that contraction of the phase space density in momentum space as it leaves the attractive mean field should be accompanied by a growth of the phase space cloud in coordinates space.

The simple cylindrical picture was expanded to include both longitudinal and collective flow [10]. For emission during the expansion stage, distortions were increased in magnitude since the boundary layer expands behind the pions, staying in the region of non-zero mean field for a longer time. Pions emitted during the collapse of the breakup surface were less affected since they spent less time near the surface.

The calculations presented here were only meant to be illustrative. The potential magnitude of mean-field effects and the ability of classical pictures to model the effects underscore the importance of incorporating mean field effects into hadronic Boltzmann descriptions of the breakup stage. Such calculations have been undertaken in the past, especially for lower-energy Fermi-velocity collisions, where the mean field is paramount [11, 12]. In this energy range, phase space points used to generate correlations are typically taken from when a particle leaves the region of mean field, rather than when they have their last collision. Given the likelihood that the dispersion relations for pions are non-trivial, it is important that calculations incorporate momentum-dependent interactions. Such an effort is crucially important for reconstructing the space-time evolution of the fireball, and for understanding the pressure in the hadronic phase.

ACKNOWLEDGMENTS

Support was provided by the U.S. Department of Energy, Grant No. DE-FG02-03ER41259.

REFERENCES

1. M. A. Lisa, S. Pratt, R. Soltz and U. Wiedemann, nucl-ex/0505014.
2. J. G. Cramer, G. A. Miller, J. M. S. Wu and J. H. S. Yoon, *Phys. Rev. Lett.* **94**, 102302 (2005).
3. G. A. Miller and J. G. Cramer, nucl-th/0507004.
4. H. W. Barz, *Phys. Rev. C* **59**, 2214-2220 (1999).
5. H. W. Barz, *Phys. Rev. C* **53**, 2536-2538 (1996).
6. F. Retiere and M. A. Lisa, *Phys. Rev. C* **70**, 044907 (2004).
7. J. I. Kapusta and Y. Li, nucl-th/0503075.
8. C. Y. Wong, *J. Phys. G* **30**, S1053-S1058 (2004).
9. M. C. Chu, S. Gardner, T. Matsui and R. Seki, *Phys. Rev. C* **50**, 3079-3087 (1994).
10. S. Pratt, nucl-th/0508029.
11. W.G. Gong, W. Bauer, C.K. Gelbke, and S. Pratt, *Phys. Rev. C* **43**, 781-800 (1991).
12. W. Bauer, C. K. Gelbke and S. Pratt, *Ann. Rev. Nucl. Part. Sci.* **42**, 77-100 (1992).

From Mach Cone to Reappeared Jet: What Do We Learn from PHENIX Results on Non-Identified Jet Correlation?

Jiangyong Jia for the PHENIX Collaboration

Columbia University, New York, NY 10027 and Nevis Laboratories, Irvington, NY 10533, USA

Abstract. Jet properties, extracted from two particle azimuth correlation, are found to be strongly modified in Au + Au collisions at $\sqrt{s_{NN}} = 200$ GeV. At intermediate p_T and in central Au + Au collisions, the modifications appear as a broadening of jet width at the near side and a cone structure at the away side. As one increase the p_T for both hadrons, the away side cone structure seems to gradually evolve into a peak structure. The interpretation of these results requires careful separation of various medium effects and surface bias.

Keywords: Mach cone, Correlation function, Elliptic flow, Reaction plane
PACS: 27.75.-q

INTRODUCTION

High p_T back-to-back jets are valuable probes for the sQGP [1] created in heavy-ion collisions at RHIC. Existing two particle jet correlation results from statistically limited RUN2 Au + Au data set revealed a strong interaction of the jets with the medium. On the one hand, jet correlation at high p_T indicates a seemingly complete disappearance of the away side jet signal [2]. On the other hand, jet correlation at low p_T shows an enhancement of the away side jet yield [3] but a broadened jet shape [4]. Qualitatively, this is consistent with the energy loss picture, where the high p_T jets are quenched by the medium and their lost energy enhances the jet multiplicity at low p_T.

Equipped with excellent statistics from RUN4 Au + Au, we would like to gain further understanding of the interaction of the jets with the medium using the non-identified charged hadron - charged hadron correlations.

JET PROPERTIES AT INTERMEDIATE P_T

Our analysis is based on 1 billion minimum bias events from Au + Au collisions at $\sqrt{s} =$ 200 GeV. The correlation function $C(\Delta\phi)$ (CF) is defined as the ratio of same event pair distribution, $dN^{\rm pairs}/d\Delta\phi$ to the mixed event pair distribution, $dN^{\rm mix}/d\Delta\phi$. $dN^{\rm mix}/d\Delta\phi$ reflects the level of combinatoric background and the geometric acceptance [5]. In heavy-ion collisions, the CF can be expressed as the sum of jets and elliptic flow,

$$C(\Delta\phi) = J(\Delta\phi) + \xi\left(1 + 2v_2^t v_2^a \cos 2\Delta\phi\right) \quad (1)$$

The superscript t and a stand for the trigger and associated particles, ξ is a normalization factor.

CP828, *Multiparticle Dynamics*
edited by V. Šimák, M. Šumbera, Š. Todorova, and B. Tomášik

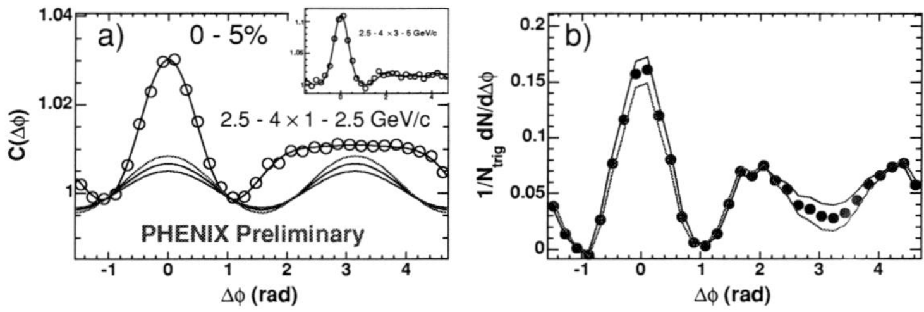

FIGURE 1. a) Correlation function in 0-5% centrality bin, the lines indicated the level of flow background and its systematic error band, the insert is CF for higher $p_{T,\mathrm{assoc}}$. b) Correspondingly background subtracted per-trigger yield.

Figure 1a shows the typical correlation function from central Au + Au collisions. The away side shape is very broad and non-gauss like. It has a plateau that expands to about 2 radians and a possible small dip at π. The subtraction of the flow contribution (shown by the curves) only makes the dip deeper (Fig. 1b). ξ is fixed by scaling the flow term to match the CF, i.e. assuming $J = 0$ at some $\Delta\phi$ (ZYAM assumption) [6]. The ZYAM procedure leads to a slight over-subtraction of jet yield, however since $2v_2^t v_2^a \approx$ few %, the over-subtraction mainly results in a vertical shift and does not affect the away side jet shape. The systematic error on $J(\Delta\phi)$ is dominated by the uncertainties on v_2.

PHENIX performed a systematic study of the jet shape and yield at intermediate p_T, as shown in Fig. 2. There is a continuous evolution of the split and the dip as function of centrality. The away side split is characterized by the split parameter D [7], which is obtained by a double gauss fit on the away side. D seems to turn on rather quickly as a function of centrality, and fall on a uniform curve as function of $\mathrm{N_{part}}$ for different collision energies and collision systems.

What is the nature of the away side dip? Energy loss models that implement the jet broadening in a random walk manner can not describe the dip or the flat jet shape [8]. Models with Cherenkov gluons [9, 10] or medium dragging effect from flow [11] predict a cone or a broadening of the away side jet, but the predicted modifications depend strongly on momentum and are expected to disappear at large p_T. Casalderrey *et. al.*

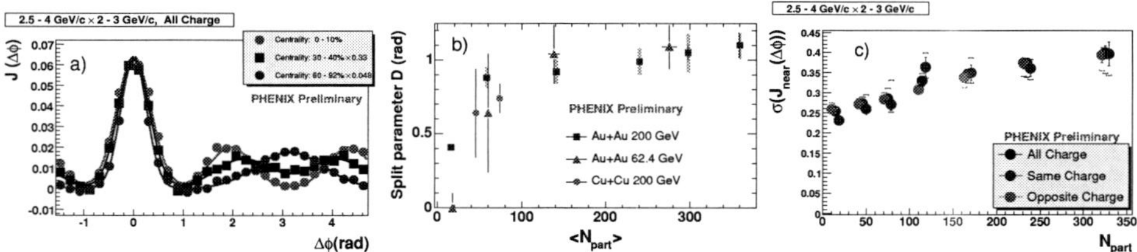

FIGURE 2. a) The correlation function for central, mid-central and peripheral bin in Au + Au. b) The away side "D" parameter as function of N_{part} for several collision systems and energies. c) the near side width as function of centrality for same charged pairs, opposite charged pairs and all pairs in Au + Au.

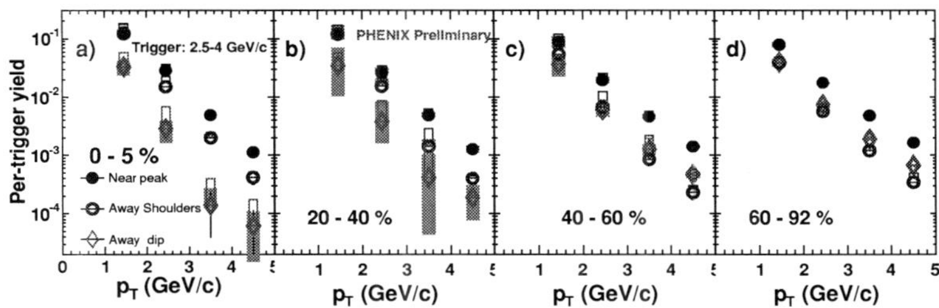

FIGURE 3. The yield for trigger 2.5-4 GeV/c plotted as function of associated hadron p_T for four different centrality bins.

[12] proposed a 'Mach cone'/'shock wave' mechanism to explain the away side jet shape. In this model, energetic jets, which travel faster than speed of sound (c_s) in the medium, excite shock waves at an angle $\theta = \cos^{-1}(c_s/c)$. The direction of the cone is independent of the p_T, but the width of the cone is predicted to narrow for higher p_T.

Figure 2 also indicates a sizable broadening of the near side jet shape in central collisions. This modification is not as dramatic as that for the away side jet, most likely due to the surface emission bias [14] in which the average distance travelled by the near side jet is much smaller than that for the away side jet. However, the relatively small amount of medium that the near side jet has to go through could already lead to some broadening. On the other hand, the baryon yield is enhanced at intermediate p_T in central collisions [15]. Since the near side jet structure could be different between baryon trigger and meson trigger [16], the broadening of the near side jet width could be a consequence of the strongly modified particle composition in central collisions.

To quantify the modifications of the jet shape, we study the jet yield in three different $\Delta\phi$ regions: near side jet region ($|\Delta\phi| < \pi/3$), the away side dip region ($|\Delta\phi - \pi| < \pi/6$), and the away side shoulder region ($|\Delta\phi - \pi \pm \pi/3| < \pi/6$). The shoulder region is sensitive to the novel medium effects, while the dip region is sensitive to the punch through jet contribution. Figure 3 plots the jet yields in the three regions as function of p_T for four centralities. In 0-5% centrality bin, there is a large separation between the yields for the dip region and near side jet region, persistent to large p_T. In more peripheral collisions, the yield of the dip region becomes closer or even exceeds that for the shoulder region, consistent with the returning of the away side jet to a normal gauss shape.

DEPENDENCE ON THE REACTION PLANE

The study of the jet yield as function of angle w.r.t. reaction plane is very important in the sense that it adds another dimension in controlling the path length dependence. It also provides additional constrains on the subtraction of the elliptic flow background. When the trigger particles are selected in a window centered around ϕ_s with a width of $\pm c$ with respect to the reaction plane, the pair distribution up to second order harmonics

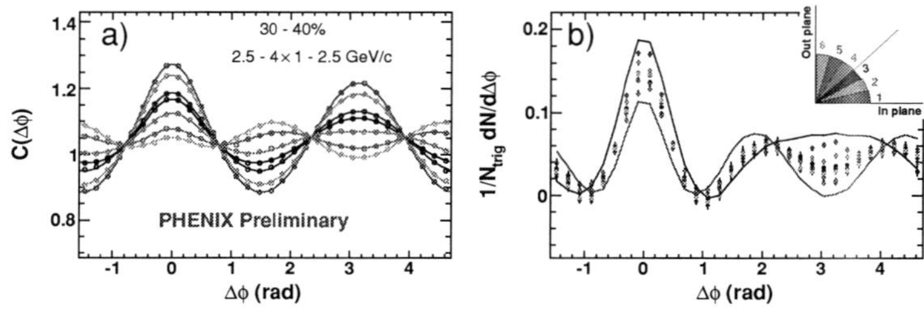

FIGURE 4. a) Correlation function for 6 trigger direction bins and the trigger integrated bin (the center curve). b) The flow subtracted per-trigger yields, the insert figure shows the 6 trigger bins.

(without jet contribution) is [13]:

$$\begin{aligned} \frac{dN^{\rm pairs}}{d\Delta\phi} &= \frac{2c}{\pi}B(a+2v_2^a b\cos 2\Delta\phi) \\ a &= 1+2{v_2}^t\cos 2\phi_s\frac{\sin 2c}{2c}\langle\cos 2\Psi\rangle \\ b &= {v_2}^t+\cos 2\phi_s\frac{\sin 2c}{2c}\langle\cos 2\Psi\rangle+{v_2}^t\cos 4\phi_s\frac{\sin 4c}{4c}\langle\cos 4\Psi\rangle \end{aligned}$$

a is the combinatoric background level, and it is proportional to the number of trigger particles in the window. The correlation function without jet contribution is:

$$C(\Delta\phi)=\frac{dN^{pairs}/d\Delta\phi}{dN^{mix}/d\Delta\phi}=\xi(1+2v_2^a b/a\cos 2\Delta\phi)=\xi(1+2v_2^a v_{2,eff}^t\cos 2\Delta\phi) \qquad (2)$$

Where $v_{2,eff}^t=b/a$ is the effective v_2 of the trigger particle in the window, and ξ is the same normalization factor as in Eq. (1). ξ does not depend on the trigger direction.

We divide the trigger range, $[0,\pi/2]$, into 6 bins. Each bin has a different flow background, which can be calculated from Eq. (2). The measured correlation functions for 30-40% centrality bin are shown Fig. 4. Several interesting features can be readily identified. The effective $v_{2,\rm eff}^t$ changes dramatically from in plane to out of plane direction, but the six CFs cross each other at $\pm\pi/4$ and $\pi\pm\pi/4$, where the harmonic contributions are zero. The away side cross points are systematically higher than those at the near side, reflecting directly the amount of jet contribution at $\pi\pm\pi/4$. The extremes of the distributions are not at $\pi/2$ where the flow influence is maximal, instead they are shifted either to the left or the right due to the jet contribution.

The jet yields in each trigger direction (Fig. 4b) are obtained by subtracting the flow contribution. The only free parameter, ξ, is fixed by the ZYAM procedure from the integrated bin, and the flow terms in all six trigger bin are automatically fixed Eq. (2). Figure 5 shows the comparison of the the measured CFs and the calculated flow contributions for 30-40% centrality bin. The systematic error bands correspond to the error of the RP v_2, propagated according to Eq. (2). The size of the systematic errors is largest for in plane bin and smallest for the out of plane bin. RP dependence study

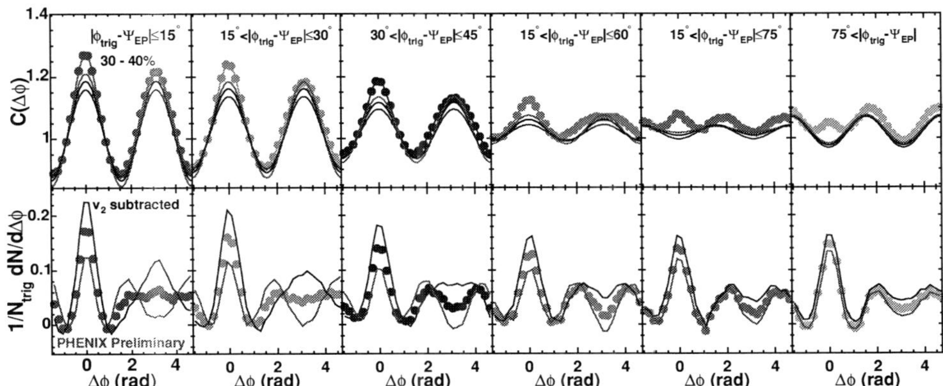

FIGURE 5. (top row) Correlation functions and (bottom row) background subtracted per-trigger yields for the 6 trigger direction bins.

helps to constrain the v_2 systematic when it is not dominated by the RP resolution[1]. In Fig. 4b, jet shapes for different bins show some subtle differences within the systematic error. We believe they are mostly due to the small v_4 terms which were not considered in current analysis.

"REAPPEARANCE" OF THE AWAY SIDE JETS AT HIGH P_T

The importance of high p_T correlation is two fold. On the one hand, high p_T jets are free from complicated intermediate p_T physics (for example recombination), thus can serve as a cleaner probe of the medium. On the other hand, studies of high p_T jets can help to disentangle normal jet fragmentation from Cherenkov gluons, shock wave or fragmentation of radiated gluons which become dominating at intermediate or low p_T.

Figure 6 shows several CFs in successively higher p_T ranges in 0-10% central Au + Au collisions. The typical away side cone structure persists to $p_T \approx 4$ GeV/c, but the edges of the cones become sharper and their magnitude drops. In $4-5\times4-5$ GeV/c bin, the relative flat away side shape does not rule out the cone shape, but it's magnitude must be significantly reduced.

As a comparison, in the most right panel of Fig. 6, we also show the hadron-hadron correlation from STAR [2]. The data are for 0-10% most central Au + Au with $4 < p_{T,\text{trig}} < 6$ GeV/c and $2 < p_{T,\text{assoc}} < p_{T,\text{trig}}$, and are comparable to the middle panel in Fig. 6. It is also almost comparable to the highest p_T point in Fig. 3a, with the p_T selection of the trigger and associated hadrons are swapped [2]. All three are qualitatively

---

[1] Since $v_2^a = v_{2,\text{raw}}^a / \langle\cos 2\Psi\rangle$, the error of v_2 from RP resolution is independent of trigger direction.

[2] When the p_T range of trigger and associated particle are swapped, the di-jet modification factor I_{AA} are connected to each other by, $I_{\text{AA}}^1 R_{\text{AA}}^1 = I_{\text{AA}}^2 R_{\text{AA}}^2 = \frac{\text{Jetpairs}_{\text{AA}}}{N_{\text{coll}}\text{Jetpairs}_{\text{pp}}}$, where R_{AA}^1 and R_{AA}^2 are the nuclear modification factor of the first and second particle, respectively.

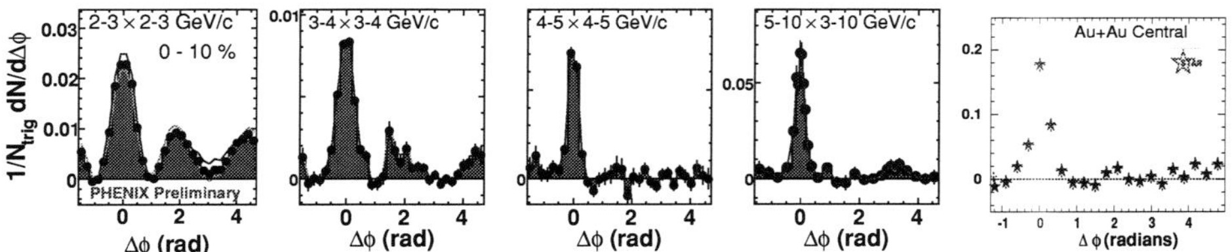

FIGURE 6. (left four panels) : per-trigger yield for different p_T selection in 0-10% centrality bin. (right panel): per-trigger yield from STAR in 0-10% centrality bin [2].

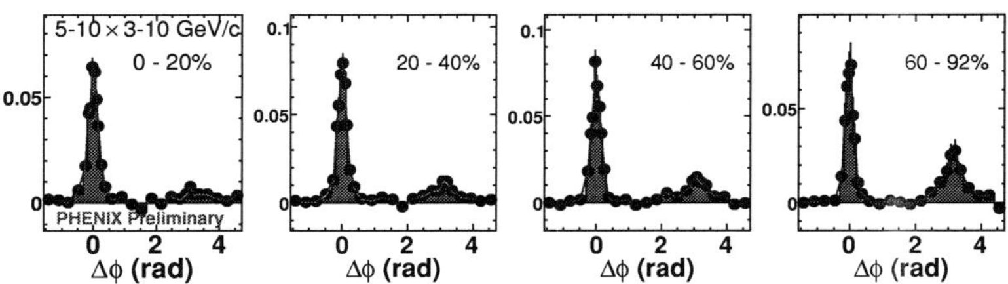

FIGURE 7. Centrality dependence of the per-trigger yield at high p_T

similar to each other. Interestingly, STAR's data are consistent with zero around π, but it seems to have a shoulder at $\pi \pm 1$ as suggested by the right panel of Fig. 6.

In the highest p_T bin of Fig. 6, a peak structure seems to reemerge around π on top of a flat background. To understand the physics behind the peak structure, we plot in Fig. 7 the centrality dependence of the CF for $5-10 \times 3-10$ GeV/c selection. The away side peak exists in all centrality bins, although its magnitude is suppressed toward central collisions. At this point, it is hard to say whether the widths of the away peaks are also broadened in central collisions. On the other hand, there seems to be little change in both the shape and magnitude of the near side jet as function of centrality. These can be compared with recent di-jet results measured at much larger p_T ($8 < p_{T,\mathrm{trig}} < 15$ GeV/c and $6 < p_{T,\mathrm{assoc}}$ GeV/c) from STAR experiment [17], where the jet width and the shape of the fragmentation function are found to be independent of p_T at fairly large z ($z > 0.4-0.5$). In energy loss picture, the large z requirement biases the detected away side jets to smaller energy loss, which biases detected jet towards surface, thus points to the picture where both jets are emitted tangential to the surface. If this scenario is true, we should recover the strong medium modification at low z (by decreasing $p_{T,\mathrm{assoc}}$). To check this, in Fig. 8a we show the CF for various associated hadron p_T with trigger p_T fixed. The $\langle z \rangle$ of the associated hadron in the four panels are approximately 0.2, 0.4, 0.6 and 1. Clearly we see a stronger distortion of the away side jet shape at smaller $p_{T,\mathrm{assoc}}$, the yields relative to the near side are also larger at smaller $p_{T,\mathrm{assoc}}$. In fact, the fragmentation functions from STAR also suggest a significant deviation from the uniform scaling shape at $z \lesssim 0.4$ as shown in Fig. 8b. It is important to measure the fragmentation function in full z range in order to separate these two competing effects.

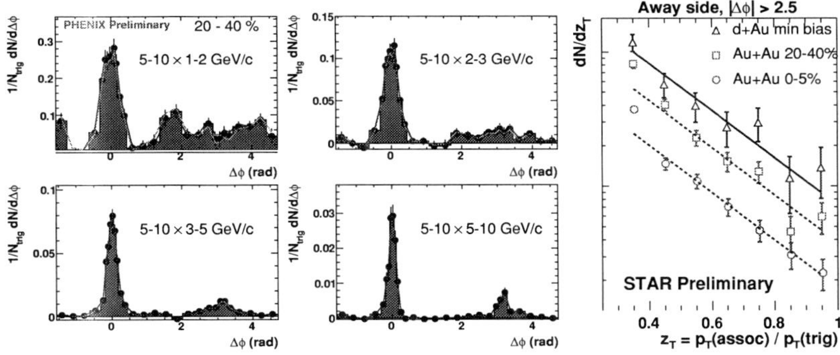

FIGURE 8. (left four panels): per-trigger yield for different $p_{T,\mathrm{assoc}}$ in 20-40% centrality bin when trigger p_T is fixed. (right panel) Away side jet fragmentation from STAR [17].

CONCLUSIONS

Jet properties from hadron-hadron correlation have been studied as function of p_T, centrality and the angle relative to the reaction plane. Precise extraction of jet signal relies on experimental control on the flow backgrounds, which can be constrained by looking their reaction plane dependence. Jet shape and yield are found to be strongly modified at intermediate and low p_T. The interpretations of these modification, however, are complicated by various competing mechanisms. By increasing the p_T for both triggering and associated hadrons, away side jet peak reappears but its yield is suppressed. This might be due to the bias effect where the detected di-jets are emitted tangentially to the surface.

REFERENCES

1. K. Adcox *et al.* [PHENIX Collaboration], *Nucl. Phys. A* **757**, 184-283 (2005); J. Adams *et al.* [STAR Collaboration], *Nucl. Phys. A* **757**, 102-183 (2005); B. B. Back *et al.* [PHOBOS Collaboration], *Nucl. Phys. A* **757**, 28-101 (2005); I. Arsene *et al.* [BRAHMS Collaboration], *Nucl. Phys. A* **757**, 1-27 (2005).
2. C. Adler *et al.* [STAR Collaboration], *Phys. Rev. Lett.* **90**, 082302 (2003).
3. J. Adams *et al.* [STAR Collaboration], *Phys. Rev. Lett.* **95**, 152301 (2005).
4. S. S. Adler *et al.* [PHENIX Collaboration], `nucl-ex/0507004`.
5. J. Jia, *J. Phys. G* **31**, S521-S531 (2005).
6. N. N. Ajitanand *et al.*, *Phys. Rev. C* **72**, 011902 (2005).
7. N. Grau [PHENIX Collaboration], proceedings of Quark Matter 2005, Aug. 4-9, Budapest, Hungary.
8. I. Vitev, `hep-ph/0506281`.
9. I. M. Dremin, *JETP Lett.* **30**, 140-144 (1979) [*Pisma Zh. Eksp. Teor. Fiz.* **30**, 152-156 (1979)].
10. V. Koch, A. Majumder and X. N. Wang, `nucl-th/0507063`.
11. N. Armesto, C. A. Salgado and U. A. Wiedemann, `hep-ph/0411341`.
12. J. Casalderrey-Solana, E. V. Shuryak and D. Teaney, `hep-ph/0411315`.
13. J. Bielcikova, S. Esumi, K. Filimonov, S. Voloshin and J. P. Wurm, *Phys. Rev. C* **69**, 021901 (2004).
14. A. Drees, H. Feng and J. Jia, *Phys. Rev. C* **71**, 034909 (2005).
15. S. S. Adler *et al.* [PHENIX Collaboration], *Phys. Rev. Lett.* **91**, 172301 (2003).
16. S. S. Adler *et al.* [PHENIX Collaboration], *Phys. Rev. C* **71**, 051902 (2005).
17. D. Magestro [STAR Collaboration], `nucl-ex/0510002`.

Femtoscopy in Heavy Ion Collisions: Wherefore, Whence, and Whither?

Mike Lisa

Physics Department, Ohio State University, 191 W. Woodruff Ave, Columbus Ohio 43210, USA

Abstract. I present a brief overview of the wealth of femtoscopic measurements from the past two decades of heavy ion experiments. Essentially every conceivable "knob" at our disposal has been turned; the response of two-particle correlations to these variations has revealed much about the space-momentum substructure of the hot source created in the collisions. I discuss the present status of the femtoscopic program and questions which remain, and point to new efforts which aim to resolve them.

Keywords: Relativistic heavy ion collisions, RHIC, HBT, interferometry, femtoscopy
PACS: 25.75.-q

The slowly crawling ants will eat our dreams.
Andrzej Białas, musing on words of Andre Breton as they might apply to femtoscopy.

Go to the ant, thou sluggard; consider her ways, and be wise. - Proverbs vi.6

WHEREFORE

High energy collisions between electrons, hadrons, or nuclei produce highly nontrivial systems. Especially in the soft (low-p_T, long spatial scale) sector, the inclusive distributions of the measured multiparticle final states are dominated by phase-space; to first order the momentum spectra and particle yields appear thermal, revealing little of the underlying physics of interest. Detailed information in this sector is obtained only through correlations; inclusive spectra tell much less than half the story.

In particular, multiparticle production is a *dynamic* process, evolving in space and time. For several decades now, small relative momentum two-particle correlations have been used to probe the space-time structure of systems at the femtometer scale. Measurements and constantly-improving techniques variously called "intensity interferometry," "HBT," "GGLP," "non-identical correlations," etc, are nowadays discussed under the common rubric of femtoscopy [1], as the title of this new workshop series reflects.

While understanding the space-time features of the system is important to both the particle and the heavy ion physicist, in the latter case it is even vital. After all, nontrivial geometrical effects *dominate* the physics of heavy ion collisions.

From the very broadest perspective, the entire heavy ion program is geared to generate and study a qualitative change in the geometric substructure of the hot system. Strongly-coupled [2] or not, the quark-gluon plasma (QGP) is a soft QCD system, in which colored degrees of freedom are relevant over large length scales. Of particular interest is the existence and nature of a deconfinement phase transition; a significant and sudden

CP828, *Multiparticle Dynamics*
edited by V. Šimák, M. Šumbera, Š. Todorova, and B. Tomášik

change in the degrees of freedom should be reflected in space-time aspects of the system [3]. Also of generic importance is the (often unasked) question of whether the "system" generated is, indeed, a system. Any discussion of "matter" or "bulk" properties relies on an affirmative answer.

More specifically, geometry defines each stage of the system's evolution. In the initial state, the entrance-channel geometry (impact parameter $\vec{b}$) determines the subsequent collective evolution and anisotropic expansion of the system [4]; the resulting "elliptic flow" [5] has been the basis of $\sim 25\%$ of the publications from the RHIC program. In the intermediate state, also, geometry dominates: quantitative understanding of exciting parton energy loss (or "jet quenching") measurements [6, 7] requires detailed information of the evolving size and anisotropic shape of the system. If coalescence is indeed the mechanism of bulk hadronization [8], space-momentum correlations in the intermediate stage induce clustering effects which must be modeled quantitatively [9].

Clearly then, for the soft (bulk) sector in heavy ion collisions, geometrical issues dominate both the physics of interest and the system with which it is probed. No surprise, then, that since the relativistic heavy ion program began roughly two decades ago, femtoscopic studies have played a major role, and a "sub-community" has developed. It was not long before the erstwhile "nuclear" physicists contributed physical and technical insights to a type of measurement initially borrowed from their particle physics colleagues. At workshops like this, such dialogue continues unabated.

Several excellent reviews of femtoscopy in heavy ion physics have very recently appeared in the literature [1, 10, 11, 12]. Together with physics discussions, the reader may find in them precise definitions of the correlation function, "homogeneity lengths," "HBT radii," "out-side-long" coordinate system etc. Here, I assume familiarity with such concepts, and very briefly review the status of heavy ion femtoscopy at present. I emphasize the breadth of systematics which has been explored so far, what (we think) it has told us, and what continues to puzzle us. I then identify a few promising directions in which the field is moving, pointing for details on these to others' contributions to these proceedings.

WHENCE

Due to their copious production and ease of detection, most femtoscopic measurements have utilized correlations between charged pions. Further, many experiments have focused on central ($|\vec{b}| = 0$) collisions, since (1) azimuthal symmetry simplifies the femtoscopic formalism [12, 13]; and (2) maximal energy densities and spatial extents are generated. The extent of measured femtoscopic systematics 15-20 years ago is represented in Fig. 1, showing that, in central collisions involving nuclei with mass number A, HBT radii scale approximately as $A^{1/3}$ [14, 15]. Apparently trivial, these data were at the same time comforting, confirming that pion correlations did indeed track with geometric scales.

Since then, femtoscopic data and techniques have evolved tremendously, generating an equally tremendous range of systematic femtoscopic studies. The first femtoscopic measurement in truly relativistic heavy ion collisions was reported almost twenty years ago by the NA35 Collaboration at the CERN SPS [16]. Similar measurements have been

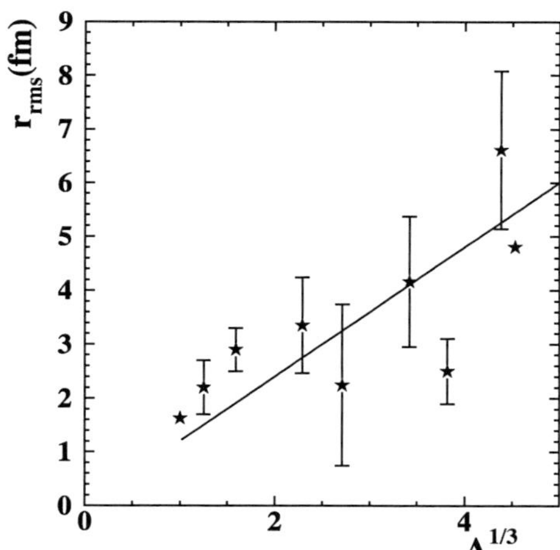

FIGURE 1. Pion HBT radius versus the mass number of colliding nuclei, from Bevalac experiments ~ 20 years ago. Compilation from [15].

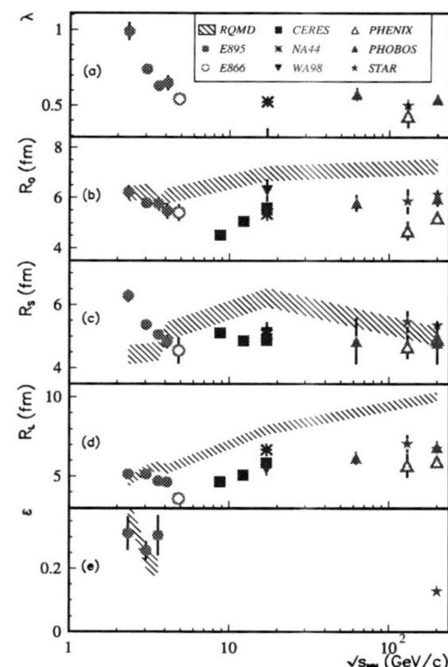

FIGURE 2. World dataset of published HBT radii from central Au+Au (Pb+Pb) collisions versus collision energy. Compilation from [12].

performed at the SPS and the BNL AGS and RHIC accelerators over the collision energy range $\sqrt{s_{NN}} \approx 2.3 - 200$ GeV. Thus, in each of the two complementary quantities– energy and time– we may consider two decades' worth of systematics [12].

The original hope was to find "anomalously" large spatial and/or temporal scales, as reflected in the HBT radii, indicating large entropy generation or a long-lived QGP state. This expectation was considered rather generic [17], and, guided by quantitative predictions from hydrodynamical models [3, 18], the most commonly-discussed systematic was the excitation function (i.e. energy dependence) of pion HBT radii. This is shown in Fig. 2, where no striking features are observed in the HBT radii at any collision energy. As I discuss later, this observed contradiction of a seemingly-generic expectation may be considered the second "HBT puzzle."

Clearly, insight into geometrically-driven physics requires more detailed systematic studies than the simple excitation function. Indeed, this has always been a generic requirement for extracting physics from *any* observable in heavy ion physics, and has required development of heavy ion programs with simultaneous, complementary, large-acceptance experiments running at dedicated machines. Especially in the crucial soft sector, more is learned by varying independent variables than by long runs at the highest possible energy a given machine can deliver.

Inspired by recent "schematic equations" [19], I denote the impressive **multi**-dimensional space explored by femtoscopic experiments as

$$\text{Heavy Ion Femtoscopy} = R\left(\sqrt{s_{NN}}; A, B, |\vec{b}|, \phi, y, m_T, m_1, m_2\right) \quad (1)$$

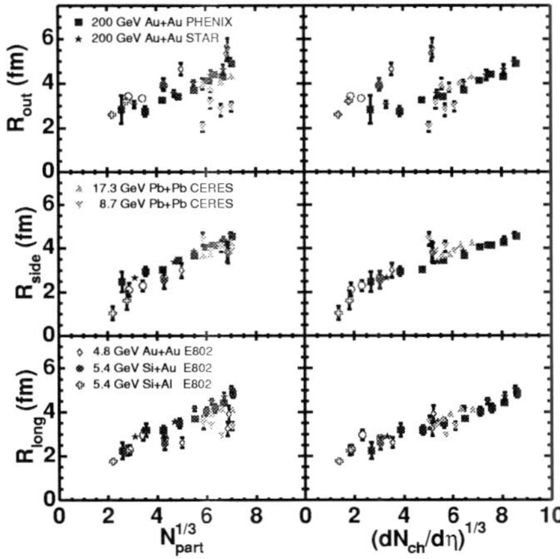
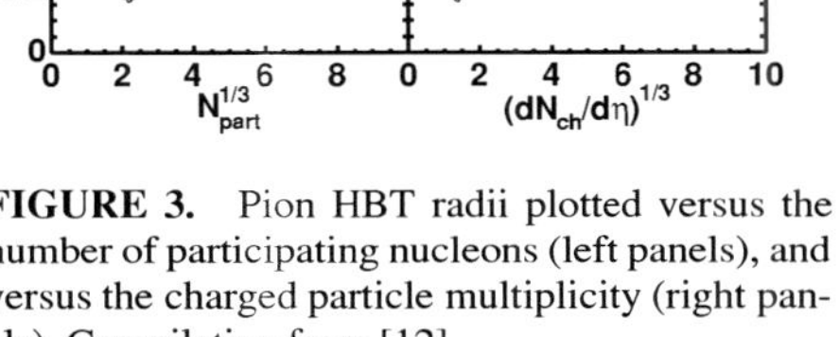

FIGURE 3. Pion HBT radii plotted versus the number of participating nucleons (left panels), and versus the charged particle multiplicity (right panels). Compilation from [12].

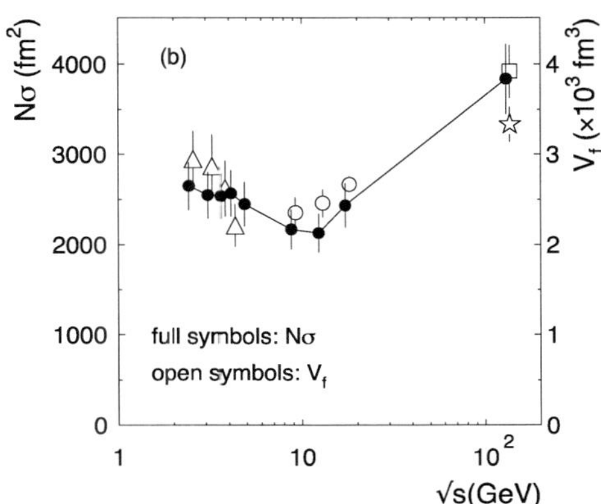

FIGURE 4. The "effective pion cross-section" $N_{\rm proton} \cdot \sigma_{p\pi} + N_{\rm pion} \cdot \sigma_{\pi\pi}$ and the "freezeout volume" $\sim R_{long} \cdot R_{side}^2$ are plotted as a function of the collision energy, for central Au+Au (Pb+Pb) collisions. Figure and further details in [21].

Global dependences Especially in light of "puzzles," we need to perform a similar study as shown in Fig. 1, checking that femtoscopic radii track with geometric collision scales to first order. We may vary the geometric scale of the reaction zone by varying the atomic numbers of the colliding nuclei, A and B, and/or by selecting events of varying impact parameter, $|\vec{b}|$. Of course, fixing only one of these parameters will not define the collision scale; instead, a natural quantity would be the number of participating nucleons N_{part} [20]. Pion HBT radii corresponding to different A, B, $|\vec{b}|$ and $\sqrt{s_{NN}}$ are collected in Fig. 3. The left panels show that these femtoscopic lengths scale similarly to those shown in Fig. 1, replacing A by N_{part}. (Note that results for central collisions, $|\vec{b}| \approx 0$, are shown in Fig. 1, so that $N_{part} \sim A$.) The HBT radius R_{out}, which mixes space and time non-trivially, may be expected to violate a pure geometrical scaling; this may explain the increased spread in the upper panels of Fig. 3.

To good approximation, at a given $\sqrt{s_{NN}}$, total multiplicity (a final-state quantity) is a function only of N_{part} (an entrance-channel quantity), independent of A, B, or $|\vec{b}|$. The relationship does, however, depend on collision energy [22]. As seen in the right panels of Fig. 3, the final-state multiplicity provides a more common scaling parameter than N_{part}; recent analyses [23, 24, 25] show that this scaling persists for different m_T values and for lighter colliding systems at RHIC.

Several observations may be made about this multiplicity scaling. Firstly, it appears that knowledge of $dN_{ch}/d\eta$ alone allows "prediction" of the HBT radii (at least R_{long} and R_{side}). This suggests that the small increase of these radii with $\sqrt{s_{NN}}$ seen in Fig. 2 is associated with increased particle production as the collision energy is raised. (Note that

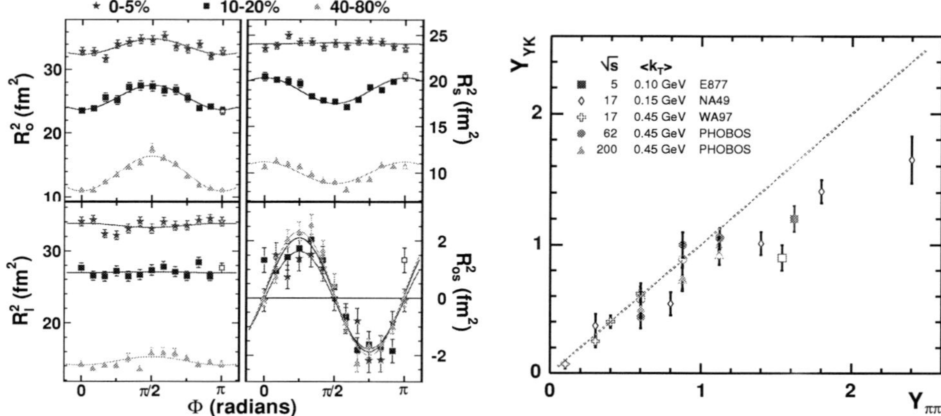

FIGURE 5. Pion HBT radii measured for Au+Au collisions at $\sqrt{s_{NN}}$, plotted as a function of azimuthal emission angle relative to the reaction plane. From [27]

FIGURE 6. The pion Yano-Koonin velocity (see text) versus pair rapidity for central Au+Au (Pb+Pb) collisions at various energies Compilation from [12].

N_{part} is approximately constant for the data in Fig. 2.) Secondly, the finite offset d in the approximately linear relationship $R_{long} \cdot R_{side}^2 = c \cdot (dN/d\eta) + d$ means that freeze-out does *not* occur at fixed density [24].

Thirdly, the scaling shown in the figure breaks down dramatically for $\sqrt{s_{NN}} \lesssim 5$ GeV, as is obvious from the non-monotonic behaviour seen in Fig. 2. As the CERES Collaboration has pointed out [21], this is likely due to the dominance of baryons at lower $\sqrt{s_{NN}}$. Indeed, a quantitative connection between the number of protons and pions, and a product of HBT radii is possible, by assuming a universal ($\sqrt{s_{NN}}$-independent) mean free path at freezeout λ_f. In Fig. 4, the "freezeout volume" $\sim R_{long} \cdot R_{side}^2$ and the "effective pion cross-section" $N_{\text{proton}} \cdot \sigma_{p\pi} + N_{\text{pion}} \cdot \sigma_{\pi\pi}$ are seen to coincide by scaling the latter by $\lambda_f = 1$ fm, apparently contradicting the standard assumption that freeze-out occurs when the mean free path becomes much larger than the system size.

HBT radii and the "freeze-out volume" may be connected only the context of a model which includes dynamical effects like flow. The analysis of [21] ignores such effects; however, its bottom line remains approximately valid, as flow effects on HBT radii are expected to be small at low p_T [26].

Kinematic dependences Insight on the dynamical evolution and geometric substructure of the emission region is gained by studying the dependence of femtoscopic lengths on the next three parameters in "Equation" 1.

In non-central collisions, the entrance-channel geometry is naturally anisotropic; the hot source geometry approximates the overlap between target and projectile, and is characterized by a "long axis" perpendicular to the impact parameter vector $\vec{b}$. At RHIC, the system expands more rapidly in-plane ($\parallel \vec{b}$) than out ($\perp \vec{b}$) [28]. If it is, indeed a collective *system* with finite lifetime, then the overall shape should evolve. Pion HBT

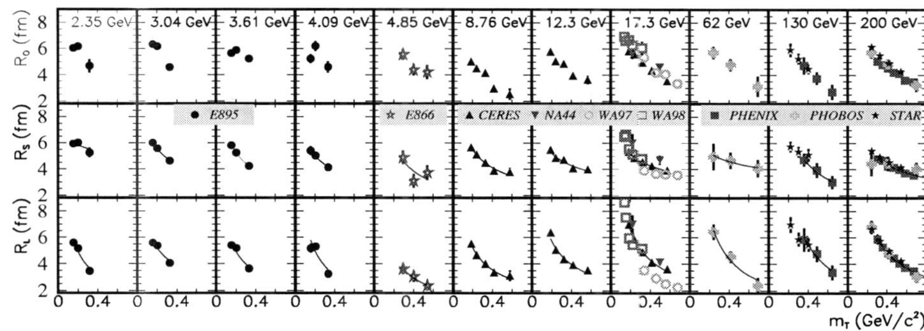

FIGURE 7. Pion HBT radii plotted versus the transverse mass m_T for all published measurements of central Au+Au (Pb+Pb) collisions over two decades in $\sqrt{s_{NN}}$. Compilation from [12].

radii have been measured as a function of their azimuthal angle $\phi_{\text{pair}} \equiv \angle\left(\vec{K},\vec{b}\right)$ for Au+Au collisions. The measurement at RHIC [27] is shown in Fig. 5. There, it is clear that as $|\vec{b}| \to 0$, the freezeout source becomes larger and rounder. In fact, there is a nice "rule of two"– the source expands to twice its original size [23, 24], and its anisotropy $\varepsilon \equiv \left(\langle y^2\rangle - \langle x^2\rangle\right) / \left(\langle y^2\rangle + \langle x^2\rangle\right)$ decreases by the same factor [27]. The relatively small change in the source shape is at least semi-quantitatively [29] consistent with short timescale estimates [26] based on the longitudinal radius, and at variance with expectations from "realistic" simulations [30].

As will become increasingly clear, the only femtoscopic systematic which might display non-trivial $\sqrt{s_{NN}}$ dependence is, in fact, the dependence on ϕ_p. This is clear from the bottom panel of Fig. 2, in which the relative paucity of such measurements is also clear. It will be especially interesting to see whether the flow and/or timescales at the LHC are sufficiently large to produce in-plane freeze-out configurations [31].

Experiments at a wide range of collision energies have mapped out the rapidity dependence of pion HBT radii. Of particular interest here is the so-called Yano-Koonin rapidity Y_{YK} [32, 33], which should approximate the rapidity of the fluid element which emits a pair of pions at some rapidity $Y_{\pi\pi}$. Fig. 6 shows an approximately "universal" behaviour $Y_{\text{YK}} \approx Y_{\pi\pi}$, independent of $\sqrt{s_{NN}}$. This is consistent with (but not proof of [12]) emission from a boost-invariant system [33].

The most extensively-studied kinematic systematic has been the p_T-dependence of pion HBT parameters. Fig. 7 shows the world dataset of published measurements for central Au+Au (Pb+Pb) collisions. The falling dependence of femtoscopic scales on transverse velocity is generally believed to arise from collective transverse and longitudinal flow (e.g. [26]). As I mentioned earlier, strong collective flow would be an indication that a real *bulk* system has been formed.

The longitudinal radius scales approximately as $R_l \sim m_T^{-0.5}$, indicating strong longitudinal flow and again consistent with expectations for emission from a boost-invariant

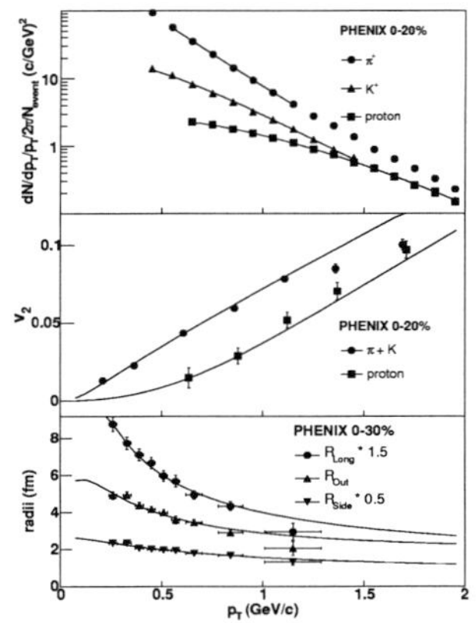

FIGURE 8. A blast-wave [26] fit reproduces several observables at RHIC. See text for details. From [34].

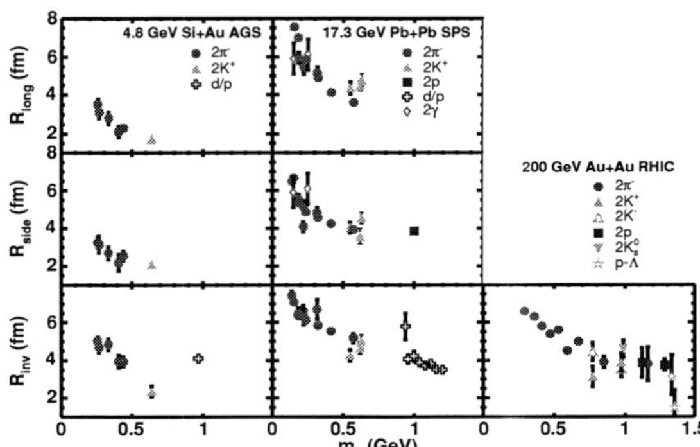

FIGURE 9. Femtoscopic radii for various similar-mass particle pairs, plotted as a function of m_T. Compilation from [12].

system [35, 26]. Decreasing transverse radii R_o and R_s may be due to collective transverse flow. The simplest flow-dominated models quantitatively interrelate these femtoscopic m_T dependences with other observations. An example is shown in Fig. 8, in which a very simple freeze-out scenario [26]—thermal motion superimposed on a collectively exploding source—can simultaneously describe a broad range of data measured at RHIC. The momentum-space distribution, quantified by the average number distribution (top panel of Fig. 8) and the number variation as a function of azimuthal angle (middle panel) give an incomplete picture by themselves. Momentum-dependent femtoscopic radii (bottom panel) probe the dynamical *sub*-structure of the collision, constraining models more stringently [12, 26, 36, 37].

Particle-species dependences Within the past several years, high-statistics datasets in experiments with good particle identification have allowed the mapping of femtoscopic systematics with the final variables in "Equation" 1—the mass (or species) of the correlated particles.

Signals of a *system's* collectivity at freeze-out should not be limited to the pions. In the simplest picture, corresponding to flow-dominated models (e.g. [26]) of Fig. 8, femtoscopic radii should approximately scale with m_T, independent of particle type. An impressive common scaling of radii from *all* measured particles is, indeed, observed at *all* energies explored, as seen in Fig. 9. The common scaling is particularly striking when one considers the quite different measurement systematics involved in charged pion correlations and, say, K_s^0 correlations. Even generalized nucleon separation scales, probed by relative yields of deuterons and protons (d/p in Fig. 9), follow the systematic, with the exception of one outlier point at the lowest energies.

Correlations between non-identical particles probe not only the sizes, but also the relative displacement of the particles' emission zones in space-time [47]. Any collective freeze-out scenario naturally implies a specific relationship between emission regions of the various particle types. In a flow-dominated picture, the emission zones for high-m_T

TABLE 1. A very incomplete table of published or ongoing femtoscopic studies at RHIC for various particle combinations. "Traditional" identical-particle interferometry lies along the lowest diagonal line of cells.

	π^+	π^-	K^+	K^-	K_s^0	p	$\bar{p}$	Λ	$\bar{\Lambda}$	Ξ	$\bar{\Xi}$
$\bar{\Xi}$	[38]	[38]									
Ξ	[38]	[38]									
$\bar{\Lambda}$						[39]	[39]				
Λ						[39]	[39]				
$\bar{p}$	[40]	[40]	[40]	[40]		[41]	[41]				
p	[40]	[40]	[40]	[40]		[41, 42]					
K_s^0					[43]						
K^-	[44]	[44]		[42]							
K^+	[44]	[44]	[42]								
π^-	[23, 41]	[23, 45, 46]									
π^+	[23, 45, 46]										

particles are not only smaller than those for low-m_T particles, but are also inevitably located further from the center of the collision region [26], as suggested by the schematic in Fig. 10. As discussed in detail in the contribution of A. Kisiel [40], available measurements of these displacements at RHIC provide further support of the flow-dominated freeze-out scenario.

Non-identical particle correlations are today a growth industry. Table 1 lists only a sampling of recently-published or ongoing analyses at RHIC energies. Similar studies have been performed at lower energies [48, 12]. The diagonal axis corresponds to identical-particle correlations, the "traditional" focus of HBT interferometry.

WHITHER

Excluding aficionados attending workshops such as this one, in the minds of most heavy ion (or high energy) physicists, the term "femtoscopy" (or, more likely "HBT") brings to mind only the single, rather uninspiring systematic plotted in Fig. 2; indeed, some may be tempted toward the dismissive view that "the measured radius is always 5 fm."

As we have just discussed, this is grossly unfair: the systematics are tremendously richer, with femtoscopic length scales varying with almost every parameter in "Equation" 1. Furthermore, these strong systematic trends are found consistently by experiments separated by decades and using quite different measurement and correction techniques; indeed, especially at RHIC, the data are almost embarrassingly consistent (c.f. Fig. 7). Yet further, it appears that these systematics may be well understood in the commonly-accepted framework of *system* evolution due to strong flow quantitatively consistent with momentum-space observables [26]. Clearly, there is much more to femtoscopy than its most notorious Figure.

On the other hand, the trends shown in Figs. 3, 6, 7, and 9 suggest that the notorious Fig. 2 quite correctly summarizes the situation after all. At any $\sqrt{s_{NN}}$, the systematics of "Equation" 1 are quite rich and may well be reconciled with a reasonable physical picture. However, in more ways than expressed by Fig. 2, those systematics are essentially

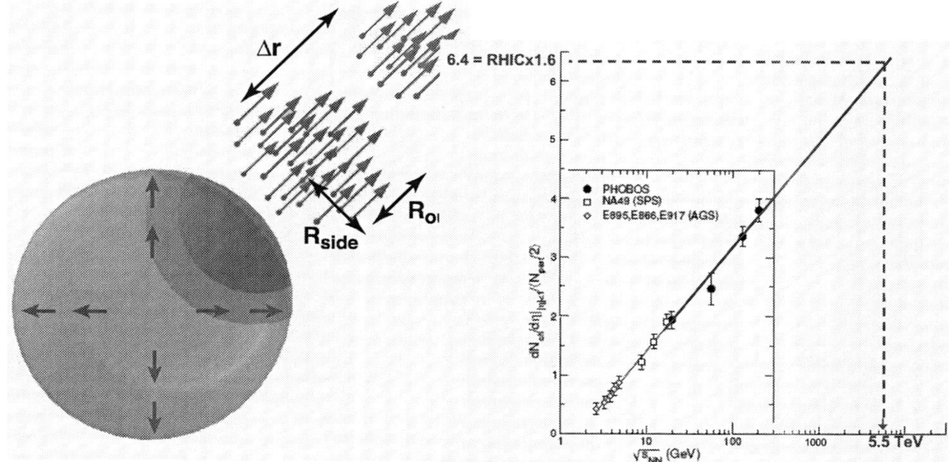

FIGURE 10. Freezeout regions for particles of different species (or different transverse masses) emitted from a common source. Two-particle correlations measure the (momentum-dependent) size, shape, and orientation of the emission regions, as well as the average displacement (Δr) in the outward direction. From [12].

FIGURE 11. Multiplicity density per participant pair, measured for Au+Au (Pb+Pb) collisions at AGS, SPS, and RHIC are shown in the lower panel, taken from [22], and naively extrapolated by the author to LHC energy.

independent of $\sqrt{s_{NN}}$! Without resorting to agreement or disagreement with particular models, this second [1] femtoscopic puzzle is startling, suggesting that the space-time consequences of the physical processes are the same at RHIC as they are near the pion production threshold. Often, "universal" behaviour is a key to deeper physical insight. Heavy ion femtoscopy, however, might display a bit *too* much universality.

Whither... or wither?

Given the prominence of nontrivial geometry to the physics of heavy ion collisions in general, and the rather generic [17] expectation of significant changes in spacetime evolution with $\sqrt{s_{NN}}$, understanding this universality remains urgent. What future efforts might shed some light?

Today, one almost reflexively points to the impending heavy ion program at the LHC for new observations generating fresh insights. While anything might happen in an unexplored energy domain, we may venture a prediction. Instead of a crystal ball, however,

---

[1] In this experimental overview, I have not discussed what has come to be known as "the" HBT puzzle [49] which, simply put, is that otherwise-successful and apparently reasonable models like hydrodynamics do not reproduce femtoscopic measurements [12]. To all but the novice heavy ion physicist, however, the initial failure of dynamical models to reproduce *diverse* observations is hardly puzzling. The experience at lower energies is that such initial failure is more the rule than the exception. In light of other, more generic puzzles, I call this problem only the "*first* HBT puzzle" [37].

we use a mirror to gaze over our shoulder at twenty years of systematics in heavy ions. The $\sqrt{s_{NN}}$-dependence of the global multiplicity (per participant pair) has been significantly extended at RHIC and summarized by the PHOBOS collaboration [22]. Boldly and probably naively extending this systematic leads to the expectation that multiplicities at the LHC will be $\sim 60\%$ higher than they are at RHIC, as shown in Fig. 11.

As discussed in the previous Section, femtoscopic length scales– for any m_T, y, N_{part}, or particle species– depend primarily on event multiplicity. Taken together, Figs. 11 and 3 suggest that radii[2] in central collisions at RHIC will simply be $\sim 17\%$ higher than they are at RHIC ($(1.6)^{1/3} = 1.17.$).

Notably, evidence is mounting that perhaps *all* soft-sector observables are determined primarily by total multipilcity, independent of $\sqrt{s_{NN}}$. Properly-scaled elliptic flow [50] and even strangeness enhancement [51] appear to show universal multiplicity scalings. Whether this is a trivial implication of entropy-driven phasespace dominance in observables sensitive to bulk medium is unclear. However, nontrivial new phases of *matter* should have signatures in the long-distance (soft momentum) sector; dependence only on multiplicity and not reaction energy would be intriguing.

So, perhaps the choice of collider facility (LHC versus RHIC) is unimportant, and heavy ion femtoscopists should focus on filling in the holes of Table 1? Most evidence thus far indicates that flow-dominated freezeout scenarios (e.g. [26]) fitted to identical π correlations essentially "predict" femtoscopic data using other particle combinations. The data is yet scant, however, so ongoing studies [40] to further explore this Table are quite important. There are even preliminary reports [38], with exotic particle combinations, of inconsistencies with these freezeout models. If confirmed, strong theoretical focus should come to bear on this result. If, on the other hand, varying the particle combination repeatedly yields results "predicted" by blast-wave models, continually filling in cells of Table 1 risks becoming a stamp-collecting exercise.

The most important recent experimental developments in heavy ion femtoscopy were presented at these workshops. In addition to those discussed above, I briefly mention here some ongoing studies which I find most promising.

Even if all of the particle combinations in Table 1 follow simple blast-wave calculations, and so reveal no new femtoscopic information, this can actually be turned to good use. In particular, one may turn around the traditional approach in which one uses the known the two-particle final state interaction (FSI) to extract geometric information, to extract the FSI itself [48, 52]. Finalized results from STAR on $p-\Lambda$ correlations [39] have extracted previously inaccessible phase shift information for low-energy baryon-antibaryon scattering. While not QGP-related physics, such studies can make a unique contribution to low-energy QCD and hadronic physics.

Ideally suited workshops like these are studies which directly compare for the first time, at a fixed energy and using identical detector and analysis techniques, correlation data from the heaviest ion collisions to that from p+p collisions [24]. As has been observed previously in high energy experiments, femtoscopic radii from identical pion

---

[2] Precise expectations R_{out} or ϕ-dependent radii at the LHC are, admittedly, less certain, as the former does not scale exactly with multiplicity (cf Fig. 3), and the multiplicity-dependence of the latter has not been extensively mapped (cf Figs. 2 and 5).

correlations measured in p+p collisions decrease with increasing p_T, qualitatively similar to the dependence shown in Fig. 7. The preliminary STAR data shows, however, that in all three HBT radii, the p_T dependence is *quantitatively identical* in p+p and A+A collisions! Since the heavy ion and high energy communities have traditionally used very different physics mechanisms to explain this dependence, this observation potentially throws the explanations of both communities into doubt. If this result is confirmed, it ranks as the "third HBT puzzle" [37].

Unexplained long-range structure in the correlation functions for the lowest-multiplicity collisions, however, presently cloud the interpretation of the HBT radii for p+p collisions [24]. Partly in an effort to understand this, a new representation of the data in terms of spherical harmonic amplitudes in $\vec{q}$-space was developed as an experimental diagnostic tool [37]. In fact, a similar harmonic decomposition method was earlier already developed by Danielewicz and collaborators [53, 54], not as a diagnostic, but as a direct link to the detailed geometry (beyond simply length scales) of the emitting source. This representation has a natural connection to source imaging [55], and, indeed, first applications to PHENIX data have been reported [56].

Harmonic decompositions as an improved representation of the correlation function and source imaging as an improved, generalized fit to the data are, in a sense, merely technical improvements, but they are quite significant ones. Just as femtoscopic studies have explored the systematic landscape of Equation 1, so should they probe the "microscape" of fine details of the measured data.

The femtoscopy of heavy ion collisions can be an addicting endevour. Systematics make sufficient sense that we are convinced that we are probing geometry at the femtometer scale. Spacetime geometry at that scale is sufficiently important to the physics that the measurements must be done well. Such measurements are sufficiently challenging that it is enjoyable to do them well and to develop improved techniques. However, for now, the overall results are sufficiently puzzling that there is plenty more to do.

ACKNOWLEDGMENTS

I express my sincere gratitude to the organizers of both ISMD and WPCF. Both actually succeeded in what is often just a stated but unattained goal of such workshops: to achieve meaningful intellectual cross-pollenation between nuclear and particle physicists. I express special congratulations to the initiators of WPCF, which itself sprang out of the earlier Warsaw meetings, for establishing a timely and most promising workshop series.

REFERENCES

1. R. Lednicky (2005), `nucl-th/0510020`.
2. M. Gyulassy, and L. McLerran, *Nucl. Phys.* **A750**, 30–63 (2005), `nucl-th/0405013`.
3. D. H. Rischke, and M. Gyulassy, *Nucl. Phys.* **A608**, 479–512 (1996), `nucl-th/9606039`.
4. P. Huovinen, P. F. Kolb, U. W. Heinz, P. V. Ruuskanen, and S. A. Voloshin, *Phys. Lett.* **B503**, 58–64 (2001), `hep-ph/0101136`.
5. J.-Y. Ollitrault, *Nucl. Phys.* **A638**, 195–206 (1998), `nucl-ex/9802005`.
6. J. Adams, et al., *Phys. Rev. Lett.* **93**, 252301 (2004), `nucl-ex/0407007`.
7. B. Cole, *plenary presentation at Quark Matter 2005, Budapest* (2005).

8. R. J. Fries, B. Muller, C. Nonaka, and S. A. Bass, *Phys. Rev. Lett.* **90**, 202303 (2003), `nucl-th/0301087`.
9. D. Molnar, *Acta Phys. Hung.* **A22**, 271–279 (2005), `nucl-th/0406066`.
10. T. Csorgo (2005), `nucl-th/0505019`.
11. S. S. Padula, *Braz. J. Phys.* **35**, 70–99 (2005), `nucl-th/0412103`.
12. M. A. Lisa, S. Pratt, R. Soltz, and U. Wiedemann, *Ann. Rev. Nucl. Part. Sci.* **55**, 311 (2005), `nucl-ex/0505014`.
13. U. W. Heinz, A. Hummel, M. A. Lisa, and U. A. Wiedemann, *Phys. Rev.* **C66**, 044903 (2002), `nucl-th/0207003`.
14. J. Bartke, *Phys. Lett.* **B174**, 32–35 (1986).
15. G. Alexander, *Rept. Prog. Phys.* **66**, 481–522 (2003), `hep-ph/0302130`.
16. T. J. Humanic, et al., *Z. Phys.* **C38**, 79–84 (1988).
17. J. W. Harris, and B. Muller, *Ann. Rev. Nucl. Part. Sci.* **46**, 71–107 (1996), `hep-ph/9602235`.
18. S. A. Bass, et al., *Nucl. Phys.* **A661**, 205–260 (1999), `nucl-th/9907090`.
19. M. Gyulassy, *J. Phys.* **G30**, S911–S918 (2004).
20. A. Bialas, M. Bleszynski, and W. Czyz, *Nucl. Phys.* **B111**, 461 (1976).
21. D. Adamova, et al., *Phys. Rev. Lett.* **90**, 022301 (2003), `nucl-ex/0207008`.
22. B. B. Back, et al., *Nucl. Phys.* **A757**, 28–101 (2005), `nucl-ex/0410022`.
23. J. Adams, et al., *Phys. Rev.* **C71**, 044906 (2005), `nucl-ex/0411036`.
24. Z. Chajecki, *contribution to these proceedings* (2005).
25. Z. Chajecki (2005), `nucl-ex/0510014`.
26. F. Retiere, and M. A. Lisa, *Phys. Rev.* **C70**, 044907 (2004), `nucl-th/0312024`.
27. J. Adams, et al., *Phys. Rev. Lett.* **93**, 012301 (2004), `nucl-ex/0312009`.
28. J. Adams, et al., *Phys. Rev. Lett.* **92**, 062301 (2004), `nucl-ex/0310029`.
29. M. A. Lisa, *Acta Phys. Polon.* **B35**, 37–46 (2004), `nucl-ex/0312012`.
30. D. Teaney, J. Lauret, and E. V. Shuryak (2001), `nucl-th/0110037`.
31. U. W. Heinz, and P. F. Kolb, *Phys. Lett.* **B542**, 216–222 (2002), `hep-ph/0206278`.
32. F. B. Yano, and S. E. Koonin, *Phys. Lett.* **B78**, 556–559 (1978).
33. Y. F. Wu, U. W. Heinz, B. Tomasik, and U. A. Wiedemann, *Eur. Phys. J.* **C1**, 599–617 (1998), `nucl-th/9607044`.
34. F. Retiere, *J. Phys.* **G30**, S827–S834 (2004), `nucl-ex/0405024`.
35. S. V. Akkelin, and Y. M. Sinyukov, *Phys. Lett.* **B356**, 525–530 (1995).
36. B. Tomasik (2005), `nucl-th/0509100`.
37. Z. Chajecki, T. D. Gutierrez, M. A. Lisa, and M. Lopez-Noriega (2005), `nucl-ex/0505009`.
38. P. Chaloupka, *contribution to these proceedings* (2005).
39. J. Adams (2005), `nucl-ex/0511003`.
40. A. Kisiel, *contribution to these proceedings* (2005).
41. H. Gos, *contribution to these proceedings* (2005).
42. M. Heffner, *J. Phys.* **G30**, S1043–S1047 (2004).
43. S. Bekele, *J. Phys.* **G30**, S229–S234 (2004).
44. J. Adams, et al., *Phys. Rev. Lett.* **91**, 262302 (2003), `nucl-ex/0307025`.
45. S. S. Adler, et al., *Phys. Rev. Lett.* **93**, 152302 (2004), `nucl-ex/0401003`.
46. B. B. Back, et al. (2004), `nucl-ex/0409001`.
47. R. Lednicky (2001), `nucl-th/0112011`.
48. R. Lednicky, *contribution to these proceedings* (2005).
49. U. W. Heinz, and P. F. Kolb (2002), `hep-ph/0204061`.
50. G. Roland, et al. (2005), `nucl-ex/0510042`.
51. H. Caines, *poster presentation at Quark Matter 2005, Budapest* (2005).
52. M. Bystersky, and F. Retiere, *contribution to these proceedings* (2005).
53. P. Danielewicz, and S. Pratt (2005), `nucl-th/0501003`.
54. P. Danielewicz, *contribution to these proceedings* (2005).
55. D. Brown, *contribution to these proceedings* (2005).
56. P. Chung, *contribution to these proceedings* (2005).

Low-Q^2 Partons in p-p and Au-Au Collisions

Thomas A. Trainor
(STAR Collaboration)

CENPA 354290, University of Washington, Seattle, WA 98195, USA

Abstract. We describe correlations of low-Q^2 parton fragments on transverse rapidity y_t and angles (η, ϕ) from p-p and Au-Au collisions at $\sqrt{s} = 130$ and 200 AGeV. Evolution of correlations on y_t from p-p to more-central Au-Au collisions shows evidence for parton dissipation. Cuts on y_t isolate angular correlations on (η, ϕ) for low-Q^2 partons which reveal a large asymmetry about the jet thrust axis in p-p collisions favoring the azimuth direction. Evolution of angular correlations with increasing Au-Au centrality reveals a rotation of the asymmetry to favor pseudorapidity. Angular correlations of transverse momentum p_t in Au-Au collisions access temperature/velocity structure resulting from low-Q^2 parton scattering. p_t autocorrelations on (η, ϕ), obtained from the scale dependence of $\langle p_t \rangle$ fluctuations, reveal a complex parton dissipation process in heavy ion collisions which includes the possibility of collective bulk-medium recoil in response to parton stopping.

Keywords: p-p collisions, p_t correlations, y_t correlations, parton scattering, minijets, angular correlations, high-p_t, mean-p_t fluctuations, heavy ion collisions, fluctuation scale dependence
PACS: 24.60.Ky, 25.75.Gz

INTRODUCTION

QCD theory predicts that many low-Q^2 scattered gluons (minijets) should be produced in relativistic nuclear collisions at RHIC energies, with rapid parton thermalization as the source of the colored medium (quark-gluon plasma) in heavy ion collisions [1, 2, 3, 4]. If so, we should discover evidence of partons with $Q/2 = 1 - 5$ GeV in the correlation structure of final-state hadrons. The discovery and analysis of nonperturbative low-Q^2 parton fragment correlations has motivated development of novel techniques, including the use of transverse rapidity y_t rather than momentum p_t and formation of angular *joint autocorrelations*. Fluctuations of event-wise mean p_t $\langle p_t \rangle$ [5, 6] may isolate fragments from low-Q^2 partons and determine the properties of the corresponding medium. A measurement of excess $\langle p_t \rangle$ fluctuations in Au-Au collisions at 130 GeV revealed a large excess of fluctuations compared to independent-particle p_t production [6].

Angular correlations of fragments from hard-scattered partons (jets) were first observed on (η, ϕ) at large p_t and with increasing $\sqrt{s}$ [7]. In a conventional high-p_t study of parton fragmentation to a hadron jet the relevant issues are the fragment distribution on p_t and the angular distribution, both relative to the parton momentum (jet thrust axis). In contrast, we adopt no *a priori* jet or factorization hypothesis. We study minimum-bias two-particle distributions on transverse rapidity space (y_{t1}, y_{t2}) to obtain fragment *distributions* (not fragmentation *functions*) and on angle space $(\eta_1, \eta_2, \phi_1, \phi_2)$ to obtain fragment angular correlations. Particle pairs are treated symmetrically, as opposed to asymmetric 'trigger' and 'associated' particle combinations. We observe that correlations obtained with this minimum-bias analysis, in contrast to the conventional trigger-

CP828, *Multiparticle Dynamics*
edited by V. Šimák, M. Šumbera, Š. Todorova, and B. Tomášik

particle approach, represent the *majority* of parton fragment pairs in nuclear collisions, those with $p_{t1} \sim p_{t2} \sim 1$ GeV/c. While we attempt to understand QCD in A-A collisions we should also revisit its manifestations in elementary collisions, where novel phenomena are still emerging.

P-P INITIAL SURVEY AND ANALYSIS METHOD

The reference system for low-Q^2 partons in A-A collisions is the *hard component* of correlations in p-p collisions. The single-particle p_t spectrum for p-p collisions at $\sqrt{s} = 200$ GeV can be decomposed into soft and hard components on the basis of event multiplicity [8]. When a fixed Lévy distribution [9] is subtracted from spectra for ten multiplicity n_{ch} classes we obtain distributions in Fig. 1 (first panel), described by a gaussian shape (solid curves) independent of n_{ch}. Transverse rapidity is defined by $y_t \equiv \ln\{(m_t + p_t)/m_0\}$, with m_0 assigned as the pion mass. That novel result motivated a study of two-particle correlations on $y_t \times y_t$ [8]. The minimum-bias distribution in Fig. 1 (second panel) exhibits separate soft and hard components interpreted as longitudinal string fragments (smaller y_t) and transverse parton fragments (larger y_t), the latter corresponding to the gaussians in the first panel.

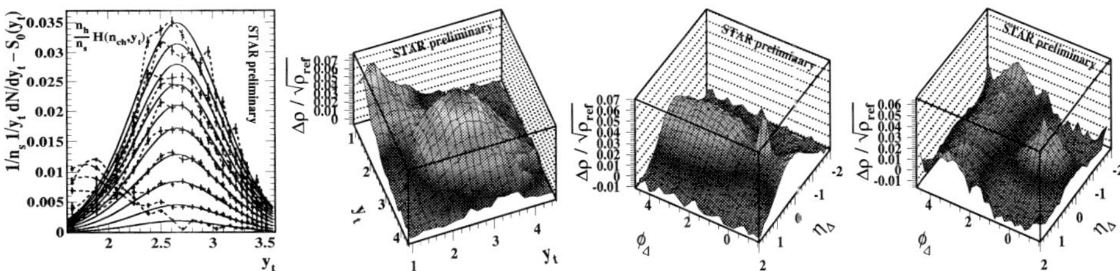

FIGURE 1. Hard components of p_t spectra *vs* n_{ch} plotted on transverse rapidity y_t; low-Q^2 parton and string fragment distributions on (y_{t1}, y_{t2}) and angular difference variables $(\eta_\Delta, \phi_\Delta)$, all for p-p collisions.

Soft and hard components on y_t produce corresponding structures in joint angular autocorrelations on $(\eta_\Delta, \phi_\Delta)$. In the third panel, string-fragment (soft) correlations for unlike-sign pairs are determined by local charge and transverse-momentum conservation (the sharp peak at the origin is conversion electrons). Minimum-bias parton fragments (hard) in the fourth panel produce classic jet correlations, with a *same-side* ($\phi_\Delta < \pi/2$) jet cone at the origin and an *away-side* ($\phi_\Delta > \pi/2$) ridge corresponding to the broad distribution of parton-pair centers of momentum. These angular *autocorrelations* do not rely on a leading or trigger particle and provide unprecedented access to low-Q^2 partons.

Density ratio $\Delta\rho/\sqrt{\rho_{ref}}$ is related to Pearson's *correlation coefficient* [10]. For event-wise particle counts n_a and n_b in histogram bins a and b on space x Pearson's coefficient is $r_{ab} \equiv \overline{(n-\bar{n})_a(n-\bar{n})_b}/\sqrt{\overline{(n-\bar{n})^2_a}\ \overline{(n-\bar{n})^2_b}}$. That suggests the form of density ratio $\Delta\rho/\sqrt{\rho_{ref}} \equiv 1/\varepsilon_x\ \overline{(n-\bar{n})_a(n-\bar{n})_b}/\sqrt{\bar{n}_a\bar{n}_b}$, where ε_x is the histogram bin size on x and Poisson values of the variances in the denominator have been substituted. $\Delta\rho/\sqrt{\rho_{ref}}$ measures correlated pairs *per particle* on (y_{t1}, y_{t2}), (η_1, η_2) and (ϕ_1, ϕ_2). For angular correlations we combine spaces (η_1, η_2) and (ϕ_1, ϕ_2) in a *joint autocorrelation*, providing projection *by averaging* to a lower-dimensional space with little information loss. An autocorrelation with index k on primary space x is obtained by averaging correlation coefficients $r_{a,a+k}$ over index a along each kth diagonal on (x_1, x_2). For space (η, ϕ) we average simultaneously along diagonals on (η_1, η_2) and (ϕ_1, ϕ_2) to obtain a *joint* autocorrelation on angular *difference variables* $\eta_\Delta \equiv \eta_1 - \eta_2$ and $\phi_\Delta \equiv \phi_1 - \phi_2$.

NUMBER CORRELATIONS ON $Y_T \times Y_T$

Transverse rapidity y_t relates to fragmentation functions on logarithmic variable $\xi_p = \ln(p_t/E_{jet})$ [11]. Particle pairs from nuclear collisions can be separated on azimuth difference variable ϕ_Δ into same-side (SS) and away-side (AS) pairs. Figure 2 (first panel) shows unlike-sign (US) SS pairs on (y_t, y_t). The US hard component is a peak at $y_t \sim 2.8$ ($p_t \sim 1$ GeV/c) elongated along $y_{t\Sigma} \equiv y_{t1} + y_{t2}$ and running into the soft (string fragment) component at small y_t.

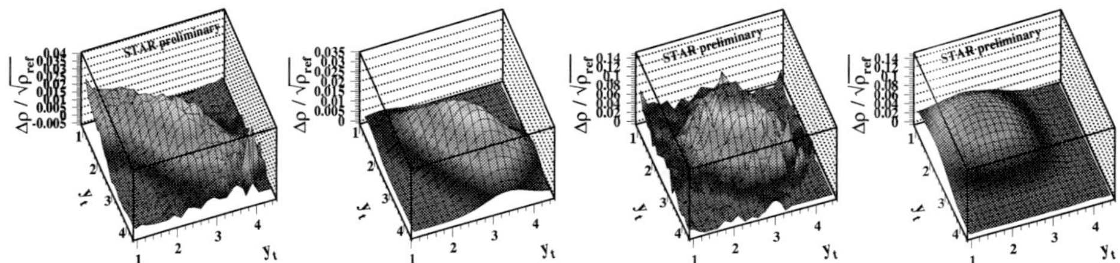

FIGURE 2. Parton fragment distributions on (y_{t1}, y_{t2}) for p-p data (US pairs), a model based on measured fragmentation functions, Au-Au mid-perpheral and Au-Au mid-central, all at $\sqrt{s_{NN}} = 200$ GeV.

The same-side unlike-sign (US) (y_t, y_t) correlations in Fig. 1 (first panel) can be interpreted as a two-particle *intra*-jet fragment distribution which we now model: we combine single-particle fragmentation functions [11] with expectations for two-particle correlations to sketch the parameterization of a two-particle fragment distribution shown in the second panel. We observe that single-particle fragmentation functions plotted on transverse rapidity y_t have a simple form represented by a beta distribution. We therefore construct a parton-fragment joint distribution on (y_t, y_t). By symmetrizing that distribution to approximate a fragment-fragment joint distribution we obtain the intra-jet fragment correlations shown in Fig. 2 (second panel). For low-Q^2 partons, which dominate the minimum-bias parton distribution, the two-particle fragment distribution is symmetric about the $y_{t\Sigma}$ diagonal. At larger $y_{t\Sigma}$ the distribution bifurcates symmetrically, the equivalent branches representing a continuum of conventional pQCD *conditional* fragmentation functions on fragment y_t *given* parton y_t.

The right two panels represent evolution with Au-Au centrality. For mid-peripheral collisions (third panel) the string fragments at smaller y_t are eliminated [12] and there is already some slight attenuation at larger y_t due to parton dissipation. For mid-central collisions (fourth panel) the fragment distribution is transported *en mass* to lower y_t, approaching the limiting case of random temperature variation and hydrodynamics [13]. The reduction with centrality at larger y_t corresponds to the variation of R_{AA} on p_t [14].

NUMBER CORRELATIONS ON $\eta \times \phi$

Jet structure is also characterized by two-particle angular correlations of hadron fragments on (η, ϕ) complementary to correlations on (y_{t1}, y_{t2}) described in the previous section. The conventional method to describe angular correlations of fragments in the absence of full jet reconstruction is as a *conditional* distribution relative to a trigger-particle momentum estimating the parton momentum. In this analysis we use no trigger condition. (y_{t1}, y_{t2}) correlations provide a cut space for study of angular autocorrelations. Figure 3 (first panel) shows two-particle correlations for p-p collisions on transverse-rapidity space (y_{t1}, y_{t2}) binned on sum and difference variables $y_{t\Sigma}$ and $y_{t\Delta} \equiv y_{t1} - y_{t2}$.

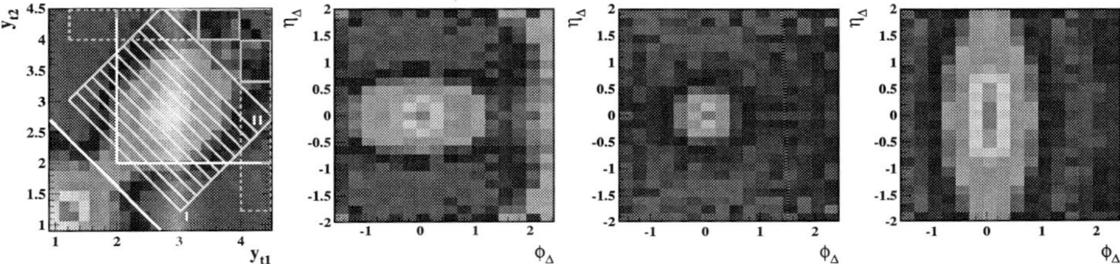

FIGURE 3. Binned (y_{t1}, y_{t2}) cut space, corresponding small-$y_{t\Sigma}$ (bin 2) and large-$y_{t\Sigma}$ (bin 11) angular correlations on $(\eta_\Delta, \phi_\Delta)$ for p-p collisions, and angular correlations for mid-central Au-Au collisions.

Hard-component fractions for angular correlation measurements are defined by the grid of bins along $y_{t\Sigma}$ (the narrow yellow lines) numbered $1, \cdots, 12$. The solid green boxes in the upper-right corner represent regions explored in *leading-particle* analyses based on a high-p_t 'trigger' particle [15]. The dashed extensions represent cuts for extended associated-particle conditions applied to heavy ion collisions [16]. Hard-component correlations on $(\eta_\Delta, \phi_\Delta)$ consist of a same-side peak at the origin and an away-side ridge. The US same-side peak represents intrajet angular correlations of parton fragments (jet cone). These hard-component (η, ϕ) systematics, fully consistent with conventional expectations for high-p_t jet angular correlations, are observed in this study for pairs of particles with *both* p_ts as low as 0.35 GeV/c ($y_t \sim 1.6$), *much lower than previously observed with leading-particle methods.*

The second and third panels of Fig. 3 show angular autocorrelations (these plots are 1:1 aspect ratio, hence exclude most or all of the away-side ridge) for bins 2 and 11 on $y_{t\Sigma}$ in Fig. 3 (first panel). In the second panel (bin 2) the most probable combination is two particles each with $p_t \sim 0.6$ GeV/c. The same-side peak (jet cone) is strongly elongated in the azimuth direction, suggesting that some aspect of the parton collision geometry is retained in these soft collisions. The third panel (bin 11) shows correlations corresponding to $p_t \sim 2.5$ GeV/c for each particle. The same-side cone is much narrower and *nearly* symmetric, more typical of a high-p_t trigger-particle analysis described by pQCD. The general trend is monotonic reduction of peak widths on $(\eta_\Delta, \phi_\Delta)$ with increasing $y_{t\Sigma}$ (Q^2). The fourth panel shows mid-central Au-Au collisions at $\sqrt{s_{NN}} = 130$ GeV. The near-side peak is strongly elongated on pseudorapidity rather than azimuth, suggesting strong coupling of the fragmenting parton to longitudinal Hubble expansion of the QCD medium [12].

$\langle P_T \rangle$ FLUCTUATIONS AND P_T CORRELATIONS ON $\eta \times \phi$

By measuring $\langle p_t \rangle$ fluctuation magnitudes as a function of bin size one can recover those aspects of the two-particle p_t correlation structure which depend on the separation of pairs of points, not on their absolute positions. Figure 4 (first panel) shows fluctuation scale dependence on bin sizes $(\delta\eta, \delta\phi)$ for 20-30% central Au-Au collisions at 200 GeV [17]. Fluctuation measurements at the full STAR acceptance [6] correspond to the single point at the apex of the distribution on scale. By inverting $\langle p_t \rangle$ fluctuation scale dependence [18], parton fragment distributions are visualized as temperature/velocity structures: joint p_t autocorrelations on angular difference variables $(\eta_\Delta, \phi_\Delta)$.

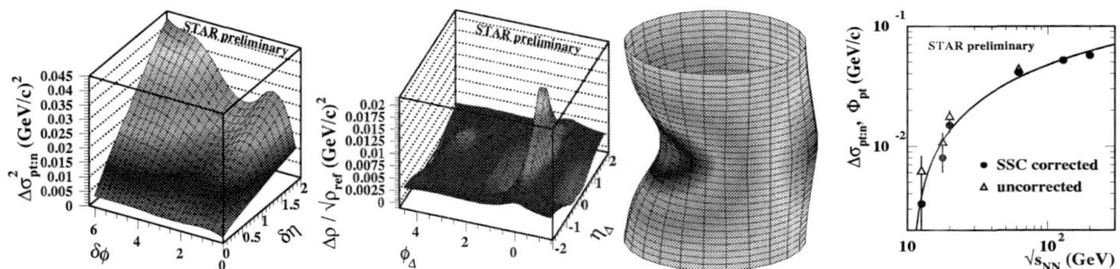

FIGURE 4. $\langle p_t \rangle$ fluctuation scale dependence for Au-Au collisions at $\sqrt{s_{NN}} = 200$ GeV, corresponding p_t autocorrelation, same with positive peak subtracted and energy dependence of $\langle p_t \rangle$ fluctuations.

Figure 4 (second panel) shows the corresponding p_t autocorrelation [17]. We have subtracted azimuth sinusoids independent of pseudorapidity (*e.g.*, elliptic flow observed for the first time as a velocity structure), revealing peak structures associated with parton scattering and fragmentation. A three-peak model of that distribution, including separate positive and negative same-side ($\phi_\Delta < \pi/2$) peaks (the two peaks have very different shapes), provides an excellent fit, with residuals at the percent level. Figure 4 (third panel) shows the result of subtracting the positive same-side model peak (representing parton fragments) from data in the second panel and plotting the difference with a cylindrical format. The negative same-side peak can be interpreted as a systematic red shift of local p_t distributions *in the neighborhood of* the positive parton fragment peak. The red shift can in turn be interpreted as the result of recoil of the bulk medium in response to stopping the parton *partner* of the observed parton (positive same-side peak). This detailed picture of parton dissipation, stopping and fragmentation is accessed for the first time with p_t autocorrelations.

Figure 4 (fourth panel) shows the energy dependence of full-acceptance $\langle p_t \rangle$ fluctuations for central heavy ion collisions at RHIC and SPS measurements by the CERES collaboration (lowest two energies) [19]. Fluctuation measure $\Delta\sigma_{p_t:n}$ is related to the variance difference by $\Delta\sigma^2_{p_t:n} \equiv 2\sigma_{\hat{p}_t}\Delta\sigma_{p_t:n}$, with $\sigma_{\hat{p}_t}$ the single-particle variance. To good approximation $\Delta\sigma_{p_t:n} \simeq \Phi_{p_t}$, the latter used for the CERES fluctuation measurements. For either measure we observe a dramatic increase in $\langle p_t \rangle$ fluctuations from SPS to RHIC energies where we have demonstrated that $\langle p_t \rangle$ fluctuations are dominated by fragments from low-Q^2 parton collisions. We observe that $\langle p_t \rangle$ fluctuations vary almost linearly with $\log\{\sqrt{s_{NN}}/10\}$ (solid curve in that panel), suggesting a threshold for *observable* parton scattering and fragmentation near 10 GeV.

SUMMARY

We have presented a survey of two-particle correlations from p-p and Au-Au collisions at RHIC. Correlations from longitudinal string fragmentation and transverse scattered parton fragmentation are clearly distinguished. The jet morphology of low-Q^2 partons requires a more general treatment of fragment p_t distributions and angular correlations. Conventional asymmetric treatments of parton fragments in terms of trigger and associated particles cannot access the low-Q^2 partons of greatest interest to us.

Using newly-devised analysis techniques we find that parton fragments are accessible down to hadron $p_t = 0.35$ GeV/c for both hadrons of a correlated pair. Fragment distributions on transverse rapidity y_t are, for p-p collisions, consistent with measured fragmentation functions in elementary collisions but reveal increasing parton dissipation with greater A-A centrality. Jet angular correlations in p-p collisions show a dramatic asymmetry about the thrust axis at low Q^2, with larger width in the azimuth direction possibly related to *nonperturbative* details of semi-hard parton collisions, rotating to elongation on pseudorapidity for central Au-Au collisions.

Inversion of the scale dependence of $\langle p_t \rangle$ fluctuations provides access to p_t angular autocorrelations, revealing a complex parton dissipation process in A-A collisions and possible evidence for bulk-medium recoil in response to parton stopping. We also observe strong energy dependence of $\langle p_t \rangle$ fluctuations, consistent with the dominant role of scattered partons in those fluctuations. Low-Q^2 partons, accessed here for the first time by novel analysis techniques including joint autocorrelations, serve as *Brownian probes* of the A-A medium at RHIC, being the softest *detectable* objects which experience QCD interactions as color charges.

REFERENCES

1. J. C. Collins and M. Perry, *Phys. Rev. Lett.* **34**, 1353-1356 (1975);
   B. Freedman and L. McLerran, *Phys. Rev. D* **17**, 1109-1122 (1978);
   Proceedings of Quark Matter 2004, *J. Phys. G* **30**, various articles.
2. K. Kajantie, P. V. Landshoff and J. Lindfors, *Phys. Rev. Lett.* **59**, 2527-2530 (1987).
3. A. H. Mueller, *Nucl. Phys. B* **572**, 227-240 (2000).
4. G. C. Nayak, A. Dumitru, L. D. McLerran and W. Greiner, *Nucl. Phys. A* **687**, 457-474 (2001).
5. K. Adcox et al. (PHENIX Collaboration), *Phys. Rev. C* **66**, 024901 (2002).
6. J. Adams et al. (STAR Collaboration), *Phys. Rev. C* **71**, 064906 (2005).
7. A. L. S. Angelis et al. (CCOR), *Phys. Lett. B* **97**, 163-168 (1980);
   C. Albajar et al. (UA1), *Nucl. Phys. B*, **309**, 405-425 (1988).
8. R. J. Porter and T. A. Trainor (STAR Collaboration), *Acta Phys. Polon. B* **36**, 353-359 (2005).
9. G. Wilk and Z. Wlodarczyk, *Phys. Rev. Lett.* **84**, 2770-2773 (2000).
10. K. Pearson, *Phil. Trans. Royal Soc.* **187**, 253-318 (1896).
11. M. Z. Akrawy et al. (OPAL Collaboration), *Phys. Lett. B* **247**, 617-628 (1990).
12. J. Adams et al. (STAR Collaboration), nucl-ex/0411003.
13. J. Adams et al. (STAR Collaboration), nucl-ex/0408012.
14. C. Adler et al. (STAR Collaboration), *Phys. Rev. Lett.* **89**, 202301 (2002).
15. C. Adler et al. (STAR Collaboration), *Phys. Rev. Lett.* **90**, 082302 (2003).
16. J. Adams et al. (STAR Collaboration), nucl-ex/0501016.
17. J. Adams et al. (STAR Collaboration), nucl-ex/0509030.
18. T. A. Trainor, R. J. Porter and D. J. Prindle, *J. Phys. G* **31**, 809-824 (2005).
19. D. Adamová et al. (CERES Collaboration), *Nucl. Phys. A* **727**, 97-119 (2003).

Two and Three Particle Flavor Dependent Correlations

N. N. Ajitanand for the PHENIX Collaboration

Dept. of Chemistry, SUNY Stony Brook, Stony Brook, NY 11794, USA

Abstract. In recent work, the PHENIX collaboration has applied a novel technique to the analysis of azimuthal correlations which extinguishes the harmonic part of the underlying event revealing the true jet shape. The extension of this method to three particle correlations allows a much more detailed study of jet topologies. Correlation functions and extracted jet landscapes are studied for wide range of p_T and centrality selections and particle flavors in Au+Au collisions at $\sqrt{s_{NN}}$=200 GeV. They reveal substantial modifications of jet properties resulting from the coupling of jets to the strongly interacting matter produced in RHIC collisions.

INTRODUCTION

The energy density achieved in Au + Au collisions at RHIC far exceeds the lattice QCD estimate for creating a de-confined phase of quarks and gluons (QGP) [1]. This energy density gives rise to large pressure gradients which are the driving force for the observed large azimuthal anisotropy (v_2) of particle emission from the collision zone [2]. The value of this anisotropy is close to the predictions of the hydrodynamic model which in turn implies the creation of a strongly interacting medium and early equilibration. In addition to the dominant soft processes giving rise to the formation of the medium, there are relatively rare hard parton-parton collisions in which the scattered partons propagate through the medium radiating gluons and interacting with the medium till they finally fragment into jet-like clusters. Thus, jets are a probe of the medium provided one can deconvolute the jet signal from the collective flow effects. Possible medium associated modifications of the jet properties are a shock wave induced conical flow or "sonic boom" [3] and "bending" induced by interactions between the propagating partons and the flowing medium [4]. Three particle azimuthal correlations can be an effective tool in the study of such modifications.

METHODOLOGY

To study jet properties we use two- and three-particle azimuthal correlation functions. These azimuthal correlation functions were built by pairing a leading hadron in a specified transverse momentum range p_T(trig), with an associated hadron also in a specified range p_T(assoc). For two-particle correlations, the correlation function $C(\Delta\phi)$ is given by:

$$C(\Delta\phi) = \frac{N_{real}(\Delta\phi)}{N_{mix}(\Delta\phi)},$$

CP828, *Multiparticle Dynamics*
edited by V. Šimák, M. Šumbera, Š. Todorova, and B. Tomášik

where $\Delta\phi$ is the difference of the azimuthal angles of the pair. The real distribution ($N_{real}(\Delta\phi)$) is built from pair members belonging to the same event and the mixed distribution ($N_{mix}(\Delta\phi)$) is made of pair members belonging to different events. Thus the correlation function is free of geometric acceptance effects and carries only the combined correlations from flow and jets. Decomposition of these correlations into their jet and flow contributions, constitute an important prerequisite for obtaining the jet function and hence, information about jet fragmentation.

Decomposition of the correlation function

Following a two source model ansatz, $C(\Delta\phi)$ can be written as;

$$C(\Delta\phi) = a_0 [H(\Delta\phi) + J(\Delta\phi)],$$

where $H(\Delta\phi)$ is a second harmonic function having an amplitude $p_2 = v_2(\text{assoc}) \times v_2(\text{trig})$ and $J(\Delta\phi)$ is the jet function [5]. Here, $v_2(\text{assoc})$ and $v_2(\text{trig})$ are the amplitudes of the harmonic distributions of trigger and associated hadrons (respectively) relative to the azimuth of the reaction plane Ψ_{RP}, i.e. $v_2(\text{assoc,trig}) = \left\langle \cos\left(2(\phi_{\text{assoc,trig}} - \Psi_{RP})\right)\right\rangle$. For these measurements, Ψ_{RP} was obtained from Beam-Beam counters having $\eta \pm 3.5$. To obtain $J(\Delta\phi)$ one simply uses the value a_0, obtained by assuming that the magnitude of the Jet function $J(\Delta\phi)$ is zero at its minimum (ZYAM assumption) [5], and subtracts the measured values of $H(\Delta\phi)$.

Extinction Method

One can also obtain $J(\Delta\phi)$ via extinction of the the harmonic function $H(\Delta\phi)$ [5]. This is done by aligning the trigger hadron within an appropriately chosen angular range $\Delta\phi_c$, perpendicular to the reaction plane. The harmonic amplitude for the trigger particle in this configuration $v_2^{out}(\text{trig})$, is given as [6]

$$v_2^{out}(\text{trig}) = \left(\frac{2v_2(\Delta\phi_c) - \sin(2\Delta\phi_c)\langle\cos(2\Delta\Psi_R)\rangle + \frac{v_2}{2}\sin(4\Delta\phi_c)\langle\cos(4\Delta\Psi_R)\rangle}{2(\Delta\phi_c) - 2v_2\sin(2\Delta\phi_c)\langle\cos(2\Delta\Psi_R)\rangle} \right).$$

Thus, it is clear that $v_2^{out}(\text{trig})$ can be made to vanish ($v_2^{out}(\text{trig}) \sim 0$) by appropriate choice of $\Delta\phi_c$ for a particular reaction plane resolution $\Delta\Psi_R$.

Figure 1 shows results from simulations in which strongly distorted away-side jets were studied. Panels (b), (c), and (d) on the left show cases for inclusive, in-plane and out-of-plane correlation functions. The figure clearly shows that our decomposition method retrieves the input jet function in detail, confirming that the decomposition procedure is robust even for unusual di-jet distributions.

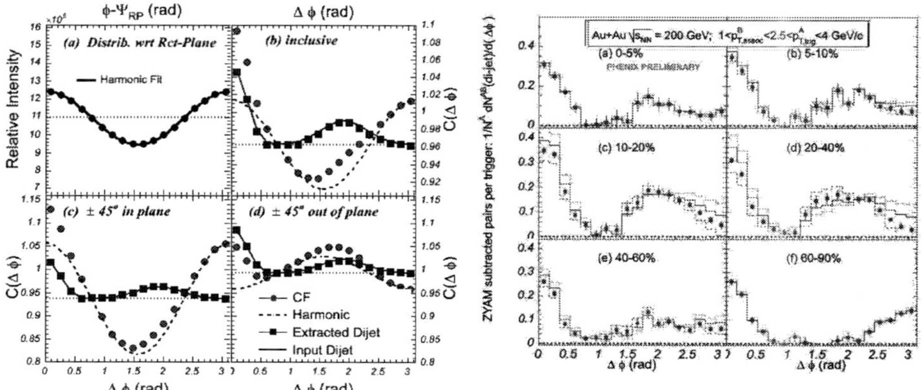

FIGURE 1. The figure on the left shows simulated data for strongly distorted away-side jets. The squares show retrieved points compared to the solid curves for the input jet functions. The figure on the right shows experimental jet-pair distributions [after flow subtraction] for several centrality selections.

RESULTS

Figure 1 show results obtained from the decomposition of the two-particle correlation measurements. The apparent shape distortions of the away-side jet is consistent with recent conjectures of a strong coupling between such jets and the high energy density matter that they traverse [3, 4].

Figure 2 show the application of these decomposition methods to meson-meson and baryon-meson correlation functions. They show that the jet-pair distributions, obtained after subtraction or extinction of the harmonic contributions, are significantly broadened on the away-side, independent of whether or not the trigger hadron is a baryon or meson. We attribute such broadening to the strong interactions between scattered partons and the high energy density medium.

It is straightforward to evaluate the conditional or per trigger yields (CY) for the near- and away-side jets via the integrated pair fractions indicated by the hatched area in Figs. 2 [5] or by building per trigger yield distributions. Figure 3 (top) compares proton and anti-proton jet yields obtained for near-side jets via the latter technique. The near-side yield is non-zero only for baryon-anti-baryon pairs, suggesting that baryon number conservation may play an important role in jet fragmentation.

One can also compare the baryonic and mesonic content of the near- and away-side jets by comparing the ratio CY(baryon)/CY(meson) for the near- and away-side jets. Figure 3 shows the double ratio

$$DBR = (CY(baryon)/CY(meson))_{away-jet} / (CY(baryon)/CY(meson))_{near-jet}$$

for central - mid-central collisions. The observed ratio of ~ 2.5 suggests that the away

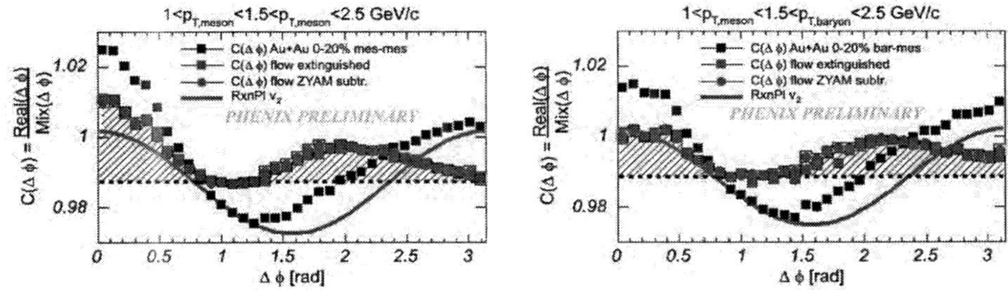

FIGURE 2. The figure on the left shows the correlation function for meson-meson pairs (black squares). The blue circles and red squares show the jet correlations which result from subtraction and harmonic extinction respectively (see text). The figure on the right shows results for baryon-meson pairs.

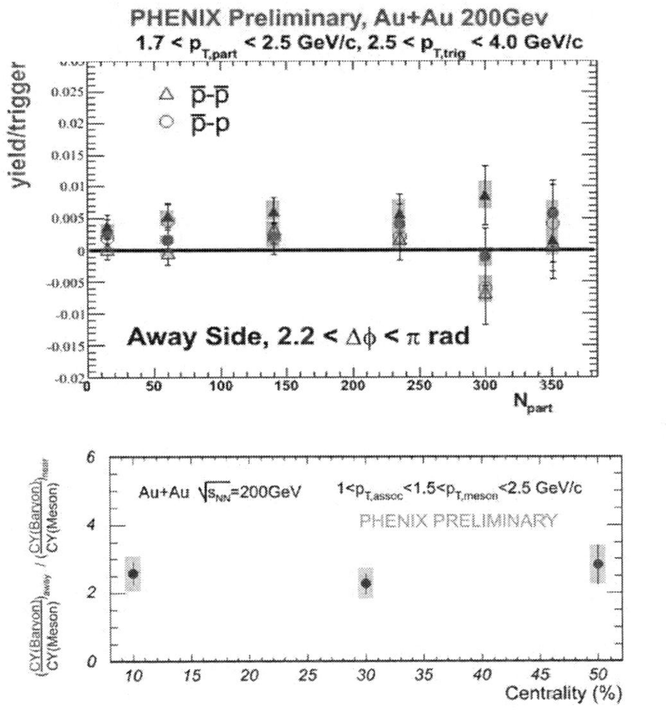

FIGURE 3. The figure on the top shows the per trigger yields vs. centrality for near-side jets obtained from p-p and p-$\bar{p}$ pairs. The figure on the bottom shows the centrality dependence of the double ratio *DBR*

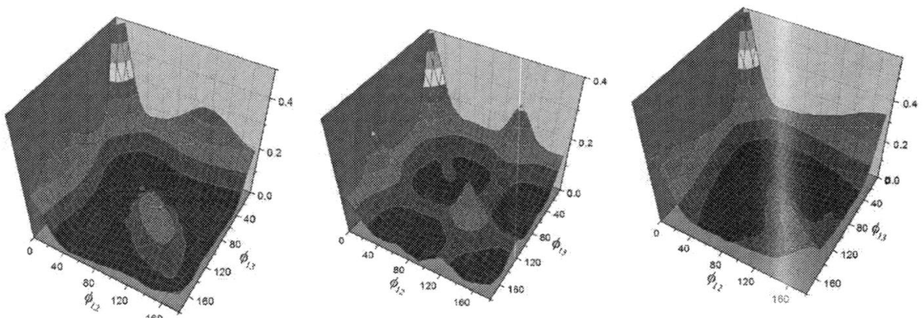

FIGURE 4. Simulated 3-particle correlations for "bent", conical and "normal" jets (left to right) respectively .

side jet is more baryon rich than the near side-side jet, possibly because of a modification to the fragmentation function of the away-side jet.

Three particle correlations

Three-particle correlation functions consisting of a trigger hadron from the range $2.5 < p_T < 4.0$ GeV/c (hadron #1) and two associated were also studied. Correlation surfaces $\Delta\phi_{1,2}$ and $\Delta\phi_{1,3}$ distributions. For these correlation functions, the harmonic extinction method was used and the mixed background was created by combining the high p_T particle from the real events with associated particles from two different events. The correlation functions therefore contain both triples and doubles contributions.

Simulated three-particle correlation surfaces ($\Delta\phi_{1,2}$ vs. $\Delta\phi_{1,3}$) are shown in Fig. 4 for three distinct away-side jet scenarios; (i) a "normal jet" in which the away-side jet axis is aligned with the leading jet axis with a spread, (ii) a "bent jet" in which the away-side jet axis is misaligned by 60°, and (iii) a "Cherenkov or conical jet" in which the leading and away-side jet axes are aligned the leading and away-side jet axes are aligned but fragmentation is confined to a very thin hollow cone with a half angle of 60°. The simulated results show relatively clear distinguishing features for the three scenarios considered.

The correlation surfaces obtained from data for the centrality selection 10-20 % are shown in Fig. 5. They show a strong dependence on the flavor (PID) of the associated particle and clearly do not follow the expected patterns for a "normal jet". We conclude that these three-particle correlation surfaces provide additional compelling evidence for strong modification of the away-side jet. Further detailed quantitative investigations are however required to firm up the signatures in the data which distinguish between a "bent jet" and a "conical jet".

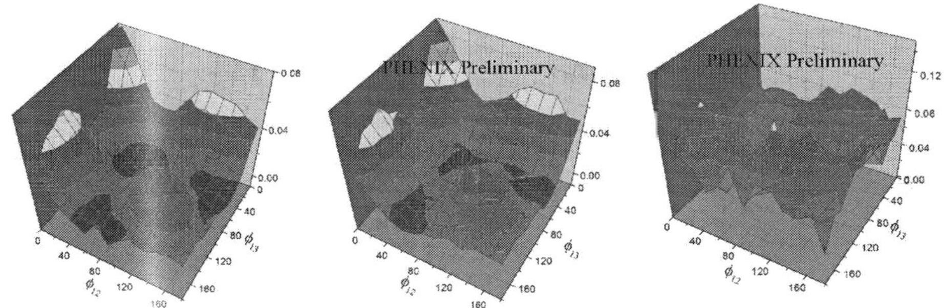

FIGURE 5. Hadron-hadron-hadron correlation function (left), Hadron-meson-meson correlation function (middle) Hadron-baryon-baryon correlation function (right)

SUMMARY

Novel methodologies have been developed to remove harmonic contributions and extract jet functions from azimuthal correlation functions. Jet function and yields show strong dependence on particle flavor. The near-side jet yield is non-zero only for baryon-anti-baryon pairs. The jet landscape for three particle correlations have been obtained as a function of particle flavor. They provide further compelling evidence for strong modification of the away-side jet.

REFERENCES

1. K. Adcox *et al*, *Phys. Rev. Lett.* **87**, 052301 (2001).
2. R. Lacey, `nucl-ex/0510029`.
3. J. Casalderrey-Solana, E. Shuryak, D. Teaney, `hep-ph/0411315`.
4. N. Armesto, C. Salgado, U.A. Wiedemann, `hep-ph/0411341`.
5. N.N. Ajitanand *et al.*, *Phys. Rev. C* **72**, 011902 (2005).
6. J. Bielcikova *et al.*, *Phys. Rev. C* **69**, 021901 (2004).

What have we learnt studying strangeness production in heavy-ion collisions at SPS ?

L. Šándor

Slovak Academy of Science, Institute of Experimental Physics, Watsonova 47, SK-04001 Košice

Abstract. A short review of main results from the strangeness production study in heavy-ion collisions at the CERN SPS is presented.

Keywords: heavy-ion collisions, strangeness production, quark-gluon plasma
PACS: 25.75.-q, 25.75.Dw, 25.75.Ld, 25.75.Nq

INTRODUCTION

The study of strange particle production in high-energy heavy-ion collisions plays an important role in understanding the nature of hadronic matter under extreme conditions and in the search for deconfined quark-gluon plasma (QGP). Strange particles are not present in the initial state systems, they are all created in the course of collision and therefore bear essential information about the physics environment developing in the reaction. As predicted more than 20 years ago [1, 2], strange and particularly multi-strange particles produced in ultrarelativistic nuclear collisions could be used as sensitive probes signalling QGP formation. Hadronic processes producing strangeness are predicted to be slow and not lead to chemical equilibrium within the typical duration of a heavy-ion reaction. On the contrary, in the QGP case the available time is sufficient to develop a flavour equilibrium among *u, d, s* quarks.

This paper provides a short report on main results from the strangeness production study in heavy-ion reactions at the CERN SPS, focusing on data from the NA49 (a large acceptance hadron spectrometer with large time projection chambers as main detectors [3]) and WA97/NA57 (utilizing a small silicon-pixel telescope as the core, covering one unit of rapidity at mid-rapidity [4]) experiments. We will briefly discuss the following topics: flavour equilibration, strangeness enhancements, possible onset of deconfinement, collective flow and thermal freeze-out scenario.

FLAVOUR EQUILIBRATION

When looking for signs of the production of a QGP one of the crucial questions is whether thermal and chemical equilibrium is achieved at some stage in the collision. Applying a statistical model which assumes equilibrium, and testing experimental data against model predictions gives one way of answering this question.

A realistic statistical model based on the use of grand canonical ensemble has been developed in [5] and compared with a set of particle ratios measured in Pb–Pb collisions

CP828, *Multiparticle Dynamics*
edited by V. Šimák, M. Šumbera, Š. Todorova, and B. Tomášik

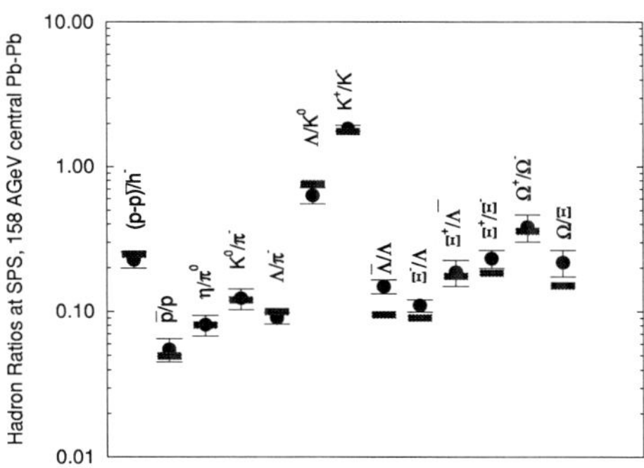

FIGURE 1. Hadron ratios: calculations in the statistical hadronization model (dashed bands) compared to experimental data (filled circles) for 158 *A* GeV central Pb–Pb collisions [7].

at top SPS energy (158 *A* GeV), including data on rare multi-strange hyperons. The overall agreement between model and data was satisfactory, indicating a close approach to chemical equilibrium with resulting temperature and chemical potential values close to the expected phase boundary between hadronic matter and the QGP. A more detailed statistical hadronisation model allowing a nonequilibrium population of strange hadrons (described by an undersaturation factor γ_s) has been developed in [6] and applied to a set of particle yields measured and extrapolated to full phase space by the NA49 experiment at laboratory energies 30 *A*, 40 *A*, 80 *A* and 158 *A* GeV. Fit results using this model indicate a significant strangeness undersaturation, however at the top SPS energy the deviation from full equilibrium is small ($\gamma_s \simeq 0.9$).

The conclusion on approaching the hadrochemical equilibrium at top SPS energy is supported by a recent comprehensive analysis of the several QGP-relevant observables in the framework of a model of the space-time evolution of the heavy-ion collision [7]. In this model all parameters of the equilibrated grand canonical ensemble are calculated from the fireball evolution leading to hadron ratios that do not depend on any free parameters. As one can see from figure 1, the results of the model calculations are in good agreement with the measured particle ratios.

STRANGENESS ENHANCEMENTS

In elementary hadron reactions the production of particles carrying strange quarks is systematically suppressed. When a QGP is created in a central heavy-ion collision at high enough energy density this suppression disappears and the colliding system undergoes the transition from a canonical ensemble to a grand canonical one. The resulting strangeness enhancement was predicted to increase with the strangeness content of the particle [2]. A comprehensive study of strangeness enhancements has been performed at the CERN SPS.

The first significant results came from the NA35 [8, 9] and WA85 [10] experiments, studying S–S and p–S (NA35) and S–W and p–W (WA85) collisions at 200 *A* GeV/c.

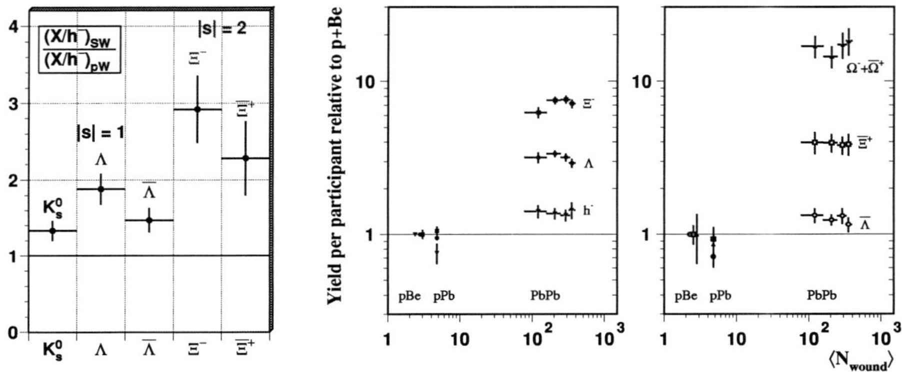

FIGURE 2. Left: a compilation of WA85 data on strangeness enhancements [11]. Right: WA97 results on the centrality dependence of strangeness enhancements in Pb–Pb with respect to p–Be collisions [13].

Both experiments observed a significant enhancement of the strange particle production with respect to the negatively charged hadron (h^-) production when going from proton-nucleus to central nucleus-nucleus interactions. A compilation of WA85 data on enhancements for particles carrying one and two units of strangeness is displayed in the left panel of figure 2 (see, e.g. [11]).

The precise measurement of enhancements in Pb–Pb reaction at the top SPS energy (158A GeV) by the WA97 experiment [12, 13] was an important piece of evidence in support of the claim that a new state of matter had been observed at the CERN SPS [14]. The WA97 results (Pb–Pb particle yields per participant relative to the p–Be ones) are shown in the right part of figure 2 as a function of collision centrality measured in number of participants. As predicted, enhancements are larger for particles with higher strangeness content, up to a factor $\sim$ 20 for triply strange Ω hyperons.

The NA57 experiment confirmed the WA97 results and extended the measurements to a wider centrality region and lower energy (40A GeV) [15, 16]. An access to the wider centrality range allowed to observe a significant centrality dependence of the Pb–Pb yields per participant for all hyperons except for $\overline{\Lambda}$. The energy dependence of the enhancements is in qualitative agreement with the prediction of canonical suppression model [17]. However, the present version of model neither reproduces the steepness of the observed centrality dependence nor the absolute values of enhancements.

THE ONSET OF DECONFINEMENT ?

The behaviour of data at AGS energies can be reproduced within hadronic transport models. In contrast, at top SPS energy there is striking evidence for new collective effects signalling the presence of a new, deconfined state of matter [14]. In order to search for the onset of deconfinement, the NA49 collaboration proposed to study the hadron production properties in central Pb–Pb collisions at different energies from the

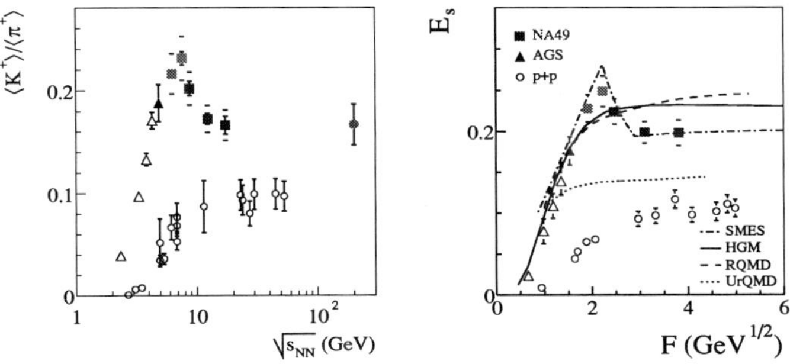

FIGURE 3. The dependence of $\langle K^+\rangle/\langle\pi^+\rangle$ (left) and E_s (right) ratios on the collision energy for central A–A (closed symbols) and inelastic p–p (open symbols) interactions [19].

top AGS (11.7A GeV) to the top SPS (158A GeV) energy. Data were taken at 20A, 30A, 40A, 80A and 158A GeV. The measurements suggest an existence of structure in energy dependence of several observables at lower SPS energies [18, 19, 20].

In figure 3 the most distinct structures observed at energy of $\sim$ 30A GeV are displayed (for details, see e.g. [18]). Both $\langle K^+\rangle/\langle\pi^+\rangle$ and $E_s = (\langle K+\overline{K}\rangle + \langle\Lambda\rangle)/\langle\pi\rangle$ (strangeness-to-pion) ratios in the Pb–Pb data show the sharp peak (horn) followed by a plateau. Such a structure is not observed in p–p interactions and is not reproduced by hadronic models. On the other hand this feature can be understood in a scenario with an onset of deconfinement around 30A GeV as proposed in the statistical model of the early stage - SMES [21]. This interesting NA49 result stimulated a lot of activity (see. e.g. [22, 23]) aiming at explanation of observed phenomena in terms of more conventional mechanisms.

COLLECTIVE FLOW AND THERMAL FREEZE–OUT

The transverse mass ($m_T = \sqrt{m^2 + p_T^2}$) spectra of particles are sensitive to the collision dynamics. The shape of the distributions is approximately exponential ($\propto \exp(m_T/T)$).

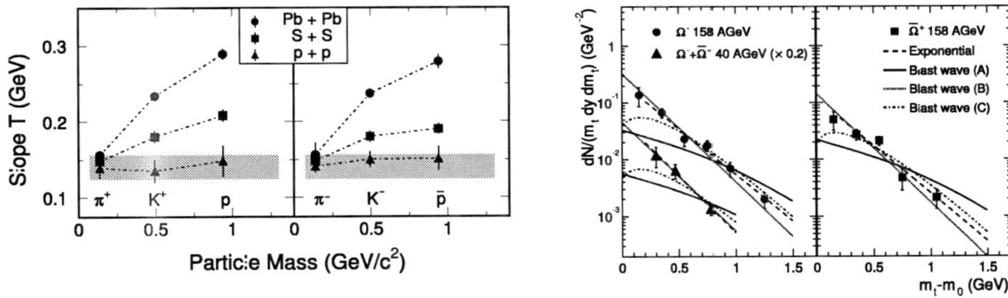

FIGURE 4. Left: Inverse slope parameter T as a function of particle mass (NA44 data at 158A GeV [24]). Right: The m_T spectra of Ω hyperons (NA49 data, [25]).

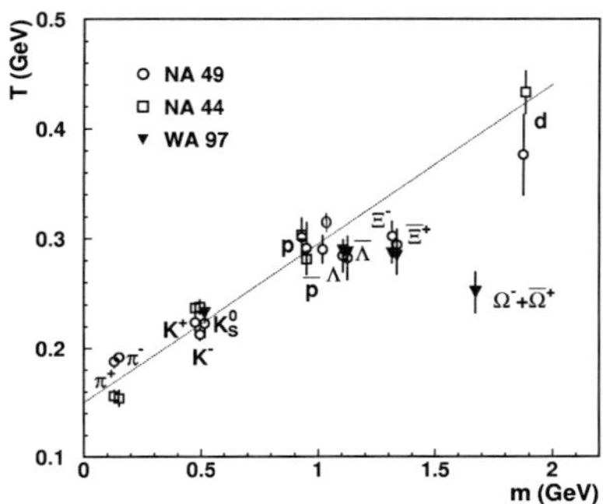

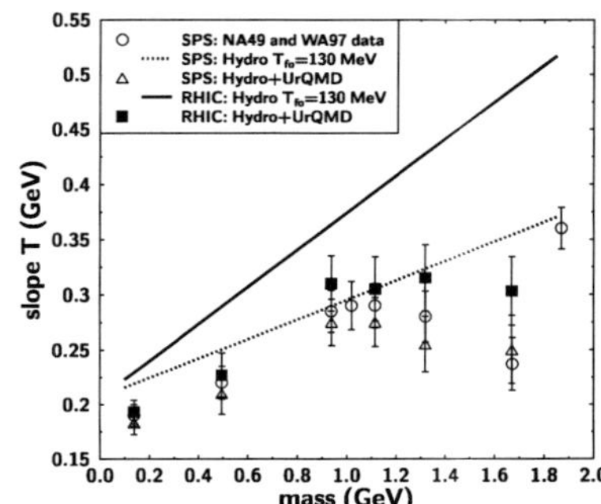

FIGURE 5. Inverse slope T as a function of particle rest mass m. Left: compilation of SPS data [26]. Right: comparison of model calculations [27] with the SPS data.

The presence of strong radial flow in Pb–Pb collisions at the top SPS energy was deduced from the systematics of experimental data suggesting approximately linear increase of the inverse slope with the particle mass. This was observed for the first time by the NA44 experiment [24] (see figure 4, left panel). Such a behaviour is expected (in a non-relativistic limit) when a thermal motion is coupled with a collective transverse expansion of the system [28]. This tendency was confirmed by hyperon spectra measurements in the NA49 [29] and WA97 [30] experiments.

A compilation of inverse slopes measured in Pb–Pb at 158A GeV (see [26] and references therein) is displayed in the left panel of figure 5. The measured slope of the Ω hyperon deviates from the common trend. This observation favours a scenario in which, in heavy-ion reaction, multi-strange hadrons freeze out early, before most of the transverse flow has developed [31]. As one can see from the right panel of figure 5, the observed pattern of SPS data, namely the "softer" spectra of multi-strange hyperons, is well described in a model [27] which combines hydrodynamic description for the early, deconfined stage of reaction with a microscopic transport UrQMD model for the later hadronic stage. The recent NA49 [32, 33] and NA57 [34, 35] measurements of m_T spectra of hadrons produced in Pb–Pb collision at different SPS energies confirm the observed pattern.

The m_T spectra of Ω hyperons measured by the NA49 experiment [25] are presented in a right panel of figure 4. The data favour a low transverse expansion velocity and high freeze-out temperature in agreement with hypothesis of early freze-out of multi-strange baryons. A similar conclusion resulted from the NA57 analysis [34].

A more complex analysis of the m_T spectra, aiming to disentangle the radial flow velocity $\beta_\perp$ from the thermal freeze-out temperature T_{fo}, is unavoidably model dependent. A hydrodynamically inspired blast-wave model [36, 37] is frequently used for this purpose. The fits of the blast-wave model with a linear velocity profile to the NA57 Pb–Pb data at 158A and 40A GeV [35] are shown in figure 6. Blast-wave model analysis performed for Pb–Pb data from 20A to 158A GeV from the NA49 and NA57 experiments [32, 33, 34, 35] revealed a significant centrality dependence of the thermal freeze-out parameters, but only week energy dependence. For central Pb–Pb collisions the temper-

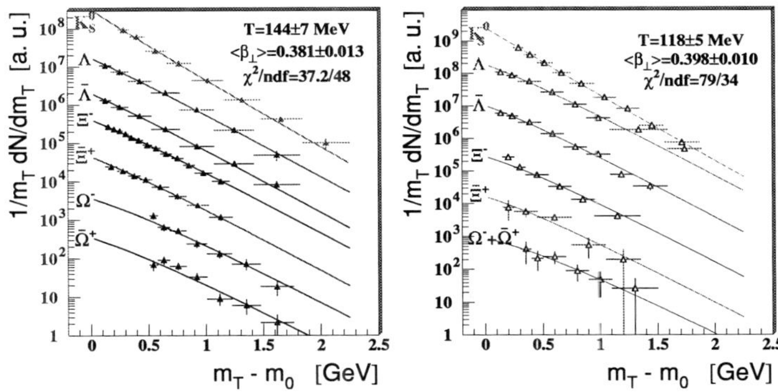

FIGURE 6. Blast-wave fits to the transverse mass spectra of strange particles for 53% most central Pb-Pb collisions at 158A GeV (left) and at 40A GeV (right) [35].

ature $T_{\rm fo}$ is about 110-130 MeV and the average transverse flow velocity $\langle\beta_\perp\rangle \simeq 0.45$.

The NA57 collaboration also studied the longitudinal flow. The value of average longitudinal velocity $\langle\beta_{\rm L}\rangle = 0.42 \pm 0.03$ has been extracted from an analysis of rapidity spectra at 158A GeV [38] indicating that collective expansion of the system is isotropic.

CONCLUSIONS

During the last decade an important advances have been made in the study of strangeness production in heavy-ion collisions at the CERN SPS. An analysis of the rich data samples is still in progress (see, e.g. [20, 16]).

Thermal model fits to data on Pb–Pb particle yields at the top SPS energy show a high degree of chemical and thermal equilibration, even for rare multi-strange baryons.

Strangeness enhancements in central Pb–Pb with respect to p–Be collisions were observed by the WA97 and NA57 experiments, which increase with the strangeness content of particle. The energy dependence of the enhancements is in qualitative agreement with the prediction of a canonical suppression model. However, more precise data on energy dependence of enhancements and more detailed model calculations are highly desirable.

Rich experimental data on Pb-Pb collisions were recorded and analysed during the SPS energy scan programme by the NA49 experiment. Several anomalies were found in the energy dependence of some observables (strangeness over pions ratio, inverse slopes of kaons, ...) indicating a possibility of the onset of deconfinement at about 30A GeV.

The analysis of transverse mass spectra of strange particles suggests that after a central Pb–Pb collision at the top SPS energy the system expands explosively and then it freezes out when the temperature is about 110-130 MeV and an average collective flow velocity is about one-half of the speed of light. An indication for the early freeze-out of multi-strange baryons has been obtained.

A discovery potential of the SPS is not yet exhausted. A possible continuation of experimenting with ion beams could answer a number of important questions.

ACKNOWLEDGMENTS

I am grateful to Emanuele Quercigh, Karel Šafařík, Boris Tomášik and Orlando Villalobos Baillie for fruitful discussions.

REFERENCES

1. J. Rafelski and B. Müller, *Phys. Rev. Lett.*, **48**, 1066-1069 (1982); erratum *ibid*, **56**, 2334 (1986).
2. P. Koch, B. Müller and J. Rafelski, *Phys. Rep.*, **142**, 167-262 (1986).
3. S. Afanasiev et al., *Nucl. Instrum. Meth. A*, **430**, 210-244 (1999).
4. NA57 collaboration, CERN-SPSC-2002-012 and references therein.
5. P. Braun-Munzinger, I. Heppe and J. Stachel, *Phys. Lett. B*, **465**, 15-20 (1999).
6. F. Becattini et al., *Phys. Rev. C*, **69**, 024905 (2004).
7. T. Renk, *J .Phys. G: Nucl. Part. Phys.*, **30**, 1495-1514 (2004).
8. J. Bartke et al., *Z. Phys. C*, **48**, 191-200 (1990).
9. R. Stock et al., *Nucl. Phys. A*, **525**, 221c-226c (1991).
10. S. Abatzis et al., *Nucl. Phys. A*, **525**, 445c-448c (1991).
11. L. Šándor et al., *Nucl. Phys. B (Proc. Suppl)*, **71**, 270-278 (1999). *Fig.1 reprinted with permission from Elsevier.*
12. E. Andersen et al., *Phys. Lett. B*, **449**, 401-406 (1999).
13. F. Antinori et al., *Nucl. Phys. A*, **661**, 130c-139c (1999). *Fig.4 reprinted with permission from Elsevier.*
14. U. Heinz and M. Jacob, *Evidence for a new state of matter: An assessment of the results from the CERN lead beam programme*, arXiv:nucl-th/0002042.
15. L. Šándor et al., *J. Phys. G: Nucl. Part. Phys.*, **31**, S919-S927 (2005).
16. R. Lietava for the NA57 collaboration: *NA57 results*, these proceedings.
17. A. Tounsi, A. Mischke and K. Redlich, *Nucl. Phys. A*, **715**, 565c-568c (2003).
18. M. Gaździcki, *J. Phys. G: Nucl. Part. Phys.*, **30**, S161-S167 (2004).
19. D. Flierl et al.: *Indications for the onset of deconfinement in nucleus nucleus collisions*, in *Quark Confinement and the Hadron Spectrum VI*, AIP Conference Proceedings **756**, 2005, pp. 433-435.
20. B. Lungwitz for the NA49 collaboration: *NA49 results on hadron production: indications of the onset of deconfinement ?*, these proceedings.
21. M. Gaździcki and M.I. Gorenstein, *Acta Phys. Pol. B*, **30**, 2705-2735 (1999).
22. J. Letessier and J. Rafelski, *Hadron production and phase changes in relativistic heavy ion collisions*, arXiv:nucl-th/0504028.
23. B. Tomášik, these proceedings.
24. I.G. Bearden et al., *Phys. Rev. Lett.*, **78**, 2080-2083 (1997).
25. C. Alt et al., *Phys. Rev. Lett.*, **94**, 192301 (2005).
26. F. Antinori et al., *Eur. Phys. J. C*, **14**, 633-641 (2000). *Fig.1 reprinted with permission of Springer Science and Business Media.*
27. S.A. Bass and A. Dumitru, *Phys. Rev. C*, **61**, 064909 (2000).
28. K.S. Lee, U. Heinz and E. Schnedermann, *Z. Phys. C*, **48**, 525-541 (1990).
29. T. Alber et al., *J. Phys. G: Nucl. Part. Phys.*, **23**, 1817-1825 (1997).
30. I. Králik at al., *Nucl. Phys. A*, **638**, 115c-124c (1988).
31. H. van Hecke, H. Sorge and Nu Xu, *Phys. Rev. Lett.*, **81**, 5764-5767 (1998).
32. M. van Leeuwen et al., *Nucl. Phys. A*, **715**, 161c-170c (2003).
33. M. Gaździcki et al., *J. Phys. G: Nucl. Part. Phys.*, **30**, S701-S708 (2004).
34. F. Antinori et al., *J. Phys. G: Nucl. Part. Phys.*, **30**, 823-840 (2004).
35. G.E. Bruno et al., *J. Phys. G: Nucl. Part. Phys.*, **31**, S127-S133 (2005).
36. E. Schnedermann, J. Sollfrank and U. Heinz, *Phys. Rev. C*, **48**, 2462-2475 (1993).
37. E. Schnedermann and U. Heinz, *Phys. Rev. C*, **50**, 1675-1683 (1994).
38. F. Antinori et al., *J. Phys. G: Nucl. Part. Phys.*, **31**, 1345-1357 (2005).

Relativistic Diffusion Model and Analysis of Large Transverse Momentum Distributions

Naomichi Suzuki* and Minoru Biyajima†

*Department of Comprehensive Management, Matsumoto University, Matsumoto 390-1295, Japan
†Department of Physics, Shinshu University, Matsumoto, 390-8621, Japan

Abstract. In order to describe large transverse momentum (p_T) distributions observed in high energy nucleus-nucleus collisions, a stochastic model in the three dimensional rapidity space is introduced. The fundamental solution of the radial symmetric diffusion equation is Gaussian-like in radial rapidity. We can also derive a p_T or radial rapidity distribution function, where a distribution of emission center is taken into account. The solution is applied to the analysis of observed large p_T distributions of charged particles. It is shown that our model approaches to a power function of p_T in the high transverse momentum limit.

Keywords: Relativistic heavy ion collisions, Stochastic model, Large p_T distributions
PACS: 25.75.-q, 02.50.Ey

INTRODUCTION

In high energy nucleus-nucleus collisions, colliding energy $\sqrt{s_{NN}}$ per nucleon grows up to $\sqrt{s_{NN}} = 200$ GeV, and secondary particles with transverse momentum p_T more than 10 GeV/c are observed. The observed p_T distributions have long tail compared with exponential distribution in p_T.

As is well known, the fundamental solution of stochastic process is Gaussian if variables in the Euclidian space are used. Therefore, as long as we consider stochastic equations in the transverse momentum space, it is very hard to describe observed p_T distributions. This fact suggests that a relativistic approach to the stochastic process would be needed.

In Ref. [1], an empirical formula for large p_T distributions at polar angle $\theta = \pi/2$,

$$\begin{aligned} E\frac{d^3\sigma}{d^3p}\Big|_{\theta=\pi/2} &= A\exp[-y_T^2/(2L_T)], \\ y_T &= \frac{1}{2}\ln\frac{E+|\mathbf{p}_T|}{E-|\mathbf{p}_T|}, \end{aligned} \tag{1}$$

was proposed from the analogy of Landau's hydrodynamical model. Polar angle θ ($0 \le \theta \le \pi$) is measured from the beam direction of colliding nuclei or incident particles. In Eq. (1), E denotes energy of an observed particle, L_T is a parameter, and y_T is called the "transverse rapidity". Equation (1) well describes large p_T distributions for $p+p \to \pi^0 + X$ and $p+p \to \pi^{\pm} + X$. However, it cannot be derived from the hydrodynamical model.

As for the relativistic approach to stochastic equation, we consider the diffusion equation in the three dimensional rapidity space or Lobachevsky space, which is non-

CP828, *Multiparticle Dynamics*
edited by V. Šimák, M. Šumbera, Š. Todorova, and B. Tomášik

Euclidean. In order to classify the longitudinal and transverse expansion, it would be appropriate to consider the diffusion equation in the geodesic cylindrical coordinate, where longitudinal rapidity y, transverse rapidity ξ and azimuthal angle ϕ are used. The longitudinal rapidity y, and the transverse rapidity ξ are defined respectively as,

$$y = \ln\frac{E+p_L}{m_T}, \quad \xi = \ln\frac{m_T+|\mathbf{p}_T|}{m},$$

where E, p_L, $\mathbf{p}_T$ and m denote energy, longitudinal momentum, transverse momentum, and mass of the observed particle, respectively, and $m_T = \sqrt{\mathbf{p}_T^2+m^2}$.

The diffusion equation in the geodesic cylindrical coordinate is given by,

$$\frac{\partial f}{\partial t} = \frac{D}{\cosh^2\xi}\frac{\partial^2 f}{\partial y^2} + \frac{D}{\sinh\xi\cosh\xi}\frac{\partial}{\partial\xi}\left(\sinh\xi\cosh\xi\frac{\partial f}{\partial\xi}\right) + \frac{D}{\sinh^2\xi}\frac{\partial^2 f}{\partial\phi^2}, \tag{2}$$

where D denotes a diffusion constant. However, we cannot solve Eq.(2) analytically at present. Therefore, we should consider somewhat simpler case.

We have proposed the relativistic diffusion model, and analyzed large p_T distributions for charged particles in $Au+Au$ collisions [2], where a radial flow effect is not included.

In section 2, the relativistic diffusion model is briefly explained. A distribution of the initial radial rapidity is taken into account. It would correspond to the distribution of radial flow. In section 3, large p_T distributions for charged particles observed at RHIC [3, 4] are analyzed. The magnitude of radial flow is estimated from p_T distributions. In section 4, high transverse momentum limit of our model is taken and relation to Hagedorn's model for p_T distribution inspired by the QCD is discussed. Final section is devoted to summary and discussions.

DIFFUSION EQUATION WITH RADIAL SYMMETRY IN THE THREE DIMENSIONAL RAPIDITY SPACE

For simplicity, we consider the diffusion equation with radial symmetry in the geodesic polar coordinate system,

$$\frac{\partial f}{\partial t} = \frac{D}{\sinh^2\rho}\frac{\partial}{\partial\rho}\left(\sinh^2\rho\frac{\partial f}{\partial\rho}\right), \tag{3}$$

with an initial condition

$$f(\rho, t=0) = \frac{\delta(\rho-\rho_0)}{4\pi\sinh^2\rho}. \tag{4}$$

In Eq. (3), ρ denotes the radial rapidity, which is written with energy E, momentum $\mathbf{p}$ and mass m of observed particle,

$$\rho = \ln\frac{E+|\mathbf{p}|}{m}. \tag{5}$$

Inversely, energy and momentum are written respectively as

$$E = m\cosh\rho, \qquad |\mathbf{p}| = \sqrt{p_L^2 + \mathbf{p}_T^2} = m\sinh\rho. \tag{6}$$

The solution of Eq. (3) with the initial condition (4) is given [5] by

$$f(\rho,\rho_0,t) = \frac{1}{2\pi\sqrt{4\pi Dt}}\mathrm{e}^{-Dt}\frac{\sinh\frac{\rho_0\rho}{2\pi Dt}}{\sinh\rho_0\sinh\rho}\exp\Big[-\frac{\rho^2+\rho_0^2}{4Dt}\Big]. \tag{7}$$

From Eq.(7), the following equation is obtained;

$$f(\rho,t) = \lim_{\rho_0\to 0} f(\rho,\rho_0,t) = \big(4\pi Dt\big)^{-3/2}\mathrm{e}^{-Dt}\frac{\rho}{\sinh\rho}\exp\Big[-\frac{\rho^2}{4Dt}\Big]. \tag{8}$$

In Eq. (7), ρ_0 denotes the radial rapidity of an emission center. It would be identified to the radial flow rapidity. We have estimated radial flow rapidity using Eq. (7) [6].

If the distribution of ρ_0 is taken into account, it would be reasonable to assume that it distributes randomly with dispersion σ_0^2;

$$f(\rho_0,t) = \big(2\pi\sigma_0^2\big)^{-3/2}\mathrm{e}^{-\sigma_0^2/2}\frac{\rho_0}{\sinh\rho_0}\exp\Big[-\frac{\rho_0^2}{2\sigma_0^2}\Big].$$

Then the distribution function of radial rapidity is given by the following equation,

$$\begin{aligned} f_c(\rho,t) &= \int_0^\infty f(\rho,\rho_0,t)f(\rho_0,t)\sinh^2\rho_0\,\sin\theta d\rho_0 d\theta d\phi \\ &= (2\pi\sigma_T^2)^{-3/2}\mathrm{e}^{-\sigma_T^2/2}\frac{\rho}{\sinh\rho}\exp\Big[-\frac{\rho^2}{2\sigma_T^2}\Big], \qquad (9) \\ \sigma_T^2 &= 2Dt + \sigma_0^2. \qquad (10) \end{aligned}$$

Equation (9) is obtained form Eq. (8), if $2Dt$ in Eq. (8) is replaced by σ_T^2. Parameter σ_T^2 is connected to the moment of ρ_0 as,

$$\langle\rho_0^2\rangle = \int_0^\infty \rho_0^2 f(\rho_0,t)\sinh^2\rho_0\,\sin\theta d\rho_0 d\theta d\phi = \sigma_0^4 + 3\sigma_0^2.$$

When $\rho << 1$, $|\mathbf{p}| = m\sinh\rho \simeq m\rho$. Then, Eq. (8) reduces to

$$f(\rho,t) \simeq \exp[-\rho^2/(2\sigma(t)^2)], \quad \sigma(t)^2 = 2Dt. \tag{11}$$

If we assume that Eq. (11) should coincide with the Maxwell-Boltzmann distribution, we have an identity, $k_BT = m\sigma(t)^2$, where k_B is the Boltzmann constant. Then, Eq. (10) reduces to

$$\sigma_T^2 = k_BT/m + \Big(-\frac{3}{2} + \frac{1}{2}\sqrt{9+4\langle\rho_0^2\rangle}\Big). \tag{12}$$

When, $\sigma_0^2 << 1$ or $\rho_0^2 << 1$, Eq. (10) is written as,

$$\sigma_T^2 \simeq k_BT/m + \langle\rho_0^2\rangle/3. \tag{13}$$

ANALYSIS OF LARGE P_T DISTRIBUTIONS OF CHARGED PARTICLES

Transverse momentum distributions of charged particles observed by STAR [3] and PHENIX [4] Collaborations are analyzed by Eq. (9). Data are taken at $\theta = \pi/2$. In this case, the identity, $\rho = y_T = \xi$, is satisfied.

The results on the data by STAR Collaboration are shown in Fig. 1 and Table 1.

Information on radial flow is extracted by the use of Eq. (12) under the assumption that the hadronization temperature is constant irrespective of centrality. The results are shown in Fig.1b. The mean radial flow rapidity, $\langle \rho_o^2 \rangle^{1/2}$, depends on centrality very weakly when it is less than 60%. At $k_B T = 0.13$ GeV, $\langle \rho_o^2 \rangle^{1/2} \simeq 0.57$ for centrality $<$ 60%.

The results on the data by PHENIX Collaboration are shown in Fig. 2 and Table 2. Estimated value of $\langle \rho_o^2 \rangle^{1/2}$ is somewhat larger than that from STAR Collaboration.

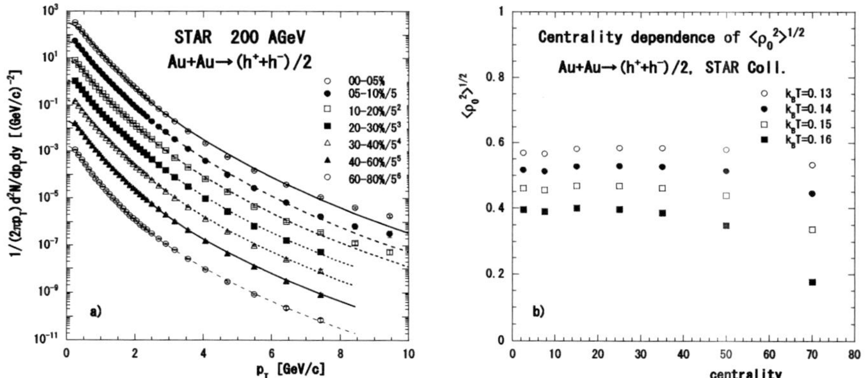

FIGURE 1. (a) p_T distribution for $Au + Au \to (h^+ + h^-)/2 + X$ at $y = 0$ [3], and (b) centrality dependence of $\langle \rho_0^2 \rangle^{1/2}$ extracted from p_T distributions

TABLE 1. Parameters on p_T distributions estimated by Eq. (9) in $Au + Au \to (h^+ + h^-)/2 + X$ at $y = 0$ at $\sqrt{s_{NN}} = 200$ GeV [3]

centrality	C	σ_T^2	m	χ^2_{min}/**n.d.f**
00-05%	1126.3 ± 51.0	0.332± 0.003	0.570± 0.007	244.9/32
05-10%	916.2 ± 47.5	0.332± 0.004	0.567± 0.009	128.8/32
10-20%	780.0 ± 36.2	0.351± 0.004	0.536± 0.008	144.2/32
20-30%	590.1 ± 28.9	0.360± 0.004	0.519± 0.008	98.1/32
30-40%	451.5 ± 22.7	0.374± 0.004	0.492± 0.008	95.4/32
40-60%	300.99 ± 14.47	0.402± 0.004	0.442± 0.006	51.7/32
60-80%	150.91 ± 7.43	0.444± 0.004	0.369± 0.006	26.5/32
pp(NSD)	17.654 ± 1.424	0.455± 0.006	0.324± 0.008	16.1/29

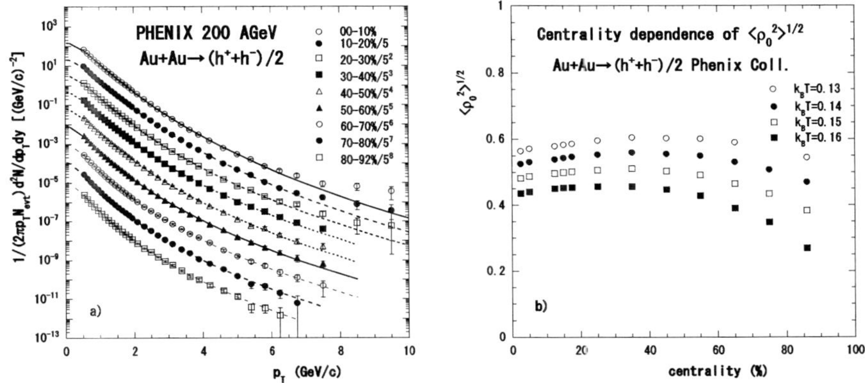

FIGURE 2. (a) p_T distributions for $Au+Au \to (h^+ + h^-)/2 + X$ at $y=0$ [4], and (b) centrality dependence of $\langle \rho_0^2 \rangle^{1/2}$ extracted from p_T distributions

TABLE 2. Parameters on p_T distributions estimated by Eq. (9) in $Au+Au \to (h^+ + h^-)/2 + X$ at $y=0$ at $\sqrt{s_{NN}} = 200$ GeV [4]

centrality	C	σ_T^2	m	χ^2_{min}/n.d.f
00-05%	410.00 ± 42.99	0.277± 0.007	0.745± 0.023	32.1/31
00-10%	408.50 ± 41.25	0.284± 0.007	0.726± 0.021	35.5/31
10-20%	305.21 ± 30.84	0.293± 0.007	0.708± 0.021	26.4/31
20-30%	265.03 ± 27.18	0.313± 0.007	0.653± 0.019	27.8/31
30-40%	209.37 ± 21.82	0.330± 0.007	0.611± 0.019	16.8/29
40-50%	143.61 ± 15.42	0.338± 0.007	0.587± 0.018	15.4/29
50-60%	103.03 ± 11.65	0.358± 0.008	0.537± 0.018	9.9/29
60-70%	70.343 ± 8.419	0.380± 0.009	0.485± 0.017	5.1/29
70-80%	40.688 ± 5.284	0.390± 0.010	0.457± 0.018	7.4/28
80-92%	22.286 ± 3.126	0.406± 0.011	0.419± 0.018	10.3/27

HIGH TRANSVERSE MOMENTUM LIMIT OF THE MODEL

In the 1970's, when large p_T distributions are observed in accelerator experiments, many models, which have power-law behavior in p_T are proposed. In Ref.[7], a model for p_T distribution, inspired by the QCD, is proposed as $(p_0/(p_T+p_0))^n$. It approaches an exponential distribution of p_T for $p_T \to 0$, and a power function of p_T for $p_T \to \infty$.

In this section, high transverse momentum limit of Eq. (9) is examined at $\theta = \pi/2$, where the identity $\rho = \ln((m_T + p_T)/m)$ holds. Radial rapidity contained in the Gaussian-like part in Eq. (9) is rewritten as,

$$\exp\left[-\frac{\rho^2}{2\sigma_T^2}\right] = \left(\frac{m_T + p_T}{m}\right)^{-\rho/(2\sigma_T^2)}. \tag{14}$$

For any positive number ε, we have,

$$\lim_{p_T\to\infty}\frac{\rho}{p_T^{\varepsilon}} = \lim_{p_T\to\infty}\frac{1}{p_T^{\varepsilon}}\ln\left(\frac{m_T+p_T}{m}\right) = 0.$$

Therefore, we can approximate ρ in the exponent in Eq. (14) as a constant, which is written by $2c_0$, within some finite transverse momentum range. Then Eq. (9) reduces to,

$$f_c(\rho,t) \sim \left(\frac{m}{m_T+p_T}\right)^{c_0/\sigma_T^2}\frac{m}{p_T} \sim \left(\frac{m}{2p_T}\right)^{c_0/\sigma_T^2+1}. \tag{15}$$

From Eq. (15), one can see that Eq. (9) shows the power-law behavior in the high transverse momentum limit, and that the power becomes smaller as σ_T^2, which should increase as the colliding energy $\sqrt{s_{NN}}$ increases.

SUMMARY AND DISCUSSIONS

In order to analyze large p_T distributions of charged particles observed at RHIC, the relativistic stochastic process in the three dimensional rapidity space, which is non-Euclidean, is introduced. The solution is Gaussian-like in radial rapidity, where the radial flow rapidity ρ_0 is included. It is very similar to the formula proposed in Ref. [1] at $\theta = \pi/2$. (See also Ref. [8].)

Transverse momentum distributions for charged particles in $Au+Au$ collisions at $y=0$ at $\sqrt{s_{NN}} = 200$ GeV are analyzed. From the observed p_T distributions, an effect of radial flow is subtracted by the use of assumption that the hadronization temperature does not depend on the centrality.

The averaged radial flow rapidity, $\langle\rho_0^2\rangle^{1/2}$, depends on the centrality very weakly from 0% up to 60%. It decreases rapidly from about 60% to higher centrality.

We have derived that our formula (9) shows power-law behavior in p_T in the high transverse momentum limit. Therefore, it behaves like the Gaussian distribution in p_T when $p_T \ll m$, and like the power-law distribution in p_T when $p_T \gg 1$.

ACKNOWLEDGMENTS

Authors would like to thank RCNP at Osaka University, Faculty of Science, Shinshu University, and Matsumoto University for financial support.

REFERENCES

1. M. Duong-van and P. Carruthers, *Phys. Rev. Lett.* **31**, 133-135 (1973)
2. N. Suzuki and M. Biyajima, *Acta Phys. Polon. B* **35**, 283-288 (2004); hep-ph/0404112
3. J. Adams, et al., STAR Collaboration, *Phys. Rev. Lett.* **91**, 172301 (2003)
4. S.S. Adler, et al., PHENIX Collaboration, *Phys. Rev. C* **69**, 034910 (2004)
5. N. Suzuki and M. Biyajima, math-ph/0406040, unpublished
6. N. Suzuki and M. Biyajima, hep-ph/0504076
7. R. Hagedorn, *CERN preprint* Ref.TH.3684-CERN (1983)
8. M. Biyajima, M. Kaneyama, T. Mizoguchi and G. Wilk, *Eur. Phys. J. C*, **40**, 243-250 (2005)

ASTROPARTICLE PHYSICS

Chairpersons: B. Shephard and G. Kozlov

The PICASSO Direct Dark Matter Search Experiment

F. Aubin[§], M. Barnabé-Heider[§], E. Behnke[‡], K. Clark[†], M. Di Marco[†], P. Doane[§], W. Feighery[ℓ], M.-H. Genest[§], R. Gornea[§], R. Guénette[§], S. Kanagalingam[¶], C.B. Krauss[†], C. Leroy[§*], L. Lessard[§], I. Levine[‡], J.- P. Martin[§], C. Muthusi[ℓ], A.J. Noble[†], R. Noulty[¶], S. Pospisil[★], J. Sodomka[★], I. Stekl[★], U. Wichoski[§] and V. Zacek[§]

[§]*Département de physique, Université de Montréal, Montréal, H3C 3J7, Canada*
[†]*Department of physics, Queens University, Kingston, K7L 3NG, Canada*
[‡]*Department of physics & astronomy, Indiana University South Bend, South Bend, Indiana, 46634, USA*
[ℓ]*Department of Chemistry, Indiana University South Bend, South Bend, Indiana, 46634, USA*
[¶]*Bubble Technology Industries, Chalk River K0J 1J0, Canada*
[★]*Institute of Experimental and Applied Physics of CTU, Prague, CZ-128 00 Prague2, Czech Republic.*

Abstract. The PICASSO experiment is searching for cold dark matter through the direct detection of weakly interacting massive particles (WIMPs), in particular neutralinos (χ) via their spin-dependent interactions with nuclei. The experiment is installed in the Sudbury Neutrino Observatory Laboratory at a depth of 2070 m (6000 mwe). PICASSO makes use of the superheated droplet technique with C_4F_{10} as the active material, and searches for χ interactions on ^{19}F. The results of these measurements are presented in terms of limits on the spin-dependent χ-proton and χ-neutron cross sections. Limits on the effective χ-proton and χ-neutron coupling strengths a_p and a_n are also reported. The results exclude new regions of the spin-dependent χ-nucleon interaction parameter space. The next phase of PICASSO is briefly discussed.

Keywords: Dark matter, Neutralino, Superheated Droplet Detectors, neutrons, alpha particles.
PACS: 24.10.Lx, 28.20.-v, 95.35, 29.40.-n

INTRODUCTION

It is commonly believed that a large fraction of the matter in the Universe is non-luminous and non-baryonic. The most recent evidence is based on the measurement of the cosmic radiative background (WMAP, [1]). The amount of cold dark matter is predicted by WMAP as $\Omega_{CDM}h^2$ = 0.095-0.129 at the 2 σ confidence level, where h ~ 0.72 is the Hubble expansion parameter. The Universe consists of ~ 4% of baryonic matter, 23% of CDM and 73% of dark energy. The lightest neutralino, χ, predicted in supersymmetric (SUSY) extensions of the Standard Model (SM) [2] emerges as the most promising candidate for dark matter. This stable spin ½ particle has no electrical

[*] Corresponding author : Bureau LRL 235, Département de physique, Université de Montréal, C.P. 6128, succ. centre-ville, Montréal (Qc), H3C 3J7, Canada. Tel.: 1-514-343-6722.
E-mail address : leroy@lps.umontreal.ca.

CP828, *Multiparticle Dynamics*
edited by V. Šimák, M. Šumbera, Š. Todorova, and B. Tomášik

charge and is a weakly interacting massive particle (WIMP) with a mass in the range from 10 GeVc^{-2} to 1 TeVc^{-2}. These neutralinos are assumed to be distributed in galactic halos[3]. In the Milky-Way galaxy, the velocity, v, of these particles is assumed to follow an isotropic Maxwellian distribution of the form $f(v) = v^2 e^{-(v+v_E)^2/v_0^2}$, where v_0 = 230 kms^{-1} is the velocity dispersion of the dark matter halo and v_E = 244 kms^{-1} is the velocity of the earth relative to the dark matter distribution (the effect of a $\pm$ 20 kms^{-1} annual variation is neglected) [3]. The local neutralino mass density at the position of the solar system is assumed to be 0.3 GeVc^{-2}cm^{-3} and the velocity distribution is truncated at the escape velocity of matter in the Milky Way, v_{esc} = 600 kms^{-1}. In a direct dark matter search experiment, the detection reaction of neutralinos is elastic scattering off detector atomic nuclei. The detection of neutralinos is through the measurement of the energy deposited by the recoiling nucleus. The rate of interaction of neutralinos with ordinary matter is of the order of a fraction of an event per day per kilogram. Therefore, it is necessary to build detectors with extremely low intrinsic background and to operate the experiment in a deep underground site.

THE EXPERIMENTAL TECHNIQUE

PICASSO detectors [4] consist of containers filled with droplets of a liquid which is in the superheated state at room temperature (presently a fluorinated halocarbon C_4F_{10}) dispersed and trapped in a polymerized gel. Liquid-to-vapour phase transitions can be induced by nuclear recoils following the neutralino interaction with a nucleus (^{19}F or ^{12}C) in the droplet. These detectors are threshold detectors, since an explosive droplet-to-bubble transition is only triggered if the energy deposited through the ionization process in the liquid by the recoiling nucleus exceeds a temperature-dependent energy threshold. The resulting shock waves are detected by piezo-electric sensors. The measurements of the detector response to monoenergetic neutrons as a function of operating temperature allow the extraction of threshold temperatures, T_{th}, for a given neutron energy and in turn the determination of the recoil threshold energy, $E_{R,th}$, as a function of temperature for various pressures of operation. For a practical range of temperatures of operation, $E_{R,th}$ follows an exponential temperature dependence:

$$E_{R,th}(T) = E_b\, e^{-K(T-T_b)}, \tag{1}$$

where K is a constant to be determined experimentally and E_b is the threshold energy at the boiling temperature T_b. From purely kinematical considerations, nuclear recoil thresholds in droplet detectors fall into the same operating temperature range for neutrons of low energy (from 10 keV up to a few MeV) and massive neutralinos (from 10 GeVc^{-2} up to 1TeVc^{-2}) at velocities which are typical for dark matter particles in our galactic halo. In this operating range the detectors are insensitive to minimally ionizing particles and γ radiation.

The probability, $P(E^i_R, E^i_{R,th}(T))$, that a recoil nucleus i at an energy near threshold will generate an explosive droplet-to-bubble transition is zero for $E^i_R < E^i_{R,th}$ and will increase gradually up to 1 for $E^i_R > E^i_{R,th}$. This probability is expressed as [4]:

$$P(E^i{}_R, E^i{}_{R,th}(T)) = 1 - \exp\left[\frac{-(1.0 \pm 0.1)(E^i{}_R - E^i{}_{R,th}(T))}{E^i{}_{R,th}(T)}\right]. \quad (2)$$

Combining Eqs. (1) and (2) (in the case of ^{19}F struck by neutrons) allows the determination of the detector efficiency as a function of the recoil energy of ^{19}F.

The neutralino induced recoil energy spectrum of ^{19}F is given by [3]:

$$\frac{dR}{dE_R} \approx c_1 \frac{R_0}{E_R} F^2(E_R) \exp\left[-c_2 \frac{E_R}{\langle E_R \rangle}\right], \quad (3)$$

where

$$R_0(counts\ kg^{-1}\ d^{-1}) = \left(\frac{403}{A_T M_\chi}\right)\left(\frac{\sigma_{SD}}{pb}\right)\left(\frac{\rho_\chi}{0.3GeVc^{-2}cm^{-3}}\right)\left(\frac{\langle v_\chi \rangle}{230kms^{-1}}\right) \quad (4)$$

is the total rate of neutralino-nucleus(A) interaction (assuming zero momentum transfer), A_T is the atomic mass of the target atoms, ρ_χ is the mass density of the neutralino, σ_{SD} is the spin-dependent neutralino-nucleus cross section, $<v_\chi>$ is the relative average neutralino velocity, $E_R = 2M_A M_\chi{}^2 (M_A + M_\chi)^{-2} \langle v_\chi{}^2 \rangle$ is the mean recoil energy and $F^2(E_R)$~1 for ^{19}F and small momentum transfer ($c_1 = c_2 = 1$ for $v_E = 0$; $c_1 = 0.75$ and $c_2 = 0.56$ for $v_E = 244$ kms^{-1}). Combining the ^{19}F recoil spectra expected from neutralino interactions (Eq. (3)) and the measured detector threshold for ^{19}F recoil energy at a given operating temperature, one can determine the neutralino detection efficiency, $\varepsilon(M_\chi,T)$, as a function of neutralino mass and operating temperature. Then, the observable neutralino count rate, R_{obs}, as a function of temperature, neutralino mass and cross section, is given by

$$R_{obs}(M_\chi, \sigma_{SD}, T) = \frac{c_1}{c_2} R_0(M_\chi, \sigma_{SD}) \cdot \varepsilon(M_\chi, T) = 1.34 \cdot R_0(M_\chi, \sigma_{SD}) \cdot \varepsilon(M_\chi, T). \quad (5)$$

Due to their large ^{19}F content (^{19}F is a spin-1/2$^+$ isotope), PICASSO detectors benefit from a very favorable spin-dependent cross section and are particularly suitable to search for spin-dependent neutralino interactions[5,6]. The general spin-dependent cross section for WIMP scattering on ^{19}F has the form [6]

$$\sigma_{SD} = 4G_F^2 \mu_F^2 C_F, \quad (6)$$

where G_F is the Fermi coupling constant and μ_F is the neutralino-^{19}F reduced mass. $C_F = (8/\pi)\left(a_p \langle S_p \rangle + a_n \langle S_n \rangle\right)^2 (J+1)/J$ is a spin-dependent enhancement factor, where a_p, a_n are the effective proton and neutron coupling strengths, $J = 1/2$ is the total nuclear spin, $\langle S_p \rangle = 0.441$ and $\langle S_n \rangle = -0.109$ are the expectation values of the proton and neutron spins within ^{19}F [6]. In order to compare our results with other experiments and predictions of supersymmetric models, the experimental cross section limits on ^{19}F, $\sigma_{\chi F}$, are converted into limits on the neutralino-proton ($\sigma_{\chi p}$) and neutralino-neutron ($\sigma_{\chi n}$) cross sections. Following the method developed in [3], we assume that all events are either due to neutralino-proton or neutralino-neutron elastic scatterings in the nucleus, i.e $a_n = 0$ or $a_p = 0$, respectively, and obtain:

$$\sigma_{\chi i} = \sigma_{\chi F}\left(\frac{\mu_i^2}{\mu_F^2}\right)\frac{C_i}{C_{i(F)}}, \quad (7)$$

where i = proton or neutron, μ_i is the χ-nucleon reduced mass (the mass difference between neutron and proton is neglected) and $C_{p(F)}$ and $C_{n(F)}$ are the proton and neutron contributions to the total enhancement factor of ^{19}F. They are related to the a_p and a_n couplings by $C_{i(F)} = (8/\pi)a_i^2 \langle S_i \rangle^2 (J+1)/J$. C_p and C_n are the enhancement factors for scattering on individual protons and neutrons. The values for the ratios $C_{p(F)}/C_p$ = 0.778 and $C_{n(F)}/C_n$ = 0.0475 are taken from [6,7]. As shown in [6], the allowed values of a_p and a_n for a given neutralino mass are constrained to fall within a region defined in our case by two straight lines with a slope given by $-\langle S_n/S_p \rangle = 0.247$:

$$a_p \leq -\frac{\langle S_n \rangle}{\langle S_p \rangle} a_n \pm \sqrt{\frac{\pi}{24 G^2{}_F \mu^2{}_p} \sigma_{\chi i}} \,. \tag{8}$$

THE SEARCH FOR DARK MATTER

The PICASSO experiment has taken data with three detectors of 1 litre volume representing a total active mass of 19.4 ± 1.0 g of ^{19}F [8]. The detectors were installed in the water purification gallery of the Sudbury Neutrino Observatory Laboratory at a depth of 2070 m (6000 mwe). The detectors were operated at an ambient pressure of 1.23 bar and at temperatures in the range from 20° to 47°C, which translates into a sensitivity to 19F recoils from 500 keV to 6 keV. The detectors were installed in an acoustically and thermally insulated container (TCS) and the temperature was regulated and controlled with a precision of 0.1°C. The setup and data taking procedure are described in [8]. There are several sources of external and internal background. The external background consists of neutrons. Neutrons at a deep underground laboratory, like SNO, are produced mainly by cosmic-ray muon interactions (photonuclear interactions of muons, muon capture) and spontaneous fission of ^{238}U and following (α,n) reactions. Water cubes surrounding the TCS slow down these external neutrons and bring them below detection threshold. The internal background consists of alpha particles [4] produced in the decay chains of uranium and thorium nuclei which are present as contaminants in the constituents of the detectors. Purification of the detector ingredients was done by chemical and mechanical means and is described in details in [4,9]. The purification achieved so far was assayed at the level of 10^{-10} gU/g. The entire detector setup is designed to be radon tight. The ^{222}Rn emanation rate for PICASSO containers was measured to be 15 ± 9 atoms/day [4].

CURRENT LIMITS AND FUTURE DEVELOPMENTS

A combined fit of the alpha background and the neutralino response to the data allowed us to rule out any positive evidence for neutralino induced nuclear recoils, translating into a 90% C.L. upper limit of 1.31 pb on $\sigma_{\chi p}$ and 21.5 pb on $\sigma_{\chi n}$ for a neutralino mass of M_χ = 29 GeVc^{-2} [8]. Figures 1 and 2 show our limits, compared with the limits obtained by some other experiments, on the spin-dependent neutralino-proton cross section ($\sigma_{\chi p}$) and neutralino-neutron cross section ($\sigma_{\chi n}$) at the 90% C.L. as a function of the neutralino mass. References to these experiments and others can be found in [10]. Figure 3 shows the excluded regions in the plane of the effective coupling strengths a_p , a_n for protons and neutrons for a neutralino mass of 50 GeVc^{-2} as obtained from Eq. (8) as well as the excluded regions obtained by other experiments. The sensitivity of PICASSO is presently limited by alpha emitting contaminants and its small active mass. The next step of the experiment is in preparation, with 2 kg of active mass, detector modules of 4.5 litre volume, hydraulic recompression, event localization capability and improved purification and fabrication techniques.

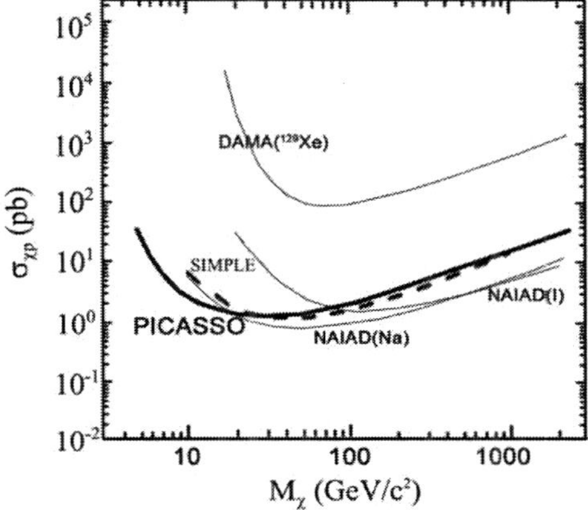

FIGURE 1. PICASSO limit on the spin-dependent neutralino-proton cross section ($\sigma_{\chi p}$) at 90% C.L. as a function of the neutralino mass compared with some other experiments.

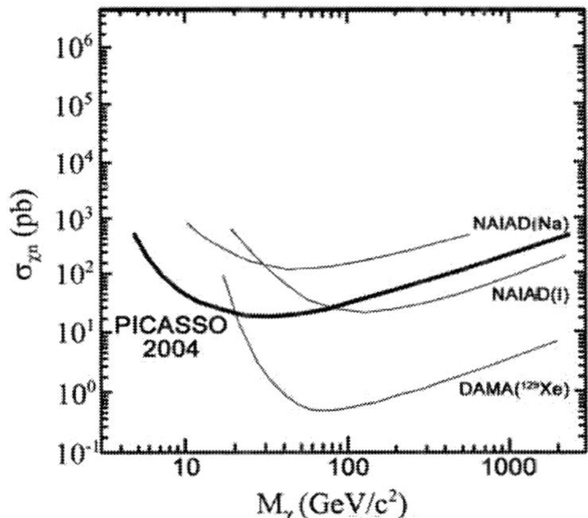

FIGURE 2. PICASSO limit on the spin-dependent neutralino-neutron cross section ($\sigma_{\chi n}$) at 90% C.L. as a function of the neutralino mass compared with some other experiments.

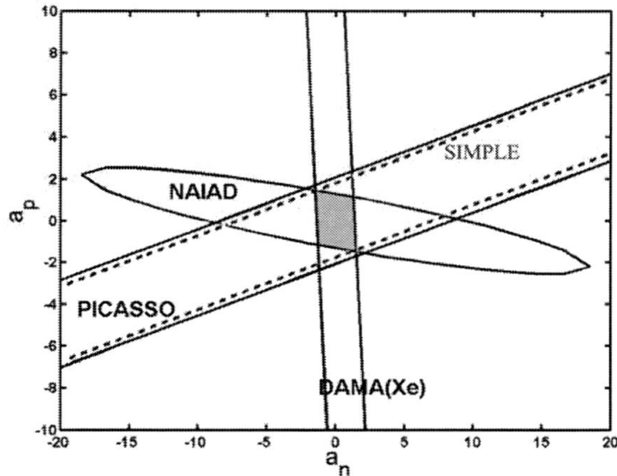

FIGURE 3. Excluded regions (outside the bands) in the plane of the effective coupling strengths a_p, a_n for protons and neutrons for a neutralino mass of 50 $GeVc^{-2}$ as obtained from Eq. (8) for PICASSO, compared with some other experiments.

REFERENCES

1. C.L. Bennett et al., *Astrophys. J. Suppl.* **148**, 1-27 (2003); D.N. Spergel et al., *Astrophys. J. Suppl.* **148**, 175-194 (2003).
2. G. Jungman, M. Kamionkowski and K. Griest, *Phys. Rept.* **267**, 195-373 (1996).
3. J.D. Lewin and P.F. Smith, *Astro. Phys.* **6**, 87-112 (1996).
4. M. Barnabé-Heider et al., PICASSO Collaboration, physics/0508098, to appear in NIM A.
5. J. Ellis and R. Flores, *Phys. Lett.* B **263,** 259 (1991)
6. D. R. Tovey et al., *Phys. Lett.* B **488**, 17-26 (2000).
7 A.F. Pacheco and D.D. Strottman, *Phys. Rev.* D **40**, 2131 (1989).
8. M. Barnabé-Heider et al., PICASSO Collaboration, hep-ex/0502028, to appear in Phys.Lett.B.
9. M. Di Marco, "Réduction du bruit de fond en vue de la détection de la matière sombre avec le projet PICASSO", Ph.D. Thesis, Université de Montréal, November 2004.
10. C. Savage et al., astro-ph/0408346v1, August 19, 2004.

Educational Cosmic Ray Arrays

R. A. Soluk[1]

Centre for Subatomic Research, University of Alberta, Edmonton, Alberta, Canada T6G 2N5

Abstract. In the last decade a great deal of interest has arisen in using sparse arrays of cosmic ray detectors located at schools as a means of doing both outreach and physics research. This approach has the unique advantage of involving grade school students in an actual ongoing experiment, rather then a simple teaching exercise, while at the same time providing researchers with the basic infrastructure for installation of cosmic ray detectors. A survey is made of projects in North America and Europe and in particular the ALTA experiment at the University of Alberta which was the first experiment operating under this paradigm.

Keywords: ALTA, NALTA, cosmic ray, outreach, high school physics
PACS: 95.55.Vj, 01.40.Ej, 01.50.Pa

INTRODUCTION

As the first source of high energy subatomic particles cosmic rays played a major role in the early exploration of elementary particles. Now almost 100 years after their discovery by Victor Hess in 1912 they are again the focus of major research efforts because they allow one to probe particle energies not achievable by terrestrial accelerators. This current interest in cosmic rays combined with the relative ease of detecting high energy cosmic ray showers, the availability of surplus detectors from decommissioned particle physics experiments and advances in GPS technology have helped to spur on the large number of educational arrays that are appearing now around the world.

A typical educational array site consists of between two and four plastic scintillator detectors connected to photomultiplier tubes and read out by custom built electronics. The detectors are placed a few meters apart on the roof of a school, college or university building along with an antenna which uses the global positioning system to provide an absolute time reference. By requiring that two or more of the detectors register particles within a narrow time window one can distinguish between the relatively high flux of individual particles from low energy cosmics and the lower rate of extensive air showers caused by higher energy primaries. With a separation on the order of 10m between detectors a primary energy of roughly 10^{14}eV is required to produce a large enough shower to cause a coincidence. A number of these detector sites are deployed to form a sparse array with distances between sites ranging from a few kilometers to hundreds of kilometers.

If the inter-site spacing is small enough these arrays can look for showers from ultra high energy particles which can be several kilometers across at the Earth's

[1] For the ALTA collaboration, mail to alta@phys.ualberta.ca

CP828, *Multiparticle Dynamics*
edited by V. Šimák, M. Šumbera, Š. Todorova, and B. Tomášik

surface. Otherwise the main physics target of these arrays is to search for a non-random component in the cosmic ray flux. Due to intergalactic magnetic fields it is expected that the arrival times and directions of cosmic ray primaries will be random. However there have been hints from experiments [1] that there may be a non-random component. Examples of possible sources for such correlated events are photodisintegration of high energy nuclei in the vicinity of the solar system or bursts of very high energy gamma rays. These can result in either changes to the average cosmic ray rate or separate but correlated showers hitting two or more sites.

From a research standpoint installing the detectors on schools has several advantages: teachers and students can be recruited to help test and maintain the equipment reducing the host institutions manpower requirements. Schools provide the infrastructure needed to maintain the site, such as power, a secure location and an internet connection thereby reducing the installation cost and complexity of the detectors. This kind of experiment also helps create publicity for physics and physics departments. From an educational standpoint enrolment in physics in most locations in North America and Europe has been decreasing during the last decade. This type of project allows students to have some exposure to a real physics experiment **before** they decide their university degree path. We've found that students can tell the difference between a real project and the usual teaching exercise and become more engaged when they realize that they are involved in actual research and have the opportunity to discover something new.

THE ALTA EXPERIMENT

The ALTA[2] experiment was the first educational cosmic ray array with its emphasis being more on research than outreach. This choice affects both the design and operation of the experiment. The roof top enclosures are installed as permanent parts of the school roofs with all cable runs in grounded metal conduit. To ensure that the direction of incoming cosmics can be related back to astronomical coordinates the detectors are fixed in position and a precision GPS compass is used to measure the alignment of the 3 detectors to north. An energy and relative timing calibration is periodically done on each site or when any component is changed. The electronics uses seven discrete crate mounted boards [2] with charge integration for energy determination, timing between detectors measured with 25ps binning and a GPS system designed specifically for precise timing [3][3]. It was decided to use a light guide to readout the scintillator since that gives the best timing and light collection uniformity. All of this ensures high quality data but increases the cost and the installation complexity of each site reducing their total number.

Critical elements in the ALTA design are monitoring, automation and remote management. For this a custom GUI based DAQ program was written with built in graphs and simple statistical tools. All electronics settings are software controlled and any changes made to the software are logged with certain values, like high voltage settings, are recorded with every event. Data is automatically uploaded to a central

[2] www.ualberta.ca/alta

[3] Relative timing between sites separated by hundreds of kilometers should be better then 16ns with the ALTA system.

computer and if a problem is detected at a site or its settings aren't correct for data taking that site is automatically reset to the correct values. If a student is conducting their own experiment they can prevent this automatic reset by simply checking a box on the screen and entering what time they will be done. In this way students are free to do what they like with the system while having the minimum impact on data collection.

Taking advantage of improvements in technology the newest electronics design reduces the number of boards from 7 to only 3 while decreasing their cost and increasing their performance. ALTA is directly working with the Silesian University in Opava and the Czech Technical University in Prague in the Czech Republic, King's college in London England and TRIUMF, the University of Victory and Saint Mary's University in Canada to create additional arrays and a more international project.

ARRAYS AROUND THE WORLD

While all of the educational cosmic ray arrays have the same general goals, research and education are to some extent in conflict with one another. For instance in research you want identical detectors with hardware settings always left at the optimum data taking values. While for education you want students to have as much hands-on experience as possible and freedom to change settings and hardware setup at will. Research demands more attention to things such as calibration and complete records of data taking conditions as well as more complex, and probably more expensive, electronics and installations. The various groups have placed different amounts of emphasis on physics and outreach which results in differences in their hardware and detector installations. Also most groups have invented their own hardware and detectors, often independently, so the same problem has been solved many times over. It should be noted that with 3 or more detectors it is possible to use the local relative timing between detectors to triangulate the incoming direction of the primary cosmic ray, whether or not this timing information is collected and used also impacts the complexity of each project's hardware. For software most groups have elected to use the commercially available Labview program.

Figure 2 shows the various projects currently operating, or under development, in North America. A brief summary of the major projects and their status follows[4]:

- ALTA (Alberta Large-area Time-coincidence Array) began detector work in 1993 with its first school sites operational in 1998. Crate mounted custom electronics is used along with a GPS system specifically designed for timing. Custom DAQ software is used to control the electronics. Each site uses 3 scintillators (60cmX60cm) in permanently installed enclosures. ALTA currently has 15 sites in Alberta and will be installing one in the Edmonton science centre this fall to create a site accessible to the general public rather then being restricted to high schools.
- CROP (Cosmic Ray Observatory Project) obtained a large number of detectors and power supplies from the decommissioned CASA array and

[4] These types of experiments are often not well publicized and it can be difficult to determine their current status. It is therefore possible that the list of projects given is not complete and some information could be out of date.

installed its first school sites in 2000 using 4 scintillators (60cmX60cm) per site in movable enclosures. They use a custom low-cost single board electronics system designed with QuarkNet and Fermilab that does rough GPS time stamping and energy measurements along with relative timing between detectors for up to 4 input channels. The board also has a built in barometric pressure sensor. CROP has 26 sites in Nebraska with a very good outreach program where students assemble their own detectors at summer workshops.

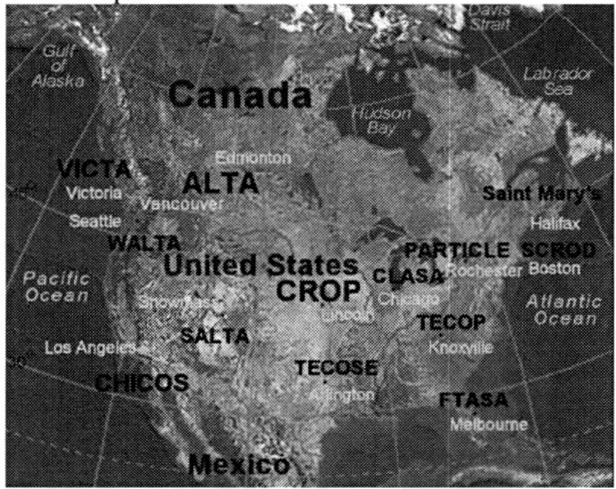

FIGURE 1. Educational cosmic ray arrays in North America

- CHICOS (California HIgh school Cosmic ray ObServatory) acquired pre-built outdoor detectors from the defunct CYGNUS array and began installing school sites in 2000. They use a simple custom built single board DAQ card with 2 scintillators (≈1m^2 each) per site. To obtain a tighter spacing and offer the possibility of observing ultra high energy showers CHICOS places detectors on colleges and middle schools as well as high schools. They now have 78 sites in the Los Angeles area making them the largest school array currently operating.
- WALTA (Washington Large area Time coincidence Array) installed its first school prototype in 1999 and currently has 16 sites using the QuarkNet DAQ card and a design similar to CROP.
- SALTA (Snowmass Area Large scale Time coincidence Array) began planning in 2001 and currently has 5 sites installed using CROP detectors.
- VICTA (VIctoria Cosmic ray Time coincidence Array) began in 2003 using the ALTA detector design and electronics. They currently have 1 test site installed with additional sites planned for this fall. In addition to schools a site will be installed at the Victoria science centre.
- TRIUMF (TRIUniversity Meson Facility) in Vancouver, BC has started an as yet unnamed project using the ALTA design and will be installing a site at the Vancouver science centre.

- Some additional projects in the United States that are planned or have started are: TECOSE (TExas Cosmic Observations by School Experimenters), PARTICLE (Physicists and Rochester Teachers Inventing CLassroom Experiments), TECOP (TEnnessee Cosmic ray Observatory Project), CLASA (Chicago Large Air Shower Array), FTASA (Florida Tech Air Shower Array) and SCROD (School Cosmic Ray Outreach Detector). Most are using the QuarkNet DAQ card.

The major groups in Canada and the United States have formed a loose collaboration known as NALTA[5] (North American Large area Time coincidence Arrays) who's goal is to share education resources and information between groups. It is also planned to have one central access point where students and researchers can make use of data from all of the NALTA sites creating in effect a single giant array.

In the past few years activity on a number of educational arrays has begun in Europe as well as can be seen in figure 3.

- HiSPARC (High School Project on Astro physics Research with Cosmics) started around 2002 and has placed pairs of detectors housed in rooftop car carrier boxes on schools in 5 cities in the Netherlands. They use their own custom electronics system and are currently up to 30 sites.

FIGURE 2. Educational cosmic ray arrays in Europe

- An unnamed project started in Finland in 2002 involving 3 universities and 3 polytechnic institutes. They have no sites operational at this time but are developing their own custom electronics using 4 detectors per site.
- SEASA (Stockholm Educational Air Shower Array) is installing 3 (30cmX100cm) detectors per site also housed in rooftop car carrier boxes. They have a custom made single board DAQ card but unlike all of the projects described so far they are not using a standard PC to connect to their data acquisition system but instead are using an embedded processor (a so called linux on a chip system) on an additional board. This processor board

[5] www.ualberta.ca/nalta

has its own network connection but no display or keyboard, instead a custom java applet is used to control the system and readout the data using TCP/IP. SEASA has two sites at AlbaNova University and plan to install their first school site this year.

- CZELTA (CZEch Large-area Time-coincidence Array) was started in 2004 by the Institute of Experimental and Applied Physics at the Czech Technical University in Prague. They currently have 2 sites using triplets of ALTA detectors and ALTA electronics installed in Prague and Opava with a third site to be installed in Pardubice later this year. Their goal is to create clusters of arrays in various cities in the Czech Republic.
- ROLAND MAZE PROJECT in Poland obtained funding in 2004 for 10 sites in the Lodz area. They plan to use a custom made PCI readout board and 4 scintillators per site ($\approx 1\text{m}^2$ each) readout using wavelength shifting fibres.
- An unnamed project in Belgium has begun using 5 scintillators (40cmX200cm) per site. They are currently developing their own custom electronics system and like SEASA will also be using a microprocessor instead of a standard PC for their readout and control computer.
- EEE (Extreme Energy Events) is a project which is now in development in Italy with an initial goal of installing 3 school sites per city in each of 7 Italian cities. This is the only array which will not be using plastic scintillator as a detector. Instead they will be using a telescope of 3 multigap resistive plate chambers (160cmX82cm) filled with a freon/sulfur hexafluoride gas mixture. Approximately 20 of these chambers have been built now at CERN.
- King's College in London is currently building detectors and will be installing 3 sites using ALTA electronics in the next few months.
- Other experiments include: RELYC in France will be using Belgium electronics with 4 detectors/site (40cmX50cm), a group in Portugal wants to start an array with a goal of 10 schools using 3 detectors/site with a custom PCI DAQ card, a project has begun in Dusseldorf Germany with one test site installed in a school, and groups in Denmark and Liverpool England are in the early stages of planning their own experiments.

Having helped found NALTA we hope that a similar collaboration can be set up between Europe and North America to share ideas, reduce the duplication of efforts and ultimately share data to allow students or researchers to treat the current collection of educational arrays as one massive global array of detectors. This would significantly extend the reach of any single local array.

REFERENCES

1. D.J. Fegan, B McBreen and C. O'Sullivan, *Phys. Rev. Lett.*, **51** (1983) 2341
   G. R. Smith, M. Ogmen, E. Buller and S. Standil, *Phys. Rev. Lett.*, **50** (1983) 2110
2. W. Brouwer, et. al., *Nucl. Instr. and Meth.* A **539** (2005) 595.
3. W. Brouwer, et. al., *Nucl. Instr. and Meth.* A **493** (2002) 79.

Astroparticle Physics and the LHC

James L. Pinfold

Physics Department, University of Alberta, Edmonton, Alberta T6G 2N5, CANADA

Abstract. Research into the fundamental nature of matter at the high energy frontier takes place in three main areas: accelerator based particle physics; high energy astrophysics; and, the cosmology of the early universe. The LHC project provides the laboratory to perform measurements of great importance for cosmic ray astrophysics and cosmology. Also, the study of astroparticle physics can have significant implications for collider physics at the LHC. This paper reviews some of the important synergistic links between astroparticle and LHC physics.

Keywords: LHC, collider physics, astroparticle physics, cosmology, cosmic rays.
PACS: 11.10.Kk, 11.25.Wx, 12.60.Jv, 13.85.-t, 13.85.Tp, 14.80.Hv, 14.80.Ly, 95.35.+d, 98.80.-k, 98.70.Vc.

INTRODUCTION

This The Large Hadron Collider (LHC) machine [1], under construction at CERN, will collide protons at the unprecedented energy centre-of-mass energy (E_{cm}) of 14 TeV and luminosity of 10^{34}cm^{-2}s^{-1}. The LHC will also run in a heavy-ion collider mode at an E_{cm} of approximately 1000 TeV using lead ions. It is currently envisaged that the first collisions will take place in July 2007.

The LHC has four main experiments: ALICE [2], ATLAS [3], CMS [4] and LHCb [5]. ATLAS and CMS are general purpose detectors with a very wide physics reach. The LHCb detector is dedicated to the study of b-physics and CP-violation in the b-sector. The TOTEM [6] detector will be integrated into the forward region of the CMS experiment. Its prime task is to measure the total p-p cross-section. It is envisaged that the CASTOR detector [7] will also deployed in the forward region of CMS in order to search for Centauro production in heavy ion collisions . ALICE is the detector dedicated to heavy ion physics, particularly the study of the quark-gluon plasma. Two experiments at the proposal stage are: MOEDAL [8], designed to search for highly ionizing particles such as magnetic monopoles; and, LHCF [9], which aims to measure the cross-section of forward π^0's to validate the MC code for very high energy cosmic ray (CR) shower development.

Recent developments in astrophysics, particle physics and cosmology are revealing the interconnectedness of these three fields. Specifically, the Wilkinson Microwave Anisotropy Probe (WMAP) measurement [10] of the Cosmic Microwave Background Radiation anisotropy to an unprecedented accuracy of 10^{-9} ^{0}K. WMAP has opened up a new era of precision cosmology that allows an accurate determination of

CP828, *Multiparticle Dynamics*
edited by V. Šimák, M. Šumbera, Š. Todorova, and B. Tomášik

cosmological parameters of relevance to particle physicists. The vastly improved precision of the WMAP data has significantly constrained the dark matter (DM) content of the Universe. This in turn strongly implies stringent model-dependent constraints on minimal SUSY. A pie graph of the energy budget of the cosmos shows: 4% ordinary matter; 23% DM and 73% dark energy.

The LHC project is in its start up phase. New neutrino oscillation results are expected and new experiments are on the way. Improved measurements of the CMB are being planned, and the next generation of large volume large area astroparticle physics experiments are in progress. The cutting edge areas of investigation of the microcosm engendered by the interplay between particle physics, astrophysics and cosmology are as follows: SUSY DM, extra dimensions (EDs) and mini-black-hole production, neutrino physics, searches for new particles such as the radion and the magnetic monopole, and ultra-high-energy cosmic rays (UHECR).

The LHC project, providing an E_{cm} equivalent roughly to a 10^{17} eV proton incident on a target proton, offers the unique possibility to perform precise measurements of the properties of high energy hadronic interactions. One application for this capability is in future studies of high energy cosmic ray (HECR) physics, since high energy particles will have central importance. The challenge here is that most of the energy flow in HECR air showers is within a few degrees of the primary particle direction.

Alternatively, CR astrophysics may point the way to new physics at accelerators. For example, many exotic phenomena such as Centauro events, observed in HECR emulsion experiments, may be harbingers of new physics at the LHC and future collider experiments. It is possible that if mini black holes can be produced in high energy particle interactions, they will first be observed in HECR interactions.

Another interesting possibility is the direct study of cosmic rays at the LHC using the very large areas of fine-grained detectors and magnetic field volumes of the ATLAS and CMS experiments. In this way, collider physics experiments can make a direct contribution to astroparticle physics.

In this article we will discuss only one example of the synergy between astroparticle physics and collider physics, that is the search for SUSY dark matter.

THE LHC PHYSICS PROGRAM

The physics reach of the LHC in the high P_T physics arena, means that we can answer, or shed light on, such fundamental open questions as: the generation of mass; the unification of fundamental interactions; new physics such as SUSY, technicolor, the signals for EDs; and, the nature of dark matter. The search for the Higgs boson is often used as an exemplar of TeV physics. The LHC detector design performance will allow us to explore the SM Higgs mass range from the LEP200 limit all the way up to 1 TeV. Entering the world of SUSY, the same detector performance enables us to largely cover the various different signatures of SUSY particle production.

Both ATLAS and CMS are developing a program of forward physics that incorporates such areas as hard and soft diffractive processes, two-photon interactions and peripheral collisions, the measurement of the p–p total cross-section, low-x

dynamics, forward physics phenomena, forward physics of proton–nucleus and nucleus–nucleus collisions, and a detailed investigation of the forward system of multiplicities, energy spectra and particle species.

SUPERSYMMETRY, COSMOLOGY AND THE LHC

Stable neutral SUSY particles such as the neutralino ($\tilde{\chi}^0$) or gravitinos are promising candidates for cold dark matter (CDM). One of the simplest SUSY models, mSUGRA, is defined by six parameters: m_0, $m_{1/2}$, A_0, $\tan\beta$, sign(μ) and the gravitino mass $m_{3/2}$. A slice through mSUGRA parameter space determines the various possibilities for the LSP. The LSP is either a stau ($\tilde{\tau}$), excluded by the limits on the abundance of charged dark matter [11], or a $\tilde{\chi}^0$ where the $\tilde{\chi}^0$ is a mixture of the gauginos, Bino ($\tilde{B}$), Wino ($\tilde{W}$) and up- and down-type Higgsinos. The favoured candidate for CDM is the lightest $\tilde{\chi}^0$ in R-parity conserving SUSY models. Although the $\tilde{\chi}^0$ could have an important Higgsino admixture for $m_0 > 1$ TeV, it is mostly pure Bino in much of the parameter space.

In the high temperature very early universe all particles were in thermal equilibrium. As the universe expanded the $\tilde{\chi}^0$ interaction rate fell behind the expansion rate and the $\tilde{\chi}^0$s were frozen out. As $\tilde{\chi}^0$s are stable their thermal relic density survives to the present and the Boltzmann equation for $\tilde{\chi}^0$s in a Friedmann–Robertson–Walker universe can be used to calculate the $\tilde{\chi}^0$ relic density $\Omega(\tilde{\chi}_1^0)\,h^2$ we see today.

Only relatively small regions of the mSUGRA model parameter space will give a low enough value of $\Omega(\tilde{\chi}_1^0)\,h^2$ to be compatible with cosmological measurements and theory. The region at low m_0–$m_{1/2}$ is called the 'bulk region'. Here the LSP has a mass of less than 200 GeV and thus this region is severely constrained by the searches for SUSY at LEP-2 and at the Tevatron. The lower limit on the Higgs boson mass obtained at LEP-2 [12] as well as the LEP-2 limits on the chargino mass [13] significantly reduce the allowed part of the bulk region.

The strip extending from the bulk region to large $m_{1/2}$, along the edge of the charged LSP region, has an increased annihilation rate as the lightest slepton and LSP are almost mass degenerate. This is called the 'co-annihilation region', where the process $\tilde{\chi}_1^0\tilde{\tau} \to \tau\gamma$ is enhanced. The cosmologically allowed co-annihilation region is now only a narrow strip which can be completely covered by LHC searches [14].

The portion of the phase space at large values of m_0 and $m_{1/2}$ is known as the 'A (rapid) annihilation funnel'. In this region, for $\tan\beta \geq 30$, the mass of the LSP is such that there is enhanced $\tilde{\chi}^0$ annihilation via a resonant intermediate (s-channel) heavy Higgs boson (A) [15]. Processes with m_0 and/or $m_{1/2}$ greater than or roughly equal to ~1 TeV can satisfy relic density constraints. The region compatible with cosmological constraints takes the form of a 'funnel' pointing towards large m_A.

Lastly, at large m_0 and along the boundary beyond which EW symmetry breaking no longer occurs there is the 'focus point' (FP) region. In the FP region the LSP acquires a significant Higgsino content [16]. The annihilation cross-section is enhanced in this region since the Higgsino component can couple to the SM gauge bosons [17]. Also, the $\tilde{\chi}_1^0$ becomes nearly degenerate in mass with the $\tilde{\chi}_2^0$ and $\tilde{\chi}_1^\pm$. Thus, additional annihilation and co-annihilation processes occur which, taken together, can reduce the neutralino relic densities to WMAP compatible values. The FP region can extend to large values of m_0 giving rise to squarks, gluinos and sleptons that are too heavy to be observed at the LHC.

Direct & Indirect Detection of Neutralino Dark Matter

If all of space is filled with relic $\tilde{\chi}^0$s it may be possible to detect them directly by searching for $\tilde{\chi}^0$-nucleus scattering in terrestrial detectors. The most promising process is elastic $\tilde{\chi}^0$–nucleon*(N)* scattering via Higgs exchange in the *t*-channel and squark exchange in the *s*-channel, with maximal recoil proton energies of ~100 keV. However, there are major uncertainties in the estimation of DM spatial and velocity distributions. The $\tilde{\chi}^0 - N$ scattering spin-independent cross-section (σ_{SI}) can vary over several orders of magnitude. In the FP region the $\tilde{\chi}^0$ is a gaugino–Higgsino mixture and the Higgs exchange diagram makes a large contribution. Cross-sections are small in the stau co-annihilation region for two reasons: squarks are relatively heavy, suppressing the squark exchange diagram; and, the $\tilde{\chi}^0$ is Bino-like, resulting in suppression of the Higgs exchange diagram. A signal has been claimed by the DAMA collaboration [18]. However, the DAMA signal has not been confirmed by ZEPLIN-1 [19] EDELWEISS [20] and CDMS-II [21]. The current generation of direct search experiments has reached a sensitivity to the normalized σ_{SI} of $\sim 4 \times 10^{-7}$ pb.

The next generation of experiments such as, ZEPLIN-2 [22], EDELWEISS-2 [23], CDMS-II (2004) [24] and CRESST-II [25] should be able to reach a σ_{SI} of 10^{-8} pb. Future tonne scale experiments are now being planned, for example, GENIUS [26], ZEPLIN4 [27], XENON [28] and CRYOARRAY [29] that are capable of probing down to $\sigma_{SI} \sim 10^{-10}$ pb. These planned detectors should be able to cover almost all of the FP region of the mSUGRA parameter space. Neutron-induced backgrounds [30] will probably define the ultimate sensitivity of direct search experiments.

Neutralino DM may be detected indirectly via $\tilde{\chi}^0$ annihilations that give rise to three potentially promising signals. The first of these signals arises from neutrinos produced by $\tilde{\chi}^0$ annihilation in the core of the sun or earth [31]. Once captured, the $\tilde{\chi}^0$s annihilate to form final states that subsequently yield neutrinos that are detected via charged current interactions in neutrino telescopes such as AMANDA, where limits on $\tilde{\chi}^0$ annihilation rates at the centre of the earth have already been obtained [32]. The planned neutrino telescopes ANTARES [33] and IceCube [34] are sensitive to $E_\mu > 10$ GeV and $E_\mu > 25$–50 GeV, respectively.

Other indirect signals arise from two main sources. Firstly, γ-rays from $\tilde{\chi}^0$ annihilations in the galactic core and halo [35]. These signals can be detected by space-based detectors [36] such as EGRET [37] or GLAST [38] with thresholds as low ~100 MeV and in atmospheric Cerenkov telescopes on the ground, with detection thresholds in the range 20→100 GeV. Secondly, hard CR positrons that are produced in the decays of leptons, heavy quarks and gauge bosons arising from $\tilde{\chi}^0$ annihilations in our galactic halo. However, an assumption of a clumpy halo is required to give a sufficient signal significance. Space-based anti-matter detectors such as AMS-2 [39] and PAMELA [40] may be able to detect a possible e^+ signal from $\tilde{\chi}^0$ annihilation.

CONCLUSION

A comprehensive description of the nature of DM can only arise as a combination of understanding achieved from both astroparticle and collider results. An analysis of the mSUGRA parameter space enables us to reach the following general conclusions.

In the stau co-annihilation region, the LHC can probe all the relevant parameter space for tan $\beta \leq 45$. Indirect searches for $\tilde{\chi}^0$ DM do not make a big contribution in this region of parameter space. The best situation arises when tan β is large and there is some overlap with the A annihilation funnel. The large $m_{1/2}$ and tan β portion of parameter space is not, at present, accessible to any planned experiments, although this region seems to be consistent with the WMAP limits.

A large part of the A annihilation funnel can be explored by the LHC. But, the large $m_{1/2}$ section might not be accessible to any search experiments. The lower part of the annihilation funnel is accessible to a 1 TeV E_{cm} linear e^+e^- collider. Indirect searches for e^+'s, $\bar{p}$'s and γ's produced by DM annihilation in the galactic core or halo can have a similar reach to that of the linear collider (LC).

In the FP region, the LHC can cover $m_{1/2}$ values ranging up to 700 GeV, corresponding to $m_{\tilde{g}} \sim 1.8$ TeV. IceCube can explore the FP region for $m_{1/2} < 1400$ GeV. Future planned direct DM search experiments should be able to access almost all the FP region .

Nearly all of the mSUGRA regions allowed by the WMAP results can be covered by combining results from all the different search experiments described above. The exceptions are a few regions in the parameter space at large $m_{1/2}$ piece of the stau co-annihilation corridor or the A annihilation funnel.

The WMAP data opened up an era of precision cosmology that has reinforced the case for cold dark matter in the universe. If the predictions for 'standard' SUSY scenarios are found to agree with observations at the LHC and are consistent with the interaction strengths and relic density as determined by astroparticle physics and cosmology, this concurrence will provide strong evidence that the DM is supersymmetric. However, if this agreement is not seen a whole host of possibilities arise. It is only through the combination of approaches in particle physics, astrophysics and cosmology that the identity of DM will be uncovered. The synergies between particle physics, astrophysics and cosmology over the next ten years should amplify our ability to make fast and deeper inroads into the terra incognita beyond the borders of our current knowledge.

REFERENCES

1. LHC Study Group, The large hadron collider conceptual design, CERN/AC/95-05.
2. ALICE Collaboration, *Technical Proposal* CERN/LHCC/95-71.
3. ATLAS Collaboration, *Detector and Physics Performance Technical Design Report* CERN/LHCC/99-15.

4. CMS Collaboration, *Technical Proposal* CERN/LHCC/94-38.
5. LHCb Collaboration, *Technical Proposal* CERN/LHCC/98-4.
6. TOTEM Collaboration, *Technical Proposal* CERN-LHCC 99-7.
7. A. L. S. Angelis *et al., Nucl. Phys.* B (Proc. Suppl.) **97** p227 (2001); A. L. S. Angelis *et al., Nuovo Cimento* C **24** p755 (2001); A. L. S. Angelis and A. D. Panagiotou , *J. Phys. G: Nucl. Part. Phys.* 23 p2069 (1997); A. Angelis *et al* CASTOR draft proposal, ALICE/97-07, Internal note/CAS.
8. J. Pinfold *et al.* (MOEDAL Collab.), CERN/LHCC 98-5, LHCC/19 http://moedal.web.cern.ch/moedal/ (1998).
9. D. A. Faus, *et al.* , CERN LHCC-2003-057; LHCC-I-012.- Geneva : CERN, 05 Nov 2003.
10. C. L. Bennett *et al., Astrophys. J. Suppl.* **148** p1 (2003) ; D. N. Spergel *et al., Astrophys. J. Suppl.* **148** p175 (2003) ; M. Tegmark *et al., Phys. Rev.* D **69** p103501 (2004).
11. J. Ellis *et al., Nucl. Phys.* B **238** p453 (1984).
12. LEP Collaboration (ALEPH, DELPHI, L3, OPAL, the LEP Electroweak Working Group and the SLD Heavy Flavour Group), CERN-EP/2002-091, *Preprint* hep-ex/0212036, (2002).
13. D. Stewart, presented at *PHENO 04: Phenomenology 2004 Symposium (April)* University of Wisconsin-Madison, Madison, (LEPSUSYWG, ALEPH, DELPHI, L3 and OPAL experiments, notes LEPSUSYWG/02-02.1, LEPSUSYWG/01-03.1, note LEPSUSYWG/02-04.1) (2004).
14. M. Battaglia, I. Hinchliffe and D. Tovey, *Preprint* hep-ph/0406147 (2004).
15. J. R. Ellis, T. Falk, G. Ganis, K. A. Olive and M. Srednicki, *Phys. Lett.* B **510** p236 (2001); A. B. Lahanas, D. V. Nanopoulos and V. C. Spanos, *Mod. Phys. Lett.* A **16** p1229 (2001); H. Baer *et al., Phys. Rev.* D **63** p015007 (2001); H. Baer and M. Brhlik, *Phys. Rev.* D **57** p567 (1998); H. Baer andM. Brhlik,. *Phys. Rev.* D **53** p597 (1996); M. Drees and M. M. Nojiri, *Phys. Rev.* D **47** p376 (1993).
16. J. L. Feng, K. T. Matchev and T. Moroi, *Phys. Rev. Lett.* **84** p2322 (2000).
17. J. L. Feng, K. T. Matchev and F. Wilczek, *Phys. Rev.* D **63** p045024 (2001).
18. R. Bernabei *et al.* (DAMA Collaboration) *Phys. Lett.* B **480** p23 (2000).
19. J. C. Barton *et al.* (ZEPLIN1 Collab.) 2003 *Proc. 4th Int. Workshop on the Identification of Dark Matter* ed N J C Spooner and A V Kudryavtsev (World Scientific); N. Spooner *et al.,* (ZEPLIN1 Collab.) 2001 *Proc. APS/DPF/DPB Summer Study on the Future of Particle Physics (Snowmass)* (*Preprint* eConf C010630 E601)
20. A. Benoit, *et al.* (EDELWEISS Collaboration), *Phys. Lett.* B **513** p15 (2001).
21. D. S. Akerib *et al,* (CDMS Collaboration), *Preprint* astro-ph/0405033 (2004).
22. R. Luscher *et al., Nucl. Phys.* B (Proc. Suppl.) **95** p233 (2001).
23. G. Chardin, *Preprint* astro-ph/0306134 (2003).
24. T. A. Perera *et al., AIP Conf. Proc.* **605** p485 (2002).
25. J. Jochum *et al.* (CRESST Collaboration), *Phys. At. Nucl.* **63** p1242 (2000).
26. H. V. Klapdor-Kleingrothaus and I. V. Krivosheina , *Found. Phys.* **33** p831 (2003).
27. N. J. Spooner (The UKDM and Boulby Collaborations) 2003 *Int. Workshop on Technique and Application of Xenon Detectors (Kashiwa, 3–4 Dec. 2001)*; D. B. Cline, *Preprint* astro-ph/0310439 (2003).
28. E. Aprile *et al., Preprint* astro-ph/0207670 (2002).
29. R. W. Schnee, D. S. Akerib, and R. J. Gaitskell, *Nucl. Phys. Proc. Suppl.* **124** p233 (2003).
30. M. J. Carson *et al., Preprint* hep-ex/0404042 (2204).
31. EG: V. Bertin, E. Nezri and J. Orloff, *J. High Energy Phys.* JHEP02(2003) p046 (2003); V. Bertin , E. Nezri and J. Orloff , *Eur. Phys. J.* C **26** 111 (2002); V. Barger *et al., Phys. Rev.* D **65** p705022 (2002); A. Coresetti and P. Nath, *Int. J. Mod. Phys.* A **15** p905 (2000); A. Bottino *et al., Astropart. Phys.* **10** p203 (1999); V. Berezinsky *et al., Astropart. Phys.* **5** p333 (1996); L. Krauss, M. Srednicki and F. Wilczek, *Phys. Rev.* D **33** p2079 (1986); J. Silk, K. Olive and M. Srednicki, *Phys. Rev. Lett.* **55** **p**257 (1985).
32. J. Ehrens *et al.* (AMANDA Collaboration), *Phys. Rev.* D **66 p**16 (2003)
33. E. Carmona *et al.* (ANTARES Collaboration) *Nucl. Phys.* B (Proc. Suppl.) **95** p161 (2001)
34. F. Halzen and D. Hooper, *J. Cosmol. Astropart. Phys.* JCAP01 p2 (2004); F. Halzen *Preprint* astro-ph/0311004 (2003); J. Ahrens J *et al., Nucl. Phys.* B (Proc. Suppl.) **118** p388 (2001).
35. P. Gondolo and J. Silk, *Rev. Lett.* **83** p1719 (1999); L. Bergstrom L *et al., Phys. Rev.* D **59** p043506 (1999); L. Bergstrom , P. Ullio and J. H. Buckley, *Astropart. Phys.* **9** p37 (1998); L. Bergstrom, J. Edsj and P. Ullio, *Phys. Rev.* D **58** p083507 (1998); V. Berezinsky, A. Bottino and G. Mignola, *Phys. Lett.* B **325** 136 (1994); V. S. Berezinsky, A. V. Gurevich and K. P. Zybin, *Phys. Lett.* B **294** p221 (1992); M. Urban M *et al., Phys. Lett.* B **293** p149 (1992).
36. P. Gondolo and J. Silk , *Phys. Rev. Lett.* **83** 1719 (1999); L. Bergstrom, P. Ullio and J. J. Buckley, *Astropart. Phys.* **9** 137 (1998); V. Berezinsky, A. Bottino and G. Mignola, *Phys. Lett.* B **325** 136 (1994); V. S. Berezinsky, A. V. Gurevich and K. P. Zybin, *Phys. Lett.* B **294** p221 (1992); M. Urban M *et al., Phys. Lett.* B **293** p149 (1992).
37. H. A. Mayer-Hasselwander *et al.* (EGRET Collaboration), *MPE-440 (1998).*
38. A. Morselli *et al.* (GLAST Collaboration), *Nucl. Phys.* B (Proc. Suppl.) **113** p213 (2002).
39. J. Casaus *et al.* (AMS Collaboration), *Nucl. Phys.* **114** (Proc. Suppl.) p59 (2003).
40. M. Pearce (PAMELA Collaboration), *Nucl. Phys.* **113** (Proc. Nucl.) p314 (2002).

HEAVY FLAVORS AND IDENTIFIED PARTICLES

Chairpersons: G. Kozlov, J. Rafelski, and W. Metzger

Heavy Flavor Production in CDF

Mario Campanelli

DPNC, University of Geneva, 24, Quai Ernest-Ansermet, 1211 Geneva 4, Switzerland

Abstract. In this paper we report on several results about heavy flavor production at the CDF experiment at Tevatron. For trigger reasons, heavy flavor production is classified in low-and high-p_t physics, and different experimental approaches are followed in the two cases. While at low-p_t it is possible thanks to the large statistics to perform exclusive reconstruction of the final state, at higher p_t particles create hadronic jets and heavy flavors are identified from reconstruction of a secondary vertex with high invariant mass.

Keywords: Hadron colliders, heavy flavors
PACS: 12.38.Qk,13.87.Ce,14.65.Fy

INTRODUCTION: WHY STUDYING HEAVY FLAVORS AT HADRON COLLIDERS

Hadron colliders are a hostile environment for studying complex events like those originating from heavy flavor decays. Dedicated accelerators and experiments have been built for this purpose, to fully reconstruct b- and c-hadrons in a clean e^+e^- environment. In hadron colliders life is more difficult, but it is possible to profit from the large cross section and luminosity that for the Tevatron parameters yields a rate for b of a few KHz and much larger for charm. The study of b production is an interesting problem in itself for perturbative QCD. Since the b mass is larger than the QCD scale, perturbative expansion is expected to work quite well in this case. State of the art calculations go above next-to-leading order and include re-summed first-order next-to-leading logarithms. New fragmentation functions taken from LEP data have significantly changed the theory predictions for b production cross section in the central region over the last years.

EXPERIMENTAL TECHNIQUES

Tevatron and CDF

Tevatron is the highest-energy particle accelerator in the world. After the upgrade, this machine is able to collide proton and anti-proton beams of 0.98 TeV each, with a peak luminosity so far of $1.2 \times 10^{32} cm^{-2} s^{-1}$. So far, over 1 fb^{-1} has been delivered to the experiments, and the expectations are for a final luminosity between 4 and 8 fb^{-1} before the start of LHC. The analyses described here are performed using an integrated luminosity between 60 and 400 pb^{-1}.

CP828, *Multiparticle Dynamics*
edited by V. Šimák, M. Šumbera, Š. Todorova, and B. Tomášik

The CDF detector has been fully upgraded for the run II of the machine. In particular, modifications involved a completely new inner silicon tracker (SVX II and ISL), a new tracking chamber (COT), similar to the previous one but much faster, an extended calorimetry range that allowed good electromagnetic calorimetry also in the end-cap region, a revolutionary device allowing on-line tracking (SVT) that can be used at trigger Level 2 to select tracks with large impact parameter, and finally a completely revised trigger and data acquisition system, that allowed the detector to cope with the much increased bunch crossing rate. Detector parameters relevant to the measurements presented in this paper are the resolution for electromagnetic showers of $13.4\%/\sqrt{E_T} \oplus 2\%$ and for hadronic showers of $75\%/\sqrt{E_T} \oplus 3\%$, a track p_t resolution of $\sigma(p_t)/p_t = 0.15\% p_t$ and an impact parameter resolution for tracks of 40 μm, including 30 μm coming from the beam spread.

Triggering and tagging

Even if large in absolute terms, heavy flavor production is anyway much smaller than ordinary QCD—a 3 or 4 orders of magnitude in the case of b production. To study heavy flavors, first we have to collect them, i.e. be able to trigger on those particles. At high p_t, we have unbiased triggers that record all events with hadronic jets with a transverse energy above a certain threshold (20, 50, 70 and 100 GeV). Since especially at low threshold the production rate largely overcomes the storing capacity of the data acquisition system, large prescale factors are applied. Jets containing heavy flavors are then identified off-line, reconstructing the secondary vertex originating from the decay of the heavy flavor. At low p_t QCD production is so high that the trigger prescale factors should be enormous and no relevant physics could be performed with those samples. In Run I heavy-flavor enriched samples were produced tagging on the presence of leptons with high p_t with respect to the jet axis, an indication of semileptonic decays of b and c quarks. In Run II, the Silicon Vertex Tracker (SVT) allows on-line measurement of track parameters, including the impact parameter, indication of a delayed decay of an heavy flavor. The SVT works in conjunction with the eXtremely Fast Tracker (XFT), that measures the p_T and ϕ of tracks in the central chamber. The SVT identifies hits in the silicon, and builds "roads" of possible track candidates. Then the hits are fed into an associative memory where thousands of possible hit configurations are stored, together with the associated track parameters. This method is needed since no fit of the hits can be performed within the time limits (13 μs) imposed by the trigger; this procedure gives anyway very competitive performances, resulting in an impact parameter resolution of 47 μm. Thanks to this device, it is now possible to have high-statistics samples of hadronic heavy flavor decays.

CHARM PRODUCTION

One of the first measurements that used the SVT capabilities was the cross section of exclusively reconstructed charm states: $D^0 \to K\pi$, $D^* \to D^0\pi$, $D^+ \to K^-\pi^+\pi^+$, $D_s^+ \to \phi\pi^+$. The analysis has been carried on with 5.8 ± 0.3 pb^{-1} of data, collected

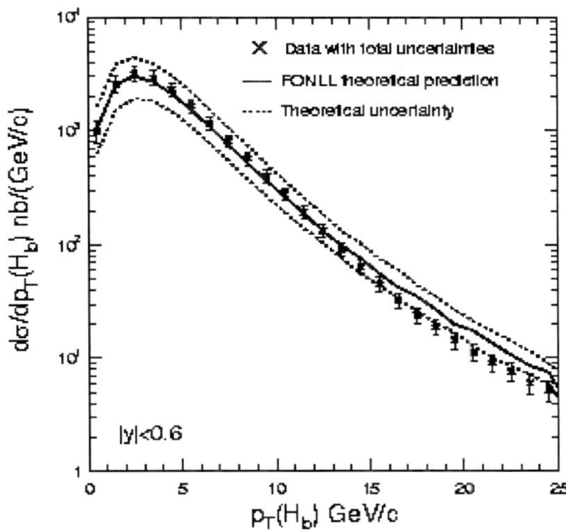

FIGURE 1. b production cross section from exclusively-reconstructed J/Ψ decays.

during the first months of Run II data taking. Events originated from prompt charm or from b decays have been distinguished fitting the distribution of the impact parameter of the fully reconstructed system. This allowed to extract direct charm fractions for the four cases: $86.4 \pm 0.4 \pm 3.5\%$ for D^0, $88.1 \pm 1.1 \pm 3.9\%$ for D^{*+}, $89.1 \pm 0.4 \pm 2.8\%$ for D^+ and $77.3 \pm 3.3 \pm 2.1$ for the D_s^+. The measured differential cross section is in good agreement with theory predictions [1], even if on the upper side of the theory error band.

SPECTROSCOPY WITH SVT DATASETS

Given the large dataset from the SVT, many channels have been studied, and CDF has the largest world sample for many exclusive channels, especially those involving heavy states like B_s, Λ_B, D^{**}, etc. CDF was also the first experiment to confirm the claim [2] from the Belle collaboration of the existence of a new resonance, so far known as X(3872), decaying into $J/\Psi\pi^+\pi^-$. The first evidence reported an excess of 730 candidates, with a mass of $3871.3 \pm 0.7 \pm 0.4$ MeV, and a width of 4.9 ± 0.7 MeV, consistent with detector resolution. X significance over background is enhanced by requiring high mass of the dipion system and this, together with an helicity analysis under investigation, is expected to shed more light on the real nature of this state.

B PRODUCTION IN J/ψ

This analysis is not using a dataset based on the SVT trigger, but on a trigger that requires two muons of 2 GeV p_t. Since the muons can also be back-to-back, the total p_t of the system can go to zero. The analysis aims to select events with J/Ψ production, and as in the case of exclusive D decays, distinguish those from J/Ψ to those coming from b decays. Opposite to the previous case, now we are interested in the secondary production; in order to do that, in every J/Ψ p_t bin the distribution of the variable $L_{xy}/p_t(J/\Psi)/M(J/\Psi)$ is fitted to get the components of prompt, b decays and background. The total b cross section is obtained de-convoluting the J/Psi branching ratio from the b fraction, and is shown in Fig. 1.

INCLUSIVE B-JET PRODUCTION

The measurements presented in the next two sections do not deal with exclusively reconstructed final states, but with hadronic jets. As already mentioned, in this case the heavy flavor production is identified from the presence of a secondary decay vertex in the event. The efficiency of this "b-tagging" algorithm is about 40%, and depends on p_t; it has been studied from Monte Carlo and cross-checked using b-enhanced samples containing isolated leptons. To merge particles in jets, CDF used during Run I the JetClu algorithm, a cone algorithm based on preclustering and including in a jet all particles inside a cone in $\eta - \phi$. This algorithm present some difficulties and was replaced in Run II analyses by Midpoint, infrared and collinear safe, that adds intermediate points to the original seeds, curing "pathological" cases often encountered by JetClu. After jets are formed and a secondary vertex is found, we still have to determine the fraction of b quarks present in our sample. In fact the tagging algorithm is not only selecting b quarks, but also charm and light quarks. The b fraction is estimated for every p_t bin making a distribution of the reconstructed invariant mass of the secondary vertex, and fitting it using templates coming from MonteCarlo for b and c+light jets. b-fractions estimated this way range from 40 to 15% of the total tagged sample.

Finally, the differential inclusive b-jet cross section is shown in Fig. 2, compared to leading-order (Pythia) MonteCarlo. The ratio between data and Pythia is about 1.4, giving an estimate of the relevance of higher orders in this measurements.

PRODUCTION OF B-JET PAIRS

A similar analysis has been carried on to measure production of $b - \bar{b}$ jets. The requirement in this case is to have two tagged jets in the central region ($|\eta| < 1.2$), and also in this case the invariant mass of the secondary vertex is used to determine the b content of the tagged jets. The measured $b - \bar{b}$ invariant mass is shown in Fig. 3, compared with two leading-order and one next-to-leading order Monte Carlo codes. We see that the agreement with the next-to-leading order code is not better than Pythia, and this is probably due to the fact that the latter gives a better description of the underlying event, absent from the next-to-leading order code.

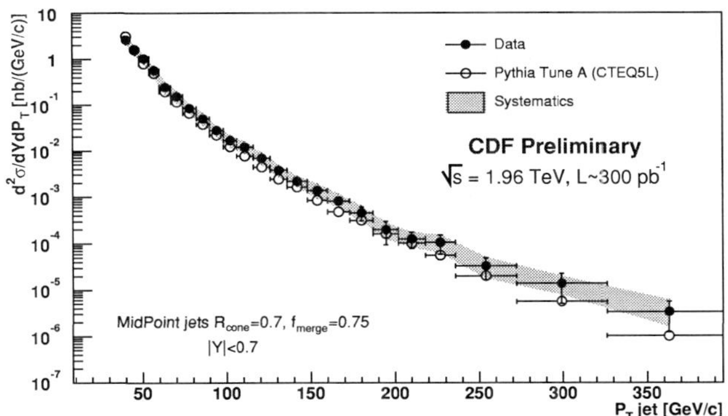

FIGURE 2. Differential cross section for b jets.

ASSOCIATED PRODUCTION OF HEAVY FLAVORS AND PHOTONS

Measuring production of b and associated with a photon is important for supersymmmetry searches, and to measure the b and c proton Pdf's. Photon production in a hadron collider is overwhelmed by a large background due to π^0 decays in $\gamma\gamma$; it is possible to distinguish the two by looking at the shape of the signal in the preshower detector in front of the calorimeter, or in the CES, a wire chamber located inside the calorimeter at the position of the maximum of the shower. It is however not possible to perform an event-by-event separation of signal and background, so only a distinction on a statistical basis is possible. Weights are given to each event according to its probability of being a

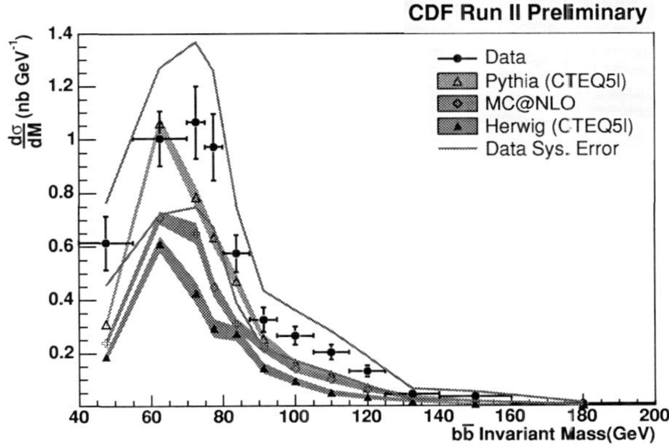

FIGURE 3. Invariant mass of the $b-\bar{b}$ system.

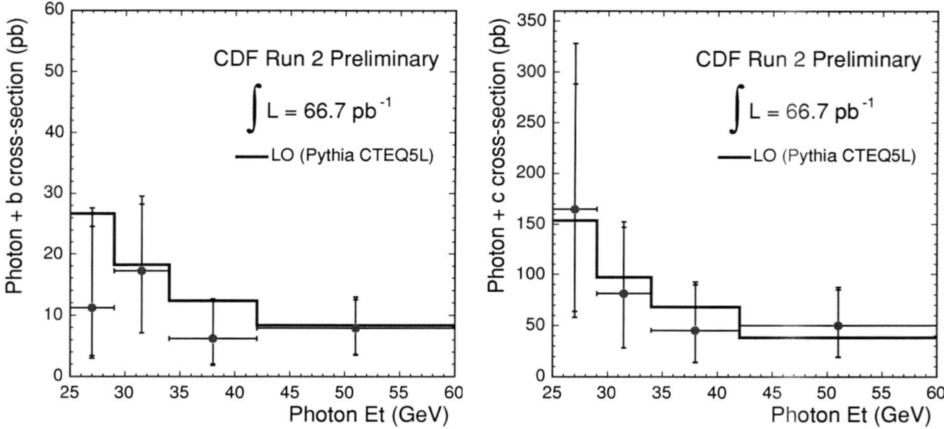

FIGURE 4. Differential cross section for b+γ and c+γ events.

real photon, and then statistical errors are properly treated to account to account for the unavoidable inclusion of background events. Also in this case the secondary vertex mass is used to identify the quark type, but instead of just separating the b from the rest, in this case a fit to three distributions, for b- c- and light-quarks is performed, to separately measure cross sections for b+γ and c+γ. Cross sections as a function of photon E_t are shown in Figs. 4, compared to LO Monte Carlo predictions (Pythia).

CONCLUSIONS

CDF has a broad program in heavy flavor production studies (not to mention decays, oscillations etc.), thanks mainly to its tracker. The main limitation is trigger and bandwidth; SVT allowed a large increase in low-p_t b physics, giving CDF the largest world samples for exclusively reconstructed hadronic final states. This allowed measuring the b cross section in the J/Ψ channel. High-p_t studies mainly used unbiased triggers. QCD analyses allowed measurements of b-and c- production for inclusive jets, multi-jets and jets in association with photons, allowing precision comparison with QCD calculations.

REFERENCES

1. M.Cacciari, P.Nason, *JHEP* 0309, 006 (2003).
2. S.K. Choi *et al.*, *Phys. Rev. Lett.* **91**, 262001 (2003).

DØ Results on Heavy Flavour Production

Isabelle Ripp-Baudot (for the DØ collaboration)

IReS, IN2P3-CNRS and Université Louis Pasteur, 23 Rue de Loess, F-67037 Strasbourg, France

Abstract. This review is focused on DØ results on hadron spectroscopy and on production cross sections: the contribution by DØ to the understanding of the $X(3872)$ object, the first observation of separated B^{**} states, measurements of B_c meson properties, and eventually measurements of the inclusive $\Upsilon(1S)$ differential cross section and of the $t\bar{t}$ cross section.

Keywords: D0, Tevatron, heavy flavour, $X(3872)$, B **, B$_c$, $\Upsilon(1S)$, top quark
PACS: 12.38.Qk, 14.40.Lb, 14.40.Nd, 14.65.Ha

INTRODUCTION

This review is focused on DØ results on hadron spectroscopy and on production cross sections, which represent only a small fraction of the DØ charm-bottom-top activities. After a general presentation of the Tevatron collider and of the DØ detector, the second section will be devoted to the contribution by DØ to the understanding of the $X(3872)$ object, to the first observation of separated B^{**} states and eventually to measurements of B_c meson properties. In the last section we will report on measurements of the inclusive $\Upsilon(1S)$ differential cross section and of the $t\bar{t}$ cross section. All these results are preliminary, unless a reference is quoted.

The Tevatron proton-antiproton collider at Fermilab operates with a center-of-mass energy of 1.96 TeV. Beams are colliding in two points instrumented by the CDF and the DØ multi-purpose detectors. The Tevatron is in its second run since 2001. In 2005 it performed better than expected, and it has already delivered 1 fb^{-1} to each experiment.

Important to heavy flavour production at the Tevatron is the considerable statistical increase over the previous datasets. B physics takes advantage of a substantial cross section, but has also to cope with a huge background, so that the trigger is an important issue. For these analyses, DØ benefits from an efficient trigger based mainly on muon and di-muon selection. All B states are produced at the Tevatron, such that it has a complementary role to B factories. In top physics, analyses depend on appropriate triggers as well, as the top quark is produced only once every ten billions events.

The DØ detector consists of three main components: the tracking system includes a high resolution silicon microstrip tracker, surrounded by a central scintillating fiber tracker, both located inside a 2 T superconducting solenoid magnet; the liquid Argon calorimeter is divided into a central part and two end cap calorimeters; eventually the muon system resides beyond the calorimeter and consists of three layers, two of them located behind a toroid magnet. All these detector systems are designed with good hermiticity and good coverage. For a detailed description of the DØ detector, see [1].

CP828, *Multiparticle Dynamics*
edited by V. Šimák, M. Šumbera, Š. Todorova, and B. Tomášik

SPECTROSCOPY RESULTS

Nature of the $X(3872)$ state

During 2003 Belle announced the observation of a new narrow hadron, temporarily called X, decaying to $J/\psi\pi^+\pi^-$ [2]. CDF [3], DØ [4] and BaBar [5] quickly confirmed this discovery. Up to now it is still unclear whether this $X(3872)$ particle is a $c\bar{c}$-state or a more complex object as a molecule-like $D\bar{D}$ resonance. DØ has observed the decays $X(3872) \to J/\psi\pi^+\pi^-$, $J/\psi \to \mu^+\mu^-$, using 230 pb^{-1} of data [4]. The signal has a statistical significance in excess of 5 sigmas and the measured mass is in agreement with those published by the other experiments. The interest in observing the $X(3872)$ at a hadronic collider resides in potentially high statistics and in the suggestion that it is not exclusively produced by weak B decays but also through the strong interaction.

The charmonium state $\psi(2S)$ with neighbouring mass and with the same decay mode provides a good benchmark for comparisons with the $X(3872)$. DØ has examined the production rate of the $X(3872)$ relative to the $\psi(2S)$ in various regions of production and decay variables: the transverse momentum p_T with respect to the beam axis, the rapidity y, a two dimensional decay-length dl, an isolation criterion with respect to all other reconstructed charged particles within a cone about its direction. Eventually the angular decay distributions of the $\pi^+\pi^-$ and $\mu^+\mu^-$ systems from the $X(3872)$ and the $\psi(2S)$ decays are investigated to check for any differences of helicities. The result of this study is shown in Fig. 1. There is no evidence for any difference between the $X(3872)$ particle and the $\psi(2S)$. Further studies include looking for other decays which could help to disentangle the different hypotheses for the nature of the $X(3872)$.

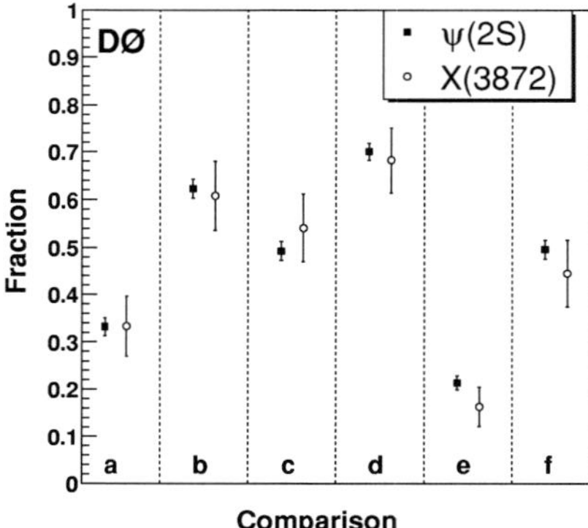

FIGURE 1. Event yield fractions in DØ for $X(3872)$ and $\psi(2S)$ in various regions of production and decay variables: (a) $p_T > 15$GeV/c; (b) $|y| < 1$; (c) $\cos(\theta_{\pi\pi}) < 0.4$; (d) effective proper decay length $dl < 0.01$ cm; (e) isolation criteria = 1; (f) $\cos(\theta_{\mu\mu}) < 0.4$.

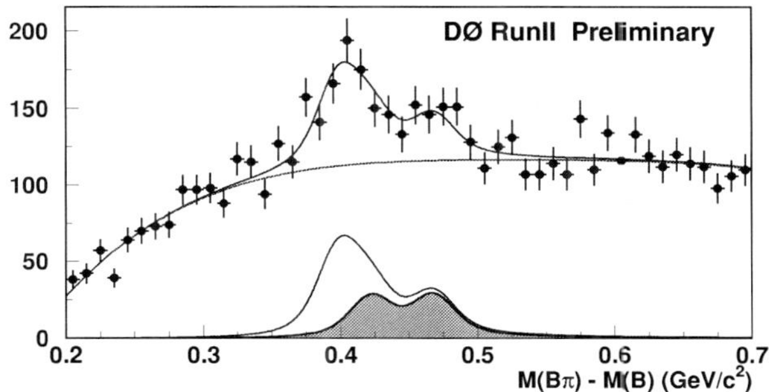

FIGURE 2. Mass difference $M(B\pi) - M(B)$ for exclusively reconstructed B mesons. The two upper curves represent the fit to the data and the estimated background. The lower curve shows the signal when the background is subtracted and the filled histogram indicates the contribution of $B_2^* \to B^*\pi$ (left peak) and $B_2 \to B\pi$ (right peak).

Observation of separate orbitally excited B mesons

For each kind of B meson, one expects four orbitally excited B states with $L = 1$, usually called B^{**}. Two of them, labelled B_1 and B_2^*, are narrow and decay to $B^{(*)}\pi$. Their properties are predicted reliably by several theoretical models [6] [7] [8] [9]. B^{**} were however up to now experimentally observed mainly at LEP through inclusive or semi-inclusive decays, preventing their separation in narrow states and precise measurements of their properties [10, 11, 12, 13].

The $B^{**} \to B^{(*)}\pi$ decays have been reconstructed in DØ without reconstructing the photon coming from the $B^* \to B\gamma$ decay, using exclusively reconstructed B mesons: B^+ mesons in the $J/\psi K^+$ channel, and B_d^0 from the $J/\psi K^{*0}$ and $J/\psi K_s^0$ final states. Using 350 pb^{-1} of data DØ obtained more than 10 000 B mesons. This led to the first observation of the two separated narrow states with a statistical significance of 7σ. Actually, due to the missing photon, one expects to find eight mass peaks, but the mass difference between B^{**+} and B_d^{**0} is negligeable relative to the experimental resolution, and the same argument applies to the mass difference between B_2^* and B_1, so that we resolve only two peaks in the invariant mass distribution, as shown in Fig. 2. The measured B_1 and B_2^* masses and average width are: $M(B_1) = 5724 \pm 4(\text{stat}) \pm 7(\text{syst})$ MeV/c^2, $M(B_2^*) - M(B_1) = 23.6 \pm 7.7(\text{stat}) \pm 3.9(\text{syst})$ MeV/c^2 and $\Gamma = 23 \pm 12(\text{stat}) \pm 9(\text{syst})$ MeV/c^2.

Properties of the B$_c$ meson decaying to $J/\psi\mu X$

The B_c^+ is the heaviest B meson, ground state of the $b\bar{c}$ system. It has been previously observed by CDF during Run I [15] but with limited statistics. Its study gives a novel

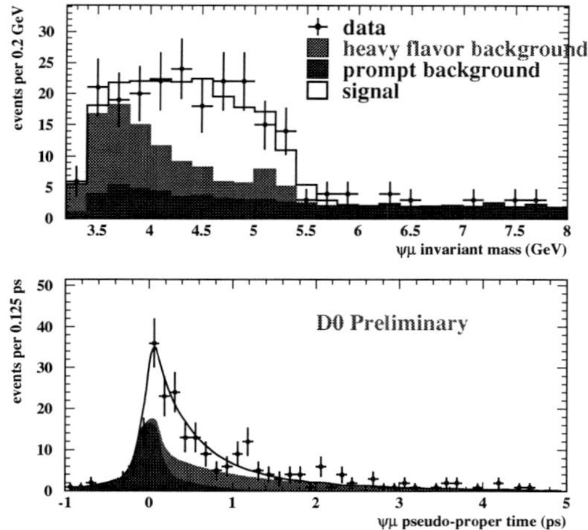

FIGURE 3. $J/\psi\mu$ invariant mass (top) and pseudo-proper time (bottom) distributions for the $B_c \to J/\psi\mu X$ candidates. The points represent the data and the solid curve shows the best combined mass and lifetime likelihood fit to the data. The filled histograms indicate the heavy flavour background (grey) and the prompt background (dark grey).

experimental insight into heavy quarks dynamics.

The $B_c \to J/\psi\mu\nu$ decays has been reconstructed in DØ, taking advantage of the presence of three muons for triggering, and also of a reduced background compared to decay channels with larger branching ratios. Using 210 pb^{-1} of data, 95 B_c with a significance greater than 5σ have been reconstructed. The mass resolution is limited, as it can be seen in Fig. 3, due to the escaping neutrino. A two-dimensional likelihood fit leads to the following mass and lifetime measurement: $M(B_c) = 5.95^{+0.14}_{-0.13}(\text{stat}) \pm 0.34(\text{syst})\text{GeV/c}^2$ and $\tau(B_c) = 0.448^{+0.123}_{-0.096}(\text{stat}) \pm 0.121(\text{syst})$ ps. As predicted by theory, the B_c lifetime is closer to a charmed meson lifetime than to that of a B meson. In conclusion, the precision of measurements of B_c properties is no longer limited by statistics. Neththeless, DØ is looking into hadronic decays, to improve the mass and lifetime precision by avoiding the missing neutrino.

CROSS SECTION MEASUREMENTS

Inclusive $\Upsilon(1S)$ differential cross section

DØ measured the inclusive $\Upsilon(1S)$ cross section as a function of the Υ transverse momentum in three rapidity regions [16]. The Υ is reconstructed from the $\mu^+\mu^-$ decay and is assumed to be unpolarized, according to the CDF Run I measurement [17]. The DØ measurement, shown in Fig. 4, is in agreement with theoretical predictions [18]

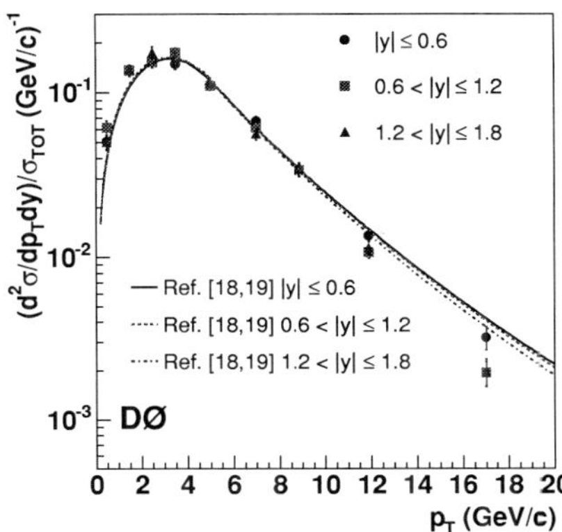

FIGURE 4. Normalized differential cross sections for $\Upsilon(1S)$ production compared to theoretical predictions [18] [19]. The p_T independent systematic uncertainties are not included.

[19]. It is compatible with the CDF Run I measurement, limited to $|y| < 0.4$ [17]. Based upon 159 pb^{-1} of data, this measurement is already limited by systematics. However, increasing statistics would benefit to the high p_T region and would improve the agreement between simulation and data.

$t\bar{t}$ production cross section

The top quark has been discovered at Tevatron in 1995 [20], the only collider able to produce it. Its study is one of the main goals of CDF and DØ. Up to now, the top quark is observed experimentally via $t\bar{t}$ pair production through strong interactions. Electroweak production of a single top is also possible, but with a smaller cross section and a larger background, so that it has not been observed yet. In the Standard Model top quarks decay to a W boson and a b quark and cross section analyses are grouped according to the W decay channel. The possible $t\bar{t}$ final states are therefore dilepton channels, lepton plus jets channels and fully hadronic final states. The more leptons present in the final state, the smaller the background, but also the lower the statistics. The background can be reduced by requiring one or two jets to be b tagged, using track impact parameters or displaced vertices. The summary of all DØ cross section measurements is shown in Fig. 5, all results are in agreement with theoretical predictions [21].

CONCLUSION

The Tevatron is a unique place to study the top quark, but also several B hadrons properties. In 2005 it has already delivered more than 1 fb^{-1} to each experiment. DØ analyses presented here are based on 150 to 370 pb^{-1}. Those limited by statistics will benefit from the expected increase in luminosity during the next years.

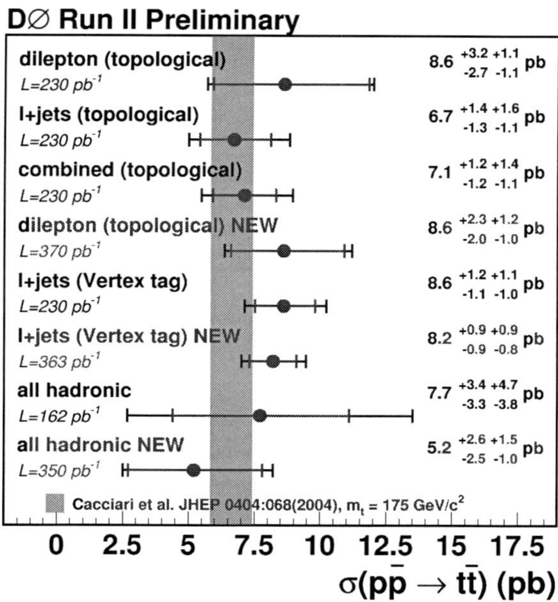

FIGURE 5. Summary of all DØ $t\bar{t}$ cross section measurements at $\sqrt{s} = 1.96$ TeV.

REFERENCES

1. V.M. Abazov *et al.*, DØ Collaboration, *FERMILAB-PUB-05-341-E*, 142 pages (2005).
2. S.K. Choi *et al.*, Belle Collaboration, *Phys. Rev. Lett.* **91**, 262001 (2003).
3. D. Acosta *et al.*, CDF Collaboration, *Phys. Rev. Lett.* **93**, 072001 (2004).
4. V.M. Abazov *et al.*, DØ Collaboration, *Phys. Rev. Lett.* **93**, 162002 (2004).
5. B. Aubert *et al.*, BABAR Collaboration, *Phys. Rev. D* **71**, 071103 (2005).
6. E.J. Eichten, C.T. Hill, C. Quigg, *Phys. Rev. Lett.* **71**, 4116-4119 (1993).
7. N. Isgur, *Phys. Rev. D* **57**, 4041-4053 (1998).
8. D. Ebert, V.O. Galkin, R.N. Faustov, *Phys. Rev. D* **57**, 5663-5669 (1998). Erratum *Phys. Rev. D* **59**, 019902 (1999).
9. A.H. Orsland, H. Hogaasen, *Eur. Phys. J. C.* **9**, 503-510 (1999).
10. R. Akers *et al.*, OPAL Collaboration, *Z. Phys. C* **66**, 19-30 (1995).
11. P. Abreu *et al.*, DELPHI Collaboration, *Phys. Lett. B* **345**, 598-608 (1995).
12. D. Buskulic *et al.*, ALEPH Collaboration, *Z. Phys. C* **69**, 393-404 (1996).
13. T. Affolder *et al.*, CDF Collaboration, *Phys. Rev. D* **64**, 072002 (2001).
14. R. Barate *et al.*, ALEPH Collaboration, *Phys. Lett. B* **425**, 215-226 (1998).
15. F. Abe *et al.*, CDF Collaboration, *Phys. Rev. Lett.* **81**, 2432-2437 (1998).
16. V.M. Abazov *et al.*, DØ Collaboration, *Phys. Rev. Lett.* **94**, 232001 (2005).
17. D. Acosta *et al.*, CDF Collaboration, *Phys. Rev. Lett.* **88**, 161802 (2002).
18. E.L. Berger, J. Qiu, Y. Wang, *Phys. Rev. D* **71**, 034007 (2005).
19. E.L. Berger, J. Qiu, Y. Wang, *Int. J. Mod. Phys. A* **20**, 3753-3755 (2005).
20. F. Abe *et al.*, CDF Collaboration, *Phys. Rev. Lett.* **74**, 2626-2631 (1995). S. Abachi *et al.*, DØ Collaboration, *Phys. Rev. Lett.* **74**, 2632-2637 (1995).
21. M. Cacciari, S. Frixione, M.L. Mangano, P. Nason, G. Ridolfi, *JHEP* **04**, 068 (2004).

Heavy Flavors in High Energy *ep* Collisions

Meng Wang
on behalf of the H1 and ZEUS collaborations

Bonn University, Institute of Physics, Nußallee 12, 53115 Bonn, Germany

Abstract. Most recent measurements of open charm and beauty production in high energy *ep* collisions at HERA are reviewed. The measurements explored the different aspects of quantum chromodynamics involved in the process of heavy flavor production. The results are compared with perturbative theoretical calculations at next-to-leading order.

Keywords: heavy flavor, structure function, fragmentation function
PACS: 13.60.Hb

INTRODUCTION

The masses of heavy quarks, charm and beauty, provide hard scales for perturbative quantum chromodynamics (QCD) calculations. Measurements of heavy flavor production therefore have been and continue to be of great interest as a rich testing ground for the reliability of perturbative QCD predictions.

HERA, which collides electrons or positron of energy 27.5 GeV with protons of energy 920 GeV (or 820 GeV before 1998) resulting in a center-of-mass energy of 318 GeV (or 300 GeV), can test heavy flavor production in a unique way. Two collider experiments, H1 and ZEUS, have accumulated approximately 135 pb^{-1} of integrated luminosity by the end of 2000. After a major upgrade during 2001 and 2002, HERA is running at much higher luminosity with the polarized electron beam and is referred to as HERA II for distinction.

Heavy flavor production at HERA is dominated by boson-gluon-fusion (BGF), as shown in Fig. 1. When the virtuality Q^2 of the exchanged boson, mainly photon, is very small, $Q^2 \ll 1\,\text{GeV}^2$, the virtual photon resembles a real one and the collisions are referred to as γp or photoproduction, in which a certain fraction of the photons can be resolved with parton contents. For large Q^2, the collisions are called deep inelastic scattering (DIS). Only the most recent measurements are presented here.

THEORY AND MODELS

In perturbative QCD (pQCD), single-particle inclusive cross sections in *ep* collisions can be factorized in the form

$$\sigma_{ep\to HX} = \sum_i \phi_{i/p} \otimes \hat{\sigma}_{Vi\to hX} \otimes D_{H/h},$$

CP828, *Multiparticle Dynamics*
edited by V. Šimák, M. Šumbera, Š. Todorova, and B. Tomášik

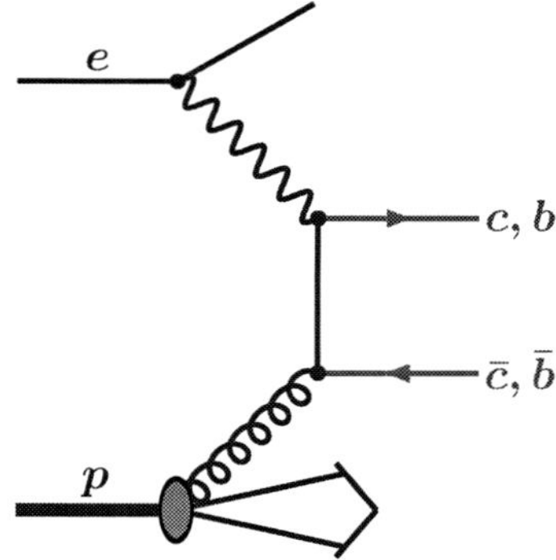

FIGURE 1. Heavy flavor production via Boson-gluon-fusion in the *ep* collision.

where $\otimes$ denotes convolution. The sum is over all relevant partons i, while $\phi_{i/p}$ is a parton density function (PDF) of the initial proton, $\hat{\sigma}_{Vi \to hX}$ is the pQCD calculable cross section of hard scattering, and $D_{H/h}$ is the fragmentation function. In resolved photoproduction, an additional PDF of the photon enters the calculation. These terms are evaluated at a renormalization scale μ_R as well as a factorization scale μ_F, which is often taken at the same value as μ_R. The PDFs and the fragmentation functions are non-perturbative and must be determined either experimentally or taken from models, but they are universal. In heavy flavor production, the mass of the heavy quark provides an additional hard scale hence yielding more accurate calculations.

Up to next-to-leading order (NLO), the fixed-flavor-number scheme (FFNS) or "massive" scheme from Frixione et al. [1] for photoproduction and from Harris and Smith [2] for DIS and the zero-mass variable-flavor-number scheme (ZMVFNS) or "massless" scheme from Cacciari and Greco [3] and from Heinrich and Kniehl [4] are compared with the data. They differ in the treatment of mass of heavy quark. In the massive scheme, the heavy quarks are non-active flavors in the proton and are produced through hard scattering such as BGF, while in the massless scheme, the heavy quarks are just the contents of the proton and can enter the reaction directly. The massive scheme works well near the threshold of heavy quark production and the massless scheme works better in the higher kinematic region. A scheme matching the two also exists but is not used for the measurements presented here.

Monte Carlo (MC) models, based on leading order QCD and parton shower approaches, are used for acceptance calculations in all measurements, and sometimes as alternative QCD predictions. They include PYTHIA [5] and HERWIG [6] which differ in the treatment of parton showers and heavy quark fragmentation.

OPEN CHARM PRODUCTION

Inclusive jet cross sections and dijet correlations in D^* photoproduction have been studied by ZEUS [7]. NLO predictions in the massive and massless schemes show reasonable

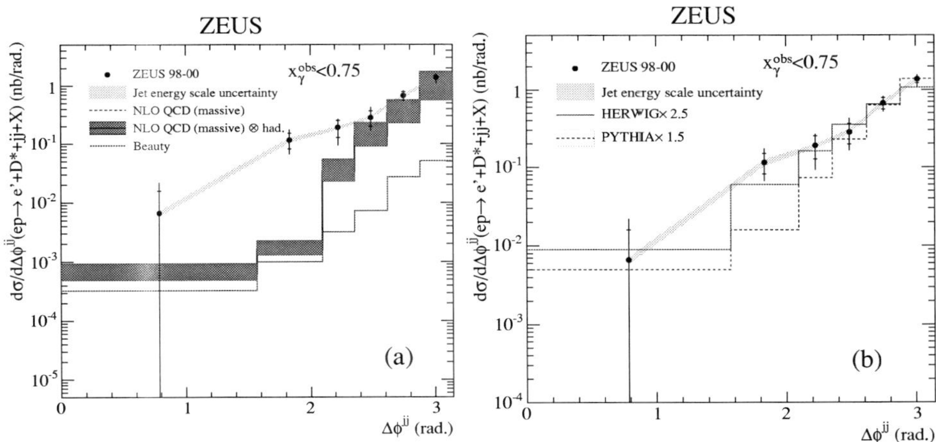

FIGURE 2. Azimuthal angular difference of the two highest E_T jets in D^* photoproduction for the resolved enriched sample, compared to (a) a NLO QCD calculation and (b) Monte Carlo models.

agreement with the data in all inclusive jet cross sections as well as some dijet correlation cross sections, but the massive prediction shows a large deviation at low $\Delta\phi^{jj}$, azimuthal angular difference of two jets, and high $(p_T^{jj})^2$, squared transverse momentum of dijet system. The discrepancy is enhanced in the resolved-enriched sample, as shown in Fig. 2a. However, the HERWIG MC model describes the shape of the data well, Fig. 2b. This indicates the necessity of higher-order calculations or additional parton showers in current NLO calculations.

Open charm production in DIS is directly sensitive to the gluon contents in the proton. Measurements [8–12] of the charm contribution to the proton structure function, $F_2^{c\bar{c}}$, are shown in Fig. 3a. The results are well described by a QCD prediction based on a QCD fit of the inclusive structure function F_2. A measurement exploring the transition region between DIS and photoproduction has also been reported recently [13] as well as the first charm measurement from HERA II data [14].

Charm fragmentation is traditionally measured at e^+e^- experiments and the results are then adapted into MC simulations or theoretical calculations. Recently, H1 and ZEUS made measurements on charm fragmentation ratios $R_{u/d}$ (the ratio of neutral to charged D-meson rates), γ_s (the strangeness suppression factor) and P_V (the fraction of D-meson in a vector state) as well as fragmentation fractions, $f(c \to D, \Lambda)$ [15–17]. The results are compared to those of e^+e^-, Fig. 3b-d, and confirm the universality of fragmentation. Measurements of the fragmentation function have also been made by H1 [18] and ZEUS [19], and are in agreement with universality. too.

OPEN BEAUTY PRODUCTION

Because of the larger mass, the pQCD calculation for open beauty production should be more reliable than for open charm. Some of the previous measurements [20, 21] have

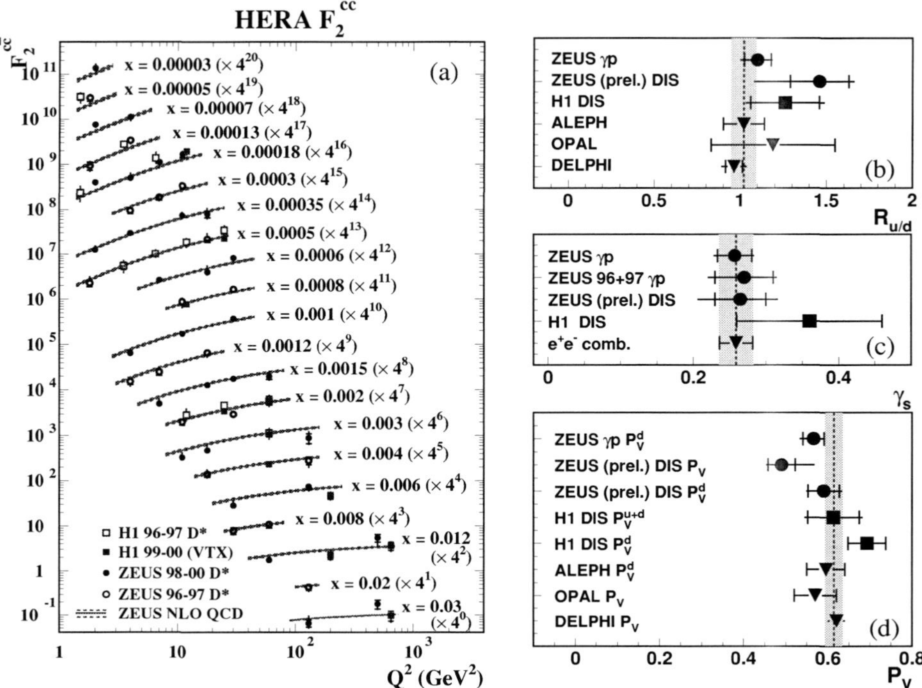

FIGURE 3. (a) $F_2^{c\bar{c}}$ as a function of Q^2 for different values of the Bjorken scaling variable x; (b-d) Comparisons of charm fragmentation ratios $R_{u/d}$, γ_s and P_V. (Comparisons of charm fragmentation fractions, $f(c \to D, \Lambda)$, are not shown for brevity.)

found the prediction to be below the data by up to a factor of 3. Recent measurements by H1 and ZEUS required events containing jets in addition to high p_T muons, which are traditionally used to tag beauty quarks [22–24]. As shown in Fig. 4a, the measurements in photoproduction agree very well between the two collaborations and are well described by the NLO calculations although the H1 data is above the calculation at low p_T^μ. In DIS, similar agreement exists, although discrepancies at low p_T^μ as well as forward η^μ are observed by both collaborations, shown in Fig. 4b as an example.

New measurements using inclusive impact parameters of tracks from decays of long lived charm and beauty hadrons have been performed in DIS by the H1 collaboration [11, 12]. They are the first measurements of $F_2^{b\bar{b}}$, Fig. 5. The data are well described by the NLO predictions and a recent calculation using NNLO structure functions [25]. The values of $F_2^{c\bar{c}}$ also confirm the previous data and have been put in Fig. 3a. Significantly improved further measurements are expected from the increased HERA II statistics and the corresponding detector improvements. First preliminary results on beauty in photoproduction using the new ZEUS micro-vertex detector have already been obtained [26].

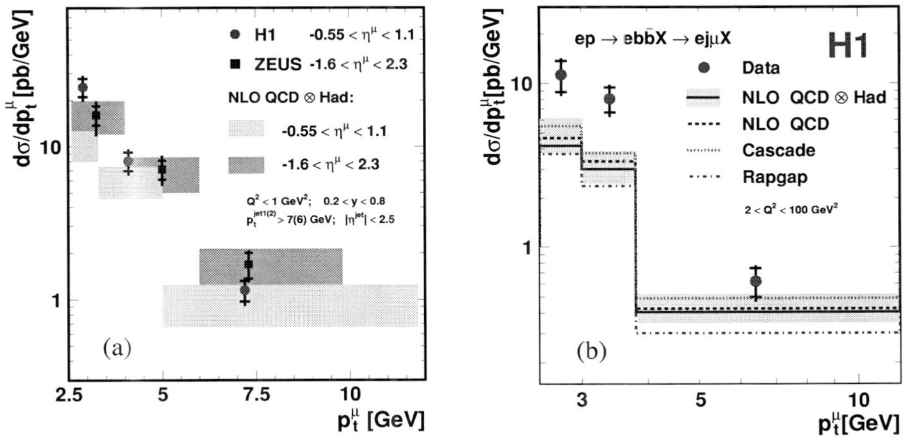

FIGURE 4. Open beauty production as a function of p_T^μ for (a) dijet photoproduction from the H1 and ZEUS experiments and (b) inclusive jet DIS from the H1 experiment. (The measurement from the ZEUS experiment for (b) has the similar result and is not shown for brevity.)

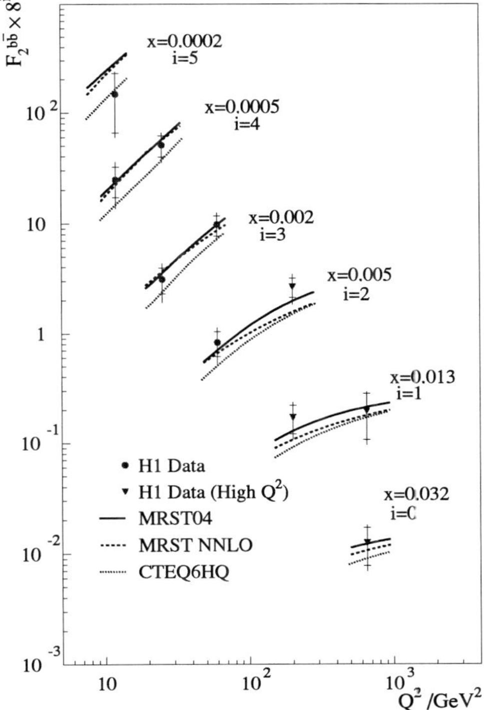

FIGURE 5. $F_2^{b\bar{b}}$ as a function of Q^2 for different x.

CONCLUSIONS

At HERA, measurements on heavy flavor production are getting more extensive and precise. New results improve the understanding of the heavy quark contributions to the proton structure function as well as the hadronic behavior of the photon. The universality of the fragmentation has also been confirmed by the recent measurements. The perturbative QCD calculation generally describes the data at next-to-leading order although discrepancies still exist in some threshold regions. Therefore higher order corrections beyond the existing models and calculations might be needed. More exciting and accurate measurements are definitely expected in the future from HERA II with higher luminosity and improved detector ability of the H1 and ZEUS experiments.

ACKNOWLEDGMENTS

I really appreciate the invitation from the conference organizers. I am also grateful to A. Geiser and M. Wing for productive discussions. This work is supported by the German Federal Ministry of Education and Research (BMBF) under Contract No. HZ4PDA.

REFERENCES

1. S. Frixione, P. Nason, and G. Ridolfi, *Nucl. Phys. B* **454**, 3–24 (1995), hep-ph/9506226.
2. B. W. Harris, and J. Smith, *Phys. Rev. D* **57**, 2806–2812 (1998), hep-ph/9706334.
3. M. Cacciari, and M. Greco, *Phys. Rev. D* **55**, 7134–7143 (1997), hep-ph/9702389.
4. G. Heinrich, and B. A. Kniehl, *Phys. Rev. D* **70**, 094035 (2004), hep-ph/0409303.
5. T. Sjostrand, et al., *Comput. Phys. Commun.* **135**, 238–259 (2001), hep-ph/0010017.
6. G. Corcella, et al., *JHEP* **01**, 010 (2001), hep-ph/0011363.
7. S. Chekanov, et al. (2005), submitted to Nucl. Phys., hep-ex/0507089.
8. J. Breitweg, et al., *Eur. Phys. J. C* **12**, 35–52 (2000), hep-ex/9908012.
9. C. Adloff, et al., *Phys. Lett. B* **528**, 199–214 (2002), hep-ex/0108039.
10. S. Chekanov, et al., *Phys. Rev. D* **69**, 012004 (2004), hep-ex/0308068.
11. A. Aktas, et al., *Eur. Phys. J. C* **40**, 349–359 (2005), hep-ex/0411046.
12. A. Aktas, et al. (2005), submitted to Eur. Phys. J., hep-ex/0507081.
13. ZEUS Collab., abstract 265 in [27].
14. ZEUS Collab., abstract 271 in [27].
15. A. Aktas, et al., *Eur. Phys. J. C* **38**, 447–459 (2005), hep-ex/0408149.
16. S. Chekanov, et al. (2005), submitted to Eur. Phys. J., hep-ex/0508019.
17. ZEUS Collab., abstract 266 in [27].
18. H1 Collab., abstract 407 in [27].
19. ZEUS Collab., abstract 778 in *XXXI International Conference on High Energy Physics*, Amsterdam, The Netherlands, 2002.
20. C. Adloff, et al., *Phys. Lett. B* **467**, 156 (1999), erratum-ibid.B518,331,2001, hep-ex/9909029.
21. J. Breitweg, et al., *Eur. Phys. J. C* **18**, 625–637 (2001), hep-ex/0011081.
22. S. Chekanov, et al., *Phys. Rev. D* **70**, 012008 (2004).
23. S. Chekanov, et al., *Phys. Lett. B* **599**, 173–189 (2004), hep-ex/0405069.
24. A. Aktas, et al., *Eur. Phys. J. C* **41**, 453–467 (2005), hep-ex/0502010.
25. R. S. Thorne, "A Variable-Flavour-Number Scheme at NNLO," in *13th International Workshop On Deep Inelastic Scattering*, Madison, Wisconsin, USA, 2005, hep-ph/0506251.
26. ZEUS Collab., abstract 359 in *International Europhysics Conference on High Energy Physics*, Lisboa, Portugal, 2005.
27. *XXII International Symposium on Lepton-Photon Interactions at High Energy*, Sweden, 2005.

In-Medium Formation of J/Psi as a Probe of Charm Quark Thermalization

R. L. Thews

Department of Physics, University of Arizona, Tucson, AZ 85721, USA

Abstract. Charmonium formation via charm quark in-medium recombination in heavy ion interactions at collider energies has the potential to probe some properties of the medium by utilizing the sensitivity of the recombination process to the momentum distribution of the quarks. We have examined the transverse momentum spectra of J/ψ, characterized by $\langle {p_T}^2 \rangle$, which result from the formation process in which the charm quark distributions are unchanged from their initial production in a pQCD process. This is contrasted with the case in which the charm quarks have completely come into thermal equilibrium with an expanding medium whose properties are determined by the spectra of produced light hadrons. We find that the resulting $\langle {p_T}^2 \rangle$ of the formed J/ψ provide a distinct signature of the underlying charm quark spectra, and that signature is essentially independent of the detailed dynamics of the in-medium formation reaction. In addition, both of these signatures are sufficiently separated from the case in which no in-medium formation takes place. Finally, utilizing a model for the fraction of J/ψ which originate from in-medium formation, we predict the centrality behavior of these signatures.

Keywords: Quark-Gluon Plasma, Color Deconfinement
PACS: 25.75i.Nq, 12.38.-t

The role of J/ψ produced in high energy heavy ion collisions as a signature for color deconfinement [1] has evolved in recent years with the realization that at collider energies an additional formation mechanism may become significant [2, 3]. This depends on the initial production of multiple $c\bar{c}$ pairs in sufficient numbers. Initial estimates [4] from extrapolation of fixed-target data put this number at about 10 for central Au-Au collisions at RHIC. Subsequently, measurements by the PHENIX and STAR experiments indicate even higher numbers, between 20 [5] and 40 [6], respectively. The in-medium formation picture we consider here uses competing formation and dissociation reactions in a Boltzmann equation to calculate the final J/ψ population. The absolute value of this formation was found to be very sensitive to the underlying charm quark momentum distributions [7]. In addition, there is significant dependence on details of the size and expansion profile of the deconfinement region, for which various model parameters must be introduced. The initial PHENIX data [8] suffered from low statistics, and was compatible with a fairly large region of model parameter space [9].

Recent work in this area concentrated on finding a signature for in-medium J/ψ formation which is independent of the detailed dynamics and magnitude of the formation. We found that the p_T spectrum of the formed J/ψ may provide such a signature [10]. Our first calculations of in-medium formation used initial charm quark momentum distributions from NLO pQCD amplitudes to generate a sample of $c\bar{c}$ pairs, supplemented

CP828, *Multiparticle Dynamics*
edited by V. Šimák, M. Šumbera, Š. Todorova, and B. Tomášik

by an initial-state transverse momentum kick to simulate confinement and nuclear effects. In the absence of in-medium formation, p_T of the J/ψ follows from that of the initially-produced individual $c\bar{c}$ pairs, but one must use a model for the magnitude of the hadronization process. For the *normalized* p_T spectrum, we start with that of the pair spectrum. In the evolution of the interacting system size from pp to pA to AA collisions, the p_T will be in general increased due to initial-state effects of interaction of constituents in the nuclei. One can express this effect as

$$\langle {p_T}^2 \rangle_{pA} - \langle {p_T}^2 \rangle_{pp} = \lambda^2 \, [\bar{n}_A - 1], \tag{1}$$

where $\bar{n}_A$ is the impact-averaged number of inelastic interactions of the projectile in nucleus A, and λ^2 is the square of the transverse momentum transfer per collision. For a nucleus-nucleus collision, the corresponding relation is

$$\langle {p_T}^2 \rangle_{AB} - \langle {p_T}^2 \rangle_{pp} = \lambda^2 \, [\bar{n}_A + \bar{n}_B - 2]. \tag{2}$$

The PHENIX measurements of J/ψ p_T spectra in pp and minimum-bias d-Au interactions [11, 12] allow us to determine the amount of initial state k_T needed to supplement our collinear pQCD events. One can then extrapolate to Au-Au and predict the spectrum of J/ψ which are produced from hadronization of the initial "diagonal" $c\bar{c}$ pairs, again for minimum bias interactions. (We use diagonal to distinguish these pairs from the "off-diagonal" combinations which contribute to in-medium J/ψ formation.) One finds $\bar{n}_A = 5.4$ for minimum bias d-Au interactions at RHIC energy (using σ_{pp} = 42 mb), which leads to $\lambda^2 = 0.35 \pm 0.14$ GeV^2. We note that the relatively large uncertainty comes entirely from the difference in p_T broadening in d-Au between positive and negative rapidity.

Our prediction for the "normal" evolution of the p_T spectrum in Au-Au interactions is shown by the triangular points in Fig.1. There is some scatter at high p_T due to the limited statistics of the number of pQCD-generated pairs. The $\langle {p_T}^2 \rangle$ of this spectrum is approximately 6.3 GeV^2.

The properties of the normalized p_T spectrum for the formation process follow from two separate effects: First, the fact that the process is dominated by the off-diagonal pairs introduces a modified initial p_T distribution. Next, one weights these pairs by a formation probability for J/ψ. We use the operator-product motivated cross section [13, 14] for $c\bar{c}$ forming J/ψ with emission of a final-state gluon, which of course is just the inverse of the dissociation process. However, any cross section which has the same general properties as this one gives essentially the same result [10]. We show by the square points in Fig.1 the prediction for the formed J/ψ. One sees that this spectrum, characterized by $\langle {p_T}^2 \rangle$ approximately 3.6 GeV^2, is substantially narrower than the one with no in-medium formation.

We next considered charm quark momentum distributions which would follow if the charm quark interaction with the medium were so strong that they come into thermal equilibrium with the expanding region of deconfinement. The parameters of temperature and maximum transverse expansion rapidity are determined by a fit to this thermal behavior of the produced light hadrons. The application to charm quarks was originally motivated in Ref. [15], who showed that the low-p_T spectrum of decay leptons from

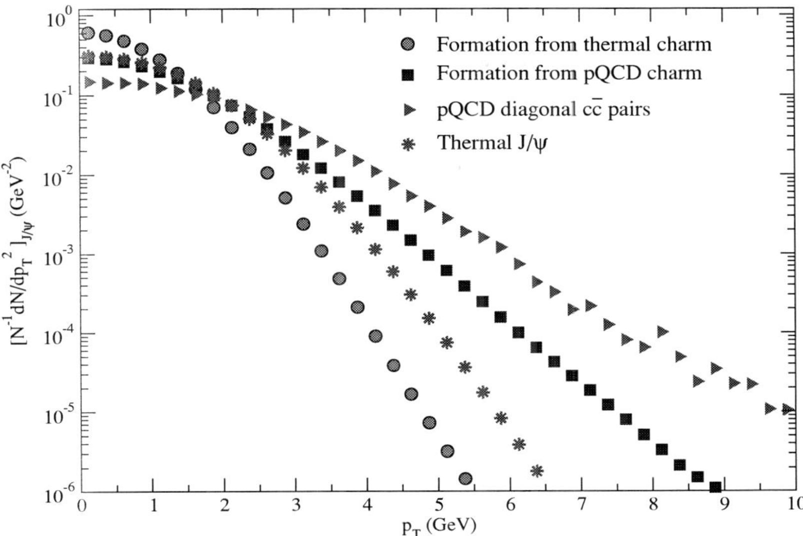

FIGURE 1. Comparison of im-medium J/ψ transverse momentum spectra predictions from various scenarios.

charmed hadron decays would not be able to differentiate between the thermal and a purely pQCD distribution. We show here, however, that the p_T spectrum of in-medium formed J/ψ is very sensitive to this distribution [16]. The circles in Fig.1 result from formation calculations using T = 170 MeV and $y_{T_{max}}$ = 0.5 for the thermal charm quarks. One sees that this p_T spectrum is narrower yet than in-medium formation from pQCD quarks, with $\langle p_T{}^2 \rangle$ approximately 1.3 GeV^2. Finally, we show by the stars the p_T spectrum of J/ψ which themselves obey this thermal distribution. The resulting spectrum falls between the in-medium formation spectra for either pQCD or thermal charm quark momentum distributions, with $\langle p_T{}^2 \rangle$ approximately 3.0 GeV^2.

We now proceed to investigate the variation of the pQCD-based results with respect to the collision centrality in Au-Au interactions. First, we use the value of λ^2 extracted from pp and pA data, together with values of $\bar{n}_A$ calculated as a function of collision centrality, to recalculate the $\langle p_T{}^2 \rangle$ values for either the initial production or the in-medium formation separately. This provides the centrality behavior of the J/ψ spectrum in the case that one or the other of these mechanisms is solely responsible for the total J/ψ population. We show these results together in Fig. 2. One sees as expected that $\langle p_T{}^2 \rangle$ is maximum for the most central collisions, but the absolute magnitudes are widely separated for initial production and in-medium formation at each centrality. One should note that the uncertainties are dominated by the difference between the p_T - broadening measurements at positive and negative rapidities in the d-Au interactions. Thus the point-to-point uncertainties are much smaller for the centrality behavior. We have also included separate values for $\langle p_T{}^2 \rangle$ in the region limited by p_T < 5 GeV, to facilitate comparison with experiment in this same range.

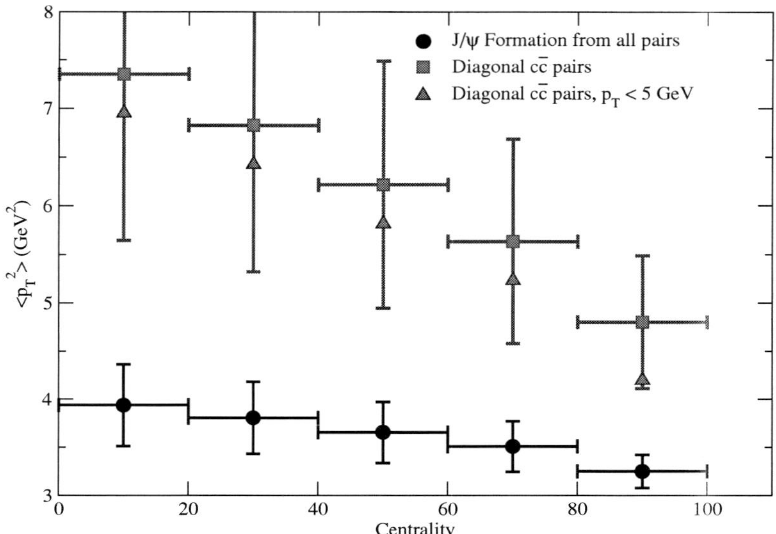

FIGURE 2. Centrality dependence of $J/\psi\langle {p_T}^2\rangle$ contrasting predictions assuming either 100 % production from initial $c\bar{c}$ pairs or 100 % in-medium formation.

In order to provide a meaningful prediction for the overall J/ψ spectrum, one should of course include both initial production and in-medium formation together as sources. This requires some estimate of the relative magnitudes of these processes, and is subject to considerable model uncertainties. What we can say, however, is that in-medium formation will be most dominant for central collisions, where the quadratic dependence on $N_{c\bar{c}}$ is enhanced. Conversely, one expects that initial production will increase in relative importance for very peripheral collisions. To get an approximate idea of how this effect will appear, we revert to our original model calculations which included the absolute magnitude results [7]. The relevant parameter is the number of initial pairs, $N_{c\bar{c}}$, parameterized by its value at zero impact parameter. These results are shown in Fig. 3 for two representative values of $N_{c\bar{c}}$ = 10 and 20. The anticipated dominance of each at opposite ends of the centrality scale is seen to be realized. In addition, since the centrality dependence of each separate contribution is in the same direction, the dependence of the total $\langle {p_T}^2\rangle$ is somewhat more flat than either of them separately.

Finally, we include a graphical representation of the evolution of $\langle {p_T}^2\rangle$ as a function of system size in Fig. 4. For simplicity of presentation, we revert back to the minimum bias case. In the absence of in-medium formation, the behavior of $\langle {p_T}^2\rangle$ is monatonic increasing with system size. Including in-medium formation as a significant component for minimum bias collisions leads to a reduced $\langle {p_T}^2\rangle$ in Au-Au interactions. The numerical results indicate that this reduction is such that the predicted Au-Au value is even below that measured in d-Au, i.e. a non-monatonic behavior. Hence we claim that observation of such a non-monatonic behavior can be taken as a signal of in-medium formation, while the absolute value of such a behavior will be correlated with the specific

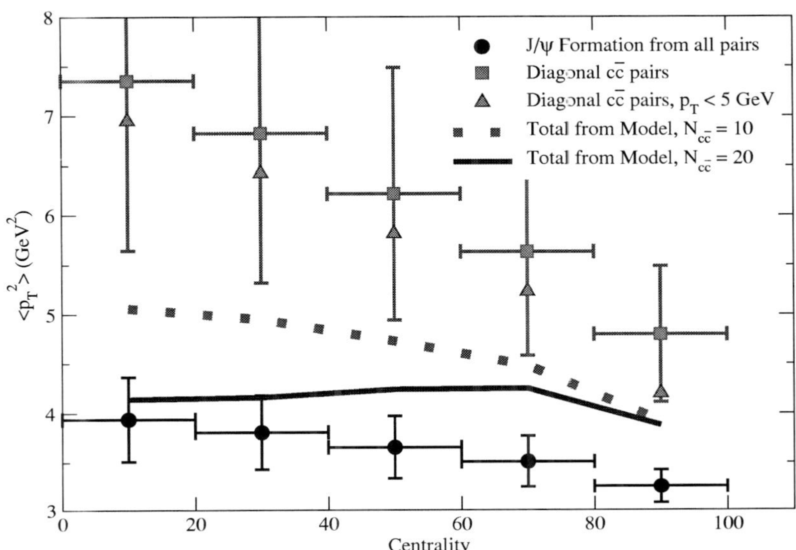

FIGURE 3. Centrality dependence of $J/\psi\langle p_T{}^2\rangle$ using a model calculation to estimate the relative contributions of initial production and in-medium formation.

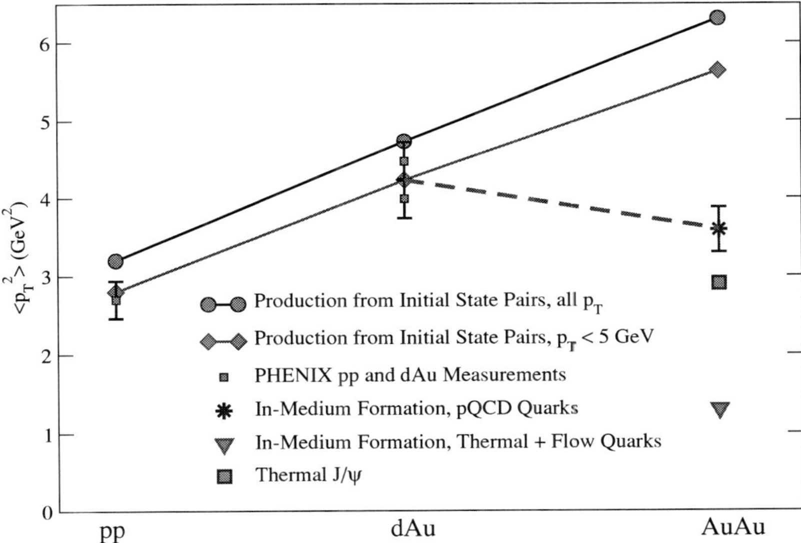

FIGURE 4. Comparison of initially-produced $J/\psi\langle p_T{}^2\rangle$ with in-medium formation predictions for minimum bias collisions.

charm quark distribution in the medium.

ACKNOWLEDGMENTS

This research was partially supported by the U.S. Department of Energy under Grant No. DE-FG02-04ER41318.

REFERENCES

1. T. Matsui and H. Satz, *Phys. Lett. B* **178**, 416-422 (1986).
2. P. Braun-Munzinger and J. Stachel, *Phys. Lett. B* **490**, 196-202 (2000).
3. R. L. Thews, M. Schroedter and J. Rafelski, *Phys. Rev. C* **63**, 054905 (2001).
4. R. Gavai, D. Kharzeev, H. Satz, G. A. Schuler, K. Sridhar and R. Vogt, *Int. J. Mod. Phys. A* **10**, 3043-3070 (1995).
5. S. S. Adler *et al.* [PHENIX Collaboration], `arXiv:nucl-ex/0409028`.
6. J. Adams *et al.* [STAR Collaboration], `arXiv:nucl-ex/0407006`.
7. R. L. Thews, *Nucl. Phys. A* **702**, 341-345 (2002).
8. S. S. Adler *et al.* [PHENIX Collaboration], *Phys. Rev. C* **69**, 014901 (2004).
9. R. L. Thews, *J. Phys. G* **30**, S369-S373 (2004).
10. R. L. Thews and M. L. Mangano, `arXiv:nucl-th/0505055`.
11. S. S. Adler *et al.* [PHENIX Collaboration], `arXiv:nucl-ex/0507032`.
12. R. G. de Cassagnac [PHENIX Collaboration], *J. Phys. G* **30**, S1341-S1346 (2004).
13. M. E. Peskin, *Nucl. Phys. B* **156**, 365-390 (1979).
14. G. Bhanot and M. E. Peskin, *Nucl. Phys. B* **156**, 391-416 (1979).
15. S. Batsouli, S. Kelly, M. Gyulassy and J. L. Nagle, *Phys. Lett. B* **557**, 26-32 (2003).
16. For the corresponding calculation in a coalescence model, see V. Greco, C. M. Ko and R. Rapp, *Phys. Lett. B* **595**, 202-208 (2004).

ϕ Production in Proton-Nucleus and Indium-Indium Collisions at the CERN SPS

M. Floris*, R. Arnaldi†, R. Averbeck**, K. Banicz‡,§, J. Castor¶, B. Chaurand||, C. Cicalo*, A. Colla†, P. Cortese†, S. Damjanovic§, A. David‡,††, A. De Falco*, A. Devaux¶, A. Drees**, L. Ducroux‡‡, H. En'yo§§, A. Ferretti†, P. Force¶, N. Guettet‡,¶, A. Guichard‡‡, H. Gulkanian¶¶, J. Heuser§§, M. Keil‡,††, L. Kluberg‡,||, J. Lozano††, C. Lourenço‡, F. Manso¶, A. Masoni*, P. Martins‡,††, A. Neves††, H. Ohnishi§§, C. Oppedisano†, P. Parracho‡, P. Pillot‡‡, G. Puddu*, E. Radermacher‡, P. Ramalhete‡, P. Rosinsky‡, E. Scomparin†, J. Seixas‡,††, S. Serci*, R. Shahoyan‡,††, P. Sonderegger††, H.J. Specht§, R. Tieulent‡‡, G. Usai*, R. Veenhof‡,†† and H.K. Wöhri‡,††

*Univ. di Cagliari and INFN, Cagliari, Italy
†Univ. di Torino and INFN, Italy
**SUNY Stony Brook, New York, USA
‡CERN, Geneva, Switzerland
§Univ. Heidelberg, Heidelberg, Germany
¶LPC, Univ. Blaise Pascal and CNRS-IN2P3, Clermont-Ferrand, France
||LLR, Ecole Polytechnique and CNRS-IN2P3, Palaiseau, France
††IST-CFTP, Lisbon, Portugal
‡‡IPN-Lyon, Univ. Claude Bernard Lyon-I and CNRS-IN2P3, Lyon, France
§§RIKEN, Wako, Saitama, Japan
¶¶YerPhI, Yerevan, Armenia

Abstract. The quality of the dimuon measurements made by NA60, in proton-nucleus and heavy-ion collisions, is much better than that reached by previous experiments, such as NA38 and NA50. The most important improvement is due to the use of a radiation-tolerant silicon vertex telescope, placed immediately downstream of the target. This allows NA60 to do a high quality measurement of ϕ meson yields and p_T distributions.

This paper presents results obtained in p-Be, p-In and p-Pb collisions at 400 GeV, from data collected in 2002, and in In-In collisions at 158 AGeV, as a function of centrality, from the 2003 running period. In particular, we show that the inverse m_T slope measured in In-In collisions, in the $\phi \to \mu\mu$ decay channel, increases with the number of nucleons participating in the collisions, rather than following a flat trend as seen in the NA50 data collected in the same decay channel but restricted to high p_T values. We also show that our measurements seem to agree with the values previously measured by NA49, using $\phi \to$ KK decays, in Pb-Pb and other collision systems.

Keywords: Heavy ion collisions, proton-nucleus collisions, ϕ meson, dimuon
PACS: 25.75.-q,25.75.Dw,14.40.Cs

NA60 is a fixed-target experiment devoted to the study of dimuon production in proton-nucleus (p-A) and nucleus-nucleus collisions at the CERN SPS. Its apparatus is composed of 4 main detectors: a muon spectrometer, a zero degree calorimeter (ZDC), a beam tracker and a vertex detector. A detailed description of the apparatus can be found

CP828, *Multiparticle Dynamics*
edited by V. Šimák, M. Šumbera, Š. Todorova, and B. Tomášik

in Ref. [1, 2]. Here we only briefly mention the detector concept.

The muon spectrometer is placed after a hadron absorber, which stops most hadrons before they can reach trigger hodoscopes and tracking chambers. The muon spectrometer also provides the main trigger to the experiment, the "dimuon trigger". To make sure that only muons can trigger the experiment, the hadron absorber is complemented by a 1.2-meter-long iron wall, placed before the last trigger station at the end of the muon spectrometer. The absorber represents the main limiting factor in the spectrometer, because of fluctuations of energy loss and multiple scattering which result in a degraded resolution of dimuon mass and of the coordinates of the interaction vertex.

The vertex detector reconstructs all charged tracks before the hadron absorber. In order to identify the muons among these, the tracks reconstructed in the muon spectrometer are extrapolated back to the vertex region and matched to the tracks reconstructed in the vertex detector. This matching is done both in coordinate and in momentum space. Once identified, the muons are refitted using the joint information of the muon spectrometer and of the vertex detector. We shall refer to these tracks as "matched muons". This technique allows to overcome the limitation due to the hadron absorber and thus results in much improved mass resolution and vertexing capability with respect to previous dimuon experiments. The mass resolution goes from around 80 MeV to 23 MeV at the ϕ, when using the information from the vertex detector. The resolution on the determination of the vertex position is $\sim 20\ \mu$m in the transverse coordinates and better than 200 μm for the longitudinal coordinate. Furthermore, the dipole field in the target region significantly increases the acceptance of low p_T and low mass dimuons (Fig. 1).

In this work, we present results on ϕ meson production in heavy ion collisions. These studies are motivated by the fact that the ϕ meson carries information about strangeness production [3]. The yield and p_T spectrum of this meson have been studied in Pb-Pb collisions at 158 AGeV incident beam energy by the NA49 (in the $\phi \to KK$ channel [4]) and the NA50 (in the $\phi \to \mu\mu$ channel [5]) experiments. They both estimated the inverse slope parameter T fitting p_T spectra with an exponential function. The T values found were in strong disagreement, both in what concerns the absolute value and the centrality dependence. NA50 values were significantly lower than NA49 ones and showed no dependence on centrality. NA49 values, on the other hand, were shown to rise as a function of the number of participants. This discrepancy is also known as the "ϕ puzzle" [6].

In this paper we report new measurements in the muon channel done by the NA60 collaboration in p-A collisions (400 GeV incident beam energy) and In-In collisions (158 AGeV incident beam energy), which can help understanding the ϕ puzzle. NA60, in fact, can measure the $\phi \to \mu\mu$ channel with very good p_T coverage and the new collision system adds further information to the general systematics. NA60 can also have access to the $\phi \to KK$ channel, by means of charged tracks reconstructed in the vertex telescope. In this work we only report on results in the dimuon channel, as the $\phi \to KK$ studies are still at a preliminary level. At present, the measurement was shown to be feasible with Monte Carlo simulations, but the background subtraction in the real data still needs to be tuned.

To extract the information on ϕ production we used spectra of matched muons. These are affected by two sources of background: the combinatorial background and the fake matches. The former is the contribution of uncorrelated muon pairs coming from the

Table 1. Centrality bins.

bins	$dN_{ch}/d\eta$ range	$\langle dN_{ch}/d\eta \rangle$	$\langle N_{\text{part}} \rangle$
1	4 – 28	16	~ 20
2	28 – 92	70	91
3	92 – 160	145	161
4	> 160	200	197

decay of pions and kaons. The latter comes from the fact that the matching procedure can fail and a muon can be associated to a wrong track in the vertex telescope. When this happens, the kinematics of the matched muon is highly degraded. The combinatorial background is subtracted with an event mixing technique [7]. The fake matches contribution is estimated via simulation, reconstructing a Monte Carlo dimuon on top of a real event ("overlay Monte Carlo"). This contribution can be alternatively estimated with an event mixing tecnique [7]. The two methods agree within 5%.

In order to extract particle yields this clean sample is fitted with the expected sources. These are (two body and dalitz) leptonic decays of low mass mesons and an underlying continuum mostly due to the simultaneous semi-muonic decay of two D mesons.

The decays included in the expected sources are: $\eta \to \mu\mu$, $\omega \to \mu\mu$, $\phi \to \mu\mu$, $\rho \to \mu\mu$, $\omega \to \mu\mu\pi^0$, $\eta \to \mu\mu\gamma$, $\eta' \to \mu\mu\gamma$.

The p-A data were collected in 2002 with a 400 GeV proton beam. Six targets were installed in the experiment: In, Pb and four Be disks, 2 mm thick each. The statistics in the sample is rather low ($\sim$ 15000 events after background subtraction and event selection). It was collected in only four days of data taking. The fit of the mass spectra with the expected sources allowed to extract the ϕ/ω cross section ratio. The results we got are 0.062 ± 0.004 for p-Be, 0.083 ± 0.007 for p-In and 0.081 ± 0.006 for p-Pb. The quoted errors are purely statistical. The p-Be result, in particular, is in agreement with a previous measurement by HELIOS-1 [8]. As can be seen from these results, ϕ production is already enhanced between p-Be and p-Pb collisions.

The indium-indium data were collected with a 158 AGeV ion beam in a 5-week-long run in 2003. In this paper we present results based on 50% of the very high collected statistics ($\sim$ 570000 events after background subtraction). The data were divided in four centrality bins as summarized in Table 1. The bins were selected using the number of charged tracks reconstructed in the vertex telescope. The corresponding number of participants was estimated with a Glauber fit to the ZDC energy spectrum. We studied the ϕ/ω cross section ratio and the p_T spectrum of the ϕ.

To extract the cross section ratios, the mass spectra were fitted independently in the four centrality bins. The parameters allowed to vary were η/ω, ρ/ω, ϕ/ω and the normalization of the continuum. Fitted spectra were arbitrarily normalized to the ω peak, in order to be compared to the data.

Figure 1 shows the acceptance of the experiment in the low mass and low p_T region. In order to check if this complex acceptance is under control and if, in general, the apparatus is well understood, the first step in the analysis was a careful study of the most peripheral bin. This was analyzed in three p_T windows. The η/ω and the ϕ/ω ratios were found to be nearly p_T independent. The value of the η/ω ratio is in good agreement

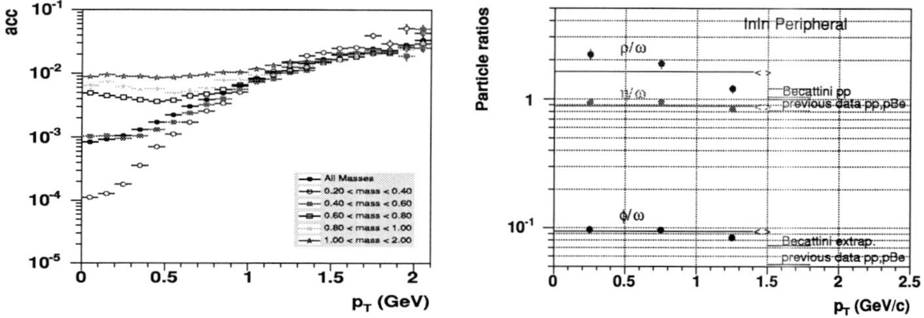

Figure 1. Dimuon acceptance as a function of p_{T} in several mass windows (left) and particle ratios in the most peripheral bin as a function of p_{T} (right).

with previous pp and p-Be measurements [9]. The ϕ/ω ratio is higher than in pp, as expected in nucleus-nucleus collisions. In general the peripheral bin is well described in terms of expected sources, indicating that the acceptances in the low mass and low p_{T} region are under control. The ρ/ω ratio, however, shows that there are "too many" ρ mesons at low p_{T}. This "excess" can be interpreted as the effect of pion annihilation in nuclear collisions. Peripheral In-In collisions, thus, are similar to C-C or O-O and not just pp-like. It can be shown that at high p_{T} ($\gtrsim$ 1 GeV) this contribution becomes negligible, as one recovers the expected ρ/ω ratio of 1.2. The effect of pion annihilation becomes dramatic in more central collisions, but NA60 can still extract a robust ω yield, thanks to its excellent mass resolution. For a detailed discussion of the modification of the ρ meson in In-In collisions see Ref. [10]. To avoid this problematic region we decided to restrict our analysis of the particle ratios to the region $\mathrm{p_T} > 1$ GeV, until the continuum below the ω is fully under control.

The ϕ/ω ratio was studied as a function of centrality and compared to previous measurements. We observe a factor 2 increase from peripheral to central collisions. Figure 2 shows the ratio compared to the $\phi/(\rho+\omega)$ ratio measured by NA50 in Pb-Pb collisions [5]. The NA50 points have a common m_{T} cut ($m_{\mathrm{T}} \geq 1.5$ GeV), so we corrected them to a common p_{T} range, assuming the T parameter measured by NA50 itself (228 MeV). The $\phi/(\rho+\omega)$ was finally converted to ϕ/ω assuming $\sigma_{\rho} = 1.2\ \sigma_{\omega}$. The trend as a function of centrality is the same in the two experiments, albeit the absolute value of NA60 points is lower. A direct comparison, however, is impossible due to the contribution of pion annihilation discussed above: this must be even higher in Pb-Pb collisions and NA50 cannot isolate it. Figure 2 shows the comparison to the ratio ϕ/π measured by the NA49 experiment. The dependence on centrality is the same also in this case, indicating that the ω/π ratio is constant, as expected in statistical models. If we use the value ω/π suggested by such models (0.07-0.08) we find that the ϕ yield in NA60 is higher than that measured by NA49.

To study the p_{T} spectrum of the ϕ we selected events in the mass window $0.98 < M_{\mu\mu} < 1.06$. The p_{T} spectrum of the continuum below the ϕ was estimated and subtracted selecting events in two side mass windows ($0.88 < M_{\mu\mu} < 0.92$ and $1.12 <$

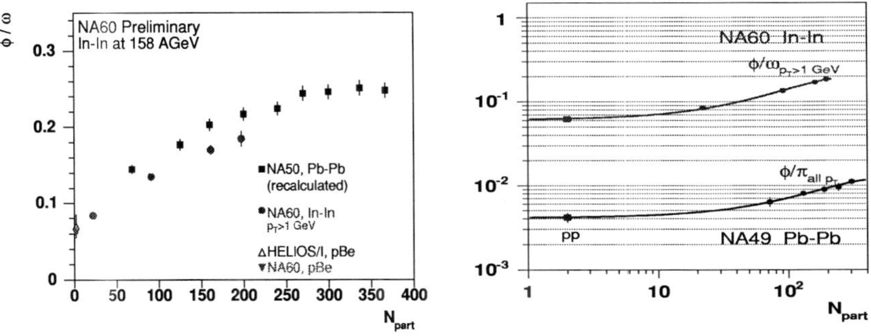

Figure 2. ϕ/ω cross section ratio as a function of centrality compared to the $\phi/(\rho+\omega)$ of NA50 (left) and to the ϕ/π ratio of NA49 (right). See text for a discussion on how NA50 points were corrected. Statistical errors only.

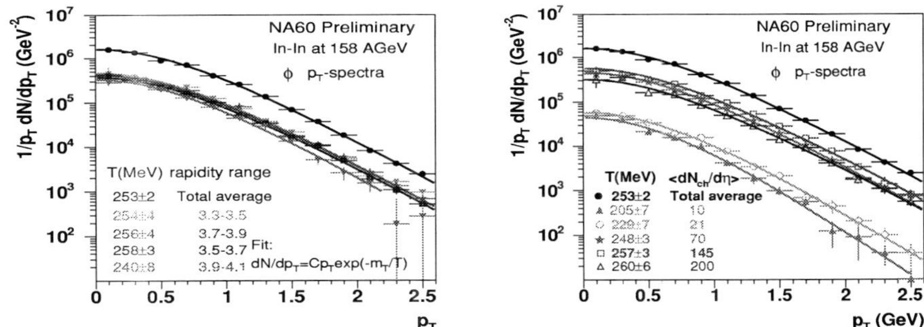

Figure 3. ϕ transverse momentum distribution in indium-indium collisions; as a function of rapidity (left) and centrality (right). Statistical errors only.

$M_{\mu\mu} < 1.16$). The p_T spectrum was then corrected for acceptance, estimated by Monte Carlo, as a 2-dimensional matrix in p_T and rapidity. The resulting distributions were studied as a function of rapidity and as a function of centrality. The study as a function of centrality was done in five bins, splitting the most peripheral one in 2 sub-bins. The spectra were fitted with an exponential function:

$$\frac{1}{\mathrm{p_T}}\frac{\mathrm{d}N}{\mathrm{dp_T}} = Ce^{-\mathrm{m_T}/T} \tag{1}$$

The results of the fit are reported in Fig. 3.

No dependence of T on rapidity is observed, while a clear increase with centrality is seen. The average value of the T slope parameter is $T = (253 \pm 2)$ MeV. Changing the fit range to the NA49 ($p_T < 1.5$ GeV) or NA50 ($m_T > 1.5$ GeV) windows produces small variations on the result: $T = (260 \pm 5)$ MeV and $T = (244 \pm 5)$ MeV, respectively.

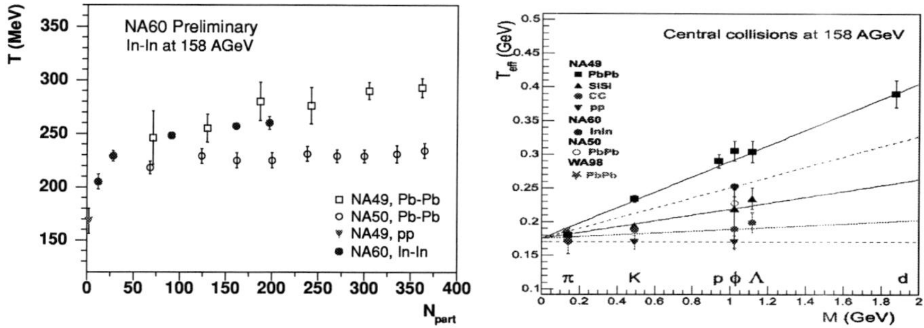

Figure 4. T slope parameter as a function of the number of participants (left) and as a function of particle mass for several collision systems (right). Statistical errors only.

Our results are in agreement with NA49. This can also be seen in Fig. 4, which shows the centrality dependence of the T slope parameter, compared to the analogous result obtained by NA49 and NA50. Our average value, moreover, fits very well in the NA49 systematics [11], the In-In point lying in between the Pb-Pb and the Si-Si points (Fig. 4). NA50 Pb-Pb result, on the other hand, is quite close to NA49 Si-Si.

We can therefore conclude that the disagreement between NA50 and NA49 was not due to the different decay channels probed.

In this paper NA60 results on ϕ production in p-A collisions at 400 GeV incident beam energy and In-In collisions at 158 GeV incident beam energy were presented. The ϕ/ω cross section ratio measured by NA60 in p-A collisions is in agreement with previous measurements. The trend with centrality in nuclear collisions agrees with previous measurements of the ϕ/π ratio by NA49 (assuming a constant ω/π) and of the $\phi/(\rho+\omega)$ ratio by NA50 (properly corrected). The absolute yield, however, seems to lie in between these two. The p_{T} spectrum of the ϕ was fitted with an exponential function and studied as a function of centrality and rapidity, giving results which agree with the previous measurement of the NA49 experiment.

REFERENCES

1. A. Baldit, et al., *CERN-SPSC-2000-010* (2000).
2. M. Floris, et al., *IEEE Trans. Nucl. Sci.* **51** (2004).
3. J. Rafelski, and B. Muller, *Phys. Rev. Lett.* **48**, 1066 (1982).
4. S. V. Afanasev, et al., *Phys. Lett. B* **491**, 59–66 (2000).
5. B. Alessandro, et al., *Phys. Lett. B* **555**, 147–155 (2003).
6. E. V. Shuryak, *Nucl. Phys. A* **661**, 119–129 (1999), `hep-ph/9906443`.
7. R. Shahoyan, et al., "Charm and intermediate mass dimuons in In+In collisions," in *Proc. Quark Matter 2005*, Budapest, 2005, in print.
8. R. J. Veenhof, Ph.D. thesis, Universiteit van Amsterdam (1993), rx-1433 (Amsterdam).
9. G. Agakishiev, et al., *Eur. Phys. J. C* **4**, 231–247 (1998).
10. S. Damjanovic, et al., "Low mass dimuon production in Indium-Indium collisions at 158 AGeV (NA60)," in *Proc. Quark Matter 2005*, Budapest, 2005, in print.
11. I. Kraus, et al., *J. Phys. G* **30**, S583–S588 (2004), `nucl-ex/0306022`.

Measurement of Identified Particle Production at RHIC

An Tai, for the STAR Collaboration

Department of Physics, University of California at Los Angeles, CA 90095, USA

Abstract. Identified particle measurements in 200 GeV Au+Au and d+Au collisions in STAR are presented. The importance of using identified particles to probe properties of medium formed at RHIC is emphasized. Magnitude of the nuclear modification factor is observed to depend on particle type (baryon vs meson) instead of particle mass at intermediate transverse momentum. This observation supports recombination/coalescence processes as dominant dynamics for particle production in this transverse momentum region. We argue that ϕ and heavy quark yields are sensitive to the medium properties at the early stage. The significant suppression of identified particle production observed at high transverse momentum in central Au+Au collisions indicates the formation of strongly-interacting medium at RHIC.

Keywords: nuclear modification factor, recombination model, jet energy loss, heavy-flavor
PACS: 25.75 Dw,14.40.Lb

INTRODUCTION

The central goal of relativistic heavy-ion collisions is to study matter properties at high temperature and high energy density. A new state of matter, the Quark-Gluon Plasma (QGP), with deconfined quarks and gluons as essential degrees of freedom is believed to be produced at the early stage of collisions. In order to investigate whether the partonic medium is actually formed, one needs to rely on probes which are sensitive to the early stage of the colliding system. Different species of identified particles are sensitive to different stage of the colliding system.

Most hadrons are formed at chemical freeze-out. Common hadrons like pions, kaons and protons continue to interact with each other and evolve until kinetic freeze-out, so they carry the information that involves the whole history of the evolution of the system and the interesting early information of the system is washed out by the final state hadronic interactions. However, some formed hadrons, like ϕ and heavy quark mesons, are believed to be sensitive to the early stage.

What is measured experimentally are colorless hadrons formed through hadronization processes of partons though we are actually interested in discovering medium with partonic degree of freedom. Therefore, it is important to understand how a hadron is formed at RHIC. In elementary collisions (e^+e^-, ep, pp etc.), the most popular hadronization scheme is parton fragmentation processes such as that implemented in Lund model [1], for example. The fragmentation functions are usually measured in e^+e^- collisions and used in other collision systems assuming universality of fragmentation functions [2]. This approach is found to work quite well reproducing measured meson spectra, however, fails to describe baryon production as shown in Fig. 1 when comparing Next-To-

CP828, *Multiparticle Dynamics*
edited by V. Šimák, M. Šumbera, Š. Todorova, and B. Tomášik

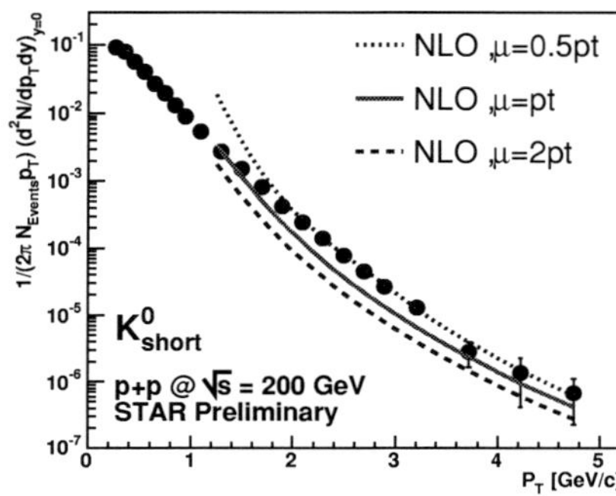

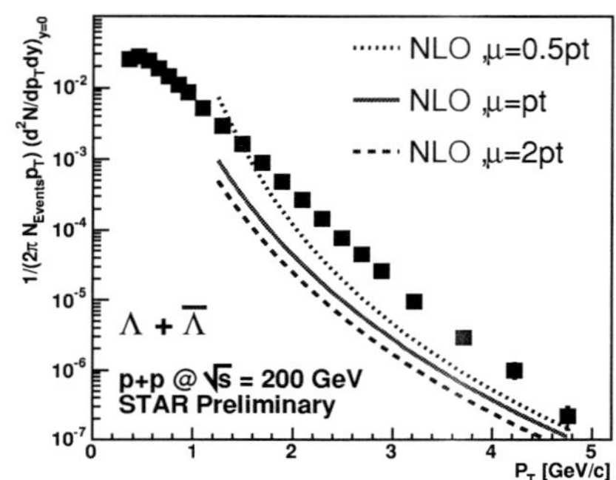

FIGURE 1. STAR measured Λ and K_s^0 spectra in 200 GeV pp collisions compared with NLO pQCD calculations with fragmentation functions from [3]

Leading-Order (NLO) pQCD predictions with STAR measured spectra for Λ and K_s^0 in 200 GeV pp collisions. Another hadronization scheme was also proposed in which hadrons are formed through recombination/coalescense of produced partons [4].

HADRON PRODUCTION IN AU+AU COLLISIONS AT RHIC

One of the most intriguing features observed at RHIC in Au+Au collisions is the enhanced production of baryons over mesons. The left panel of Fig. 2 shows STAR measured Λ/K_s^0 ratios for pp collisions and Au+Au collisions of different centralities. The ratios increase with increasing centrality at intermediate transverse momentum, $1.0<p_T<5.0$ GeV/c. This feature can be easily understood in a recombination/coalescense model. In a heavy-ion collision, the parton density at low p_T is high, therefore it is much more efficient to form a hadron into the intermediate p_T region by combining quarks than through fragmentation of a high p_T parton which has a much smaller production cross section [5]. Because a baryon contains one more constitute quark than a meson, baryon production would be enhanced over meson production at intermediate p_T.

In order to investigate nuclear effects of hadron production, nuclear modification factor, R_{cp}, is introduced, which is defined by

$$R_{cp}(p_T) = \frac{N_{coll}^{peripheral}}{N_{coll}^{central}} \frac{dn_{central}/dp_T dy}{dn_{peripheral}/dp_T dy}, \tag{1}$$

where $N_{coll}^{peripheral}$ and $N_{coll}^{central}$ are number of binary nucleon-nucleon collisions for peripheral and central Au+Au collisions, respectively. Another quantity, R_{AA}, is also used in literature, which is obtained by replacing $N_{coll}^{peripheral}$ and $dn_{peripheral}/dp_T dy$ in Eq. (1) by one and the yield of pp collisions, respectively. R_{cp} for different particle species are shown in the right panel of Fig. 2. At intermediate p_T, magnitude of R_{cp} is different

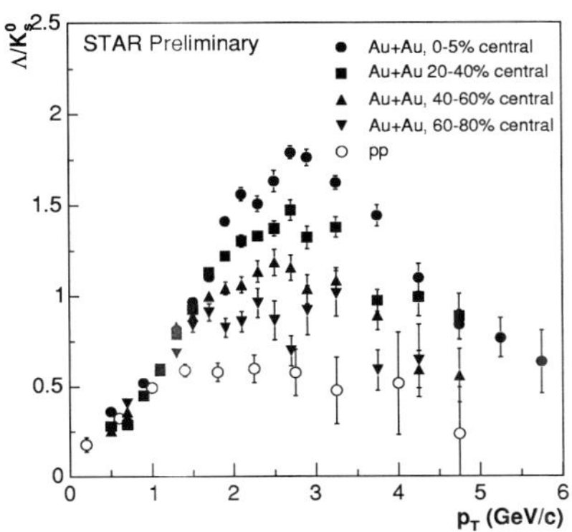

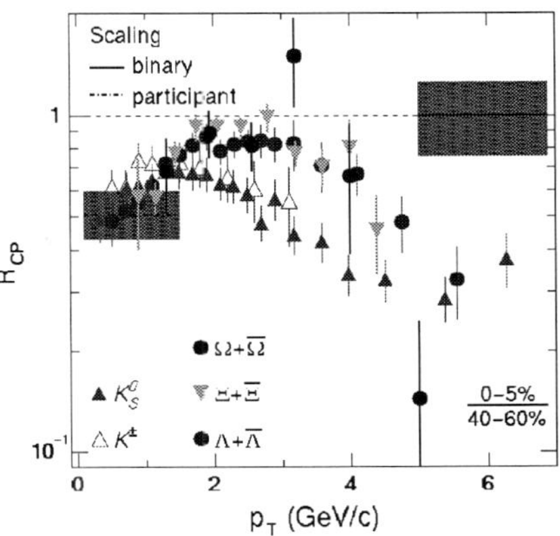

FIGURE 2. STAR measured Λ/K_s^0 ratios in 200 GeV pp and Au+Au collisions (left) and R_{cp} for strange hadrons in 200 GeV Au+Au collisions (right)

for baryons (Λ, Ξ and Ω) and mesons ($K^{\pm}$ and K_s^0) while at high p_T, R_{cp} is smaller than one, indicating a suppression of particle production in central Au+Au collisions with respect to number of binary collision scaling. Parton energy loss in QGP medium (jet quenching) was proposed to be responsible for the suppression of high p_T hadron production [6].

Is the separation of R_{cp} between baryons and mesons due to the particle mass or particle type ? How do we test the jet energy loss scenario ? Could the high p_T hadron suppression result from modification of parton distribution functions (PDF) of nucleons in a Au nucleus [7] ? These are the questions that we are going to address in the following sections.

HADRON PRODUCTION IN D+AU COLLISIONS AT 200 GEV

d+Au collisions set a critical reference to understand the observed nuclear effects in Au+Au collisions. For example, if the high p_T hadron suppression is due to the modification of parton distribution functions, this effect should also be seen in d+Au collisions. On the other hand, if the energy loss in the final state is responsible for the high p_T suppression, it will not be observed in d+Au collisions because small transverse dimension of the d+Au system would not cause significant energy loss for partons traversing the medium .

The spectra of K_s^0, ϕ, Λ and Ξ measured in 200 GeV d+Au collisions are shown in Fig. 3, plotted as a function of $m_T - mass$ ($m_T \equiv \sqrt{p_T^2 + mass^2}$). The spectra are fit with a double exponential function reflecting contributions of hard and soft processes in particle production.

R_{cp} of these particles in d+Au collisions are plotted in the left panel of Fig. 4. Unlike Au+Au collisions, suppression of hadron production at high p_T is not observed in d+Au

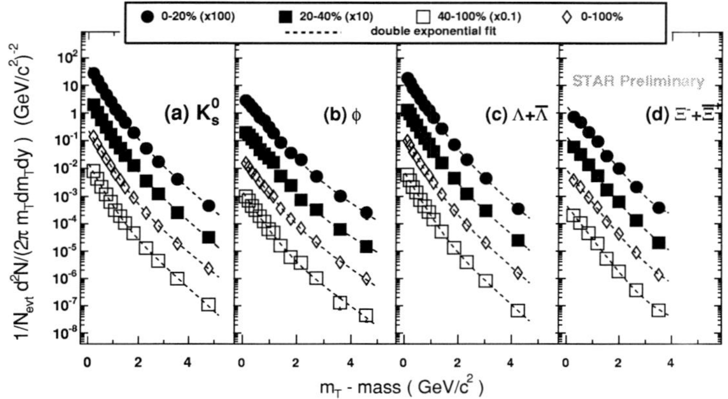

FIGURE 3. Spectra of K_s^0, ϕ, Λ and Ξ measured in 200 GeV d+Au collisions

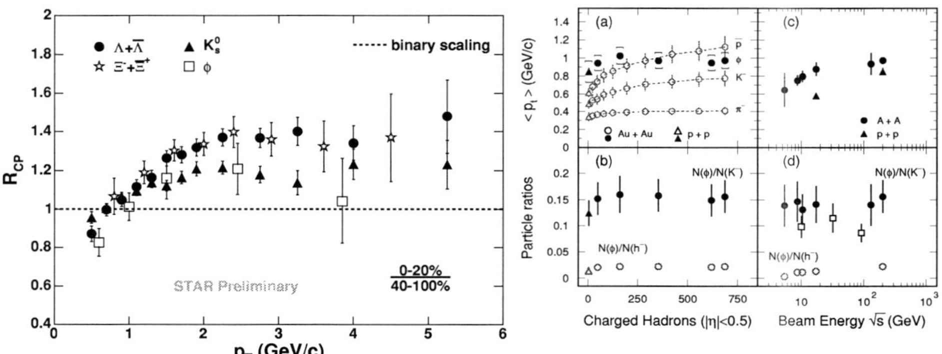

FIGURE 4. Left: R_{cp} of K_s^0, ϕ, Λ and Ξ in 200 GeV d+Au collisions. Right: mean p_T of ϕ plotted as a function of charged hadrons comparing with mean p_T of negative pions, kaons and anti-protons (a); ratios of ϕ yields to yields of negative kaons (filled symbols) and negatively charged hadrons (open symbols) (b); ϕ mean p_T vs center of mass beam energy for central nucleus-nucleus and pp collisions (c); ratios of ϕ yields to yields of negative kaons (filled symbols) and negatively charged hadrons (open symbols) in central nucleus-nucleus collisions vs center of mass beam energy; data from e^+e^- (open squares) are also shown (d).

collisions. Instead, R_{cp} in d+Au collisions is higher than number of binary collision scaling for p_T >1.0 GeV/c, which is conventionally called as Cronin effect [8]. This observation demonstrates that high p_T suppression in Au+Au collisions is a final state effect. In addition, it is worth pointing out that the results in the left panel of Fig. 4 also show the separation of baryon and meson R_{cp}, similar to that in Au+Au collisions.

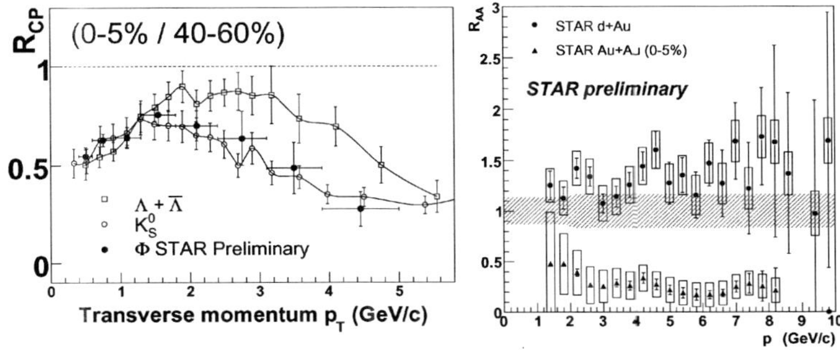

FIGURE 5. R_{cp} of ϕ comparing to Λ and K_s^0 in 200 GeV Au+Au collisions (left) and R_{AA} of single electrons for d+Au and central Au+Au collisions (right)

ϕ MESON AND HEAVY FLAVOR PRODUCTION

The ϕ meson is an important probe of particle production dynamics in heavy-ion collisions. ϕ is a meson and has a mass of 1019 MeV/c^2, close to a Λ baryon with a mass of 1116 MeV/c^2. Therefore, ϕ can be used to test whether R_{cp} depends on particle mass or particle type. In addition, it is believed that ϕ production is insensitive to the later hadronic interactions. The right panel of Fig. 4 is a compilation of ϕ data from various experiments [9] in which ϕ/K^-, ϕ/h^- ratios and mean p_T of ϕ from different collision systems are studied as a function of centralities characterized by charged hadron multiplicity $(dn/d\eta)$ and beam energy. $< p_T >$ of ϕ shows little dependence on centrality in contrast to antiproton whose $< p_T >$ increases with centrality reflecting an increase of transverse flow in Au+Au collisions due to final state interactions of antiprotons with medium. A ϕ meson is heavier than an antiproton. If ϕ mesons flow with the same velocity as antiprotons, $< p_T >$ of ϕ would increase even faster than that of antiprotons. The observation indicates that ϕ should have a small hadronic cross section with medium. Furthermore, ϕ/K^- ratios show no dependence on centrality, indicating that production of ϕ through coalescence of K^+K^- pairs is not a dominant channel. Otherwise an increase of this ratio would be expected due to increasing kaon density as a function of centrality. We then conclude that ϕ should primarily be produced through $ggg\rightarrow\phi$ and $s\bar{s}\rightarrow\phi$. These features make ϕ production a valuable probe of medium properties at the prehadronic stage.

R_{cp} of ϕ is plotted in the left panel of Fig. 5 in comparison with R_{cp} of K_s^0 and Λ. It clearly shows that R_{cp} of a hadron at intermediate p_T depends on whether the hadron is a baryon or a meson, but not on its mass. This observation supports the recombination/coalescence model where the particle production yield is determined by number of constituent quarks in a hadron, not by its mass. In addition, since ϕ has a small hadronic cross section, it may also indicate that the final state interactions which cause the suppression of ϕ production at high p_T occur in the partonic stage.

Measurements of heavy flavor production at RHIC take us a step further forward in our effect of probing matter properties at the early stage of relativistic heavy ion collisions using identified particles since heavy quarks are produced primarily through gluon fusion. The right panel of Fig. 5 shows the STAR measured R_{AA} for single electrons using the EMC detector at 200 GeV. The single electron spectra were obtained after the electrons from photon conversion in the TPC, π^0 Dalitz decay etc were subtracted out (dubbed as non-photonic electrons) and presumably represent electrons from decays of heavy quarks. Contrary to the theoretical expectation that a heavy quark would lose a smaller fraction of its energy in the QGP medium comparing to light quarks due to its heavier mass (the dead-cone effect) [10], it is found that the nuclear modification factor of non-photonic electrons is suppressed as much as that for pions. The elliptic flow of single electrons are also measured by STAR, which provides us another important tool to study properties of produced medium. These exciting results are stimulating a lot of discussions [11].

In summary, STAR is carrying on a vigorous program to investigate properties of medium with probes sensitive to the early stage information of nucleus-nucleus collisoins. STAR data reveal that a dense and strongly interacting medium is formed at the early stage of nucleus-nucleus collisions at RHIC, whose properties can not be understood based on hadronic interaction models where partons are still confined in hadrons.

ACKNOWLEDGMENTS

I would like to thank conference organizers for all their efforts making this meeting very fruitful and pleasant.

REFERENCES

1. B. Andersson *et al.*, *Phys. Rept.* **97**, 31-145 (1983).
2. B. A. Kniehl, G. Kramer and B. Pötter, *Nucl. Phys. B* **582**, 514-536 (2000); D. de Florian and W. Vogelsang *Phys. Rev. D* **71**, 114004 (2005).
3. B. A. Kniehl, G. Kramer and B. Pötter, *Nucl. Phys. B* **597**, 337-369 (2001); D. de Florian, M. Stratmann and W. Vogelsang *Phys. Rev. D* **57**, 5811-5824 (1998).
4. R. C. Hwa, *Phys. Rev. D* **22**, 1593-1608 (1980).
5. R. J. Fries *et al.*, *Phys. Rev. C* **68**, 044902 (2003).
6. X. N. Wang, *Phys. Rev. C* **61**, 064910 (2000).
7. D. Kharzeev and M . Nardi , *Phys. Lett. B* **507**, 121-128 (2001).
8. J. W. Cronin *et al.*, *Phys. Rev. D* **11**, 3105-3123 (1975).
9. J. Adams *et al.* STAR Collaboration *Phys. Lett. B* **612**, 181-189 (2005).
10. Y. L. Dokshitzer and D. E. Kharzeev, *Phys. Lett. B* **519**, 199-206 (2001).
11. X. Dong, *these proceedings*.

NA49 Results on Hadron Production: Indications of the Onset of Deconfinement?

Benjamin Lungwitz for the NA49 collaboration

Institut für Kernphysik, Johann-Wolfgang Goethe Universität Frankfurt, Max-von-Laue-Str. 1, 60438 Frankfurt am Main, Germany

Abstract. The NA49 experiment at the CERN SPS measured the energy and system size dependence of particle production in A+A collisions. A change of the energy dependence of several hadron production properties at low SPS energies is observed which suggests a scenario requiring the onset of deconfinement.

INTRODUCTION

In heavy ion interactions at sufficiently high collision energy the creation of a deconfined state of matter, the quark gluon plasma (QGP), is expected [1, 2]. In order to look for the onset of deconfinement the NA49 experiment at the CERN SPS studied the hadronic final state of Pb+Pb collisions in the energy range 20*A* – 158*A* GeV.

In addition to the energy scan program, NA49 also measured the dependence of hadron production properties on the size of the colliding nuclei and the centrality of collisions.

THE NA49 EXPERIMENT

The NA49 detector system [3] is a large acceptance fixed target hadron spectrometer. Its main devices are four large time projection chambers (TPCs). Two of them, called vertex TPCs, are located in two superconducting dipole magnets. The other two TPCs are installed behind the magnets left and right of the beam line allowing precise particle tracking in the high density region of heavy ion collisions. The accuracy of momentum determination is $\Delta p/p^2 \approx (0.3-7)\cdot 10^{-4}$ (GeV/c)$^{-1}$.

The measurement of the energy loss dE/dx allows identification of pions, protons, kaons and electrons for momenta $p > 4$ GeV/c. The NA49 experiment is equipped with two time of flight walls which, together with the energy loss measurement, allow a good separation of the particle species at midrapidity. Decaying particles can be identified with a good precision via the reconstruction of the decay vertex and / or the invariant mass method.

The downstream veto calorimeter allows a determination of the centrality of a collision by measuring the energy in the projectile spectator region.

A large variety of different colliding systems was studied. Beams of p and Pb are available at the CERN SPS directly, C and Si beams were produced via fragmentation

CP828, *Multiparticle Dynamics*
edited by V. Šimák, M. Šumbera, Š. Todorova, and B. Tomášik

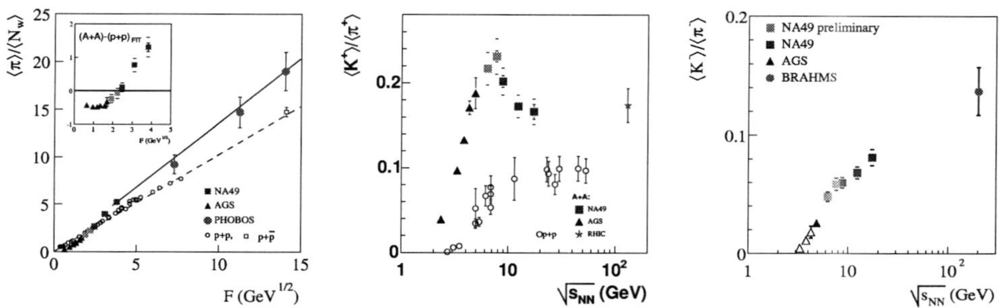

FIGURE 1. Energy dependence of pion multiplicity $\langle\pi\rangle$ to the nubmer of wounded nucleons $\langle N_W\rangle$ [4] (left), of the $\langle K^+\rangle/\langle\pi^+\rangle$ ratio (middle) and the $\langle K^-\rangle/\langle\pi^-\rangle$ ratio [4] (right) in Pb+Pb (Au+Au) (full symbols) and p+p($\bar{\text{p}}$) (open symbols) collisions.

of the primary Pb beam. Targets of liquid hydrogen or solid foils of different materials were used.

In this paper results of the energy scan, which covers Pb+Pb collisions at 20*A*, 30*A*, 40*A*, 80*A* and 158*A* GeV, and results of the system size scan at 40*A* and 158*A* GeV are shown. The data should be considered preliminary unless a reference to the corresponding publication is given.

RESULTS

Pions:

Figure 1 shows the pion yield per wounded nucleon in central Pb+Pb (Au+Au) collisions and p+p interactions as a function of the Fermi variable $F = \left(\sqrt{s_{NN}} - 2m_N\right)^{3/4} / \left(\sqrt{s_{NN}}\right)^{1/4} \approx \sqrt{\sqrt{s_{NN}}}$. The ratio $\langle\pi\rangle/\langle N_W\rangle$ rises linearly with F for p+p interactions. For heavy ion collisions at low energies $\langle\pi\rangle/\langle N_W\rangle$ is smaller than for p+p by a constant amount which may be attributed [5] to pion absorption in the hadronic medium. At low SPS energies the pion multiplicity in Pb+Pb collisions starts to increase faster with energy than in p+p interactions. This enhancement of pion production may indicate an increase of the number of degrees of freedom in the QGP phase and suggests the onset of deconfinement at low SPS energies.

The system size dependence of pion production for various energies is shown in Fig. 2. Both the suppression at low energies and the enhancement at high energies start already in small systems.

Kaons:

At SPS energies s-quarks are carried mainly by Λs and anti-kaons, while most $\bar{\text{s}}$-quarks are carried by kaons. Therefore the K^+ meson multiplicity is approximately equal to half to the total number of $\bar{\text{s}}$ (and therefore also s) quarks produced in the collision. In order to take out the influence of center of mass energy on particle production the strange hadron yields are divided by the number of pions.

The energy dependence of the $\langle K^+\rangle/\langle\pi^+\rangle$ ratio shows (Fig. 1) the famous "horn" which can not be reproduced by hadron gas models assuming $\gamma_S = 1$ or string hadronic

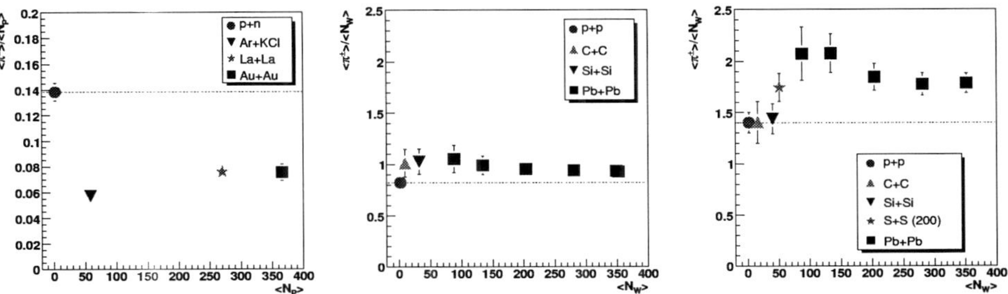

FIGURE 2. System size dependence of pion yield for 2*A* GeV [6, 7] (left), 40*A* GeV [8] (middle) and 158*A* GeV [9, 10] (right).

models. The increase of the $\langle K^- \rangle / \langle \pi^- \rangle$ ratio slows down at low SPS energies. Both features are consistent with a model assuming a first order phase transition [5] in which they are explained by a lower fraction of strange to non-strange degrees of freedom in the QGP than in the hadron gas.

For central collisions the $\langle K^+ \rangle / \langle \pi^+ \rangle$ ratio rises quickly with system size for all energies as shown in Fig. 3. This effect can be attributed to the strong volume dependence of relative strangeness production, in statistical models known as canonical strangeness suppression [9, 12]. The relative kaon production increases with centrality of the collision, the increase is faster and it saturates earlier for higher energies. The number of wounded (or participant) nucleons N_W (or N_P) is not a good scaling parameter. The $\langle K^+ \rangle / \langle \pi^+ \rangle$ ratio for central collisions of small systems is larger, especially at lower energies, than the ratio for peripheral collisions of large systems with the same number of wounded nucleons $\langle N_W \rangle$. This may be due to low collision density in peripheral reactions.

The $\langle K^- \rangle / \langle \pi^- \rangle$ ratio, not shown here, behaves qualitatively similar.

ϕ-mesons:

The energy dependence of the $\langle \phi \rangle / \langle \pi \rangle$ ratio (Fig. 4) shows indications of a non-monotonic behavior at the low SPS energies. The system size dependence shows a

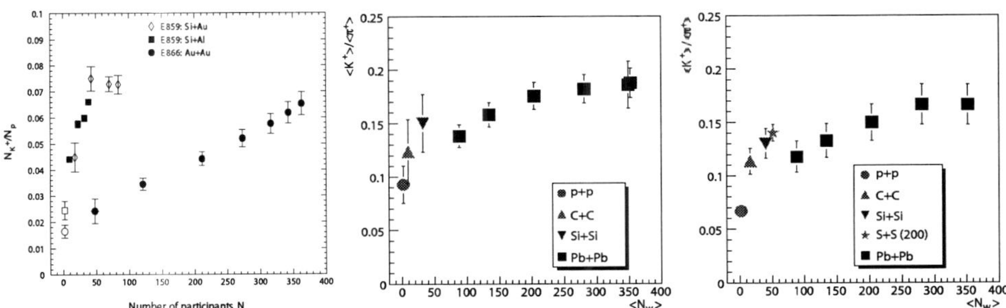

FIGURE 3. System size dependence of the $\langle K^+ \rangle / \langle \pi^+ \rangle$ ratio at 5*A* GeV [11], 40*A* GeV [8] and 158*A* GeV [9, 10].

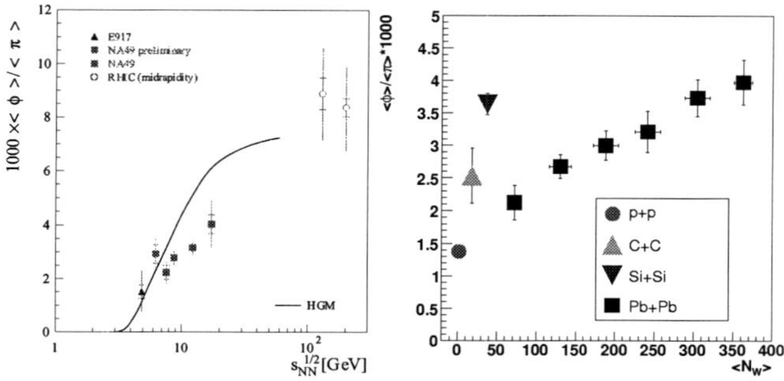

FIGURE 4. Left: Energy dependence of the $\langle\phi\rangle / \langle\pi\rangle$ ratio for central Pb+Pb (Au+Au) collisions [13, 14]. Right: System size dependence of the $\langle\phi\rangle / \langle\pi\rangle$ ratio at 158*A* GeV [9, 15].

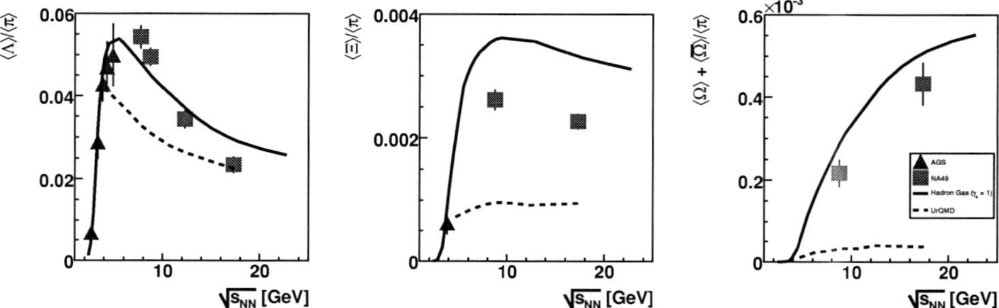

FIGURE 5. Energy dependence of the $\langle\Lambda\rangle / \langle\pi\rangle$ [9, 14], $\langle\Xi\rangle / \langle\pi\rangle$ [14] and $\langle\Omega+\bar{\Omega}\rangle / \langle\pi\rangle$ [17] ratios in central Pb+Pb (Au+Au) collisions.

similar structure as it was observed for the $\langle K^+\rangle / \langle\pi^+\rangle$ ratio. This indicates that ϕ production is sensitive to total strangeness production and favors the picture of the ϕ production via coalescence of a strange and an anti-strange quark. In contrast the statistical hadron gas model predicts no correlation of ϕ production to strange quark density because of its zero net strageness.

Hyperons:

The energy dependence of the $\langle\Lambda\rangle / \langle\pi\rangle$ ratio shows a maximum at high AGS or low SPS energies. The position of that maximum is consistent with the position of the maximum in the $\langle K^+\rangle / \langle\pi^+\rangle$ ratio. The energy dependence of $\langle\Lambda\rangle / \langle\pi\rangle$ can be approximately reproduced by a hadron gas model assuming $\gamma_S = 1$ [16] (see solid curve in Fig. 5).

The $\langle\Xi\rangle / \langle\pi\rangle$ ratio seems to have a maximum at low SPS energies like the $\langle K^+\rangle / \langle\pi^+\rangle$ and the $\langle\Lambda\rangle / \langle\pi\rangle$ ratios. The maximum is absent in the $\langle\Omega\rangle / \langle\pi\rangle$ ratio. The analysis of multi-strange hyperons is in progress and future results might clarify the situation.

Both the $\langle\Xi\rangle / \langle\pi\rangle$ and the $\langle\Omega\rangle / \langle\pi\rangle$ ratios at SPS energies are underestimated by the string hadronic UrQMD model [18] (see curves in Fig. 5).

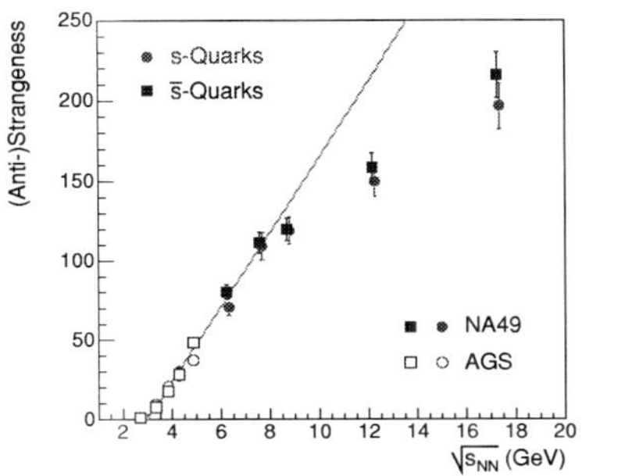

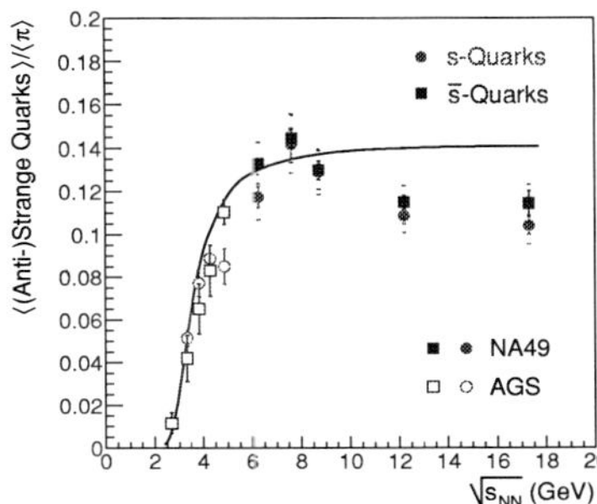

FIGURE 6. Energy dependence of constituent s and s̄ quark multiplicities (left) and their ratios to the pion yield (right).

s and s̄ yield:
The yields of K^-, Λ, Ξ and Ω were used to calculate the total number of constituent s-quarks produced in the reaction. The corresponding antiparticles were used to estimate the s̄-quark yield. The correction for the missing measurements was calculated based on the hadron gas model [19].

Figure 6 shows that both NA49 and AGS data fulfill strangeness conservation ($\langle \mathrm{s} \rangle = \langle \bar{\mathrm{s}} \rangle$). The "horn" which was observed in the $\langle K^+ \rangle / \langle \pi^+ \rangle$ ratio is also seen in the $\langle s \rangle / \langle \pi \rangle$ and $\langle \bar{s} \rangle / \langle \pi \rangle$ ratios. The energy dependence of $\langle \mathrm{s} \rangle$ and $\langle \bar{\mathrm{s}} \rangle$ also shows an anomaly at low SPS energies: the increase of strangeness with energy is getting weaker at the low SPS energies.

Transverse mass spectra:
The transverse mass spectra can be parametrized by the inverse slope parameter T obtained by fitting the function $d^2n/(m_T dy dm_T) = C \cdot \exp(-(m_T)/\mathrm{T})$. Due to flow effects and power law behavior at high m_T the fit range has to be limited to intermediate m_T.

Another characterization of the shape of transverse mass spectra is the mean transverse mass $\langle m_T \rangle$. Its advantage is that it can be used even for non-exponential spectra. Figure 7 shows that the inverse slope parameter of kaons increases in the AGS and RHIC energy domains but it stays constant at SPS energies. This feature, which is not seen in p+p interactions, might be attributed to the latent heat of a phase transition [5] and is in fact consistent with hydrodynamic model calculations assuming a first order phase transition [21].

Similar energy dependence (the "step") is evident for $\langle m_T \rangle - m_0$. It seems to be present for pions, kaons and protons.

CONCLUSION

Several anomalies in the energy dependence of hadron production properties were observed by the NA49 experiment. These are the "kink" for pion multiplicities, the "horn" for the strangeness to pion ratio and the "step" for the mean transverse mass of various

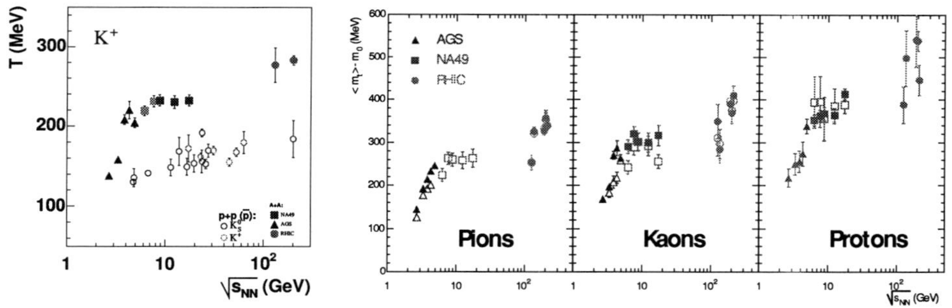

FIGURE 7. Energy dependence of the inverse slope parameter of kaons for central Pb+Pb (Au+Au) and p+p [20] collisions (left) and of the mean transverse mass of various hadrons produced in central Pb+Pb (Au+Au) collisions (right). Positive (negative) particles are represented by full (open) symbols.

hadrons. These observations can not be described by present hadronic models but are consistent with the onset of deconfinement at low SPS energies [5, 22].

The system size dependence of various hadronic observables, e.g. the $\langle K \rangle / \langle \pi \rangle$ ratio, shows early saturation in central collisions of different size nuclei. Peripheral collisions of large nuclei behave quite differently from central collisions of small nuclei with the same number of wounded nucleons. This difference is larger for smaller energies.

REFERENCES

1. F. Karsch, E. Laermann, and A. Peikert, *Nucl. Phys. A* **605**, 579–599 (2001), hep-lat/0012023.
2. Z. Fodor, and S. D. Katz, *JHEP* **04**, 050 (2004), hep-lat/0402006.
3. S. Afanasev, et al., *Nucl. Instrum. Meth. A* **430**, 210–244 (1999).
4. M. Gazdzicki, et al., *J. Phys. G* **30**, S701–S708 (2004), nucl-ex/0403023.
5. M. Gazdzicki, and M. I. Gorenstein, *Acta Phys. Polon. B* **30**, 2705 (1999), hep-ph/9803462.
6. M. Gazdzicki, and D. Roehrich, *Z. Phys. C* **65**, 215–223 (1995).
7. J. L. Klay, et al., *Phys. Rev. C* **68**, 054905 (2003), nucl-ex/0306033.
8. P. Dinkelaker, *J. Phys. G* **31**, S1131–S1136 (2005).
9. C. Alt, et al., *Phys. Rev. Lett.* **94**, 052301 (2005), nucl-ex/0406031.
10. J. Bachler, et al., *Nucl. Phys. A* **661**, 45–54 (1999).
11. F. Wang, *J. Phys. G* **27**, 283–300 (2001), nucl-ex/0010002.
12. S. Hamieh, K. Redlich, and A. Tounsi, *Phys. Lett. B* **486**, 61–66 (2000), hep-ph/0006024.
13. S. V. Afanasev, et al., *Phys. Lett. B* **491**, 59–66 (2000).
14. V. Friese, *J. Phys. G* **31**, S911–S918 (2005), nucl-ex/0412013.
15. V. Friese, *Nucl. Phys. A* **698**, 487–490 (2002).
16. J. Cleymans, and K. Redlich, *Phys. Rev. C* **60**, 054908 (1999), nucl-th/9903063.
17. C. Alt, et al., *Phys. Rev. Lett.* **94**, 192301 (2005), nucl-ex/0409004.
18. S. A. Bass, et al., *Prog. Part. Nucl. Phys.* **41**, 225–370 (1998), nucl-th/9803035.
19. F. Becattini, M. Gazdzicki, A. Keranen, J. Manninen, and R. Stock, *Phys. Rev. C* **69**, 024905 (2004), hep-ph/0310049.
20. M. Kliemant, B. Lungwitz, and M. Gazdzicki, *Phys. Rev. C* **69**, 044903 (2004), hep-ex/0308002.
21. M. Gazdzicki, et al., *Braz. J. Phys.* **34**, 322–325 (2004), hep-ph/0309192.
22. J. Letessier, and J. Rafelski (2005), nucl-th/0504028.

Stopping and the $\langle K \rangle / \langle \pi \rangle$ Horn

Boris Tomášik*,† and Evgeni E. Kolomeitsev**

*The Niels Bohr Institute, Blegdamsvej 17, 2100 Copenhagen Ø, Denmark
†Ústav jaderné fyziky AVČR, 25068 Řež, Czech Republic
**School of Physics and Astronomy, University of Minnesota, 116 Church Street S.E., Minneapolis, Minnesota 55455, USA

Abstract. We propose a non-equilibrium hadronic model to interpret the observed excitation function of kaon-to-pion ratios in nuclear collisions at AGS and SPS energies. The crucial assumption of our model is that due to stronger stopping at lower energies the lifetime of the fireball is prolonged because the system has to build up the longitudinal expansion from internal pressure.

Keywords: strangeness production, stopping, fireball expansion
PACS: 25.75.-q, 25.75.Dw, 24.10.Pa

THE HORN

The excitation function of the ratio $\langle K^+ \rangle / \langle \pi^+ \rangle$ which exhibits a sharp peak (a "horn") at projectile energies of 30 AGeV is certainly one of the most intriguing experimental results from the energy scan performed at the CERN SPS [1]. Statistical hadronisation model expects a maximum connected with the transition from baryon to meson dominated energy regime but cannot reproduce the observed sharpness of the peak [2]. The sharp maximum together with the excitation functions of $\langle K^- \rangle / \langle \pi^- \rangle$ and $\langle \Lambda \rangle / \langle \pi \rangle$ is not reproduced in hadronic transport codes either [3, 4]. The only successful, though schematic, interpretation of the data uses the framework of the so-called *Statistical model of the early stage (SMES)* [5]. The model is based on the assumption that the primordial particle production is realized according to grand-canonical equilibrium distribution and the steep descent of the peak corresponds to transition to deconfined phase via a mixed phase. A kinetic description of the excitation function, also including phase transition in the region of the peak, has been proposed recently [6].

Concerning the SMES, a question appears whether it is realistic to assume a chemical equilibrium thus early in the collision? Moreover, if the peak is to be regarded as a signature for deconfinement, are all hadronic scenarios safely excluded? We shall explore the possibility of reproducing the "horn" in a hadronic non-equilibrium model.

Let us recall the two important quantities which regulate the final amount of produced strangeness. Firstly, it is the energy density, because the rates of reactions producing strangeness depend on the energy of particles in the incoming channel. Secondly, it is the total lifespan of the fireball since strange particles are out of chemical equilibrium and the relative strangeness content, most probably, grows with time. Thus we propose the following scenario potentially leading to the $\langle K^+ \rangle / \langle \pi^+ \rangle$ "horn": at lower energies ($\lesssim 30A$GeV) the increase of the excitation function is due to an increase in the energy available for the strangeness production. With the further increase of the collision energy

CP828, *Multiparticle Dynamics*
edited by V. Šimák, M. Šumbera, Š. Todorova, and B. Tomášik

the nuclear stopping power decreases. Since we know from M_T-dependence of HBT radii that the longitudinal expansion pattern at freeze-out looks roughly the same at all energies, there must be accelerated longitudinal expansion and it must last longer at lower energies with stronger stopping. We put the sharp decrease of the ratio (on the right-hand side of the "horn") into connection with shorter lifetime of the fireball in more energetic collisions. The fireball just has less time to "cook up" strangeness and therefore the relative strangeness yield is lower. This hypothesis will be tested with a kinetic model.

THE MODEL

As we only want to calculate the ratios of total yields, and not the yields themselves, it is enough to study the evolution of the *spatially averaged densities* of individual species only. The spatial distribution is inessential here. We will in particular investigate the evolution of *kaon* density. Assuming that kaons are kept in thermal equilibrium with a fireball medium, the variation of the kaon density follows from the simple relation

$$\frac{dn_K}{d\tau} = \frac{d}{d\tau}\frac{N_K}{V} = -\frac{N_K}{V}\frac{1}{V}\frac{dV}{d\tau} + \frac{1}{V}\frac{dN_K}{d\tau}, \tag{1}$$

where N_K and V are the total number of kaons and the volume of the system, respectively, and τ is the time in the co-moving frame. The second term on the right-hand side of this equation expresses the change of density due to chemical reactions. It can be split into gain term and loss term

$$\frac{dn_K}{d\tau} = n_K\left(-\frac{1}{V}\frac{dV}{d\tau}\right) + \sum_{ij}\langle v_{ij}\sigma_{ij}^{+}\rangle\frac{1}{1+\delta_{ij}}n_i n_j - \sum_{j}\langle v_{Kj}\sigma_{Kj}^{-}\rangle\frac{1}{1+\delta_{Kj}}n_K n_j. \tag{2}$$

In the gain term we sum over all processes producing a kaon, in the loss term these are the processes destroying a kaon. The relative-velocity-averaged cross-sections are multiplied with densities of the incoming species.

The first term on the right-hand-sides of eqs. (1) and (2) includes the expansion rate and stands for the density change due to expansion. In a simulation of the collision (hydrodynamic or transport), the expansion rate is obtained naturally. Here, we shall adopt a *parametrisation* of the expansion. Though in this way we do not make a direct contact to the underlying microscopic equation of state, it gives us a possibility to explore many different evolution scenarios and their impact on data.

The ansatz for the energy density and the baryon density which will be used reads

$$\varepsilon(\tau) = \begin{cases} \varepsilon_0(1-a\tau-b\tau^2) & \text{for} \quad \tau<\tau_s \\ \frac{\varepsilon_0'}{(\tau-\tau_0)^{\alpha/\delta}} & \text{for} \quad \tau>\tau_s \end{cases} \tag{3a}$$

$$\rho_B(\tau) = \begin{cases} \rho_{B,0}(1-a\tau-b\tau^2)^{\delta} & \text{for} \quad \tau<\tau_s \\ \frac{\rho_{B,0}'}{(\tau-\tau_0)^{\alpha}} & \text{for} \quad \tau>\tau_s \end{cases} \tag{3b}$$

In the above expressions the parameters can be tuned. Two of them are constrained by the requirements that the functions are continuous together with their first time derivatives.

Initially ($\tau < \tau_s$) the fireball expands acceleratedly with a quadratic dependence on time. Then, at $\tau > \tau_s$, the power-law expansion is dictated by the $M_\perp$-dependence of HBT radii. Note the shift by τ_0: it allows to delay this type of expansion. Such a delay would be unobservable in the $M_\perp$-dependence of R_{long} which is standardly used to argue for a short fireball lifetime [7]. The "lifetime" measured by R_{long} would correspond in our case to the difference $(\tau_{\text{total}} - \tau_0)$. Note also that the two parameterisations for energy and baryon density have similar forms and differ essentially by the exponent δ which is given by the assumed equation of state.

By tuning the parameters a and b we specify the initial rate of density decrease and the pace on which the decrease accelerates. This way we quantify the stopping and re-acceleration. We can construct a large class of models "between Landau and Bjorken scenarios".

Densities of kaons (i.e. K^+, K^0, K^{*+}, K^{*0}) are calculated according to eq. (2). We include the following processes:

$$
\begin{array}{llll}
\pi N \leftrightarrow KY, & \pi N \rightarrow NK\bar{K}, & \pi\Delta \leftrightarrow KY, & \pi\Delta \rightarrow NK\bar{K} \\
NN \rightarrow KNY, & NN \rightarrow NNK\bar{K}, & NN \rightarrow \Delta KY & \\
N\Delta \rightarrow NYK, & N\Delta \rightarrow NNK\bar{K}, & N\Delta \rightarrow \Delta KY & \\
\Delta\Delta \rightarrow \Delta YK, & \Delta\Delta \rightarrow NNK\bar{K} & & \\
\pi\pi \leftrightarrow K\bar{K}, & \pi\rho \leftrightarrow K\bar{K}, & \rho\rho \leftrightarrow K\bar{K}, & \\
K^* \leftrightarrow K\pi, & \pi Y \leftrightarrow K\Xi. & &
\end{array}
$$

For those processes with two particles in final state also inverse reactions have been included which annihilate kaons.

Species with negative strangeness must balance the total strangeness of the system to zero. Reactions which just swap the strange quark between them are quick. Therefore, we assume that these species are in chemical equilibrium with respect to each other, while the strangeness-weighted sum of their densities is given by the requirement of strangeness neutrality.

All non-strange species are assumed to be chemically equilibrated. We assume no antibaryons at these energies. Their influence would be highest at the top SPS energy $\sqrt{s_{NN}} = 17A\text{GeV}$ where we expect 10% error due to their neglect. In a schematic model like ours this is acceptable.

In a model set up in this way, the total produced amount of strangeness is mainly controlled by the lifetime (as we shall show) and slightly less by the temperature. The rate at which strange quarks are distributed among K^- and Λ is fixed by the final temperature.

We try to choose the parameterisations (3) in such a way that the final state resulting from fits to chemical composition at different energies [8] is reached. The initial amount of strangeness is estimated from pp, pn, and nn collisions [9].

RESULTS

We show an example of our results in Figure 1. Each point in that figure corresponds to

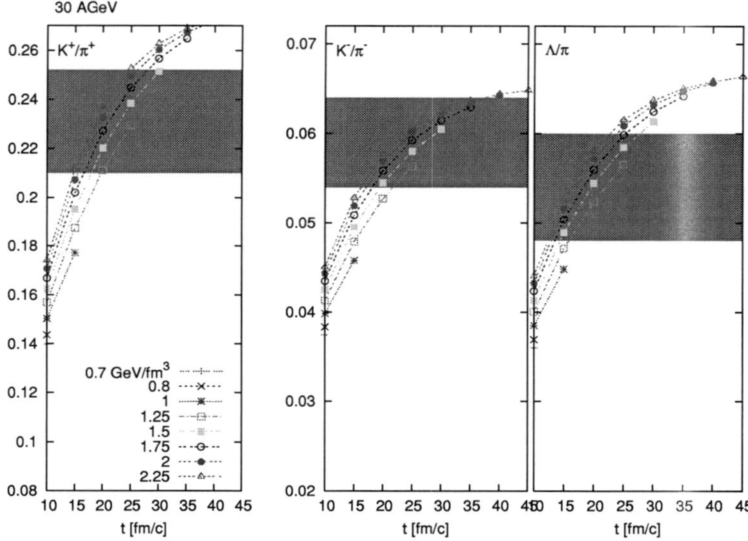

FIGURE 1. Calculated ratios $\langle K^+\rangle/\langle\pi^+\rangle$, $\langle K^-\rangle/\langle\pi^-\rangle$, $\langle\Lambda\rangle/\langle\pi\rangle$ as functions of the total lifetime of the fireball. Different curves correspond to different initial energy densities. Calculation for Pb+Pb collisions at projectile energy of 30 *A*GeV. The bands show values accepted by data [1].

a different evolution scenario; they differ by total lifetime and initial energy density. Though the presented results are obtained for Pb+Pb collisions at projectile energy of 30 *A*GeV, it applies generally that dependence on lifetime is crucial, while the dependence on initial energy density is less important. The latter gains more weight at lower energies and is completely irrelevant at the highest SPS energy. It mainly stems from the temperature dependence of reactions including nucleons like $\pi + N \rightarrow \Lambda + K$ which make up a bigger share of the total production rate at lower energies where the baryon density is higher.

Summary of the values of total lifetime and initial energy density allowed by comparison to data is presented in Figure 2. The more important limitations are put on the lifetime. In general, for SPS energies our hypothesis is confirmed that with increasing energy of the collision lifetime of the fireball decreases. Only at the lowest studied energy, 11.6 *A*GeV (AGS), we miss the $\langle\Lambda\rangle/\langle\pi\rangle$ ratio by about 3 standard deviations, see Figure 3. This is caused by much too high temperature in the final state, which can be fixed by re-parameterising the time dependence of the energy density. Otherwise, we reproduce the data well.

CONCLUSIONS

The excitation function of the ratios $\langle K^+\rangle/\langle\pi^+\rangle$, $\langle K^-\rangle/\langle\pi^-\rangle$ and $\langle\Lambda\rangle/\langle\pi\rangle$ can be reproduced in a non-equilibrium hadronic scenario. The crucial assumption is that the total

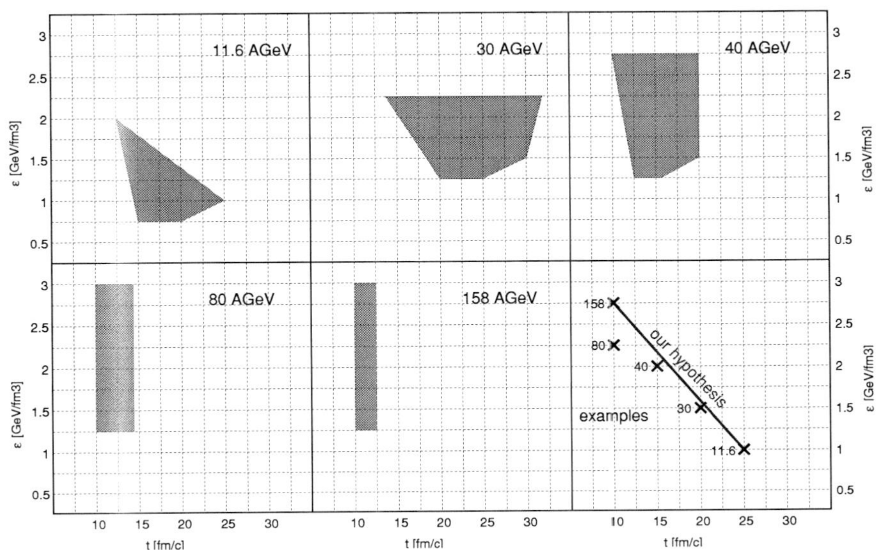

FIGURE 2. Values of initial energy density and total lifetime accepted by data [1].

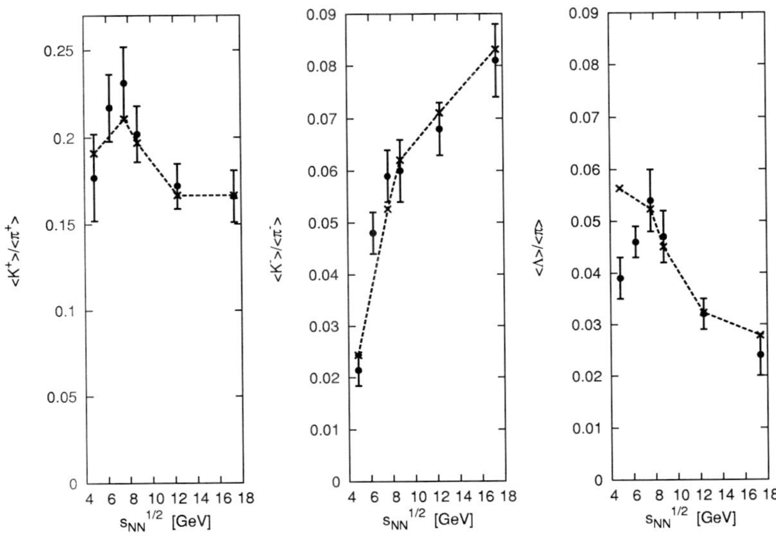

FIGURE 3. Results obtained with parameter sets indicated in lower right panel of Figure 2 compared to data.

lifetime of the fireball is a decreasing function of the collision energy.

Thus we have made a specific prediction for the evolution dynamics of the fireball. Such a prediction must now be checked against other observables. Although we were lead by the knowledge of all particle abundances, transverse momentum spectra and HBT radii, a careful analysis of these data in the framework of the proposed model must be performed. Furthermore, dilepton spectra gained an unprecedented accuracy recently [10, 11, 12]. They are sensitive to whole fireball history [13] and could provide the crucial test of our hypothesis.

ACKNOWLEDGEMENTS

The work of BT was supported by a Marie Curie Intra-European Fellowship within the 6th European Community Framework Programme. The work of EEK was supported by the US Department of Energy under contract No. DE-FG02-87ER40328.

REFERENCES

1. B. Lungwitz, *this volume*, nucl-ex/0509041.
2. J. Cleymans, H. Oeschler, K. Redlich and S. Wheaton, *Phys. Lett. B* **615**, 50–54 (2005).
3. M. Bleicher, *this volume*.
4. H. Weber, E.L. Bratkovskaya, W. Cassing, H. Stöcker, *Phys. Rev. C* **67**, 014904 (2003).
5. M. Gaździcki and M.I. Gorenstein, *Acta Physica Polonica B* **30**, 2705–2735 (1999).
6. J. K. Nayak, J. Alam, P. Roy, A.K. Dutt-Mazumder, B. Mohanty, "Kaon to pion ratio in heavy ion collisions", *poster at the conference Quark Matter 2005, Aug.4-9, 2005, Budapest, Hungary.*
7. D. Adamová *et al.* [CERES/NA45 Collaboration], *Nucl. Phys. A* **714**, 124–144 (2003).
8. F. Becattini *et al.*, *Phys. Rev. C* **69**, 024905 (2004).
9. B. Tomášik and E.E. Kolomeitsev, in preparation.
10. G. Agakichiev *et al.* [CERES/NA45 collaboration], *Phys. Lett. B* **422**, 405–412 (1998).
11. S. Damjanovic and K. Filimonov [CERES/NA45 Collaboration], *Pramana* **60**, 1067–1072 (2002), [arXiv:nucl-ex/0111009].
12. E. Scomparin for the NA60 collaboration, *to be published in the proceedings of Quark Matter 2005, Aug.4-9, 2005, Budapest, Hungary.*
13. T. Renk, *J. Phys. G* **30**, 1495–1514 (2004).

NA57 Results

F. Antinori[a], P. Bacon[b], A. Badalà[c], R. Barbera[c], A. Belogianni[d], I. Bloodworth[b], M. Bombara[e], G. E. Bruno[f], S.A. Bull[b], R. Caliandro[f], M. Campbell[g], W. Carena[g], N. Carrer[g], R.F. Clarke[b], A. Dainese[a], D. Di Bari[f], S. Di Liberto[h], R. Divià[g], D. Elia[f], D. Evans[b], G. Feofilov[i], R.A. Fini[f], P. Ganoti[d], B. Ghidini[f], G. Grella[j], H. Helstrup[k], K.F. Hetland[k], A.K. Holme[l], A. Jacholkowski[c], G.T. Jones[b], P. Jovanovic[b], A. Jusko[b], R. Kamermans[m], J.B. Kinson[b], K. Knudson[g], V. Kondratiev[i], I. Králik[e], A. Kravčáková[e], P. Kuijer[m], V. Lenti[f], R. Lietava[b], G. Løvhøiden[l], V. Manzari[f], M. A. Mazzoni[h], F. Meddi[h], A. Michalon[n], M. Morando[a], P.I. Norman[b], A. Palmeri[c], G.S. Pappalardo[c], B. Pastirčák[e], R.J. Platt[b], E. Quercigh[a], F. Riggi[c], D. Röhrich[o], G. Romano[j], K. Šafařík[g], L. Šándor[e], E. Schillings[m], G. Segato[a], M. Sené[p], R. Sené[p], W. Snoeys[g], F. Soramel[a], M. Spyropoulou-Stassinaki[d], P. Staroba[q], R. Turrisi[a], T.S. Tveter[l], J. Urbán[r], P. van de Ven[m], P. Vande Vyvre[g], A. Vascotto[g], T. Vik[l], O. Villalobos-Baillie[b], L. Vinogradov[i], T. Virgili[j], M.F. Votruba[b], J. Vrláková[r] and P. Závada[q]

[a]*University of Padua and INFN, Padua, Italy*
[b]*University of Birmingham, Birmingham, UK*
[c]*University of Catania and INFN, Catania, Italy*
[d]*Physics Department, University of Athens, Athens, Greece*
[e]*Institute of Experimental Physics, Slovak Academy of Science, Košice, Slovakia*
[f]*Dipartiemnto IA di Fisica dell' Università e del Politecnico di Bari and INFN, Bari, Italy*
[g]*CERN, European Laboratory for Particle Physics, Geneva, Switzerland*
[h]*University "La Sapienza" and INFN, Rome, Italy*
[i]*State University of St. Petersburg, St. Petersburg, Russia*
[j]*Dipartiemnto di Scienze Fisiche "E.R.Caianiello" dell' Università and INFN, Salerno, Italy*
[k]*Høgskolen i Bergen, Bergen, Norway*
[l]*Fysisk institutt, Universitetet i Oslo, Oslo, Norway*
[m]*Utrecht University and NIKHEF, Utrecht, The Netherlands*
[n]*IReS/ULP, Strasbourg, France*
[o]*Institutt for fysikk og teknologi, Universitetet i Bergen, Bergen, Norway*
[p]*Collège de France, Paris, France*
[q]*Institute of Physics, Prague, Czech Republic*
[r]*P.J.Šafárik University, Košice, Slovakia*

Abstract. Hyperon enhancements at 160 and 40 A Gev/c are presented and compared. The momentum spectra are analysed in the framework of the blast wave model and freeze-out temperature, transverse and longitudinal flows are extracted. Nuclear modification factors are presented and discussed.

Keywords: heavy ions, QGP, strangeness, flow, nuclear modification factor
PACS: 12.38Mh, 25.75Dw, 25.75Ld, 25.75Nq

CP828, *Multiparticle Dynamics*
edited by V. Šimák, M. Šumbera, Š. Todorova, and B. Tomášik

INTRODUCTION

The study of relativistic heavy ion collisions provides a unique opportunity to search for a new predicted state of matter—the quark-gluon plasma (QGP) [1]. A number of experimental signatures which could signal the QGP have been proposed (for a recent review see [2]).

Strange particles have proved over the past years to be a powerful tool for the study of reaction dynamics in high-energy heavy-ion collisions. NA57 is a dedicated experiment at the CERN SPS for the study of the production of strange and multi-strange particles in Pb-Pb collisions [3]. It continues and extends the study initiated by its predecessor the WA97 experiment, by (i) enlarging the trigger fraction of the inelastic cross section, thus extending the centrality range towards less central collisions and (ii) collecting data also at lower (40 A GeV/c) beam momentum in order to study the energy dependence of the strangeness enhancement.

In this contribution the NA57 results on enhancements and transverse flow at both SPS energies are summarised, and the longitudinal flow and nuclear modification factors at 160 A GeV/c are presented.

ENHANCEMENT AT 160 AND 40 A GEV/C

The enhanced yield of strange and multi-strange particles in nucleus-nucleus reactions with respect to proton-nucleus interactions has been suggested as one of the sensitive signatures for a phase transition to a QGP state [4, 5]. It is expected that the enhancement should be more pronounced for multi-strange than for singly strange particles [6]. The NA57 experiment measures *enhancement* as the particle yield per wounded nucleon in Pb-Pb centrality class $'Class'$ relative to the particle yield per wounded nucleon in p-Be collisions:

$$E = \frac{\langle N_{wound} \rangle_{p-Be}}{\langle N_{wound} \rangle_{Class}} \times \frac{dN^{Class}_{Pb-Pb}/dy}{dN_{p-Be}/dy}. \qquad (1)$$

The Pb-Pb data are divided into five centrality classes [7] corresponding to 0-4.5%, 4.5-11%, 11-23%, 23-40% and 40-53% most central collisions of the total inelastic Pb-Pb cross section. At 160 A GeV/c in Fig. 1 the hierarchy of hyperon enhancements $E(\Lambda) < E(\Xi) < E(\Omega)$ is observed in agreement with WA97 measurement [8]. The comparison of Λ, $\bar{\Lambda}$ and Ξ^- enhancements at two energies is shown in Fig. 2.

We do not observe an indication of threshold behaviour at the lower energy suggesting that the properties of the medium created in collisions at both energies are similar.

The centrality dependence of the enhancement at 40 A GeV/c is steeper than at 160 A GeV/c. This is in qualitative agreement with canonical suppression models although the size of the effect is larger in data [9, 10].

TRANSVERSE MASS AND RAPIDITY SPECTRA

The momentum spectra of produced particles are sensitive to the details of the production dynamics [11]. For the fireball in local thermal equilibrium the shapes of the m_T and

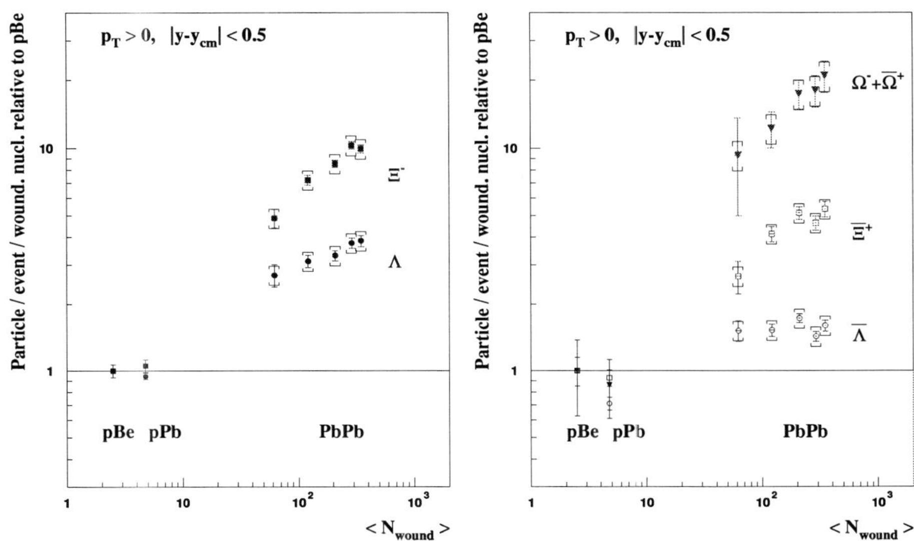

FIGURE 1. The centrality dependence of hyperon enhancements at 160 A GeV/c. The error bars inidcate the statistical errors only, while the bracket symbols represent the systematic errors.

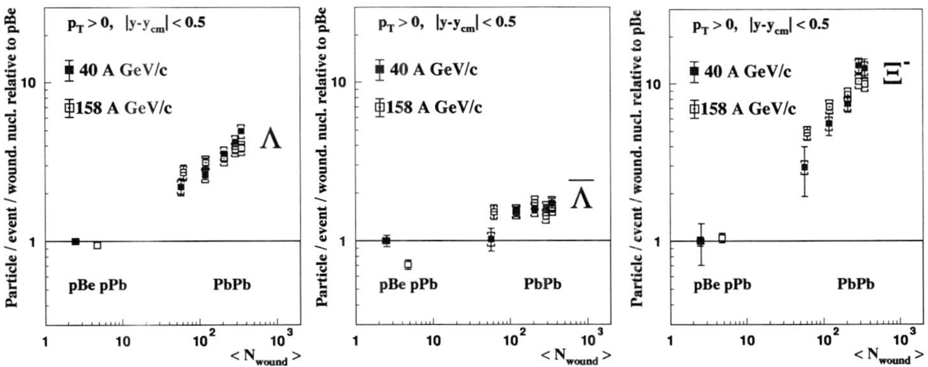

FIGURE 2. Centrality dependence of Λ, $\bar{\Lambda}$ and Ξ^- enhancements at both 160 and 40 A GeV/c. The error bars indicate the statistical errors only, while the bracket symbols represent the systematic errors.

rapidity spectra depend both on the thermal motion of the particles and on the collective flow driven by pressure. To disentangle the two contributions, namely the thermal motion and the transverse and longitudinal flows, one has to rely on models. In the present analysis we consider the blast wave model [11]. The detailed procedure of the NA57 analysis can be found in [12].

The parameters which characterise the expanding fireball are local temperature T, average transverse expansion velocity $\langle \beta_\perp \rangle$ and average longitudinal velocity $\langle \beta_L \rangle$.

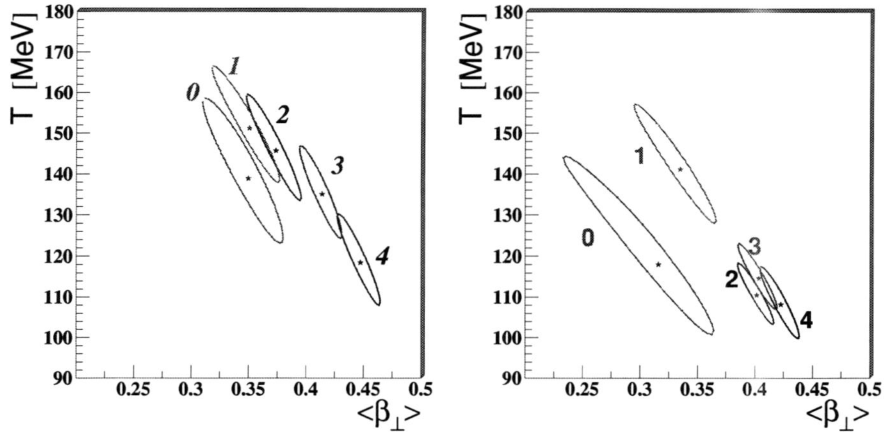

FIGURE 3. The 1 σ confidence level contours at 160 A GeV/c (left) and 40 A GeV/c (right). The data are divided into 5 centrality classes [12], most peripheral labelled 0 to most central labelled 4.

Figure 3 shows the centrality dependence of the temperature T and velocity $\langle\beta_\perp\rangle$ which are extracted from the transverse mass spectra [12].

We do not observe a different pattern of centrality dependence at two energies. At both energies we observe a decrease of T and an increase of $\langle\beta_\perp\rangle$ with centrality, indicating earlier freeze-out in peripheral collisions.

The longitudinal velocity $\langle\beta_L\rangle = 0.42 \pm 0.03$ is extracted from the rapidity spectra [13]. The centrality dependence of the rapidity spectra and longitudinal flow is not observed in data [13].

NUCLEAR MODIFICATION FACTORS

The idea of jet quenching by parton energy loss was introduced in [14] and later revisited in [15]. The quenching of high transverse momentum p_T particles in central heavy ion collisions has indeed been observed at BNL RHIC. The effect is quantified using the nuclear modification factor:

$$R_{CP}(p_T) = \frac{\langle N_{coll}\rangle_P}{\langle N_{coll}\rangle_C} \times \frac{d^2N^C_{AA}/dp_Tdy}{d^2N^P_{AA}/dp_Tdy} \tag{2}$$

The observation of the effect at RHIC [16] triggered the reanalysis of SPS data by the WA98, NA57 and NA49 collaborations [17, 18, 19]. Although hard processes play a minor role at SPS energies, the nuclear modification factors may shed light on a modification of particle production in a dense medium.

The NA57 experiment calculates $R_{CP}(p_T)$ for h^-, K^0_S, Λ and $\bar{\Lambda}$ using uncorrected p_T distributions for all reconstructed strange particles [18]. The acceptance, selection

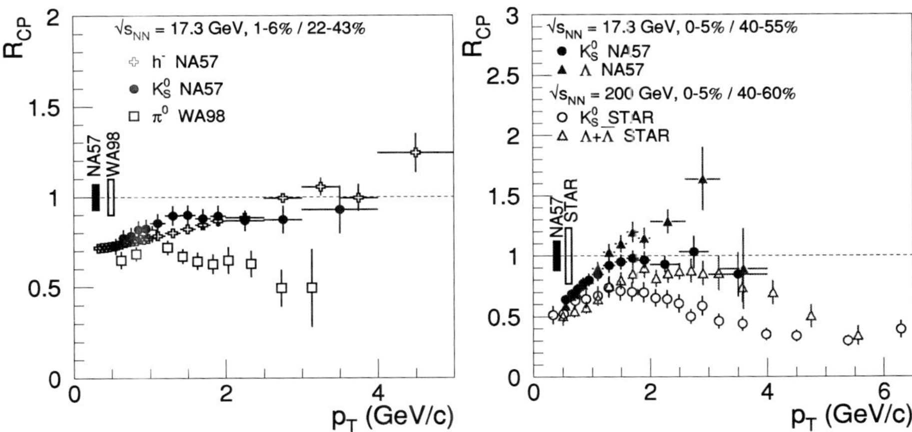

FIGURE 4. Left: $R_{CP}(p_T)$ for h^- and K^0_S from NA57 and π^0 from WA98 [17] in Pb-Pb collisions at $\sqrt{s}$=17.3 GeV. Right: $R_{CP}(p_T)$ for K^0_S and Λ at $\sqrt{s} = 17.3$ GeV (NA57) and in Au-Au at $\sqrt{s} = 200$ GeV (STAR) [20]: The bars centred at $R_{CP} = 1$ represent the normalisation errors; the point-by-point bars are the quadratic sum of statistical and systematic errors.

criteria and reconstruction biases are cancelled out in the ratio R_{CP} as has been checked explicitly [18].

In Fig.4 we compare our results to other R_{CP} measurements at the SPS and at RHIC. In the left panel, the WA98 π^0 data [17] in Pb-Pb collisions are plotted together with the NA57 h^- and K^0_S data for the same centrality classes. The R_{CP} of K^0_S is approximately constant at 0.8 for $p_T > 1$ GeV/c and is significantly larger than that measured by the WA98 collaboration for π^0, even when taking into account the normalisation systematic errors, which are independent for the two experiments. The h^- data from NA57 are compatible, within the systematic errors, with the π^0 data from WA98 for $p_T < 1.5$ GeV/c, where the h^- sample is expected to be dominated by π^-. For higher p_T, h^- have a larger R_{CP} than π^0; this may be due to increasing contributions from K^- and $\bar{p}$ in the h^- sample.

The NA49/NA57 comparison is at the moment not straight-forward, since different definitions of the reference peripheral class are used by the two experiments.

A comparison for K^0_S and Λ data at SPS and RHIC is presented in the right hand panel of Fig. 4. In the p_T range covered by our data, the relative pattern for K^0_S and Λ is similar at the two energies, while the absolute values are higher at the SPS than at RHIC, where parton energy loss is believed to have strong effect. The larger R_{CP} for Λ than for K^0_S at p_T around $2-3$ GeV/c can be interpreted at both energies as due to parton coalescence, although it may be explained also in terms of larger Cronin effect for Λ with respect to kaons.

CONCLUSIONS

The NA57 experiment has measured:

- A hierarchical pattern of hyperon enhancements with respect to their strangeness content: $E(\Lambda) < E(\Xi) < E(\Omega)$. This pattern is a consequence of strangeness phase space saturation in Pb-Pb collisions [9, 21, 22] due to fast strange quark creation in the QGP phase [4].
- Strange particle transverse mass and rapidity distributions: The spectra are consistent with the blast wave model parametrisation. The centrality dependence results indicate that with increasing centrality the transverse flow velocity increases and freeze-out temperature decreases at both energies. At top SPS energy collective flow with similar transverse and longitudinal average velocities is observed.
- A central to peripheral R_{CP} pattern qualitatively similar to that observed at RHIC energy, although higher in absolute value. The difference between K_S^0 and Λ suggests recombination effects at medium transverse momenta.

ACKNOWLEDGMENTS

We thank T. Csörgő, U. Heinz and B. Tomášik for fruitful discussions on the blast wave model.

REFERENCES

1. N. Cabibbo and G. Parisi, *Phys. Lett. B* **59**, 67–69 (1975).
2. U. Heinz, "Concepts of Heavy Ion Physics", hep-ph/0407360.
3. K. Fanebust et al. (NA57 collaboration), *J. Phys. G* **28**, 1607–1614 (2002).
4. J. Rafelski and B. Müller, *Phys. Rev. Lett.* **48**, 1066–1069 (1982), Erratum *ibid.* **56**, 2334 (1986).
5. P. Koch, B. Müller and J. Rafelski, *Phys. Rep.* **142**, 167–262 (1986).
6. J. Rafelski, *Phys. Lett. B* **262**, 333–340 (1991).
7. F. Antinori et al. (NA57 collaboration), *J. Phys. G* **31**, 321–335 (2005).
8. E. Andersen et al. (WA97 collaboration), *Phys. Lett. B* **449**, 401–406 (1999).
9. A. Tounsi et al., *Nucl. Phys. A* **715**, 565c–568c (2003).
10. F. Antinori, *J. Phys. G* **30**, S725–S734 (2004).
11. E.Schnedermann, J.Sollfrank and U.Heinz, *Phys. Rev. C* **48**, 2462–2475 (1993).
12. G. E. Bruno et al. (NA57 collaboration), *J. Phys. G* **31**, S127–S133 (2005),
    F. Antinori et al. (NA57 collaboration), *J. Phys. G* **30**, 823–840 (2004).
13. F. Antinori et al. (NA57 collaboration), *J. Phys. G* **31**, 1345–1357 (2005).
14. J. D. Bjorken, Fermilab-Pub-82/59/THY (1982) and erratum (unpublished).
15. M. Gyulassy and M. Plümer, *Phys. Lett. B* **243**, 432–438 (1990).
16. S.S. Adler et al. (PHENIX collaboration), *Phys. Rev. C* **69**, 034910 (2004),
    J. Adams et al. (STAR collaboration), *Phys. Rev. Lett.* **91**, 172302 (2003).
17. M.M. Aggarwal et al. (WA98 collaboration), *Eur. Phys. J. C* **23**, 225–236 (2002).
18. F. Antinori et al. (NA57 collaboration), *Phys. Lett. B* **623**, 17–25 (2005).
19. C. Höhne et al. (NA49 collaboration), A. Lásló et al. (NA49 collaboration), *Proceedings of Quark Matter 05*, Budapest, August 2005.
20. J. Adams et al. (STAR collaboration), *Phys. Rev. Lett.* **92**, 052302 (2004).
21. F. Becattini, *Nucl. Phys. A* **702**, 336–340 (2002).
22. A. Bialas, *Phys. Lett. B* **442**, 449–452 (1998).

SMALL X-PHYSICS AND DIFFRACTION

Chairpersons: A. Valkárová, N. Schmitz, and Y. Hama

Factorization and Factorization Breaking in Diffraction at HERA

S. Levonian

DESY, Notkestrasse 85, 22607 Hamburg, Germany

Abstract. A recent progress is reviewed in our understanding of diffractive processes at HERA. New precision measurements are now available from both H1 and ZEUS collaborations, from which diffractive parton distribution functions are determined. They are used to test QCD factorization in diffractive production of charm and dijets.

Keywords: Diffraction, Factorization, HERA
PACS: 12.38.Aw, 12.39.St, 13.60.Hb, 13.85.Hd

INTRODUCTION

A significant progress has been achieved over the last decade in understanding the nature of *diffractive phenomena* at high energies. This is to a large extent due to the electron-proton collider HERA, which for the first time offers the possibility for the partonic structure of colour singlet exchange to be probed in hard diffractive deep-inelastic scattering (DIS).

At the hadronic level diffraction is best described in the framework of Regge formalism [1, 2] as a t-channel exchange of a leading trajectory with the vacuum quantum numbers, named *Pomeron*. Specific interest to diffraction as a 'physics of Pomeron' is related to the fact that the Pomeron exchange asymptotically dominates over all other contributions to the scattering amplitude, and thus represents the essence of strong interactions in high energy limit. Translating to the modern partonic language this reveals that colourless exchange is important in low-x regime (which is high energy limit of QCD) where gluons are expected to dominate. Since the diffractive cross section, being proportional to the gluon density squared, rises with energy faster than the total cross section unitarity correction effects, e.g. in a form of gluon saturation, are expected to be first seen in diffraction. Hence an interesting question is: can this be observed already at HERA?

Another aspect in which diffraction may play a key rôle is an interplay between soft and hard processes. Since both short distance and long distance physics contribute to diffractive DIS, HERA has high potential to provide new insight into non-perturbative QCD phenomena, as well as to bridge soft (Regge) and hard (pQCD) domains.

The central problem in hard diffraction is the question of *QCD factorization*, i.e. the question whether it is possible to describe diffractive cross section by a convolution of the diffractive parton distribution functions (DPDFs) with universal partonic cross sections. For diffractive DIS QCD factorization has been proven by Collins [3] while it fails for hadron induced reactions. In addition it is important to test a conjecture of *Regge factorization* [4] which assumes that the diffractive cross section can be expressed as a

CP828, *Multiparticle Dynamics*
edited by V. Šimák, M. Šumbera, Š. Todorova, and B. Tomášik

product of Pomeron flux and its structure function. In this paper both hard QCD and Regge factorization hypotheses are confronted with recent HERA data [5, 6, 7, 8, 9].

TESTING REGGE FACTORIZATION IN INCLUSIVE DIFFRACTION

Several experimental techniques are used to measure diffractive processes $ep \to eXY$ at HERA, where a single proton or low mass excitation system Y is separated from the final state X by large rapidity gap. They have different advantages and drawbacks and are complementary to each other. Most clean, but statistically limited sample is provided by so called LPS method in which scattered proton in the forward direction is detected directly by *Roman Pot* technique. To obtain high statistics sample one can use characteristic properties of the hadronic final state. H1 and ZEUS follow slightly different approaches trying to to fully utilize their detector capabilities.

H1 method is based on the requirement of the large rapidity gap separating the leading baryon system Y from the photon dissociation system X. The rapidity gap is positively identified by the absence of activity in pseudorapidity interval $3.2 < \eta < 7.5$, covered by various detector components. In this approach non-diffractive background is small and can be estimated using Monte Carlo simulation.

The extraction of the diffractive contribution in ZEUS is performed using so called M_X method, which is based on the fact that diffractive and non-diffractive final states have very different $\ln M_X^2$ distributions. The non-diffractive contribution exhibiting exponential fall-off towards lower M_X values is extrapolated underneath the diffractive plateau and statistically subtracted from the total M_X distribution.

Inclusive diffractive cross sections as a function of longitudinal momentum fraction of the colour singlet exchange relative to the incoming proton, $x_{I\!P}$, as measured by different techniques in the intermediate Q^2 range are compared in Fig. 1. Data show fair agreement in the bulk of the phase space. However some systematic differences are present at low M_X. Moreover, less steep Q^2 dependence is observed in ZEUS M_X sample as compared to H1 LRG data.

One of the basic predictions in Regge model is a specific asymptotic behaviour of the total, elastic and diffractive cross sections with energy, W, which is controlled by the value of intercept of the universal Pomeron trajectory, $\alpha_{I\!P}(t)$ at $t = 0$. In addition, if Regge factorization holds the Pomeron flux $f_{I\!P}(x_{I\!P},t) = x_{I\!P}^{1-2\alpha_{I\!P}(t)} e^{bt}$ (and hence $\alpha_{I\!P}(0)$) should not depend on $\beta = x/x_{I\!P}$ and Q^2. Figure 2 summarizes the Q^2 dependence of $\alpha_{I\!P}(0)$ as extracted using Regge motivated fits from diffractive and inclusive DIS data. In the inclusive case proton structure function can be parameterised at low $x < 0.01$ in form of $F_2(x,Q^2) = cx^{-\lambda(Q^2)}$ with $\lambda = \alpha_{I\!P}(0) - 1$. In diffractive DIS $\alpha_{I\!P}(0)$ can be obtained from W, or equivalently from $x_{I\!P}$-dependence of the diffractive cross section. One can see, that the data cannot be described by the universal factorizable Pomeron.

NLO DGLAP QCD fit [10] is performed to inclusive diffractive cross sections as measured by H1 and ZEUS. Resulting DPDFs are shown in Fig. 3. While singlet component is similar at low Q^2, H1 and ZEUS gluon distributions differ in normalization by factor of two, reflecting different Q^2 dependence in the original cross sections.

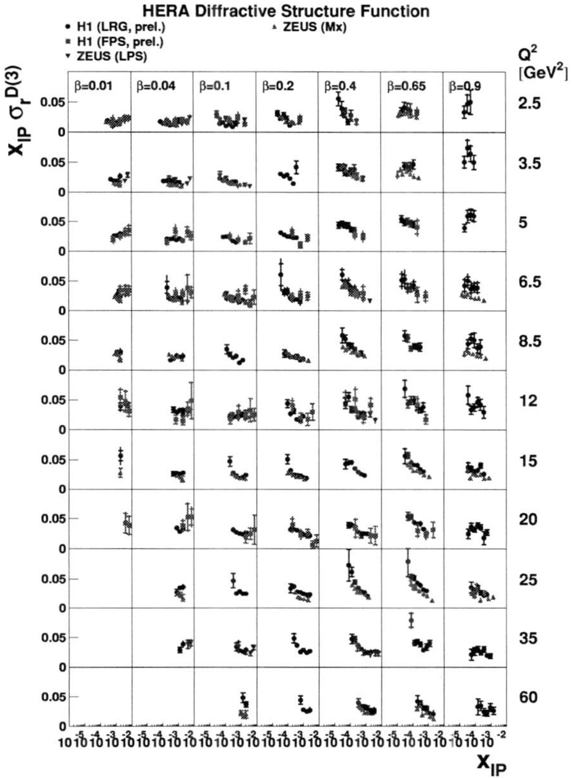

FIGURE 1. Diffractive reduced cross sections, measured in four different samples, as a function of $x_{I\!P}$ in bins of β and Q^2. All data are scaled to $M_Y < 1.6$ GeV range and transported to H1 LRG bin centers.

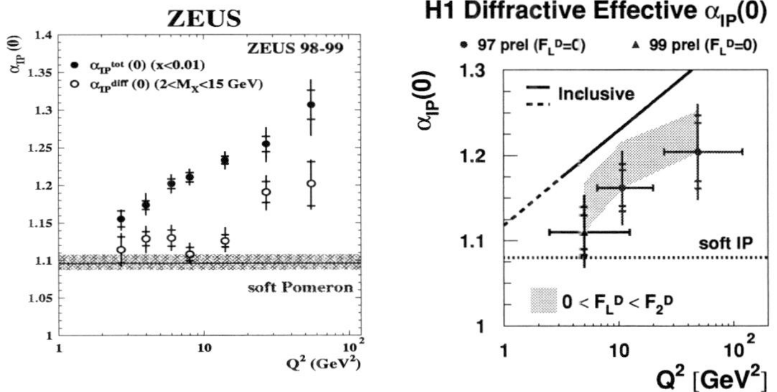

FIGURE 2. The intercept of the Pomeron trajectory, $\alpha_{I\!P}(0)$, as a function of Q^2 in diffractive and inclusive DIS, as determined by ZEUS(left) and H1 (right).

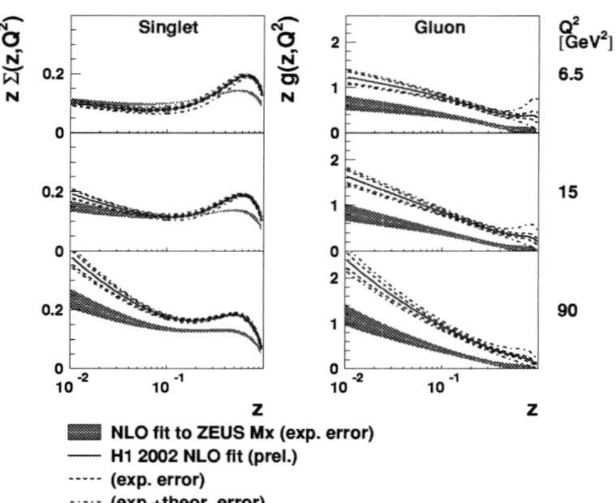

FIGURE 3. Diffractive parton densities obtained from the NLO QCD fits [10] and normalized such that the Pomeron flux is unity at $x_{I\!P} = 0.003$.

TESTING QCD FACTORIZATION IN DIFFRACTIVE FINAL STATES

In the following the H1 diffractive PDFs [10] are used to describe more exclusive final states, such as charm and dijets production, in which hard scale makes partonic cross sections perturbatively calculable and hence allows to test QCD factorization and the universality of the extracted DPDFs.

Charm production is tagged via $D^{*\pm}$ in diffractive DIS. The cross section is determined for the kinematic region $2 < Q^2 < 100$ GeV^2, $0.05 < y < 0.7$ and $p_{T,D^*} > 2$ GeV. The NLO QCD prediction is calculated using program HVQDIS [11] interfaced to the H1 DPDFs, with renormalisation and factorization scales set to $\mu_f^2 = \mu_r^2 = Q^2 + 4m_c^2$ and $m_c = 1.5$ GeV. Figure 4 (left) shows that H1 and ZEUS measurements are in agreement and both are well described by NLO QCD calculations. This supports the validity of QCD factorization in diffractive DIS. Same conclusion can be drawn from the diffractive dijets production in DIS regime [8] which is compared to NLO QCD calculations using DISENT [12] program on Fig. 4 (right).

Note, that in spite of overall good agreement withing systemetic uncertainties the calculations overestimate the cross section at high $z_{I\!P}^{jets}$. This indicates a potential of the data to further constrain diffractive gluon density.

Contrary to DIS a suppression factor of ~ 2 is observed when comparing H1 diffractive dijets in photoproduction to NLO predictions [13]. Such a supprerssion (sometimes also called the *gap survival probability*) is expected theoretically in resolved photoproduction and can be estimated using eikonal models based on s-channel unitarity [14].

In order to study the behaviour of direct and resolved component separately, the dijet cross section is measured as a function of $x_\gamma^{obs} = \sum_{jets}(E_T^{jet} e^{-\eta^{jet}})/2yE_e$. The result as shown in Fig. 6 disfavours the suppression of only the resolved contribution by a factor

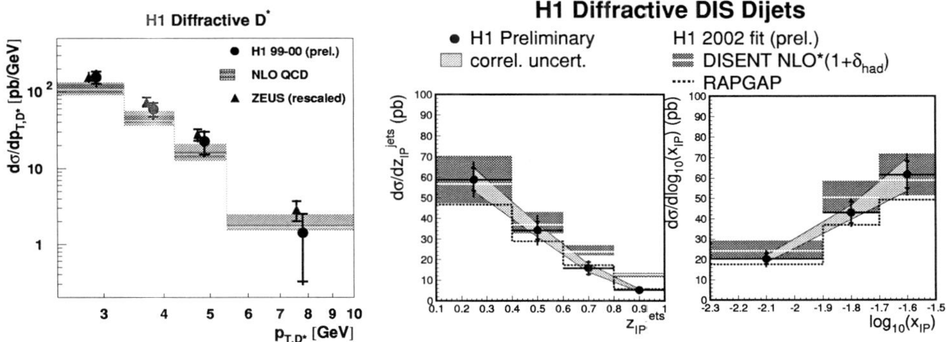

FIGURE 4. Differential $D^{*\pm}$ cross section at $x_{I\!P} < 0.04$ in diffractive DIS as a function of p_{T,D^*}, compared to NLO QCD predictions (left). Differential dijet cross sections in diffractive DIS compared to NLO QCD predictions (right).

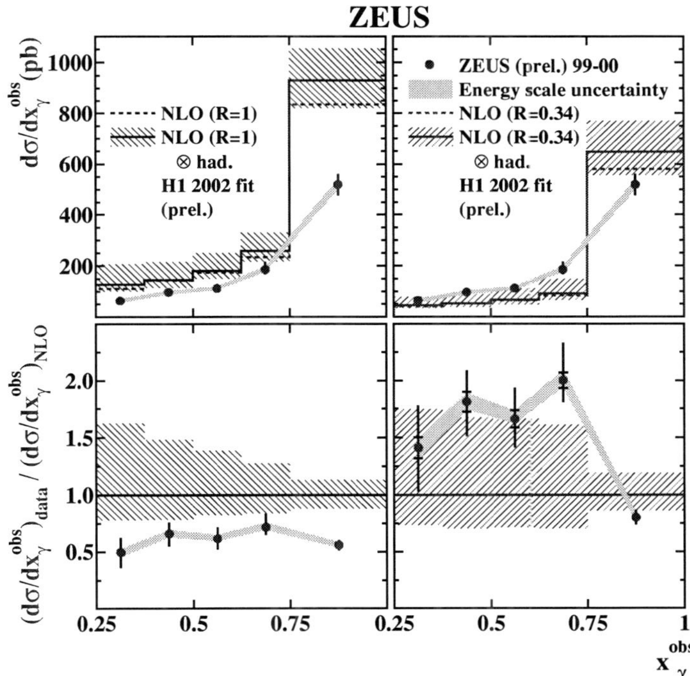

FIGURE 5. Differential dijet cross sections and their ratios to NLO QCD predictions [15] in diffractive photoproduction as a function of x_γ^{obs} when applying resolved component suppression by factor of R.

of $R = 0.34$ [14, 15] but rather suggests a global x_γ^{obs} independent suppression by a factor of ~ 2. Note however that this conclusion strongly depends upon the choice of DPDFs [16].

SUMMARY

New generation of diffractive parton distributions together with their uncertainties are determined from HERA data in NLO QCD framework. They are used to test factorization properties of diffraction in different reactions. Despite measured F_2^D are in fair agreement between the two HERA experiments remaining differences lead to approximately two times different gluon in diffractive PDFs. This has to be clarified with new data.

Q^2 dependence of the Pomeron intercept suggests Regge factorization breaking in DIS regime. Energy dependence of the diffractive cross section has no simple explanation at the moment. It may indicate the onset of unitarity corrections (or saturation effect) is already seen in diffraction at HERA. Alternatively, it could be a manifestation of a complicated interplay between soft and hard phenomena in diffractive DIS.

NLO predictions for diffractive final states strongly depend upon specific choice of diffractive parton density functions. When using H1 DPDFs both diffractive charm and dijets production in DIS support QCD factorization. In diffractive photoproduction QCD factorization is broken, showing at the moment global x_γ independent suppression factor of ~ 2, contrary to the most plausible theoretical expectation.

In spite of the recent progress a complete understanding of the nature of colour singlet exchange remains a major challenge in QCD.

REFERENCES

1. T. Regge, *Nuovo Cimento* **14**, 951 (1959); *ibid.* **18**, 947 (1960); G. Chew and S. Frautchi, *Phys. Rev. Lett.*, **7**, 394-397 (1961); G. Chew, S. Frautchi and S. Mandelstam, *Phys. Rev.* **126**, 1202-1208 (1962).
2. V.N. Gribov, *Sov. Phys. JETP* **26**, 414-422 (1968).
3. J. Collins, *Phys. Rev. D* **57**, 3051-3056 (1998); erratum *ibid.* **61**, 019902 (2000).
4. G. Ingelman and P. Schlein, *Phys. Lett. B* **152**, 256-260 (1985).
5. M. Kapishin, "Measurements of inclusive diffraction at HERA", Proc. 32nd ICHEP, Beijing, 2004, pp. 724–727.
6. ZEUS Coll., S.Chekanov *et al.*, Preprint DESY-05-011 (2005).
7. O. Gutsche, "Diffractive charm and jet production", Proc. 32nd ICHEP, Beijing, 2004, pp. 728–731.
8. H1 Coll., paper 6-0177 submitted to 32nd ICHEP, Beijing (2004).
9. ZEUS Coll., paper 293 subm. to LP05, Uppsala, Sweeden (2005).
10. H1 Coll., paper 89 subm. to EPS 2003; P. Newmann and F.-P. Schilling, "Diffractive structure function data and PDF fits", in *HERA and the LHC Workshop* Proceedings (2005).
11. B.W. Harris and J. Smith, *Nucl. Phys. B* **452**, 109-160 (1995).
12. S. Catani and M.H. Seymour, *Nucl. Phys. B* **485**, 291-419 (1997); [*erratum* ibid. **510**, 503-504 (1997)].
13. S. Frixione, Z. Kunszt and A. Signer, *Nucl. Phys. B* **467**, 399-442 (1996); S. Frixione, *Nucl. Phys. B* **507**, 295-314 (1997).
14. A.B. Kaidalov, V.A. Khoze, A.D. Martin and M.G. Ryskin, *Phys. Lett. B* **567**, 61-68 (2003).
15. M. Klasen and G. Kramer, Preprint DESY-04-011, `hep-ph/0401202`, (2004).
16. T. Tawara, "Jet production in diffractive processes", EPS 2005, Lisbon, Portugal (2005).

Hard Diffractive Results and Prospects at the Tevatron

Krisztian Peters

Department of Physics & Astronomy, University of Manchester, Manchester M13 9PL, UK

Abstract. We review hard diffractive results and prospects at the Tevatron with an emphasis on factorization breaking in diffractive processes. Upper limits on the exclusive di-jet and χ_c^0 production cross sections at CDF and the status of the DØ Forward Proton Detectors are discussed.

Keywords: Diffraction, Factorization, Structure Functions, CDF, DØ
PACS: 12.38.Qk, 13.85.-t, 29.40Vj

DIFFRACTION AT THE TEVATRON

Diffractive events are mediated by the exchange of color singlets with vacuum quantum numbers and have clear experimental signatures. These are on one hand rapidity gaps: the absence of particles in some regions of rapidity in contrast to non-diffractive events where gaps are filled by additional soft parton interactions which yields an exponential suppression of rapidity gaps. On the other hand tagged protons or anti-protons: p or $\bar{p}$ scattered at small angle and measured in Roman Pots far away from the interaction point. Depending on these rapidity gaps and tags, the main diffractive event topologies at the Tevatron are: single diffraction (SD), double diffraction (DD) or double Pomeron exchange (DPE). Single diffraction is characterized by a leading proton or anti-proton which escapes the collision intact and the presence of a further particle or a di-jet resulting from a hard scattering separated by a rapidity gap. Double diffraction is characterized by a gap in the central region and dissociated protons and anti-protons. In DPE there is a gap on both, the proton and anti-proton side, with a central produced di-jet or other particles. Proton and anti-proton remains intact.

In RUN II the CDF Forward Proton Detectors (FPDs) include Roman Pot detectors approximately 57m from the interaction point. These consist of three stations and each station comprises one scintillation fiber detector and one trigger counter. Beam Shower Counters, a set of scintillation counters around the beam pipe, are used to reject non-diffractive (ND) background at the trigger level. This makes it possible to collect diffractive data at high luminosities. The energy flow in the event in the very forward direction is measured by Miniplug Calorimeters. These consist of alternating layers of lead plates and liquid scintillator readout. It has a towerless geometry without dead regions due to the lack of internal mechanical boundaries. The DØ FPDs are described in the following.

CP828, *Multiparticle Dynamics*
edited by V. Šimák, M. Šumbera, Š. Todorova, and B. Tomášik

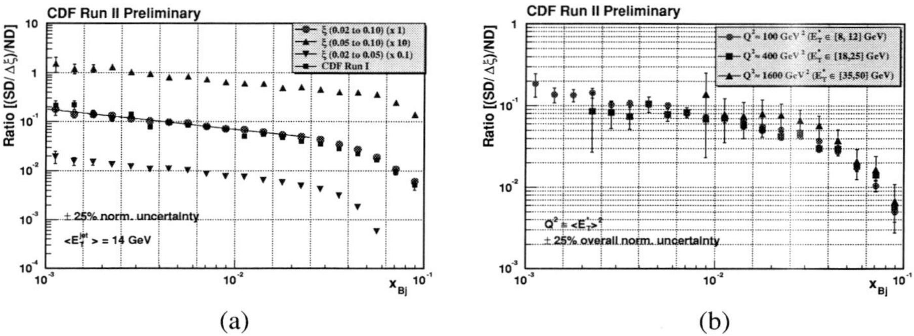

(a) (b)

FIGURE 1. Ratio of diffractive to non-diffractive di-jet event rates as a function of x_{B_j} at CDF for different ξ ranges and compared to Run I (a) and for different Q^2 values.

FACTORIZATION IN DIFFRACTION

One of the central issues of diffraction is whether hard diffractive processes obey QCD factorization. As Collins proved [1] for the general class of diffractive DIS processes, the cross section can be described as the convolution of a hard scattering matrix element (process dependent) and parton density functions (process independent). The question arises if this factorization theorem is more general, are parton densities really universal in diffractive exchange? Can we use them for different collider processes and energies? To answer this question is fundamental for the understanding of diffraction and it is also important to extrapolate Tevatron results to the LHC. The general strategy to prove factorization is to extract parton density functions (PDFs) and compare predictions to measurements of other processes and experiments.

Before the extraction of PDFs, diffractive fractions already yield some insight. Both, CDF and DØ, measured for different processes the fractions of events with one gap to all events. It was found that all ratios are at the order of 1% at the Tevatron which would support a more general factorization theorem. However the ratio of 1% yields an uniform gap suppression w.r.t. HERA where the diffractive rates are approximately 10 times higher. This discrepancy indicates already the breakdown of QCD factorization.

CDF measured in Run I the diffractive structure function of the anti-proton from SD di-jets and the result was compared with expectations from diffractive DIS measurements of H1 [2]. These events can be described in terms of a Pomeron emitted from the anti-proton and scattering with a parton from the proton. The diffractive structure function was obtained by measuring the ratio of the diffractive and non-diffractive cross section. The product of this ratio with the known non-diffractive structure function gives the result on the diffractive structure function. Although the shapes of the structure functions from HERA and Tevatron have similar shapes, there is a normalization discrepancy of a factor of 10. This result again confirms the breakdown of QCD factorization between the Tevatron and HERA as already expected from the result on the diffractive fractions. One possibility to explain these observations is that, since there are more spectator partons in $p\bar{p}$ collisions w.r.t. $\gamma^* p$ collisions, the rate of gap destructions due to soft partonic

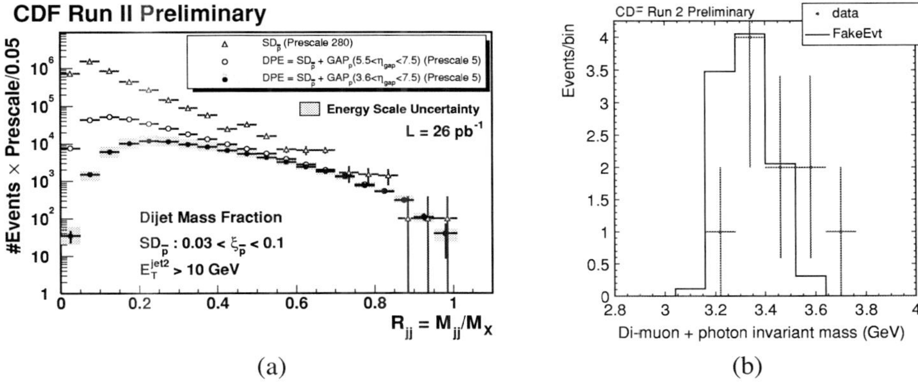

FIGURE 2. Di-jet mass fraction for different rapidity gap selections (a) and di-muon plus photon invariant mass in the exclusive sample compared to Monte Carlo predictions (b) at CDF.

interactions is larger at the Tevatron. One attempt to describe this rate is made with the introduction of the concept of the "gap survival probability" $|S|^2$ [3, 4]. The observation that the shapes of the two structure functions from HERA and the Tevatron are similar supports this concept. The gap survival probability factor corrects the normalization discrepancy.

In Run II the diffractive di-jet sample was collected with a dedicated trigger which selects events with at least one calorimeter tower above the 5 GeV E_{T} threshold and a threefold Roman Pot Spectrometer coincidence. Calorimeter information is used to determine the momentum loss of the anti-proton,

$$\xi_{\bar{p}} = \frac{\sum E_T e^{-\eta}}{\sqrt{s}}. \tag{1}$$

SD and background regions are selected according to the measured $\xi_{\bar{p}}$ values. A large number of events are at $\xi_{\bar{p}} \sim 1$ which are due to the overlap of at least one ND contribution. A plateau is observed in the $\xi_{\bar{p}}$ distribution which results from a distribution proportional to $1/\xi_{\bar{p}}$ as expected for diffractive production.

In Fig. 1 the ratio of SD to ND event rates is plotted versus Bjorken x_{B_j}. In Fig. 1(a) this ratio is integrated over three different ξ regions and compared to the Run I result. There is no ξ dependence observable between 0.03 and 0.1. Furthermore the slope and normalization agrees and thus confirms the Run I result. In Fig. 1(b) the same ratio is plotted for three different Q^2 values, where Q^2 is the mean jet transverse energy $\langle (E_{\mathrm{T}}^1 + E_{\mathrm{T}}^2)/2 \rangle^2$. There is no appreciable Q^2 dependence in the observable region of $100 < Q^2 < 1600$ GeV^2. Thus both structure functions seem to have a similar Q^2 evolution which does not disfavor any of the two mechanisms of hard diffraction, namely the existence of a hard Pomeron (exchange of a colorless object) or a soft color rearrangement in the final state.

Using the di-jet sample CDF extracted also DPE di-jet production events. These events are characterized by a leading anti-proton, two jets in the central pseudorapidity region and a large rapidity gap on the outgoing proton side. If factorization holds the ratio of DPE to SD, $R_{SD}^{DPE}(x_p, \xi_p)$ has to be equal to the ratio of SD to ND, $R_{ND}^{SD}(x_{\bar{p}}, \xi_{\bar{p}})$ (in LO QCD) for a fixed x_{B_j} and ξ value. Although the collected events have different ξ ranges for the proton and anti-proton, the weighted average of $R_{ND}^{SD}(x_{\bar{p}}, \xi_{\bar{p}})$ is flat in ξ and the ratio was extrapolated to $\xi = 0.02$. At this ξ values the above mentioned ratios differ by a factor of 5 [5]. The deviation from unity yields again a breakdown of factorization. Since the formation of a second gap is less suppressed this result is coherent with the concept of the gap survival probability. The number of spectator partons does not change with the formation of a second gap, i.e. one does not have to pay the price for the gap two times.

In the same manner as was previously done by the SD/ND ratio, the diffractive structure function can be extracted from the DPE/SD ratio. The obtained result now approximately agrees with expectations from H1 leading again to the above mentioned conclusions.

SEARCH FOR EXCLUSIVE EVENTS

Since the CDF Roman Pots have been installed at the end of Run I, CDF collected in Run II two orders of magnitude more di-jet data which made a study of exclusive di-jet production in DPE possible. The strategy was to obtain an inclusive DPE di-jet event sample and look for exclusive signature using the di-jet mass fraction $R_{jj} = M_{jj}/M_X$ where the di-jet mass is divided by the mass of the rest of the system excluding the proton and anti-proton. In Fig. 2(a) the obtained number of events is plotted versus the di-jet mass fraction where no gap, a narrow gap and a wide gap was required on the proton side. The result is a smoothly falling spectrum all the way down to $R_{jj} = 1$ and a similar event yield at high mass fraction regardless of the gap requirements. From this it follows that no significant excess due to exclusive di-jets is seen at high R_{jj}. An upper limit on the exclusive di-jet production cross section is calculated based on events with $R_{jj} > 0.8$. For example requiring a minimum jet transverse energy of 10 GeV, this upper limit is

$$E_{\mathrm{T}}^{\min} = 10\,\mathrm{GeV}: \quad \sigma(R_{jj} > 0.8) < 1.1 \pm 0.1(stat) \pm 0.5(sys)\ \mathrm{nb}. \qquad (2)$$

There are also other production channels available in DPE, we mention here the exclusive χ_c^0 production. This process is of particular interest, since the χ_c^0 quantum numbers are very similar to the ones of the Higgs boson, thus it can be used to test and normalize the predictions for exclusive Higgs production at the LHC. CDF did an analysis in Run II where the χ_c^0 further decays in a $J/\psi + \gamma$. 93 pb^{-1} of di-muon triggered data was used and events have been selected in the J/ψ mass window. A large rapidity gap on both the proton and anti-proton side was required. Ten events have been found which are exclusive $\chi_c^0(\rightarrow J/\psi + \gamma)$ candidates. In Fig. 2(b) the di-muon plus photon invariant mass of the 10 events is compared with a sample of generated χ_c^0 events passed through a detector simulation. The invariant mass is consistent with that

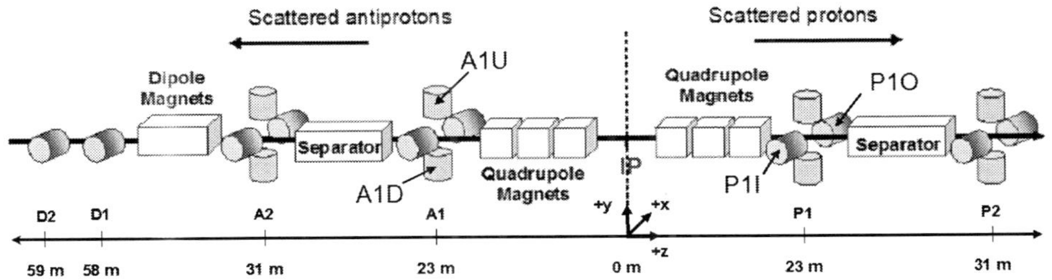

FIGURE 3. The Forward Proton Detector at DØ. Quadrupole Pots are named P or A when placed on the proton or the anti-proton side, respectively. Dipole Pots are named D.

of the χ_c^0, although the mean mass is higher and the distribution broader in the data than in the simulation. This may be due to the simulation or there may be contributions from cosmic events, higher mass χ_c mesons or $J/\psi+\pi^0$ events. Since it is very difficult to fully understand the background one may calculate an "upper limit" on exclusive χ_c^0 production assuming that the observed 10 events are all $J/\psi+\gamma$ events. This upper limit is:

$$\sigma = 49 \pm 18(stat) \pm 39(sys) \text{ pb}. \tag{3}$$

DØ FORWARD PROTON DETECTORS

The DØ Forward Proton Spectrometers have been installed and recently commissioned. These are 9 momentum spectrometers with 2 scintillating fiber detectors each, as schematically shown in Fig. 3. There is a dipole spectrometer located on the scattered anti-proton side behind the dipole magnets approximately 58 m away from the interaction point. They have in the range of $|t| \approx 0-1$ GeV^2 and $\xi \approx 0.03-0.07$ a very good acceptance. Eight quadrupole spectrometers are on both side of the main detector approximately 23 and 31 m away from the interaction point, behind the quadrupole magnets and the separators. They have a very good acceptance in the region of: $|t| \approx 0.8-3.0$ GeV^2 and $\xi \approx 10^{-3}-0.05$. The position detectors housed inside Roman Pots operate millimeters away from the beam, however outside of the ultra high vacuum. They enable the reconstruction of the high energy protons and anti-protons directly, thus providing a first time possibility of tagging both the protons and anti-protons and measuring their ξ and t dependence at the Tevatron.

The DØ scintillating fiber detectors have 6 layers of scintillating fiber channels and one trigger scintillator layer. The fibers are oriented within $\pm 45^o$ to reconstruct hits and obtain redundancy. Furthermore every second channel is offset by 2/3 fiber for a finer hit resolution. All 18 detectors regularly brought close to the beamline and diffractive samples being collected.

In a small dedicated test run of the FPDs in a stand alone mode the slope of the elastic cross section of proton anti-proton scattering was measured. In Fig. 4 the result (not normalized) is plotted and compared with theory predictions [6]. DØ access a new kinematic domain with these measurements and the result of the slope agrees well

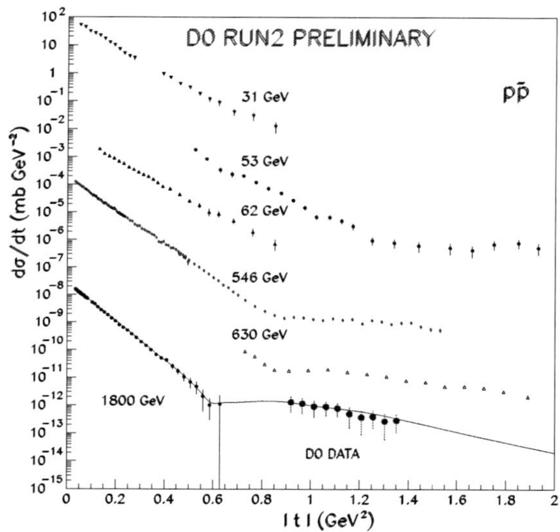

FIGURE 4. The slope of the elastic $p\bar{p}$ cross section measured at DØ and compared to predictions of [6] (solid line).

with the model of Bock *et al.* [6]. This measurement is being redone using the fully integrated FPDs. The alignment and detector understanding of the Roman Pot detectors is in progress and more physics results are expected soon.

REFERENCES

1. J. C. Collins, *Phys. Rev. D* **57**, 3051–3056 (1998), hep-ph/9709499.
2. T. Affolder, et al., *Phys. Rev. Lett.* **84**, 5043–5048 (2000).
3. J. D. Bjorken, *Phys. Rev. D* **47**, 101–113 (1993).
4. E. Gotsman, E. M. Levin, and U. Maor, *Phys. Lett. B* **309**, 199–204 (1993), hep-ph/9302248.
5. T. Affolder, et al., *Phys. Rev. Lett.* **85**, 4215–4220 (2000).
6. M. Block, R. Fletcher, F. Halzen, B. Margolis, and P. Valin, *Phys. Rev. D* **41**, 978–981 (1990).

High p_T Suppression in Au+Au at $\sqrt{s_{NN}} = 200$ GeV Measured with BRAHMS

Catalin Ristea* for the BRAHMS Collaboration

*The Niels Bohr Institute, University of Copenhagen, Blegdamsvej 17, 2100 Copenhagen Ø, Denmark

Abstract. We present the pseudorapidity evolution of high p_T suppression for unidentified charged hadrons, from $\eta \sim 0$ to $\eta \sim 4$. The study provides a test of the correlation between high p_T suppression and the total multiplicity of charged particles. High p_T spectra and nuclear modification factors R_{AA} and R_{CP} for inclusive charged hadrons will be discussed.

PACS: 25.75.-q

INTRODUCTION

From the study of nucleon-nucleon interactions it is known that when two partons undergo a scattering with large momentum transfer Q^2 in the early stages of the collision, the hard-scattered partons fragment into jets of hadrons with high transverse momentum ($p_T > 2$GeV/c) [1]. When the hard scattered partons traverse the hot and dense nuclear matter that is created in a high energy heavy ion collision, they may lose energy through gluon bremsstrahlung. The energy loss depends on the density of color charges in the matter which parton propagates [2, 3]. This effect is called jet-quenching or high p_T suppression, termed as final state effect, whereas modifications of the parton distribution functions, gluon shadowing or saturation are initial state effects. The most directly measurable consequence is the suppression of high transverse momentum hadrons in the final state.

BRAHMS has shown that in d-Au collisions at 200 GeV the nuclear modification factors are less than unity in the forward region, suggesting that initial state effects are important for $\eta > 0$, in contrast to the situation for $\eta \sim 0$ [4]. In the Au-Au system, we observe that the suppression of high p_T particles persists to higher rapidities suggesting that the medium causing the nuclear modification is longitudinally extended [5, 6]. In order to disentangle initial and final state effects in Au+Au collisions, we use a high statistics data set from 200 GeV to study charged hadron yields.

DATA ANALYSIS

The data presented here were collected with BRAHMS detector system [7]. BRAHMS consists of a set of global detectors for event characterization and two magnetic spectrometers, the mid-rapidity spectrometer (MRS) and the forward spectrometer (FS), which identify charged hadrons over a broad range in rapidity and transverse momen-

CP828, *Multiparticle Dynamics*
edited by V. Šimák, M. Šumbera, Š. Todorova, and B. Tomášik

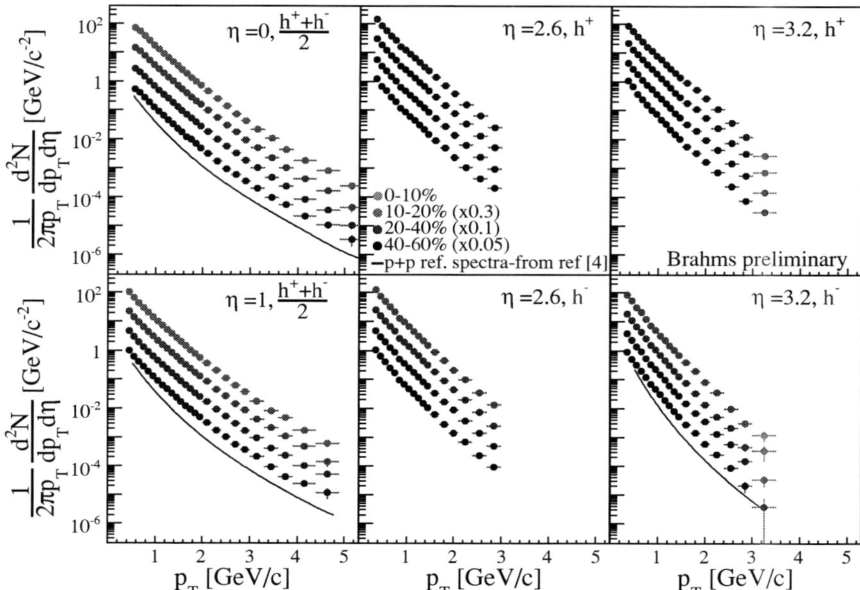

FIGURE 1. Invariant p_T spectra for charged hadrons produced in Au+Au collisions at $\sqrt{s_{NN}} = 200$ GeV GeV at $\eta = 0$, $\eta = 1$, $\eta = 2.6$, $\eta = 3.2$ for 0-10%, 10-20%, 20-40% and 40-60% centrality. The p+p reference spectra (from [4]) used for calculating R_{AA} are shown with solid lines. For clarity, some spectra have been divided by the indicated factors.

tum. Collision centrality is determined from the charged particle multiplicity measured by a set of global detectors.

Since BRAHMS is a small solid angle device, the charged particle spectra are obtained by mapping out the particle phase space by collecting data with many different spectrometer settings. BRAHMS is the only experiment from RHIC to perform detailed measurements away from midrapidity.

When combining different settings, the deviation of the single settings spectra to the final spectra are within 20% in the low p_T regions. In the high p_T region, the different settings are consistent within the statistical errors.

RESULTS

Spectra

For the Au-Au data which are presented here, the midrapidity spectrometer was positioned at 90 and 40 degrees relative to the beam axis, and measured charged

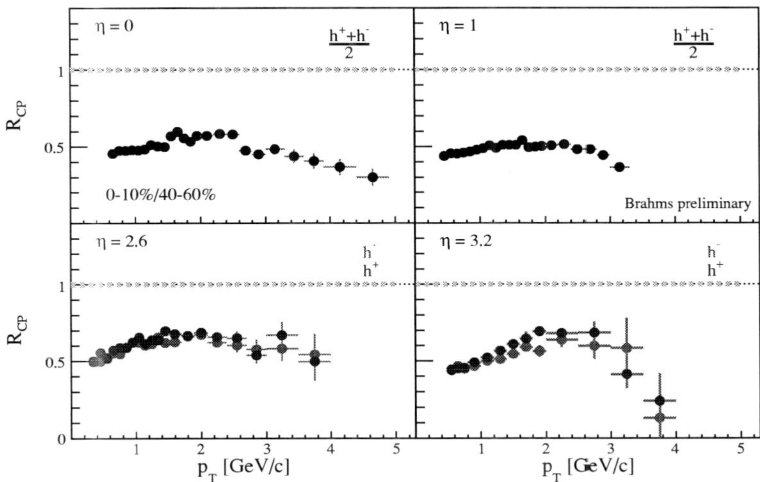

FIGURE 2. R_{CP} ratio for charged hadrons in Au+Au collisions at $\sqrt{s_{NN}}$=200GeV for pseudorapidities $\eta = 0, 1, 2.6, 3.2$ as a function of transverse momentum p_T.

hadrons at pseudorapidities in the range η ∈[-0.1, 0.1] and η ∈[0.9, 1.1] respectively ($\eta = -\ln(\tan(\theta/2))$, where θ is the angle of emission relative to the beam direction). The forward spectrometer (FS) was placed at 8 and 4 degrees, for the ranges in pseudorapidity [2.4, 2.8] and [3.0, 3.5] respectively. For the results presented here only the front part of the forward spectrometer (FFS) was used. At the most forward angle, the consistency with the full FS spectra has been investigated in order to identify the effects of increased background. The global detectors were used for the minimum bias trigger and event characterization. This trigger selects approximately 95% of the Au-Au interaction cross section. Spectrometer triggers are also used to enhance the track sample. The IP position is determined with a precision $\sigma < 0.85cm$ by the use of beam counters (BB) placed at $z = \pm 2.2m$.

Figure 1 shows the measured invariant p_T spectra for inclusive charged hadrons $(h^+ + h^-)/2$ at 90^0 and 40^0 (left panel), for negative hadrons (h^-) and positive hadrons (h^+) at 8° (middle panel) and at 4° (right panel), corresponding to $\eta = 0$, $\eta = 1$, $\eta = 2.6$ and $\eta = 3.2$. The displayed spectra correspond to centralities of 0-10%, 10-20%, 20-40% and 40-60% of the total interaction cross section. The spectra are from measurements at various magnetic fields (mostly high magnetic field chosen in order to increase the statistics at high p_T) and have been corrected for the acceptance of the spectrometers and for centrality dependent tracking efficiencies. No corrections for feed-down, decay or absorption have been applied for the FFS data.

All the charged hadron spectra exhibit a power law shape. At forward angles 90% of the particles are emitted in the region $p_T < 2GeV/c$.

The $dN/d\eta$ values extracted from fits to these data are consistent with the measured multiplicities for charged particles [8] extracted from the global detectors, for all the centrality cuts.

Nuclear Modifications

In order to determine the high p_T hadron suppression in nucleus-nucleus collisions, the hadron p_T spectra have to be compared to the reference data from nucleon-nucleon collisions at the same collision energy. The nuclear modification factor is defined as:

$$R_{AA}(p_T) = \frac{d^2N/dp_T d\eta}{T_{AA} d^2\sigma^{NN}_{inel}/dp_T d\eta} \tag{1}$$

where $T_{AA} = \langle N_{bin} \rangle / \sigma^{NN}_{inel}$ accounts for the collision geometry, averaged over the event centrality class. $\langle N_{bin} \rangle$, the number of binary NN collisions, is calculated using the Glauber model. σ^{NN}_{inel} and $d^2\sigma^{NN}_{inel}/dp_T d\eta$ are the cross section and differential cross section for inelastic nucleon-nucleon (NN) collisions, respectively.

In the absence of nuclear medium effects such as shadowing, the Cronin effect or gluon saturation, hard processes are expected to scale with $\langle N_{bin} \rangle$ and consequently, R_{AA}=1. Any deviation from unity indicate nuclear medium effects, especially for the high p_T region ($p_T > 2GeV/c$) where the hard production dominates.

In order to remove the systematic errors introduced by the comparison of the measurements of nucleus-nucleus and p+p collisions, we construct the ratio of central to peripheral collisions, R_{CP}, defined as:

$$R_{CP} = \frac{1/\langle N^C_{bin} \rangle}{1/\langle N^P_{bin} \rangle} \frac{d^2N_C/dp_T d\eta}{d^2N_P/dp_T d\eta} \tag{2}$$

where $dN^2_{C(P)}/dp_T d\eta$ are the differential yields in a central (peripheral) collision, respectively. Nuclear medium effects are expected to be much stronger in central relative to peripheral collisions, which makes R_{CP} another measure of these effects. If the yield of the process scales with the number of binary collisions, R_{CP}=1.

Figure 2 shows the pseudorapidity dependence of the R_{CP} ratio in Au-Au collisions, at $\eta = 0$, $\eta = 1$, $\eta = 2.6$ and $\eta = 3.2$. The observed suppression is similar at forward rapidities as compared to midrapidity. This result may indicate that quenching extends also in the longitudinal direction.

It has been proposed that this suppression at forward rapidity might be related to the initial conditions of the colliding nuclei, in particular to the possible formation of the Color Glass Condensate (CGC) in the initial state at RHIC [9]. This effect has been observed in d-Au collisions at $\eta \sim 3$ [4].

Figure 3 shows the R_{AA}, as a function of p_T for two centrality cuts for the Au-Au measurements at $\eta = 0$, $\eta = 1$ and $\eta = 3.2$. For the most central $(0-10\%)$ bin we use $N_{bin} = 897 \pm 117$, and for the most peripheral $(40-60\%)$ $N_{bin} = 78 \pm 26$. All the $p+p$ reference spectra used are from ref. [4]. The R_{AA} rise from values of 0.2–0.4 at low p_T to a maximum at $p_T \approx 2$ GeV/c. The low p_T part of the spectrum is associated with soft

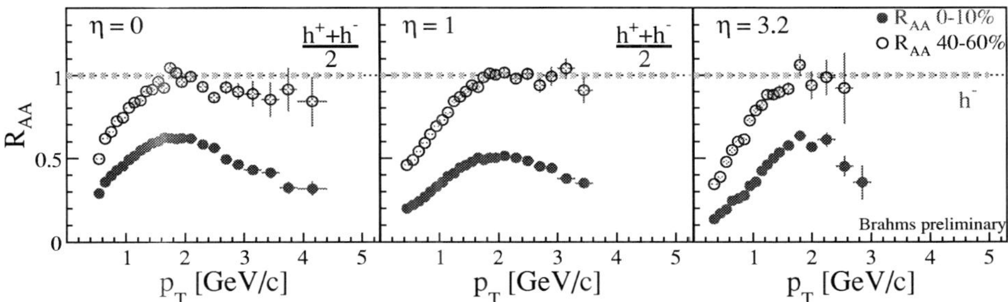

FIGURE 3. Nuclear modification factor for charged hadrons at pseudorapidities $\eta = 0, 1, 3.2$. Red points are the R_{AA} for the most central collisions (0-10% centrality) and blue points are the R_{AA} for semi-peripheral collisions (40-60% centrality).

collisions and should therefore scale with the number of participants. Thus the applied scaling with the (larger) N_{bin} value reduces R_{AA} at the lower p_T.

For central collisions, above $p_T \approx 2$ GeV/c, the R_{AA} distributions decrease and are systematically lower than unity. We observe that the suppression persists over a wide range in pseudorapidity and it is almost constant for all the studied angles. As expected, the semi-central events follow the pp spectra, and the R_{AA} reach the unity in the p_T range [1.5, 2.0] GeV/c for all data. The observed behavior of nuclear modification factors is consistent with jet surface emission [10, 11], modeled in the frame of PQM.

The study of nuclear modification factors for the identified particle performed over the BRAHMS wide rapidity coverage could bring additional information on the high p_T suppression phenomenon observed at RHIC.

CONCLUSIONS

BRAHMS has measured the pseudorapidity dependence of nuclear modification factors in Au-Au collisions. In Au-Au collisions the suppression persists over 3 units in pseudo-rapidity, indicating that the hot and dense partonic matter may be extended also along the beam direction. The similarity observed in the forward direction both in Au-Au and d-Au collisions at large pseudorapidities suggest that initial state effects play a significant role in this region.

REFERENCES

1. J.F.Owens *et al.*, *Phys. Rev. D* **18**, 1501-1514 (1978).
2. M. Gyulassy, M. Plumer, *Phys. Lett. B* **243**, 432-438 (1990).
3. X. N. Wang, M. Gyulassy, M. Plumer, *Phys. Rev. D* **51**, 3436-3446.
4. BRAHMS Collaboration, I. Arsene *et al.*, *Phys. Rev. Lett.* **93**, 242303 (2004).
5. C. E. Jørgensen, Ph.D. thesis, The Niels Bohr Institute, september 2004.

6. BRAHMS Collaboration, I. Arsene *et al.*, *Phys. Rev. Lett.* **91**, 072305 (2003).
7. BRAHMS Collaboration, M. Adamczyk *et al.*, *Nucl. Instr. Meth. A* **499**, 437-468 (2003).
8. BRAHMS Collaboration, I. Bearden *et al.*, *Phys. Rev. Lett.* **88**, 202301 (2002).
9. L. McLerran, R. Venugopalan, *Phys. Rev. D* **49**, 2233-2241 (1994); *Phys. Rev. D* **49**, 3352-3355 (1994); *Phys. Rev. D* **50**, 2225-2233 (1994).
10. K.J. Eskola, H. Honkanen, C.A. Salgado, U.A. Wiedemann, *Nucl. Phys.* **A747**, 511-529 (2005).
11. A. Dainese, C. Loizides, G. Paic, *Eur. Phys. J. C* **38**, 461-474 (2005).

Factorization in Hard $\gamma - p$, $\gamma^* - p$ and $p - p$ Scattering

A. Bialas

M. Smoluchowski Institute of Physics, Jagellonian University, Reymonta 4, 30-059 Kraków, Poland

Abstract. Starting from the idea that the diffractive collisions reflect the absorption of the incident particle wave, it is argued that one should expect a strong factorization breaking between $\gamma - p$ and $p - p$ diffractive cross-sections, as well as between two-gap, one-gap and no-gap cross-sections in $p - p$ collisions. One the other hand, there are no "absorptive" corrections which would destroy factorization of $\gamma - p$ and $\gamma^* - p$ diffractive cross-sections.

Keywords: Factorizaton, hard scattering, dijets
PACS: 13.15.+g,13.87.-a

INTRODUCTION

This presentation is based on the paradigm that the diffractive dissociation processes are *driven* by absorption of the incident particle wave [1], elegantly formulated by Good and Walker [2]. The idea is that the state vector of the incident particle can be decomposed into quantum mechanical superposition of "diffractive states" having the same additive quantum numbers

$$|\psi> = a_1|\hat{\psi}_1> + a_2|\hat{\psi}_2> + ... \tag{1}$$

In the following I shall work under the assumption that the diffractive states are the partonic states with a fixed number and transverse positions of partons [3].

A diffractive state is characterized by the condition that, when interacting with the target, it does not produce any diffractive dissociation: it can only scatter elastically or be absorbed, i.e. change into a state with quantum numbers different from those of the original state:

$$T|\hat{\psi}> = t|\hat{\psi}> + \omega|\phi> \tag{2}$$

where t is the elastic amplitude while $|\omega|^2$ represents the probability of absorption (state $|\phi>$ is orthogonal to all diffractive states). In QCD, the diffractive states are states without colour (as is the incident particle), while $|\phi>$ is a state with a non-vanishing colour quantum number (thus the transition $|\hat{\psi}> \rightarrow |\phi>$ involves an exchange of colour between the projectile and target).

The expansion (1) is supposed to be *coherent* during the whole interaction process and therefore the masses of the diffractive states must satisfy the inequality

$$\frac{2E_{lab}}{M^2 - M_0^2 - Q^2} \gg d \tag{3}$$

CP828, *Multiparticle Dynamics*
edited by V. Šimák, M. Šumbera, Š. Todorova, and B. Tomášik

where d is the length of the target particle (in its rest frame). This condition implies the "rapidity gap" between the final state particles and the target.

Unitarity of the S-matrix implies that

$$(1-t)^2 = 1 - |\omega|^2 \;\rightarrow\; 2t = |t|^2 + |\omega|^2 \tag{4}$$

which simply expresses the fact that the total cross-section of a diffractive state ($2t$) is the sum of elastic (t^2) and non-diffractive ($|\omega|^2$) cross-sections (there is no diffractive dissociation, therefore non-diffractive and inelastic cross-sections are identical).

CROSS-SECTIONS FOR HARD PROCESSES

We restrict ourselves to discussion of *hard* processes, like production of two hard jets[1]. Writing the general Good-Walker expansion as

$$|m> = \sum_n <n|m> |\hat{n}> \tag{5}$$

where the diffractive states are denoted by $|\,\hat{}>$, it is not too difficult to derive the formulae for cross-sections of non-diffractive and diffractive interactions which lead to states with two hard jets.

Denoting by $|H_k>$ the collection of all *coloured* states containing the two hard jets we obtain

$$\sum_k \sigma(1 \rightarrow H_k)_{nondif} = \sum_{n^*} |\omega_{n^*}|^2 | <\hat{n}^*|1> |^2 \tag{6}$$

where the sum on the R.H.S. runs over all diffractive states which contain two hard jets[2]. The physical meaning of (6) is clear: each term in the sum on the R.H.S. is a product of the probability of a fluctuation of the initial particle into a diffractive state with two jets and the probability of the non-diffractive interaction of this state with the target.

Denoting by $|k>$ the *non-coloured* hadronic states containing two jets we have

$$\sigma(1 \rightarrow k)_{dif} = \sum_k | \sum_n <k|\hat{n}> t_n <\hat{n}|1> |^2. \tag{7}$$

At this point it is important to note that the expansion coefficients must satisfy the equality

$$\sum_n <k|\hat{n}><\hat{n}|1> = 0 \tag{8}$$

because the state $|k>$ is orthogonal to $|1>$.

---

[1] One may also consider production of heavy flavours or other hard probes.

[2] This is a shorthand. A better formulation is that these states contain one dipole of a very small transverse size which decays into two hard jets when partons are tranformed into hadrons.

REGGE FACTORIZATION AND FACTORIZATION OF THE PARTON DISTRIBUTION

To discuss factorization property of the amplitudes and of the cross-sections one should not mix up the Regge factorization and factorization of parton distributions.

In its simplest form, Regge factorization states that the elastic scattering amplitude can be written as a product of three factors: two vertices describing coupling of the Pomeron to the colliding objects and the Pomeron propagator. On the other hand, the factorization of parton densities refers to the analogous property of non-diffractive cross-sections.

In the Good-Walker language, these requirements should apply to the interaction of the diffractive states which play the fundamental role. Thus Regge factorization demands that each t_n factorizes into a product of a three factors: a vertex describing the couplig of the Pomeron to the target, a vertex decribing its coupling to the diffractive state $|\hat{n}>$ and the Pomeron propagator. The factorization of the parton distribution implies analogous requirement for the probability of inelastic interaction $|\omega_n|^2$.

In view of the non-linear unitarity relation (4) they cannot be both strictly valid at the same time, although—depending on the details of the dynamics—they both can represent a reasonable approximation.

It is clear that these properties of the elastic amplitudes and/or probabilities of inelastic interaction do not lead, in general, to a straightforward factorization of the cross-sections for different processes, as predicted by the simple Regge model. Indeed, the formulae (7) and (8) express the cross-sections as sums of many terms. Even if factorization holds for each of them, the total sum will not, in general, factorize.

Therefore the application of the factorization rules demands a good knowledge of the expansion (1) and of the properties of diffractive states. For this reason, in practice they can only be used for hard processes where one may hope that perturbative QCD provides a method to estimate these quantities with sufficient accuracy.

"ABSORPTIVE" CORRECTIONS

Assuming the factorization property of the parton distributions, one sees from (6) that it is applicable to all hard non-diffractive processes. It is of course necessary to estimate correctly the contributions from the projectile and therefore the procedure is far from trivial but the structure of Eq. (6) does not lead to any surprizes.

Thus our first conclusion is that when the proper fluctuation probabilities of the incident particle are taken into account, the factorization of the target parton distributions should hold for all non-diffractive processes. This is, of course, in perfect agreement with the factorization theorems derived from QCD.

The Good-Walker picture gives, however, very different results for diffractive processes. The main reason is that, as seen from (7), the diffractive processes are described by a sum of coherent amplitudes rather than by a sum of probabilities. The orthonormality of the diffractive states and of the produced hadronic states implies strong constraints on the amplitudes (c.f. (8)), leading to drastic violations of the naive factorization predictions.

The simplest illustration of this phenomenon is provided by comparison of the photon induced with hadron induced hard diffractive processes. Denoting by $|k>$ a colourless hadron state with two hard jets we obtain for the amplitudes $p \to k$ and $\gamma \to k$

$$T_{dif}(p \to k) = \sum_n <k|\hat{n}> t_n <\hat{n}|p>;$$
$$T_{dif}(\gamma \to k) = \sum_n <k|\hat{n}> t_n <\hat{n}|\gamma> \tag{9}$$

where the sum runs over all diffractive (parton) states $|\hat{n}>$ of the projectile. These two formulae seem identical but their content is different because the expansion in the diffractive states is rather different for a hadron and for the photon. The photon expansion is largely dominated by the diffractive state $|\hat{\gamma}>$, i.e., the state with no partons in it (and thus vanishing cross-section).

Expressing the amplitudes t_n as

$$t_n = <t> + \delta_n \tag{10}$$

we obtain

$$T_{dif}(p \to k) = \sum_n <k|\hat{n}> \delta_n <\hat{n}|p>;$$
$$T_{dif}(\gamma \to k) \approx - <k|\gamma><t> + \sum_n <k|\hat{n}> \delta_n <\hat{n}|\gamma> \tag{11}$$

where we have used the orthogonality relation (8) and the approximate equality

$$<k|\hat{\gamma}><\hat{\gamma}|\gamma> \approx <k|\gamma> \tag{12}$$

which follows from the condition $<\hat{\gamma}|\gamma> \approx 1$.

One sees from the Eq. (9) that both the average $<t>$ and fluctuations δ_n contribute to the photon induced processes. On the other hand, only fluctuations of the elastic amplitude around the average contribute to the hadronic diffractive dissociation cross-section [2]. Consequently, if the fluctuations of the scattering amplitudes δ_n are not large as compared to the average value $<t>$, one obtains a drastic reduction of the hadronic cross-section with respect to the one extrapolated naively from measurements of the photon induced reactions. This effect is often called "absorption" of the term $<k|p><t>$ which is apparently missing on the R.H.S. of the first line in (11)[3]. Several simple implementations [4, 5] of this general formulation were used to explain the strong factorization breaking observed between HERA and FERMILAB measurements [6].

Similar, although more complicated, phenomena are also present if one compares the no-gap and multi-gap hadron-induced processes measured recently by CDF collaboration [7]. A detailed explanation can be found in [8, 9]. For lack of space we shall not discuss them here[4].

---

[3] Actually it has nothing to do with any absorption. As we have seen, it is the effect of coherence of the interaction and orthogonality of the states.

[4] A more general treatment is being prepared [10].

DIFFRACTIVE $\gamma - P$, $\gamma^* - P$ TWO-JET PRODUCTION

Finally, let us consider the controversial issue of factorization in diffractive production of two hard jets by real photons and by deeply virtual photons. Since the formula (11) is clearly valid for both real and virtual photons, the Good-Walker approach does not predict any strong violation of factorization. Of course the cross-sections may differ because the diffractive states of real and virtual photons are not identical but there should not be any "absorptive" corrections of the type seen between the first and second line of (11).

This statement is in apparent contradiction with the results of the NLO calculations of Klasen and Kramer [11]. Their calculated cross-section at low Q^2 must be reduced by factor of about $1/3$ to obtain agreement with data, thus suggesting the necessity of an important "absorptive" correction. This argument is hardly convincing, however, because the difference between the LO and NLO cross-sections is very large [11] indicating that, most likely, a higher order calculation is needed to obtain a reliable estimate of the cross-section[5].

SUMMARY

To summarize, we have argued that the absorption picture of hard diffractive dissociation processes implies (i) validity of the factorization of the parton distributions in hard non-diffractive scattering; (ii) strong violation of the factorization between diffractive hadron-induced and photon-induced processes; (iii) strong violation of the Regge factorization between hard no-gap and multigap diffractive hadron-induced process; (iv) no important factorization breaking in dijet production by real and virtual photons. The qualitative features of this picture seem to be confirmed by existing data. The detailed quantitative description based on the available fits to parton distributions also does exist [4, 8].

ACKNOWLEDGMENTS

Most of the ideas presented in this note were developped in collaboration with Robi Peschanski. Discussions with Hannes Jung and Leszek Motyka were very helpful and are highly appreciated. Thanks are due to Vlád'a Šimák for the kind hospitality at Kroměříž.

---

[5] In absence of such high-order calculations (which is probably very difficult to perform), perhaps a better estimate can be obtained from the leading order formulae corrected by energy-momentum conservation (which, in any case, is responsible for a substantial part of the NLO contribution). This may be a possible justification of the recent result of the H1 collaboration, showing that the data are well described by a Monte-Carlo code employing the leading order formulae and including parton showering [12].

REFERENCES

1. I.Ya. Pomeranchuk and E.L. Feibnerg, *Doklady AN SSSR* **93**, 439 (1953).
2. M.L. Good and W.D. Walker, *Phys. Rev.* **120**, 1857-1860 (1960).
3. K. Fialkowski and L. van Hove, *Nucl. Phys. B* **107**, 211-220 (1976); H. Miettinen and J. Pumplin, *Phys.Rev. D* **18**, 1696-1708 (1978).
4. A.B. Kaidalov, V.A. Khoze, A.D. Martin and M.G. Ryskin, *Europ. Phys. J. C* **21** 521-529 (2001).
5. A. Bialas, *Acta Phys. Pol. B* **33**, 2635-2642 (2002).
6. T. Affolder et al.,CDF coll., *Phys. Rev. Lett.* **85**, 4215-4220 (2000).
7. T. Affolder et al.,CDF coll., *Phys. Rev. Lett.* **84**, 5043-5048 (2000).
8. A.B. Kaidalov, V.A. Khoze, A.D. Martin and M.G. Ryskin, *Phys. Lett. B* **559**, 235-238 (2003).
9. A. Bialas and R. Peschanski, *Phys. Lett. B* **575**, 30-36 (2003).
10. A. Bialas and R. Peschanski, in preparation.
11. M. Klasen and G. Kramer, *Phys. Rev. Lett.* **93**, 232002 (2004).
12. H1 coll., Paper submitted to 32 ICHEP04, Beijing (2004) and private communication from H.Jung.

Possible Saturation Effects at HERA and LHC in the k_T−Factorization Approach

A.V. Kotikov*, A.V. Lipatov† and N.P. Zotov †[1]

*N.N. Bogoliubov Laboratory of Theor. Physics, Joint Institute for Nuclear Research, 141980 Dubna, Russia

†D.V. Skobeltsyn Institute of Nuclear Physics, M.V. Lomonosov Moscow State University, 119992 Moscow, Russia

Abstract. We consider possible saturation effects in the structure function F_L at fixed W in HERA energy range and in $b\bar{b}$−production at LHC energies in the framework of k_T−factorization approach.

Keywords: QCD, small x, BFKL evolution, k_T−factorization
PACS: 12.38.-t, 13.60.-r, 13.87.Ce

INTRODUCTION

Remarkable progress in recent years has resulted from the observation that the gluon density in a proton at small x grows as x decreases. Both DGLAP and BFKL evolution equations predict this rapid growth of the parton densities, thus demonstrating the triumph of perturbative QCD. It is, however, clear that this growth cannot continue for ever, because it would violate the unitarity constraint [1]. Consequently, the parton evolution dynamics must change at some point, and a new phenomenon must come into play. As the gluon density increases, non-linear parton interactions are expected to become more and more important, resulting eventually in the slowdown of the parton density growth (known under the name of "saturation effect") [1, 2]. The underlying physics can be described by the non-linear Balitsky-Kovchgov (BK) equation [3]. It is expected that these nonlinear interactions lead to an equilibrium-like system of partons with some definite value of the average transverse momentum k_T and the corresponding saturation scale $Q_s(x)$. This equilibrium-like system is the so called Color Glass Condensate (CGC) [4]. Since the saturation scale increases with decreasing of x, one may expect that the saturation effects will be more clear at the LHC energies. In the preasymptotic region ($k_T \geq Q_s(x)$) of energies the heavy quark production is described more adequately [5] by the so called k_T−factorization approach [1, 6]. At $k_T < Q_s(x)$ the k_T−factorization approach gives us a chance to account for saturation effects (if they are under control the BK equation) using scale properties of the dipole model, which is equivalent the k_T factorization [7].

Here we demonstrate a description of possible saturation effects in the framework of k_T−factorization approach. As example of that we consider the structure function F_L at

[1] speaker at the conference

CP828, *Multiparticle Dynamics*
edited by V. Šimák, M. Šumbera, Š. Todorova, and B. Tomášik

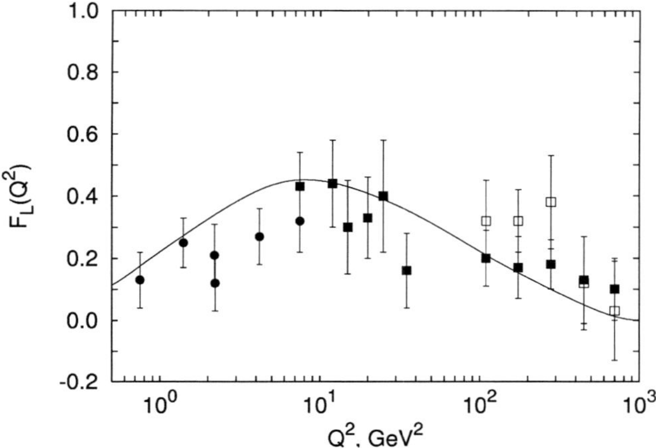

FIGURE 1. The Q^2 dependence of $F_L(x,Q^2)$ (at fixed $W = 276$ GeV). The experimental points are from [8]. Solid curve is the result of the k_T-factorization approach with the GLLM unintegrated gluon distribution from [9].

fixed W and low Q^2 in HERA energy range and $b\bar{b}-$quark production at LHC energies.

THEORETICAL FRAMEWORK AND NUMERICAL RESULTS

The structure function F_L at fixed W in HERA range

The SF $F_L(x,Q^2)$ is driven at small x primarily by gluons and in the k_T-factorization approach it is related in the following way to the unintegrated distribution $\Phi_g(x,k_\perp^2)$:

$$F_L(x,Q^2) = \int_x^1 \frac{dz}{z} \int^{Q^2} dk_\perp^2 \sum_{i=u,d,s,c} e_i^2 \cdot \hat{C}_L^g(x/z,Q^2,m_i^2,k_\perp^2)\, \Phi_g(z,k_\perp^2), \qquad (1)$$

where e_i^2 are charge squares of active quarks. The function $\hat{C}_L^g(x,Q^2,m_i^2,k_\perp^2)$ can be regarded as SF of the off-shell gluons with virtuality $k_\perp^2$ (hereafter we call them *hard structure function* by analogy with similar relations between cross-section and hard cross-section). They are described by the sum of the quark box (and crossed box) diagram contribution to the photon-gluon interaction (see, for example, Fig. 1 in [10]). To calculate the longitudinal SF $F_L(x,Q^2)$ we used the hard SF $\hat{C}_L^g(x,Q^2,m_i^2,k_\perp^2)$ obtained in Ref. [10] and different unintegrated gluon distributions. [2] Notice that the $k_\perp^2$-integral in Eq. (1) can be divergent at lower limit, at least for some parameterizations of $\Phi_g(x,k_\perp^2)$.

[2] Here we are interested in the parameterizations of unintegrated gluon functions which take into account saturations effects.

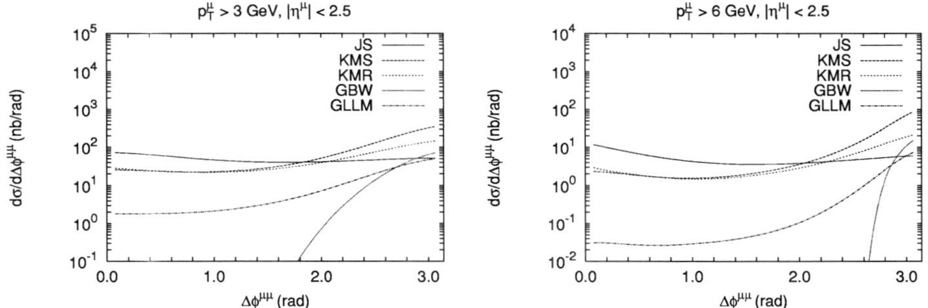

FIGURE 2. Azimuthal muon-muon correlations at LHC with different unintegrated gluon distributions

To overcome the problem we change the low Q^2 asymptotics of the QCD coupling constant within hard structure functions. We applied the so called "soft" version of "freezing" procedure [11],which contains the shift $Q^2 \to Q^2 + M^2$, where M is an additional scale, which strongly modifies the infrared α_s properties. For massless produced quarks, ρ-meson mass m_ρ is usually taken as the M value, i.e. $M = m_\rho$. In the case of massive quarks with mass m_i, the $M = 2m_i$ value is usually used. We calculate the SF F_L as the sum of two types of contributions - the charm quark one F_L^c and the light quark one F_L^l:

$$F_L \; = \; F_L^l + F_L^c. \tag{2}$$

For the F_L^l part we used the massless limit of hard SF and restricted ourselves to the modification of the argument in the strong coupling constant of the hard SF only. We have shown [12] that our k_T-factorization results are in good agreement with the data for large and small part of the Q^2 range. However, there was some disagreement between the data and theoretical predictions at $Q^2 \sim 3$ GeV^2. The disagreement comes from two possible reasons: additional higher-twist contributions, which are important at low Q^2 values[3], or/and NLO QCD corrections.

It was shown that the saturation (non-linear QCD) approaches contain information of all orders in $1/Q^2$, they resum higher-twist contributions [13]. The analysis of the behavior of the longitudinal structure function $F_L(x,Q^2)$ in the saturation models was done in Ref. [14][4]. In Fig.1 we demonstrate our k_T-factorization description of $F_L(Q^2)$ at fixed W with the unintegrated gluon distribution proposed in Ref. [9] which takes into account non-linear (saturation) effects. We see very well description of H1 experimental data at fixed W in all Q^2 region.

[3] Some part of higher-twist contributions was took into account by the 'freezing" procedure.
[4] N.Z. thanks M.V.T. Machado for useful discussion of this problem.

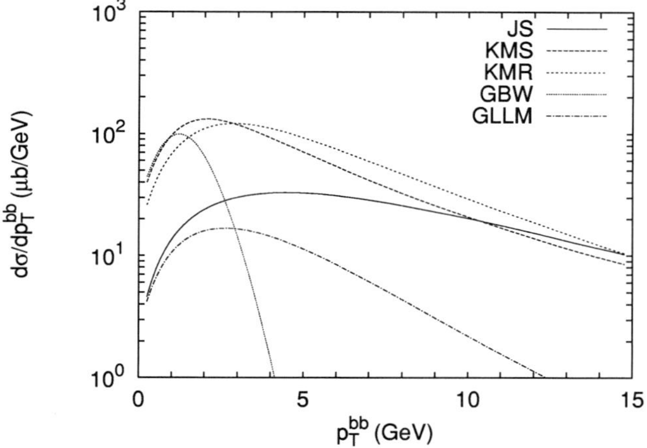

FIGURE 3. The total transverse momentum distribution at LHC with different unintegrated gluon distributions

Saturation effects in $b\bar{b}-$production at LHC

It was shown that the data on $b\bar{b}$ azimuthal correlations at Tevatron (measured as the decay muon ones, $d\sigma/d\Delta\phi^{\mu\mu}$) are much more informative to distinguish different unintegrated gluon distributions [15]. In Fig. 2 we show our predictions for the $b\bar{b}$ azimuthal correlations at LHC obtained with different unintegrated gluon distributions (see, for example, [16]). The GBW and GLLM ones take into account saturation effects. We see that the latter two unintegrated gluon densities reduce to different behavior of differential cross section $d\sigma/d\Delta\phi^{\mu\mu}$. Figure 3 displays the total transverse momentum distribution, $d\sigma/dp_T^{b\bar{b}}$, where $p_T^{b\bar{b}} = p_T^b + p_T^{\bar{b}}$, at LHC energy obtained with the same unintegrated gluon densities. In this case the difference between the results with usual unintegrated gluon densities and ones obtained with account of the non-linear (saturation) effects is more dramatic.

CONCLUSIONS

We considered in the framework of k_T-factorization approach possible manifestation of the saturation effects at HERA and LHC. We shown that account of these effects (in particle, by the GLLM unintegrated gluon distribution) improves to a marked degree description of the F_L data at low Q^2 at HERA energy. We demonstrated also that the experimental data for $b\bar{b}$ azimuthal correlations and $p_T^{b\bar{b}}-$distribution at LHC will give us additional possibilities to study non-linear (saturation) effects.

ACKNOWLEDGMENTS

N.Z. thanks DESY directorate for financial support and Organizing Committee for hostly and friendly atmosphere during the Symposium.

REFERENCES

1. L.V. Gribov, E.M. Levin and M.G. Ryskin, *Phys. Rep.* **100**, 1–150 (1983).
2. A.H. Mueller and J. Qiu, *Nucl. Phys. B* **268**, 427–452 (1986);
   L.D. Mclerran and R. Venugopalan, *Phys. Rev. D* **49**, 2233–2241, 3352–3355 (1994); **50**, 2225–2233 (1994); **53**, 458–475 (1996); **59**, 094002 (1999);
   K. Golec-Biernat and M. Wusthoff, *Phys. Rev. D* **59**, 014017 (1999); **60**, 114023 (1999);
   A.H. Mueller, hep-ph/9911289.
3. I.I. Balitsky, *Nucl. Phys. B* **463**, 99–160 (1996);
   Y.V. Kovchegov, *Phys. Rev. D* **60**, 034008 (1999).
4. M. Gyulassy and L. McLerran, *Nucl. Phys. A* **750**, 30–63 (2005).
5. F. Gelis, R. Venugopalan, *Phys. Rev. D* **69**, 014019 (2004);
   D. Kharzeev, K. Tuchin, *Nucl. Phys. A* **735**, 248–266 (2004);
   R. Baier, A.H. Mueller, D. Schiff, *Nucl. Phys. A* **741**, 358–380 (2004).
6. E.M. Levin, M.G. Ryskin, Yu.M. Shabelsky and A.G. Shuvaev, *Yad. Fiz.* **53**, 1059–1076 (1991);
   S. Catani, M. Ciafaloni and F. Hautmann, *Nucl. Phys. B* **366**, 135–188 (1991);
   J.C. Collins and R.K. Ellis, *Nucl. Phys. B* **360**, 3–30 (1991).
7. V. Barone, M. Genovese, N.N. Nikolaev et al., *Phys. Lett. B* **326**, 161–167 (1994);
   A.Bialas, H. Navelet and R. Peschanski, *Nucl. Phys. B* **593**, 438–450 (2001).
8. E.M. Lobodzinska, *Acta Phys. Polon. B* **35**, 223–227 (2004); hep-ph/0311180.
9. G.Gotsman, E. Levin, M. Lublinsky and U. Maor, *Eur. Phys. J. C* **27**, 411–425 (2003).
10. A.V. Kotikov, A.V. Lipatov, G. Parente, N.P. Zotov, *Eur. Phys. J. C* **26**, 51–66 (2002).
11. N.N. Nikolaev and B.M. Zakharov, *Z. Phys. C*, **49**, 607–618 (1991); **53**, 331–346 (1992).
12. A.V. Kotikov, A.V. Lipatov and N.P. Zotov, *JETP*, **101**, 811–816 (2005).
13. J. Bartels, K. Golec-Biernat, K. Peters, *Eur. Phys. J. C*, **17**, 121–128 (2001);
    E. Gotsman, E. Levin, U. Maor, L.D. McLerran, K.Tuchin, *Nucl. Phys. A*, **683**, 383–405 (2001).
14. V.P. Goncalves, M.V.T. Machado, *Eur. Phys. J. C* **37**, 299–305 (2004).
15. S.P. Baranov, N.P. Zotov, A.V. Lipatov, *Phys. Atom. Nucl.* **67**, 837–845 (2004).
16. B. Andersson et al. (Small x Collaboration), *Eur. Phys. J. C* **25**, 77–101 (2002).

$k_\perp$ Factorization and Quark Production from the Color Glass Condensate

H. Fujii*, F. Gelis† and R. Venugopalan**

*Institute of Physics, University of Tokyo, Komaba, Tokyo 153-8902, Japan
†CEA/DSM/SPhT, 91191 Gif-sur-Yvette cedex, France
**Physics Department, Brookhaven National Laboratory, Upton, New York, 11973, USA

Abstract. We examine the violation of the $k_\perp$ factorization approximation for quark production in high-energy proton-nucleus collisions. We comment on its implications for the open charm and quarkonium production in collider experiments.

Keywords: small-x, classical field, multiple scattering, factorization
PACS: 12.38.-t, 11.15.Kc, 11.80.La, 12.39.St

INTRODUCTION

Semi-hard processes, where $\sqrt{s} \gg m_{q_\perp} \gg \Lambda_{\rm QCD}$, contribute significantly to particle production in high-energy collider experiments due to the large density of the small-x gluons. The $k_\perp$ factorization formalism [1] systematically resums corrections of $(\alpha_s \ln(s/q_\perp^2))^n$ from gluon branchings in perturbative QCD. In this framework, the particle production cross-section is expressed as a convolution of a hard matrix element and *unintegrated* distributions of gluons in the hadrons with definite transverse momentum $k_{i\perp}$ and longitudinal fraction x_i in each projectile hadron (i=1, 2).

Multiple-scattering (higher twist) effects become important at small x due to the large density of the small-x gluons. It is expected to be the origin of the Cronin enhancement and $p_\perp$ broadening of hadrons observed in nuclear experiments. It is also relevant for the nuclear suppression of quarkonium production.

The simplest situation for studying the impact of higher twist effects on $k_\perp$ factorization is in proton-nucleus (pA) collisions, wherein the proton is dilute and the nucleus is dense. The $k_\perp$ factorization formalism was examined in the color glass condensate framework [2]. It is shown that the factorization is recovered when one keeps only the terms that are of the lowest order in the charge sources $\rho_{p,A}$ of the projectiles [3]. The cross-sections at the leading order in ρ_p, but at all orders in the dense source ρ_A of the nucleus are obtained analytically. Gluon production by the "2-to-1" processes is shown to be $k_\perp$-factorizable [4, 5, 6] whereas the quark production is generally not [7, 8, 9].

Here we report the numerical estimates for the $k_\perp$ factorization breaking in quark production within the McLerran-Venugopalan (MV) model [10]. We briefly discuss open charm production and quarkonium suppression in pA collisions in this framework.

CP828, *Multiparticle Dynamics*
edited by V. Šimák, M. Šumbera, Š. Todorova, and B. Tomášik

FACTORIZATION BREAKING IN QUARK PAIR PRODUCTION

The cross-section of the quark pairs produced in pA collisions is obtained as [7]:

$$\frac{d\sigma}{d^2p_\perp d^2q_\perp dy_p dy_q} = \frac{\alpha_s^2 N}{8\pi^4(N^2-1)} \int_{k_{1\perp},k_{2\perp}} \frac{\delta(p_\perp + q_\perp - k_{1\perp} - k_{2\perp})}{k_{1\perp}^2 k_{2\perp}^2}$$
$$\times \left\{ \int_{k_\perp,k'_\perp} \mathrm{tr_d}\left[(\not{q}+m)T_{q\bar{q}}(\not{p}-m)\gamma^0 T'^{\dagger}_{q\bar{q}}\gamma^0\right] \phi_A^{q\bar{q},q\bar{q}}(k_{2\perp};k_\perp,k'_\perp) \right.$$
$$+ \int_{k_\perp} \mathrm{tr_d}\left[(\not{q}+m)T_{q\bar{q}}(\not{p}-m)\gamma^0 T_g^{\dagger}\gamma^0 + \mathrm{h.c.}\right] \phi_A^{q\bar{q},g}(k_{2\perp};k_\perp)$$
$$\left. + \mathrm{tr_d}\left[(\not{q}+m)T_g(\not{p}-m)\gamma^0 T_g^{\dagger}\gamma^0\right] \phi_A^{g,g}(k_{2\perp}) \right\} \varphi_p(k_{1\perp}) , \quad (1)$$

where the explicit forms for the Dirac matrices $T_{q\bar{q}}(k_{1\perp},k_\perp)$ and $T_g(k_{1\perp})$ are given in [7]. Here $\varphi_p(l_\perp) \equiv (\pi^2 R_p^2\, g^2/l_\perp^2)$ F.T. $\langle \rho_p^a(0)\rho_p^a(x_\perp)\rangle$ is the unintegrated gluon distribution for the proton, and F.T. denotes the Fourier transformation. On the nucleus side, however, one needs *three* nuclear distributions defined as (see Eqs. (42), (43) and (45) in [7])

$$\phi_A^{g,g}(l_\perp) \equiv \frac{\pi^2 R_A^2 l_\perp^2}{g^2 N} \text{ F.T. tr} \left\langle U(0)U^\dagger(x_\perp)\right\rangle ,$$
$$\phi_A^{q\bar{q},g}(l_\perp;k_\perp) \equiv \frac{2\pi^2 R_A^2 l_\perp^2}{g^2 N} \text{ F.T. tr}\left\langle \widetilde{U}(x_\perp)\, t^a \widetilde{U}^\dagger(y_\perp)\, t^b U_{ba}(0)\right\rangle ,$$
$$\phi_A^{q\bar{q};q\bar{q}}(l_\perp;k_\perp,k'_\perp) \equiv \frac{2\pi^2 R_A^2 l_\perp^2}{g^2 N} \text{ F.T. tr}\left\langle \widetilde{U}(0)\, t^a \widetilde{U}^\dagger(y_\perp)\, \widetilde{U}(x'_\perp)\, t^a \widetilde{U}^\dagger(y'_\perp)\right\rangle , \quad (2)$$

where U and $\widetilde{U}$ denote the path-ordered exponentials of the gauge fields in the nucleus in the adjoint and fundamental representations, respectively, and describe the multiple scatterings of the gluon and the quarks. The average $\langle\cdots\rangle$ is taken over the Gaussian distribution of the color charge sources characterized by the saturation scale Q_s^2.

$k_\perp$ factorization is violated by the transverse structure of the quark pair probed by the momentum $k_\perp{}^{(\prime)}$ from the nucleus since each quark from the pair can resolve and interact with several gluons from the nucleus. If any of the transverse masses $m_{q_\perp}$ and $m_{p_\perp}$ of the produced quarks is large compared with the typical rescattering scale, Q_s, we can neglect the $k_\perp^{(\prime)}$-dependence in $T_{q\bar{q}}(k_{1\perp},k_\perp^{(\prime)})$ and recover the $k_\perp$ factorized formula thanks to the sum rule for ϕ_A's; $\int_{k_\perp,k'_\perp}\phi_A^{q\bar{q},q\bar{q}} = \int_{k_\perp}\phi_A^{q\bar{q},g} = \phi_A^{g,g}$.

In the subsequent numerical calculation we adopt the large N approximation, which brings two simplifications: (i) $\phi_A^{q\bar{q};q\bar{q}}(l_\perp;k_\perp,k'_\perp) = (2\pi)^2\delta(k_\perp - k'_\perp)\phi_A^{q\bar{q},g}(l_\perp;k_\perp)$ and (ii) $\phi_A^{q\bar{q},g}(l_\perp;k_\perp)$ becomes a product of two 2-point functions and positive definite. This approximation is reasonably good and reduces the numerical task [10]. At this stage

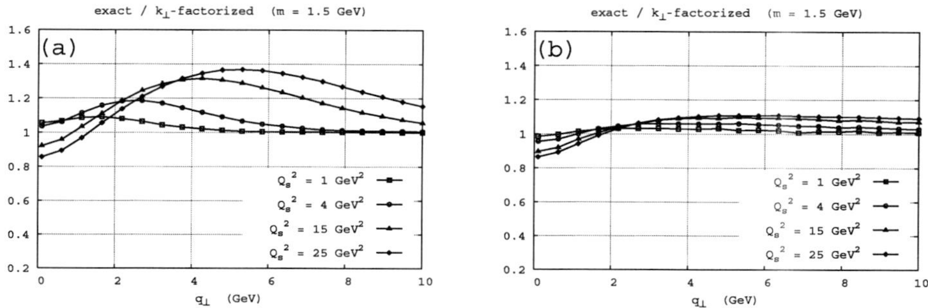

FIGURE 1. Breaking of $k_\perp$ factorization in single charm quark production in (a) the MV model and in (b) the non-local Gaussian model.

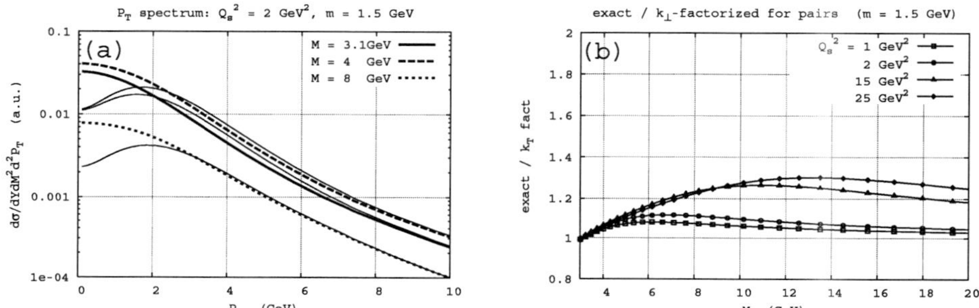

FIGURE 2. (a) $P_\perp$ spectrum of the quark pair with fixed invariant mass M. (b) Breaking of $k_\perp$ factorization in charm quark pair production in the MV model.

we can revive the $k_\perp$-factorized expression if we replace the full $\phi_A^{q\bar{q},g}(l_\perp;k_\perp)$ to $(1/2)(2\pi)^2[\delta(k_\perp)+\delta(k_\perp-l_\perp)]\phi_A^{g,g}(l_\perp)$. In this approximation either quark or anti-quark exchanges all the momentum from the nucleus.

In Fig. 1 (a) we compare the exact result (but in the large N) with the $k_\perp$ factorized approximation for single charm quark production. The breaking is relatively small for the saturation momentum Q_s^2=1 GeV^2, which may be the relevant scale for RHIC at central rapidity. At Q_s^2=15, 25 GeV^2 (corresponding to very forward rapidities in the proton fragmentation region at RHIC and LHC) the correction can be as large as 40% at $q_\perp \sim Q_s$. Because of the multiple scatterings with the momentum scale Q_s, the quarks are less likely to be produced at smaller momentum $q_\perp$ but are enhanced at the scale of Q_s. The breaking effect disappears as $q_\perp \to \infty$ as it should. For the bottom quark production the violation is smaller. To assess the model-dependence of our results, we compute them now, shown in Fig. 1 (b), with a non-local Gaussian model known to be the asymptotic solution of renormalization equations for x evolution [11]; non-linear evolution effects reduce the magnitude of the violation of $k_\perp$ factorization.

In Fig. 2 (a) shown is the total $P_\perp$ distribution of the charm quark pair with the fixed invariant masses M=3.1, 4, 8 GeV. In the $k_\perp$ factorized approximation (thin curves),

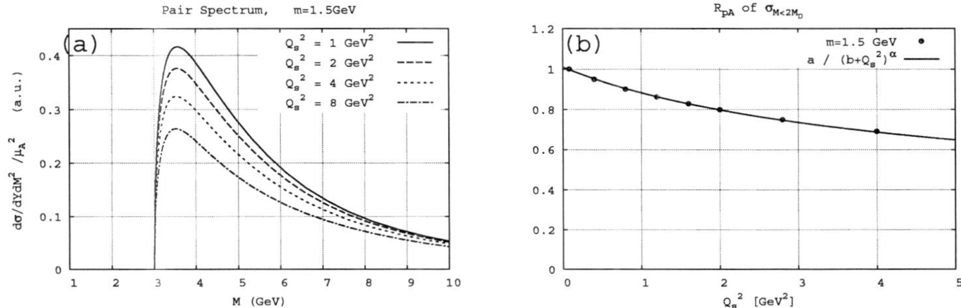

FIGURE 3. (a) Q_s^2 dependence of the charm pair production. (b) Suppression of low mass pairs in pA collisions.

either quark or antiquark exchanges all the momentum from the nucleus and we see the bump structure near Q_S, reflecting the gluon distribution of the nucleus. The bump is smeared out due to multiple scatterings of both the quark and antiquark in the full formula, which enhances the pair production at small $P_\perp$. This trend is opposite to the single quark case. Integrating over $P_\perp$, we show in Fig. 2 (b), the magnitude of factorization breaking in the invariant mass spectrum of the pair.

PHENOMENOLOGY

We study the importance of small-x distributions in D meson production by convoluting the single quark spectrum with an appropriate fragmentation function [12]. We find, however, the production spectrum is determined not by the quark distribution with $q_\perp \lesssim Q_S$, but largely by the tail part $\propto 1/q_\perp^4$ of the MV model. Moreover, in order to assess the rapidity dependence of open charm production, the x-dependence of the unintegrated gluon distributions should be taken into account, which requires going beyond the MV model. Our results on open charm production will be reported elsewhere [13].

The Q_s^2-dependence of the pair spectrum (divided by the charge density μ_A^2) is diplayed in Fig. 3 (a). At larger M, where the high-density effects are diminished, all curves converge to a single one. The multiple scatterings of the pair quarks suppress the yield in the low M region. (The overall cross-section is of course enhanced with increasing Q_s^2.) One can get an idea about the normal suppression of the quarkonium production in the pA collisions, relying on the color evaporation picture. We show the nuclear modification ratio, R_{pA}, for the pairs with M less than the open charm threshold $2M_D$, as a function of Q_s^2 in Fig. 3 (b). The suppression pattern fits the form $1/(Q_s^2)^\alpha$ with $\alpha \sim 0.42$, and not the frequently assumed exponential form [14]. One should note here that $Q_s^2 \sim A^{1/3}$ in the MV model.

ACKNOWLEDGMENTS

HF's work was supported by the Grants-in-Aid for Scientific Research No. 16740132. RV's research was supported in part by DOE Contract No. DE-AC02-98CH10886.

REFERENCES

1. S. Catani, M. Ciafaloni, F. Hautmann, *Nucl. Phys. B* **366**, 135-188 (1991); J.C. Collins, R.K. Ellis, *Nucl. Phys. B* **360**, 3-30 (1991).
2. E. Iancu, R. Venugopalan, hep-ph/0303204; E. Iancu, A. Leonidov, L.D. McLerran, hep-ph/0202270; A.H. Mueller, hep-ph/9911289.
3. F. Gelis, R. Venugopalan, *Phys. Rev. D* **69**, 014019 (2004).
4. Yu.V. Kovchegov, A.H. Mueller, *Nucl. Phys. B* **529**, 451-479 (1998); A. Dumitru, L.D. McLerran, *Nucl. Phys. A* **700**, 492-508 (2002).
5. J.P. Blaizot, F. Gelis, R. Venugopalan, *Nucl. Phys. A* **743**, 13-56 (2004).
6. Yu.V. Kovchegov, K. Tuchin, *Phys. Rev. D* **65**, 074026 (2002); D. Kharzeev, Yu. Kovchegov, K. Tuchin, *Phys. Rev. D* **68**, 094013 (2003).
7. J.P. Blaizot, F. Gelis, R. Venugopalan, *Nucl. Phys. A* **743**, 57-91 (2004).
8. K. Tuchin, *Phys. Lett. B* **593**, 66-74 (2004).
9. N. N. Nikolaev and W. Schafer, *Phys. Rev. D* **71**, 014023 (2005).
10. H. Fujii, F. Gelis and R. Venugopalan, Phys. Rev. Lett. **95**, 162002 (2005).
11. E. Iancu, K. Itakura, L.D. McLerran, *Nucl. Phys. A* **724**, 181-222 (2003).
12. D. Kharzeev and K. Tuchin, *Nucl. Phys. A* **735**, 248-266 (2004).
13. H. Fujii, F. Gelis and R. Venugopalan, in preparation.
14. H. Fujii and T. Matsui, *Phys. Lett. B* **545**, 82-90 (2002); H. Fujii, *Phys. Rev. C* **67**, 031901 (2003).

Nonlinear $k_\perp$-factorization: a New Paradigm for Hard Processes in a Nuclear Medium

N.N. Nikolaev*,†, W. Schäfer*, B.G. Zakharov† and V.R. Zoller**

*Institut für Kernphysik, Forschungszentrum Jülich, D-52425 Jülich, Germany
†L.D. Landau Institute for Theoretical Physics, Moscow 117940, Russia
**Instiute for Theoretical and Experimental Physics, 117259 Moscow, Russia

Abstract. We review the origin, and salient features, of the breaking of the conventional linear $k_\perp$-factorization for hard processes in a nuclear environment. A realization of the nonlinear $k_\perp$-factorization which emerges instead is shown to depend on color properties of the underlying pQCD subprocesses. We discuss the emerging universality classes and extend nonlinear $k_\perp$-factorization to AGK unitarity rules for the excitation of the target nucleus.

Keywords: Hard pQCD processes, $k_\perp$ factorization breaking, AGK rules
PACS: 13.97.-a, 11.80La,12.38.Bx, 13.85.-t

INTRODUCTION

An extension of factorization theorems to nuclear targets is of burning urgency - the notion of nuclear gluon densities can be made meaningful only if they furnish a unified description of the whole variety of nuclear hard processes. We recall the 1975 observation [1] that the Lorentz contraction of ultrarelativistic nucleus entails a spatial overlap/fusion/screening of partons from nucleons at the same impact parameter if $x \lesssim x_A = 1/R_A m_N$. In the laboratory frame, this amounsts to a coherency over the whole nucleus of the dijet excitation $a \to bc$, where $a = \gamma^*, q, g,\ b, c = q, \bar{q}, g, \gamma$. Excitation of the perturbative $|bc\rangle$ Fock state of the physical projectile $|a\rangle$ by one-gluon exchange leaves the target nucleon debris in the color excited state. For nuclear targets one has to deal with multiple gluon exchanges which are enhanced by a large target thickness.

Within the fusion reinterpretation of multiple gluon exchanges, the nuclear glue will be a nonlinear functional of the free nucleon glue: the same sea and glue will be shared by many nucleons. Specifically, the collective nuclear unintegrated glue $\phi(b,\kappa)$, per unit area in the impact parameter plane, can be defined in terms of the coherent diffractive dijet production off nuclei [2, 3]. In can be expanded as

$$\phi(b,\kappa) = \frac{1}{\sigma_0} \sum_{j=1}^{\infty} w_j(b) f^{(j)}(\kappa), \tag{1}$$

where the collective glue $f^{(j)}(\kappa)$ of j overlapping nucleons satisfies the convolution representation $f^{(j)}(\kappa) = \frac{1}{\sigma_0}(f^{(j-1)} \otimes f)(\kappa)$, a probability to find j overlapping nucleons equals

$$w_j(b) = \frac{\nu_A^j(b)}{j!} \exp\left[-\nu_A(b)\right], \quad \nu_A(b) = \frac{1}{2}\sigma_0 T(b), \tag{2}$$

CP828, *Multiparticle Dynamics*
edited by V. Šimák, M. Šumbera, Š. Todorova, and B. Tomášik

$\sigma_0 = \int d^2\kappa f(\kappa)$ is the dipole cross section $\sigma(x,r)$ for large dipoles r, $T(b)$ is the optical thickness of the nucleus at the imapct parameter b and $f(\kappa)$ is the unintegrated glue in the target nucleon,

$$f(\kappa) = \frac{4\pi\alpha_S}{N_c} \cdot \frac{1}{\kappa^4} \cdot \frac{\partial G_N(x,\kappa)}{\partial \log \kappa^2},$$

Subsequently we will use also $\Phi(b,\kappa) = \phi(b,\kappa) + w_0(b)\delta(\kappa)$. The antishadowing properties [2, 3] of the so-defined $\phi(b,\kappa)$ are responsible for the familiar Cronin effect [4]. It furnishes the linear $k_\perp$-factorization for the nuclear structure function $F_{2A}(x,Q^2)$ and the forward single jets in DIS off nuclei, precisely the same as for the free-nucleon target in terms of $f(\kappa)$ – this linear $k_\perp$-factorization is, though, rather an exception because of the special Abelian features of these observables. Dijets in DIS, and dijet and single-jet spectra in hA collisions, however, prove to be highly nonlinear functionals of $\Phi(b,\kappa)$ [3, 4, 5, 6, 7, 8]. Our findings have recently been corroborated in [9, 10]. Any application of linear $k_\perp$–factorization to nuclei is entirely unwarranted.

Here we review the origin, and universality properties, of nonlinear $k_\perp$-factorization. The crucial point is that the calculation of the single-jet and dijet cross sections can be reduced to an interaction with the nucleus of color-singlet multiparton systems [11]. The interaction cross-section for such systems of n partons is a (matrix) superposition of elementary dipole cross-sections. Consequently, their nuclear S-matrix is a (matrix) product of nuclear S-matrices for elementary dipoles. We derive explicit quadratures for nuclear cross sections in terms of the collective glue $\Phi(b,\kappa)$ and establish universality classes for nonlinear $k_\perp$-factorization depending on the color properties of the relevant pQCD subprocesses. Our important finding is that hard processes in a nuclear environment can not be described entirely in terms of the classical gluon field of a whole nucleus [12]. Specifically, we will see that the description of final states requires the nuclear glue defined for the slice $[0,\beta]$ of the nucleus ($0 \leq \beta \leq 1$)

$$\exp\left[-\frac{1}{2}\beta\sigma(x,r)T(b)\right] = \int d^2\kappa \Phi(\beta;b,\kappa)\exp(i\kappa r), \tag{3}$$

and the corresponding overlap probabilities $w_j(\beta;b)$. We will need also the wave function of bc Fock states coherently distorted in this slice, $\Psi(\beta;z,p) \equiv \int d^2\kappa\Phi(\beta;b,x,\kappa)\Psi(z,p+\kappa)$.

We recall first the linear $k_\perp$-factorization for forward $q\bar{q}$ dijets in DIS,

$$\frac{2(2\pi)^2 d\sigma_N(\gamma^* \to q\bar{q})}{dzd^2pd^2\Delta} = f(x,\Delta)\,|\Psi(z,p) - \Psi(z,p-\Delta)|^2\,, \tag{4}$$

where $\Delta = p_q + p_{\bar{q}}$ is the jet-jet decorrelation momentum, and $p \equiv p_{\bar{q}}$, $z \equiv z_{\bar{Q}}$ refer to the $\bar{q}$-jet. The master formula, which sums all multiple gluon exchanges in a nuclear target, is derived in [4] based on [11, 3]:

$$\frac{d\sigma(a^* \to bc)}{dz_b d^2p_b d^2p_c} = \frac{1}{(2\pi)^4}\int d^2b_b d^2b_c d^2b'_b d^2b'_c$$
$$\times \exp[-ip_b(b_b - b'_b) - ip_c(b_c - b'_c)]\Psi(z_b, b_b - b_c) \times \Psi^*(z_b, b'_b - b'_c)$$
$$\left\{S^{(4)}_{\bar{b}\bar{c}cb}(b'_b,b'_c,b_b,b_c) + S^{(2)}_{\bar{a}a}(b',b) - S^{(3)}_{\bar{b}\bar{c}a}(b,b'_b,b'_c) - S^{(3)}_{\bar{a}bc}(b',b_b,b_c)\right\}. \tag{5}$$

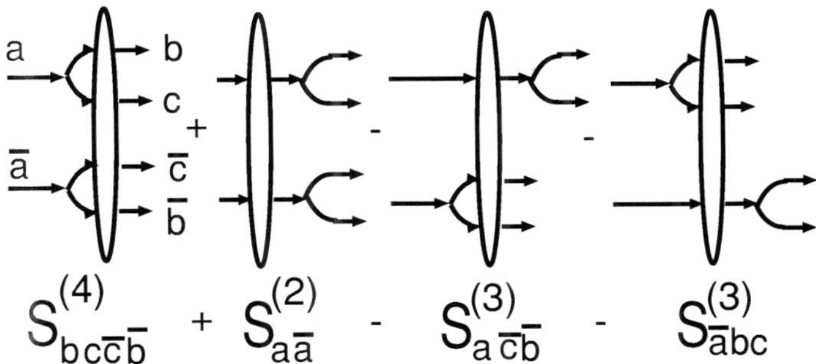

FIGURE 1. The S-matrix structure of the two-body density matrix for the excitation $a \to bc$.

The connection between the wave function of $|bc\rangle$ dipoles $\Psi(z,r)$ and its momentum space version $\Psi(z,p)$ and the familiar parton-splitting functions is found in [4]. All $S^{(n)}$ describe a scattering of color-singlet systems of n partons, as indicated in Fig. 1. For the dilute-gas nucleus

$$S_i(b_c{}',b_b{}',b_c,b_b) = \exp\{-\frac{1}{2}\Sigma_i(b_c{}',b_b{}',b_c,b_b)T(b)\}. \tag{6}$$

The 2-parton and 3-parton dipole cross sections are well known [11, 13], the major nontrivial task is a calculation of the coupled-channel Σ_4 [3, 6, 7]. The relevant basis of color states depends on the pQCD subprocess, $\gamma^* \to q\bar{q}:\ 1_1 + 8_{N_c^2},\quad g \to q\bar{q}:\ 1_1 + 8_{N_c^2},\quad q \to qg:\ 3_{N_c} + \{6+15\}_{N_c^3},\quad g \to gg:\ 1_1 + \{8_A + 8_S\}_{N_c^2} + \{10 + \overline{10} + 27 + R_7\}_{N_c^4}$, where the subscripts indicate the N_c-dependence of the size of the relevant color multiplets. For the description of the multiplet R_7 which exists for $N_c > 3$ only, see [7]. In the case of nuclear targets one must distinguish truly inelastic processes and coherent diffraction $aA \to (bc)A$ with retention of the target nucleus in the ground state.

UNIVERSALITY CLASSES OF NONLINEAR $K_\perp$-FACTORIZATION

The origin of coherent diffractive (CD) DIS is a difference between the in-vacuum and intranuclear-distorted wave function of the $q\bar{q}$ state of the photon:

$$\frac{(2\pi)^2 d\sigma_A(\gamma^* \to q\bar{q})}{d^2b dz d^2p d^2\Delta} = \delta^{(2)}(\Delta)\Big|\Psi(1;z,p) - \Psi(z,p)\Big|^2. \tag{7}$$

CD makes $\approx 50\%$ of total DIS off heavy nuclei [14]. In the case of $q \to qg$ the incident partons are colored, and intranuclear attenuation of the incident quark wave entails the suppression of CD by the factor $w_0^2(b)$, a similar gap survival probability is found for $g \to gg$ [7]. In all above cases CD is of leading order in $1/N_c$, while CD $g \to q\bar{q}$ is suppressed.

The universality class of dijets in higher color multiplets excited from partons in lower multiplet is the most interesting one. As an example we cite the large-N_c result [6]

$$\begin{aligned}
&\frac{d\sigma\left(q^* \to qg(6+15)\right)}{d^2bdzd^2\Delta d^2p} = \frac{1}{2(2\pi)^2}T(b)\int_0^1 d\beta \\
\times \; &\int d^2\kappa d^2\kappa_1 d^2\kappa_2 d^2\kappa_3 \delta(\kappa+\kappa_1+\kappa_2+\kappa_3-\Delta) \\
\times \; &\underbrace{\Phi(\beta;b,\kappa_3)}_{Quark\ \ ISI}\underbrace{f(\kappa)\left|\Psi(\beta;z,p-\kappa_2-\kappa_3)-\Psi(\beta;z,p-\kappa_2-\kappa_3-\kappa)\right|^2}_{Hard\ \ Excitation} \\
\times \; &\underbrace{\Phi(1-\beta;b,\kappa_1)}_{Quark\ \ FSI}\underbrace{\Phi(\frac{C_A}{C_F}(1-\beta);b,\kappa_2)}_{Gluon\ \ FSI}, \qquad (8)
\end{aligned}$$

where we indicated the rôle of different factors in the integrand. Notice the sixth order nonlinearity of the dijet spectrum (8) in nuclear and free-nucleon glue. The most nontrivial point here is how incoherent distortions of the incident quark wave come along with the coherent distortions of the qg wave function in the same slice $[0,\beta]$ of the nucleus.

The ratio of Casimirs C_A/C_F in $\Phi(\frac{C_A}{C_F}(1-\beta),b,x,\kappa_2)$ for FSI of gluons is a reminder that the collective nuclear glue derives from a density matrix in color space [3, 4]. In DIS, $\gamma^* \to q\bar{q}(8)$, ISI of the photon vanishes and $\Phi(\beta;b,\kappa_3) = \delta^{(2)}(\kappa_3)$, for the incident gluons, $g \to gg(10+\overline{10}+27+R_7)$, the quark FSI factor must be swapped for the gluon FSI factor, and there will be a slight modification of incoherent ISI distortions - at large N_c the gluon behaves like an uncorrelated quark-antiquark pair. [4, 7]. The striking distinction between ISI and FSI requires the collective nuclear glue for slices of the nucleus—nonlinear $k_\perp$-factorization can not be described by the classical gluon field of the whole nucleus. The nonlinear $k_\perp$-factorization entails a nuclear enhancement of the decorrelation of dijets [3, 6, 7].

The universality class of dijets in the same lower color multiplet as the beam parton has its own unique features. For instance, in the case of color-triplet qg dijets [6]

$$\frac{d\sigma\left(q^*A \to qg(3)\right)}{d^2bdzd^2\Delta d^2p} \;=\; \frac{1}{(2\pi)^2}\phi(b,\Delta)\left|\Psi(1;z,p-\Delta)-\Psi(z,p-z\Delta)\right|^2 \qquad (9)$$

The free-nucleon cross section has exactly the same form in terms of $f(\Delta)$ and the in-vacuum wave functions. Recalling that $\Psi(z,p-z\Delta)$ is the probability amplitude for the qg state in the physical quark of transverse momentum Δ, we reinterpret this result as a fragmentation of the quark, scattered quasielastically with the differential cross section $\propto \phi(b,\Delta)$, see also below. The change from the free-nucleon to a nuclear target, $|\Psi(z,p-\Delta)-\Psi(z,p-z\Delta)|^2 \Longrightarrow |\Psi(1;z,p-\Delta)-\Psi(z,p-z\Delta)|^2$, must be interpreted as a nuclear modification of the fragmentation function.

Apart from the fact that ISI of incident gluons looks like an ISI of the uncorrelated quark-antiquark pair, the nonlinear $k_\perp$-factorization for $g \to q\bar{q}(8)$, $g \to gg(8_A+8_S)$, $g \to gg(8_S)$ is very similar to that of $q \to qg(3)$ [7].

UNITARITY CUTS AND AGK RULES FOR COLOR EXCITATION OF THE TARGET NUCLEUS

Our technique can readily be extended [15] to partial cross sections $d\sigma_j$ for final states with j color-excited nucleons of the target nucleus – in the language of the so-called AGK unitarity rules that corresponds to j cut pomerons [16]. This multiplicity j controls the hadronic multiproduction in the nucleus hemisphere, i.e., the collision centrality, as well as the nonperturbative energy-loss contribution to the quenching of forward jets. The simplest case is the AGK rule for the universality class of coherent diffraction: $d\sigma_j = \delta_{j0} d\sigma_D$.

Now we notice that the differential cross section of quark-nucleon quasielastic scattering $qN \to q'N^*$ equals $d\sigma_{qN}/d^2\kappa = \frac{1}{2}f(\kappa)$. Here the target debris N^* is in the color-excited state. The collective glue $f^{(j)}(\kappa)$ is similarly related to the differential cross section of j-fold incoherent quasielastic scattering,

$$\frac{d\sigma^{(j)}}{d^2\kappa} = \frac{1}{2} f^{(j)}(\kappa) \tag{10}$$

which, evidently, is simply a j-fold convolution of single scattering cross sections. This suggest a simple interpretation of $f^{(j)}(\kappa)$ in terms of j cut pomerons.

We illustrate the emerging AGK rule for the universality class of dijets in the same lower color representation as the beam parton on the example of $q \to qg(3)$:

$$\begin{aligned}\frac{d\sigma_j(q^*A \to qg(3))}{d^2b dz d^2\Delta d^2p} &= \frac{1}{(2\pi)^2\sigma_0} w_j(b) f^{(j)}(\Delta) \left|\Psi(1;z,p-\Delta) - \Psi(z,p-z\Delta)\right|^2 \\ &= \frac{2}{(2\pi)^2\sigma_0} w_j(b) \frac{d\sigma^{(j)}}{d^2\Delta} \left|\Psi(1;z,p-\Delta) - \Psi(z,p-z\Delta)\right|^2 . \end{aligned} \tag{11}$$

It makes the interpretation of this final state as a result of fragmentation of the quasielastically scattered quark an obvious one—the nuclear-distorted fragmentation function does not depend on j. Notice the uncut pomerons which enter the intranuclear distortion of $\Psi(1;z,p-\Delta)$.

As an example of AGK rules from the universality class of dijets in higher color multiplet excited from partons in lower multiplet we show the large-N_c result for $q \to qg(6+15)$ with j color-excited nucleons:

$$\begin{aligned}\frac{d\sigma_j(q^* \to qg(6+15))}{d^2b dz d^2\Delta d^2p} &= \frac{1}{(2\pi)^2} T(b) \int_0^1 d\beta \\ \times \int d^2\kappa d^2\kappa_1 d^2\kappa_2 d^2\kappa_3 &\delta(\kappa+\kappa_1+\kappa_2+\kappa_3-\Delta) \sum_{n,k,m} \delta(j-n-k-m-1) \\ \times \underbrace{w_m(\beta;b)\frac{f^{(m)}(\kappa_3)}{\sigma_0}}_{Quark\ ISI} &\underbrace{f(\kappa)\left|\Psi(\beta;z,p-\kappa_3-\kappa_2) - \Psi(\beta;z,p-\kappa_3-\kappa_2-\kappa)\right|^2}_{Hard\ excitation}\end{aligned}$$

$$\times \underbrace{w_k\left(\frac{C_A}{C_F}(1-\beta);b\right)\frac{f^{(k)}(\kappa_2)}{\sigma_0}}_{Gluon\ FSI} \times \underbrace{w_n\left((1-\beta);b\right)\frac{f^{(n)}(\kappa_1)}{\sigma_0}}_{Quark\ FSI} \tag{12}$$

It corresponds to the following counting of cut pomerons: one for hard excitation at the depth β; m for m-fold quasielastic intranuclear scattering of the incident quark in the slice $[0,\beta]$; k for k-fold quasielastic intranuclear scattering of the final-state gluon in the slice $[\beta,1]$ and n for n-fold quasielastic intranuclear scattering of the final-state quark in the slice $[\beta,1]$ of the nucleus. There are more uncut pomerons which describe the coherent distortion of the qg wave function in the slice $[0,\beta]$.

We summarize: Hard processes in a nuclear environment must be described by nonlinear $k_\perp$-factorization in terms of the collective nuclear glue defined through the coherent diffractive dijet production. We have explicit quadratures for single-jet to dijet spectra from all pQCD subprocesses. An indispensable virtue of nonlinear $k_\perp$-factorization is the nontrivial interplay of incoherent rescatterings and coherent distortions of the dijet wave function. Still another important virtue is that it requires the collective glue for slices of a nucleus—nonlinear $k_\perp$-factorization can not be described by a classical gluon field of the whole nucleus. Our connection between the collective glue for ovelapping nucleons and the cross section of multiple quasielastic scattering of partons gives a simple interpretation of the AGK unitarity rules for excitation of the target nucleus. These unitarity rules can be applied to evaluation of the energy loss and quenching of forward jets.

REFERENCES

1. N.N. Nikolaev and V.I. Zakharov, *Sov. J. Nucl. Phys.* **21**, 227 (1975); *Phys. Lett. B* **55**, 397-399 (1975).
2. N. N. Nikolaev, W. Schäfer and G. Schwiete, *Phys. Rev. D* **63**, 014020 (2001); *JETP Lett.* **72**, 405-409 (2000).
3. N. N. Nikolaev, W. Schäfer, B. G. Zakharov and V. R. Zoller, *J. Exp. Theor. Phys.* **97**, 441-465 (2003).
4. N. N. Nikolaev and W. Schäfer, *Phys. Rev. D* **71**, 14023 (2005).
5. N. N. Nikolaev, W. Schäfer, B. G. Zakharov and V. R. Zoller, *Phys. At. Nucl.* **68**, 661-676 (2005).
6. N. N. Nikolaev, W. Schafer, B. G. Zakharov and V. R. Zoller, *Phys. Rev. D* **72**, 034033 (2005).
7. N. N. Nikolaev, W. Schafer and B. G. Zakharov, `arXiv:hep-ph/0508310`.
8. N. N. Nikolaev, W. Schafer and B. G. Zakharov, `arXiv:hep-ph/0502018`, accepted for publication in Phys. Rev. Lett. (2005).
9. J. P. Blaizot, F. Gelis and R. Venugopalan, *Nucl. Phys. A* **743**, 57-91 (2004).
10. J. Jalilian-Marian and Y. V. Kovchegov, *Phys. Rev. D* **70**, 114017 (2004).
11. N. N. Nikolaev, G. Piller and B. G. Zakharov, *J. Exp. Theor. Phys.* **81**, 851-859 (1995); *Z. Phys. A* **354**, 99-105 (1996).
12. L. D. McLerran and R. Venugopalan, *Phys. Rev. D* **49**, 2233-2241 (1994); R. Venugopalan, `arXiv:hep-ph/0412396`. and references therein.
13. N. N. Nikolaev and B. G. Zakharov, *Z. Phys. C* **49**, 607-618 (1991); *Z. Phys. C* **64**, 631-652 (1994); *J. Exp. Theor. Phys.* **78**, 598-618 (1994).
14. N. N. Nikolaev, B. G. Zakharov and V. R. Zoller, *Z. Phys. A* **351**, 435-446 (1995).
15. N. N. Nikolaev, W. Schafer and B. G. Zakharov, paper in preparation.
16. V. A. Abramovsky, V. N. Gribov and O. V. Kancheli, *Yad. Fiz.* **18**, 595-616 (1973).

Multiple Collisions and Final State Properties

Gösta Gustafson

Dept. of Theoretical Physics, Lund University, Sölvegatan 14A, S-22362 Lund, Sweden

Abstract. In the collinear factorization scheme the minijet cross section diverges for small $k_\perp$. In the $k_\perp$-factorization formalism this divergence is avoided, and it is possible to calculate the number of minijets and the $E_\perp$-flow exclusively from HERA data, without reliance on a phenomenological soft cutoff. The hadron multiplicity is very sensitive to non-perturbative effects. Analyses of minimum bias and underlying events at the Tevatron show that colours appear to combine to give strings, which are as short as possible. This leads to important questions about the properties of the pomeron and the applicability of the AGK cutting rules.

Keywords: Multiple collisions, minijets, underlying events, minimum bias events
PACS: 12.38.Aw, 13.85.Hd

INTRODUCTION

In high energy collisions multiple interactions, saturation, and multipomeron exchange are essential problems, which still are largely unsolved. In high energy pp or p$\bar{\text{p}}$ collisions the inclusive cross section for minijet production is much larger than the total cross section. This implies that multiple hard subcollisions must be common, and this feature is clearly demonstrated in data from the Tevatron [1]. The more detailed properties of the underlying event and the relation to unitarity and diffraction are, however, still very unclear. Models differ by factors 3-4 in their predictions for the hadron multiplicity at LHC [2]. In this talk I want first to discuss minijet production and $E_\perp$-flow, and in the subsequent section the hadron multiplicity.

MINIJETS

In hadron-hadron collisions *collinear factorization* works well for calculations of high-$p_\perp$ jets. However, in this formalism the minijet cross section diverges with $\sigma_{jet} \sim 1/p_\perp^4$, which implies that also the total $E_\perp$ diverges. This implies the need for a soft cutoff, and in PYTHIA [3] fits to experimental data give a cutoff $p_{\perp 0} \sim 2$ GeV. The cutoff is also growing with energy, which makes it difficult to extrapolate safely to the high energies at LHC.

In the $k_\perp$*-factorization* formalism the off shell matrix element does not blow up when the exchanged transverse momentum $k_\perp \to 0$. Assume that two incoming partons, with momenta k_1 and k_2, scatter producing the outgoing partons (jets) q_1 and q_2. When the momentum exchange $k_\perp^2$ is smaller than the incoming virtualities $k_{\perp 1}^2$ and $k_{\perp 2}^2$, the cross section does not diverge, and the total $E_\perp$ stays finite. The result is an "effective cutoff", $p_{\perp 0}$, which increases with energy [4, 5]. The increase is less steep for larger s, and in the leading logarithmic approximation it saturates for asymptotically high energies.

CP828, *Multiparticle Dynamics*
edited by V. Šimák, M. Šumbera, Š. Todorova, and B. Tomášik

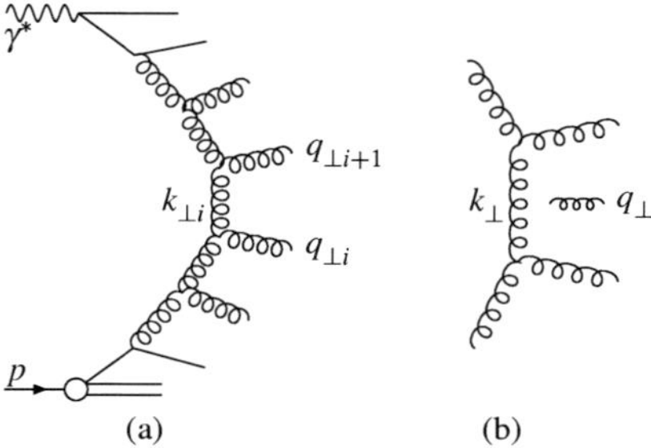

FIGURE 1. (a) A large transverse momentum, $k_{\perp i}$, in the middle of the chain corresponds to a hard subcollision between a proton and a resolved photon. (b) In the BFKL and LDC formalisms a soft emitted gluon is treated as final state radiation, while in the CCFM model some of these gluons are treated as initial state radiation.

At high energies $\sigma_{\text{jet}} \gg \sigma_{\text{tot}}$. Thus at Tevatron energies we have $\sigma_{\text{jet}} \sim 300$mb, to be compared to $\sigma_{\text{inel,non-diff}} \sim 40$mb. This implies that there are normally several hard subcollisions in each event. For the experimental evidence for multiple collisions see the talk by R. Field in these proceedings [1]. The large average number of minijets implies that correlations are important. The observed "pedestal effect" shows that the hard subcollisions are not independent, indicating that central collisions have many minijets, while peripheral collisions have fewer minijets. In the model by Sjöstrand and van Zijl [6], implemented in the PYTHIA MC, this feature is described by a parton distribution in impact parameter space in form of a double Gaussian with a denser central core.

In the small x (BFKL) region non-$k_\perp$-ordered parton chains are important. We let k_i denote the virtual links and q_i the (quasireal) emitted partons in the chain, as indicated in Fig. 1a. Then, if the link with transverse momentum $k_{\perp i}$ is a local maximum, it corresponds to a hard subcollision, where $q_{\perp i} \approx k_{\perp i} \approx q_{\perp i+1}$. A single local maximum corresponds to a resolved photon interaction, and with several local maxima there are several correlated hard subcollisions.

The BFKL equation describes only *inclusive* cross sections. A link in the ladder corresponds to a Reggeized gluon, where soft emissions are compensated by virtual corrections (*cf* Fig. 1b). Therefore such soft emissions do *not* contribute to parton distribution functions. They *do* contribute, however, to the final state properties, where they have to be added associated with appropriate Sudakov form factors.

The CCFM model interpolates between BFKL and DGLAP evolution. Here soft emissions are included in the initial state radiation, with an extra suppression from non-eikonal form factors. The *Linked Dipole Chain* (LDC) model [7, 8] (implemented in the LDCMC Monte Carlo [9]) is a reformulation and generalisation of CCFM. The separation between initial and final state radiation is here more similar to the BFKL formalism. Thus the parton with momentum q in Fig. 1b corresponds to final state radiation if $q_\perp < k_\perp$, and should then not be included in the initial state ladder.

An important consequence is that in LDC the ISR chain is fully symmetric between the photon end and the proton end of the ladder. This symmetry implies that the LDC

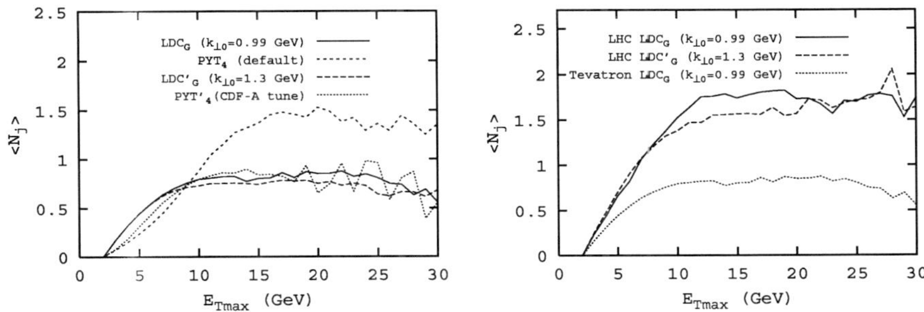

FIGURE 2. The average number of minijets in the "minimum azimuth region" for $|\eta| < 2.5$ *vs.* $E_\perp$ for the hardest jet. (a) For $\sqrt{s} = 1.8$ TeV. (b) For 14 TeV.

model can also be applied to hadronic collisions. Thus from a fit to HERA data for F_2 it is possible to predict the cross section for a chain in pp collisions, which means that the effective cutoff $p_{\perp 0}$ can be fixed. If one also assumes that the correlations are given by a double Gaussian distribution in the impact parameter, as in PYTHIA, one arrives at the results presented in Fig. 2 [5]. Fig. 2a shows the number of minijets in the "minimum azimuth region" $60° < \phi < 120°$ at $\sqrt{s} = 1.8$ TeV. The two LDC curves are obtained from two different fits to HERA data [10]. The two PYTHIA curves correspond to default parameter values (thin dashed curve), and Field's tune A (dotted curve), which gives a good fit to the CDF data [1]. We note that the two LDC result are very similar and agree well with the tuned PYTHIA result. Fig. 2b shows corresponding results for LHC. Also here the two curves correspond to different HERA fits. For comparison the result for 1.8 TeV is also indicated (dotted curve). We see that the activity increases by a little more than a factor of 2 between the two energies.

HADRON MULTIPLICITIES

In pp collisions the multiplicity of final state hadrons depends strongly on the colour connections between the produced partons. This implies that the result is very sensitive to soft non-perturbative effects. Multiple interactions are related to multiple pomeron exchange, which is expected to obey the Abramovskiĭ-Gribov-Kancheli (AGK) cutting rules [11]. These rules were derived from Regge theory in a multiperipheral model, but such a multiperipheral chain has important similarities with a gluonic chain in QCD. An essential feature is the dominance of small momentum transfers at each vertex. The colour structure of QCD gives, however, some extra complications, as discussed by J. Bartels at this symposium [12, 13].

The pomeron is identified with a gluonic ladder, and a cut pomeron to a gluon chain as in Fig. 1a. Multi-pomeron exchange corresponds to multiple ladders, and as an example the AGK rules for two pomeron exchange give the relative weights $1 : -4 : 2$ for cutting 0, 1 or 2 pomerons. Similar cutting rules apply to a diagram with two pomerons attached to one of the colliding protons and one pomeron to the other, connected by a central

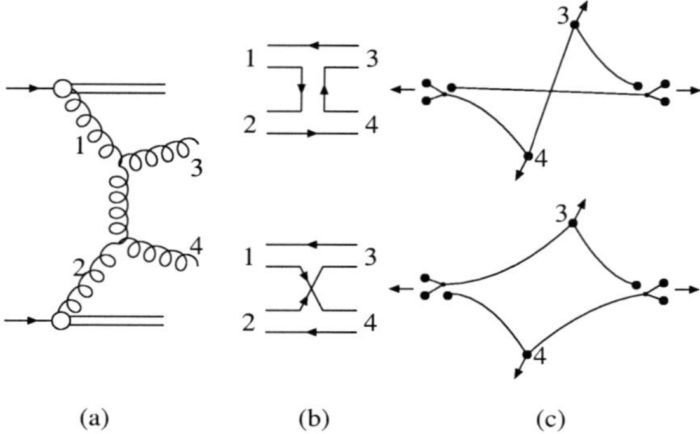

FIGURE 3. For the gluon-gluon subcollision in Fig. (a), the different colour flow patterns give four different colour connecting string configurations. Two possibilities are shown in Figs. (b) and (c), and the other two are obtained by interchanging colour and anticolour.

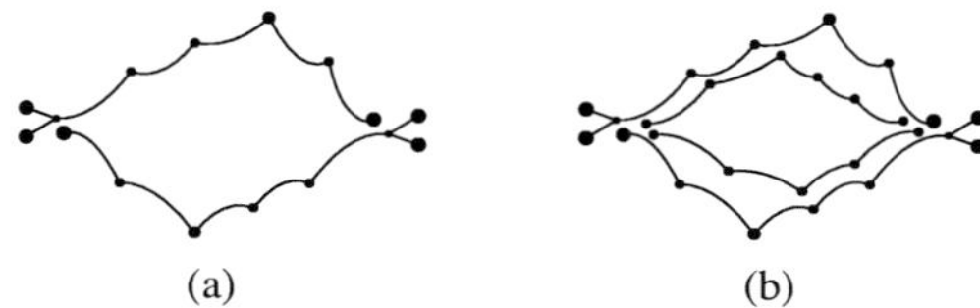

FIGURE 4. (a) Initial state radiation corresponds to more kinks on the strings. Due to colour coherence these kinks are connected so that the total string length is minimal. (b) Two cut pomerons with two hard subcollisions would naively be expected to result in four strings and approximately doubled multiplicity.

triple-pomeron coupling. In Ref. [11] this and similar diagrams are, however, expected to give smaller contributions.

A hard gg $\rightarrow$ gg subcollision will imply that the two proton remnants carry colour octet charges. This is expected to give two colour triplet strings, or two cluster chains, connecting the two remnants and the two final state gluons. In the string model the strings are stretched between the remnants, with the gluons acting as kinks on the strings. As indicated in Fig. 3 these kinks can either be on different strings or both on one of the strings, depending on the colour flow in the gluon-gluon scattering subprocess [14]. (There are also interference terms for which the colour connections cannot be determined from perturbative QCD. These terms are, however, suppressed by factors $\sim 1/N_c^2$ [14].) Including initial state radiation will give extra kinks, which due to colour coherence will be connected so as to result in minimal extra string length, as schematically indicated in Fig. 4a.

Multiple collisions with two independent gg $\rightarrow$ gg scatterings would be expected to correspond to two cut pomerons, with four triplet strings stretched between the proton remnants as indicated in Fig. 4b. This would give approximately a doubled multiplicity, in accordance with the AGK cutting rules. However, the CDF data show that this is *far from reality.*

Rick Field's successful tune A is a fit using an early PYTHIA version based on a model by Sjöstrand and van Zijl [6]. Already in their early analysis it was realized that four strings as in Fig. 4b would give too high multiplicity. Therefore, in this PYTHIA version there are three possible string connections for a secondary hard subcollision. 1) An extra closed string loop between the two final state gluons. 2) A single string between the scattered partons, which are then treated as a $q\bar{q}$ system. 3) The new hard gluons are inserted as extra kinks among the initial state radiations, in a way which corresponds to minimum extra string length. In the successful tune A the last possibility is chosen in 90% of the cases, which corresponds to *minimal extra multiplicity*. The default PYTHIA tune, which contained equal probabilities for the three cases, does not give a good fit. A more advanced treatment of pp collisions presented in Refs. [15, 16] is implemented in a new PYTHIA version (PYTHIA 6.3 [17]). This model does, however, not work as well as Field's tune A of the older model.

Thus we have the following conclusions from the CDF data:

i) Multiple interactions are definitely needed.

ii) The data are fitted if colours rearrange so that secondary hard scatterings give minimum extra string length, *i.e.* minimum extra multiplicity.

Consequently two independent hard collisions do not correspond to two cut pomeron ladders stretched between the proton remnants. It also does not correspond to a cut pomeron loop in the centre. Instead it looks like a single ladder, with a higher density of gluon rungs in the central region. How can this be understood? It raises important questions:

– Do rescattering and unitarity constraints (and AGK) work in the initial perturbative phase?

– If so, does this correspond to an initial hard collision inside a confining bag, with the final state partons colour connected in a later non-perturbative phase?

We can compare with the situation in e^+e^--annihilation. If two gluons are emitted from the quark or antiquark legs, these gluons form a colour singlet with probability $\sim 1/N_c^2$. They could then hadronize as a separate system. Analyses of data from LEP indicate that such isolated systems are suppressed, possibly even more than by a factor $\sim 1/N_c^2$. This gives further questions:

– Why do the strings make the shortest connections in $\approx 100\%$ in pp and almost never in e^+e^-?

– What is the relation between high multiplicity events, diffraction, and gap survival probability, and how does this reflect features of AGK in ep, γp, and pp?

– Is the pomeron a much more complicated phenomenon than the simple ladder envisaged by Abramovskiĭ-Gribov-Kancheli?

– How could this be tested experimentally?

If these questions can be answered at HERA and the Tevatron, it would be of great value for the analyses of both new and old physics at the LHC.

SUMMARY

In high energy pp collisions perturbative QCD and PDFs from HERA can be used to describe the production of jets with large $p_\perp$. In the collinear factorization scheme the cross section for minijets grows $\sim 1/p_\perp^4$ for small $p_\perp$. This implies that also the total $E_\perp$-flow is divergent, and it is necessary to introduce a phenomenological soft cutoff. In the $k_\perp$-factorization scheme the minijet production is suppressed for small $k_\perp$. This implies that the soft singularity is avoided, and in the Linked Dipole Chain model it is possible to calculate an "effective cutoff", which fixes the number of minijets and the $E_\perp$-flow. This effective cutoff increases with energy, but saturates when the energy goes towards infinity. This implies that the minijets and the $E_\perp$-flow can be determined exclusively from HERA data for DIS, without reliance on a phenomenological soft cutoff and uncertain extrapolations from lower energies to LHC conditions.

The hadron multiplicity distributions are very sensitive to non-perturbative effects. Field's PYTHIA tune A reproduces the essential features of minimum bias and underlying events at the Tevatron. This shows that colours appear to combine to give strings which are as short as possible, in a way very different from e^+e^--annihilation. This leads to a set of questions: What is the relation between high multiplicity events, diffraction, and gap survival probability? Do unitarity effects and AGK cutting rules work as expected in an initial perturbative phase, followed by colour recombination in a later soft phase? Or is the pomeron a much more complicated phenomenon than the simple multiperipheral ladder? Can analyses of data from HERA and the Tevatron improve predictions for, and analyses of, results at the LHC?

REFERENCES

1. R. Field, "Jet physics at Tevatron", Talk presented at this symposium (2005).
2. A. Moraes, Contribution to the workshop HERA and the LHC, CERN Oct. 11-13, 2004 (2004).
3. T. Sjostrand, et al., *Comput. Phys. Commun.* **135**, 238–259 (2001), hep-ph/0010017.
4. G. Gustafson, and G. Miu, *Phys. Rev. D* **63**, 034004 (2001), hep-ph/0002278.
5. G. Gustafson, L. Lonnblad, and G. Miu, *Phys. Rev. D* **67**, 034020 (2003), hep-ph/0209186.
6. T. Sjostrand, and M. van Zijl, *Phys. Rev. D* **36**, 2019–2041 (1987).
7. B. Andersson, G. Gustafson, and J. Samuelsson, *Nucl. Phys. B* **467**, 443–478 (1996).
8. B. Andersson, G. Gustafson, H. Kharraziha, and J. Samuelsson, *Z. Phys. C* **71**, 613–624 (1996).
9. H. Kharraziha, and L. Lonnblad, *JHEP* **03**, 006 (1998), hep-ph/9709424.
10. G. Gustafson, L. Lonnblad, and G. Miu, *JHEP* **09**, 005 (2002), hep-ph/0206195.
11. V. A. Abramovsky, V. N. Gribov, and O. V. Kancheli, *Yad. Fiz.* **18**, 595–616 (1973).
12. J. Bartels, "Multiple scatterings in ep/pp and AGK rules", Talk presented at this symposium (2005).
13. J. Bartels, M. Salvadore, and G. P. Vacca, *Eur. Phys. J. C* **42**, 53–71 (2005), hep-ph/0503049.
14. G. Gustafson, *Z. Phys. C* **15**, 155–160 (1982).
15. T. Sjostrand, and P. Z. Skands, *JHEP* **03**, 053 (2004), hep-ph/0402078.
16. T. Sjostrand, and P. Z. Skands, *Eur. Phys. J. C* **39**, 129–154 (2005), hep-ph/0408302.
17. T. Sjostrand, L. Lonnblad, S. Mrenna, and P. Skands (2003), hep-ph/0308153.

Traveling Waves in High Energy QCD

Robi Peschanski

Service de physique théorique, CEA/Saclay, 91191 Gif-sur-Yvette cedex, France
URA 2306, unité de recherche associée au CNRS

Abstract. Saturation is expected to occur when a high density of partons (mainly gluons)—or equivalently strong fields in Quantum Chromodynamics (QCD)—is realized in the weak coupling regime. A way to reach saturation is through the high-energy evolution of an extended target probed at a fixed hard scale. In this case, the transition to saturation is expected to occur from nonlinear perturbative QCD dynamics. We discuss this approach to saturation, which is mathematically characterized by the appearance of traveling wave patterns in a suitable kinematical representation. A short review on traveling waves in high energy QCD and a first evidence of this phenomenon in deep-inelastic proton scattering is presented.

INTRODUCTION

One of the most intringuing empirical observations about deep-inelastic scattering (DIS) on a proton is the geometric scaling property [1], which states that the γ^*-p total cross-section data can be approximately ploted on a one-dimensional curve $\sigma^{\gamma*-p}(Q^2/Q_s^2(Y))$ where (Q^2,Y) span the virtuality-rapidity kinematical domain of DIS. One finds $Q_s^2(Y) \approx e^{\lambda Y}$.

The question we want to address is whether the observed geometric scaling can be explained in terms of QCD evolution equations describing *saturation*.

Saturation of parton (mainly gluon) densities at high rapidity may be at the origin of the geometric scaling property. Saturation is expected to occur when a high density of partons is created in the limited geometry of the target. In the expected scenario, the parton wave functions overlap and eventually create a new state of matter, the Color Glass Condensate (CGC) [2]. In the case of DIS on a proton, increasing densities are obtained by high-energy evolution at fixed virtuality Q^2, when the corresponding linear QCD evolution, named BFKL after Balitsky, Kuraev, Lipatov, and Fadin [3], gives rise to an exponential growth of the number of gluons. At high enough energy the BFKL formulation gets modified. In the approximation of uncorrelated hard probes, one deals with the Balitsky-Kovchegov (BK) equation [5], where the corrections due to the high density of partons are expressed in terms of a non-linear damping term. We shall investigate whether and how geometric scaling could be due to the nonlinear QCD evolution of the gluon distribution in the target.

CP828, *Multiparticle Dynamics*
edited by V. Šimák, M. Šumbera, Š. Todorova, and B. Tomášik

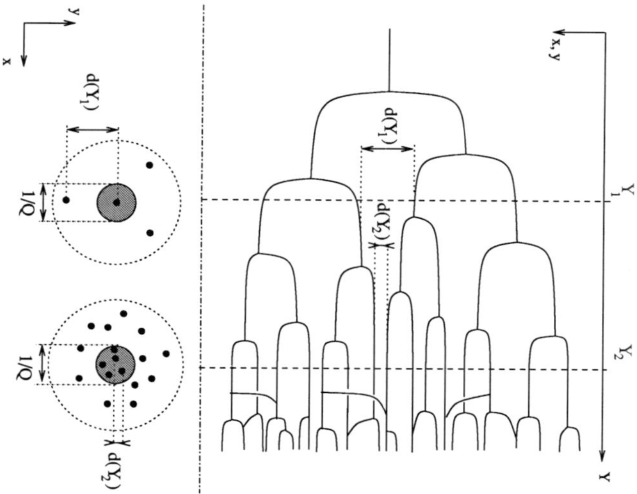

FIGURE 1. Schematic picture of the transition to saturation with rapidity. In a region $Y \sim Y_1$: the projectile of size $\sim 1/Q$ is able to probe the number of partons in exponential growth; when $Y \sim Y_2$: the probe counts partons by groups, leading to a nonlinear damping evolution factor in the probed parton density. For $Y > Y_2$: A new phase of partonic matter (CGC) may appear through parton correlations.

GEOMETRIC SCALING AS TRAVELING WAVES

Let us start from the BK equation expressed in momentum space (using the impact-parameter independent formulation)

$$\partial_Y \mathcal{N} = \bar{\alpha}\chi(-\partial_L)\mathcal{N} - \bar{\alpha}\mathcal{N}^2$$

where $\mathcal{N}(Y, L = 2\log k)$ is related to the tranverse-momentum k distribution of gluons in the target. The (leading order) BFKL kernel is

$$\chi(-\partial_L) = 2\psi(1) - \psi(-\partial_L) - \psi(1+\partial_L)\ .$$

The coupling constant will be considered either constant, or running like $\bar{\alpha} = 1/bL$.

From a mathematical point of view, the BK equation can be related [6] to an "universality class" of nonlinear equations for which characteristic properties can be rigorously derived. "Universality" means here that the solutions have general characteristics which are essentially independent from initial conditions. In the case of the BK equation (and also for some extensions beyond BK) this generic feature is the formation of *traveling wave* patterns when the energy increases. As we shall see, traveling waves are intimately related to geometric scaling. Let us define these traveling waves and give a qualitative explanation of their formation during the rapidity evolution.

Consider first the "diffusive approximation" of the BFKL kernel

$$\bar{\chi}(-\partial_L) \sim \chi\left(\tfrac{1}{2}\right) + \tfrac{1}{2}\,\chi''\left(\tfrac{1}{2}\right) \times \left(\partial_L + \tfrac{1}{2}\right)^2\ .$$

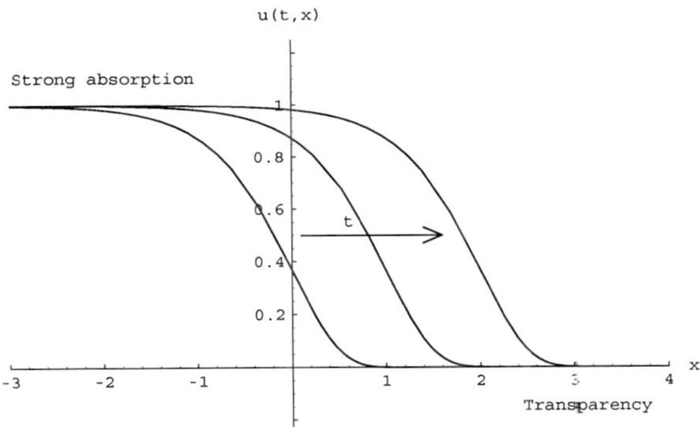

FIGURE 2. Typical traveling wave pattern. The wave front (in QCD: amplitude) connecting the regions $u = 1$ (in QCD: strong absorption) and $u = 0$ (in QCD: transparency) travels from the left to the right as t (in QCD: Y) increases.

By a suitable redefinition of the function and variables one can map the BK equation in the diffusive approximation onto the Fisher Kolmogorov-Petrovski-Piskounov (F-KPP) equation [7] describing a reaction-diffusion process in space and time:

$$\partial_t u(t,x) = \partial_x^2 u(t,x) + u(t,x) - u^2(t,x) \; ,$$

In reaction-diffusion language, $\partial_x^2 u(t,x)$ is the diffusion term, $u(t,x)$ is responsible for the exponential growth and $u^2(t,x)$, the damping term.

The "dictionnary" between F-KPP and BK can be written as follows

$$\begin{array}{rcl} Time\ t & \rightarrow & Y \\ Space\ x & \rightarrow & L + \frac{1}{2}\,\bar{\alpha}\chi''\left(\frac{1}{2}\right) Y \\ Wave\ Front\ \ u(x-ct) & \rightarrow & \mathcal{N}(L - vY) \\ Traveling\ Waves & \rightarrow & Geometric\ Scaling. \end{array}$$

The key feature of the solutions of the F-KPP equation (and of equations belonging to the same universality class) is the formation of traveling wave patterns, see Fig. 2. It can be rigorously proven that, at large times ($\Rightarrow$ large rapidities), the solutions verify

$$u(t,x) \xrightarrow[t\rightarrow\infty]{} u(x - m(t)) \Rightarrow \mathcal{N}(Y,L) \xrightarrow[Y\rightarrow\infty]{} \mathcal{N}(L - L_s(Y))$$

with

$$L_s(Y) = \log Q_s^2(Y) = \bar{\alpha}\,\chi'(\bar{\gamma})Y - \frac{3}{2\bar{\gamma}}\log Y - \frac{3}{(\bar{\gamma})^2}\sqrt{\frac{2\pi}{\bar{\alpha}\chi''(\bar{\gamma})}}\frac{1}{\sqrt{Y}} + \mathcal{O}(1/Y)\; .$$

$\bar{\gamma}$ is the implicit solution of the characteristic equation $\chi(\bar{\gamma})/\bar{\gamma} = \chi'(\bar{\gamma})$ which leads to the asymptotic wave speed $v = \bar{\alpha}\chi'(\bar{\gamma})$. It plays the rôle of a critical QCD anomalous

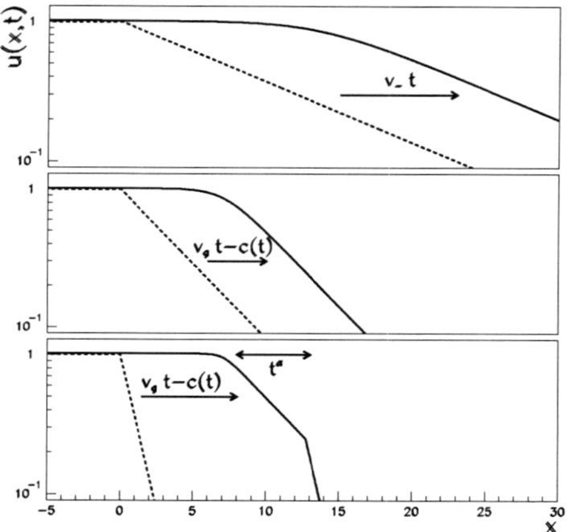

FIGURE 3. "Pulled vs. Pushed fronts". The function $u(t,x)$ is represented for the three different classes of initial conditions. Top: "Pushed fronts," the wave front keeps the initial front profile with a non-universal speed v_-; Middle: "Pulled = Pushed"; Bottom: "Pulled front:" both the wave front and the speed acquire universal values at large t ($\sim Y$) in an intermediate kinematical region called the "wave interior".

dimension. One immediately recognizes the equivalence between geometric scaling and traveling waves solutions of BK. Moreover, quite a detailed information can be obtained from the sole knowledge of the linear kernel, as for the three first terms [6] of the asymptotic expansion of the saturation scale $Q_s^2(Y)$.

ASYMPTOTIC TRAVELING WAVES

In fact the results using the diffusive approximation of BK and its mapping to the F-KPP equation are more general. Instead of entering here in refined (and technically useful) mathematical arguments [8], let us qualitatively illustrate the reason why traveling wave patterns generically develop during the rapidity evolution from a given initial condition, using the concepts of "pulled fronts".

Universality features are obtained in the "pulled" front situation, see the bottom plot of Fig. 3. In this case, one starts with initial conditions corresponding to an anomalous dimension $\gamma_0 > \bar{\gamma}$, where $\bar{\gamma} = .6275...$ is for the (leading order) BFKL kernel [6]. The formation of the traveling wave comes from the competition between the exponential growth at small values of the solution u, where the linear term dominates the evolution, with the nonlinear damping exercised at larger values of u. Constrained by these two opposite trends, the solution is "forced" to adopt a universal behaviour. Fortunately enough, the perturbative QCD initial conditions, through the transparency property ($\gamma_0 = 1 > \bar{\gamma}$,), fall into the universal "pulled" front regime [6].

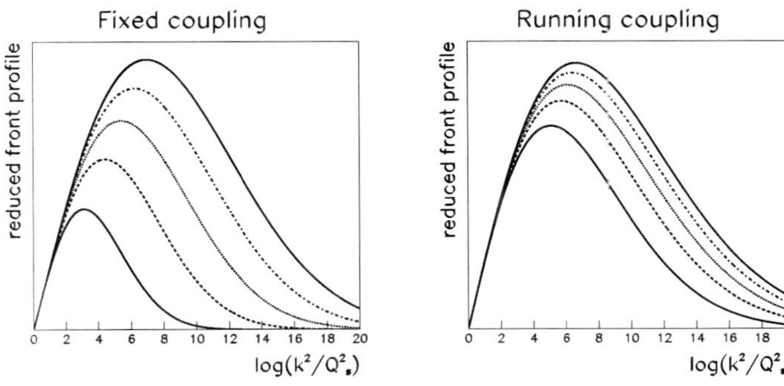

FIGURE 4. The "Reduced" Front profile.

In Fig. 4 the "reduced front" $(k^2/Q_s(Y)^2)^{\bar{\gamma}}\mathcal{N}(L,Y)$, obtained by factorizing out the wave propagation, is represented for fixed (left) and running (right) coupling constant. It shows the scaling straight line, which envelopes the curves obtained for increasing rapidity. Hence geometric scaling violations are also predicted due to diffusion. One may notice that the diffusive approach to scaling is slower for running coupling, corresponding to "anomalous diffusion" in $1/t^{1/3} \sim 1/Y^{1/6}$ (for running coupling $t \sim Y^{1/2}$). Normal diffusion $1/t^{1/2} \sim 1/Y^{1/2}$ is characteristic of the F-KPP universality class or BK with fixed coupling.

PARAMETRIC TRAVELING WAVES

The abovementionned mathematical properties have the major interest of revealing the link between geomeric scaling and the nonlinear dynamics of QCD evolution equations at high energy. However the path from phenomenology to theory is not yet accomplished. On the phenomenological ground, the simulations of BK equation solutions [9] have shown that pre-asymptotics effects may be important and endenger geometrical scaling predictions from QCD in the physical region where it is observed. On the theory side, it has been recently realized [10] that the effect of parton fluctuations and correlations, especially in the dilute (transparency) region which is the driving force of the pulled front, may be responsible for a breaking of geometric scaling. In fact, traveling wave patterns still exist [11] but their event-by-event fluctuations tends to break geometric scaling in the average.

Despite these problems, the numerical simulations [9] also show that patterns behaving like traveling waves exist in a preasymptotic region in energy. Moreover the mean field BK equation seems to be still valid in this region. In order to explore this possibility and understand its origin, we proposed [12] to attack the problem of traveling wave solutions of the BK equation in a new way.

The initial mathematical idea [13] is to assume and directly insert a traveling wave solution $u(x,t) \to U(s = x/c - t)$ into the F-KPP equation, or equivalently to impose a geometric scaling form to a solution of the BK equation. Keeping for simplicity the

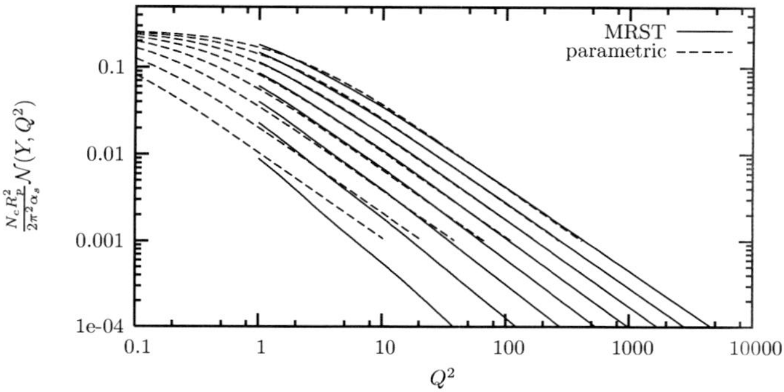

FIGURE 5. The amplitude $\mathcal{N}(L,Y)$ as a fonction of L for different values of $Y = 3, 4, .., 10$. Full lines: MRST gluon parametrization. Dashed lines: Parametric QCD traveling waves.

space-time language and the FKPP equation (the dictionnary to BK being still easy), one gets

$$U(1-U) + \frac{dU}{ds} + \frac{1}{c^2}\frac{d^2U}{(ds)^2} = 0 .$$

taking into account the previous study showing [7] that the wave speed $c \geq 2$, we obtain an iterative solution

$$U(s) = U_0 + \frac{1}{c^2}U_1 + \sum_{p \geq 2} \frac{1}{c^{2p}} U_{2p}$$

obeying an exactly solvable [12] hierarchy of equations for the U_{2p}'s. The first one only (for U_0) is nonlinear and can be exactly solved, the other boiling down to a rather simple linear algebra. This mathematical result translates into a *parametric* form of a geometric scale invariant gluon amplitude $\mathcal{N}(L - L_s(Y))$ which can be used in phenomenology. $\mathcal{N}$ depends only on few parameters determined by the linear kernel. One obtains

$$\mathcal{N} \propto \frac{1}{1+\left[\frac{k^2}{Q_s^2(Y)}\right]^{\mu}} - \frac{1}{c^2} \frac{\left[\frac{k^2}{Q_s^2(Y)}\right]^{\mu}}{\left(1+\left[\frac{k^2}{Q_s^2(Y)}\right]^{\mu}\right)^2} \log \frac{\left(1+\left[\frac{k^2}{Q_s^2(Y)}\right]^{\mu}\right)^2}{4\left[\frac{k^2}{Q_s^2(Y)}\right]^{\mu}} .$$

This solution (valid also for the running case with an appropriate mapping [12]) is mathematically valid [8] in the "wave interior", see Fig. 3. Remarkably, we observed that the corresponding region has a large overlap with the physical region where geometrical scaling is valid at HERA. In order to check this opportunity, we determined the phenomenological observable related to $\mathcal{N}(L,Y)$ from a MRST parametrization of F_2 data. The obtained kernel parameters are not far but different from those expected from the leading order BFKL kernel one and thus could require higher order contributions [12]. Interestingly, higher order contributions plays an important rôle in a different but related problem: the existence of "backward" traveling waves appearing [14] from the energy evolution starting from an inital high density (and not initial low density as in DIS) state, which could be the case in heavy-ion reactions.

CONCLUSION

Traveling wave patterns have been shown to result from nonlinear QCD equations. The phenomenologically observed geometric scaling in DIS on a proton at large energy is consistent with the traveling wave patterns. It remains to be found whether one can relate the geometric scaling curve with the QCD kernel and more generally to a complete solution of high-energy QCD, including correlation and fluctuation contributions.

ACKNOWLEDGMENTS

I wish to take the opportunity of the kind hospitality from the organizers of ISMD 2005 to acknowledge the long-term interactions with Andrzej Bialas, Jean-Louis Meunier and Bernard Derrida, which inspired me to look for properties of nonlinear equations such as F-KPP. I also acknowledge the fruitful collaboration with Stéphane Munier during the first stage of the present study and, more recently, a stimulating cooperation with Rikard Enberg, Cyrille Marquet and Gregory Soyez. Lively discussions with Edmond Iancu, Al Mueller and many other colleagues have to be mentionned.

REFERENCES

1. A. M. Staśto, K. Golec-Biernat, and J. Kwiecinski, *Phys. Rev. Lett.* **86**, 596-599 (2001).
2. For a recent review and references, see K. Itakura, "Recent results on saturation and CGC", hep-ph/0511031.
3. L. N. Lipatov, *Sov. J. Nucl. Phys.* **23**, 338-345 (1976); E. A. Kuraev, L. N. Lipatov, and V. S. Fadin, *Sov. Phys. JETP* **45**, 199 (1977); I. I. Balitskiĭ and L. N. Lipatov, *Sov. J. Nucl. Phys.* **28**, 822-829 (1978).
4. E. Iancu, K. Itakura and S. Munier, *Phys. Lett. B* **590**, 199-208 (2004).
5. I. Balitsky, *Nucl. Phys. B* **463**, 99-157 (1996); Y.V. Kovchegov, *Phys. Rev. D* **60**, 034008 (1999).
6. S. Munier and R. Peschanski, *Phys. Rev. Lett.* **91**, 232001 (2003); *Phys. Rev.* **D69** (2004) 034008; *Phys. Rev. D* **70**, 077503 (2004).
7. R. A. Fisher, *Ann. Eugenics* **7**, 355 (1937); A. Kolmogorov, I. Petrovsky, and N. Piscounov, Moscou *Univ. Bull. Math.* **A1**, 1 (1937); M. Bramson, *Memoirs of the American Mathematical Society* **285** (1983).
8. E. Brunet and B. Derrida, *Phys. Rev. E* **56**, 2597-2604 (1997). U. Ebert, W. van Saarloos, *Physica D* **146**, 1-99 (2000). For a review, see W. van Saarloos, *Phys. Rep.* **386**, 29-222 (2003).
9. For most recent numerical studies see: G. Soyez, "Fluctuation effects in high-energy QCD," hep-ph/0509138; R. Enberg, K. Golec-Biernat and S. Munier, "The high energy asymptotics of scattering processes in QCD," hep-ph/0505101.
10. For a first evidence see, A. H. Mueller and A. I. Shoshi, *Nucl. Phys. B* **692**, 175-208 (2004).
11. E. Iancu, A. H. Mueller and S. Munier, *Phys. Lett. B* **606**, 342-350 (2005).
12. R. Peschanski, *Phys. Lett. B* **622**, 178-182 (2005); C. Marquet, R. Peschanski and G. Soyez, *Phys. Lett. B* **628**, 239-249 (2005).
13. J. David Logan, "An Introduction to Nonlinear Partial Differential Equations", John Wiley and sons eds., New York, 1994.
14. R. Enberg and R. Peschanski, "Infrared instability from nonlinear QCD evolution," hep-ph/0510352.

Pentaquarks – a brief update

Michal Praszalowicz

M. Smoluchowski Institute of Physics, Jagellonian University, Kraków, Poland

Abstract. We argue that recently reported light and narrow $uudd\bar{s}$ pentaquark state requires modification of a naive quark model. On the other hand such a state is naturally accommodated in chiral soliton models. We briefly describe the logic behind such models and underline differences between Skyrme model and chiral quark soliton models. We discuss phenomenological and theoretical consequences of Θ^+. Finally we comment on recent experimental reports from CLAS, LEPS and SVD-2.

Keywords: pentaquarks, chiral models, quark models
PACS: 12.39.Jh, 12.39.Dc, 12.40.Yx, 14.20.-c

INTRODUCTION

There is still a lack of consensus whether the lightest member of the exotic antidecuplet has been discovered. After almost two years of excitement [1, 2] the results from high statistics G11 and G10 dedicated experiments at CLAS were presented in April and June 2005 respectively, with negative result for the photoproduction of Θ^+ on proton [3] and deuteron [4]. On the other hand the high statistics run at Spring-8 [5] and recent analysis by SVD [6] confirm previous results announced by these groups. There is also a number of other reports from different groups that conducted searches for Θ^+ and found nothing (for review see Ref.[7]). The reasons why some experiments see Θ^+ while the others do not, maybe either of experimental nature or a peculiar production mechanism or both. Therefore the present confusion concerning exotics calls for a new high precision KN experiment in the interesting energy range.

The sighting of the heaviest members of $\overline{10}$ that were seen only by NA49 experiment at CERN [2] is even more problematic. Nevertheless the positive evidence of 11 experiments that reported the existence of Θ^+ cannot be simply ignored.

MODELS FOR EXOTICS

The founders of the quark model had no clue why the physical states consisted only of 3-quark and quark–antiquark configurations. Indeed, Gell-Mann already in the 60-ies mentioned a possible four-quark–antiquark configuration forming an antidecuplet of SU(3) flavor [8]. With an introduction of color and the concept of confinement some exotic combinations like one-quark, two-quark, four-quark were excluded from the possible spectrum, however, pentaquark states remained. Quark model in its most naive nonrelativistic version predicts that mass of the 5-quark ($uudd\bar{s}$) state should be approximately $5 \times 330 + 150 = 1800$ MeV, where it is assumed that the mass of the constituent quark is 330 MeV and the strange (anti)quark is 150 MeV heavier. This

CP828, *Multiparticle Dynamics*
edited by V. Šimák, M. Šumbera, Š. Todorova, and B. Tomášik

state was believed to be rather wide for the following reason. For the allowed decay $\Theta^+ \to KN$ the momentum of the outgoing Kaon in the Θ^+ rest frame $p_K \simeq 530$ MeV (for $M_{\Theta^+} = 1800$ MeV) *i.e.* two times more than the momentum of the outgoing pion in the decay $\Delta \to \pi N$. However, the phase-space argument is not enough to explain the expected large width of the pentaquark state, since in the quark model Θ^+ has negative parity and therefore can decay in the $s-$wave, while Δ decays in the $p-$wave and the phase space dependence is p_K and p_π^3 respectively. It was then argued that the decay of Δ proceeds by a creation of $q\bar{q}-$pair, while Θ^+ has a so-called *fall apart* mode, resulting in a much larger coupling that in the Δ case.

The states with such properties were looked for in the 70's and 80's in the KN scattering and nothing was found. Recent experimental reports mentioned in the Introduction find that the mass $\Theta^+$is in the range of 1520 – 1560 MeV and the width is very small, presumably of the order of 1 MeV. If true, these properties of the pentaquark states require modification of the naive nonrelativistic quark model. Such modifications were indeed proposed in the literature [9].

An alternative approach to baryons based on the old idea of Skyrme has been developed over the years under the name chiral soliton models (χSM). Chiral soliton models are often regarded as orthogonal to the quark picture. Very often they are generally referred to as Skyrme type models where only mesonic degrees are present. In fact they are deeply rooted in QCD, take into account quark degrees of freedom maybe even in a more complete way than the quark models themselves, and that they are fully operative providing predictions of static baryon properties, structure functions, skewed and off-forward amplitudes and light-cone distribution amplitudes for baryons (for review see *e.g.* Refs. [10, 11, 12]), not to mention properties of pseudoscalar mesons [13]. That of course does not mean that they capture all physics, since — for example — they do not posses confinement. They rely on large N_c limit and chiral symmetry breaking.

Imagine QCD in which gluons have been integrated out. The resulting theory involves only quarks that interact nonlocally in a chirally invariant way. A simplified version of such a theory is a local Nambu-Jona–Lasinio model:

$$\mathscr{L} = \overline{\psi}(i\partial\!\!\!/ - MU^{\gamma_5}[\varphi])\psi \tag{1}$$

which looks like a Dirac Lagrangian density for a massive fermion ψ if not for matrix U. In fact ψ is a 3-vector in flavor space and also in color. Unitary matrix

$$U^{\gamma_5} = \exp\{\frac{i}{F_\varphi}\vec{\lambda}\cdot\vec{\varphi}\,\gamma_5\} \tag{2}$$

is parameterized by a set of eight pseudoscalar fields $\vec{\varphi}$ that, however, do not propagate since there is no meson kinetic term in (1). When we integrate out the quark fields the pseudoscalar fields appear as physical states. The resulting effective action contains only meson fields and can be organized in terms of a derivative expansion

$$S_{\text{eff}}[\varphi] = \frac{F_\varphi^2}{4}\int \text{Tr}\left(\partial_\mu U\,\partial^\mu U^\dagger\right) + \frac{1}{32e^2}\int \text{Tr}\left(\left[\partial_\mu U\,U^\dagger, \partial_\nu U U^\dagger\right]^2\right) + \Gamma_{\text{WZ}} + \ldots \tag{3}$$

where constants F_φ and e can be calculated from (1) with an appropriate cut-off. Γ_{WZ} is the Witten Wess-Zumino term which takes into account axial anomaly. Perhaps the most important part are the ellipses which encode an infinite set of terms that are effectively summed up by the fermionic model of Eq.(1). The truncated series of Eq.(3) is the basis of the Skyrme model. Hence the Skyrme model is (a somewhat arbitrary, because it does not include another possible 4 derivative term) approximation to (1).

Baryons are introduced in two steps, following large N_c strategy described by Witten [14]. First, one constructs a soliton solution, *i.e.* solution to the classical equations of motion that corresponds to matrix U_0 which cannot be expanded in a power series around unity. Second, since the classical soliton has no quantum numbers (except baryon number), one has to quantize the system. Perhaps this quantization procedure, which reduces both models to the nonrelativistic quantum system analogous to the symmetric top [15] with two moments of inertia $I_{1,2}$, makes chiral-soliton models look odd and counterintuitive.

In chiral quark soliton models stabilization of the soliton occurs due to the valence quark level which also provides the baryon number. In the Skyrme model where no quarks are present the soliton is stable due to the specific choice of the 4-derivative term in (3) and the baryon number is given as a charge of the conserved topological current. The quantization on the other hand proceeds in both models almost identically [15], the only difference being that some model parameters dominated by the valence level in the quark soliton model are exactly zero in the Skyrme model.

Chiral soliton models predict that positive parity baryons fall into SU(3) representations that contain hypercharge $Y = N_c/3$ which is 1 in the real world. Therefore the lowest lying multiplets are octet and decuplet, exactly as in the quark models. Moreover, chiral models predict a tower of exotic rotational states starting with $\overline{10}_{1/2}$, $27_{3/2,1/2}$, $\overline{35}_{5/2,3/2}$ (subscripts refer to spin) etc. The splittings between the centers of the lowest-lying octet, decuplet and antidecuplet baryons are given in the χSM by

$$\Delta M_{10-8} = 3/(2I_1), \; \Delta M_{\overline{10}-8} = N_c/(2I_2) = 3/(2I_2) \tag{4}$$

where $I_{1,2}$ are two soliton moments of inertia that depend on details of the chiral Lagrangian. Since $I_1, I_2 \sim \mathcal{O}(N_c)$, this means that $\Delta M_{\overline{10}-8} \sim \mathcal{O}(N_c^0)$, whereas ΔM_{10-8} is $\mathcal{O}(1/N_c)$. This N_c dependence of the splittings has triggered some arguments [16, 17] and counter-arguments [18], regarding the applicability of collective coordinate quantization to the $\overline{10}$.

Early estimates of the masses of exotics were done in the 80's. In 1984 Biedenharn and Dothan estimated that $\Delta M_{\overline{10}-8} \sim 600$ MeV in a specific modification of the Skyrme model [19]. Phenomenological estimates of *both* Θ^+ and $\Xi_{\overline{10}}$ masses obtained in the Skyrme model in 1987 are in a surprising agreement with present experimental findings [20]. Estimates of the antidecuplet mass have been recently reviewed in [21]. The bottom line is that antidecuplet is much lighter than in the quark models. Therefore χSM predicts light exotic baryons belonging to antideucuplet of positive parity.

In 1997 Diakonov Petrov and Polyakov [22] used the quark-soliton model based on (1) and obtained again a small mass $M_{\Theta^+} \simeq 1530$ MeV. However, the most intriguing result of this paper was the prediction that width of Θ^+ and of other members of $\overline{10}$ should be very small. The decay width is calculated by means of the formula for the decay width for $B \to B' + \varphi$:

$$\Gamma_{B\to B'+\varphi} = \frac{1}{8\pi}\frac{p_\varphi}{M_B M_{B'}}\overline{\mathscr{M}^2} = \frac{1}{8\pi}\frac{p_\varphi^3}{M M'}\overline{\mathscr{A}^2} \tag{5}$$

up to linear order in m_s. The "bar"over the amplitude squared denotes averaging over initial and summing over final spin (and, if explicitly indicated, over isospin). Anticipating linear momentum dependence of the decay amplitude $\mathscr{M}$ we have introduced reduced amplitude $\mathscr{A}$ which does not depend on the meson momentum p_φ.

Soliton models can be used to calculate the matrix element $\mathscr{M}$. Explicitly

$$\Gamma_{B\to B'+\varphi} = \frac{3G_{\mathscr{R}}^2}{8\pi M_B M_{B'}} C^{\mathscr{R}}_{B\to B'+\varphi}\, p_\varphi^3.$$

For antidecuplet decays ($\mathscr{R} = \overline{10}$):

$$G_{\overline{10}} = G_0 - G_1 - 1/2\, G_2, \quad C^{\overline{10}}_{\Theta^+\to N+K} = 1/5, \tag{6}$$

whereas for decuplet ($\mathscr{R} = 10$):

$$G_{10} = G_0 + 1/2\, G_2, \quad C^{10}_{\Theta\to N+\pi} = 1/5. \tag{7}$$

In the nonrelativistic small soliton limit [23] in which chiral quark soliton model reproduces many results of the nonrelativistic quark model $G_1/G_0 = 4/5$, $G_2/G_0 = 2/5$ and $G_{\overline{10}} \equiv 0$! This nonintuitive cancellation [22] explains the small width of antidecuplet as compared to the one of 10 for example. Another argument explaining the small width can be found in [24].

CONSEQUENCES OF Θ^+

The existence of ligth and narrow Θ^+ has far reaching theoretical and phenomelogical consequences. On phenomenological side not only Θ^+ but also other members of antidecuplet should exist. Indeed NA49 reported another exotic state that can be interpreted as $\Xi_{\overline{10}}$ at 1860 MeV, however, this state has not been seen so far by any other experiment.

If Θ^+ mass is 1539 MeV and $\Xi_{\overline{10}}$ 1862 MeV then equal spacing within the antidecuplet requires additional cryptoexotic nucleon-like and Σ-like states with masses 1648 MeV and 1757 MeV respectively. These states should be in principle narrow with the decay widths related to Γ_{Θ^+} by the SU(3) symmetry. However, these states will mix with other nonexotic and/or exotic octet states and this mixing will modify these relations [25]. Recent data on photoexcitation of nucleon resonances from GRAAL [26] may be interpreted as a new narrow antidecuplet N* resonance at 1680 MeV. GRAAL sees resonant structure only on neutron but not on proton. This can be understood in terms of magnetic transition moment [27] $\mu_{8\to\overline{10}}$ which is proportional to $Q-1$. Similarly modified PWA [28] of πN scattering indicates that such a state migth exist, also STAR data show some structure in the same energy range.

Spin and parity of Θ^+ are at present unknown. While almost all theorists agree that spin should be 1/2 the parity distinguishes between different models. Chiral models predict positive parity, similarly quark models with flavor dependent forces and correlated quark models predict $\mathscr{P}=+$. In uncorrelated quark models and sum rules $\mathscr{P}=-$.

Unlike model calculations lattice simulations (see S. Sasaki [29] or Kovacs [30] for review) should give clean theoretical answer whether pentaquarks exists and what their quantum numbers are. However, since pentaquarks are excited QCD states, lattice simulations are difficult and give ambiguous message: either there is no bound Θ^+ state, or there is one but with negative parity. One simulation indicates $\mathscr{P}=+$.

Let us stress that, unlike in the case of Ω^- whose spin and parity are not measured but *assumed* after the quark model, the parity of Θ^+ is of utmost importance to discriminate between various models and to understand how QCD binds quarks.

Are there any exotics beyond the $\overline{10}$? Chiral models predict a tower of exotic rotational states starting with $\overline{10}_{1/2}$, $27_{3/2,1/2}$, $\overline{35}_{5/2,3/2}$ (subscripts refer to spin) etc. The lowest excitation of Θ^+ is an isospin triplet of spin 3/2 belonging to flavor 27. The mass of Θ_{27} is only slightly larger than the mass of Θ^+ [21].

In the correlated quark models additional states are also unavoidable. In the diquark model [9] the spin-orbit interaction splits spin 1/2 and 3/2 states by a tiny amount of a $\Delta E \sim 35 \div 65$ MeV [31]. Similarly in the diqaurk-triquark scenario [9], the mass splitting would be of the order of 40 MeV. Hence a nearby isosinglet Θ^+ state of spin $3/2$ is expected in these models. This is a distinguishing feature, since the soliton models do not accommodate spin $3/2$ antidecuplet.

Although there are no more exotics in the minimal diquark model [9], the tensor diquarks in 6 of SU(3) flavor are almost unavoidable. They lead to further exotics like 27 which in the schematic model of Shuryak and Zahed [32] is even lighter than $\overline{10}$.

CONCLUDING REMARKS

The answer whether ligth and narrow Θ^+ does really exist is an experimental issue. Recent reports from CLAS do not confirm previously seen narrow structures in reactions [3, 4]

$$\begin{aligned} \text{G11} &: \quad \gamma+p \rightarrow \Theta^+ + K^0, \\ \text{G10} &: \quad \gamma+d \rightarrow \Theta^+ + K^- + p. \end{aligned} \tag{8}$$

Although discouraging, these results may in fact indicate that – given Θ^+ does exist – the production mechanism is such that the cross-section for these reactions is suppressed. Indeed Lipkin and Karliner [33] indicated that the simplest possible mechanism in G11 case is photon fluctuation into $\bar{K}^0 K^0$ which is indeed suppressed. In the case of G10 photon hits the neutron in the deuteron but the proton from the deutron is also observed. That requires rescattering so that the proton gets enough energy to be seen in the detector. The rescattering cross section is unknown and may be as well suppressed. The question, however, remains why the low statistics experiments: SAPHIR and G2 have found peaks that are not seen in G11 and G10 respectively.

One can replace $\bar{s}$ quark in Θ^+ by $\bar{c}$ or even $\bar{b}$ getting in this way heavy pentaquarks. Here also the situation is not clear. H1 at HERA reported Θ_c at 3.1 GeV [34], however, sister experiment ZEUS did not find the similar peak.

There is no strong theoretical argument against pentaquarks except its unnaturally small width. But – as a large number of recent theoretical papers shows – theoretical explanation may be found in many different models. So if high precision experiments will not find Θ^+ and its partners, this may be even more difficult to understand than the small width and the small mass.

ACKNOWLEDGMENTS

The author would like to thank the organizers of the ISMD 2005 for support and kind hospitality. The present work was partially supported by the Polish State Committee for Scientific Research (KBN) under grant 2 P03B 043 24.

REFERENCES

1. T. Nakano *et al.* (LEPS Collaboration), *Phys. Rev. Lett.* **91**, 012002 (2003); V.V. Barmin *et al.* (DIANA Collaboration), *Phys. Atom. Nucl.* **66**, 1715 (2003); S. Stepanyan *et al.* (CLAS Collaboration), *Phys. Rev. Lett.* **91**, 252001 (2003); J. Barth *et al.* (SAPHIR Collaboration), *Phys. Lett.* B **572** 127 (2003); V. Kubarovsky *et al.* (CLAS Collaboration) *Phys. Rev. Lett.* **92** 032001 (2004); A.E. Asratyan, A.G. Dolgolenko, M.A. Kubantsev, *Phys. Atom. Nucl.* **67**, 682 (2004); A. Airapetian *et al.* (HERMES Collaboration), *Phys. Lett.* B **585**, 213 (2003); The ZEUS collaboration, *Phys. Lett.* B **591** 7 (2004); R. Togoo *et al.*, *Proc. Mongolian Acad. Sci.* **4**, 2 (2003); M. Abdel-Bary, *et al.* (COSY-TOF), Phys. Lett. B **595**, 127 (2004); A. Aleev *et al.* (SVD Collaboration), hep-ex/0401024.
2. C. Alt *et al.* (NA49 Collaboration), *Phys. Rev. Lett.* **92**, 042003 (2004).
3. R. De Vita (CLAS Collaboration), talk at 2005 APS meeting, Tampa, http://www.phy.ohiou.edu/~hicks/thplus/New/RDeVita-APS05.pdf.
4. Lei Guo (CLAS Collaboration), talk at 2005 JLab Users Meeting, Newport News, http://www.jlab.org/ugm05/LGuo1.pdf.
5. T. Nakano (LEPS Collaboration), talk at Int. Conf. on QCD and Hadronic Physics, Beijing, 2005, http://www.phy.pku.edu.cn/~qcd/transparency/20-plen-m/Nakano.pdf; T. Hotta, (LEPS Collaboration), *Acta Phys. Polon.* B **36**, 2173 (2005).
6. A. Aleev *et al.* (SVD Collaboration), hep-ex/0509033.
7. K. H. Hicks, *Prog. Part. Nucl. Phys.* **55**, 647 (2005).
8. M. Gell-Mann, *Phys. Rev.* **125**, 1067 (1962); *Phys. Lett.* **8**, 214 (1964).
9. R. L. Jaffe, F. Wilczek, *Phys. Rev. Lett.* **91**, 232003 (2003); M. Karliner, H. J. Lipkin, *Phys. Lett.* B **575**, 249 (2003).
10. D. Diakonov, V.Y. Petrov, hep-ph/0009006.
11. H. Weigel, *Int. J. Mod. Phys.* A **11**, 2419 (1996).
12. C.V. Christov *et al.*, *Prog. Part. Nucl. Phys.* **37**, 91 (1996).
13. V.Y. Petrov, P.V. Pobylitsa, hep-ph/9712203; V.Y. Petrov, M.V. Polyakov, R. Ruskov, C. Weiss, K. Goeke, *Phys. Rev.* D **59**, 114018 (1999); M. Praszalowicz, A. Rostworowski, *Phys. Rev.* D **64**, 074003 (2001).
14. E. Witten, *Nucl. Phys.* B **160**, 57 (1979) and **223**, 422, 433 (1983).
15. P.O. Mazur, M.A. Nowak, M. Praszałowicz, *Phys. Lett.* B **147**, 137 (1984); A.V. Manohar, *Nucl. Phys.* B **248**, 19 (1984); M. Chemtob, *Nucl. Phys.* B **256**, 600 (1985); S. Jain, S.R. Wadia, *Nucl. Phys.* B **258**, 713 (1985); M.P. Mattis, M. Karliner, *Phys. Rev.* D **31**, 2833 (1985); M. Karliner, M.P. Mattis, *Phys. Rev.* D **34**, 1991 (1986).
16. T.D. Cohen, *Phys. Lett.* B **581**, 175 (2004); *Phys. Rev.* D **70**, 014011 (2004); A. Cherman, T.D. Cohen, A. Nellore, *Phys. Rev.* D **70**, 096003 (2004).

17. P.V. Pobylitsa, *Phys. Rev.* D **69**, 074030 (2004).
18. D. Diakonov, V. Petrov, *Phys.Rev.* D **69**, 056002 (2004).
19. L. C. Biedenharn, Y. Dothan, Print-84-1039 (DUKE), in Gotsman, E., Tauber, G. (eds.): *From SU(3) To Gravity*, pages 15-34.
20. M. Praszalowicz, *Proc. of the Workshop on Skyrmions and Anomalies*, Kraków, 1987, eds. M Jeżabek and M. Praszalowicz, World Scientific, Singapore 1987, p. 531; M. Praszalowicz, *Phys. Lett.* B **575**, 234 (2003).
21. J. Ellis, M. Karliner, M. Praszalowicz, *JHEP* **0405**, 002 (2004).
22. D. Diakonov, V. Petrov, M.V. Polyakov, *Z. Phys.* A **359**, 305 (1997).
23. M. Praszalowicz, A. Blotz, K. Goeke, *Phys. Lett.* B **354** (1995) 415; M. Praszalowicz, T. Watabe, K. Goeke, *Nucl. Phys.* A **647** (1999) 49.
24. D. Diakonov, V. Petrov, hep-ph/0505201.
25. D. Diakonov, V. Petrov, *Phys. Rev.* D **69**, 094011 (2004); S. Pakvasa, M. Suzuki, *Phys. Rev.* D **70**, 036002 (2004); M. Praszalowicz, *Acta Phys. Polon.* B **35**, 1625 (2004); V. Guzey, M. V. Polyakov, hep-ph/0501010.
26. V. Kouznetsov (GRAAL), talk at NSTAR2004, http://lpsc.in2p3.fr/congres/nstar2004/talks/slava_nstar2004.pdf.
27. M. V. Polyakov, A. Rathke, Eur. Phys. J. A **18**, 691 (2003); H. C. Kim, M. Polyakov, M. Praszalowicz, G. S. Yang, K. Goeke, *Phys. Rev.* D **71**, 094023 (2005).
28. R.A. Arndt, Y.I. Azimov, M.V. Polyakov, I.I. Strakovsky, R.L. Workman, *Phys. Rev.* C **69**, 035208 (2004).
29. S. Sasaki, *AIP Conf. Proc.* **717**, 416 (2004).
30. F. Csikor, S. D. Katz, Z. Fodor, T. G. Kovacs, *Acta Phys. Polon.* B **36**, 2271 (2005).
31. J. J. Dudek, F. E. Close, Phys. Lett. B **583**, 278 (2004),
32. E. Shuryak, I. Zahed, Phys. Lett. B **589**, 21 (2004), D. K. Hong, Y. J. Sohn, I. Zahed, *Phys. Lett.* B **596**, 191 (2004).
33. M. Karliner, H. J. Lipkin, hep-ph/0506084.
34. A. Aktas *et al.* (H1 Collaboration), *Phys. Lett.* B **588**, 17 (2004).

Diffractive Higgs boson production at LHC

Marek Taševský

Institute of Physics, Academy of Sciences of the Czech Rep., Na Slovance 2, 182 21 Prague, Czech Rep.
and
Physics Department of the Antwerp University, Universiteitsplein 1, B-2610 Antwerp, Belgium

Abstract. The exclusive double Pomeron exchange production of Higgs boson in diffraction is briefly discussed. Three dedicated Monte Carlo event generators are described and their predictions compared. Two decay channels, namely $\mathrm{H} \rightarrow \mathrm{b}\bar{\mathrm{b}}$ and $\mathrm{H} \rightarrow \mathrm{W}^{+}\mathrm{W}^{-}$, are discussed in detail.

Keywords: Diffraction, Higgs boson, LHC
PACS: 13.85.-t, 14.80.Bn

INTRODUCTION

Data from the proton-proton collisions which will be produced at the LHC at the highest-ever centre-of-mass energy, $\sqrt{s}$, of 14 TeV will provide information about unexplored phase space regions and new physics domains. The luminosity at startup, planned for the year 2007, is expected to be $10^{33}\mathrm{cm}^{-2}\mathrm{s}^{-1}$. In the high luminosity mode, it will reach values of $10^{34}\mathrm{cm}^{-2}\mathrm{s}^{-1}$ leading to an integrated luminosity of 100 fb^{-1} per year. At these highest luminosities, it is expected to see on average 23 overlapping (mostly soft) hadronic interactions per bunch crossing.

The forward physics project includes the CMS [1], TOTEM [2] and CASTOR [3] detectors. The CMS detector is a general purpose detector with an acceptance of $|\eta| < 3$ for tracking and of $|\eta| < 5$ for calorimetry. The pseudorapidity is defined as $\eta = -\ln\tan(\theta/2)$ where θ is the polar angle of a particle with respect to the beam axis. The TOTEM experiment will use the same interaction point (IP) as CMS and is designed to measure the total and elastic pp cross sections, and the diffraction dissociation. It will use two telescopes to detect inelastic events, namely T1 with an acceptance of $3 < |\eta| < 5$ and T2 with an acceptance of $5.3 < |\eta| < 6.6$ (Fig. 1), and three Roman Pot (RP) stations, placed symmetrically (at 147, 180 and 215 m) from the IP to measure protons scattered under very small θ angles. The CASTOR calorimeter is designed to cover the same region as T2. A combination of these detectors thus provides the largest acceptance detector ever built at a hadron collider and enables a study of a large variety of processes in detail. The issue of a common usage of the TOTEM and CASTOR detectors an their integration into the CMS detector and trigger/DAQ system is being investigated in a common study group established in 2002.

CP828, *Multiparticle Dynamics*
edited by V. Šimák, M. Šumbera, Š. Todorova, and B. Tomášik

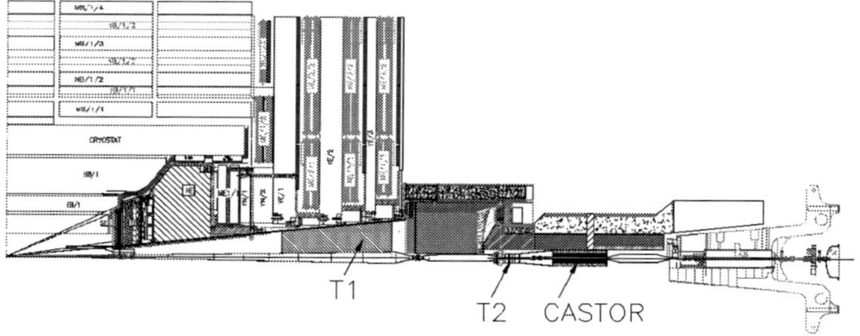

FIGURE 1. Positions of the T1 and T2 TOTEM telescopes and the CASTOR calorimeter, integrated into CMS.

COMPARISON OF MONTE CARLO GENERATORS

The exclusive diffraction mechanism is defined as pp $\rightarrow$ p+X+p where no radiation is emitted between the outgoing beam hadrons and the central system X. The three generators that are examined here are DPEMC [4], EDDE [5] and ExHuMe [6]. ExHuMe is an implementation of the perturbative calculation of Khoze, Martin and Ryskin [7] in which two gluons couple perturbatively to the off-diagonal unintegrated gluon distribution in the proton. This approach includes a Sudakov factor to suppress radiation between the outgoing protons and the central system X. The bare cross section is suppressed by a rapidity gap survival probability, S^2, that accounts for additional momentum transfer between the proton lines which could fill in the gap. ExHuMe takes $S^2 = 0.03$ to be at the LHC.

In contrast, DPEMC and EDDE treat the proton vertices non-perturbatively using Pomeron emission from each of the protons. DPEMC follows the Bialas-Landshoff approach [8] of parameterising the Pomeron flux within the proton. DPEMC also sets S^2 to 0.03 at the LHC. EDDE uses an improved Regge-eikonal approach [9] to calculate the soft proton vertices and includes the Sudakov suppression factor. Using the default settings at the LHC energy of 14 TeV, the total cross sections for production of a 120 GeV Higgs boson are 3.0, 1.9 and 2.8 fb for DPEMC, EDDE and ExHuMe respectively. However, despite these similar values, the physics reach of the central exclusive process is predicted to differ significantly between the generators [10]. Fig. 2a) shows ExHuMe and EDDE predicting that the cross section for the exclusive production of the Higgs boson decreases with the Higgs boson mass (M_H) much faster than in DPEMC. This is a direct effect of the presence of the Sudakov suppression factors growing as the available phase space for gluon emission increases with an increasing mass of the central system. Figure 2b) shows different behaviour of the generators concerning the collision energy dependence of an exclusively produced 120 GeV Higgs boson. Since with a fixed central mass, an increase in collision energy is identical to a decrease in $\xi_1 * \xi_2$, the flatter $\sqrt{s}$ dependence of DPEMC and EDDE is reflected in the flatter ξ distributions compared to ExHuMe (Fig. 3a)). The distributions of ξ (ξ_1 and ξ_2 are the momentum losses of both protons measured in the RPs) as well as of the central system rapidity (Fig. 3b)) have a direct impact on the physics potential. The more central rapidity distribution of ExHuMe is due to the gluon distribution falling more sharply than the Pomeron flux parameterisation present in DPEMC. It should, however, also be noted that the selection of the PDF parameterisation has a non-negligible effect. It was observed in ExHuMe that going from the default MRST 2002 set to the CTEQ6M set gives rise a flatter ξ distribution which leads to a broader peak and sharper fall in rapidity distribution and hence a larger cross section of 3.7 fb. Predictions of the cross sections for two decay modes are shown in Fig. 2c). The increasing branching ratio to WW as M_H increases compensates for the falling central exclusive cross section.

The acceptance of any forward proton taggers that might be installed at the LHC are sensitive to the rapidity distribution of the central system. The differences seen in Fig. 3b) are reflected in different acceptance curves shown in Fig. 4. The predicted acceptances using Roman Pots at 420 and 220 m as a function of the mass of the central system were obtained using a fast simulation of the CMS detector. The fast simulation includes a parameterisation of the responses of the Roman Pots based on a detailed simulation of the detectors [11].

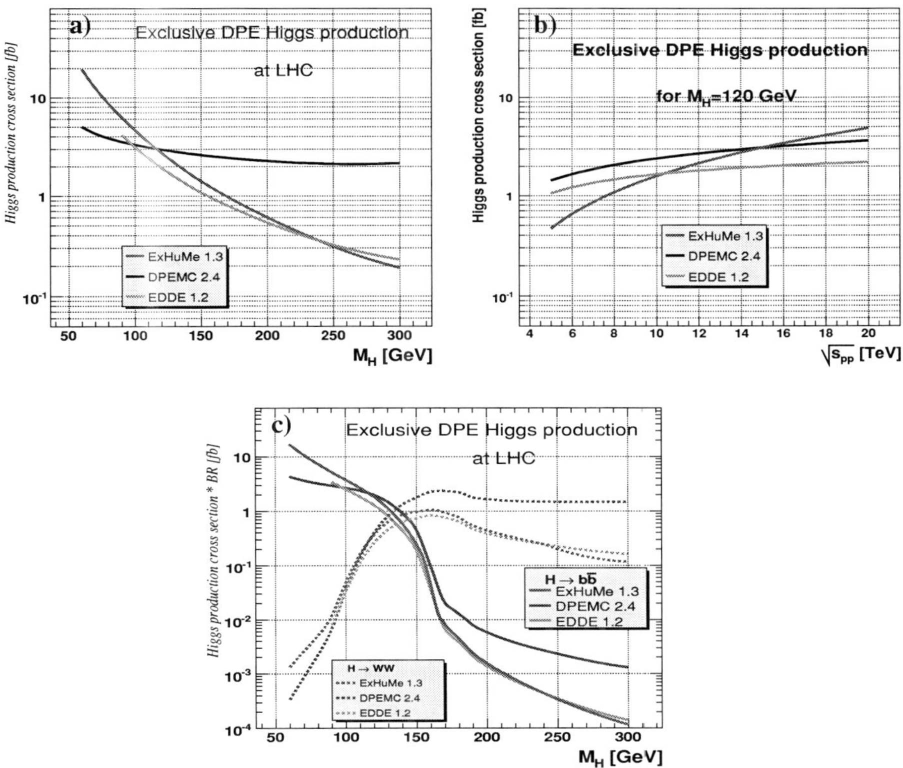

FIGURE 2. The cross section for the exclusive production of the Higgs boson as a function of (a) the Higgs boson mass, (b) the collision energy, (c) the Higgs boson mass (for H→WW and H→ b$\bar{b}$).

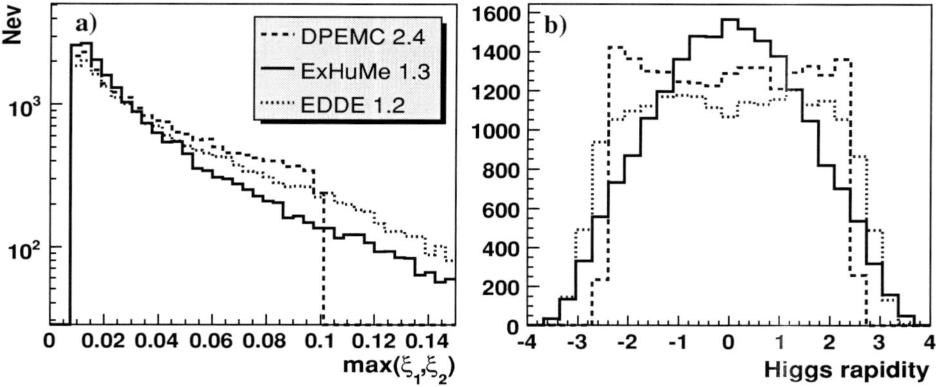

FIGURE 3. The distribution of the a) $\max(\xi_1, \xi_2)$ and b) the Higgs boson rapidity. A cut $\xi_{\max} = 0.1$ is applied in DPEMC, as required by the Bialas-Landshoff approach. Nev is the number of events.

As seen in Fig. 4, as the mass of the central system increases the combined acceptance using detectors at both 220 and 420 m (420+220 m) increases, with the relative difference between the predictions from the three generators decreasing (from about 40% down to 15% for the most extreme relative differences). At $M_H = 120$ GeV the acceptances are predicted to be 46, 50 and 57% for EDDE, DPEMC and ExHuMe, respectively. Corresponding acceptances for the 420 m detector alone are 24, 25 and 32%, respectively.

Experimental data with which the generators could be confronted are very scarce. The only existing set of results comes from the CDF collaboration [12]. They have experimentally defined exclusive events to be those where $R_{jj} = M_{jj}/\sqrt{\hat{s}} > 0.8$ with M_{jj} defined as the mass of two jets with the highest transverse energy and $\sqrt{\hat{s}}$ is the invariant mass of the central system. It should be noted that M_{jj} depends on the particular jet algorithm used and the $\sqrt{\hat{s}}$ measurement relies on the central calorimeter rather than on forward detectors.

H→B$\bar{\text{B}}$

This decay channel is challenging. The b-jet background is controlled by a combination of the $J_z = 0$ selection rule, which strongly suppresses central $b\bar{b}$ production at leading order, and a very good mass resolution from the Roman Pots using the missing mass method. This method gives resolutions of the order of 1% for a 140 GeV central system, if both protons are detected at 420 m from the IP. For the 420+220 m combination the resolution deteriorates to approximately 6%. The detectors at 220 m alone can accept only central systems with masses larger than 200 GeV. There are three issues in this channel. Firstly, the estimate of the whole background turns out to be very complex [13]. The LO contributions are fully suppressed by the $J_z = 0$ selection rule in the

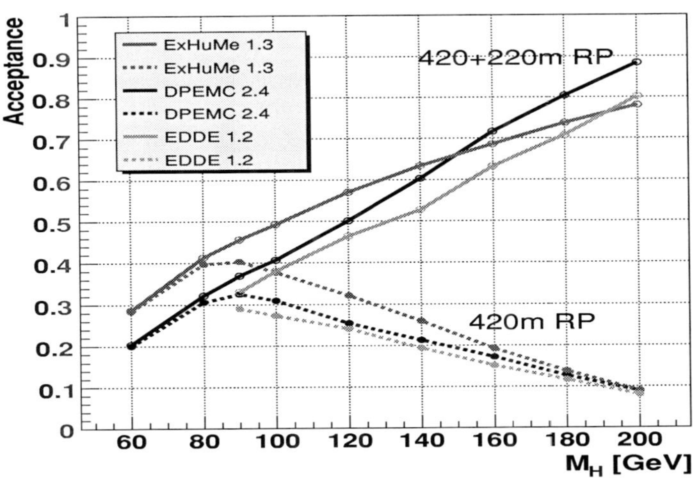

FIGURE 4. The predicted acceptances for the proposed forward taggers at 420 m and for a combination of taggers at 220 and 420 m from the central detector.

limit of massless quarks and forward going protons. There is an admixture of $|J_z| = 2$ production, arising from non-forward going protons and also contributions of the order of m_b^2/E_T^2 from massive quarks. There are also indications that the NLO contributions are non-negligible, however, they still need to be calculated. Another problem is the acceptance-resolution compromise. If we wish to detect a 120 Higgs boson we can choose from two detecting modes, namely the combination of 220 and 420 m or the 420 m alone. While the former gives a reasonable acceptance and a rather moderate resolution, the latter offers a rather low acceptance and the excellent resolution leading to the efficient background suppression. The third problem is the level-1 triggering. Due to latency requirements, the signal from the 220 m RP arrives in time for a level-1 trigger decision, while that from the 420 m RP does not. The cuts used to trigger diffractively produced H$\rightarrow$b$\bar{\text{b}}$ events (with $M_H = 120$ GeV) with a bandwidth of 1 kHz lead to a selection efficiency of 15–20% [14].

H$\rightarrow$WW

This channel does not suffer from any of the above problems: suppression of the dominant backgrounds does not rely primarily on the mass resolution of the RPs, it is calculated with a sufficient precision and certainly in the semi-leptonic decay channel, level-1 triggering is not a problem. From experimental point of view, there are three main categories of events with two W bosons in the final state. Events in which at least one of the W bosons decays in either the e or μ are the simplest and will usually pass the level-1 trigger thresholds due to the hight p_T final state lepton. If one of the W bosons does not decay in the e or μ channel, the event can still pass the level-1 trigger thresholds if a W decays in the τ channel, with τ subsequently decaying leptonically. The 4-jet decay mode occurs approximately half the time, but is unlikely that this signature will pass the level-1 trigger thresholds without information from the RPs. The CMS level-1 trigger has a single (double) electron threshold of 29 (17) GeV with a pseudorapidity coverage of $|\eta| < 2.5$, falling to 14 (3) GeV for a single (double) muon, with a coverage of $|\eta| < 2.1$. In the semi-leptonic decay channel, two quarks are required to have $p_T > 25$ GeV and $|\eta| < 2.5$ and at least one e (μ) to have p_T >20 (10) GeV. In table 1 we show the event yields per luminosity of 30 fb^{-1} as obtained with ExHuMe. Similar results were obtained for the ATLAS level-1 trigger thresholds [15].

TABLE 1. Cross sections, acceptances and event yields per luminosity of 30 fb^{-1} for H→WW decay channel as obtained with ExHuMe 1.3.

M_H[GeV]	σ[fb]	Acc[%]	semi-lept.[Nev]	fully-lept.[Nev]	Total[Nev]
120	0.37	57	1.2	0.2	1.3
135	0.77	62	3.1	0.6	3.4
140	0.87	63	3.5	0.6	3.8
150	1.00	66	4.9	1.0	5.3
160	1.08	69	6.0	1.0	6.6
170	0.94	71	5.4	1.0	5.9
180	0.76	74	4.5	0.8	4.9
200	0.44	78	2.9	0.6	3.2

ACKNOWLEDGMENTS

The work was supported by Interuniversity Attraction Poles Programme - Belgian Science Policy.

REFERENCES

1. CMS Coll., *The Compact Muon Solenoid, Technical Proposal*, **CERN-LHCC-1994-038.**
2. TOTEM Coll., *TOTEM Technical Design Report*, **CERN-LHCC-2004-022.**
3. A.L.S. Angelis *et al., Nuovo Cim.* **24C** (2001) 755; A. Panagiotou, " Progress with CASTOR", talk at CMS/TOTEM meeting on Diffraction, CERN, 12.12.2003.
4. M. Boonekamp and T. Kucs, *Comp. Phys. Commun.* **167** (2005) 167.
5. V.A. Petrov and R. Ryutin, *JHEP* **0408** (2004), 13.
6. J. Monk and A. Pilkington, hep-ph/0502077.
7. V.A. Khoze, A.D. Martin and M.G. Ryskin, *Eur. Phys. J.* **C23** (2002) 311.
8. A. Bialas and P.V. Landshoff, *Phys. Lett.* **B256** (1991) 540.
9. V.A. Petrov and R. Ryutin, *Eur. Phys. J.* **C36** (2004) 509.
10. M. Boonekamp et al., "Diffractive Higgs: Monte Carlo generators for central exclusive diffraction", Proceedings of the HERA-LHC workshop.
11. V. Avati and K. Österberg, "TOTEM forward measurements: leading proton acceptance", Proceedings of the HERA-LHC workshop.
12. T. Affolder et al., *Phys.Rev.Lett.* **85**, (2000) 4215.
13. V.A. Khoze et al., hep-ph/0507040.
14. M. Arneodo et al., "Diffractive Higgs:CMS/TOTEM level-1 trigger studies", Proceedings of the HERA-LHC workshop.
15. B.Cox et al., hep-ph/0505240.
16. M. Carena et al., hep-ph0202167.

SPECIAL SESSION

Chairperson: Y. Hama

Summary of ISMD 2005

A.Bialas

M.Smoluchowski Institute of Physics, Jagellonian University, Reymonta 4, 30-059 Krakow, Poland
e-mail:bialas@th.if.uj.edu.pl

Abstract. A short review of the presentations at the International Symposium of Multipaticle Dynamics held at Kromeriz from 9th till 15th of August 2005 is given.

Keywords: Multiparticle dynamics, correlations, hadronization
PACS: 12.38-t,12.40-y,25.75-q

INTRODUCTION AND APOLOGIES

This year, discussions at the International Symposium on Multiparticle Dynamics covered a particularly large number of different topics. Although I tried my best to give justice to most of them, I found very soon that there is no possible way I can do it. Some subject are even entirely omitted for the simple reason of my ignorance. This is, in particular, the fate of the very interesting lectures on astroparticle physics. Also the talks about the future experiments are not covered in this summary. I am really sorry for that. I should also apologize to all who will find their work inadequately presented and, even worse, misunderstood.

I grouped the subjects, somewhat arbitrarily, into 8 sections which do not fully coincide with the structure adopted by organizers. This has the disadvantage that some talks are mentioned several times (in different sections), but I did not know how to do it better.

STUDIES OF QUARK-GLUON PLASMA

Arguments supporting existence of QGP and investigation of its properties were presented by many authors, most of them representing the experimental collaborations. I shall be brief here, mostly because this subject is also covered at other conferences (recently at QM2005 in Budapest) and the interested reader can easily consult the relevant material. I will thus mention only few recent results.

(i) In the studies of the large transverse momentum jets, two new effects were observed [1, 2]. First, very high statistics of these experiments allowed to study jets at transverse momenta ranging up to 14 GeV. At such high transverse momenta, the away-side jet, although significantly broadened, survives the passage through the dense matter. This gives hope for a future successful investigation of the interaction inside the QGP and thus for detailed studies of its properties. Second, at lower transverse momenta it was found that the remnants of the away-side jet show a broad structure in the azimuthal angle, possibly with a dip in the back-to-back position. Several hypotheses were invoked

CP828, *Multiparticle Dynamics*
edited by V. Šimák, M. Šumbera, Š. Todorova, and B. Tomášik

to explain the structure, most exotic ones being those of Cerenkov-like radiation [3] and of the Mach cone caused by the blast wave propagating through QGP with the velocity of sound [4]. A simpler possibility is a deflection of the away-jet supplemented by coalescence with the remnants of QGP [5]. The preliminary data from STAR support the jet deflection [1], those from PHENIX coll.- the Mach cone [6].

We have also heard an interesting expose by R.Field [7] showing the difficulties in obtaining a well-defined experimental determination of jets. It turns out that the backround is rather substantial, that its understanding is badly needed for the correct interpretation of the present experiments and is absolutely necessary for future (LHC) ones.

(ii) Several talks were devoted to the measurement of the flow and its consequences for the properties of the QGP [8]-[11]. The most important new observation is that the flow in QGP turns out indeed universal: all particles seem to flow, even charm (at transverse momentum above 2 GeV there is still controversy between STAR and PHENIX data in this respect). This shows that the QGP must interact rather strongly, that the equilibrium is attained rather early (probably within 1 f after collision) and that the hadronization is happening rather quickly. All these properties support the Blast Wave picture [12] of the collision.

(iii) On the problem of strangeness enhancement, the final data from WA97 and NA57 [13, 14] were reported. They confirm large enhancement for multistrange baryons, suggesting presence of deconfinement [15]. Rather strong dependence on centrality of the collision was found in qualitative agreement with the idea of canonical enhancement [16]. The observed effect is, however, much stronger than predicted by the theory.

(iv) The collection of data about the "Horn" observed at SPS energies was presented [17] and interpreted as a possible evidence for phase transition [18]. The subject remains controversial: an explanation in terms of hadronic degrees of freedom was also shown [19].

(v) The data on ψ suppression were discussed by A.Tai [20].On theoretical side, R.Thews argued that these data cannot be explained by a simple suppression of ψ in QGP [21] but that some recombination effects are also necessary [22].

HADRONIZATION

The coalescence model [23], which was a great issue at the previous ISMD (cf. the summary by W.Ochs [24]), was only marginally mentioned this year: M.Shao [25] argued that the Cronin effect observed at RHIC cannot be explained without invoking coalescence at the final state of the collision.

Testing the coalescence model remains, however, a rather important issue because its confirmation would give a strong support for the color deconfinement inside QGP. At this point, I would like to make two remarks.

(a) The coalescence model is often applied in two rather different ways. For estimate of rates and fluctuations, the momentum measurements are not necessary and thus the predictions [23, 26] may be valid at all transverse momenta, even very small. On the other hand, the arguments which involve momenta of the constituents [27], may be expected to hold only at relatively large transverse momenta where the precise mechanism of coalescence is not essential[1], say above 1 GeV. Finally, at very large transverse momenta the coalescence model has no chance to work since the density of constituents is simply too low in this region. It is necessary to keep these restrictions in mind when the validity of the coalescence model is discussed.

(b) In its simplest version, the coalescence model does not take into account possible correlatons between the constituents. Such correlations cannot be excluded [28], however, and their investigation may open a new and rather broad field of research.

Another interesting topic related to the mechanism of hadronization was reported by G.Torrieri [29]. He argued that simultaneous measurements of rates *and* fluctuations of the occurence of various hadrons in the final state may distinguish between statistical models of hadronization which assume or do not assume chemical equilibrium. Since establishment of the chemical equilibrium demands a fairly long time, these investigations give another possibility of testing the current models of QGP, favouring a short hadronization time [12].

HBT CORRELATIONS

Several new results were presented.

(i) New tools.

(a) Harmonic decomposition of angular distribution of the HBT correlations [30] was presented by M.Lisa [31]. This new method of analysis represents, in my opinion, a real breakthrough: giving much more systematic information than the traditional representation of HBT data in terms of projections on orthogonal axes, it has a good chance to become a standard tool in all experiments.

(b) A new idea of implementation of quantum interference into the Monte Carlo codes was presented by O.Utyuzh [32].

(ii) New data.

Measurements of $\pi\pi$ correlations in Cu-Cu collisions at RHIC were presented by S.Panitkin [33]. They are in line with the general ideas of the blast wave model. The recent data on K^0K^0 correlations at HERA were presented by L.Zawiejski [34]. They confirm former observations [35] that the HBT radii for pions and kaons are by a large factor higher than those for protons and Λ's. This effect is normally attributed to the strong momentum-position correlations at freeze-out [36, 37]. The origin of such correlations remains, however, unclear. In this respect, the data shown by M.Lisa [31] were indeed striking: it turns out that the measured HBT radii depend on the transverse mass almost identically for nuclear and *pp* collisions [38]. This observation provides yet another serious "HBT puzzle", this time for "elementary" collisions. I personally feel that its solution may provide a new isight into the hadronization process of the QCD cascade [39].

(iii) New theoretical approaches.

A novel approach to the estimate of the HBT effect in heavy nuclei and its application

[1] Extension of these arguments to low transverse momenta may be considered a challenge for theory: It would require understanding the details of the binding of the constituents inside a hadron.

to Au-Au data was presented by J.Cramer [40]. The starting point is the observation that a pion created inside the dense, deconfined medium must struggle before it can escape into the vacuum and thus its wave functon can hardly be a simple plane wave. This effect was described in terms of an optical potential whose imaginary part was responsible for the absorption in medium and the real part approximated the effect of the chiral phase transition (the jump from the chiral symmetry inside the medium to the broken phase outside). Fitting the 11 parameters to describe the data, it was possible to obtain information not only on space-time structure of the system (showing substantial deviations from the blast wave fits) but also about the parameters of the chiral potential. This last feature seems -potentially- very promising.

Similar ideas were presented by S.Pratt [41], who showed that quantum calculation can be replaced by consideration of classical trajectories, giving a more intuitive picture of the phenomenon and saving much of the computer time.

PARTICLE DISTRIBUTIONS

New measurements of in pseudorapidity spectra from CuCu and pp collisions were presented by PHOBOS coll. [42]. They confirm earlier observations that the spectrum is dominated by the large region of limiting fragmentation which increases with increasing energy of the collision. It should be emphasized that this observation invalidates to a large extent the standard picture of boost-invariant system created in high-energy collisions [43]. In my opinion, the consequences of this important change of the paradigm are rather fundamental but they were not yet fully absorbed and digested by the high-energy community.

The analysis of energy and centrality dependence of these spectra [42] showed that these two variables are independent of each other. This factorization is yet another empirical fact which was not anticipated by the theory.

Another unexpected effect in CuCu data was also reported by the PHOBOS collaboration [42]. It turns out that, at a given number of participants, the particle yield in CuCu collisions is identical to that observed in AuAu collisions. Remembering that this rule does not work for *pp* collisions [44], one may be led to the conclusion that the particle production per one wounded nucleon undergoes a discontinuous change somewhere between Copper and Hydrogen collisions. It may be very interesting to investigate this point in more detail, also as function of centrality. I am convinced that we shall hear still more about this in future.

An interesting observation concerning this problem was made by R.Noucier [42]. He compared the spectra not at a fixed number of wounded nucleons but at a fixed number of wounded constituent quarks. It turns out that particle production per one wounded quark is universal ! Similar effects were also reported by E.Sarkisyan [45], who has shown that the wounded quark model provides the correct initial conditions for the Landau hydrodynamical model.

I was particularly pleased by these results because about 25 years ago I was myself an ardent supporter of the wounded quark model [46, 47]. It was thus very nice to hear that this old idea can be successfully applied to the new data at high energies. If this turns out to be indeed the case, we shall have to admit that the observed spectra seem to remember, at least to a large extent, the initial conditions of the collision. Very strange indeed.

GENERAL STUDIES OF CORRELATIONS

As was shown some time ago [48], correlations between non-identical particles carry information about the particle emission profiles in space-time. A large program for studying such correlations is being organized by STAR collaboration and was reported by M.Lisa [31]. This is certainly a promising enterprise and I am sure it will provide substantial new information about the development of the system.

Measurements of charge balance functions by STAR, NA49 and NA22 collaborations were reported by M.Lisa [31], P.Christakoglou [49] and N.Li [50]. They confirm the observation that the width of the balance functions measured in nuclear collisions is substantially narrower than that in hadronic collisions. Furthermore, N.Li, studying hadronic collisions, found that the width of the charge balance function decreases with increasing multiplicity of the collision. I think it would be interesting to compare quantitatively this effect with that found in heavy ion collisions where it is attributed to centrality dependence.

The new data do not change the conclusion that the electric charge is created at the late stage of the development of the QGP [51]. At this moment it may be useful to point out that, apart from "clocking" hadronization [52], measurements of balance functions can provide interesting quantitative information about the process of hadronization. For example, in coalescence model balance functions give information about the correlations between the constituents (before they coalesce into observed hadrons) [28]. In the statistical model, which ignores all correlations except those which arize from resonance decay, the balance function for baryon-antibaryon system should be rather small and very broad, as there are practically no resonances decaying into $B\bar{B}$ pairs [53]. I am thus convinced that this tool can and should be more intensively exploited.

Forward-backward correlations were measured by PHOBOS and reported by G.Roland [54]. These measurements gave evidence for particle production in clusters containg -on average- 2-2.5 charged particles, a result similar to that obtained in early studies of hadron-hadron collisions [55].

Correlatons in trasverse momentum were studied by NA22 collaboration. Y.Huang [56] studied rapidity dependence: it turns out that the properly defined correlation is almost independent of rapidity. J.Fu [57] investigated the dependence of transverse momentum correlations on multiplicity, showing that they must be disentangled before the physical conclusions are drawn from the data.

The results from STAR collaboration on autocorrelations among the produced particles in η -ϕ space were presented by T.Trainor [58]. This interesting and elaborate work (using novel methods of analysis) shows that the character and importance of the observed correlations changes rather dramatically (from one-dimesional to two-dimensional) with inreasing centrality of the collision. The shear volume of the presented data and conclusions makes it impossible, however, to describe them here. I am thus forced to refer the reader directly to the original contribution.

Event by event fluctuations were studied by NA22 coll.[60], aiming at the direct measurement of entropy of the multiparticle system[61], the problem started almost 20 years ago by the organizers of this Symposium [59]. The first, pioneering measurement of entropy of the multipartile system created in π^+- and K^+-p collisions, using the coincidence method, was reported by Z.Li [60]. Also recent progress in theoretical understanding of this method was presented [62].

STUDIES OF QCD

Many contributions were presented in this group. Some of them are listed below.

E. De Wolf presented the data from Opal coll. showing that the correlations among the hadrons forming a jet both at small $|p|$ and at small $p_\perp$ with respect to jet axis do not follow the predictions of perturbative QCD and parton-hadron duality [63]. In my opinion this result should finally close the long discussions about this controversial subject.

K.Fialkowski demostrated that the high order multiplicity moments cannot be used for testing the perturbative QCD. Their complicated behaviour seems to be a reflection of the properties of the tail of the multiplicty distribution and of experimental procedures rather than reflection of fundamental physics [64].

F.Fabbri discussed the difference between multiplicity of jets originated from heavy and light quarks. He showed that the data agree almost perfectly with predictions from perturbative QCD [65].

Production of the hard jets at HERA was discussed in 4 contributions. K.Sedlak have shown the H1 results from small Q^2 data and indicated the possibility of testing the relative weight of various diagrams representing different patters of colour flow by studying the angular distribution of jets [66]. I find this possibility rather interesting and hope that the results will be soon available. L.Joensson discussed jets produced at high Q^2 [67]. The data seem to agree reasonably well with QCD, although the best description is obtained in terms of the Colour Dipole Model. M.Sutton have described how the information on jet production may help in improving the accuracy of determination of the gluon distributions [68]. The results are indeed impressive, although not very surprizing, as the gluon distribution enters directly in the formulae for jet production. Finally, B.Straub have shown that the QCD factorization is satisfied in the data on inlusive production of jets [69], whereas S.Levonian reported violation of factorization in diffractive production [70]

Dijet production with a leading neutron was discussed in contribution by W.Yan [71]. The dependence on dijet variables (E_T and x_γ) is reasonably described by the pion-exchange model. But no model can describe the neutron transverse momentum dependence in the full range of longitudinal momentum fraction.

K.Peters decribed in detail the present status of the factorization breaking in jet production at FERMILAB [72]. The theoretical description of this effect was presented in a comprehensive report by A.Kupco [73].

An interesting contribution was presented by N.Nikolaev, who described how one can derive parton distribution in a nucleus [74]. This work, which demanded rather impressive calculations of colour flow inside the nuclear target, may have, in my opinion, an important impact on further studies of hadron-nucleus interactions.

J.Bartels reported about the progress in his investigation of the AGK rules in perturbative QCD [75]. This work also requires rather involved calculations of complicated diagrams. From what I understood, it seems that the AGK rules are not applicable in pQCD. The work is in progress, however, and it is not clear how they may be eventually modified.

Finally G.Gustafson [76] described the progress in the investigation of the multipartonic collisions which, very likely, will dominate the hadron-hadron interactions at LHC energies. From his description I gather that the problem is awfully complicated. Nevertheless, given the number of previous achievements, the great experience of the Lund group and incredible dexterity with which Gosta presented his argument, I believe they will be able to provide a workable solution at the time it will be needed.

SATURATION

There were only few contributions on this subject but they carried the weight. L.McLerran summarized the present status of the Colour Glass Condensate and indicated interesting remaining problems [77]. I understood from this talk that the idea of CGC came now out of age and attained a real self-confidence. This was clearly seen in the attitude of Larry, who did not try to seduce the audience but simply stated that CGC *is* the long searched high-energy solution of QCD and that its understanding will provide answer to all pertinent questions one may ask in this context. This understanding seems to him to be within reach of the present generation. I am personally not so optimistic (particularly when I hear that "it is enough to sum the loop diagrams") but it is difficult not to agree that the idea is powerful and attractive.

C.Marquet presented a review of successful applications of the ideas of saturation and CGC to several processes, particularly the diffractive ones [78].

H.Fujii have shown a calculation of $q\bar{q}$ production from Colour Glass Condensate [79].

In the final talk of this group, R.Peschanski presented a new analysis of data using the travelling wave solution [81] of the Balitsky-Kovchegov equation. I must say that I was really impressed seeing how the data follow the travelling wave and how they approach geometrical scaling. At the same time, one could see from this picture that the data barely approach the saturation region. Thus saturation, although certainly suggested by this analysis, seems still beyond reach of the present experiments.

MISCELLANEA

Here I would like to call attention to two contributions which are somewhat outside the main stream of the discussion at the Symposium.

M.Praszalowicz [82] gave a top class review of the present status of search for pentaquarks. The subject was put by organizers among "Future projects", probably to emphasize that the status of pentaquarks is still uncertain. But several new experiments are planned (or just running) and soon the situation will be clarified. Anyway, I was under impression of the amount of work already performed and theoretical consequences it implies.

Finally, let me mention an intriguing presentation by M.Tokarev [83] which is difficult to classify in any of the subjects listed till now. He advocated a mysterious z-scaling and have shown a number of plots demonstrating that this idea works in many situations. Since I did not understand what it is all about, it is better to refrain from comments, although I must admit that the results were rather impressive.

ACKNOWLEDGMENTS

This summary could not be realized without the unconditional help and expertise I received from Šárka Todorova-Nová and without a friendly but nevertheless coercive pressure from Vláda Šimák whom I was not able to resist.

REFERENCES

1. T. Hallman, these proceedings. See also J.Adams et al., STAR coll, *Phys.Rev.Lett.* 93 (2004) 252301; nucl-ex/0501009.
2. J.Jia, these proceedings; See also, B.Jacak, nucl-ex/0508036 and references quoted there.
3. I.Dremin, *Pisma v ZhETF* **30** (1979) 152; *Yad.Fiz.* **33** (1981)1357; hep-ph/0507167; V.Koch, A.Majumder and X.-N. Wang, nucl-th/0507063.
4. J.Casalderrey-Solana, E.Shuryak and D.Teaney, hep-ph/0411315; H.Stoecker, *Nucl.Phys.* **A750** (2005) 121; J.Ruppert and B.Muller, *Phys. Lett.* **B618** (2005) 123; hep-ph/0503158.
5. R.Hwa and Z.Tan, nucl-th/0503060; N.Armesto, C.Salgado and U.Wiedemann, hep-ph/0411341.
6. N.Ajitanand, these proceedings.
7. R.Field, these proceedings.
8. R.Lacey, these proceedings; PHENIX coll., S.S.Adler et al., *Phys.Rev.Lett.* **94** (2005) 082302.
9. H.Huang, these proceedings.
10. X.Dong, these proceedings.
11. K.Schweda, these proceedings.
12. F.Retiere and M.Lisa, *Phys.Rev.* **C70** (2004) 044907 and references quoted there.
13. L.Sandor, these proceedings.
14. R.Lietava, these proceedings.
15. P.Koch, B.Muller and J.Rafelski, *Phys.Rept.* **142** (1986) 167; *Z.Phys.* **A324** (1986) 453.
16. A.Tounsi and K.Redlich, *J.Phys.* **G28** (2002) 2095; A.Tounsi, A.Mischke and K.Redlich, *Nucl.Phys.* **A715** (2003) 565.
17. B.Lungwitz, these proceedings.
18. M.Gazdzicki and G.Gorenstein,*Acta Phys.Pol.* **B30** (1999) 2705.
19. B.Tomasik, these proceedings.
20. A.Tai, these proceedings.
21. T.Matsui and H.Satz, *Phys.Lett.* **B178** (1986) 416.
22. M.Schroedter, R.L.Thews and J.Rafelski, *J.Phys.* **G27** (2001) 691; R.L.Thews, M.Schroedter and J.Rafelski, *J.Phys.* **G27** (2001) 715; R.L.Thews and J.Rafelski, *Nucl. Phys.* **A698** (2002) 575; R.L.Thews, these proceedings.
23. J.Zimanyi and P.Levai, nucl-th/0404060; J.Zimanyi, P.Levai and T.Biro, *Heavy Ion Phys.* **17** (2003) 205; J.Zimanyi, T.Biro, T.Csorgo and P.Levai, *Heavy Ion Phys.* **4** (1996) 15; T.Biro, P.Levai, J.Zimanyi, *Phys.Lett.* **B347** (1995)6.
24. W.Ochs, *Acta Phys.Pol.* **B36** (2005) 761.
25. M.Shao, these proceedings.
26. A.Bialas, *Phys.Lett.* **B442** (1998) 449; J.Zimanyi et al., *Phys.Lett.* **B472** (2000) 243.
27. See, e.g., S.Voloshin, *Nucl.Phys.* **A715** (2003) 379; *Acta Phys.Pol.* **B36** (2005) 551.

28. A.Bialas, *Phys.Lett.* **B579** (2004) 31; A.Bialas and R.Rafelski, hep-ph/0508084.
29. G.Torrieri, S-Y Jeon and J.Rafelski, nucl-th/0503046 and G.Torrieri, these proceedings.
30. P.Danielewicz and S.Pratt, *Phys.Lett.* **B618** (2005) 60; M.Lisa, S.Pratt, R.Soltz and U.Wiedemann, nucl-ex/0505014 and references quoted there.
31. M.Lisa, these proceedings.
32. O.Utyuzh, G.Wilk and Z.Wlodarczyk, hep-ph/0503046 and O.Utyuzh, these proceedings.
33. S.Panitkin, these proceedings.
34. L.Zawiejski, these proceedings.
35. For a recent review, see G.Alexander, *Rep.Prog.Phys.* **66** (2003) 481.
36. T.Csorgo and J.Zimanyi,*Nucl.Phys.* **A517** (1990) 588.
37. A.Bialas and K.Zalewski,*Acta Phys.Pol.* **B30** (1999) 359; A.Bialas, M.Kucharczyk, H.Palka and K.Zalewski, *Phys.Rev.* **D62** (2000) 114007; *Acta Phys.Pol.* **B32** (2001) 2901.
38. STAR coll., Z Chajecki et al., nucl-ex/0505009.
39. An interesting possibility of explaining this "puzzle" may exist in the Lund model. G.Gustafson, private communication.
40. J.G.Cramer, G.A.Miller, J.M.S.Wu and J.H.Joon, *Phys.Rev.Lett.* **94** (2005) 102302; G.A.Miller and J.G.Cramer, nucl-th/0507004; J.G.Cramer, these proceedings.
41. S.Pratt, nucl-th/0508029 and these proceedings.
42. R.Noucier, these proceedings and private communication.
43. J.Bjorken, *Phys.Rev.* **D27** (1983) 140.
44. B.B.Back et al., *Phys.Rev.* **C65** (2002) 31901R; *Phys.Rev.* **C**0201005; G.S.F. Stephans et al., *Acta Phys.Pol.* **B**33 (2002) 1419.
45. E.Sarkisyan and A.Sakharov, hep-ph/0410324 and E.Sarkisyan, these proceedings.
46. A.Bialas, W.Czyz and W.Furmanski, *Acta Phys.Pol.* **B8** (1977) 585; A.Bialas, *Proc. ISMD Kayersberg* (1977), p.
47. De Marzo et al.,*Phys.Rev.* **D26** (1982) 1019.
48. R.Lednicky and V.L.Lyubishits, *Heavy Ion Phys.* **3** (1996) 93; R.Lednicky and V.L.Lyubishits, B.Erasmus and D.Nouais, *Phys.Lett.* **B373** (1996) 30.
49. P.Christakoglou, A.Petridis and M.Vassiliou (for NA49 coll.), these proceedings.
50. N.Li (for NA22 coll.), these proceedings.
51. S.Pratt, *Nucl.Phys.* **A715** (2003) 389c.
52. S.A.Bass, P.Danielewicz and S.Pratt, *Phys.Rev.Lett.* **85** (2000) 2689.
53. W.Broniowski, private communication.
54. G.Roland, these proceedings.
55. For a review, see L. Foa, *Phys.Rep.* **22** (1975) 1.
56. Y.Huang, these proceedings.
57. J.Fu, these proceedings.
58. T.Trainor, these proceedings. See also D.Prindle and T.Trainor, hep-ph/0506173; STAR coll., J.Adams et al., hep-ph/0506172; *Acta Phys.Pol.* **B36** (2005) 353; nucl-ex/0411003; nucl-ex/0408012; nucl-ex/0406035.
59. V.Simak and M.Sumbera and I.Zborovsky, *Phys.Lett.* **B206** (1988) 159.
60. M.Atayan et al., *Acta Phys.Pol.* **B36** (2005) 2969 and Z.Li, these proceedings.
61. A.Bialas and W.Czyz, *Phys.Rev.* **D61** (2000) 074021; *Acta Phys.Pol.* **B31** (2000) 687.
62. A.Bialas, W.Czyz and K.Zalewski, hep-ph/0506233; *Acta Phys.Pol.* **B36** (2005) 3109 [hep-ph/0508289] and A.Bialas, these proceedings.
63. E.DeWolf, these proceedings and Note OPAL-PN528 (July 2005).
64. K.Fialkowski and R.Wit, hep-ph/0507008 and K.Fialkowski, these proceedings.
65. Yu. Dokshitzer, F.Fabbri, V.Khoze and W.Ochs, hep-ph/0508074 and F.Fabbri, these proceedings.
66. K.Sedlak, these proceedings.
67. L.Joensson, these proceedings.
68. M.Sutton, these proceedings.
69. B.Straub, these proceedings.
70. S.Levonian, these proceedings.
71. W.Yan, these proceedings. See also H1 coll., *Eur.Phys.J.* **C41** (2005) 273.
72. K.Peters, these proceedings.
73. A.Kupco, these proceedings.

74. N.Nikolaev and W.Schafer, *Phys.Rev.* **D71** (2005) 014023; N.Nikolaev,W.Schafer and B.Zakharov, hep-ph/0508310; N.Nikolaev, these proceedings.
75. J.Bartels, these proceedings.
76. G.Gustafson, these proceedings.
77. L.Mc Lerran, these proceedings.
78. C.Marquet, these proceedings.
79. H.Fujii, F.Gelis and R.Venugopalan, hep-ph/0502204 and H.Fujii, these proceedings.
80. R.Peschanski, these proceedings.
81. S.Munier and R.Peschanski, *Phys.Rev.Lett.* **91** (2003) 232001; *Phys.Rev.* **D69** (2004) 034008.
82. M.Praszalowicz, these proceedings.
83. M.Tokarev, these proceedings.

WORKSHOP ON PARTICLE CORRELATIONS AND FEMTOSCOPY

FEMTOSCOPY IN HEAVY ION COLLISIONS

Chairperson: J. Cramer

Femtoscopy: Theory

Richard Lednický

Institute of Physics ASCR, Na Slovance 2, 18221 Prague 8, Czech Republic
Joint Institute for Nuclear Research, Dubna, Moscow Region, 141980, Russia

Abstract. The theoretical basics of correlation femtoscopy, recent results from femtoscopy in relativistic heavy ion collisions and their consequences are shortly reviewed.

Keywords: correlations, quantum statistics, final state interaction
PACS: 25.75.Gz

INTRODUCTION

The momentum correlations of two or more particles at small relative momenta in their center-of-mass (c.m.) system are widely used to study space-time characteristics of the production processes on a level of fm $= 10^{-15}$ m, so serving as a correlation femtoscopy tool. Particularly, for non-interacting identical particles, like photons or, to some extent, pions, these correlations result from the interference of the production amplitudes due to the symmetrization requirement of quantum statistics (QS) [1-5].[1] The momentum QS correlations were first observed as an enhanced production of the pairs of identical pions with small opening angles (GGLP effect [1]). Later on, Kopylov and Podgoretsky settled the basics of correlation femtoscopy in more than 20 papers (see a review [5]) and developed it as a practical tool; particularly, they suggested to study the interference effect in terms of the correlation function, proposed the mixing techniques to construct the uncorrelated reference sample and clarified the role of the space–time characteristics of particle production in various physical situations. The momentum correlations of particles emitted at nuclear distances are also influenced by the effect of final state interaction (FSI) [10-14]. Though the FSI effect complicates the correlation analysis, it is an important source of information allowing for the coalescence femtoscopy [15-18], the correlation femtoscopy with unlike particles [12-14] including the access to the relative space–time asymmetries in particle production [19-30] and a study of particle strong interaction [26, 29, 30]. In this review, I will concentrate on the assumptions behind the correlation femtoscopy formalism and discuss the recent results obtained from the femtoscopy analysis of like and unlike particle correlations in relativistic heavy ion collisions. One can inspect recent reviews [30-32] for a number of important topics

[1] There is [2-8] an analogy of the momentum QS correlations of photons with the space–time correlations of the intensities of classical electromagnetic fields used in astronomy to measure the angular radii of stellar objects based on the superposition principle (HBT effect) [9]. This analogy is sometimes misunderstood and the momentum correlations are mixed up with the space-time (HBT) correlations although their orthogonal character and thus the absence of the former in astronomy measurements due to extremely large space-time extent of stellar objects (and vice versa) was already pointed out early in [8].

CP828, *Multiparticle Dynamics*
edited by V. Šimák, M. Šumbera, Š. Todorova, and B. Tomášik

that are not touched here, such as non-Gaussian tails, imaging techniques, correlations of penetrating probes, comparison of different colliding systems or spin correlations.

ASSUMPTIONS

The two-particle correlation function $\mathscr{R}(p_1, p_2)$ is usually defined as a ratio of the measured two-particle distribution to the reference one obtained by mixing particles from different events of a given class, normalized to unity at sufficiently large relative momenta. The space-time information contained in the momentum correlations is usually extracted based on the following assumptions:

(i) The mean freeze-out phase space density $\langle f \rangle$ is assumed sufficiently small so that only the mutual QS and FSI effects can be considered when calculating the correlation function of two particles emitted with a small relative momentum $Q = 2k^*$ in their c.m. system.[2] Note that $\langle f \rangle$ increases with energy and for central collisions seems to saturate at the highest SPS energy [31, 43] (see, however, [44]); $\langle f \rangle \ll 1$ for pions with $p_t > 0.2$ Gev/c so pointing to negligible multiboson effects (see, e.g., [45]) in this p_t-region.

(ii) The momentum dependence of the one-particle emission probabilities is assumed inessential when varying the particle four-momenta p_1 and p_2 by the amount characteristic for the correlations due to QS and FSI. This *smoothness assumption*, requiring the components of the mean space-time distance between particle emitters much larger than those of the space-time extent of the emitters, is well justified for heavy ion collisions.

(iii) An independent or incoherent particle emission is assumed. This assumption is reasonable for most of particle pairs produced in heavy ion collisions and is consistent with the observed strength of two- and three-pion correlations (see, e.g., [46, 47]).

(iv) To simplify the calculation of the FSI effect, the Bethe-Salpeter amplitude describing two particles emitted at space-time points $x_i = \{t_i, \mathbf{r}_i\}$ and detected with four-momenta p_i is usually calculated at equal emission times in the pair c.m. system.[3] Then, similar to the Fermi factor in $\beta-$decay, the correlation function can be calculated as a square of the properly symmetrized stationary solution of the scattering problem, $\psi^{S(+)}_{-\mathbf{k}^*}(\mathbf{r}^*)$, averaged over the relative distance $\mathbf{r}^* = \mathbf{r}_1^* - \mathbf{r}_2^*$ of the emitters in the pair c.m. system ($\mathbf{k}^* = \mathbf{p}_1^* = -\mathbf{p}_2^*$) and over the total pair spin S.[4]

(v) The FSI separation in the elastic transitions $1+2 \rightarrow 1+2$ implies a long FSI time as compared with the characteristic production time, i.e. k^* much less than typical

---

[2] This *two-particle approximation* may not be justified for the rare pairs associated with large phase-space density fluctuations and also in low energy heavy ion reactions when the particles are produced in a mean field of residual nuclei. To deal with this field a quantum adiabatic approach [33, 34] or transport simulations [35-37] can be used. At high energies, there are also attempts to account for the mean field effects in eikonal [38-40] and optical potential [41, 42] approaches, the latter giving some tens of percent change of the outgoing phase space density profiles in heavy ion collisions at RHIC assuming however unrealistic in-medium masses at the kinetic freeze-out and neglecting time dependence of the potential.

[3] The *equal time* approximation is usually valid for heavy particles like kaons or nucleons. But even for pions, it merely leads to a slight overestimation (typically $< 5\%$) of the strong FSI effect [12, 48] and, it doesn't influence the leading zero–distance effect of the Coulomb FSI.

[4] Note that in the case of small k^*, we are interested in, the short-range interaction is dominated by central forces and s-waves so the two-particle amplitude depends only on the total spin S.

production momentum transfer of hundreds MeV/c. In fact, the long-time FSI can be separated also in the inelastic transitions, $1+2 \to 3+4$, characterized by a slow relative motion in both entrance and exit channels [49]. Then, one has to solve the two-channel scattering problem and take into account the contributions of both transitions.[5]

FEMTOSCOPY WITH IDENTICAL PARTICLES

It is well known that the directional and velocity dependence of the correlation function can be used to estimate both the duration of the emission and the form of the emission region [2-5], as well as—to reveal the details of the production dynamics (such as collective flows; see, e.g., [50-52] and reviews [53, 54]). For this, the correlation functions can be analyzed in terms of the out (x), side (y) and longitudinal (z) components of the relative momentum vector $\mathbf{q} = \{q_x, q_y, q_z\}$ [55-58]; the out and side denote the transverse components of the vector $\mathbf{q}$, the out direction is parallel to the transverse component of the pair three–momentum. The corresponding correlation widths are usually parameterized in terms of the Gaussian correlation radii and their dependence on pair rapidity and transverse momentum is studied. It appears that with the increasing energy of heavy ion collisions from AGS and SPS up to the highest energies at RHIC, the correlation data show rather weak energy dependence [60] and point to the kinetic freeze-out temperature somewhat below the pion mass, a strong transverse flow (with the mean transverse flow velocity at RHIC exceeding half the velocity of light), a short evolution time of 8-10 fm/c and a very short emission duration of about 2-3 fm/c (see, e.g., a recent review [31]). The short evolution and emission duration at RHIC are also supported by the correlation analysis with respect to the reaction plane [60]. The small time scales at RHIC were not expected in traditional transport and hydrodynamic models (see, e.g., [61, 62]) and may indicate an explosive character of particle production [63, 64] or, large parton cross sections [65] and early hadronization [66] as in the corresponding successful multi-phase and hadron transport models. In view of a successful hydrodynamical description of the elliptic flow at RHIC (see, however, [67]) the failure of hydrodynamics to explain the femtoscopy data is often considered as a puzzle. However, it is still possible that the overestimation of the outward and longitudinal radii will be cured in the full three-dimensional hydrodynamic calculations with the modified initial conditions (earlier thermalization time or initial transverse flow [68, 69]), tuned equation of state with the account for chemical non-equilibrium after hadronization [68-71] and a better modeling of the freeze-out process (including continuous emission during the hydrodynamic stage [72, 73] and coupling to the hadron cascade [74, 75]). This hope is supported by a successful description of the SPS and RHIC data in a number of hydro-motivated parametrizations (see a brief review [76]). The success of these simple parametrizations can be caused by the formation of particle spectra and correlation radii during an early stage of the kinetic freeze-out (see, e.g., [75, 65, 66]). It was argued [73] that such an early spectra formation can be related with the fact that the solution

---

[5] In practice, the particles 1,3 and 2,4 are members of the same isomultiplets (as, e.g., in the transition $\pi^- p \leftrightarrow \pi^0 n$) so one can assume the same $\mathbf{r}^*$-distributions in both channels.

of the non-relativistic Boltzmann equation at spherically symmetric initial conditions exactly coincides with free streaming. Indeed, numerical solutions of relativistic Boltzmann equation at anisotropic initial conditions showed only a slight deviation from free streaming even in the case of a large number of rescatterings [77].

FEMTOSCOPY WITH UNLIKE PARTICLES

The complicated dynamics of particle production, including resonance decays and particle rescatterings, leads to essentially non-Gaussian tail of the r^*-distribution. Therefore, due to different r^*-sensitivity of the QS, strong and Coulomb FSI effects, one has to be careful when analyzing the correlation functions in terms of simple models. Thus, the QS and strong FSI effects are influenced by the r^*-tail mainly through the correlation strength parameter λ while, the shape of the Coulomb FSI is sensitive to the distances as large as the pair Bohr radius $|a|$ (hundreds fm for the pairs containing pions). These problems can be at least partially overcome with the help of transport code simulations accounting for the dynamical evolution of the emission process and providing the phase space information required to calculate the QS and FSI correlation effects. Thus, in a preliminary analysis of the NA49 correlation data from central Pb+Pb 158 AGeV collisions [26], the transport RQMD v.2.3 code was used. To account for a possible mismatch in $\langle r^* \rangle$, the correlation functions were calculated with the space-time coordinates of the emission points scaled by 0.7, 0.8 and 1. The scale parameter was then fitted using the quadratic interpolation. The fits of the $\pi^+\pi^-$, $\pi^+ p$ and $\pi^- p$ correlation function indicate that RQMD overestimates the distances r^* by 10-20% thus indicating an underestimation of the collective flow in this model. Recently, there appeared data on $p\Lambda$ correlation functions from Au+Au experiments E985 at AGS [78], STAR at RHIC [29] and Pb+Pb experiment NA49 at SPS CERN [27]. As the Coulomb FSI is absent in $p\Lambda$ system, one avoids here the problem of its sensitivity to the r^*-tail. Also, the absence of the Coulomb suppression of small relative momenta makes this system more sensitive to the radius parameters as compared with pp correlations [79]. In spite of rather large statistical errors, a significant enhancement is seen at low relative momentum, consistent with the known singlet and triplet $p\Lambda$ s-wave scattering lengths. The fitted correlation radii of 3-4 fm are in agreement with the radii obtained from pp correlations in the same experiments. These radii are smaller than those obtained from two-pion and two-kaon correlation functions at the same transverse momenta and are in qualitative agreement with the approximate m_t scaling expected in the case of the collective expansion.

ACCESSING PARTICLE STRONG INTERACTION

In case of a poor knowledge of the two-particle strong interaction, which is the case for meson-meson, meson-hyperon or hyperon-hyperon systems, it can be improved with the help of correlation measurements. In heavy ion collisions, the effective radius r_0 of the emission region can be considered much larger than the range of the strong interaction potential. The FSI contribution to the correlation function is then independent of the actual potential form and, at small $Q = 2k^*$, it is determined by the s-wave scattering amplitudes $f^S(k^*)$ at a given total spin S [12]. In case of $|f^S| > r_0$, this contribution is of the order of $|f^S/r_0|^2$ and dominates over the effect of QS. In the opposite case, the sensitivity of the correlation function to the scattering amplitude is determined by the linear term f^S/r_0. The possibility of the correlation measurement of the scattering amplitudes has been demonstrated [26] in a preliminary analysis of the NA49 $\pi^+\pi^-$ correlation data within the RQMD model. The fitted strong interaction scale, redefining the original s-wave scattering length of 0.23 fm, appeared to be significantly lower than unity: 0.63 ± 0.08. To a similar shift ($\sim$ 20%) point also the recent BNL data on K_{l4} decays in agreement with the two-loop calculation in the chiral perturbation theory with a standard value of the quark condensate. Also the singlet $\Lambda\Lambda$ s-wave scattering length has been estimated [26, 27] based on the fits of the NA49 data from Pb+Pb collisions at 158 AGeV. Though the fit results are not very restrictive, they likely exclude the possibility of a large singlet scattering length comparable to that of $\sim$20 fm for the two-nucleon system. The STAR experiment measured for the first time the $p\bar{\Lambda}$ and $\bar{p}\Lambda$ correlation functions and performed simultaneous fit of the correlation radius and the spin-averaged s-wave scattering length. The fitted imaginary part of the scattering length of $\sim$ 1 fm is in agreement with the $\bar{p}p$ results while the real part appears to be more negative [29].

ACCESSING RELATIVE SPACE-TIME ASYMMETRIES

The correlation function of two non-identical particles, compared with the identical ones, contains a principally new piece of information on the relative space-time asymmetries in particle emission [19]. Since this information enters in the two-particle amplitude $\psi^{S(+)}_{-\mathbf{k}^*}(\mathbf{r}^*)$ through the terms odd in $\mathbf{k}^*\mathbf{r}^* \equiv \mathbf{p}_1^*(\mathbf{r}_1^* - \mathbf{r}_2^*)$, it can be accessed studying the correlation functions $\mathscr{R}_{+i}$ and $\mathscr{R}_{-i}$ with positive and negative projection k_i^* on a given direction $\mathbf{i}$ or, - the ratio $\mathscr{R}_{+i}/\mathscr{R}_{-i}$. For example, $\mathbf{i}$ can be the direction of the pair velocity or, any of the out (x), side (y), longitudinal (z) directions. In longitudinally comoving system (LCMS), $r_i^* = r_i$ except for $r_x^* \equiv \Delta x^* = \gamma_t(\Delta x - v_t\Delta t)$, where γ_t and v_t are the pair LCMS Lorentz factor and velocity. One may see that the asymmetry in the out (x) direction depends on both space and time asymmetries $\langle\Delta x\rangle$ and $\langle\Delta t\rangle$. In case of a dominant Coulomb FSI, the intercept of the correlation function ratio is directly related with the asymmetry $\langle r_i^*\rangle$ scaled by the Bohr radius $a = (\mu z_1 z_2 e^2)^{-1}$: $\mathscr{R}_{+i}/\mathscr{R}_{-i} \approx 1 + 2\langle r_i^*\rangle/a$. The out correlation asymmetries between pions, kaons and protons observed in heavy ion collisions at CERN SPS and BNL RHIC are in agreement with practically charge independent meson production and a negative $\langle\Delta x\rangle$ (for $m_1 < m_2$) and/or positive $c\langle\Delta t\rangle$ on the level of several fm [26,30]. In fact they are in quantitative agreement with the RQMD transport model as well as with the hydro-motivated blast wave parametrization, both predicting the dominance of the spatial part of the asymmetries generated by large transverse flows. Particularly, the RQMD simulation of central Pb+Pb collisions in conditions of the experiment NA49 at 158 AGeV yields practically zero asymmetries for $\pi^+\pi^-$ system while, for $\pi^\pm p$ systems, $\langle\Delta x\rangle \doteq -5.2$ fm, $\langle\Delta t\rangle \doteq 2.9$ fm/c, $\langle\Delta x^*\rangle \doteq -8.5$ fm. Besides, it predicts $\langle x\rangle$ increasing with particle p_t or $v_t = p_t/m_t$, starting from zero due to kinematic reasons. The asymmetry arises because of a faster increase with v_t for heavier particle. In fact, the hierarchy $\langle x_\pi\rangle < \langle x_K\rangle < \langle x_p\rangle$ is a signal of a universal transversal collective flow [26]; the mean thermal velocity is smaller for heavier particle and thus washes out the positive shift due to the flow to a lesser extent. It should be noted that the shift $\langle x \rangle$ and its mass dependence predicted by the blast wave model are weakened when introducing the transverse temperature gradient. The large correlation asymmetries at RHIC may thus point against a hot emitting central zone surrounded by a cooler hadronic matter indicated by the analysis within the Buda-Lund model [80].

CONCLUSIONS

Wealth of data on momentum correlations of various particle species ($\pi^{\pm}, K^{\pm 0}, p^{\pm}, \Lambda, \Xi$) is available and gives unique space-time information on production characteristics including collective flows. Rather direct evidence for a strong transverse flow in heavy ion collisions at SPS and RHIC is coming from unlike particle correlation asymmetries. Being sensitive to relative time delays and collective flows, the correlation asymmetries can be especially useful to study the effects of phase transitions. Weak energy dependence of correlation radii contradicts to traditional hydrodynamic and transport calculations which strongly overestimate outward and longitudinal radii at RHIC. It remains to be clarified whether this can be cured in full three-dimensional hydrodynamic calculations with a more refined treatment of initial conditions, equation of state and freeze-out process. A number of succesful hydro motivated parametrizations give useful hints in this direction. The momentum correlations yield also a valuable information on particle strong interaction hardly accessible by other means.

REFERENCES

1. G. Goldhaber et al., *Phys. Rev.* **120**, 300 (1960).
2. V.G. Grishin, G.I. Kopylov and M.I. Podgoretsky, *Sov. J. Nucl. Phys.* **13**, 638 (1971).
3. G.I. Kopylov and M.I. Podgoretsky, *Sov. J. Nucl. Phys.* **15**, 219 (1972).
4. G.I. Kopylov, *Phys. Lett. B* **50**, 472 (1974).
5. M.I. Podgoretsky, *Sov. J. Part. Nucl.* **20**, 266 (1989).
6. E.V. Shuryak, *Phys. Lett. B* **44**, 387 (1973).
7. G. Cocconi, *Phys. Lett.* **49**, 459 (1974).
8. G.I.Kopylov and M.I.Podgoretsky, *Sov. Physics JETP* **42**, 211 (1975).
9. R. Hanbury-Brown and R.Q. Twiss, *Nature* **178**, 1046 (1956).
10. S.E. Koonin, *Phys. Lett. B* **70**, 43 (1977).
11. M. Gyulassy, S.K. Kauffmann and L.W. Wilson, *Phys. Rev. C* **20**, 2267 (1979).
12. R. Lednicky and V.L. Lyuboshitz, *Sov. J. Nucl. Phys.* **35**, 770 (1982).
13. D.H. Boal and J.C. Shillcock, *Phys. Rev. C* **33**, 549 (1986).
14. D.H. Boal, C.-K. Gelbke and B.K. Jennings, *Rev. Mod. Phys.* **62**, 553 (1990).
15. H. Sato and K. Yazaki *Phys. Lett. B*, **98**, 153 (1981).
16. V.L. Lyuboshitz, *Sov. J. Nucl. Phys.* **48**, 956 (1988).
17. S. Mrowczynski, *Phys. Lett. B* **277**, 43 (1992); *ibid.* **308**, 216 (1993).
18. R. Scheibl and U. Heinz, *Phys. Rev. C* **59**, 1585 (1999).
19. R. Lednicky et al., *Phys. Lett. B* **373**, 30 (1996).
20. B. Erazmus et al., CERN Note ALICE-INT-1995-43.
21. S. Voloshin et al., *Phys. Rev. Lett.* **79**, 4766 (1997).
22. D. Ardouin et al., *Phys. Lett. B* **446**, 191 (1999).
23. R. Lednicky, S. Panitkin and N. Xu, arXiv:nucl-th/0304062.
24. R. Lednicky, arXiv:nucl-th/0304063, 0304064.

25. D. Miskowiec et al. (E877), arXiv:nucl-ex/9808003.
26. R. Lednicky, NA49 Note number 210 (1999); arXiv:nucl-th/0112011, 0212089.
27. Ch. Blume (NA49), *Nucl. Phys. A* **715** (2003) 55c.
28. J. Adams et al. (STAR), *Phys. Rev. Lett.* **91**, 262302 (2003).
29. A. Kisiel (STAR), *J. Phys. G*, **30**, S1059 (2004).
30. R. Lednicky, arXiv:nucl-th/0305027; *Phys. Atom. Nucl.* **67**, 72 (2004).
31. M. Lisa et al., arXiv:nucl-ex/0505014.
32. T. Csörgő, arXiv:nucl-th/0505019.
33. L. Martin et al., *Nucl. Phys. A* **604**, 69 (1996).
34. H.W. Barz, *Phys. Rev. C* **59**, 2214 (1999); *ibid.* **53**, 2536 (1996).
35. H. Sorge et al., *Phys. Lett. B* **243**, 7 (1990).
36. W.G. Gong et al., *Phys. Rev. C* **43**, 781 (1991).
37. M. Isse et al., arXiv:nucl-th/0502058.
38. C.Y. Wong, *J. Phys. G* **30**, S1053 (2004).
39. W.N. Zhang et al., *Chin. Phys. Lett.* **21**, 1918 (2004).
40. J.I. Kapusta and Y. Li, arXiv:nucl-th/0503075.
41. J.G. Cramer et al., *Phys. Rev. Lett.* **94** 102302 (2005).
42. S. Pratt, arXiv:nucl-th/0508029.
43. S.V. Akkelin and Yu.M. Sinyukov, arXiv:nucl-th/0505045.
44. D. Ferenc et al., *Phys. Lett. B* **457**, 347 (1999).
45. R. Lednicky et al., *Phys. Rev. C* **61**, 034901 (2000).
46. K. Morita, S. Muroya and H. Nakamura, arXiv:nucl-th/0310057.
47. C. Adams et al. (STAR), *Phys. Rev. Lett.* **91**, 262301 (2003).
48. R. Lednicky, arXiv:nucl-th/0501065.
49. R. Lednicky, V.V. Lyuboshitz and V.L. Lyuboshitz, *Phys. of Atomic Nuclei* **61**, 2050 (1998).
50. S. Pratt, *Phys. Rev. Lett.* **53**, 1219 (1984); *Phys. Rev. D* **33**, 1314 (1986).
51. Y. Hama and S. S. Padula, *Phys. Rev. D* **37**, 3237 (1988).
52. A.N. Makhlin and Yu.M. Sinyukov, *Z. Phys. C* **39**, 69 (1988).
53. U. Wiedemann and U. Heinz, *Phys. Rep.* **319**, 145 (1999).
54. T. Csörgő, *Heavy Ion Phys.* **15**, 1 (2002).
55. M.I. Podgoretsky, *Sov. J. Nucl. Phys.* **37**, 272 (1983).
56. G.F. Bertsch, P. Danielewicz and M. Herrmann, *Phys. Rev. C* **49**, 442 (1994).
57. S. Pratt, in *Quark Gluon Plasma 2* ed. by R.C. Hwa, World Scientific, Singapore, 1995, pp. 700-748.
58. S. Chapman, P. Scotto and U. Heinz, *Phys. Rev. Lett.* **74**, 4400 (1995).
59. S.S. Adler et al. (PHENIX), *Phys. Rev. Lett.* **93**, 152302 (2004).
60. J. Adams et al. (STAR), *Phys. Rev. C*, **71**, 044906 (2005); *Phys. Rev. Lett.* **93**, 012301 (2004).
61. S. Soff, S.A. Bass and A. Dumitru, *Phys. Rev. Lett.* **86**, 3981 (2001).
62. U. Heinz and P.Kolb, *Nucl. Phys. A*, **702**, 269 (2002).
63. T. Csörgő and L.P. Csernai, *Phys. Lett. B* **333**, 494 (1994).
64. A. Dumitru and R.D. Pisarski, *Nucl. Phys. A* **698**, 444 (2002).
65. Z.W. Lin and C.M. Ko, *Phys. Rev. Lett.* **89**, 152301 (2002).
66. T. Humanic, *Nucl. Phys. A* **715**, 641 (2003).
67. N. Borghini, arXiv:nucl-th/0509092.
68. P.F. Kolb and R. Rapp, *Phys. Rev. C* **67**, 044903 (2003).
69. D. d'Enterria and D. Peressounko, arXiv:nucl-th/0503054.
70. R. Rapp, *Phys. Rev. C* **66**, 017901 (2002).
71. T. Hirano and K. Tsuda, *Phys. Rev. C* **66**, 054905 (2002).
72. F. Grassi, Y. Hama and T. Kodama, *Phys. Lett. B* **355**, 9 (1996).
73. Yu.M. Sinyukov, S.V. Akkelin and Y. Hama, *Phys. Rev. Lett.* **89**, 052301 (2002).
74. S.A. Bass and A. Dumitru, *Phys. Rev. C* **61**, 064909 (2000).
75. D. Teaney, J. Lauret and E.V. Shuryak, arXiv:nucl-th/0110037.
76. W. Florkowski, arXiv:nucl-th/0509039.
77. N. Amelin et al., arXiv:nucl-th/0507040.
78. M. Lisa et al. (E895), *Nucl. Phys. A* **698**, 185 (2002).
79. F. Wang and S. Pratt, *Phys. Rev. Lett.* **83**, 3138 (1999).
80. M. Csanád et al., *J. Phys. G: Nucl. Part. Phys.* **30**, S1079 (2004).

Is HBT Really Puzzling?

Scott Pratt and Daniel Schindel

Department of Physics and Astronomy, Michigan State University
East Lansing, Michigan 48824

Abstract. Two-particle correlations from RHIC have provided a surprising snapshot of the final state at RHIC. In this talk I discuss the nature of the HBT puzzle and attempt to delineate several factors which might ultimately resolve the issue.

Keywords: Correlations, Femtoscopy, Heavy Ion Collisions
PACS: 25.70.Nq

Correlation, or HBT, measurements provide our best insight into the space-time development of relativistic heavy-ion collisions [1]. Immediately after the first two-pion correlations from RHIC were presented in 2001, the term "HBT puzzle" came into common usage. The term was inspired by the failure of some of our most sophisticated dynamical models of the collision. However, there is no problem in describing the data with simple parameterizations of the break-up space-time profile. In this talk, I will review one of these simple parameterizations, the blast-wave model, to illustrate the unsettling aspects of the inferred parameters. I then discuss several factors which might ultimately lead to a more satisfying interpretation of the data.

Numerous parameterizations of the breakup space-time geometry can be applied to fit correlation data from RHIC. Among these are the Buda-Lund model [2] and numerous variations of the blast-wave model [3, 4, 5]. Minimal blast wave models incorporate four common parameters: the breakup temperature T, the radius of the fireball surface R, the collective velocity at the surface $v_\perp$, and the breakup time τ, which is related to the collective-velocity gradient along the beam direction by the relation $v_z = z/t$ by virtue of assuming boost invariance. Additionally, chemical potentials can be added to normalize the spectra. Parameters for the surface diffuseness or the temporal duration of the emission $\Delta\tau$ are often included. The Buda-Lund model also allows for a temperature gradient.

Pionic observables do not tightly constrain the parameters by themselves [6], but after fitting the spectra of heavier particles the following parameters emerge from blast-wave fits: $T \sim 110$ MeV, $v_\perp \sim 0.7c$, $R \sim 12 - 13$ fm, and $\tau \sim 9$ fm/c. The duration of the emission appears to be sudden, less than 5 fm/c, and roughly consistent with zero. The quality of the fits with data are illustrated in Fig. 1.

On inspection, there are two puzzling aspects of these numbers. First, the breakup density is remarkably high. If one takes the 1000 hadrons per unit of rapidity observed at RHIC and divide it by the volume $V = \tau\pi R^2$, one obtains densities higher than nuclear matter density, in the range of 0.18 fm^{-3}. Assuming cross sections of 20 mb, estimates of the mean free path are in the neighborhood of 2.5 fm, which is much smaller than the 25 fm diameter of the fireball. The second surprising aspect of the fit is that it

CP828, *Multiparticle Dynamics*
edited by V. Šimák, M. Šumbera, Š. Todorova, and B. Tomášik

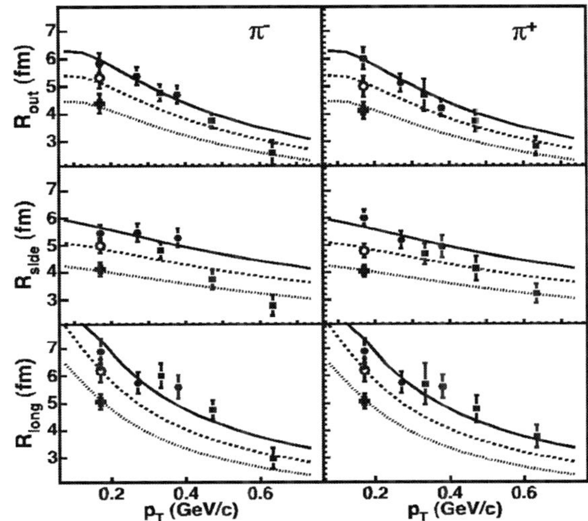

FIGURE 1. Blast wave fits to source sizes from Lisa and Retière [3].

implies the fireball, whose original diameter was 6 fm, expanded by 7 fm in 10 fm/c. Given the surface velocity was 0.7c, this implies the surface unphysically accelerates instantaneously to its final velocity.

The density, though surprising, can be readily explained. First, due to collective expansion, the density will have fallen substantially by the time particles move a few fm. Secondly, at the time of breakup, the expansion has become isotropic. For non-relativistic Gaussian-like expansions of particles of the same mass, collisions do not alter the outgoing phase space density, and thus cease to have an effect on the overall source size [7]. Thus, the phase space distribution tends to effectively freeze out before particles have their last collisions. In fact, microscopic simulations incorporating cross sections of tens of mb often yield source sizes smaller than those measured at RHIC. The freezing-out of the source size can be thought of as a competition between the cooling, which pushes the system toward smaller source sizes and the growth of the overall source. For the qualifiers mentioned above, these two balance perfectly. In practice, collisions of pions with heavier particles tend to reduce the pion source and increase the apparent source size of heavier particles such as protons, as entropy moves from pions to protons. In fact, source sizes from microscopic models usually vary remarkably little as a function of the assumed cross section [8].

Source sizes are constrained by entropy, which can be calculated from final-state phase space densities, which can in turn be extracted from experiment given measured radii,

$$\bar{f}(\mathbf{p}) \equiv \frac{\int d^3r[f(\mathbf{p},\mathbf{r})]^2}{\int d^3r f(\mathbf{p},\mathbf{r})} \quad (1)$$

$$
\begin{aligned}
&= \frac{\pi^{3/2}}{(2S+1)} \cdot \frac{dN/d^3p}{R_{\rm out}R_{\rm long}R_{\rm side}} \\
\frac{dS}{dy} &\approx 2\pi \int p_t dp_t\, E\frac{E}{d^3p}\left[\frac{5}{2} - \frac{3}{2}\log(2) - \log(\bar{f}(p_t)) \pm \frac{1}{2^{3/2}}\bar{f}(p_t)\right]
\end{aligned}
\quad (2)
$$

These expressions were applied in [9] to extract entropy and phase space density in central 130*A* GeV collisions at RHIC. Phase space densities at this and lower energies from [1] are displayed in Fig. 2. Note that the phase space densities tend to rise, then saturate near the top SPS energies. At this point the phase space density, which is approximately frozen after chemical freezeout in an isentropic expasion, saturates when the entirety of the fireball surpasses temperatures of 170 MeV [10], the temperature for which particle ratios suggest chemical compositions have frozen out [11].

Pionic source volumes are a factor of two smaller than those predicted by hydrodynamic models based on lattice-gauge-theory-inspired equations of state. This represents a deficit of ln(2) units of entropy per pion, which might suggest that the equation of state used for the hydrodynamic calculations in [12] is excluded. However, an analysis of the entropy carried by all the particles is consistent with entropy expected from lattice calculations [9] as shown in Fig. 2. Since baryons tend to carry twice as much entropy per particle as pions, a modest increase in the population of baryons is able to absorb the $\sim 20\%$ shortfall in pionic entropy. If one knows $s(\varepsilon)$, the entropy density as a function of the energy density, one can also derive the pressure $P(\varepsilon)$ from thermodynamic identities. A higher-entropy equation of state, i.e., one where the effective number of degrees of freedom is higher, must also have lower pressure than an equation of state whose entropy rises more slowly with energy. Since entropy can increase during the expansion, the fact that the measured entropy is consistent with the lattice equation of state places a lower-bound on the stiffness of the equation of state. A very stiff equation of state, like the pion gas example illustrated in Fig. 2, could in principle be consistent with the final-state entropy, but would require the entropy to double during the expansion. Most mechanisms for generating entropy such as viscosity and shocks are expected to provide $\sim 10\%$ increases in entropy.

Unfortunately, there are no systematic studies of the dependence of source radii as a function of the equation of state. However, a compilation of several hydrodynamic and microscopic transport models presented in Fig. 3 shows a trend of stiffer equations of state resulting in smaller radii, as expected. By incorporating more resonances, or by including strings, microscopic models effectively lower the pressure. The largest source sizes resulted from hydrodynamic calculations using an equation of state with a large latent heat. Although it is premature, one might conclude that the equation of state appears somewhat stiffer than some of the lattice models, but not nearly so stiff as that of a pion gas. Taken as a whole the model calculations are encouraging as they illustrate the sensitivity of correlation meausurements to the equation of state.

Even though the average source sizes can be reproduced with some of the transport models represented in Fig. 3, none of the models does a good job of reproducing the p_t dependence for all three dimensions. Part of this difficulty lies in the fact that, as stated earlier, models tend to require some time to accelerate transversely and thus have difficulty reproducing the final-state geometry described by the blast-wave model. This

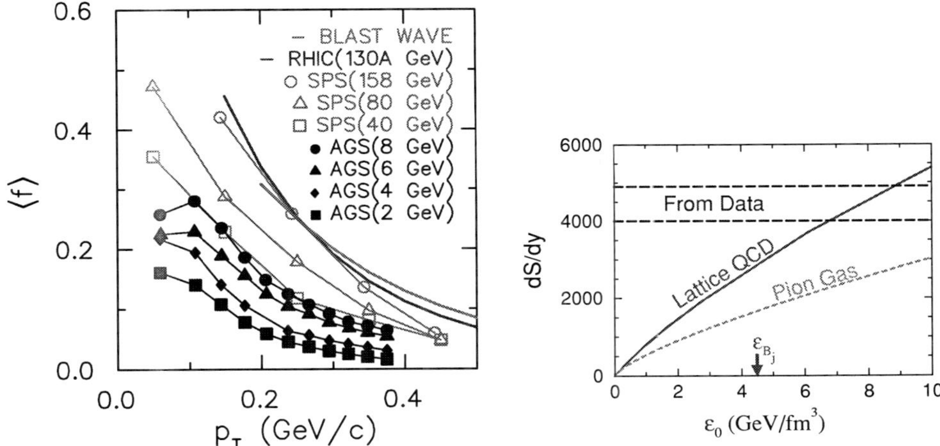

FIGURE 2. Average pionic phase space densities as a function of p_t (left panel) rise with increasing beam energy, but saturate at upper SPS energies. As a function of the initial energy density at 1 fm/c, expected entropies per unit rapidity can be calculated given the equation of state (right panel). A blast-wave model with a chemical potential of 70 MeV for pions roughly reproduces the RHIC results. The Bjorken estimate probably underestimates the true energy density by $\sim 50\%$, thus the lattice equation of state seems consistent with the experimentally extracted dS/dy which is slightly above 4000.

aspect of the HBT puzzle might be resolved by some combination of the following improvements or alterations of dynamic models:

1. Reduced emissivity. Sequential evaporation from the surface tends to increase $R_{\mathrm{out}}/R_{\mathrm{side}}$ beyond the value of unity observed. Since pions with $p_t > 100$ MeV/c move faster than the surface, those pions emitted earlier get ahead of other pions and extend the shape of the phase space distribution. Any mechanism to reduce the emissivity of the surface, such as super-cooling [13], would help the system reach the sudden freeze-out picture of a blast-wave model.
2. Non-infinite longitudinal extent. Many of the hydrodynamic models assume boost invariance. A finite extent along the beam direction lowers R_{long}, and since R_{long} is related to the inferred lifetime, accounting for the finiteness should lead to longer lifetimes which gives the matter more time to expand transversely to the large 13 fm size.
3. Longitudinal acceleration. Boost invariant models also neglect acceleration along the beam axis. Accounting for this acceleration alters the connection between the velocity gradient and the lifetime, $dv_z/dz \sim 1/\tau$, and should result in somewhat longer times [14].
4. Shear viscosity. The high velocity gradient along the beam axis at early times can lead to shear effects which lower the pressure along the beam axis while raising the transverse pressure [15, 16]. This effect could be especially significant if the early stage is well described by longitudinal classical fields. For non-interacting electric

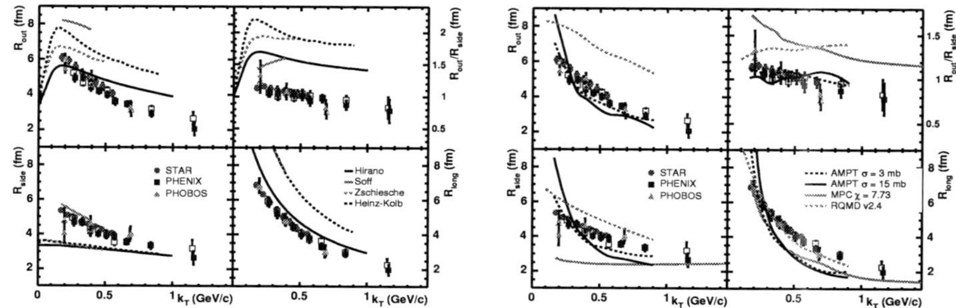

FIGURE 3. Dimensions of several models from identical-pion correlations compared to RHIC data as a function of the average pair momentum k_t. A basic trend is apparent that those models with soft equations of state overestimate the source sizes while the stiffest example, a pion gas (MPC) grossly underestimates the source sizes. None of the models quantitatively describe the k_t dependence of all the dimensions.

fields along the z axis, the components of the stress energy tensor are $T_{xx} = T_{yy} = \varepsilon$, $T_{zz} = -\varepsilon$.

5. Refraction through the mean field. An attractive mean field for pions could refract trajectories and lead to phase space distributions whose R_{out} dimension exceeds the physical size of the emitting source [17, 18, 19], especially the R_{side} projection.

None of these effects are expected to represent more than a 10% correction, except for the refractive effects at very low p_t, but they all push the analysis of the data toward a more physical interpretation. Thus, the HBT puzzle might well have originated from a conspiracy of effects, each of which borders on being negligible when considered by itself.

Unraveling the space-time picture of these collisions will undoubtedly benefit from improved measurement, especially the measurement of other correlations besides pion correlations [20, 21]. However, the main impediment toward resolving correlation measurements remains the lack of a cohesive well-tested modeling framework. For instance, there is the need of a detailed comparison of hydrodynamic and Boltzmann approaches using the same effective equations of state, so that the role of viscosity, and the sensitivity to breakup conditions might be quantified. The random comparison of different models, each run with different initial conditions, unknown equations of state and viscosities, untested breakup criteria, and varying chemistries, precludes robust or rigorous scientific conclusions being generated from comparison with data. But, despite the daunting modeling challenge that lies ahead, the analyses performed thus far have demonstrated a keen sensitivity between the bulk properties used in the calculations and experimental correlation analyses.

ACKNOWLEDGMENTS

Support was provided by the U.S. Department of Energy, Grant No. DE-FG02-03ER41259.

REFERENCES

1. M. A. Lisa, S. Pratt, R. Soltz and U. Wiedemann, nucl-ex/0505014.
2. M. Csanád, T. Csörgő and B. Lörstad, *Nucl. Phys. A* **742**, 80-94 (2004).
3. F. Retière and M. A. Lisa, *Phys. Rev. C* **70**, 044907 (2004).
4. E. Schnedermann, J. Sollfrank and U. W. Heinz, *Phys. Rev. C* **48**, 2462-2475 (1993).
5. A. Kisiel, T. Taluc, W. Broniowski and W. Florkowski, nucl-th/0504047.
6. B. Tomasik, nucl-th/0304079.
7. M. Chojnacki, W. Florkowski and T. Csörgő, *Phys. Rev. C* **71**, 044902 (2005).
8. S. Pratt and J. Murray, *Phys. Rev. C* **57**, 1907-1919 (1998).
9. S. Pal and S. Pratt, *Phys. Lett. B* **578**, 310-317 (2004).
10. S.V. Akkelin and Y.M. Sinyukov, nucl-th/0505045.
11. P. Braun-Munzinger, D. Magestro, K. Redlich and J. Stachel, *Phys. Lett. B* **518**, 41-46 (2001).
12. S. Soff, S. A. Bass and A. Dumitru, *Phys. Rev. Lett.* **86**, 3981-3984 (2001).
13. T. Csörgő and L. P. Csernai, *Phys. Lett. B* **333**, 494-499 (1994).
14. T. Renk, these proceedings, hep-ph/0509053.
15. D. A. Teaney, *J. Phys. G* **30**, S1247-S1250 (2004).
16. A. Muronga and D. H. Rischke, nucl-th/0407114.
17. J. G. Cramer, G. A. Miller, J. M. S. Wu and J. H. S. Yoon, *Phys. Rev. Lett.* **94**, 102302 (2005).
18. G. A. Miller and J. G. Cramer, arXiv:nucl-th/0507004.
19. S. Pratt, nucl-th/0508029.
20. S. Pratt and S. Petriconi, *Phys. Rev. C* **68**, 054901 (2003).
21. P. Danielewicz, these proceedings.

HOW CORRELATIONS REFLECT DYNAMICS

Chairperson: Y. Hama

Transport Model Study of HBT at RHIC

Che Ming Ko

Cyclotron Institute and Physics Department, Texas A&M University, College Station, Texas 77843-3366, USA

Abstract. The pion interferometry at the Relativistic Heavy Ion Collider (RHIC) has been studied in a multiphase transport (AMPT) model that includes both initial partonic and final hadronic interactions. It was found that the two-pion correlation function was sensitive to the magnitude of the cross sections for parton scattering in the partonic matter. Also, the emission source of pions from the AMPT model is non-Gaussian with strong space-time and momentum correlations as well as a large halo around the central core, leading to source radii that can be more than twice larger than the radius parameters extracted from a Gaussian fit to the correlation function.

Keywords: Transport model, HBT, RHIC
PACS: 25.75.Gz, 25.75.-q, 24.10.Lx

INTRODUCTION

Two-particle interferometry based on the Hanbury-Brown Twiss (HBT) effect [1] has been used in nuclear collisions at RHIC to extract information about the emission source [2, 3, 4]. Because of the long emission time as a result of the phase transition from the quark-gluon plasma to hadronic matter, the emission source in relativistic heavy ion collisions is expected to have a much larger radius in the direction of the total transverse momentum of detected two particles than that perpendicular to both this direction and the beam direction [5, 6, 7]. The extracted ratio of these two radii from a Gaussian fit to the measured two-pion correlation function in Au+Au collisions at $\sqrt{s} =$ 130 AGeV is, however, close to one [2, 3, 4]. Also, the extracted radius parameters are small compared to theoretical predictions based on the hydrodynamic model [8]. These puzzling results have been attributed to strong space-time and momentum correlations in the emission source [9]. In this talk, they are addressed in the framework of a multiphase transport model (AMPT) [10, 11, 12] that includes both initial partonic and final hadronic interactions.

A MULTIPHASE TRANSPORT MODEL

The AMPT model is a hybrid model that uses minijet partons from hard processes and strings from soft processes in the HIJING model [13] as the initial conditions for modelling heavy ion collisions at ultra-relativistic energies. In the default version, time evolution of resulting minijet partons is described by the ZPC parton cascade model [14] with an in-medium cross section derived from the lowest-order Born diagram with an effective gluon screening mass taken as a parameter for fixing the magnitude and angular distribution of parton scattering cross section. After minijet partons stop interacting, they

CP828, *Multiparticle Dynamics*
edited by V. Šimák, M. Šumbera, Š. Todorova, and B. Tomášik

are combined with their parent strings, as in the HIJING model with jet quenching, to fragment into hadrons using the Lund string fragmentation model as implemented in the PYTHIA program [15]. The final-state hadronic scatterings are then modelled by the ART hadronic transport model [16]. In an extended string melting version of the AMPT model [12], hadrons that would have been produced from string fragmentation are converted to valence quarks and/or antiquarks in order to model the initially formed partonic matter. Interactions among these partons are again described by the ZPC model. The transition from the partonic matter to the hadronic matter is achieved using a simple coalescence model, which combines two nearest quark and antiquark into mesons and three nearest quarks or antiquarks into baryons or anti-baryons that are close to the invariant mass of these partons.

HBT CORRELATION

From the AMPT model, the source of emitted particles is obtained from their space-time coordinate x and momentum $\mathbf{p}$ at freeze-out, i.e., at their last interactions. Denoting the single-particle emission function for particles by $S(x,\mathbf{p})$, the HBT correlation function for two identical particles of momenta $\mathbf{p}_1$ and $\mathbf{p}_2$ in the absence of final-state interactions, such as the Coulomb interaction, is then given by [5, 17]:

$$C_2(\mathbf{Q},\mathbf{K}) \approx 1+\frac{\int d^4x_1 d^4x_2 S(x_1,\mathbf{p}_1)S(x_2,\mathbf{p}_2)\cos\left[Q\cdot(x_1-x_2)\right]}{\int d^4x_1 S(x_1,\mathbf{p}_1)\int d^4x_2 S(x_2,\mathbf{p}_2)}, \tag{1}$$

where $\mathbf{K}=(\mathbf{p}_1+\mathbf{p}_2)/2$ and $Q=(\mathbf{p}_1-\mathbf{p}_2,E_1-E_2)$. The three-dimensional correlation function in $\mathbf{Q}$ is usually shown as a function of the invariant relative momentum ($Q_{\rm inv}=\sqrt{-Q^2}$) or as a function of the projection of the relative momentum $\mathbf{Q}$ in the "out-side-long" (*osl*) system [6, 7], defined by the beam direction ($Q_{\rm long}$), the direction along the total momentum of the two particles in the transverse plane ($Q_{\rm out}$), and the direction orthogonal to the above two directions ($Q_{\rm side}$).

TWO-PION CORRELATION FUNCTIONS

With the emission function obtained from the AMPT model for central ($b=0$ fm) Au+Au collisions at $\sqrt{s}=130$ *A*GeV, the correlation function $C_2(\mathbf{Q},\mathbf{K})$ has been evaluated using the program Correlation After Burner [18]. Figure 1 shows the one-dimensional projections of the calculated correlation function including final-state Coulomb interactions for mid-rapidity ($-0.5<y<0.5$) charged pions with transverse momentum $125<p_{\rm T}<225$ MeV$/c$ for both the default AMPT model and the extended one with string melting. In evaluating the one-dimensional projections of the correlation function onto one of the $Q_{\rm out},Q_{\rm side}$ and $Q_{\rm long}$ axes, integrations of the other two $\mathbf{Q}$ components over the range $0-35$ MeV/c have been carried out. To reproduce measured one-dimensional correlation functions by the STAR collaboration [2], shown by filled circles, requires a parton scattering cross section of about 10 mb in the extended AMPT model with string melting. Such a large parton scattering cross section was also needed

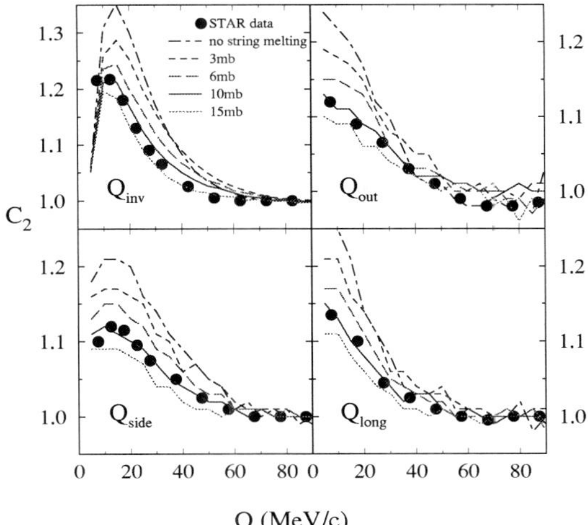

FIGURE 1. Correlation functions for mid-rapidity charged pions with $125 < p_T < 225$ MeV/c in central ($b = 0$ fm) Au+Au collisions at $\sqrt{s} = 130A$ GeV. Theoretical results with the Coulomb correction are shown for the default AMPT model (dash-dotted curves) and for the extended AMPT model with string melting and various values for the parton scattering cross section. Coulomb-uncorrected correlation functions from the STAR collaboration [2] are shown by filled circles.

in the AMPT model to describe the observed anisotropic flows of charged hadrons in heavy ion collisions at RHIC [12, 19, 20].

EMISSION SOURCE RADII

The size of the emission source can be determined from the emission function via the curvature of the correlation function at $\mathbf{Q} = 0$:

$$R_{ij}(K)^2 = -\frac{1}{2}\frac{\partial^2 C_2(\mathbf{Q},\mathbf{K})}{\partial Q_i \partial Q_j}\bigg|_{\mathbf{Q}=0} = D_{x_i,x_j}(K) - D_{x_i,\beta_j t}(K) - D_{\beta_i t,x_j}(K) + D_{\beta_i t,\beta_j t}(K). \quad (2)$$

In the above, $x_i (i = 1-3)$ denotes the projections of the particle position at freeze-out in the *osl* system, i.e., $x_{\rm out}$, $x_{\rm side}$ and $x_{\rm long}$; $\beta = \mathbf{K}/K_0$ with K_0 being the average energy of the two particles; and $D_{x,y} = \langle x \cdot y \rangle - \langle x \rangle \langle y \rangle$, with $\langle x \rangle$ denoting the average value of x.

Empirically, the size of emission source is usually estimated by fitting the measured correlation function $C_2(\mathbf{Q},\mathbf{K})$, that has been corrected for effects due to final-state Coulomb interactions, with a four-parameter Gaussian function:

$$C_2(\mathbf{Q},\mathbf{K}) = 1 + \lambda \exp\left(-\sum_{i=1}^{3} R_{ii}^2(K) Q_i^2\right). \quad (3)$$

If the emission source is Gaussian in space and time, then for central heavy ion collisions considered here the radius parameters obtained from the above Gaussian fit to the

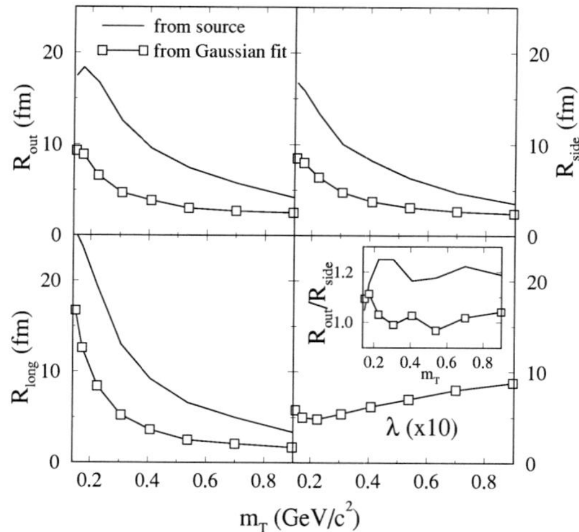

FIGURE 2. Source radii from the emission function (solid curves) as well as the fitted radii and λ parameters from the Gaussian fit to the correlation function (curves with squares) of mid-rapidity pions obtained from the AMPT model with string melting and a parton scattering cross section of 10 mb for central ($b = 0$ fm) Au+Au collisions at $\sqrt{s} = 130$ AGeV as functions of pion transverse mass m_T. The inset shows the corresponding ratio R_{out}/R_{side}.

correlation function would be the same as those determined directly from the emission function of the source via Eq. (2).

In Fig. 2, solid lines are the transverse mass (m_T) dependence of the source radii R_{out} (upper-left panel), R_{side} (upper-right panel), and R_{long} (lower-left panel) of mid-rapidity charged pions in central Au+Au collisions at $\sqrt{s} = 130$ AGeV, that are determined from the emission function given by the AMPT model with string melting and parton scattering cross section of 10 mb. The radius parameters determined from the Gaussian fit to the three-dimensional correlation function in **Q** obtained from the emission function without including final-state Coulomb interactions are shown by curves with squares. It is seen that they are about a factor of 2 to 3 smaller than the source radii obtained directly from the emission function but similar to those extracted from a Gaussian fit to the measured correlation function after correcting for the final-state Coulomb interactions [2]. However, the ratio R_{out}/R_{side} obtained from the emission function (solid curve in the inset of the lower-right panel of Fig. 2) has a value between 1.0 and 1.3, and is not much larger than that (curve with squares) from the radius parameters obtained from the Gaussian fit to the correlation function. The latter is similar to the experimental values extracted from the measured correlation function [2].

DISCUSSIONS

The large difference between the source radii obtained directly from the emission function and from the Gaussian fit to the correlation function can be understood from the

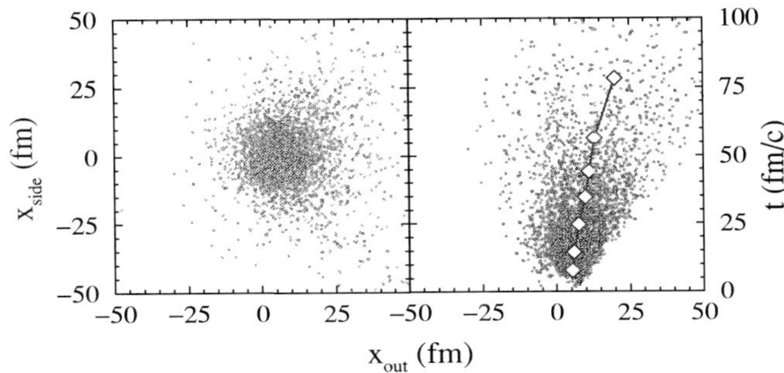

FIGURE 3. $x_{\text{out}} - x_{\text{side}}$ (left panel) and $x_{\text{out}} - t$ (right panel) distributions at freeze-out for mid-rapidity charged pions with $125 < p_{\text{T}} < 225$ MeV$/c$ from the AMPT model with string melting and a parton scattering cross section of 10 mb for central ($b = 0$ fm) Au+Au collisions at $\sqrt{s} = 130$ AGeV. The curve with open diamonds represents $\langle x_{\text{out}} \rangle$ as a function of t.

$x_{\text{out}} - x_{\text{side}}$ (left panel) and $x_{\text{out}} - t$ (right panel) distributions in the emission function shown in Fig. 3 for mid-rapidity pions with $125 < p_{\text{T}} < 225$ MeV$/c$ from the AMPT model with string melting and a parton scattering cross section of 10 mb. It is seen that the emission source is shifted in the direction of the pion transverse momentum, i.e, $\langle x_{\text{out}} \rangle > 0$. This positive shift in x_{out} results from the collective expansion of the emission source [12]. The emission source also shows a large halo around a central core, which consists not only of pions from decays of long-lived resonances such as the ω but also of thermal pions. Furthermore, there exists a strong positive $x_{\text{out}} - t$ correlation as clearly seen from the solid curve with open diamonds, which shows that the average value $\langle x_{\text{out}} \rangle$ increases with the freeze-out time t. Since Eq. (2) gives $R_{\text{out}}^2 = D_{x_{\text{out}},x_{\text{out}}} - 2\ D_{x_{\text{out}},\beta_\perp t} + D_{\beta_\perp t,\beta_\perp t}$ and $R_{\text{side}}^2 = D_{x_{\text{side}},x_{\text{side}}}$, the $x_{\text{out}} - t$ correlation makes it difficult to extract information about the duration of emission from the ratio $R_{\text{out}}/R_{\text{side}}$ as it leads to a large positive value for the $x_{\text{out}} - t$ correlation term $D_{x_{\text{out}},\beta_\perp t}$. For pions included in generating Fig. 3, the value of $D_{x_{\text{out}},\beta_\perp t}$ is 168 fm^2 and is indeed appreciable compared to 185 and 431 fm^2, respectively, for the first and last terms in above equation. Because of the large halo and the space-time correlations, the emission source from the AMPT model thus deviates appreciably from a Gaussian one. The radius parameters obtained from a simple Gaussian fit to the measured correlation functions thus may not give the true source radii. The imaging method [21], developed for extracting the emission function of a source from the correlation function, will be very useful for verifying the non-Gaussian features of the emission source in high-energy heavy-ion collisions.

SUMMARY

Within a multiphase transport model that includes both initial partonic and final hadronic interactions, the two-pion correlation function is found to be sensitive to the parton scattering cross section, which controls the density at which the parton-to-hadron transition occurs in the AMPT transport model. To reproduce the measured correlation function in central Au+Au collisions at $\sqrt{s} = 130\ A\mathrm{GeV}$ requires both the melting of initial strings to partons and a large parton scattering cross section. Furthermore, the emission source is non-Gaussian in space and time. It not only shifts significantly to the direction along the pion transverse momentum but also has a strong correlation between this displacement and the freeze-out time. Consequently, the source radii extracted directly from the emission function are about a factor of 2 to 3 larger than the radius parameters extracted from a Gaussian fit to the three-dimensional correlation function. The ratio $R_{\mathrm{out}}/R_{\mathrm{side}}$ obtained from the emission function is, however, not much larger than that extracted from a Gaussian fit to the correlation function, which is close to one. Although to evaluate the correlation function requires only the space-time information of pions at freeze-out, it is shown to be sensitive to the partonic dynamics during the early stage of heavy ion collisions. The study of pion interferometry as well as that of kaons [22] thus helps to confirm the formation of the partonic matter at RHIC and to study its properties.

ACKNOWLEDGEMENTS

This talk is based on work in collaboration with Zi-Wei Lin and Subrata Pal, and supported by the U.S. National Science Foundation under Grant No. PHY-0457265 and the Welch Foundation under Grant No. A-1358.

REFERENCES

1. R. Hanbury Brown and R.Q. Twiss, *Nature (London)* **178**, 1046 (1956).
2. C. Adler *et al.*, STAR Collaboration, *Phys. Rev. Lett.* **87**, 082301 (2001).
3. S. C. Johnson, PHENIX Collaboration, *Nucl. Phys. A* **698**, 603-606 (2002).
4. K. Adcox *et al.* [PHENIX Collaboration], *Phys. Rev. Lett.* **88**, 192302 (2002)
5. S. Pratt, *Phys. Rev. Lett.* **53**, 1219-1221 (1984).
6. G. Bertsch, M. Gong and M. Tohyama, *Phys. Rev. C* **37**, 1896-1900 (1988).
7. S. Pratt, T. Csörgö and J. Zimanyi, *Phys. Rev. C* **42**, 2646-2652 (1990).
8. S. Soff, S. A. Bass and A. Dumitru, *Phys. Rev. Lett.* **86**, 3981-3984 (2001).
9. B. Tomasik and U. W. Heinz, *Eur. Phys. J. C* **4**, 327-338 (1998); `nucl-th/9805016`.
10. B. Zhang, C. M. Ko, B. A. Li and Z. W. Lin, *Phys. Rev. C* **61**, 067901 (2000).
11. Z. W. Lin, S. Pal, C. M. Ko, B. A. Li and B. Zhang, *Phys. Rev. C* **64**, 011902 (2001); *Nucl. Phys. A* **698**, 375c-378c (2002); `nucl-th/0411110`.
12. Z. W. Lin and C. M. Ko, *Phys. Rev. C* **65**, 034904 (2002).
13. X. N. Wang and M. Gyulassy, *Phys. Rev. D* **44**, 3501-3516 (1991).
14. B. Zhang, *Comput. Phys. Commun.* **109**, 193-206 (1998). .
15. T. Sjostrand, *Comput. Phys. Commun.* **82**, 74-90 (1994).
16. B. A. Li and C. M. Ko, *Phys. Rev. C* **52**, 2037-2063 (1995). .
17. U. A. Wiedemann and U. W. Heinz, *Phys. Rept.* **319**, 145-230 (1999). .
18. S. Pratt *et al.*, *Nucl. Phys. A* **566**, 103c-114c (1994).
19. L. W. Chen, C. M. Ko and Z. W. Lin, *Phys. Rev. C* **69**, 031901 (2004).
20. L. W. Chen, V. Greco, C. M. Ko and P. F. Kolb, *Phys. Lett. B* **605**, 95-100 (2005).
21. D. A. Brown and P. Danielewicz, *Phys. Lett. B* **398**, 252-258 (1997); *Phys. Rev. C* **57**, 2474-2483 (1998); *ibid* **64**, 014902 (2001); S. Y. Panitkin and D. A. Brown, *Phys. Rev. C* **61**, 021901 (2000).
22. Z. W. Lin and C. M. Ko, *J. Phys. G* **30**, S263-S270 (2004)

Evolution of Observables in Hydrodynamic and Kinetic Models of A+A Collisions

Yu.M. Sinyukov

Bogolyubov Institute for Theoretical Physics, Metrologichna st. 14b, 03143 Kiev, Ukraine

Abstract. Time evolution of the momentum spectra, Bose-Einstein correlation functions and averaged phase-space densities (APSD) is analyzed within hydrodynamic and kinetic models. The results shed light on behavior of the interferometry volumes and APSD in A+A collisions at different energies and for different nuclei.

Keywords: *relativistic heavy ion collisions, phase-space density, hydrodynamic evolution, Boltzmann equations, inclusive spectra, Bose-Einstein correlations.*
PACS: *24.10.Nz, 24.10.Pa, 25.75.-q, 25.75.Gz, 25.75.Ld.*

INTRODUCTION

This is a brief review of recent results [1, 2, 3] that were obtained in collaboration with S. Akkelin, N. Amelin, R. Lednicky, L. Malinina and T. Pocheptsov as for the time behavior of the observables in hydrodynamic and numerical kinetic models of heavy ion collisions. It is known that the bulk of hadronic observables which belong to the "soft physics" are related to the very last period of the matter evolution, so called kinetic freeze-out when the hadronic system decays. We argue for the particle transverse momentum spectra, interferometry volumes and averaged phase-space densities (APSD), if they were measured at any time of the expansion of hadron-resonance gas, are approximately conserved upon isentropic and chemically frozen evolution [1]. This is an important result for a study of the hadronic matter at the early stages of the evolution, in particular, near the hadronization point. We discuss the treatment of experimental data for pions that accounts for such a behavior of above values.

"CONSERVED OBSERVABLES" IN HYDRODYNAMIC AND KINETIC MODELS

As it is well known the momentum spectra in multi-particle production can be expressed through the Wigner functions $f(x,p)$ which are associated with the phase-space densities in classical case (see, e.g. [4]). For locally equilibrated weekly interacting gas the Wigner functions are just the local Bose-Einstein or Fermi-Dirac distribution functions if the Compton length of quanta in the gas is much less than the typical homogeneity length in expanding system [4]. Supposing such distributions or, if it is possible, the Boltzmann approximation to them we analyze temporal behavior of the "observables" in different hydrodynamic models.

CP828, *Multiparticle Dynamics*
edited by V. Šimák, M. Šumbera, Š. Todorova, and B. Tomášik

Let us start from analytic results. The exact non-relativistic solution [5] of the ideal hydrodynamics with longitudinal (L) velocity $v_L = x_L/t$, initially zero transverse (T) velocity $v_T = 0$ and the Gaussian transverse density profile (with radii R) leads to a temporal behavior of the observables as it follows [1].

At each moment t the longitudinal spectrum is $m\frac{dN}{dp_L} = const$, similarly to $\frac{dN}{dy} = const$ in the relativistic boost-invariant model, and the momentum spectrum is

$$\frac{d^3N}{d^3p} = \frac{n_0 R_0^2 t_0}{m^2 T_{eff}(t)} \exp\left(-\frac{p_T^2}{2mT_{eff}(t)}\right), \tag{1}$$

where n_0 is the initial particle density at $\mathbf{x}_T = 0$, R_0 is the initial Gaussian radius of the fireball and the effective temperature, T_{eff}, is

$$T_{eff} = m\dot{R}^2(t) + T(t). \tag{2}$$

The value of $T_{eff}(t)$ does not change much since a decrease of the temperature $T(t)$ during the system expansion is accompanied by an increase in time of the transverse flows described by $\dot{R}(t)$. The typical behaviors of the temperature and effective temperature in this model are presented in the left panel of Fig. 1.

The interferometry radii, *out-*, *side-* and *long-* take the forms

$$R_O^2 = R_S^2 = \frac{T(t)}{T_{eff}(t)} R^2(t),\ R_L^2 = t^2 \frac{T(t)}{m} \tag{3}$$

and are similar to the ones in the relativistic case in midrapidity when $m \to m_T \equiv \sqrt{m^2 + p_T^2}$. The interferometry volume is

$$V_{int} \equiv R_O R_S R_L = \frac{R_0^2 t_0 (T_0)^{3/2}}{T_{eff}(t)\sqrt{m}} \tag{4}$$

and does not change significantly since it depends on T_{eff} only. The average momentum phase-space density is

$$\langle f(t,\mathbf{p})\rangle = \frac{\int \left(f^{l.eq.}(t,\mathbf{r},\mathbf{p})\right)^2 d^3r}{\int f^{l.eq.}(t,\mathbf{r},p) d^3r} = \frac{d^3N}{d^3p} / (8\pi^{3/2} V_{int}) \tag{5}$$

This quantity as well as the interferometry volume V_{int} will be conserved in time if and when $T_{eff}(t)$ is time constant at some stage of the evolution.

In the middle panel of Fig. 1 we demonstrate results for the evolution of the averaged pion phase-space density $\langle f(t,\mathbf{p})\rangle$ within realistic model of Pb+Pb collisions at CERN energies 158 AGeV [5]. The model describes hydrodynamically the evolution of the chemically frozen hadron-resonance gas from the hadronization point to thermal, or kinetic freeze-out. As one can see, the values of $\langle f(t,\mathbf{p})\rangle$ are changed by less than 10% during the evolution and are even higher at the final than at initial time at small transverse momenta. Contrary to this the particle densities of the thermal pions drop rather quickly in the central part of the system as is shown in the right panel of Fig. 1. Note, when

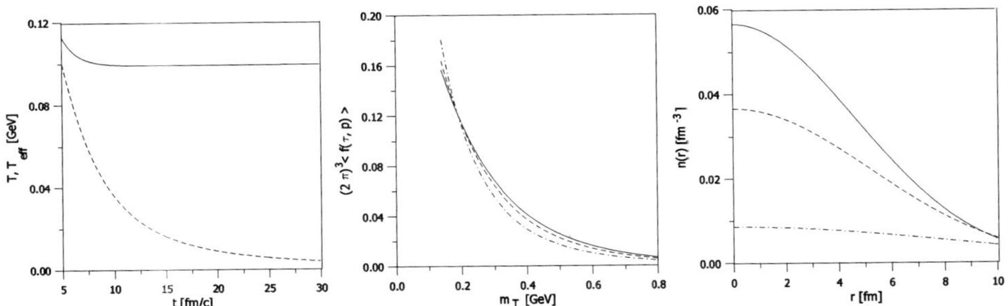

FIGURE 1. Left: Evolution of the pion effective temperature T_{eff} (solid line) versus the temperature of the system T (dashed line) in the non-relativistic one-component Bjorken-like model with transverse expansion. The plots correspond to initial conditions: $T_0 = 0.1$ GeV, time $t_0 = 5$ fm/c, transverse radius $R_0 = 5$ fm and the initial transverse flow $\dot{R}(t_0) = 0.3$. Middle and right: The pion APSD $\langle f(\tau, p)\rangle$ (middle) and the pion densities $n(r)$ (right) taken at typical proper times: at hadronization, $\tau = 7.24$ fm/c, (solid line) and at kinetic freeze-out $\tau = 8.9$ fm/c (dashed line). The dot-dashed line corresponds to the "asymptotic" time $\tau = 15$ fm/c of hydrodynamic evolution. The initial conditions of hydrodynamic expansion of hadron-resonance gas are taken from Ref. [5].

densities becomes sufficiently small the hydrodynamic picture destroys, nevertheless, formally calculated for this asymptotic regime the slope parameters, interferometry radii and other observables do not depend on time [6, 7] in complete accordance with analogous (and obvious) results for a free streaming.

An analysis of a temporal behavior of the observables has been also done in Ref. [3] within Universal Kinetic Model UKM-R, a Monte Carlo event generator, realized as a numerical code written in object oriented C++ language. First, the known results based on analytic solutions of the non-relativistic Boltzmann equation (BE) were reproduced. As it is shown in Fig. 2 neither momentum spectra nor interferometry radii of expanding fireball depend on the evolution time despite a huge number of collisions, if the initial state of a gas corresponds to Maxwell-Boltzmann thermal function with a spacial spherically symmetric Gaussian distribution. It can be easily understood from the result [6] that the difference between the initial and final momentum spectra $\Delta(p)$ is defined by the dissipative effects:

$$\Delta(p) = \int_{\sigma_0}^{\sigma_{\text{out}}} d^4x \left[C_{\text{gain}}(x,p) - C_{\text{loss}}(x,p)\right], \tag{6}$$

(and similar for Bose-Einstein correlation functions) in BE:

$$p^{\mu}\frac{\partial f(x,p)}{\partial x^{\mu}} = C_{\text{gain}}(x,p) - C_{\text{loss}}(x,p). \tag{7}$$

Therefore if $C_{\text{gain}} \approx C_{\text{loss}}$ at least in integral sense, it leads to a conservation of the above observables. In the described case *gain-* and *loss-* terms are coincided locally .

Further, the initial conditions have been modified and it was found that with the increasing elastic cross section the system of nonrelativistic particles more and more recovers spherical symmetry and thermal momentum distribution. At the same time, sim-

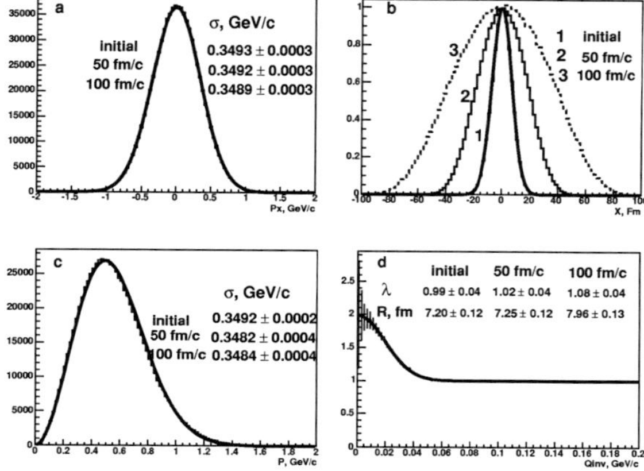

FIGURE 2. Distributions of the x-components of particle three-momentum (**a**) and three-coordinate (**b**), particle momentum (**c**) and the corresponding CF's as functions of q_{inv} (**d**) obtained in the UKM-R simulation of the kinetic evolution of $N = 400$ heavy spin-0 bosons of mass $m = 0.938$ GeV/c^2 at the evolution time $t = 0$, 50 and 100 fm/c. The elastic cross section $\sigma_{el} = 400$ mb, the initial Gaussian radius $R_0 = 7$ fm and the initial temperature $T_0 = 0.130$ GeV. The shown results of Gaussian fits of the momentum distributions and the CF's agree with the input initial values $\sigma = (mT_0)^{1/2} = 0.3492$ GeV/c and $\lambda = 1, R_0 = 7$ fm respectively.

ilar to the above case, the momentum dispersion and the interferometry volume is practically conserved during the evolution [3]. The conservation effect takes place also in the evolution of the system of relativistic particles, except for some increase ($\sim 30\%$ at 400 mb) of the final interferometry volume. However, since the mean mass of hadron-resonance gas is fairly large, $\bar{m} = 0.662$ [5, 1], as compare with typical temperatures, 100-170 MeV, one can expect the interferometry volume should not change noticeably during the evolution. Thus UKM-R studies support similar results obtained in hydrodynamic approximation.

ANALYSIS AND TREATMENT OF EXPERIMENTAL DATA

As it was shown in Ref. [1] the phase-space density of thermal pions *totally averaged* over space (over freeze-out hypersurface σ) and over momenta except the longitudinal one (rapidity is fixed, e.g., $y = 0$),

$$(2\pi)^3 \langle f \rangle_{y=0} = \frac{\int d^3p \overline{f}^2_{eq}}{\int d^3p \overline{f}_{eq}} = \kappa \frac{2\pi^{5/2} \int \left(\frac{1}{R_O R_S R_L} \left(\frac{d^2N}{2\pi m_T dm_T dy} \right)^2 \right) dm_T}{dN/dy}, \quad (8)$$

is an approximate integral of motion. Here $\overline{f}_{eq} \equiv (\exp(\beta(p_0 - \mu) - 1)^{-1}$, and β and μ coincide with the inverse of the temperature and chemical potential which are supposed

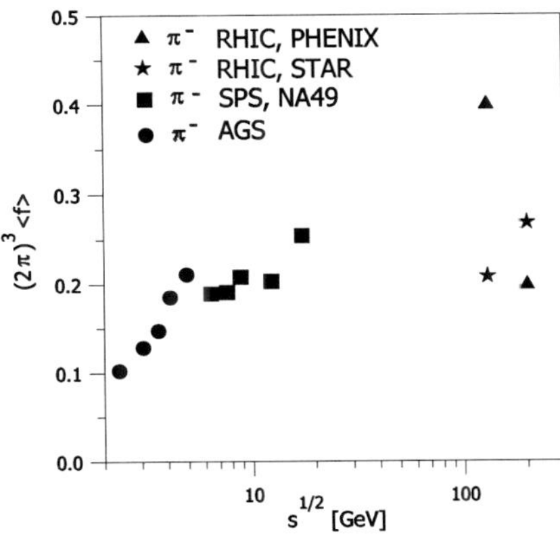

FIGURE 3. The average phase-space density of all negative pions at mid-rapidity, $(2\pi)^3\langle f\rangle$, as function of c.m. energy per nucleon in heavy ion central collisions.

to be uniform on freeze-out hypersurface. The $\kappa = 1$ if one ignores resonance decays. Roughly, $\frac{dN}{m_T dm_T dy} \propto \exp(-m_T/T_{eff})$. Then, estimating the APSD, $\langle f\rangle$, and assuming that integral C over dimensionless variable m_T/T_{eff} depends on energies of collisions fairly smoothly, one can write

$$V_{int} \simeq C\frac{dN/dy}{\langle f\rangle T_{eff}^3} \quad (9)$$

It is easy to see then that at any *fixed* energy $\sqrt{s_{NN}}$ the V_{int} is nearly constant in time since the values dN/dy, APSD $\langle f\rangle$ and effective temperature T_{eff} in r.h.s. of Eq. (9) are approximately conserved for the thermal pions during the chemically frozen hydro-evolution. As the result, the HBT microscope at diverse energies "measures" the radii that are similar to the sizes of colliding nuclei.

The conservation of the APSD allows one to study the hadronization stage of the matter evolution based on the Eq. (8). The results for the APSD at mid-rapidity for *all* negative pions ($\kappa = 1$) at the AGS, SPS, RHIC energies are presented in Fig. 3 [2]. One can see that the APSD grows significantly with energy at the AGS energies, then has the plateau starting from the lowest SPS energy, 20 AGeV, till 80 AGeV and then begins to grow again, apparently very slowly at RHIC as one can conclude from the non quite compatible experimental data of the STAR and PHENIX Collaborations. Since the conserved APSD values of *thermal* pions are almost proportional to the presented total values (with coefficients $\kappa(\sqrt{s}) \approx 0.6-0.7$ in (8) accounting for decays of long- and short- lived resonances [1]) they follow the same tendencies which are presented in Fig. 3. Then a plateau at low SPS energies indicates, apparently, a transformation of an excess of initial energy to non-hadronic forms of matter: the pure hadronic stage with densities smaller than initial ones appears later (after some expansion) at the hadronization temperature T_c defining the APSD of thermal pions. A saturation of the APSD at RHIC energies can be treated as an existence of the limiting Hagedorn temperature of hadronic matter, or maximal temperature of deconfinement.

The RHIC experiments show clearly that there is no proportionality law between V_{int} and dN^{π^-}/dy: the later value grows with energy significantly faster than V_{int}. This fact is the main component of the HBT puzzle. According to Eq. (9) a proportionality between V_{int} and the particle numbers dN/dy may be destroyed by a factor $\langle f \rangle T_{eff}^3$. So, if the APSD and V_{int} only slightly grow with energy, mostly an increase of T_{eff}^3 could compensate a growth of dN/dy in Eq. (9). One can see that it is the case: for example, the ratio of cube of effective temperatures of negative pions at $\sqrt{s_{NN}} = 200$ GeV (RHIC) to one at 40 AGeV (CERN SPS) gives approximately 2, while the ratio of correspondent mid-rapidity densities is approximately equal to 3. It can be only in the case of an increase of the pion transverse flows in A+A collisions with energy. If the intensity of flows grows, it leads to a reduction of the corresponding homogeneity lengths which contribute to the interferometry radii. This effect can almost compensate a contribution to observed interferometry volumes of the geometrical system sizes that grow with energy. The question is then: why does the intensity of flow grow? It is clear that an increase of collision energy $\sqrt{s}$ results in a rise of initial energy density ε and hence of (maximal) initial pressure p_{max}. At the same time the initial transverse acceleration $a = grad(p)/\varepsilon \propto p_{max}/\varepsilon$ does not change significantly. So the HBT puzzle transforms into puzzling increase with energy of transverse pion flows.

CONCLUSIONS

In ultrarelativistic A+A colisions the effective temperature of transverse spectra does not change much since heat energy transforms into collective flows of expansion [1, 3].

The interferometry volumes, if they were measured at any time of the expansion of hadron-resonance gas, also do not grow significantly during the system evolution even if initial conditions are quite asymmetric [2, 3].

The APSD is an approximate integral of motion upon a chemically frozen evolution [1]. This makes it possible to analyse the hadronization stage. Then the plateau in APSD $\sqrt{s}$-behavior can be associated with deconfinement phase transition at low SPS energies; a saturation of this quantity at RHIC energies indicates, apparently, the limiting Hagedorn temperature for hadronic matter. Also, an experimental behavior of the interferometry volumes at different energies and for different nuclei can be explained [2].

ACKNOWLEDGMENTS

The research described in this publication was made possible in part by Award No. UKP1-2613-KV-04 of the U.S. Civilian Research & Development Foundation for the Independent States of the Former Soviet Union (CRDF) and Ministry of Education and Sciences of Ukraine, also Fundamental Researches State Fund of Ukraine, Agreement No. F7/209-2004. Research carried out within the scope of the ERG (GDRE): Heavy ions at ultrarelativistic energies - a European Research Group comprising IN2P3/CNRS, EMN, Universite de Nantes, Warsaw University of Technology, JINR Dubna, ITEP Moscow and Bogolyubov Institute for Theoretical Physics NAS of Ukraine.

REFERENCES

1. S.V. Akkelin, Yu.M. Sinyukov, *Phys. Rev. C* **70**, 064901 (2004).
2. S.V. Akkelin, Yu.M. Sinyukov, nucl-th/0505045.
3. N.S. Amelin, R. Lednicky, L.V. Malinina, T.A. Pocheptsov, Yu.M. Sinyukov, nucl-th/0507040.
4. Yu.M. Sinyukov, *Nucl. Phys. A* **566**, 589c-592c (1994); *Heavy Ion Phys.* **10**, 113-136 (1999).
5. S.V. Akkelin, P. Braun-Munzinger, Yu.M. Sinyukov, *Nucl. Phys. A* **710**, 439-465 (2002).
6. Yu.M. Sinyukov, S.V. Akkelin and Y. Hama, *Phys. Rev. Lett.* **89**, 052301 (2002).
7. T. Csorgo, J. Zimanyi, *Heavy Ion Phys.* **17**, 281-293 (2003).

Invariance Group Important for the Interpretation of Bose-Einstein Correlations

Kacper Zalewski

M. Smoluchowski Institute of Physics, Jagellonian University, ul. Reymonta 4, 30-059 Cracow, Poland
and The Henryk Niewodniczański Institute of Nuclear Physics, Polish Academy of Sciences, 152 Radzikowskiego St, 31-342 Cracow, Poland

Abstract. A group of transformations changing the phases of the elements of the single-particle density matrix, but leaving unchanged the predictions for identical particles concerning the momentum distributions, momentum correlations etc., is identified. Its implications for the determinations of the interaction regions from studies of Bose-Einstein correlations are discussed.

Keywords: Bose-Einstein correlations, interaction region determination.
PACS: 25.75.Gz, 13.65.+i

INTRODUCTION

Bose-Einstein correlations are helpful when trying to derive properties of the interaction regions from the measured momentum distributions. In this report we point out an ambiguity inherent in such derivations [1]. There are many ways from the data to the inferred properties of the interaction regions (cf. e.g. [2] and references given there). One can use density matrices, Wigner functions, emission functions, distances between the pairs of points where the identical particles are produced etc. The ambiguity seems to be common to all of them.

We will not discuss here the evolution of the density matrix during the freeze-out period. Formally, one can introduce the assumption that all the hadrons have been produced instantaneously and simultaneously at some time which we may choose as $t = 0$. Then what one measures is the density matrix (in the interaction representation) at freeze-out.

As is well known, the diagonal elements of the k-particle density matrix in the momentum representation give, and are unambiguously given by, the k particle momentum distribution. On the other hand, in most models these elements can be expressed as symmetrized products of the single particle density matrix elements [3]:

$$\rho(\mathbf{p}_1,\ldots,\mathbf{p}_k;\mathbf{p}_1,\ldots,\mathbf{p}_k) = \sum_P \prod_{j=1}^{k} \rho_1(\mathbf{p}_j;\mathbf{p}_{Pj}), \tag{1}$$

CP828, *Multiparticle Dynamics*
edited by V. Šimák, M. Šumbera, Š. Todorova, and B. Tomášik

where the summation is over all the $k!$ permutations of the momenta $\mathbf{p}_1, \ldots, \mathbf{p}_k$. Thus, all the momentum distributions are unambiguously determined when the single particle density matrix $\rho_1(\mathbf{p}_1;\mathbf{p}_2)$ is known.

Our main observation [1] is that the converse is not true. Given the momentum distributions for all the sets of $k = 1, 2, \ldots$ particles, it is not possible to find unambiguously the matrix ρ_1. This trivial observation will be seen to have very non trivial consequences. Since the matrix ρ_1 is further used to derive conclusions about the interaction region, the ambiguity affects our capacity for making unambiguous statements about such regions.

INVARIANCE GROUP

Consider the transformation

$$\rho_1(\mathbf{p};\mathbf{p'}) \to \rho_{1\alpha}(\mathbf{p};\mathbf{p'}) \equiv e^{i\alpha(\mathbf{p})}\rho_1(\mathbf{p};\mathbf{p}')e^{-i\alpha(\mathbf{p'})}, \tag{2}$$

where α is any real-valued function of momentum. According to formula (1) for $k = 1$, ρ_1 is a single particle density matrix. Therefore, it must be hermitian and have trace one. Also $\rho_{1\alpha}$, as seen from its definition, is hermitian and has trace one. Consequently, it can be substituted for ρ_1 on the right-hand side of formula (1). This introduces on the right-hand side, for every p_i a term $e^{i\alpha(\mathbf{p}_i)}$ and a term $e^{-i\alpha(\mathbf{p}_i)}$ which cancels it. Thus, the diagonal matrix element on the left-hand side does not change. Experimentally, the substitution of $\rho_{1\alpha}$ for ρ_1 is invisible. The transformations from ρ_1 to $\rho_{1\alpha}$ form a local (in momentum space) U(1) invariance group.

There are several quantities related to the single particle density matrix in the momentum representation and yielding information about the interaction region. In order to get the space distribution of the sources one can use the diagonal elements of the single particle density matrix in the coordinate representation

$$\tilde{\rho}_1(\mathbf{x};\mathbf{x}) = \int \frac{d^3K d^3q}{(2\pi)^3}\, e^{i\mathbf{q}\cdot\mathbf{x}}\rho_1(\mathbf{p};\mathbf{p'}), \tag{3}$$

where

$$\mathbf{K} = \frac{1}{2}(\mathbf{p}+\mathbf{p'}); \qquad \mathbf{q} = \mathbf{p}-\mathbf{p'}. \tag{4}$$

In order to get an approximate phase-space distribution one can use the Wigner function

$$W_1(\mathbf{K},\mathbf{X}) = \int \frac{d^3q}{(2\pi)^3} e^{i\mathbf{q}\cdot\mathbf{X}}\rho_1(\mathbf{p};\mathbf{p'}), \tag{5}$$

or the emission function

$$\rho_1(\mathbf{p};\mathbf{p'}) = \int d^4X S(K,X)e^{iqX}. \tag{6}$$

In the last two formulae $X = \frac{1}{2}(x + x')$ and the approximation consists in interpreting K and X as the energy-momentum and space-time position of the particle. This approximation is not always good, but in a well-defined sense [4] this is the best one can have without contradicting the principles of quantum mechanics. In the last formula the four-vector K has the same value on both sides of the equality. Since the momenta $\mathbf{p}$ and $\mathbf{p'}$ are on mass shell, for any $\mathbf{q}$ the value of q_0 is fixed by the condition $Kq = 0$. Thus, it is not possible to invert the Fourier transform and to express the emission function S in terms of the density matrix ρ_1. In fact, there is an infinity of different emission functions, corresponding to various off mass shell continuations of a given on mass shell function ρ_1. We conclude that the ambiguities when using the emission function formalism are more severe than when using Wigner functions.

The point is that the transition from ρ_1 to $\rho_{1\alpha}$, which has no effect on the momentum distributions, changes the functions $\tilde{\rho}_1$, W_1 and S and consequently changes the conclusions concerning the interaction region. In the following section we will illustrate this fact by some examples.

EXAMPLES

The class of transformations (2) is very rich. We will just discuss three simple examples. Consider

$$\alpha(\mathbf{p}) = \mathbf{b} \cdot \mathbf{p} \qquad \Rightarrow \qquad \alpha(\mathbf{p}) - \alpha(\mathbf{p'}) = \mathbf{b} \cdot \mathbf{q}, \tag{7}$$

where $\mathbf{b}$ is any vector. This gives

$$\tilde{\rho}_1(\mathbf{x};\mathbf{x}) = \int dK dq \, e^{i\mathbf{q}\cdot(\mathbf{x}+\mathbf{b})} \rho_1(\mathbf{p};\mathbf{p'}). \tag{8}$$

The interaction region gets shifted by $\mathbf{b}$. Similarly, replacing $\mathbf{b}$ and $\mathbf{p}$ by four-vectors and using the emission function one can generate an arbitrary shift in space-time. This result is of little interest. It is obvious that the momentum distributions do not depend on where and when the experiment was done.

As our second example consider

$$\alpha(\mathbf{p}) = \frac{1}{2} c \mathbf{p}^2 \qquad \Rightarrow \qquad \alpha(\mathbf{p}) - \alpha(\mathbf{p'}) = c\mathbf{K} \cdot \mathbf{q}, \tag{9}$$

where c is any real number. Then

$$W_{1\alpha}(\mathbf{K},\mathbf{X}) = \int \frac{d^3 q}{(2\pi)^3} \, e^{i\mathbf{q}\cdot(\mathbf{X}+c\mathbf{K})} \rho_1(\mathbf{p};\mathbf{p'}). \tag{10}$$

This time the shift is proportional to $\mathbf{K}$ with a proportionality coefficient which is unconstrained by momentum measurements. The corresponding distribution in space is

$$\tilde{\rho}_{1\alpha}(\mathbf{x};\mathbf{x}) = \int d^3K W_{1\alpha}(\mathbf{K},\mathbf{x}) \tag{11}$$

Suppose now that for $c = 0$ there are no position-momentum correlations. Then for each $\mathbf{K}$ the interaction region occupies the same portion of space and when averaged over $\mathbf{K}$ coincides with that for any given $\mathbf{K}$. For $c \neq 0$, the interaction regions corresponding to different values of $\mathbf{K}$ are shifted with respect to each other and the averaged size of the interaction region gets bigger. Its increase with respect to the situation at $c = 0$ depends on the value of $|c|$. Since this is unconstrained by the data on momentum distributions, the radius can be made as large as one wishes.

For instance, for the Gaussian

$$\rho_1(\mathbf{p};\mathbf{p'}) = \frac{1}{(\sqrt{2\pi\Delta^2})^3} \exp\left[-\frac{\mathbf{K}^2}{2\Delta^2} - \frac{1}{2}R^2\mathbf{q}^2\right] \tag{12}$$

one obtains $\tilde{\rho}_{1\alpha}(\mathbf{x};\mathbf{x})$ also Gaussian with root mean square width

$$R_\alpha^2 = R^2 + c^2\Delta^2. \tag{13}$$

In order to see that this broadening is due to the averaging over $\mathbf{K}$ it is enough to have a look at the corresponding Wigner function

$$W_{1\alpha}(\mathbf{K},\mathbf{X}) = \frac{1}{(2\pi R\Delta)^3} \exp\left[-\frac{\mathbf{K}^2}{2\Delta^2} - \frac{(\mathbf{X} + c\mathbf{K})^2}{2R^2}\right]. \tag{14}$$

The conclusion from this example is that without further assumptions one can at best obtain a lower limit for the radius of the interaction region. In practice, everybody uses, more or less consciously, a model which supplies the necessary additional assumptions. The problem how to choose the best model among those which give exactly the same fit to all the data is an interesting open problem. In order to show that this problem is not purely academic let us consider the following example from the literature.

Some models [5], [6] assume correlations between position in space-time and energy-momentum of the type[1]

$$K^\mu = \lambda X^\mu. \tag{15}$$

A convenient notation is

$$X_0^2 - X_{||}^2 = \tau^2; \qquad K_0^2 - K_{||}^2 = M_T^2 \qquad \Rightarrow \qquad \lambda = \frac{M_T}{\tau}. \tag{16}$$

Note that here M_T is a temporal-longitudinal variable. Let us assume that [6]

$$S = S_{||}S_T \tag{17}$$

[1] Sometimes called Hubble flow

where

$$S_T = \exp\left[-\frac{\mathbf{X}_T^2}{2r_T^2} - \frac{(\mathbf{K}_T - \lambda\mathbf{X}_T)^2}{2\delta_T^2}\right]. \tag{18}$$

This is a possible quantum-mechanical rendering of the transverse part of the classical relation (15). The only information about $S_\parallel$ important for our purpose is that it depends neither on $\mathbf{K}_T$ nor on $\mathbf{X}_T$. The distribution of sources in the transverse plane is obtained by integrating S_T over $\mathbf{K}_T$. The result is a Gaussian with constant root mean square width r_T. On the other hand S_T can be rewritten in the form

$$S_T = \exp\left[-\frac{\phi_T^2}{2R_D^2} - \frac{(\mathbf{X}_T - \phi_T)^2}{2R_\phi^2}\right], \tag{19}$$

where

$$R_\phi = \frac{r_T}{\sqrt{1+\mu^2}}; \qquad R_D = \mu R_\phi; \qquad \mu = \frac{r_T}{\tau\delta_T}M_T; \qquad \phi_T = r_T\frac{\mu}{1+\mu^2}\frac{\mathbf{K}_T}{\delta_T}. \tag{20}$$

This can be considered[2] as a transform with $\alpha(\mathbf{p}) = \frac{1}{2}c\mathbf{p}^2$ and

$$c = \frac{r_T\mu}{\delta_T(1+\mu^2)} \tag{21}$$

of

$$S_{\alpha T} = \exp\left[-\frac{\mathbf{K}_T^2}{2\delta_T^2} - \frac{\mathbf{X}_T^2}{2R_\phi^2}\right]. \tag{22}$$

Performing the integration of $S_{\alpha T}$ over $\mathbf{K}_T$ one finds for the distribution of sources in the transverse plane a Gaussian with root mean square width

$$R_{T\alpha}^2 = R_\phi^2 = \frac{r_T^2}{1+\mu^2}, \tag{23}$$

which exhibits the familiar decrease of the transverse radius with the transverse mass as reported by so many experimental papers. Let us summarize the situation: if our prejudice is in favor of a Hubble flow as interpreted in [6], we conclude that the transverse radius does not depend on M_T; if our prejudice is against correlations between position in the transverse plane and transverse momentum, we conclude that the transverse radius decreases with increasing M_T; experimental data on momentum distributions will not help us to decide which of these two prejudices is the right one.

---

[2] This is an approximation, because c depends on M_T, it is, however, good enough for our qualitative discussion; compare [6], where the full calculation can be found.

Our last example is one dimensional. It could e.g. apply to one transverse component. We choose the Gaussian density matrix (12) and

$$\alpha(p) = \frac{4}{3a^3}p^3 \qquad \Rightarrow \qquad \alpha(p) - \alpha(p') = \frac{4}{3a^3}q\left(K^2 + \frac{q^3}{3}\right). \tag{24}$$

A simple calculation [1] yields

$$S_\alpha = \frac{a}{\sqrt{2\pi\Delta}}\exp\left[-\frac{K^2}{2\Delta^2} + B\right]Ai(A), \tag{25}$$

where $Ai(\ldots)$ is the Airy function and

$$A = a\tilde{X} + \frac{\omega^4}{4}; \qquad B = \frac{\omega^2}{2}\left[A - \frac{\omega^4}{12}\right]; \qquad \omega = aR; \qquad \tilde{X} = X - \frac{4}{a^3}K^2. \tag{26}$$

For large values of a the emission function is almost Gaussian. A numerical calculation shows that $a = 2/R$ is already large enough. For smaller values of a, however, at negative $\tilde{X}$ the emission function develops big wiggles, oscillating between positive and negative values. Its shape in the positive $\tilde{X}$ region also significantly changes [1]. This example shows how the transformations discussed in the present report can lead to changes of the interaction region which are much more complicated than just momentum dependent shifts.

CONCLUSIONS

The experimental data about momentum distributions tell us little about the interaction regions, unless additional assumptions are made. The usual recommendations: reduce the experimental errors, include more particle correlations etc. are not enough. Exactly the same fit can be obtained from widely different models, differing in these additional assumptions and giving conflicting information about the interaction region. The caveat for model users is that only some of the assumptions of the model are being tested by comparison with the data, while others, which may be very important for drawing inferences about the interaction region, are unconstrained by the data.

REFERENCES

1. A. Bialas and K. Zalewski, hep-ph/0501017.
2. U.A. Wiedemann and U. Heinz, *Phys. Rep.* **319**, 145–230 (1999).
3. J. Karczmarczuk, *Nucl. Phys. B* **78**, 370–380 (1974).
4. M. Hillery et al., *Phys. Rep.* **106**, 121–167 (1984).
5. T. Csörgő and J Zimányi, *Nucl. Phys. A* **517**, 588–598 (1990).
6. A. Bialas et al., *Phys. Rev. D* **62**, 114007 (2000).

The Particle Interferometry Method as a Tool Reflecting Evolution of Hadron Source

Hanna Paulina Gos

Warsaw University of Technology, Faculty of Physics, ul. Koszykowa 75, 00-662 Warsaw, Poland
and
SUBATECH, Laboratoire de Physique Subatomique et des Technologies Associees, EMN-IN2P3/CNRS-Universite, Nantes, F-44307, France

Abstract. The particle interferometry method provides a powerful tool to study the properties of the matter produced in heavy ion collisions at ultra-relativistic energies. Applied to identical and nonidentical hadron pairs it makes the study of space-time evolution of the source possible. Identical baryon interferometry allows to extract the HBT radii of produced sources, which can be compared to those deduced from the identical pions study providing more complete information about the source characteristics. In the case of nonidentical baryon systems we are able to explore the space-time differences in the emission process of considered particles. Such measurements may reveal any asymmetry occurring in emission due to a specific behaviour of the matter e.g. some collective phenomena. Theoretical predictions from EPOS model will be presented as well, helping us to understand the behaviour of matter created in heavy ion better.

Keywords: femtoscopy, hadron source, HBT, interferometry, nonidentical particles
PACS: 25.70.Mn, 25.70.Pq, 25.75.-q, 25.75.Dw, 25.75.Gz, 25.75.Ld.

INTRODUCTION

The correlation effect is determined by the distance separating emission points of particles in space and time and by their relative momentum [1, 2]. By analysing momentum correlations it is possible to access information about source which cannot be measured directly. The paper contains three different parts: the first one concerns baryon correlations and shows results for various proton and antiprotons combinations, the second introduces nonidentical particle correlation method which allows to measure asymmetry in emission process and the third part presents results obtained via simulations of p+p and d+Au collisions. The preliminary results from EPOS model are presented, which illustrate two particle distributions of the emission points.

BARYON-BARYON CORRELATIONS

Identical baryon correlations exhibit the properties of the quantum statistics [3, 4] and of the final state interactions [5, 6]. Nonidentical baryon correlations are sensitive to final state interaction only, dependent on type of the pair: Coulomb in case of charged particles and the nuclear one. Nonidentical correlations for proton-antiproton pairs exhibit both. For proton-antiproton system the annihilation channel occurs. The correlations are measured in terms of Qinv or k* [1]. Measured particles come from Au+Au collisions

CP828, *Multiparticle Dynamics*
edited by V. Šimák, M. Šumbera, Š. Todorova, and B. Tomášik

TABLE 1. Results of fit for identical baryon combinations

Centrality	$p-p$ @ 62GeV	$p-p$ @ 200GeV	$\overline{p}-\overline{p}$ @ 200GeV
Minimum bias	$3.1^{+0.2}_{-0.1}$	$3.4^{+0.1}_{-0.2}$	$3.4^{+0.2}_{-0.2}$
Central	$3.6^{+0.2}_{-0.2}$	$4.0^{+0.1}_{-0.3}$	$4.2^{+0.2}_{-0.3}$
Midcentral	$3.1^{+0.2}_{-0.2}$	$3.3^{+0.2}_{-0.1}$	$3.3^{+0.2}_{-0.2}$
Peripheral	$2.3^{+0.2}_{-0.3}$	$2.5^{+0.2}_{-0.1}$	$2.3^{+0.3}_{-0.2}$

TABLE 2. Results of fit for nonidentical baryon combinations

Centrality	$p-\overline{p}$ @ 62GeV	$p-\overline{p}$ @ 200GeV
Minimum bias	$2.0^{+0.1}_{-0.1}$	$2.1^{+0.1}_{-0.1}$
Central	$2.3^{+0.1}_{-0.2}$	$2.5^{+0.2}_{-0.1}$
Midcentral	$1.9^{+0.2}_{-0.1}$	$2.0^{+0.1}_{-0.1}$
Peripheral	$1.6^{+0.2}_{-0.2}$	$1.6^{+0.3}_{-0.2}$

at $\sqrt{s_{NN}} = 62$GeV and $\sqrt{sNN} = 200$GeV. Minimum bias data samples are selected and divided according to three centralities of collision: central, midcentral, peripheral.

Antiproton-antiproton and proton-antiproton correlation functions for Au+Au collision at $\sqrt{sNN} = 200$GeV for different centrality bins are shown in Fig.1. For all systems the correlation effect decreases with increasing centrality. For fitting procedure the Gaussian source is assumed. The correlation functions fitted by CorrFit [7] tool give consistent description within all centralities and energies for all systems. The errors presented in the table are statistical ones, however systematic uncertainties which take into account the stability of purity correction are estimated to be 0.1fm. Other sources of errors are still under study.

During fitting procedure we exclude the first bin in all systems as we do not expect changes in fitting sizes when taking into account its value. It decreases the number of necessary calculations. Smaller sizes for all combinations with annihilation channel within all centrality bins are observed and some explanations could be proposed: the existence of two separate sources with different location of emitting points for different types of particles, neglecting the p-wave in the description of strong interaction which might influence more in case of systems with annihilation channel, and neglecting residual correlations. These three factors still need to be studied in detail.

NONIDENTICAL PARTICLE CORRELATIONS

The nonidentical particle correlation method of analysis [8, 9, 10] allows to study space-time asymmetries in the emission of two types of particles (e.g. $\pi^+\pi^-$ system). In the classical picture the correlation strength, due to Coulomb interaction depends on whether two particles move towards each other or away from each other in the pair rest frame. The distribution of space-time emission points is assumed to be the same for both particles. We use the decomposition of k* into out, side and long components. In our calculations positive pion is always consequently taken as a first particle in the pair. One

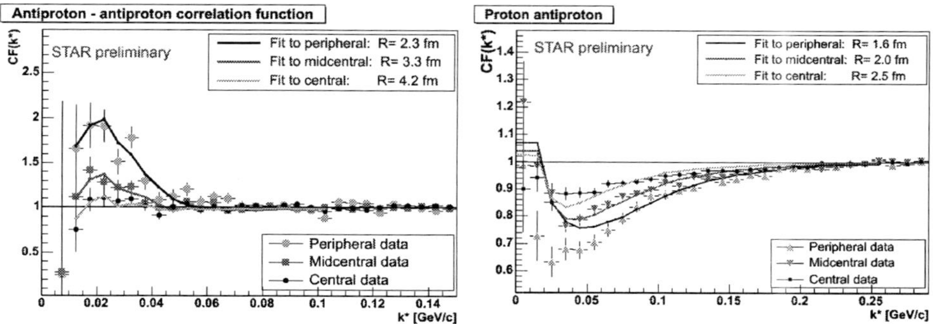

FIGURE 1. Antiproton-antiproton and proton-antiproton correlation functions for $\sqrt{s_{NN}} = 200\text{GeV}$

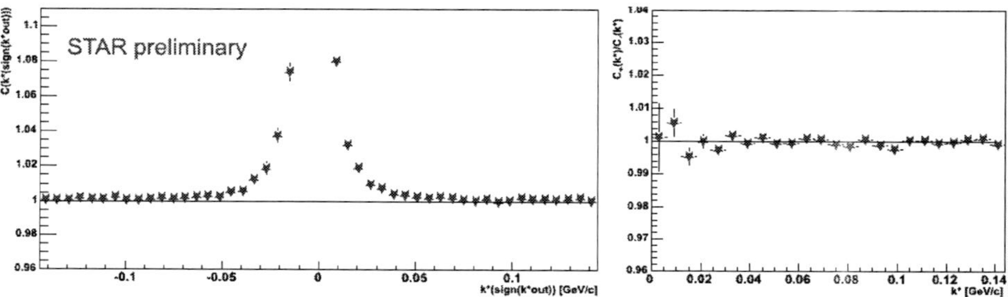

FIGURE 2. $\pi+\pi-$ correlation functions for $\sqrt{s_{NN}} = 130\text{GeV}$ is presented in left figure. The ratio of two functions dependent on k*out sign shown in right figure equals the unity for every value of k*, implies symmetry in emission process of charged pions.

can study asymmetries in different directions by studying the correlations dependent on negative and positive components of k* projections in three directions [9]. No deviation from unity seen in Fig. 2 in out projections, which implies no asymmetry in emission process between positive and negative pions.

EPOS

EPOS [11] is an improved parton model with remnants and means: Energy conserving quantum mechanical multiple scattering approach based on

- partons (parton ladder)
- off-shell remnants
- splitting of parton ladder.

It is based on elementary parton-parton scattering, where the hard scattering is preceded by parton emissions and when these partons emit further partons. EPOS takes into consideration not only hard but soft processes as well, both exist at RHIC energies.

Two-particle distributions defined as a differences of time and space coordinates of freeze-out show how the source looks like. The distributions in Fig. 3 illustrate positive pion pairs produced in p+p reaction. The distributions wider in space and time correspond to all pairs of pions. The RMS values are calculated. The distributions of pairs coming from correlated region which have momenta differences not bigger than 300MeV/c are also shown. All RMS values of pairs from correlated region comparing to ones from the whole space are smaller what indicates that they are emitted closer to each other.

$\pi^+\pi^+$ correlation functions for p+p collisions are shown below. The correlation effect is estimated to occur up to 300MeV/c. Presented in the Fig.4 the projections of three-dimensional correlation function for different ranges of cut of the other momenta components do not indicate sensitivity of the radii to the range of the projection. The height of function is sensitive to that but it does not affect the extracted sizes.

The proton-proton one-dimensional correlation function for d+Au reaction is calculated and presented in Fig.5. The width of interaction is smaller than in $\pi^+\pi^+$ system. In order to estimate the radii of proton source, we compare the correlation function calculated using space-time emission points taken from EPOS to functions calculated with assumption of Gaussian distributions of freeze-out coordinates [6]. Using Gaussian source parametrisation we are not able to reproduce both: width and height of the correlation function obtained from EPOS. The width of the function calculated with assumption of Gaussian parametrisation of the source with Rinv= 2fm is comparable to one obtained from EPOS. The correlation function with Gaussian source calculated for Rinv= 2.5fm has a height which is much closer to that from EPOS than the function for Rinv= 2fm. This figure indicates that the assumption of the Gaussian source emitting protons may cause the discrepancy.

SUMMARY

In the first part the measured results of baryon correlations are presented. A good agreement of experimental results with theoretical predictions assuming the Gaussian source are observed. On the other hand, baryon study shows different radii depending on the existence or the absence of the annihilation channel. All studies indicate smaller sizes for systems where annihilation occurs. As most results confirm good agreement between the fit and experimental function, the estimated Gaussian shape of source is not suspected to change the radii. Nonidentical correlation method which refers to dynamics of two-particle systems allows to study the pions space-time asymmetry in emission process. Result presented in this paper shows that this method is useful to conclude about symmetry in emission process as well. The last topic concerns predictions from EPOS model. Various two-particle distributions illustrating that the source is not a Gaussian are shown. Two-particle distributions of space-time freeze-out coordinates illustrate the geometry of the source which can not be described by any analytical formula. In order to estimate the size of emitting area, the correlation function from EPOS is compared to ones calculated with the assumption that the source distribution is a Gaussian. Presented results suggest that this assumption might be wrong.

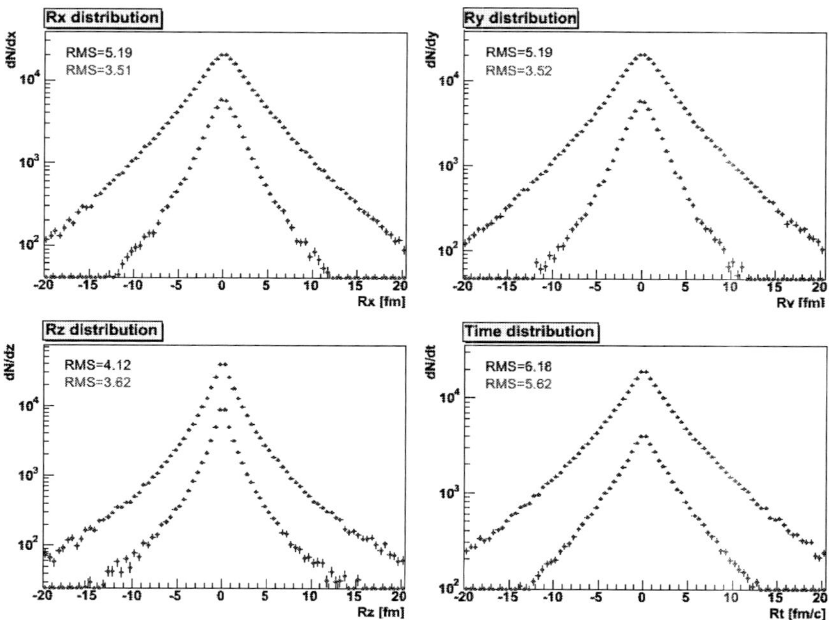

FIGURE 3. Two-particle distributions for positive pion pairs. Two studies are shown: first including all pairs of positive pions and the second illustrating only pairs from correlated region (up to $Q_{inv} = 300\text{MeV}/c$)

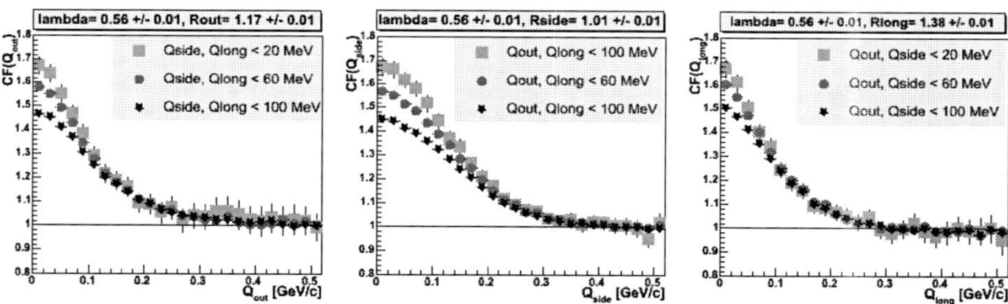

FIGURE 4. The projections of three-dimensional $\pi+\pi+$ correlation functions

ACKNOWLEDGMENTS

We thank the RHIC and RCF at BNL, the NERSC Center at LBNL for their support. Research carried out within the scope of the ERG (GDRE): Heavy ions at ultrarelativistic energies- an European Research Group comprising IN2P3/CNRS, Ecole des Mines de Nantes, Universite de Nantes, Warsaw University of Technology. We acknowledge SUBATECH department in France and Faculty of Physics in Poland.

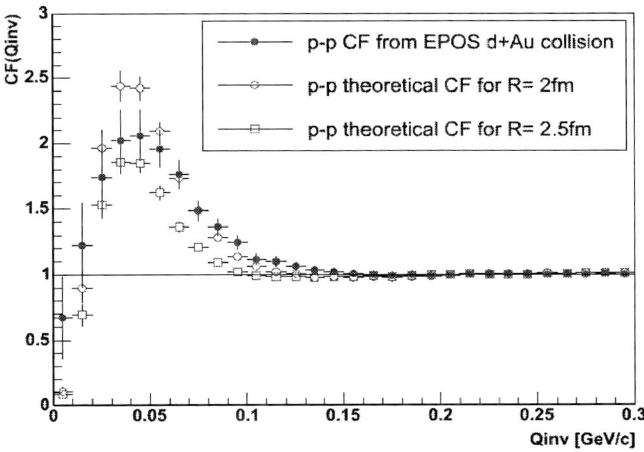

FIGURE 5. Proton-proton correlation for d+Au reaction compared to two theoretical calculated functions.

REFERENCES

1. R. Lednicky, *Phys. Atom. Nucl.* **67**, 72-82 (2004).
2. M. A. Lisa, S. Pratt, R. Soltz and U. Wiedemann, Submitted to *Ann.Rev.Nucl.Part.Sci Appl*, (2005).
3. G. Goldhaber, S. Goldhaber, W.Y. Lee and G. Pais, *Phys. Rev.* **120**, 300-312 (1960).
4. G. I. Kopylov and M. I. Podgoretsky, *Sov. J. Nucl. Phys.* **15**, 219-223 (1972).
5. M. Gyulassy, S.K. Kauffamann and L.W. Willson, *Phys. Rev. C* **20**, 2267-2292 (1979).
6. R. Lednicky, V.L. Lyuboshitz, *Sov. J. Nucl. Phys.* **35**, 770(1972).
7. A. Kisiel, "CorrFit - a program to fit arbitrary two-particle correlation functions" in *2nd Warsaw Meeting On Particle Correlations and Resonannces in Heavy Ion Collisions*, Nukleonika Conference Proceedings,**49** suppl. 2, pp.81-83 (2004).
8. R. Lednicky, V.L. Lyuboshitz, B. Erazmus and D. Nouais, *Phys. Lett. B* **373**, 30 (1996)
9. A. Kisiel, *J. Phys. G* **30**, S1059-S1064 (2004).
10. A. Kisiel, "Non-identical particle femtoscopy in heavy ion collisions", these proceedings.
11. K. Werner, F.-M. Liu, T. Pierog, hep-th/0506232.

Azimuthally Sensitive Femtoscopy and v_2

Boris Tomášik

The Niels Bohr Institute, Blegdamsvej 17, 2100 Copenhagen Ø, Denmark
and Ústav jaderné fyziky AVČR, 25068 Řež, Czech Republic

Abstract. I investigate the correlation between spatial and flow anisotropy in determining the elliptic flow and azimuthal dependence of the HBT correlation radii in non-central nuclear collisions. It is shown that the correlation radii are in most cases dominantly sensitive to the anisotropy in space. In case of v_2, the correlation depends strongly on particle species. A procedure for disentangling the spatial and the flow anisotropy is proposed.

Keywords: HBT, non-central collisions, anisotropy, v_2
PACS: 25.75.-q, 25.75.Ld

MOTIVATION

In non-central nuclear collisions at RHIC energies, the resulting fireball can exhibit anisotropy in both spatial shape and transverse expansion velocity profile. They both influence the measured "elliptic flow" coefficient v_2 [1]. A question arises: how are they correlated in the determination of v_2, i.e., which combinations of spatial and flow anisotropy lead to the same elliptic flow?

On the other hand, dependence of HBT correlation radii on the azimuthal angle is also shaped by the two mentioned anisotropies. Therefore, the same question can be asked: how do the spatial anisotropy and transverse expansion flow anisotropy combine in the ϕ-dependence of correlation radii?

An analogical situation appears in determining the slopes of single-particle p_T-spectra. It is well known that they are determined by temperature and transverse expansion velocity and that it is impossible to disentangle these two quantities from a single measured spectrum. There is, however, also the M_T-dependence of HBT radii in which the correlation of temperature and transverse flow is qualitatively different from that in the determination of spectra. Temperature and transverse flow velocity then can be unambiguously measured from analysing both spectra and HBT radii.

A similar solution shall be sought here: can we disentangle spatial and flow anisotropy in non-central collisions by analysing both v_2 and the azimuthally sensitive HBT radii?

Note that several statements have been made in literature which are related to this programme. In [2] the STAR collaboration concluded that it was impossible to determine spatial anisotropy just from the measurement of v_2 and a conjecture was made that HBT analysis would be able to gain such result. Two qualitatively different final states resulted from hydrodynamic simulations by Heinz and Kolb [3] and the authors demonstrated the possibility to distinguish these states by HBT interferometry. Here I report on a systematic study of the interplay between spatial and flow anisotropy in framework of generalisations of the blast-wave model.

CP828, *Multiparticle Dynamics*
edited by V. Šimák, M. Šumbera, Š. Todorova, and B. Tomášik

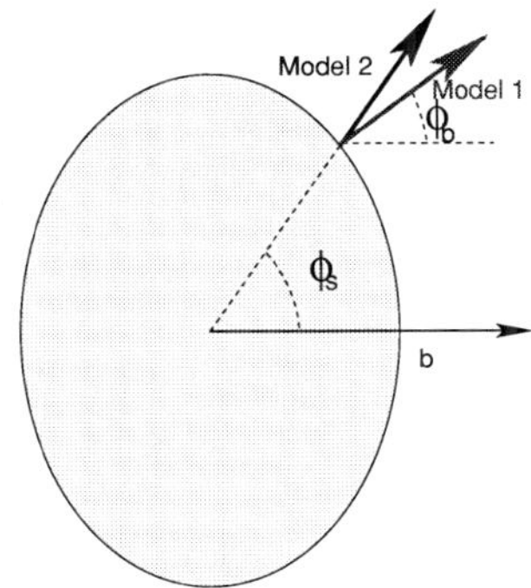

FIGURE 1. The two different models for transverse expansion velocity used here.

AN AZIMUTHALLY ANISOTROPIC BLAST-WAVE MODEL

Instead of fully describing the used model I will just focus on those features which are important for this work and refer the reader to literature for more detailed discussion [5, 4]. Suffice it to say that the fireball is thermalised with a temperature T and exhibits longitudinally boost-invariant expansion. Its transverse profile is ellipsoidal and the emission function is

$$S(x,p) \propto \Theta(1-\tilde{r}), \qquad \tilde{r} = \sqrt{\frac{x^2}{R_x^2} + \frac{y^2}{R_y^2}}, \tag{1}$$

where R_x and R_y are the two transverse radii, in and out-of-plane, respectively. They can be parametrised with the help of a spatial anisotropy parameter a

$$R_x = aR, \qquad R_y = \frac{R}{a}. \tag{2}$$

Thus an out-of-plane elongated source is characterised by $a < 1$, whereas for an in-plane elongated source we have $a > 1$.

The transverse expansion velocity also depends on the azimuthal angle. The velocity is given as

$$v_\perp = \tanh \rho(\tilde{r}, \phi). \tag{3}$$

We shall have a closer look at two models which differ in the azimuthal variation of the velocity. In *Model 1* [5] the velocity is always perpendicular to a surface given by $\tilde{r} = \text{const}$. This direction together with the reaction plane defines the azimuthal angle ϕ_b, as illustrated in Figure 1. The transverse rapidity

$$\rho(\tilde{r}, \phi) = \tilde{r}\rho_0 (1 + \rho_2 \cos(2\phi_b)), \tag{4}$$

where the parameter ρ_0 measures the radial flow and ρ_2 is the flow anisotropy parameter. As the velocity is perpendicular to the surface of the fireball, this model resembles the expansion profile early in the fireball evolution: the direction of velocity coincides with acceleration which in turn is given by the pressure gradient.

In *Model 2* the transverse expansion velocity is directed radially and varies with the usual azimuthal angle, which is denoted as ϕ_s here

$$\rho(\tilde{r},\phi) = \tilde{r}\,\rho_0\,(1+\rho_2\cos(2\phi_s))\,. \tag{5}$$

THE ELLIPTIC FLOW

Recall that v_2 is defined as the second Fourier coefficient of the azimuthal dependence of spectrum

$$P_1(p_T,\phi) = \left.\frac{d^3N}{p_T\,dp_T\,dy\,d\phi}\right|_{y=0} = \frac{1}{2\pi}\left.\frac{d^2N}{p_T\,dp_T\,dy}\right|_{y=0}(1+2v_2(p_T)\cos(2\phi)+\dots)\,. \tag{6}$$

It can be calculated in the two used models and the result reads [4]

$$v_2 = \frac{\int_0^1 d\tilde{r}\,\tilde{r}\int_0^{2\pi} d\phi\,\cos(2\phi)\,J(\phi)\,K_1(a)\,I_2(b)}{\int_0^1 d\tilde{r}\,\tilde{r}\int_0^{2\pi} d\phi\,J(\phi)\,K_1(a)\,I_0(b)}\,, \tag{7}$$

where the arguments of the Bessel functions are $a = m_T\cosh\rho(\tilde{r},\phi)/T$ and $b = p_T\sinh\rho(\tilde{r},\phi)/T$. The *only* difference between the two models appears in the Jacobian $J(\phi)$

$$\text{Model 1:}\quad J(\phi) = (a^2\cos^2\phi + a^{-2}\sin^2\phi)\,, \tag{8a}$$

$$\text{Model 2:}\quad J(\phi) = (a^{-2}\cos^2\phi + a^2\sin^2\phi)\,. \tag{8b}$$

From these relations it is obvious that the two models lead to the same v_2 if they are related by transformation $a \to a^{-1}$. In other words, one in-plane and another out-of-plane source give the same v_2. This is an *analytic illustration of the claim that it is impossible to determine even the qualitative type of spatial anisotropy just from measurement of* v_2.

Now we can look on the correlation between flow and spatial anisotropy and study it only for Model 1, since results for the other Model are obtained simply by substitution $a \to a^{-1}$. In Figure 2 we see that the correlation between a and ρ_2 strongly depends on the particle species. Hence, here is a strategy for determining both a and ρ_2: first determine the temperature and radial flow coefficient ρ_0 from azimuthally integrated spectra. Their dependence on azimuthal anisotropies was shown to be small [5]. Then measure v_2 for at least two particle species and obtain a and ρ_2. Of course, this procedure assumes that we know which model to use for the analysis. This leaves an open question which is to be answered by correlation measurement.

AZIMUTHALLY SENSITIVE HBT

In non-central collisions, the HBT correlation radii can be measured as a function of the azimuthal angle ϕ. We shall focus mainly on the two transverse radii R_o and R_s and

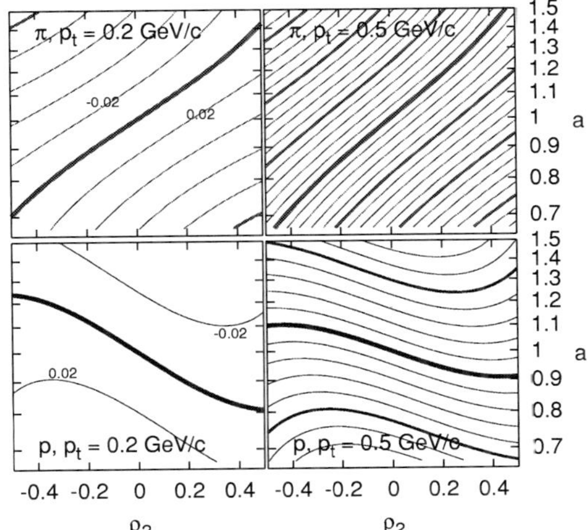

FIGURE 2. Elliptic flow v_2 calculated with Model 1 for pions (upper row) and protons (lower row) with $p_T = 0.2\,\mathrm{GeV}/c$ (left column) and 0.5 GeV/c (right column). The used values $T = 100\,\mathrm{MeV}$ and $\rho_0 = 0.88$. Thickest contour lines show where v_2 vanishes; consecutive lines correspond to steps by 0.02.

decompose their azimuthal angle dependence as [6, 7]

$$R_o^2(\phi) = R_{o,0}^2 + 2R_{o,2}^2 \cos 2\phi + \dots \tag{9a}$$

$$R_s^2(\phi) = R_{s,0}^2 + 2R_{s,2}^2 \cos 2\phi + \dots . \tag{9b}$$

The individual terms of these decompositions are obtained as various combinations of space-time variances taken with the emission function [6, 7]. Because we are rather interested in the oscillation of the radii and not so much in their absolute size, we shall look at the normalised oscillation amplitudes $R_{i,2}^2/R_{i,0}^2$ [5][1]. They are sensitive to a and ρ_2, but less sensitive to R and ρ_0.

From Figure 3 we conclude that the azimuthal oscillations of the HBT correlation radii are mainly shaped by the *spatial* anisotropy parameter a. Dependence on flow anisotropy is weaker, with the only exception of R_s^2 at high K_t in Model 2 which is determined mainly by flow. This confirms the statement that the azimuthal dependence of correlation radii follows mainly the spatial anisotropy, especially at low K_t. This has been shown here in framework of two models. It would be natural to expect this behaviour to be valid in general. It can be spoilt by very strong flow gradients which differ by much in in-plane and out-of-plane directions. A questions arises, however, whether large enough difference of the flow gradients is realistic.

---

[1] In fact, Retière and Lisa realised in [5] that because it also includes time contributions, $R_{o,0}^2$ is not a good normalisation quantity and so used $R_{s,0}^2$ to normalise *all* of $R_{o,2}^2$, $R_{s,2}^2$, and $R_{ol,2}^2$. This is not done here.

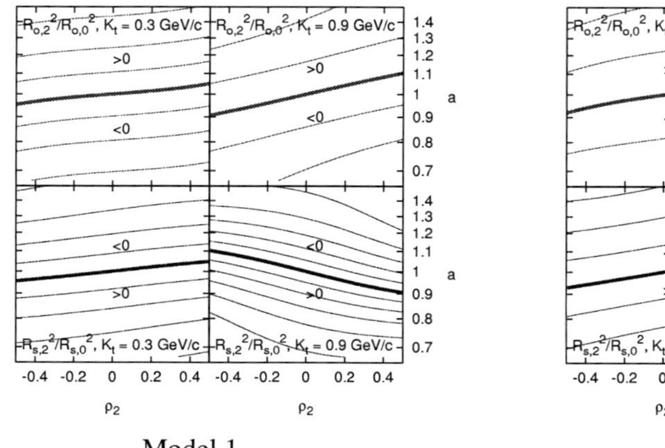

Model 1

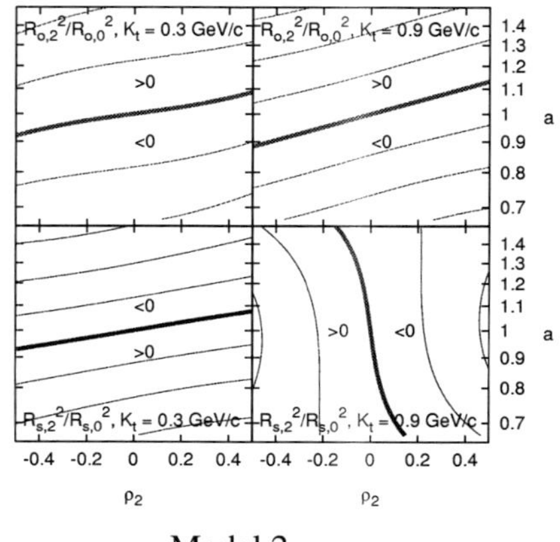

Model 2

FIGURE 3. Normalised oscillation amplitudes $R_{o,2}^2/R_{o,0}^2$ (upper row) and $R_{s,2}^2/R_{s,0}^2$ (lower rows) as functions of a and ρ_2 calculated for $T = 0.1\,\text{GeV}$, $\rho_0 = 0.88$, $R = 9.41\,\text{fm}$, $\tau_0 = 9\,\text{fm}/c$, $\Delta\tau = 1\,\text{fm}/c$ (see [4] for definitions of all parameters). Left columns show results for $K_T = 0.3\,\text{GeV}/c$, right columns correspond to $K_T = 0.9\,\text{GeV}/c$. Thickest lines show where the second-order oscillation terms vanish, other contours are set in steps of 0.1.

The two investigated models exhibit similar dependence on the spatial anisotropy parameter a when focusing on the oscillation of HBT radii. Recall, however, that they were related by transformation $a \to a^{-1}$ when reproducing the same v_2. Therefore, two different models which both reproduce v_2 measurement will behave differently when fitting the azimuthal dependence of HBT radii. This is illustrated in Figure 4. Both models used in this figure fit measured the v_2 for pions and protons well. However, while Model 1 reproduces the RHIC data qualitatively well, Model 2 leads to the phase of oscillation just opposite to data [8].

Thus we conclude that among the two models used in this study, Model 1 seems to correspond to RHIC data, whereas Model 2 is clearly ruled out. This does not disqualify it, however, from future applications at the LHC where possibly longer lived fireballs could be produced which will develop a different transverse flow pattern.

CONCLUSIONS

It has been demonstrated analytically that one cannot disentangle spatial and flow anisotropy of the fireball just from a measurement of v_2. I also demonstrated that, at least for two classes of models, the azimuthal dependence of correlation radii reflects the type of spatial anisotropy the source actually exhibits.

Thus I can propose the following (schematic) procedure for disentangling a and ρ_2: first measure the azimuthal dependence of HBT radii and determine the spatial anisotropy a. Then, with that a try to reproduce v_2 for more species. Since for different species a and ρ_2 are correlated in different ways, this should lead to unique pair of the

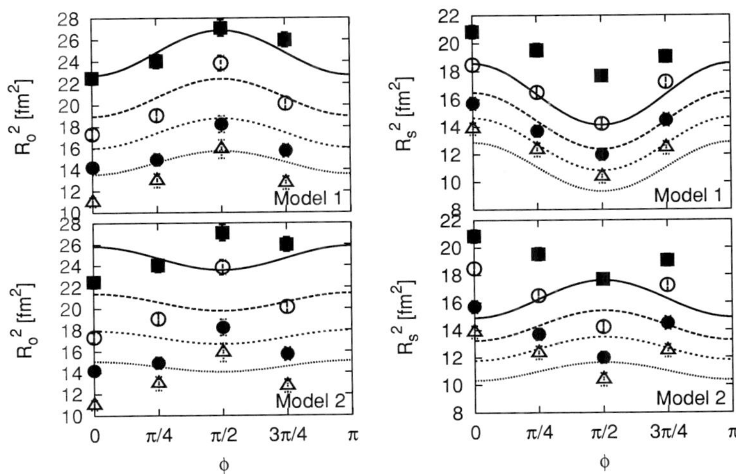

FIGURE 4. Comparison of the dependence of R_o^2 and R_s^2 on azimuthal angle ϕ with the data measured by STAR collaboration (Au+Au at 200 *A*GeV, centrality class 20–30%) [8]. The curves and data points correspond from top to bottom to $K_T = 0.2, 0.3, 0.4, 0.52\,\mathrm{GeV}/c$. Parameters of the models are: $T = 0.12\,\mathrm{GeV}$, $\rho_0 = 0.99$, $\rho_2 = 0.035$, $R = 9.41\,\mathrm{fm}$, $\tau_0 = 5.02\,\mathrm{fm}/c$, $\Delta\tau = 2.9\,\mathrm{fm}/c$, and $a = 0.946$ (Model 1) or $a = 1.057$ (Model 2). Both models were tuned to fit $v_2(p_T)$ [9].

anisotropy parameters.

ACKNOWLEDGEMENTS

This research and presentation were supported by a Marie Curie Intra-European Fellowship within the 6th European Community Framework Programme.

REFERENCES

1. S. Voloshin and Y. Zhang, *Z. Phys. C* **70**, 665–672 (1996).
2. C. Adler *at al.* [STAR Collaboration], *Phys. Rev. Lett.* **87**, 182301 (2001).
3. U. Heinz and P.F. Kolb, *Phys. Lett. B* **542**, 216–222 (2002).
4. B. Tomášik, *Acta Physica Polonica B* **36**, 2087–2104 (2005).
5. F. Retière and M.A. Lisa, *Phys. Rev. C* **70**, 044907 (2004).
6. U. Heinz, A. Hummel, M. A. Lisa and U. A. Wiedemann, *Phys. Rev. C* **66**, 044903 (2002).
7. B. Tomášik and U. A. Wiedemann, "Central and non-central HBT from AGS to RHIC", in *Quark Gluon Plasma 3*, edited by R.C. Hwa and X.N. Wang, World Scientific, Singapore, 2004, pp. 715-777, [arXiv:hep-ph/0210250].
8. J. Adams *et al.* [STAR Collaboration], *Phys. Rev. Lett.* **93**, 012301 (2004).
9. B. Tomášik, *Nucl. Phys. A* **749**, 209–218 (2005).

RAPIDITY DEPENDENCE, CONTRAINTS FROM FLOW, v_2 ...

Chairperson: S. Manly

Rapidity Dependence of Bose-Einstein Correlations at SPS energies

Stefan Kniege for the NA49 collaboration

Institut für Kernphysik Frankfurt (IKF), Universität Frankfurt, Max-von-Laue-Str. 1 60438 Frankfurt am Main, Germany

Abstract. This article is devoted to results on π^--π^--Bose-Einstein correlations in central Pb+Pb collisions measured by the NA49 experiment at the CERN SPS. Rapidity as well as transverse momentum dependences of the correlation lengths will be shown for collisions at 20*A*, 30*A*, 40*A*, 80*A*, and 158*A* GeV beam energy. Only a weak energy dependence of the radii is observed at SPS energies. The k_t-dependence of the correlation lengths as well as the single particle m_t-spectra will be compared to model calculations. The rapidity dependence is analysed in a range of 2.5 units of rapidity starting at the center of mass rapidity at each beam energy. The correlation lengths measured in the longitudinally comoving system show only a weak dependence on rapidity.

Keywords: Bose-Einstein correlations, HBT
PACS: 25.75.Gz

1. INTRODUCTION

The measurement of correlations of identical bosons in heavy ion collisions provides a unique tool to investigate the space time evolution of the particle emitting source. Bose-Einstein correlations are observed as an enhancement of the yield of pairs of particles with small relative momenta. Measurements of the range and strength of the correlations in momentum space allow to derive the extension of the source in coordinate space. Due to space-momentum correlations in expanding sources the correlations do not reflect the whole extensions of the source. In such a scenario, the study of the correlations in different regions of phase space helps to understand the evolution of the source. While the dependence of the correlation lengths on the mean transverse momentum $k_t = \frac{1}{2}|\vec{p}_{t,1} + \vec{p}_{t,2}|$ of the pairs reflects the transverse expansion dynamics of the source the dependence on the pair rapidity $Y = \frac{1}{2}\log\left(\frac{E_1+E_2+p_{z,1}+p_{z,2}}{E_1+E_2-p_{z,1}-p_{z,2}}\right)$, which is measured in the center of mass system, should shed light on the profile of the source in longitudinal direction. The large acceptance of the NA49 experiment allows us to obtain a comprehensive picture of the dynamical evolution of the source.

The article is organized as follows: A brief survey of the experiment, the construction of the correlation function and the fit method are presented in section 2. In section 3, crucial systematic uncertainties in the analysis due to detector effects are discussed. The results on the k_t- as well as the Y- dependence of the correlation lengths are presented in section 4 and compared to model calculations.

CP828, *Multiparticle Dynamics*
edited by V. Šimák, M. Šumbera, Š. Todorova, and B. Tomášik

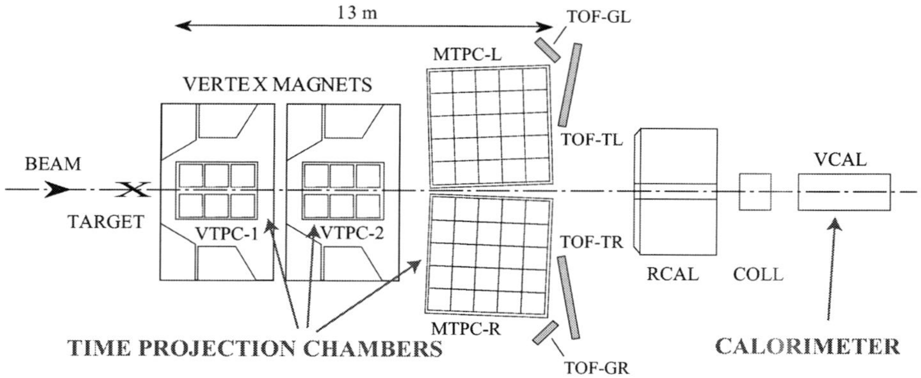

FIGURE 1. NA49 detector setup.

2. EXPERIMENTAL SETUP AND ANALYSIS

NA49 [1] is a fixed target experiment located at the CERN SPS comprising four large-volume Time Projection Chambers (TPC), two of which are located inside the magnetic field of two superconducting dipole magnets (Fig. 1). The TPCs are read out at 90 (MTPC) and 72 (VTPC) pad rows resulting in a very good determination of the momentum of the traversing particles. A zero degree calorimeter at the downstream end of the experiment is used to trigger on the centrality of the collisions. The data presented here correspond to the 7.2% most central events for data samples taken at 20*A*, 30*A*, 40*A*, 80*A*, and 158*A* GeV beam energy. By scaling the magnetic field it was possible to obtain a similar coverage of phase space relative to the center of mass rapidity for the different beam energies. Measuring the specific energy loss dE/dx of charged particles in the gas of the TPCs with a resolution of 3-4% allows particle identification. However, due to ambiguities in particle identification by specific energy loss measurements in certain regions of phase space, negative hadrons rather than identified negative pions are studied in this analysis.

The correlation function is constructed as the ratio of a distribution of the momentum difference of pairs from the same event (signal) and a mixed event background distribution (background). Following the approach of Pratt and Bertsch [2, 3] the momentum difference is decomposed into a component parallel to the beam axis q_{long} and two components in the transverse plane q_{out} and q_{side} with q_{out} defined parallel, and q_{side} perpendicular to k_t. The correlation function is parameterised by a Gaussian function

$$C_2(q)_{BP} = 1 + \lambda \cdot \exp(-R^2_{out}q^2_{out} - R^2_{side}q^2_{side} - R^2_{long}q^2_{long} - 2R^2_{outlong}q_{out}q_{long}) \qquad (1)$$

and the parameters R_{out}, R_{side}, R_{long}, $R_{outlong}$, and λ are determined by a fit to the measured correlation function. The Coulomb repulsion of the particles is accounted for by weighting the theoretical Bose-Einstein correlation function $C_2(q)_{BP}$ by a factor $F(q_{inv}, \langle r \rangle)$ [4] in the fit procedure. The weight is determined by the invariant momentum difference q_{inv} of the pair and the mean pair separation $\langle r \rangle$ of the particles in the source. According to [4] this quantity can be derived from the extracted radii. We therefore determine the source parameters as well as the value of $<r>$ in an iterative fit-procedure. The following fit function was used:

$$C_2(q)_f = n\{p \cdot (C_2(q)_{BP} \cdot F(q, <r>)) + (1-p)\}. \tag{2}$$

The contamination of the sample with pairs of non-identical particles and pairs of pions from long lived resonances or weak decays is accounted for by a purity factor p which is determined by a VENUS/GEANT simulation. The fit parameter n is introduced to account for the different statistics in signal and background distributions. Beside the uncertainties which arise due to the construction of the correlation function and the fit formalism there are further detector related effects which will be discussed in the next section.

3. SYSTEMATIC STUDIES

3.1. Two track resolution

The momentum difference of a pair is closely related to the distance of the tracks traversing the detector. Bose-Einstein correlations are restricted to a narrow window in momentum difference, hence it is crucial to understand the two track resolution of the detector. The overlap of charge clusters induced at the pad planes of the TPCs can lead to an assignment of points to the wrong track or in an extreme case to complete merging of two tracks. In this case, a pair with small relative momentum will be lost in the signal distribution and the observable Bose-Einstein enhancement will be reduced. To study the impact of the limited two track resolution on the extracted radii, the distance of the tracks was measured at each pad row where both tracks lie in the sensitive volume of the TPCs. Starting from the downstream end of the TPCs the distance of closest approach (dca) of two tracks after a given number of passed pad rows n_{rows} was determined. Pairs with a dca smaller than a given cut value dca_{cut} were rejected both from the signal

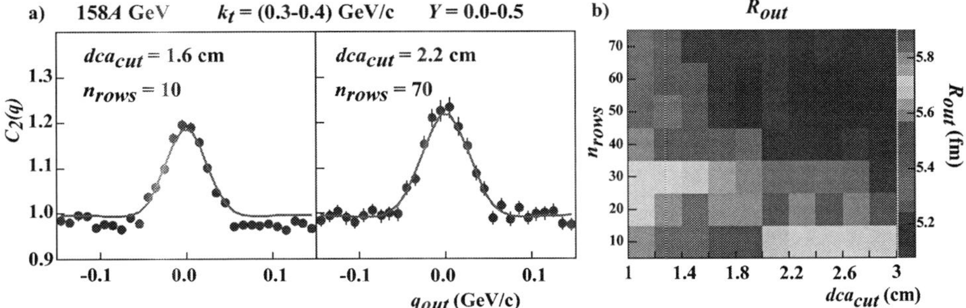

FIGURE 2. a) Impact of two track resolution inefficiencies on the shape of the correlation function and b) fit results for R_{out} for different combinations of the cut parameters dca_{cut} and n_{rows}.

and the background distribution. The impact of a variation of the two parameters dca_{cut} and n_{rows} on the correlation function is shown in Fig. 2. Requiring only small values of dca_{cut} and n_{rows}, tracks can approach each other very closely over a considerable part of the track length in the TPCs. In this case, track merging effects can lead to a significant loss of pairs in the signal. This effect is very pronounced at high transverse momenta and shows up in an undershoot of the projection of the correlation function onto q_{out} (Fig. 2a). Increasing the cut parameters, the influence of merging effects is reduced, the undershoot of the correlation function vanishes and the extracted radii vary only by less than 0.2 fm (Fig. 2b). This can serve as an estimate of the systematic error on the radii due to the specific treatment of the two track inefficiencies in this analysis.

Further systematic errors on the radii arise due to uncertainties concerning the treatment of the Coulomb interaction (which can significantly influence the parameters R_{out} and λ), the missing particle identification, the finite momentum resolution of the detector and the normalisation of the correlation function. The overall systematic error on the radii is not specified for each k_t-Y-bin but estimated to be smaller than 1 fm for all extracted radii.

4. RESULTS

4.1. The k_t-dependence

Figure 3 presents the k_t-dependence of the radii at midrapidity. Radius and λ parameters were obtained from fits of (2) to the correlation function in bins of Y and k_t. Since the influence of the finite momentum resolution was found to be small, no corrections were applied for this effect. The bin width in k_t was chosen to be 0.1 GeV/c and increased to 0.2 GeV/c for the last bin ((0.4-0.6) GeV/c) to obtain sufficient statistics at all energies. A strong decrease of R_{long} with k_t is observed. R_{out} is slightly larger than R_{side} for all energies indicating a finite duration of particle emission [3]. Also shown in this picture is a fit of the k_t-dependence of the radii according to a blast wave parameterisation [5] of the source. In this model the source is treated as boost invariant in the longitudinal direction. In the transverse direction a box-shaped density profile and a linearly increasing flow profile is assumed. Space momentum correlations induced by flow reduce the measured correlation lengths. This effect is partly compensated in case of a superimposed thermal velocity field. Therefore ambiguities arise in the two model parameters temperature and flow, which can not be resolved by only analysing the k_t-dependence of the radii. To resolve these ambiguities the single particle p_t-spectra of protons and negatively charged pions measured by NA49 were fitted to a parameterisation derived from the same model. The lines in Fig. 3 correspond to a combined fit to the radii and the particle p_t-spectra. As expected from the weak energy dependence of the radii only small variations of the extracted source parameters were observed. The extracted parameters were the temperature T, the maximum transverse flow rapidity ρ, the transverse geometrical radius R, the emission time τ and the emission duration $\Delta\tau$. The fit results are inserted in Fig. 3. The temperature T slightly increases with the beam energy, transverse flow and geometrical radius stay approximately constant over the ob-

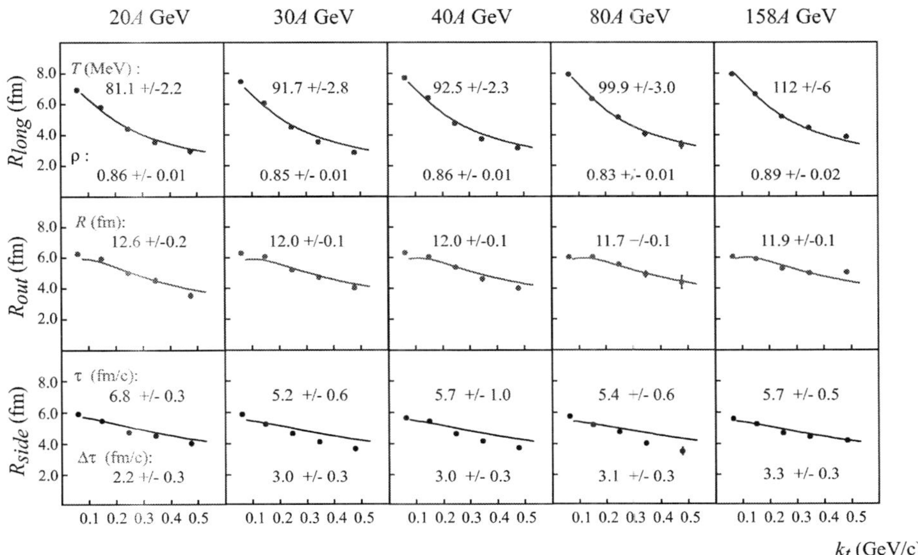

FIGURE 3. The k_t-dependence of R_{side}, R_{out}, and R_{long} at midrapidity ($0.0 < Y < 0.5$) for the different data sets (dots). The lines correspond to a combined fit of the blast wave model to the radii and the single particle p_t-spectra.

served energy range. For the emission time we obtain values of 5.4 to 6.8 fm/c. The fit slightly overpredicts R_{side} at high k_t but still results in a finite emission duration of 2.2-3.2 fm/c. The fit might be further constrained by adding more particle spectra or by including contributions from resonance decays in the model.

4.2. Y-dependence

The model described in section 4.1 is only applicable in case of a longitudinally boost invariant source. Under such conditions it is expected that the cross term $R_{outlong}$ vanishes [6]. Considering the systematic error of 1 fm this condition is fulfilled to good approximation at midrapidity. In Fig. 4 the rapidity dependence of R_{long}, R_{side}, R_{out}, and $R_{outlong}$ is shown at k_t=(0.0-0.1) GeV/c for the different beam energies. In [7] the impact of a non-boost invariant expansion on the parameter $R_{outlong}$ is studied. An increase of $R_{outlong}$ with increasing rapidity is predicted due to the decrease of the inclusive pion yields and is in agreement with the results for all energies. However a change in the longitudinal expansion dynamics which is indicated by the change in $R_{outlong}$ is not reflected in the rapidity dependence of the other observables. R_{side}, which is supposed to determine the geometrical size of the source [8] does not change significantly with rapidity. R_{out} is approximately constant over the investigated rapidity region. Only slight

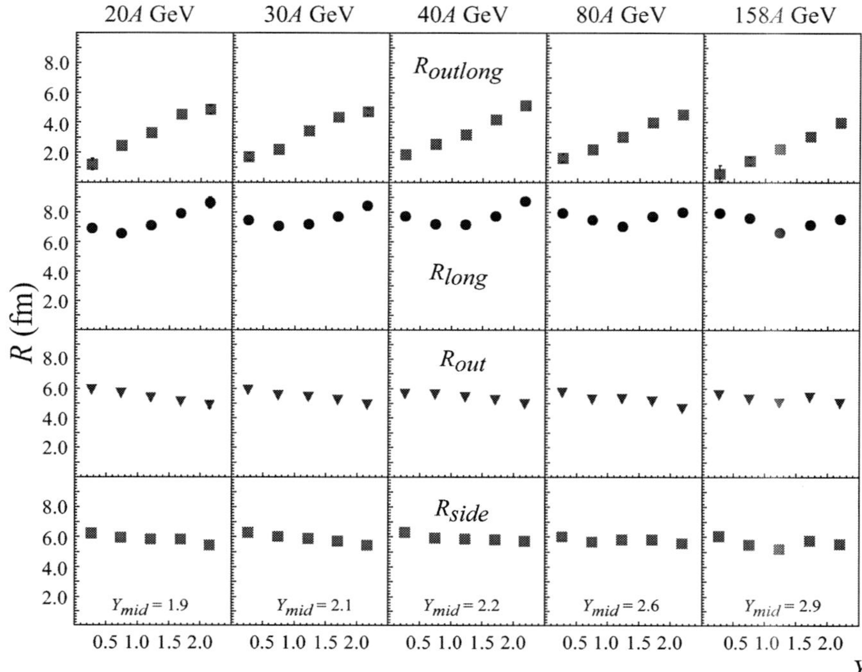

FIGURE 4. Rapidity dependence of R_{side}, R_{out}, R_{long}, and $R_{outlong}$ at k_t=(0.0-0.1) GeV/c for the different beam energies. Shown are as well the midrapidity values Y_{mid} for the different beam energies.

changes in R_{long} are observed. These are even less pronounced at higher transverse momenta.

In summary, a distinct energy dependence of the radii is not observed even though there is a dramatic change in the energy dependence of other hadronic observables like e.g. the kaon to pion ratio [9]. Furthermore the radii do not show a pronounced rapidity dependence, in contrast to the particle yields which decrease strongly with rapidity.

REFERENCES

1. S. Afanasiev et al., *Nucl. Instr. Meth. A* **430**, 210-244 (1999).
2. S. Pratt, *Phys. Rev. D* **33**, 1314-1327 (1986).
3. G. F. Bertsch, *Nucl. Phys. A* **498**, 173c-180c (1989).
4. Yu. M. Sinyukov et al., *Phys. Lett. B* **432**, 248-257 (1998).
5. F. Retiere, M. Lisa, *Phys. Rev. C* **70**, 044907 (2004).
6. S. Chapman, P. Scotto und U. Heinz, *Phys. Rev. Lett.* **74**, 4400-4403 (1995).
7. S. Chapman, P. Scotto und U. Heinz, *Nucl. Phys. A* **590**, 449c-452c (1995).
8. S. Chapman, J. Nix and U. Heinz, *Phys. Rev. C* **52**, 2694-2703 (1995).
9. M. Gazdzicki, *J. Phys. G: Nucl. Part. Phys.* **30**, 701-708 (2004).

Understanding the Rapidity Dependence of the Elliptic Flow and the HBT Radii at RHIC

M. Csanád*, T. Csörgő†, B. Lörstad** and A. Ster†

*Department of Atomic Physics, ELTE, Pázmány P. 1/A, H-1117 Budapest, Hungary
†MTA KFKI RMKI, H - 1525 Budapest 114, P.O.Box 49, Hungary
**Department of Physics, University of Lund, S-22362 Lund, Sweden

Abstract. The pseudo-rapidity dependence of the elliptic flow at various excitation energies measured by the PHOBOS Collaboration in Au+Au collisions at RHIC is one of the surprising results that has not been explained before in terms of hydrodynamical models. Here we show that these data are in agreement with theoretical predictions and satisfy the universal scaling relation predicted by the Buda-Lund hydrodynamical model, based on exact solutions of perfect fluid hydrodynamics. We also show a theoretical prediction on the rapidity and transverse momentum scaling of the HBT radii measured in heavy ion collisions, based on the Buda-Lund model.

Keywords: elliptic flow, correlation radii
PACS: 25.75.Gz,25.75.Ld

INTRODUCTION

One of the unexpected results from experiments at the Relativistic Heavy Ion Collider (RHIC) is the relatively strong second harmonic moment of the transverse momentum distribution, referred to as the elliptic flow. Measurements of the elliptic flow by the PHENIX, PHOBOS and STAR collaborations (see Refs. [1, 2, 3, 4]) reveal rich details in terms of its dependence on particle type, transverse and longitudinal momentum variables, on the centrality and the bombarding energy of the collision. In the soft transverse momentum region, these measurements at mid-rapidity are reasonably well described by hydrodynamical models [5, 6]. However, the dependence of the elliptic flow on the longitudinal momentum variable pseudo-rapidity and its excitation function has resisted descriptions in terms of hydrodynamical models (but see their new description by the SPHERIO model [7]).

Here we show that these data are consistent with the theoretical and analytic predictions that are based on Eqs. (1-6) of Ref. [8], that is, on perfect fluid hydrodynamics.

We furthermore calculate rapidity dependent HBT (Bose-Einstein) radii in the framework of the model and make prediction on the universal scaling of these observables.

Our tool in describing the pseudorapidity-dependent elliptic flow and HBT radii is the Buda-Lund hydrodynamical model. The Buda-Lund hydro model [9] is successful in describing the BRAHMS, PHENIX, PHOBOS and STAR data on identified single particle spectra and the transverse mass dependent Bose-Einstein or HBT radii as well as the pseudorapidity distribution of charged particles in Au + Au collisions both at $\sqrt{s_{\rm NN}} = 130$ GeV [10] and at $\sqrt{s_{\rm NN}} = 200$ GeV [11]. However the elliptic flow would be zero in an axially symmetric case, so we developed the ellipsoidal generalization of

CP828, *Multiparticle Dynamics*
edited by V. Šimák, M. Šumbera, Š. Todorova, and B. Tomášik

the model that describes an expanding ellipsoid with principal axes X, Y and Z. Their derivatives with respect to proper-time (expansion rates) are denoted by $\dot{X}, \dot{Y}$ and $\dot{Z}$.

The generalization goes back to the original one, if the transverse directed principal axes of the ellipsoid are equal, ie $X = Y$ (and also $\dot{X} = \dot{Y}$).

The deviation from axial symmetry can be measured by the momentum-space eccentricity,

$$\varepsilon_p = \frac{\dot{X}^2 - \dot{Y}^2}{\dot{X}^2 + \dot{Y}^2}. \tag{1}$$

The exact analytic solutions of hydrodynamics (see Ref. [8, 12, 13]), which form the basis of the Buda-Lund hydro model, develop Hubble-flow for late times, i.e. $X \underset{\tau\to\infty}{\longrightarrow} \dot{X}\tau$, so the momentum-space eccentricity ε_p nearly equals space-time eccentricity ε.

Let us introduce $\Delta\eta$ additionally. It represents the elongation of the source expressed in units of space-time rapidity. Let us consider furthermore that at the freeze-out $\tau\Delta\eta = Z$ and $Z \approx \dot{Z}\tau$, and so $\Delta\eta \approx \dot{Z}$

Hence, in this paper we extract space-time eccentricity (ε), average transverse flow (u_t) and longitudinal elongation ($\Delta\eta$) from the data, instead of $\dot{X}, \dot{Y}$ and $\dot{Z}$.

In the time dependent hydrodynamical solutions, these values evolve in time, however, it was show in Ref. [14] that $\dot{X}, \dot{Y}$ and $\dot{Z}$, and so ε, u_t and $\Delta\eta$ become constants of the motion in the late stages of the expansion.

RAPIDITY DEPENDENT ELLIPTIC FLOW

The result for the elliptic flow (under certain conditions detailed in Ref. [15]) is the following simple universal scaling law:

$$v_2 = \frac{I_1(w)}{I_0(w)}. \tag{2}$$

The model predicts an *universal scaling:* every v_2 measurement is predicted to fall on the same *universal* scaling curve I_1/I_0 when plotted against w.

This means, that v_2 depends on any physical parameter (transverse or longitudinal momentum, center of mass energy, centrality, type of the colliding nucleus etc.) only through the (universal) scaling paremeter w.

Here w is the scaling variable, defined by

$$w = \frac{p_t^2}{4\overline{m}_t}\left(\frac{1}{T_{*,y}} - \frac{1}{T_{*,x}}\right), \tag{3}$$

and

$$T_{*,x} = T_0 + \overline{m}_t \dot{X}^2 \frac{T_0}{T_0 + \overline{m}_t a^2}, \tag{4}$$

$$T_{*,y} = T_0 + \overline{m}_t \dot{Y}^2 \frac{T_0}{T_0 + \overline{m}_t a^2}, \tag{5}$$

and

$$\overline{m}_t = m_t \cosh(\eta_s - y). \tag{6}$$

Table 1. Results of fits to PHOBOS data of Ref. [1]. Both space-time eccentricity (ε) and longitudinal elongation ($\Delta\eta$) increase with increasing $\sqrt{s_{NN}}$. Remaining parameters were fixed as follows: $T_0 = 175$ MeV, $a = 1.19$ and $u_t = 1.64$.

	19.6 GeV	62.4 GeV	130 GeV	200 GeV
ε	0.294 ± 0.029	0.349 ± 0.008	0.376 ± 0.005	0.394 ± 0.006
$\Delta\eta$	1.70 ± 0.25	2.16 ± 0.05	2.46 ± 0.04	2.56 ± 0.04
χ^2/NDF	1.84/11	20.1/13	34.8/15	27.5/15
conf. level	100%	21.4%	1.00%	7.03%

Here $a = \langle \Delta T/T \rangle_t$ measures the temperature gradient in the transverse direction, at the freeze-out, m_t is the transverse mass, T_0 the central temperature at the freeze-out, while η_s is the space-time rapidity of the saddle-point (point of maximal emittivity). This saddlepoint depends on the rapidity, the longitudinal expansion, the transverse mass and on the central freeze-out temperature:

$$\eta_s - y = \frac{y}{1 + \Delta\eta \frac{m_t}{T_0}}, \tag{7}$$

where $y = 0.5\log(\frac{E+p_z}{E-p_z})$ is the rapidity.

More details about the ellipsoidally symmetric model and its result on $v_2(\eta)$ can be found in Ref. [15].

Equation 2 depends, for a given centrality class, on rapidity y and transverse mass m_t. Before comparing our result to the $v_2(\eta)$ data of PHOBOS, we thus performed a saddle point integration in the transverse momentum variable and performed a change of variables to the pseudo-rapidity $\eta = 0.5\log(\frac{|p|+p_z}{|p|-p_z})$, similarly to Ref. [16]. This way, we have evaluated the single-particle invariant spectra in terms of the variables η and ϕ, and calculated $v_2(\eta)$ from this distribution, a procedure corresponding to the PHOBOS measurement decribed in Ref. [1].

We have found that the essential fit parameters are ε and $\Delta\eta$, and the quality of the fit is insensitive to the precise value of T_0, a and u_t. These parameters dominate the azimuthal-averaged single particle spectra as well as the HBT (Bose-Einstein) radii, however they only marginally influence v_2. Their precise value is irrelevant in a broad region of values and does not influence the confidence level of the $v_2(\eta)$ fits. Hence we have fixed their values as given in the caption of Table 1. We also excluded points with large rapidity from the fits in case of lower center of mass energies.

Fits to PHOBOS data of Ref. [1] and its 1-3σ error contours are shown on the top two panels of Fig. 1. The fitting package is available at Ref. [17]. Bottom panel of Fig. 1 demonstrates that the investigated PHOBOS data points follow the theoretically predicted scaling law.

RAPIDITY DEPENDENT HBT RADII

In the framework of the model we can also calculate the HBT radii. In the simplest case, where system of ellipsoidal expansion equals the out-side-longitudinal coordinate

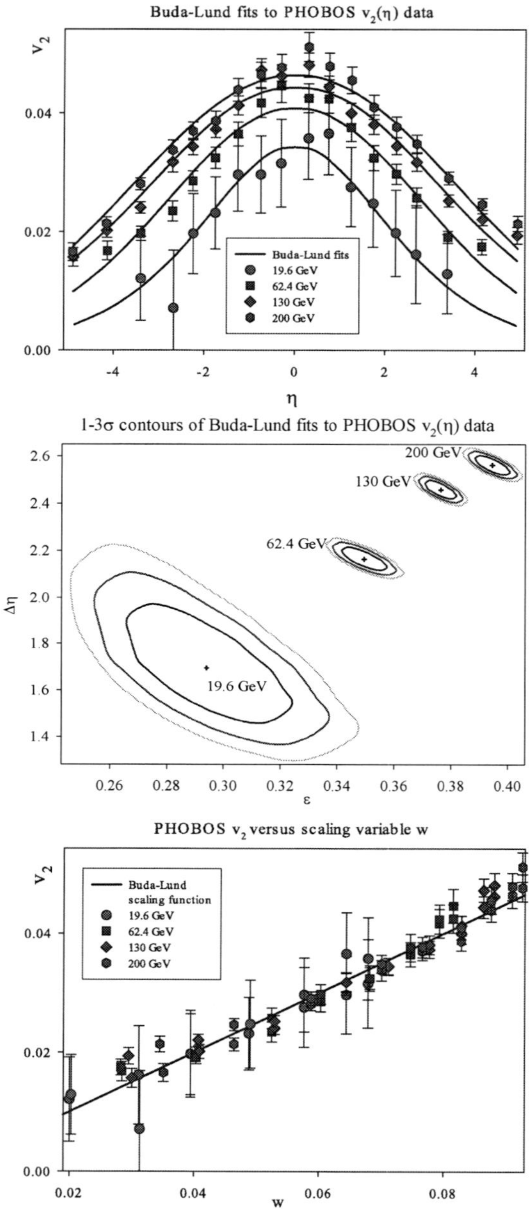

Figure 1. Top: PHOBOS data on the pseudorapidity dependence of the elliptic flow [1], at various center of mass energies, with Buda-Lund fits. Middle: Error contours of the fits. Bottom: Elliptic flow versus scaling variable w is plotted. The data points of Ref. [1] show theoretically predicted [15]) universal scaling, when plotted against the universal scaling variable w.

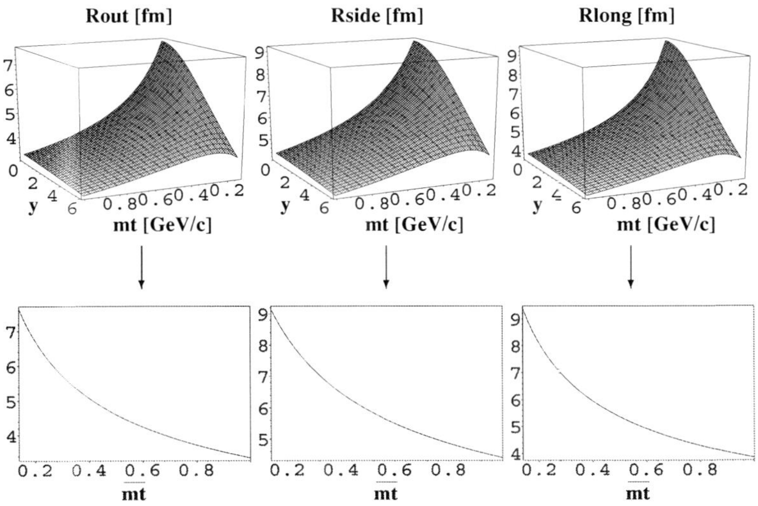

Figure 2. Upper panel: R_{out}, R_{side} and R_{long} as a function of rapidity y and transverse mass m_t. Lower panel: The two-dimensional $R(m_t, y)$ functions are predicted to show a scaling behavior, insofar as they depend only on scaling variable $\overline{m_t}$.

system:

$$R_{\text{out}}^2 = X^2 \left(1 + \frac{\overline{m}_t \left(a^2 + \dot{X}^2\right)}{T_0}\right)^{-1}, \tag{8}$$

$$R_{\text{side}}^2 = Y^2 \left(1 + \frac{\overline{m}_t \left(a^2 + \dot{Y}^2\right)}{T_0}\right)^{-1}, \tag{9}$$

$$R_{\text{long}}^2 = Z^2 \left(1 + \frac{\overline{m}_t \left(a^2 + \dot{Z}^2\right)}{T_0}\right)^{-1}. \tag{10}$$

This means, that the HBT radii depend on transverse mass and rapidity only through the scaling paremeter $\overline{m_t}$, as illustrated on Fig. 2. This behavior could easily be checked by measurement of rapidity and transverse momentum dependence of the HBT radii and comparing this data to the present prediction of the Buda-Lund model. Such a comparision could be a further test of perfect fluid hydrodynamics.

CONCLUSIONS

In summary, we have shown that the excitation function of the pseudorapidity dependence of the elliptic flow in Au+Au collisions is well described with the formulas that are predicted by the Buda-Lund type of hydrodynamical calculations.

We have provided a quantitative evidence of the validity of the perfect fluid picture of soft particle production in Au+Au collisions at RHIC but also show here that this perfect fluid extends far away from mid-rapidity.

We also suggest a further test of perfect fluid hydrodynamics at large rapidities, expressed by Eqs. 8-10 and illustrated by Fig. 2.

The universal scaling of PHOBOS $v_2(\eta)$, expressed by Eq. 2 and illustrated by Fig. 1 provides a successful quantitative as well as qualitative test for the appearence of a perfect fluid in Au+Au collisions at various colliding energies at RHIC.

ACKNOWLEDGMENTS

It is our pleasure to acknowledge the inspiring discussions with Roy Lacey, Stephen Manly and Gábor Veres. This research was supported by the NATO Collaborative Linkage Grant PST.CLG.980086, by the Hungarian - US MTA OTKA NSF grant INT0089462 and by the OTKA grant T038406 .

REFERENCES

1. B. B. Back *et al.* [PHOBOS Collaboration], *Phys. Rev. Lett.* **94**, 122303 (2005).
2. S. S. Adler *et al.* [PHENIX Collaboration], *Phys. Rev. Lett.* **91**, 182301 (2003).
3. C. Adler *et al.* [STAR Collaboration], *Phys. Rev. Lett.* **87**, 182301 (2001).
4. P. Sorensen [STAR Collaboration], *J. Phys. G* **30**, S217-S222 (2004).
5. K. Adcox *et al.* [PHENIX Collaboration], *Nucl. Phys. A* **757**, 184-283 (2005).
6. J. Adams *et al.* [STAR Collaboration], *Nucl. Phys. A* **757**, 102-183 (2005).
7. F. Grassi, Y. Hama, O. Socolowski and T. Kodama, *J. Phys. G* **31**, S1041-S1044 (2005).
8. T. Csörgő, L. P. Csernai, Y. Hama and T. Kodama, *Heavy Ion Physics A* **21**, 73-84 (2004).
9. T. Csörgő and B. Lörstad, *Phys. Rev. C* **54**, 1390-1403 (1996).
10. M. Csanád, T. Csörgő, B. Lörstad, A. Ster, *Acta Phys. Pol. B* **35**, 191-196 (2004).
11. M. Csanád, T. Csörgő, B. Lörstad and A. Ster, *Nukleonika* **49**, S45-S48 (2004).
12. S. V. Akkelin, T. Csörgő, B. Lukács, Y. M. Sinyukov, M. Weiner, *Phys. Lett. B* **505**, 64-70 (2001).
13. T. Csörgő, hep-ph/0111139.
14. T. Csörgő and J. Zimányi, *Heavy Ion Phys.* **17**, 281-293 (2003).
15. M. Csanád, T. Csörgő and B. Lörstad, *Nucl. Phys. A* **742**, 80-94 (2004).
16. D. E. Kharzeev, Y. V. Kovchegov and E. Levin, *Nucl. Phys. A* **699**, 745-769 (2002).
17. `http://www.phenix.bnl.gov/viewcvs/offline/analysis/budalund/`

Effects of LatticeQCD EoS and Continuous Emission on Some Observables

Y. Hama*, R. Andrade*, F. Grassi*, O. Socolowski†, T. Kodama**, B. Tavares** and S.S. Padula‡

*Instituto de Física, Universidade de São Paulo, C.P. 66318, 05315-970 São Paulo-SP, Brazil
†Instituto Tecnológico da Aeronáutica, Praça Marechal Eduardo Gomes, 50 - Vila das Acácias, CEP 12228-900 São José dos Campos-SP, Brazil
**Instituto de Física, Universidade Federal do Rio de Janeiro, C.P. 68528, 21945-970 Rio de Janeiro-RJ, Brazil
‡Instituto de Física Teórica, Universidade Estadual Paulista, Rua Pamplona 145, Bela Vista - CEP 01405-000 São Paulo-SP, Brazil

Abstract. Effects of lattice-QCD-inspired equations of state and continuous emission on some observables are discussed, by solving a 3D hydrodynamics. The particle multiplicity as well v_2 are found to increase in the mid-rapidity. We also discuss the effects of the initial-condition fluctuations.

Keywords: LatticeQCD equations of state, hydrodynamic model
PACS: 24.10.Nz, 25.75.-q, 25.75.Ld

1. HYDRODYNAMIC MODELS

Hydrodynamics is one of the main tools for studying the collective flow in high-energy nuclear collisions. Here, we shall examine some of the main ingredients of such a description and see how likely more realistic treatment of these elements may affect some of the observable quantities. The main components of any hydrodynamic model are the initial conditions, the equations of motion, equations of state and some decoupling prescription. We shall discuss how these elements are chosen in our studies.

Initial Conditions: In usual hydrodynamic approach, one assumes some highly symmetric and smooth initial conditions (IC). However, since our systems are small, large event-by-event fluctuations are expected in real collisions, so this effect should be taken into account. We introduce such IC fluctuations by using an event simulator. As an example, we show here the energy density for central Au+Au collisions at 130A GeV,

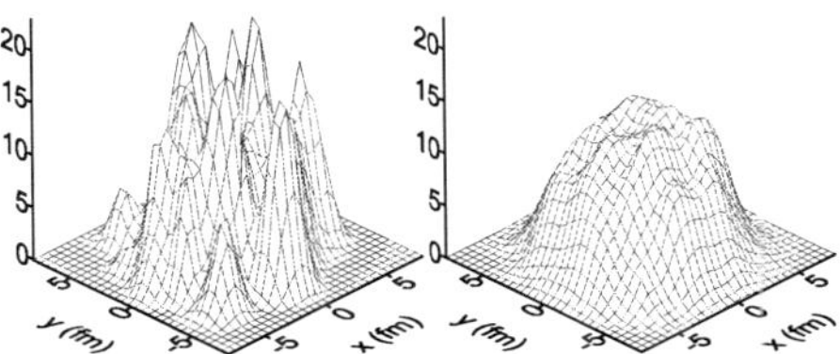

FIGURE 1. The initial energy density at $\eta = 0$ is plotted in units of GeV/fm^3. One random event is shown vs. average over 30 random events ($\simeq$ smooth initial conditions in the usual hydro approach).

CP828, *Multiparticle Dynamics*
edited by V. Šimák, M. Šumbera, Š. Todorova, and B. Tomášik

given by NeXuS[1] [1]. Some consequences of such fluctuations have been discussed elsewhere[5, 6, 7]. We shall discuss some others in Sec. 2.

Equations of Motion: In hydrodynamics, the flow is governed by the continuity equations expressing the conservation of energy-momentum, baryon-number and other conserved charges. Here, for simplicity, we shall consider only the energy-momentum and the baryon number. Since our systems have no symmetry as discussed above, we developed a special numerical code called SPheRIO (**S**moothed **P**article hydrodynamic **e**volution of **R**elativistic heavy **IO**n collisions) [8], based on the so called Smoothed-Paricle Hydrodynamics (SPH) algorithm [9]. The main characteristic of SPH is the parametrization of the flow in terms of discrete Lagrangian coordinates attached to small volumes (called "particles") with some conserved quantities.

Equations of State: In high-energy collisions, one often uses equations of state (EoS) with a first-order phase transition, connecting a high-temperature QGP phase with a low-temperature hadron phase. A detailed account of such EoS may be found, for instance, in [7]. We shall denote them 1OPT EoS. However, lattice QCD showed that the transition line has a critical end point and for small net baryon surplus the transition is of crossover type [10]. The following parametrization may reproduce this behavior, in practice:

$$P = \lambda P_H + (1-\lambda)P_Q + 2\delta/\sqrt{(P_Q-P_H)^2+4\delta}\,, \tag{1}$$

$$s = \lambda s_H + (1-\lambda)s_Q\,, \tag{2}$$

$$\varepsilon = \lambda \varepsilon_H + (1-\lambda)\varepsilon_Q - 2\left[1+(\mu/\mu_c)^2\right]\delta/\sqrt{(P_Q-P_H)^2+4\delta}\,, \tag{3}$$

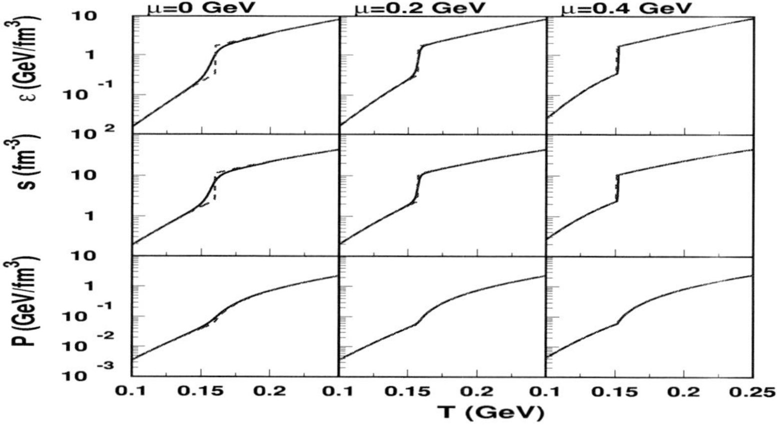

FIGURE 2. A comparison of $\varepsilon(T)$, $s(T)$ and $P(T)$ as given by our parametrization with a critical point (solid lines) and those with a first-order phase transition (dashed lines).

[1] Many other simulators, based on microscopic models, *e.g.* HIJING [2], VNI [3], URASiMA [4], $\cdots$, show such event-by-event fluctuations.

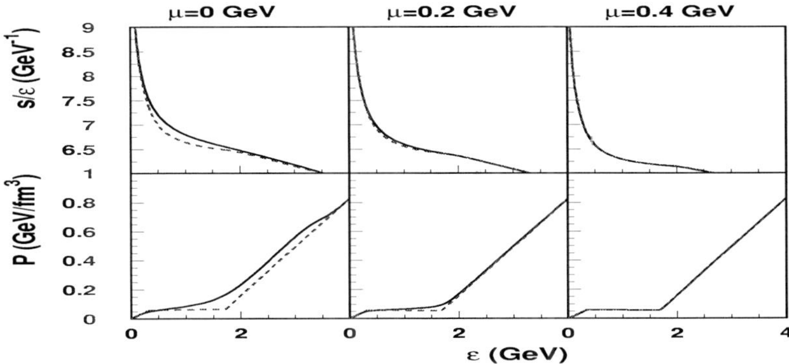

FIGURE 3. Plots of s/ε and P as function of ε for the two EoS shown in Fig. 2.

where $\lambda \equiv [1-(P_Q-P_H)/\sqrt{(P_Q-P_H)^2+4\delta}\ ]/2$ and suffixes Q and H denote those quantities given by the MIT bag model and the hadronic resonance gas, respectively, and $\delta \equiv \delta(\mu_b) = \delta_0 \exp[-(\mu_b/\mu_c)^2]$, with μ_c =const. As is seen, when $\delta(\mu_b) \neq 0$, the transition from hadron phase to QGP is smooth. We could choose $\delta(\mu_b)$ so to make it exactly 0 when $\mu_b > \mu_c$, to guarantee the first-order phase transition there. However, in practice our choice above showed to be enough. We shall denote the EoS given above, with $\delta_0 \neq 0$, CP EoS. Let us compare, in Fig. 2, $\varepsilon(T)$, $s(T)$ and $P(T)$, given by the two sets of EoS. one can see that the crossover behavior is correctly reproduced by our parametrization for CP EoS, while finite jumps in ε and s are exhibited by 1OPT EoS, at the transition temperature. It is also seen, as mentioned above, that at $\mu_b \sim 0.4\,\mathrm{GeV}$ the two EoS are indistinguishable. Now, since in a real collision what is directly given is the energy distribution at a certain initial time (besides n_b, s, *etc.*), whereas T is defined with the use of the former, we plotted some quantities as function of ε in Fig. 3. One immediately sees there some remarkable differences between the two sets of EoS: naturally p is not constant for CP EoS in the crossover region; moreover, s is larger. We will see in Sec.2 that these features affect the observables in non-negligible way.

Decoupling Prescription: Usually, one assumes decoupling on a sharply defined hypersurface. We call this *Sudden Freeze Out* (FO). However, since our systems are small, particles may escape from a layer with thickness comparable with the systems' sizes. We proposed an alternative description called *Continuous Emission* (CE) [11] which, as compared to FO, we believe is closer to what happens in the actual collisions. In CE, particles escape from any space-time point x^μ, according to a momentum-dependent escaping probability $\mathscr{P}(x,k) = \exp[-\int_\tau^\infty \rho(x')\,\sigma v\,\mathrm{d}\tau']$. To implement CE in SPheRIO code, we had to approximate it to make the computation practicable. We took $\mathscr{P}$ on the average, *i.e.*,

$$\mathscr{P}(x,k) \to \langle \mathscr{P}(x,k) \rangle \equiv \mathscr{P}(x) = \exp\left(-\kappa\, s^2/|\mathrm{d}s/\mathrm{d}\tau|\right). \tag{4}$$

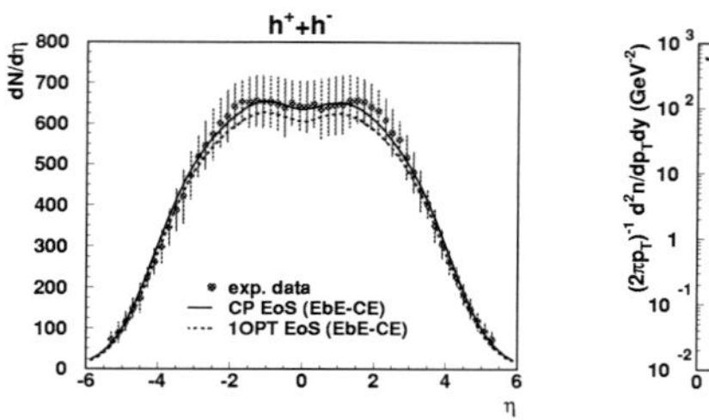

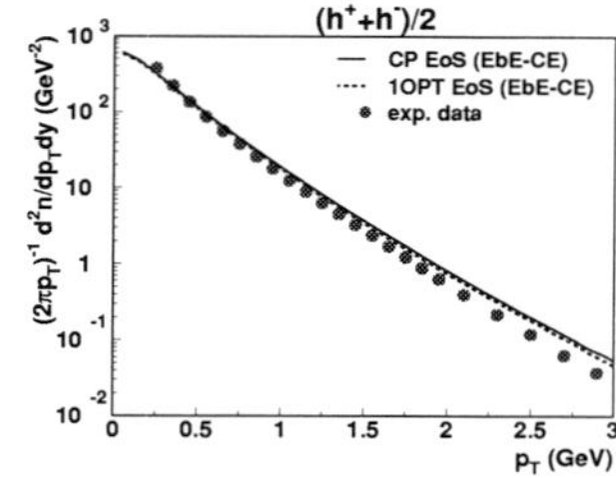

FIGURE 4. η and p_T distributions for the most central Au+Au at 200A GeV. Results of CP EoS and 1OPT EoS are compared. The data are from PHOBOS Collaboration[12].

The last equality has been obtained by making a linear approximation of the density $\rho(x') = \alpha s(x')$ and $\kappa = 0.5\alpha\langle\sigma v\rangle$ is estimated to be 0.3, corresponding to $\langle\sigma v\rangle \approx$ 2 fm^2. It will be shown in Sec. 2 that CE gives important changes in some observables.

2. RESULTS

Let us now show results of computation of some observables, as described above, for Au+Au at 200A GeV. We start computing η and p_T distributions for charged particles, to fix the parameters. Then, v_2 and HBT radii are computed free of parameters.

Pseudo-rapidity distribution: Figure 3 shows that the inclusion of a critical end point increases the entropy per energy. This means that, given the same total energy, CP EoS produces larger multiplicity, which is clearly shown in the left panel of Fig. 4, especially in the mid-rapidity region. Now, we shall mention that, once the equations of state are chosen, fluctuating IC produce smaller multiplicity, for the same decoupling prescription, as compared with the case of smooth averaged IC [7].

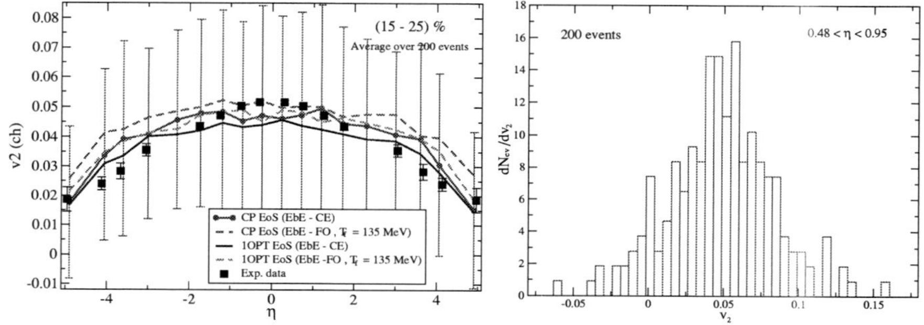

FIGURE 5. Left: η distribution of v_2 for charged particles in the centrality $(15-25)\%$ Au+Au at 200A GeV, computed with fluctuating IC. The vertical bars indicate dispersions. The data are from PHOBOS Collaboration [13]. Right: v_2 distribution in the interval $0.48 < \eta < 0.95$, corresponding to CP EoS and CE.

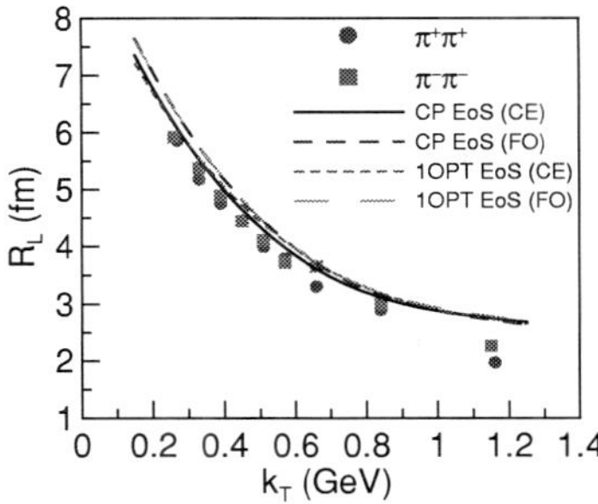

FIGURE 6. k_T dependence of HBT radius R_L for π in the most central Au+Au at 200A GeV, computed with fluctuating IC. The data are from PHENIX Collaboration [14].

Transverse-Momentum Distribution: As discussed in Sec. 1, since the pressure does not remain constant in the crossover region, we expect that the transverse acceleration is larger for CP EoS, as compared with 1OPT EoS case. In effect, the right panel of Fig. 4 does show that p_T distribution is flatter for CP EoS, but the difference is small. The freezeout temperature suggested by η and p_T distributions turned out to be $T_f \simeq 135 - 140\,\text{MeV}$.

Elliptic-Flow Parameter v_2: We show, in Fig. 5, results for the η distribution of v_2 for Au+Au collisions at 200A GeV. As seen, CP EoS gives larger v_2, as a consequence of larger acceleration in this case as discussed in Sec.1. Notice that CE makes the curves narrower, as a consequence of earlier emission of particles, so with smaller acceleration, at large-$|\eta|$ regions. Due to the IC fluctuations, the resulting fluctuations of v_2 are large, as seen in Figs. 5. It would be nice to measure such a v_2 distribution, which would discriminate among several microscopic models for the initial stage of nuclear collisions.

HBT Radii: Here, we show our results for the HBT radii, in Gaussian approximation as often used, for the most central Au+Au collisions at 200A GeV. As seen in Figs. 6 and

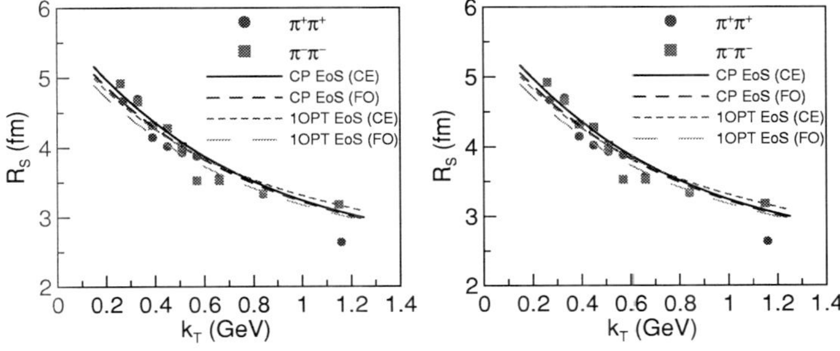

FIGURE 7. k_T dependence of HBT radii R_s and R_o for pions in the most central Au+Au at 200A GeV, computed with event-by-event fluctuating IC. The data are from PHENIX Collaboration [14].

7, the differences between CP EoS results and those for 1OPT EoS are small. For R_s, and especially for R_o, one sees that CP EoS combined with continuous emission gives steeper k_T dependence, closer to the data. However, there is still numerical discrepancy in this case.

3. CONCLUSIONS AND OUTLOOKS

In this work, we introduced a parametrization of lattice-QCD EoS, with a first-order phase transition at large μ_b and a crossover behavior at smaller μ_b. By solving the hydrodynamic equations, we studied the effects of such EoS and the continuous emission. Some conclusions are: *i*) The multiplicity increases for these EoS in the mid-rapidity; *ii*) The p_T distribution becomes flatter, although the difference is small; *iii*) v_2 increases; CE makes the η distribution narrower; *iv*) HBT radii slightly closer to data.

In our calculations, the effect of the continuous emission on the interacting component has not been taken into account. A more realistic treatment of this effect probably makes R_o smaller, since the duration for particle emission becomes smaller in this case. Another improvement we should make is the approximations we used for $\mathscr{P}(x,p)$.

ACKNOWLEDGMENTS

We acknowledge financial support by FAPESP (04/10619-9, 04/15560-2, 04/13309-0), CAPES/PROBRAL, CNPq, FAPERJ and PRONEX.

REFERENCES

1. H.J. Drescher, F.M. Liu, S. Ostapchenko, T. Pierog and K. Werner, *Phys. Rev. C* **65**, 054902 (2002).
2. M. Gyulassy, D.H. Rischke and B. Zhang, *Nucl. Phys. A* **613**, 397-434 (1997).
3. B.R. Schlei and D. Strotman, *Phys. Rev. C* **59**, 9-12 (1999).
4. S. Daté, K. Kumagai, O. Miyamura, H. Sumiyoshi and Xiao-Ze Zhang, *J. Phys. Soc. Japan* **64**, 766-776 (1995).
5. Y. Hama, F. Grassi, O. Socolowski Jr., C.E. Aguiar, T. Kodama, L.L.S. Portugal, B.M. Tavares and T. Osada, in *Proc. of 32nd. ISMD*, eds. A. Sissakian *et al.* (World Sci. – Singapore, 2003) pp.65–68.
6. O. Socolowski Jr., F. Grassi, Y. Hama and T. Kodama, *Phys. Rev. Lett.* **93**, 182301 (2004).
7. Y. Hama, T. Kodama and O. Socolowski Jr., *Braz. J. Phys.* **35**, 24-51 (2005).
8. C.E. Aguiar, T. Kodama, T. Osada and Y. Hama, *J. Phys. G* **27**, 75-94 (2001); T. Kodama, C.E. Aguiar, T. Osada and Y. Hama, *J. Phys. G* **27**, 557-560 (2001).
9. L.B. Lucy, *Astrophys. J.* **82**, 1013 (1977); R.A. Gingold and J.J. Monaghan, *Mon. Not. R. Astro. Soc.* **181**, 375 (1977).
10. Z. Fodor and S.D. Katz, *J. High Energy Phys.* **03**, 014 (2002); F. Karsh, *Nucl. Phys. A* **698**, 199-208 (2002);
11. F. Grassi, Y. Hama and T. Kodama, *Phys. Lett. B* **355**, 9-14 (1996); *Z. Phys. C* **73**, 153-160 (1996).
12. PHOBOS Collab., B.B. Back *et al.*, *Phys. Rev. C* **65**, 054902 (2002); *Phys. Rev. Lett.* **93**, 052303 (2004).
13. PHOBOS Collab., B.B. Back *et al.*, `nucl-ex/0407012`.
14. PHENIX Collab., S.S. Adler *et al.*, *Phys. Rev. Lett.* **93**, 152302 (2004).

Rapidity Dependence of HBT Correlation Radii in Non-Boost Invariant Models

Thorsten Renk

Department of Physics, Duke University, PO Box 90305, Durham, NC 27708 , USA

Abstract. Hanbury-Brown Twiss (HBT) correlation measurements provide valuable information about the phase space distribution of matter in ultrarelativistic heavy-ion collisions. The rapidity dependence of HBT radii arises from a nontrivial interplay between longitudinal and transverse expansion and the time dependence of the freeze-out pattern. For a non-accelerating longitudinal expansion the dependence primarily arises from the amount of radiating matter per unit rapidity $dN/d\eta$, but for a scenario with strong longitudinal acceleration additional complications occur. In this paper I explore schematically what type of dependence can be expected for RHIC conditions under different model assumptions for the dynamics of spacetime expansion and freeze-out.

Keywords: HBT, correlations, rapidity dependence
PACS: 25.75.-q

INTRODUCTION

While the final state of a ultrarelativistic heavy-ion collision is to a large degree reflected in the measured distribution of hadrons there are different possible evolutions leading to this final state which leave a more subtle imprint on observables such as transverse mass $m_T = \sqrt{m^2 + p_T^2}$ spectra and two particle correlations.

The two extreme assumptions are the boost-invariant model proposed by Bjorken [1] in which the incoming nuclei are nearly transparent and the Landau scenario [2] in which the incoming nuclei are completely stopped and the resulting extremely hot and dense system is assumed to undergo strongly accelerated expansion. In a Bjorken expansion matter moves on free streaming trajectories from a source with negligible longitudinal extension (the initial long. nuclear overlap), leading to rapidity $\eta = \frac{1}{2}\frac{p_0+p_z}{p_0-p_z}$ being equal to spacetime rapidity $\eta_s = \frac{1}{2}\frac{t+z}{t-z}$. In this scenario a plateau in the multiplicity per unit rapidity $dN/d\eta$ is expected around $\eta = 0$. In contrast, it has been argued that $dN/d\eta$ at RHIC can reasonably well be described with Landau hydrodynamics [3]. In this scenario, $\eta_s \neq \eta$, no plateau is expected and $dN/d\eta$ changes strongly in time as the system undergoes longitudinal acceleration.

Most 3d hydrodynamic calculations start with initial conditions close to Bjorken-like free streaming (albeit without a plateau in rapidity) and consequently do not alter $dN/d\eta$ significantly during the evolution [4]. In contrast, in [5] I have suggested that a significant longitudinal acceleration seems to be necessary if a simultaneous description of transverse mass spectra and HBT correlation radii is required.

In a longitudinal free-flow scenario (with $dN/d\eta$ approximately unchanged), the effect dominating the rapidity dependence of Hanbury-Brown-Twiss (HBT) radii and

CP828, *Multiparticle Dynamics*
edited by V. Šimák, M. Šumbera, Š. Todorova, and B. Tomášik

m_T-spectra is the shape of $dN/d\eta$ — the amount of thermalised matter at given η governs the amount of expansion and flow that can be reached before decoupling.

In contrast, in a scenario in which $dN/d\eta$ changes strongly over time, radiation from the expanding system will only contribute to the yield at forward rapidities in later evolution stages as matter accelerates outward. Thus, e.g. the duration of the presence of thermalised matter at forward rapidities will be potentially smaller than at midrapidity quite independent from variations in $dN/d\eta$. However, spectra and HBT radii will only reflect this if emission of particles during the evolution time is a significant fraction of the total particle emission including final breakup. It is the aim of this paper to study to what degree the rapidity dependence of HBT correlation radii is capable of reflecting the evolution history of the system.

MODEL FRAMEWORK

I use the model described in [5, 6] for the analysis. Its main assumption is that an equilibrated system is formed a short time τ_0 after the onset of the collision. This thermal fireball subsequently expands isentropically until the mean free path of particles exceeds the dimensions of the system and particles decouple at a timescale τ_f.

For the entropy density at a given proper time we make the ansatz

$$s(\tau,\eta_s,r) = NR(r,\tau)\cdot H(\eta_s,\tau) \tag{1}$$

with τ the proper time as measured in a frame co-moving with a given volume element and $R(r,\tau), H(\eta_s,\tau)$ two functions describing the shape of the distribution and N a normalization factor. Woods-Saxon distributions

$$R(r,\tau) = 1/\left(1+\exp\left[\frac{r-R_c(\tau)}{d_{\rm ws}}\right]\right), \quad H(\eta_s,\tau) = 1/\left(1+\exp\left[\frac{\eta_s-H_c(\tau)}{\eta_{\rm ws}}\right]\right). \tag{2}$$

are used to describe the shapes for given τ. Thus, the ingredients of the model are the skin thickness parameters $d_{\rm ws}$ and $\eta_{\rm ws}$ and the parametrizations of the expansion of the spatial scales $R_c(\tau), H_c(\tau)$ as a function of τ. For the radial expansion assuming constant acceleration one finds $R_c(\tau) = R_0 + \frac{a_\perp}{2}\tau^2$. $H_c(\tau)$ is obtained by integrating forward in τ a trajectory originating from the collision center which is characterized by a rapidity $\eta_c(\tau) = \eta_0 + a_\eta\tau$ with $\eta_c = \text{atanh}\ v_z^c$ where v_z^c is the longitudinal velocity of that trajectory. Since the relation between proper time as measured in the co-moving frame and lab time is determined by the rapidity at a given time, the resulting integral is in general non-trivial and solved numerically (see [6] for details). R_0 is determined in overlap calculations using Glauber theory, the initial size of the rapidity interval occupied by the fireball matter. The set of model parameters then includes initial rapidity η_0 transverse velocity $v_\perp^f = a_\perp\tau_f$ and rapidity at decoupling proper time $\eta^f = \eta_0 + a_\eta\tau_f$. Thus, specifying $\eta_0, \eta_f, v_\perp^f$ and τ_f sets the scales of the spacetime evolution and $d_{\rm ws}$ and $\eta_{\rm ws}$ specify the detailed distribution of entropy density.

For transverse flow a linear relation between radius r and transverse rapidity $\rho = \text{atanh}\ v_\perp(\tau) = r/R_c(\tau)\cdot\rho_c(\tau)$ is assumed with $\rho_c(\tau) = \text{atanh}\ a_\perp\tau$. The model allows for the possibility of accelerated longitudinal expansion which in general implies $\eta \neq \eta_s$

[6]. This mismatch between spacetime and momentum rapidity can be parametrized as a local $\Delta\eta = \eta - \eta_s$ which is a function of τ and η_s.

With the help of a quasiparticle equation of state (EOS), one can find the local temperature $T(\eta_s, r, \tau)$ from the entropy density $s(\eta_s, r, \tau)$ and net baryon density $\rho_B(\eta_s, r, \tau)$. Particle emission is calculated throughout the lifetime of the fireball by selecting a freeze-out temperature T_f, finding the hypersurface characterized by $T(\eta_s, r, \tau) = T_f$ and evaluating the Cooper-Frye formula

$$E\frac{d^3N}{d^3p} = \frac{g}{(2\pi)^3}\int d\sigma_\mu p^\mu \exp\left[\frac{p^\mu u_\mu - \mu_i}{T_f}\right] = \int d^4x S(x,p) \tag{3}$$

with p^μ the momentum of the emitted particle and g its degeneracy factor.

HBT correlation radii are determined as averages of coordinates with the emission function [7]

$$R^2_{side}(\mathbf{K}) = \langle\tilde{y}^2\rangle(\mathbf{K}), \quad R^2_{out}(\mathbf{K}) = \langle(\tilde{x} - \beta_\perp\tilde{t})^2\rangle(\mathbf{K}) \quad \text{and} \quad R^2_{long}(\mathbf{K}) = \langle\tilde{z}^2\rangle(\mathbf{K}) \tag{4}$$

where

$$\tilde{x}^\mu(K) = x^\mu - \langle x^\mu\rangle(K) \quad \text{with} \quad \langle f\rangle(K) = \frac{\int d^4x f(x) S(x,K)}{\int d^4x S(x,K)}.$$

By choosing η_0 different from the measured width of $dN/d\eta$ it is possible to introduce longitudinal acceleration into the model. Note that the model description cannot be assumed to be valid in the target/projectile fragmentation — as one probes forward rapidities, thermal physics is less relevant, evident from the fact that a large fraction of matter is below T_F *ab initio*. Thus, the model prediction beyond $\eta = 2$ should be taken with some caution.

R_{OUT} AND THE FREEZE-OUT PATTERN

We expect that the spacetime evolution of the emission source is reflected in the η dependence of HBT radii only if the system is not dominated by the final breakup. Since freeze-out is determined dynamically, there is no apparent parameter to adjust amount of hadrons released by the final breakup vs. the amount of particles continuously emitted before. From Fig. 1 it is however evident that selecting the skin thickness d_{ws} in the radial distribution of entropy density has a visible impact on the radial position $R_{CF}(\tau)$ of the Cooper-Frye surface.

R_{CF} shrinks for a large skin thickness (almost a Gaussian) whereas it grows for a small skin thickness (a box-like distribution). Since particles must move faster than the Cooper-Frye surface in order to be released, an inward-moving surface implies much more early emission than an outward moving one. This is confirmed in the right panel of Fig. 1 where we present the number of particles being released as a function of τ. The width of the final peak in this figure is connected with the emission time, i.e. the difference in R_{out} and R_{side}. Note that the data seem to prefer the solution with expanding R_{CF}, i.e. a sharp surface, at least in the last stages of the evolution. Shockwaves in the

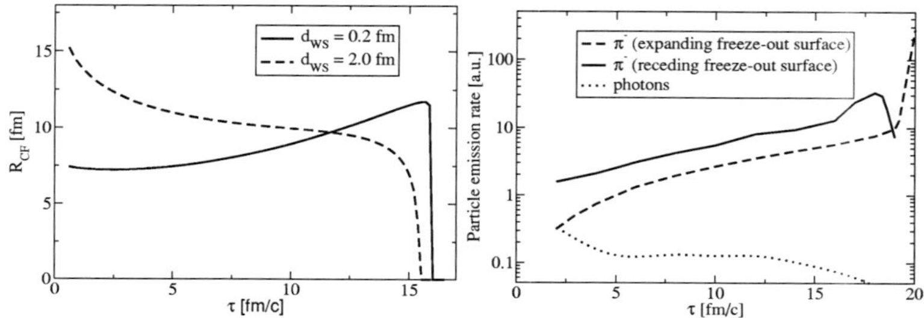

FIGURE 1. Left panel: Evolution of the Cooper-Frye surface radius R_{CF} at $\eta = 0$ for $T_F = 110$ MeV as a function of τ for different values of radial skin thickness parameter. Right panel: Particle emission (in arbitrary units) into midrapidity as a function of τ for pions as a function of $R_{CF}(\tau)$ and for photons.

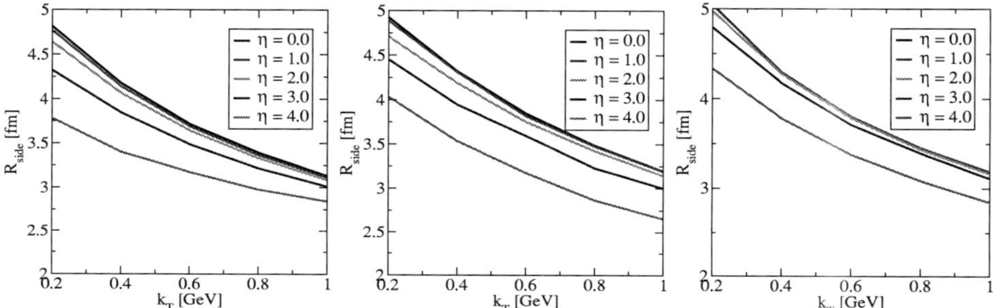

FIGURE 2. R_{side} as a function of pair momentum k_T for different rapidities η in an approximate boost invariant expansion (left), a longitudinally accelerating scenario with continuous particle emission (middle) and a longitudinally accelerating scenario with sudden breakup (right).

medium excited by hard parton energy loss as observed in two particle correlations [8, 9] could be a potential mechanism for creating such a sharply defined surface.

RESULTS

I investigate rapidity dependence in three different scenarios designed to show the essential effects to be expected: 1) a longitudinal free-flow expansion where $dN/d\eta$ remains approximately constant during the evolution 2) a longitudinally accelerated scenario where the width of $dN/d\eta$ changes from 1.7 to 3.8 during the evolution and continuous particle emission occurs and 3) a longitudinally accelerated expansion with the same parameters as above but a sudden breakup in the end (note that case 1 is also characterized by a sudden breakup). In all three cases single particle m_T spectra describe the data at midrapidity and the measured $dN/d\eta$ is reproduced, though only 3) describes the measured HBT correlation radii at midrapidity correctly.

Figure 2 shows R_{side} in the three different scenarios as a function of pair transverse momentum for different rapidities η.

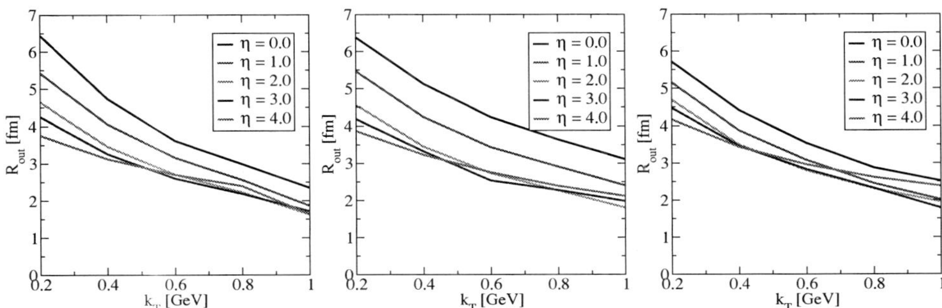

FIGURE 3. R_{out} as a function of pair momentum k_T for different rapidities η in an approximate boost invariant expansion (left), a longitudinally accelerating scenario with continuous particle emission (middle) and a longitudinally accelerating scenario with sudden breakup (right).

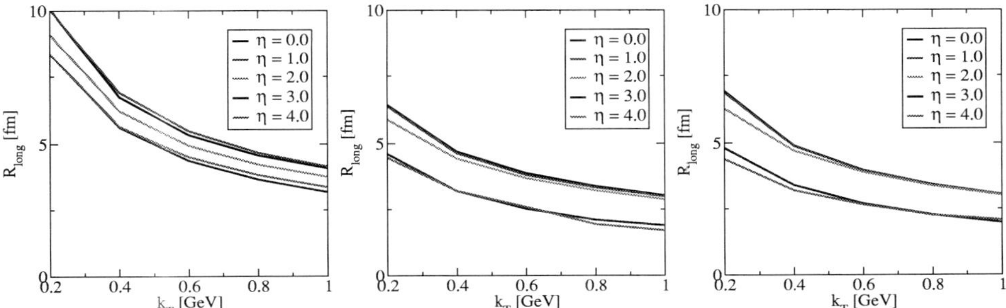

FIGURE 4. R_{long} as a function of pair momentum k_T for different rapidities η in an approximate boost invariant expansion (left), a longitudinally accelerating scenario with continuous particle emission (middle) and a longitudinally accelerating scenario with sudden breakup (right).

The differences between the three scenarios is generally small, however the geometrical radius in the free-flow expansion is systematically smaller than in the accelerated cases. This is due to the fact that breakup occurs roughly at a fixed volume (determined by the total entropy), hence the stronger longitudinal expansion in the Bjorken case (which expands with η_0 at all times) implies less transverse radius expansion and flow.

Fig. 3 compares R_{out} for the three different scenarios. The most striking effect is the general increase in the free flow and the continuous emission scenario as compared to the sudden breakup one. The reason can easily be identified: with an expanding Cooper-Frye surface and a large fraction of energy going into transverse expansion, the sudden breakup scenario has the strongest positive $x-t$ correlation and the least release of particles prior to final breakup. While sudden breakup also occurs in the approximate Bjorken solution, the stronger longitudinal expansion leaves less energy for transverse expansion, significantly lessening this correlation. Finally, since the Cooper-Frye surface is moving inward in the continuous emission scenario, there is no such correlation and R_{out} significantly exceeds R_{side}.

Finally, in Fig. 4 I compare the three different scenarios with respect to R_{long}. Due to $\eta \approx \eta_s$ in this scenario but $\eta > \eta_s$ in the presence of longitudinal acceleration, the scaled falloff with k_T is the same in all three cases (since the final $dN/d\eta$ is fixed), the correlation radius normalization however is given by η_s and this reduces the overall normalization. The radii in the sudden breakup scenario are systematically larger than in the continuous emission one, reflecting the fact that freeze-out happens on average later and hence at larger spatial extension. The sudden drop at $\eta = 3$ is probably an artefact of the model getting sensitive to the target/projectile fragmentation region which is not well described by thermal physics.

CONCLUSIONS

The most remarkable finding is that to first order there is hardly any influence of the expansion and freeze-out scenario on the rapidity dependence of HBT correlations. The reason for this is that the measured single particle spectra and $dN/d\eta$ already provide rather stringent constraints on the expansion pattern. The most pronounced effect of a longitudinally accelerated expansion is the deviation from the scaling $\eta \approx \eta_s$, ultimately leading to a smaller R_{long} in the presence of acceleration while keeping the correct falloff in k_T. For the most part changes in the sideward and outward correlation radii are minor. Since the data practially rule out strong continuous emission throughout the lifetime by $R_{out}/R_{side} \approx 1$, the correlation radii are dominated only by the state of the system at breakup, not by the evolution path. It seems that for reasonable evolution scenarios which are in agreement with other hadronic data the η dependence of HBT radii is primarily determined by the amount of matter at given η. Thus, photons might be a more promising tool in order to study differences in the evolution history [10].

ACKNOWLEDGMENTS

I would like to thank B. Müller and J. Ruppert for helpful discussions during the preparation of this paper. This work was supported by the DOE grant DE-FG02-96ER40945 and a Feodor Lynen Fellowship of the Alexander von Humboldt Foundation.

REFERENCES

1. J. D. Bjorken, *Phys. Rev. D* **27**, 140-151 (1983).
2. L. D. Landau, *Izv. Akad. Nauk SSSR, Physics Series* **17**, 51 (1953).
3. P. Steinberg, nucl-ex/0405022.
4. see e.g. T. Hirano and Y. Nara, *J. Phys. G*, **31** S1-S14 (2005), also C. Nonaka, private communication.
5. T. Renk, *Phys. Rev. C* **70**, 021903 (2004).
6. T. Renk, *J. Phys. G* **30**, 1495-1514 (2004).
7. U. A. Wiedemann and U. W. Heinz, *Phys. Rept.*, **319**, 145-230 (1999).
8. S. S. Adler et al. [PHENIX collaboration], nucl-ex/0507004.
9. T. Renk and J. Ruppert, hep-ph/0509036.
10. T. Renk, *Phys. Rev. C* **71**, 064905 (2005).

SOURCE IMAGING

Chairperson: R. Lacey

Femtoscopy in PHENIX: Evidence for a Long Range Structure in the Pion Emission Source in Au+Au Collisions at RHIC

P. Chung*, P. Danielewicz†, W. Holzmann*, R. Lacey* and J. Alexander*

*Dept of Chemistry, SUNY Stony Brook, Stony Brook, NY 11794, USA
†National Superconducting Cyclotron Laboratory and Department of Physics and Astronomy, Michigan State University, East Lansing, MI 48824-1321, USA

Abstract. The PHENIX experiment has acquired ~1 billion minimum bias Au+Au events at $\sqrt{s} = 200$AGeV during the year 2004. This high statistics data set, coupled with a state-of-the-art analysis technique, allows for the extraction of 3D emission sources for various particle types. These 3D sources lend fresh insight into the nature of a long-range source previously reported by PHENIX. The new results indicate an anisotropic pion emission source in the pair center of mass system (PCMS) having an extended space-time extent in the outward direction. The two-proton emission source from the same data set is essentially isotropic in the PCMS. These results provide information on the evolution dynamics of the high energy density nuclear matter created at RHIC.

Keywords: correlation, interferometry, intensity interferometry, cartesian harmonics, anisotropic correlation
PACS: 25.70.Pq, 25.75.Gz

INTRODUCTION

A de-confined phase of nuclear matter is expected to be formed at the high energy densities created in Au+Au collisions at RHIC [1]. An observation of the presence (or absence) of an emission source of large space-time extent for particle emission can provide important constraints for understanding the nature of this phase transition.

In recent measurements, the PHENIX Collaboration has observed a long range structure in the 1D two-pion emission source function for Au+Au collisions at $\sqrt{s} = 200$ AGeV [2]. This long range structure was resolved using the 1D Source Imaging technique of Brown and Danielewicz [3, 4]. Hence, it yielded no directional information. Thus, a detailed analysis of the dynamical origin of this structure was not possible. To address this shortcoming, Danielewicz and Pratt introduced the more powerful technique of decomposing correlation functions into correlation moments using a cartesian surface-spherical harmonics basis [5]. In this representation, each moment (of a particular order) corresponds to a specific deformation (i.e dipole, quadrupole etc) of the 3D source function. They provide detailed 3D information about the emission source.

In this paper, we present the first application of the moment decomposition technique for the analysis of $\pi^+\pi^+$ and pp pairs produced in Au+Au collisions at $\sqrt{s} = 200$ AGeV.

CP828, *Multiparticle Dynamics*
edited by V. Šimák, M. Šumbera, Š. Todorova, and B. Tomášik

EXPERIMENTAL SETUP AND DATA ANALYSIS

The data presented here were taken by the PHENIX Collaboration during the 2004 run. The colliding beams ($\sqrt{s} = 200$ AGeV) were provided by the RHIC accelerator. Charged tracks were detected in the two central arms of PHENIX [6], each of which subtends 90 degrees in azimuth and ± 0.35 units of pseudo-rapidity. Tracking information was provided by a drift chamber followed by two layers of pad chambers. Particle identification was performed by an electromagnetic calorimeter and a time-of-flight wall.

3D correlation functions , C(**q**), were obtained as the ratio of foreground to background distributions in relative momentum **q** for $\pi^+\pi^+$ and pp pairs. Here, $\mathbf{q} = \frac{(\mathbf{p_1}-\mathbf{p_2})}{2}$ is half of the relative momentum between the two particles in the PCMS frame. The foreground distribution was obtained using pairs of particles from the same event and the background was obtained by pairing particles from different events. The events used have a z-vertex position within ± 30cm from the center of the PHENIX spectrometer. Track merging and splitting effects were removed by appropriate cuts in the relevant coordinate space on both the foreground and background distributions. There was no significant effect on the correlation functions due to the momentum resolution of 0.7%.

In the cartesian harmonic decomposition, the 3D correlation function is expressed as

$$C(\mathbf{q}) = \sum_{l} \sum_{\alpha_1 \ldots \alpha_l} C^l_{\alpha_1 \ldots \alpha_l}(q) A^l_{\alpha_1 \ldots \alpha_l}(\Omega_{\mathbf{q}}) \qquad (1)$$

where $l = 0, 1, 2, \ldots$, $\alpha_i = x, y$ or z, $A^l_{\alpha_1 \ldots \alpha_l}(\Omega_{\mathbf{q}})$ are cartesian harmonic basis elements ($\Omega_{\mathbf{q}}$ is solid angle in **q** space) and $C^l_{\alpha_1 \ldots \alpha_l}(q)$ are cartesian correlation moments given by

$$C^l_{\alpha_1 \ldots \alpha_l}(q) = \frac{(2l+1)!!}{l!} \int \frac{d\Omega_{\mathbf{q}}}{4\pi} A^l_{\alpha_1 \ldots \alpha_l}(\Omega_{\mathbf{q}}) C(\mathbf{q}) \qquad (2)$$

The cartesian coordinate system is oriented such that z is parallel to the beam (longitudinal), x points in the direction of the total momentum of the pair in the PCMS frame (outward) and y is perpendicular to the other two axes (sidewards).

RESULTS

Figure 1 (left panel) shows the $l = 0$ moment C^0 (solid stars) and 1D correlation function $C(q)$ (open stars) for $\pi^+\pi^+$ pairs with $0.20 < k_T < 0.36$ GeV/c, from Au+Au collisions in the centrality range 0-30% of total cross section. Here, $k_T = \frac{(\mathbf{p_{1T}}+\mathbf{p_{2T}})}{2}$ is the mean tranverse momentum of the two particles in the pair. The $l = 0$ moment is essentially identical to the 1D correlation function. Approximately, the two distributions should be the same. Hence, Fig. 1 serves as a good consistency check of the moment calculation procedure.

Figures 1(a), (b) and (c) show the $l = 1$ moments C^1_x, C^1_y and C^1_z respectively for $\pi^+\pi^+$ pairs with $0.20 < k_T < 0.36$ GeV/c, from Au+Au collisions in the centrality range 0-30%. All 3 dipole moments vanish, indicating absence of any dipole deformation in the source function, as expected from symmetry requirements of like particle pair correlations.

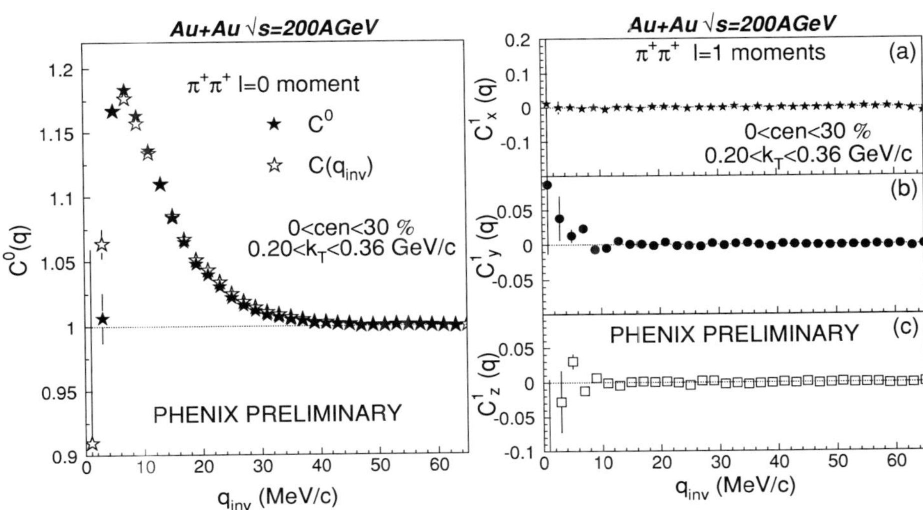

FIGURE 1. Left panel: $l = 0$ moment (solid stars) and 1D correlation function (open stars) for $\pi^+\pi^+$ pairs with $0.20 < k_T < 0.36$ GeV/c from Au+Au collisions in centrality range 0-30%. Right panel: $l = 1$ moments (a) C_x^1 (b) C_y^1 and (c) C_z^1 for same k_T and centrality selection.

Figures 2 (a), (b) and (c) show the $l = 2$ moments C_{xx}^2, C_{yy}^2 and C_{zz}^2, respectively for $\pi^+\pi^+$ pairs with $0.20 < k_T < 0.36$ GeV/c, from Au+Au collisions in the centrality range 0–30%. The non-zero values for the $l = 2$ moments represent source anisotropies of a quadrupole nature and hence represent specific deformations in the direction specified by the moment. C_{xx}^2 is negative for all q_{inv} values representing a negative contribution to the correlation function in the x (outward) direction. In contrast, both C_{yy}^2 and C_{zz}^2 are positive and hence represent positive contributions to the correlation function in the y (sideward) and z (longitudinal) directions. Consequently, the overall correlation function (obtained by adding the $l = 0$ and $l = 2$ moments) is narrower in x and broader in y and z directions, indicating the 3D source size is larger in x and smaller in y and z directions, as compared to the angle-averaged source size.

Figures 2 (d),(e) and (f) show the $l = 4$ moments $C_{x^4}^4$, $C_{y^4}^4$ and $C_{x^2y^2}^4$, respectively for the same centrality and k_T selection. These 3 moments are independent of each other and their non-zero values indicate octupole deformation structure in the source function. The magnitude of this anisotropy is smaller compared to the quadrupole deformation which therefore represents the main shape deformation for the pion source function.

Figure 3 (left panel) shows the $l = 0$ moment C^0 (solid stars) and 1D correlation function $C(q)$ (open stars) for pp pairs with $0.4 < k_T < 2.0$ GeV/c, from Au+Au collisions in

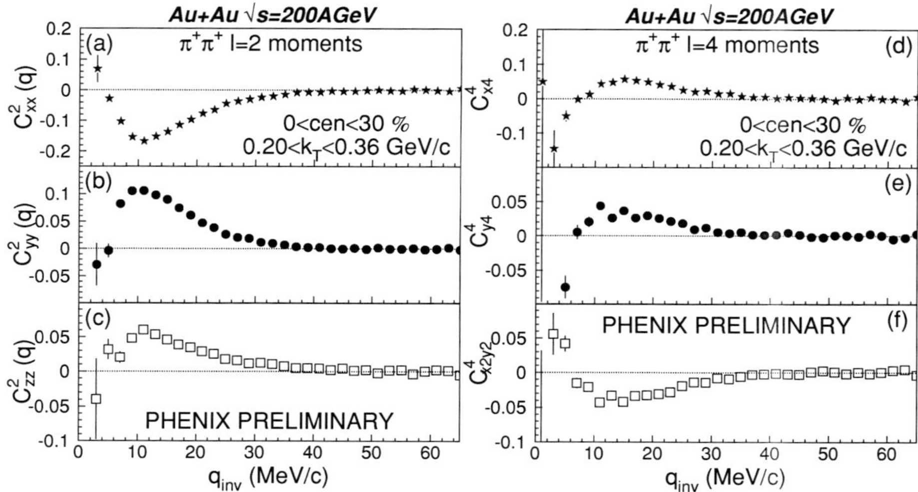

FIGURE 2. Left panel: $l = 2$ moments (a) C^2_{xx} (b) C^2_{yy} and (c) C^2_{zz} for $\pi^+\pi^+$ pairs with $0.20 < k_T < 0.36$ GeV/c from Au+Au collisions in centrality range 0-30%. Right panel: $l = 4$ moments (d) $C^4_{x^4}$ (e) $C^4_{y^4}$ and (f) $C^4_{x^2y^2}$ for same k_T and centrality selection.

the centrality range 0-90%. Again, the very good agreement between the $l = 0$ moment and the 1D correlation function attests to the reliability of the moment calculation procedure. Figures 3 (a),(b) and (c) show the $l = 1$ moments C^1_x, C^1_y and C^1_z, respectively. As with the pions, the dipole moments vanish due to symmetry requirements of identical particle pair correlations.

Figures 4 (a),(b) and (c) show the $l = 2$ moments C^2_{xx}, C^2_{yy} and C^2_{zz} respectively for pp pairs, with $0.4 < k_T < 2.0$ GeV/c, from Au+Au collisions in the centrality range 0-90%. In contrast to the $\pi^+\pi^+$ moments, the $l = 2$ moments for the pp correlation function are all consistent with 0 within statistical fluctuations. Likewise, the $l = 4$ moments shown in Figs. 4 (d) $C^4_{x^4}$,(e) $C^4_{y^4}$ and (f) $C^4_{x^2y^2}$ all vanish. Hence, the overall pp correlation function is the same in all 3 directions, namely the angle-averaged $l = 0$ moment C^0.

DISCUSSION

Analysis of the $\pi^+\pi^+$ 1D correlation function indicates a long range structure in the angle-averaged pion source function [2]. The cartesian moments, shown in Fig. 2,

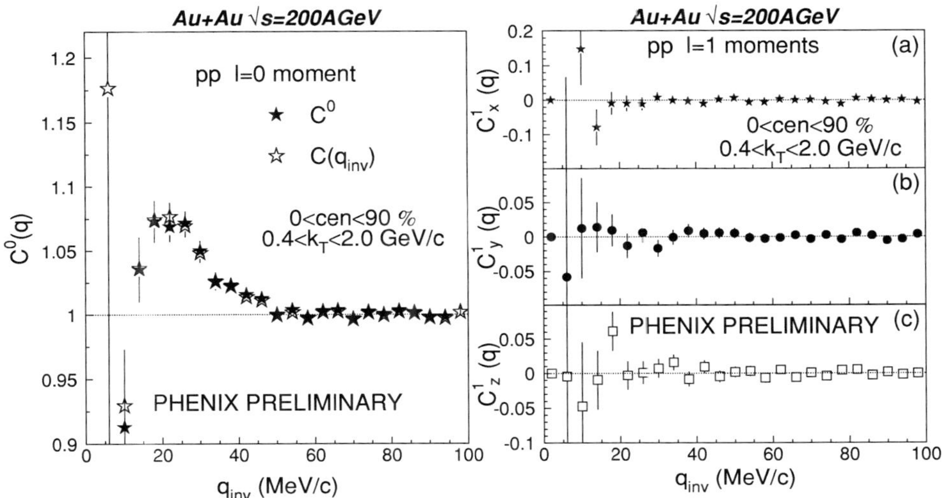

FIGURE 3. Left panel: $l = 0$ moment (solid stars) and 1D correlation function (open stars) for pp pairs with $0.4 < k_T < 2.0$ GeV/c from Au+Au collisions in centrality range 0-90%. Right panel: $l = 1$ moments (a) C_x^1 (b) C_y^1 and (c) C_z^1 for same k_T and centrality selection.

indicate that the correlation function is narrower in the outward direction compared to the sideward and longitudinal directions. This results in a source function which is elongated in the outward direction in the PCMS frame.

The elongation must be necessarily due to a prolonged emission in the pair c.m. frame. A source isotropic in its rest frame, breaking up instantaneously in the pair c.m. frame, would have been shortened in the outward direction by Lorentz factor γ_{boost}. On the other hand, a source breaking up instantaneously in its own rest frame would lead to a source elongated in the pair c.m. frame by the factor γ_{boost} in the outward direction, with nonsimultaneity of emission and movement of the source overcompensating the Lorentz contraction. The most extreme Lorentz elongation factor would be obtained under the assumption of a source freezing out instantaneously in the locally co-moving frame. However, such is an unlikely source. Rather a plausible source is one with a radial velocity along the pair momentum, that freezes out over a finite time in its frame. Specific conclusions, however, must be based on a model, either global parametrization of the emission or dynamic.

In contrast to the pions, the cartesian moments for pp pairs, shown in Figs. 3 and 4, indicate that the proton source function is isotropic in the pair c.m. frame. This suggests

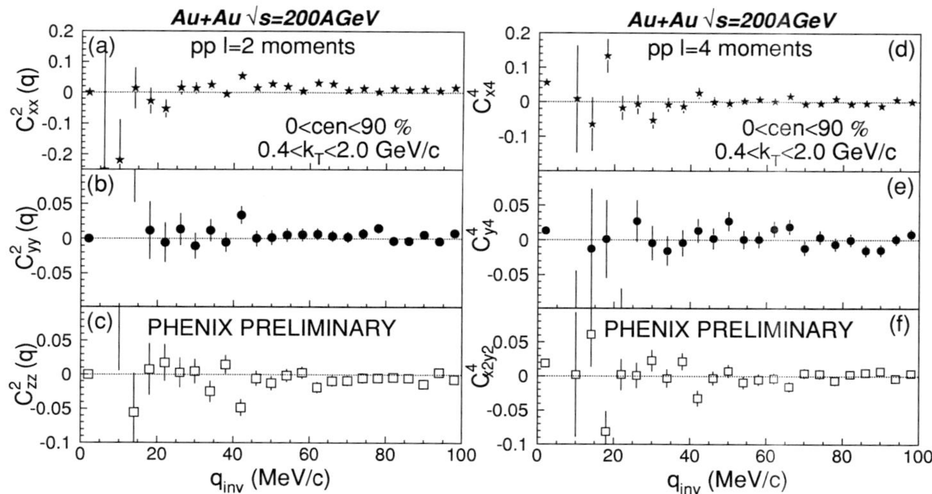

FIGURE 4. Left panel: $l = 2$ moments (a) C^2_{xx} (b) C^2_{yy} and (c) $C^2_{zz}$for pp pairs with $0.4 < k_T < 2.0$ GeV/c from Au+Au collisions in centrality range 0-90%. Right panel: $l = 4$ moments (d) $C^4_{x^4}$ (e) $C^4_{y^4}$ and (f) $C^4_{x^2y^2}$ for same k_T and centrality selection.

an proton emission source of velocity close to the pair velocity.

A model explaining the sources would need to explain the pion and proton emission differences and associated correlation anisotropies simultaneously.

ACKNOWLEDGMENTS

The authors would like to thank Dr Scott Pratt for fruitful discussions.

REFERENCES

1. QM2002, *Nucl. Phys. A* **715**, 1c-930c (2003).
2. P. Chung et al, *Nucl. Phys. A* **749**, 275c-278c (2005).
3. D.A. Brown and P. Danielewicz, *Phys. Lett. B* **398**, 252-258 (1997).
4. D.A. Brown and P. Danielewicz, *Phys. Rev. C* **57**, 2474-2483 (1998).
5. P. Danielewicz and S. Pratt, `nucl-th/0501003`.
6. K. Adcox et al., *Nucl. Instrum. Meth. A* **499**, 469-479 (2003).

Understanding the Emission Duration through Femtoscopy and Imaging

D.A. Brown*, A. Enokizono*, M. Heffner* and R. Soltz*

*Lawrence Livermore National Laboratory, Livermore, California 94550, USA

Abstract. We investigate the role of lifetime effects in femtoscopy at RHIC. We find the long emission durations do not necessarily lead to large outward radii. Within the Core-Halo model, we show that resonance induced halos lead to non-Gaussian tails in the two particle source. We note that the emission duration of the system can mask this tail. We illustrate the strong differences between the two-particle source shape in the side and outwards directions in the presence of a lifetime effect. We comment on the observability of these lifetime induced tails through source imaging.

Keywords: femtoscopy, HBT, source imaging, freeze-out
PACS: 25.75.-q, 25.75.Gz

INTRODUCTION

One of the most perplexing results from RHIC are those from the like-pion femtoscopic measurements of the size of the exploding source. These results pose several mysteries that get to the heart of the dynamical evolution of the system. First, the extracted pion $R_{\rm out}$ and $R_{\rm side}$ parameters suggest an instantaneous freeze-out of particles from the reaction zone [1]. This goes against all expectations of a long-lived source [2]. Second, all of the extracted source sizes have a weak dependence on bombarding energy over several decades of bombarding energy, from the AGS to RHIC [1]. Finally, all systems from pp, to dAu, to CuCu to AuAu exhibit the same general scaling with the number of charged particles, but otherwise behave similarly [1].

In these proceedings, we will examine the role of finite lifetime effects on the source within a simple model. With this model, we will show that the expected long-lived source may in fact have been realized at RHIC. Furthermore, we suggest that a simple interpretation of the two-particle correlations in terms the source sizes $R_{\rm side}$, $R_{\rm out}$, $R_{\rm long}$ and λ are insufficient to characterize the complex RHIC sources. These parameters only serve to characterize the gross "homogeneity length" as suggested by Mahklin, Sinyukov and others [1, 3, 4] and so are only sensitive to the overall dynamics of the system.

To move beyond this gross characterization of sources, we will need to move toward three-dimensional imaging analyses of the femtoscopic results using the tools of Ref. [5]. Source imaging allows us to cleanly separate effects due to final state interactions and symmetrization from effects due to the source function itself. Because no assumption is made about the functional form of the source function, the imaging is model independent. Thus, it is possible to image non-Gaussian sources such as those suggested in the Core-Halo model [6]. In our model studies, we find two particle sources that strongly deviate from the naive Gaussian expectations in the Core-Halo model. While we focus on like-pion sources, our observations have a more general applicability.

CP828, *Multiparticle Dynamics*
edited by V. Šimák, M. Šumbera, Š. Todorova, and B. Tomášik

OUR MODEL

To explore the consequences of a finite source lifetime, we introduce a variant of the Core-Halo model [6]. In the original model, the core of the single particle source does not evolve dynamically, but does emit resonances which decay with finite time. These resonances create a non-Gaussian tail. In reality, the exploding system at RHIC does have a finite lifetime, so one must deal with the interplay of the system lifetime and resonance effects. Furthermore, one needs a model in which there is some dynamics since, as we describe later, one needs a source of moving particles coupled with a finite lifetime to generate the effects we find in our model.

The building blocks for our two-particle source are the normalized particle emission rates:

$$D(r,\mathbf{p}) = \frac{Ed^7N}{d^4rd^3p} \bigg/ \frac{Ed^3N}{d^3p}. \tag{1}$$

We decompose this single particle source into a core, with fraction f, and a halo with fraction $(1-f)$:

$$D(\mathbf{r},t,\mathbf{p}) = fD_{\text{core}}(\mathbf{r},t,\mathbf{p}) + (1-f)D_{\text{halo}}(\mathbf{r},t,\mathbf{p}) \tag{2}$$

The core portion of this single-particle source has a simple Gaussian spacial part, exponential time decay, and momentum dependence arising from hydro-like Boltzman factor:

$$D_{\text{core}}(\mathbf{r},t,\mathbf{p}) \propto e^{-(E-\mathbf{p}\cdot\mathbf{v}_{\text{flow}})/T} D_{\text{gauss}}(\mathbf{r}) e^{-t/\tau_{fo}} \tag{3}$$

Here, f, T, and τ_{fo} are all adjustable and are specified in the lab frame. Typically we choose $T = 175$ MeV, $\tau_{fo} = 10$ fm/c, $f = 0.5$ and set the Gaussian's radii $R_x = R_y \equiv R_T = 4$ fm to ensure radial symmetry of the emitting system. In fact, we go further and set $R_T = R_z$, making the spacial part of our source spherical. Here, the flow profile is given by

$$\mathbf{v}_{\text{flow}} = \begin{cases} \alpha\mathbf{r} & \text{if } |\mathbf{r}| < R_{\text{max}} = 6.98 \text{ fm} \\ \frac{1}{\sqrt{3}}c\hat{\mathbf{r}} & \text{otherwise} \end{cases} \tag{4}$$

and $\alpha = \frac{1}{\sqrt{3}}c/R_{\text{max}}$, giving a maximal flow speed of $\frac{1}{\sqrt{3}}c$, corresponding to the speed of sound in a free relativistic ideal gas. While the flow profile does have a dramatic role in shaping the core part of the source function [7], it has only a minor role in shaping the lifetime effects because the characteristic length scales associated with lifetime effects are so much larger than the core size. We comment that there are no adjustable parameters in this or the remainder of the model single particle source.

We assume that the halo is dominated by the decay of the ω resonance, with lifetime $\tau_\omega = 23$ fm/c. Other potential candidate resonances have decay times that are too short, e.g. the ρ with lifetime $\tau_\rho = 1.3$ fm/c, or too long, e.g. the η' with lifetime $\tau_{\eta'} = 975$ fm/c or have charged pionic decay modes with small branching fractions. Given this, we create the ω's from the same core (with the ω's mass in the Boltzman factor), but we allow it to propagate classically for some distance before decaying into pions:

$$D_{\text{halo}}(\mathbf{r},t,\mathbf{p}) \propto \int d\Delta t\, d^3p_\omega P(\mathbf{p}_\omega,\mathbf{p}) e^{-\Delta t/\tau_\omega} \times D_{\text{core}}(\mathbf{r}-\frac{\mathbf{p}_\omega}{E_\omega}\Delta t, t-\Delta t, \mathbf{p}_\omega). \tag{5}$$

Here, since the ω decays via the three-body reaction $\omega \to \pi^+\pi^-\pi^0$ with a branching fraction of 88.8%, we define the probability, P, for finding a π with momentum $\mathbf{p}$ from the decay of the ω with momentum $\mathbf{p}_\omega$, using standard three-body decay kinematics. Since we are working with charged pions, the only other charged pion decay mode of the ω is $\omega \to \pi^+\pi^-$ with a branching fraction of 2.2%, which is negligible.

The single particle sources can be converted into the two-particle source, such as one might image in a heavy-ion collision, through [8]

$$S(\mathbf{r}') = \int dr'_0 \int d^4R D_1(R+r/2,\mathbf{P}/2)D_2(R-r/2,\mathbf{P}/2). \tag{6}$$

This source function gives the probability to emit a pair with a separation of $\mathbf{r}'$ in the pair CM frame. The time integral here takes place in the pair CM frame and is a consequence of the lack of time dependence of the pair relative wavefunction in the pair CM frame. This integral has some subtle but important consequences. Because each pair has a different total momentum and hence a different boost to the pair CM frame, the time integral is some line integral in along some space-time direction, in any frame other than the pair CM. Therefore, this time integral serves to mix the space and time effects in the source function.

We construct the source function from this single particle source by Monte-Carlo integrating the emission function in Eq. (6). We work in Bertsch-Pratt coordinates [9], so that the time integral in Eq. (6) serves to move the time effects into the outward and longitudinal directions. We also make one optional modification to the model source function: we discard all pairs in which one of the pions has a pseudo-rapidity $\eta > 0.35$, corresponding to the pseudo-rapidity window of the PHENIX detector. In the following, we explore briefly the consequences of this rapidity cut. We will see that this cut is responsible for eliminating a non-Gaussian tail in the longitudinal direction that would otherwise mirror that in the outwards direction.

When examining the "core" and "halo" parts of our source functions, it is important to keep in mind that these do not directly correspond to the "core" and "halo" in the single particle source. First, since there are two emission functions convoluted together in Eq. (6), there are actually *three* terms in the source function: a core piece convoluted with another core piece, a halo piece convoluted with another halo piece, and then the core-halo cross terms. Because the halo in the emission function is so dilute, the halo-halo piece is essentially unobservable. Because of the lifetime effects, below we show that the halo of the source is a combination of an effect of the time integral in the pair CM frame in Eq. (6) and the core-halo cross term. However, the core-core term gives rise to the core of the source function.

On the left in Fig. 1 we show our model source function sliced along the various directions in the Bertsch-Pratt coordinate system. Clearly the outwards and longitudinal

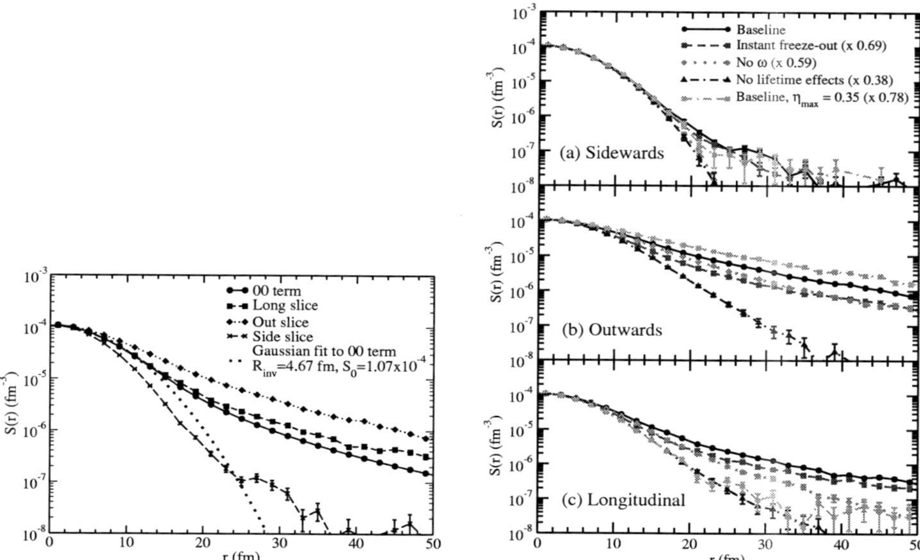

FIGURE 1. Left: Comparison of slices of default source function along the sidewards, outwards and longitudinal directions and of the angle-averaged source (the ℓm=00 term in a spherical harmonic expansion). We also show the best fit Gaussian source fit to the angle-averaged source. Right: Comparison of the sources sliced in the side, out and long directions for each of the cases we consider. The upper panel corresponds to the sidewards direction, the middle panel to the outwards direction and the lower panel to the longitudinal direction.

slices show a large non-Gaussian tail. As we will show, this tail is a direct result of the lifetime effects built into the model. The sidewards direction also shows a small non-Gaussian tail and this is a result of the ω mesons propagating some distance before decaying, creating a geometrical halo. In this figure, we also show the angle average source (or ℓm=00 term in a spherical harmonic expansion of the source) along with a Gaussian fit to the core part of the source.

We now characterize the effects of lifetime on our source function in more detail. To do this, we examine several modifications of our model (called the "Baseline" model in the right in Fig. 1). In the first modification, we set the freeze-out duration to 0 fm/c ("Instant freeze-out"). In the second, we set the core fraction to 100%, turning off the ω resonance ("No ω"). Third, we turn off all lifetime effects, setting $\tau_{f/o} = 0$ fm/c and $f = 1$ ("No lifetime effects"). Finally, we apply the PHENIX pseudo-rapidity acceptance to the source ("Baseline, $\eta_{max} = 0.35$).

On the right in Fig. 1 we show slices of our model source function in the three panels, for the different variations of our model. In all cases, the sources are normalized to match the "Baseline" source in order to emphasis the differences in source shape.

In panel 1(a), we show plots of the slices in the sidewards direction. In all cases, the core shape in this direction is clearly Gaussian with only a small deviation from the

Gaussian behavior caused by the propagation of the ω before its decay, as noted above.

In panel 1(b), we show the outwards direction. Here we see the most dramatic effects of lifetime. In the curve labeled "Instant freeze-out" we see that part of the non-Gaussian tail is caused by the finite lifetime of the core in the single particle source. The size and shape of this portion of the tail is comparable to the tail caused by the ω resonance as seen in the curve labeled "No ω." Unfortunately this means that we cannot uniquely attribute an origin to any non-Gaussian tail we observe. Finally, only by turning off both the freeze-out duration and the ω do we eliminate both the non-Gaussian tail and all lifetime effects. Note that the best Gaussian fits will only ever resolve the Gaussian core of the source function [5] as the non-Gaussian tail corresponds to the low-q fall off in the Coulomb hole in the measured correlation functions.

Interestingly, while the non-Gaussian tail dramatically changes when we alter the lifetime effects, the Gaussian core of the source function does not change very much at all. We have fit the core part of the source functions with Gaussians. As expected, the sideward radii do not vary much from the mean 4.1 fm. Surprisingly, the outward radius only changes at the 20% level from 5.6 fm when we turn on and off the lifetime effects. This has profound implications for the first RHIC HBT puzzle: because the source function core is only weakly dependent on the lifetime effects encoded in the emission function, R_{out} does not change dramatically from R_{side} and hence the ratio stays close to 1, as observed at RHIC [1].

Finally, in panel 1(c), we show the longitudinal direction. The slices in this direction are very similar to the shape in the outwards direction, with the exception of the curve corresponding to the PHENIX pseudo-rapidity acceptance. We see that the finite acceptance essentially turns off the non-Gaussian tail in the "Baseline" model. We attribute this to low-r pairs flowing with a high longitudinal momentum, and thus high rapidity, being cut out of the source.

IMAGING ANALYSIS

We now describe our initial attempts to observe these lifetime effects in like-pion correlations. To do this, we use the CRAB code [10] to generated raw $\pi^-\pi^-$ correlations for all of the cases above. We then proceed as in Ref. [5] and expand the correlations in spherical harmonics. We image these following the Ref. [5] and, as a double-check, we restore the correlation. Typical results from this process are shown in Fig. 2, for the "Baseline" model only. On the left, we show the original and restored correlation terms. Clearly we can reproduce the shape of the large q fall off, but have some trouble at low q. On the right, we show the source function reconstructed from these correlation terms. In this first attempt, we are able to reproduce the core piece, but struggle to image the halo. There are clearly some problems, such as the low r bump in the longitudinal and outwards directions and a corresponding dip in the sidewards direction. This artifact is caused by an artificially high concentration of knots in the basis spline representation used. We hope to fix this with future tuning. Finally, we have not imaged the $\ell = 4$ terms. Clearly these terms are important as we see in the lower set of panels in the left. Until these technical problems are resolved, we can not address the central question of whether we can characterize the effects of lifetime on the source using source imaging.

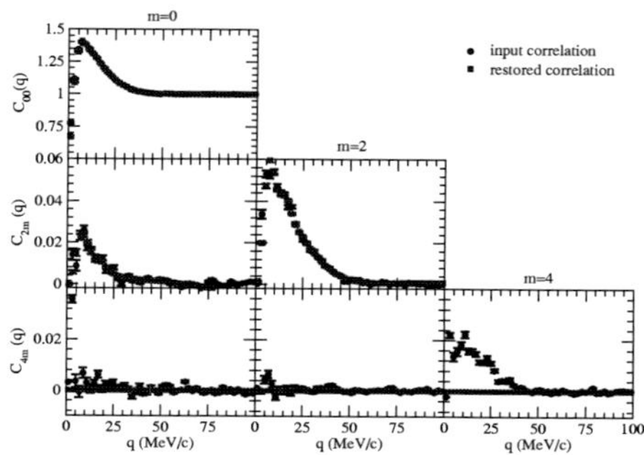

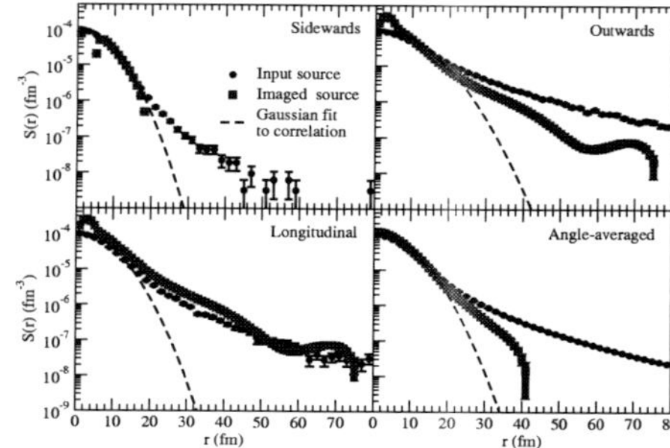

FIGURE 2. Typical imaging results for the "Baseline" model. Left: terms of the spherical harmonic expansion of the model correlation functions, compared to the correlations restored from source images of each term. We only imaged the $\ell = 0, 2$ terms. Right: slices of the imaged source in the sidewards, outwards and longitudinal directions and a plot of the angle-averaged source, compared to the input model source. The images have not yet been optimized.

CONCLUSIONS

We have demonstrated that the first RHIC HBT puzzle can be accounted for in a model that contains realistic and long-lived single particle sources. The reason is simple: the lifetime information can be encoded in the non-Gaussian tails of the source function rather than the HBT radii R_{side}, R_{out}, and R_{long}. In the near future, we hope to show that these tails can be extracted from the measured two-particle correlations using the 3d imaging technique described in Ref. [5].

ACKNOWLEDGMENTS

This work was performed under the auspices of the U.S. Department of Energy by Lawrence Livermore National Laboratory under Contract W-7405-Eng-48.

REFERENCES

1. M. Lisa, R. Soltz, S. Pratt, U. Weidemann, to appear *Ann. Rev. Nucl. Part. Phys.* (2005).
2. D. Rischke, M. Gyulassy, *Nucl. Phys. A* **608**, 479-512 (1996) .
3. A.N. Makhlin and Yu.M. Sinyukov, *Zeit. Phys. C* **39**, 69-82 (1988).
4. U. Heinz and B. Jacak, *Ann. Rev. Nucl. Part. Sci.* **49**, 529-579 (1999).
5. D. Brown *et al.* to appear *Phys. Rev. C* (2005).
6. S. Nickerson, T. Csörgő, and D. Kiang, *Phys. Rev. C* **57**, 3251-3262 (1998); T. Csörgő, B. Lörstad, J. Zimányi, *Zeit. Phys. C* **71**, 491-497 (1996).
7. S.Y. Panitkin and D.A. Brown, *Phys. Rev. C* **61**, 021901 (2000).
8. D.A. Brown and P. Danielewicz, *Phys. Lett. B* **398**, 252-258 (1997).
9. S. Pratt, T. Csörgő and T. Zimányi, *Phys. Rev. C* **42**, 2646-2652 (1990).
10. S. Pratt, *CoRrelation AfterBurner*, `http://www.nscl.msu.edu/~pratt/freecodes/crab/home.html`.

BEYOND THE GAUSSIAN APPROXIMATION

Chairperson: H. Eggers

Intermittency, fractal sources, Levy distributions

A.Bialas

M.Smoluchowski Institute of Physics, Jagellonian University, Reymonta 4, 30-059 Krakow, Poland
e-mail:bialas@th.if.uj.edu.pl

Abstract.
A theoretical introduction to intermittency and its relation to HBT correlations is presented.

Keywords: Intermittency, fractals, cascades, Levy distributions
PACS: 05.45.-a,05.45.Df,05.30.-d

INTERMITTENCY

An explanation is needed: I did not work on these problems since about 10 years. Therefore my views are essentially frozen at that time and I feel somewhat embarrased to talk about all this today. If I nevertheless accepted this kind invitation, it is for two reasons. First, I still believe that our early work on the subject may be helpful in understanding the new developments. Second, ten years is a long time and I am convinced that most of you never heard of it and those who did may have already forgotten. I thus decided to update the talk I gave at the XXVII Rochester Conference (Glasgow) in 1994 [1].

Let me start by reminding what the intermittency in particle physics [2, 3] is all about. In short, it is a search for *self-similarity* of the spectra. Of course the spectra cannot be *exactly* self-similar; the problem was to define self similarity and look for it in the statistical sense. It was soon realized that the key point is the study of the spectra as function of the resolution: power law behaviour of factorial moments [2, 3] and cumulants [4] indicates self-similarity.

The obvious consequence of self-similarity is that the spectrum has no scale and therefore observed fluctuations are not characterized by a definite correlation length. On the contrary: there must be fluctuations at all scales. This helped to understand the presence of some unusual events found in high-energy collisions [5, 6] and was, actually, the starting point of the whole story [2].

PARTON CASCADE

The experimental discovery of the power law dependence versus resolution [7] opened a way to speculations about its possible origin. We have started our studies using the model of a general self-similar cascade which is a prototype of a self-similar statistical system. Among such cascades the most natural seemed the parton cascade [8] which had

CP828, *Multiparticle Dynamics*
edited by V. Šimák, M. Šumbera, Š. Todorova, and B. Tomášik

two advantages: (i) it is the approximate consequence of perturbative QCD[1] and, (ii) it was known to describe correctly the data through the widely used Monte Carlo codes.

This idea run, however, into two serious difficulties. First, self-similarity was observed in the region of up to and even below 50 MeV momentum difference. This is a rather small scale indeed, and one may doubt if the perturbative calculations are applicable in this region. Second, more important: even if the self-similar parton cascading is still operating at these small momentum scales, it is hard to believe that the transision from partons to hadrons[2] does not destroy the subtle structure of the correlations in this region. This was in fact the reason for which I myself was rather reluctant to accept the parton cascade as an explanation.

HBT CORRELATIONS

The situation changed dramatically with the discovery of the close relation between the observed power law in momentum space at short distances and quantum interference between the identical bosons. The idea that HBT correlations may be at the origin of these short-range correlations was suggested almost immediately after the first experimental data were released [9]. It took several years, however, before the quality of the data was good enough to prove that, indeed, same-sign spectra show a much stronger effect that those of unlike-sign particles.

This observation implied a fundamental change in the way of thinking about the problem.

(i) It became clear that not the momentum space but the configuration space must be examined in order to understand the origin of self-similarity[3]. Perhaps even the combination of both: momentum space at large momentum scales and configuration space at small momentum scales.

(ii) It also removed the difficulty I mentioned in the previous section: in this picture the short-range correlations in momentum have nothing to do with parton cascade in momentum space but rather reflect the shape of the interaction region at large distances. Thus the problem of parton-hadron transition simply disappears.

All this led some people to the conclusion that "intermittency is explained by HBT correlations", the statement I find blatantly wrong. HBT correlations explained, indeed, the presence of the short-range correlations in momentum space. They are not sufficient, however, to explain the essence of intermittency, i.e. self-similarity, the power law behaviour of the correlation functions. To this end one must explain why the shape of the phase-space region in configuration space is such that it leads to the power-law correlations in momentum space.

This question was posed from the very beginning [11] and discussed more thoroughly later [13]. I will summarize briefly the results.

---

[1] Running coupling constant destroys the exact self-similarity.

[2] We do not observe partons but hadrons.

[3] This is a simplification: a residual effect is also present in the unlike-sign spectra [7, 10]. To my knowledge, it still not fully understood.

POWER LAW IN CONFIGURATION SPACE

First, it was fairly clear that to obtain power law in momentum space, a sort of power law behavior in configuration space is also needed. There is, however, a technical problem. The power law cannot hold in the full phase-space, however: such behaviour leads to non-normalized distributions. It is therefore necessary to introduce modifications which, in turn, introduce ambiguities.

In [12] I have tried a simple cut at large distances in configuration space. Here I will show a mathematically more elegant method which modifies the distribution at large distances by an exponential factor.

Consider the (two-dimensional, for simplicity) correlation function is momentum space of the form:

$$C(q^2) = \frac{1}{(1+q^2L^2/4)^{2\lambda}} \tag{1}$$

which describes the power law as long as $|q| \gg 1/L$ ($|q|$ is the momentum difference and L is a parameter). By choosing L large enough we can extend the power law arbitrarily close to the point $|q| = 0$.

According to the standard algorithm, the Fourier transform of $[C(q^2)]^{1/2}$ is the distribution $W(X)$ of particle sources in the configuration space. A simple algebra gives

$$W(X) = \frac{1}{\Gamma(\lambda)} \int_0^\infty \frac{dR^2}{L^2} \left(\frac{R^2}{L^2}\right)^{\lambda-1} e^{-R^2/L^2} \frac{e^{-X^2/R^2}}{\pi R^2} \tag{2}$$

which we recognize as a superposition of Gaussian sources with radii distributed according to the power law, modified at large distances by the factor e^{-R^2/L^2}. It turns out that the integration in (2) can be completed [14] with the result

$$W(X) = \frac{2}{\pi L^2 \Gamma(\lambda)} \left(\frac{|X|}{L}\right)^{\lambda-1} K_{\lambda-1}(2|X|/L) \tag{3}$$

In the limit $|X|/L \to 0$ we thus have

$$W(X) \sim \left(\frac{|X|^2}{L^2}\right)^{\lambda-1} \tag{4}$$

and in the limit $|X|/L \to \infty$

$$W(X) \sim \left(\frac{|X|}{L}\right)^{\lambda-3/2} e^{-2|X|/L} \tag{5}$$

This example illustrates, I think, the point: to obtain a power law in momentum space modified at small $|q|$, a power law in configuration space, modified at large X, is needed.

HIGHER ORDER CORRELATIONS: SELF-SIMILAR CASCADE IN CONFIGURATION SPACE

The formula (2), taken literally, suggests that the power law may be not necessarily present in individual events but is rather an effect of averaging over many events. To verify this possibility, we have investigated [13] higher order correlations in this framework. Generalizing (2) to multiparticle distributions

$$W(X_1,...,X_k) = \frac{1}{\Gamma(\lambda)} \int_0^\infty \frac{dR^2}{L^2} \left(\frac{R^2}{L^2}\right)^{\lambda-1} e^{-R^2/L^2} \frac{e^{-(X_1^2+...+X_k^2)/R^2}}{(\pi R^2)^k} \tag{6}$$

one obtains

$$C(q_{12},...,q_{k1}) = \frac{1}{\left[1+(q_{12}^2+...+q_{k1}^2)L^2/4\right]^{2\lambda}} \tag{7}$$

showing that the exponent in the power law is independent of k, in a blatant contradition with experiment [7].

We have also investigated the possibility of independent production, each particle having distribution described by (2). This leads to a linear dependence of intermittency exponents on k (i.e. simple fractal) [7], also in disagreement with data[4].

Therefore we finally concluded that the most natural interpretation is a self-similar cascade in configuration space leading to the fractal source of particles extending to rather large distances [13]. It is not clear if perturbative QCD can be used for evaluation of what really happens because, if L is very large as required by the data, the distances are indeed large. On the other hand, the cascade itself is a natural framework for colour interactions (since gluons carry colour) and therefore it seemed likely that an effort in this direction may be worth to try.

A step in this direction was undertaken by R. Peschanski and myself [16]. We have attempted to evaluate the transverse size of the dipole cascade introduced by Mueller [17] as a configuration space representation of the BFKL approximation. It was found that, in the central rapidity region, the emitted gluons fill a region in transverse space which has a power-law tail

$$W(X) \sim |X_\perp|^{\gamma_m - 2} \tag{8}$$

with $\gamma_M \approx 0.37$ being a solution of the equation $2\psi(1) - \psi(1-\gamma_m/2) - \psi(1+\gamma_M/2) = 8\log 2$. Therefore one expects the power law in the transverse momentum space

$$C(q_\perp) \sim \frac{1}{|q_\perp|^{2\gamma_M}}. \tag{9}$$

The measurements performed by NA22 coll. [18] showed indeed that, for $|Y| \leq 0.5$, data agree with the power law and $\gamma_M^{exp} = 0.4 \pm 0.04$, in good agreement with the theoretical expectation.

---

[4] This was definitely established later by the high-statistics data on factorial cumulants [15].

The paper [16] was criticized for not including virtual corrections in the calculation [19], but I still believe that the main result (8) is correct. It would be certainly interesting to perform a new calculation to verify this important point.

LEVY DISTRIBUTIONS

Recently, a new and very interesting idea was suggested [20]. The point is to use the Levy stable distribution in configuration space. The Levy distribution is a generalization of the Gaussian: similarly as a Gaussian, it remains stable under the convolution. This property means that after convolution of two such distributions, depending on two independent variables X_1 and X_2, the resulting one (which is a function of $X = X_1 + X_2$) is of the identical form, except of a possible shift and rescaling of X. It should be noted that this is a much stronger requirement than the well-known infinitely divisible distributions[5].

The first application of the Levy distributions to particle spectra was suggested by Brax and Peschanski [21] who applied it to momentum space and derived several relations for the intermittency exponents. The new idea [20], is that the configuration space is described by the stable distribution. The authors suggest that this may be a consequence of the QCD cascade being controlled by the renormalization group constraints. I find this a very stimulating suggestion, although I am not sure how it may actually work in practice. Nevertheless it is clear that this new road is certainly worth to pursue.

The Levy distributions are fairly complicated and most of them do not have an expression in a closed (analytical) form. However, their Fourier transforms (called characteristic functions in probability theory) are rather simple. Thus the prediction for the correlation function is indeed straightforward. In the simplest case one obtains

$$C(q) \sim e^{-(|q|R)^{\alpha}} \tag{10}$$

where α is a parameter. The authors of [20] found that the present data do not contradict this form. Moreover, using the ideas of the Lund model[6], they argue that the parameter α obtained from the fit to data of UA1 and NA22 coll. [22] is consistent with the expectations from QCD. This very exciting speculation demands, naturally, more work but one cannot deny that it opens a new way of thinking about the problem.

SUMMARY AND CONCLUSIONS

(i) It should be emphasized that the approximate self-similarity of the spectra is a *natural* consequence of QCD and thus should be actively pursued in the data.

---

[5] The well-know example of infinitely divisible distribution is the Γ distribution. One can easily see that it is not invariant under the convolution (although it always remains the Γ distribution): after many convolutions the Γ distribution approaches $\delta(X-1)$.

[6] Observing, in particular, that the Lund cascade can be interpreted as well in the momentum as in the cofiguration space !

(ii) It is likely that the origin of the approximate scaling of the *hadron* spectra at high resolution is the consequence of a self similar structure of the *parton* spectra at large distances in configuration space.

(iii) By the very nature of fractal structures, the investigation of scaling behaviour is more effective in a dilute system than in a dense one. Therefore the heavy ion collisions, being dominated by the large fixed scale, the nuclear radius, is not the best place to look for the effects we are discussing here. On the other hand, the pp data collected at RHIC may become the important source of new information, particularly as they are tuned to obtain good accuracy at very high resolution in momentum space.

(iv) Exciting new possibility of Levy stable distribution in configuration space gives a new boost to the search of scaling and its relation to fundamental theory.

ACKNOWLEDGMENTS

I would like to thank Hans Eggers and Wolfram Kittel for encouragement, Robi Peschanski for discussions, Edward Sarkisyan for correspondence and Vlada Simak for the kind hospitality at Kromeriz.

REFERENCES

1. A.Bialas, *Proc. XXVII ICHEP*, Glasgow (1994), p.1287. IOP Publishing, London (1995).
2. A.Bialas and R.Peschanski, *Nucl.Phys.* **B273** (1986) 703.
3. A.Bialas and R.Peschanski, *Nucl.Phys.* **B308** (1988) 847.
4. P.Lipa, P.Carruthers, H.Eggers, and B.Buschbeck, *Phys.Lett.* **B285** (1992) 300; H.Eggers, P.Lipa, P.Carruthers and B.Buschbeck, *Phys.Rev* **D48** (1993) 2040.
5. JACEE coll., T.H.Burnett et al., *Phys. Rev. Lett.* **50** (1983) 2062.
6. NA22 coll., M.Adamus et al., *Phys. Lett.* **B185** (1987) 200.
7. For a comprehensive review of the experimental data, see E.DeWolf, I.Dremin and W.Kittel, *Phys. Rep.* **270** (1996) 1.
8. B.Buschbeck, P.Lipa and R.Peschanski, *Phys. Lett.* **B215** (1988) 788; I.Sarcevicz and H.Satz, *Phys. Lett.* **B233** (1989) 251.
9. P.Carruthers et al., *Phys. Lett.* **B222** (1989) 487; M.Gyulassy, Festschrift L.Van Hove, p.479; Eds. A.Giovannini and W.Kittel (World Scientific, Singapore 1990).
10. OPAL coll., Abbiendi et al., *Eur. Phys. J.* **C11** (1999) 239.
11. A.Bialas, *Nucl. Phys.* **A525** (1991) 345c; **A545** (1992) 285c.
12. A.Bialas, *Acta Phys. Pol.* **B23** (1992) 561.
13. A.Bialas and B.Ziaja, *Acta Phys. Pol.* **B24** (1993) 1509.
14. I.S.Gradstein and I.M.Ryzhik, *Tablitsy Integralov, summ, riadov i proizvedenii*, p.354, Moskva (1962).
15. OPAL coll., G.Abbiendi et al., *Phys. Lett.* **B523** (2001) 35.
16. A.Bialas and R.Peschanski, *Phys. Letters* **B355** (1995) 301.
17. A.H.Mueller, *Nucl. Phys.* **B415** (1994) 471.
18. EHS/NA22 coll., N.M.Agababayan et al., *Phys. Lett.* **B393** (1997) 205.
19. J.Bjorken, private communication.
20. T.Csorgo, S.Hegyi and W.A.Zajc, *Eur. Phys. J.* **C36** (2004) 67; nucl-th/0402035 T.Csorgo, S.Hegyi, T.Novak and W.A.Zajc, *Acta Phys. Pol.* **B36** (2005) 329.
21. Ph.Brax and R.Peschanski, *Phys. Letters* **B253** (1991) 225.
22. N.Neumeister et al., *Z.Phys.* **C60** (633) 1993; N.M.Agababian et al., *Z.Phys.* **C59** (1993) 405;

Beyond the Gaussian Approximation (Experimental Review)

W. Kittel

Radboud University Nijmegen, Toernooiveld 1, 6525ED Nijmegen, The Netherlands

Abstract. Evidence for non-Gaussian behavior of the BE correlation function at small four-momentum difference is reviewed. Recent data lead to the class of Lévy stable distributions as the most likely candidate for a parametrization.

Keywords: Bose-Einstein correlations, power-law scaling, Lévy stable distributions
PACS: 12.38.Qk, 13.66.Bc, 13.85.Hd, 13.87.Fh

More important than the parameters extracted from "forcing" the two-particle correlation function into a fit by a pre-selected (often merely Gaussian) parametrization, is the actual experimentally observed shape of this distribution itself.

The simple geometrical interpretation of the interference pattern based on the optical analogy is invalid when emitters move relativistically with respect to each other, leading to strong correlations between the space-time and momentum-energy coordinates of emitted particles [1, 2, 3]. Correlations of this type e.g. arise due to the nature of inside-outside cascade dynamics [4] as e.g. in color-string fragmentation [5]. In the interpretation of BEC by Andersson and Hofmann [6] in the string model, the length scale measured by BEC is therefore not related to the size of the total pion emitting source, but to the space-time separation between production points for which the momentum distributions still overlap. This distance is, in turn, related to the string tension. The model predicts an approximately exponential shape of the correlation function with the "radius" parameter r independent of the total interaction energy.

Scale invariance implies that multiparticle correlation functions exhibit power-law behavior over a considerable range of the relevant relative distance measure (such as Q^2) in phase space [7, 8]. As such, BEC from a static source do not exhibit power-law behavior. However, a power law is obtained if the size of the particle source fluctuates event-by-event, and/or, if the source itself is a self-similar (fractal-like) object extending over a large volume [9]. In fact, is has been demonstrated [10] that already the interplay of resonance decays and some 50% directly produced pions can give such a power-law scaling at small Q, provided that the formation time of resonance and direct pions is short ($\tau_{\mathrm{f}} \approx 0.2$ fm/c).

The usually "reasonable" χ^2 values of the single-Gaussian fits hide the fact that the Gaussian parametrization in general undershoots the data at low values of Q^2. For the case of two-particle correlations, this has been clearly demonstrated by NA22 [11] and UA1 [12] (see Figs. 1), but deviations from a Gaussian are also observed in lepton-hadron [13, 14], e^+e^- [15], and even heavy-ion [16] collisions.

CP828, *Multiparticle Dynamics*
edited by V. Šimák, M. Šumbera, Š. Todorova, and B. Tomášik

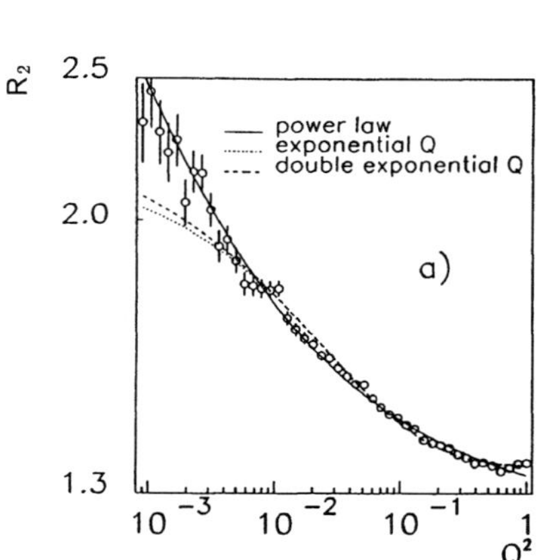

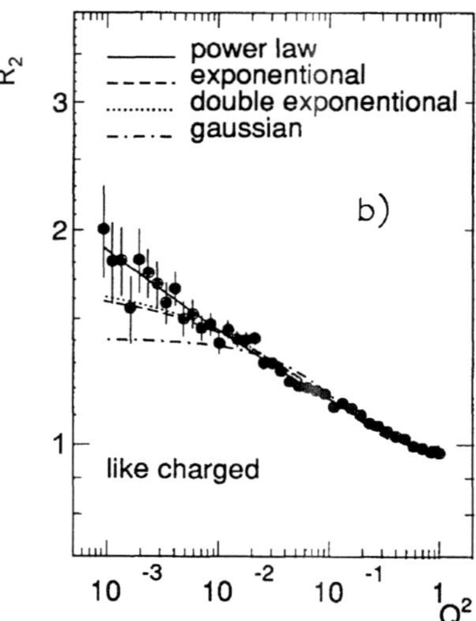

FIGURE 1. Bose-Einstein correlation function R_2 as a function of Q^2 for like-charged pairs in UA1 [12] and NA22 [11], compared to power-law, exponential, double-exponential and Gaussian fits, as indicated.

Deviations from a (multi-) Gaussian form are also observed for higher-order correlations. In Fig. 2a, the NA22 data [17] on BE correlations of order $q = 2$ to 4 are plotted as a function of Q^2 in conventional linear scale. The curves are the fits by a q-fold Gaussian parametrization [18]. In Fig. 2b the same data and the same fits are repeated for $Q^2 < 1$ GeV^2 on ln-ln scale. Even though the statistical errors at small Q^2 are large (the very reason why small Q^2 does not contribute much to χ^2), it is obvious that the small-Q^2 points *systematically* lie above the multi-Gaussian fit, thus supporting a power-law behavior. This effect is even enhanced when the data are corrected for Coulomb repulsion.

Fig. 3a shows the second-order cumulant $K_2 = R_2 - 1$ as a function of Q (on log-scale) [19] compared to a general quantum statistical model by Andreev, Plümer and Weiner (APW), based on a classical source current formalism applied successfully in quantum optics [20]. It includes as special cases more specific models such as [21] and [18]. The APW normalized cumulant predictions are built from normalized correlators d_{ij}, the on-shell Fourier transforms of classical space-time current correlators. The specific parametrizations tested in Fig. 3 are

$$\begin{aligned} \text{Gaussian}: &\quad d_{ij} = \exp(-r^2 Q_{ij}^2) \\ \text{exponential}: &\quad d_{ij} = \exp(-r Q_{ij}) \\ \text{power law}: &\quad d_{ij} = Q_{ij}^{-2\beta} \ . \end{aligned} \tag{1}$$

For constant chaoticity λ and real-valued currents, the APW model predicts

$$K_2(Q_{12}) = 2\lambda(1-\lambda)d_{12} + \lambda^2 d_{12}^2 \ , \tag{2}$$

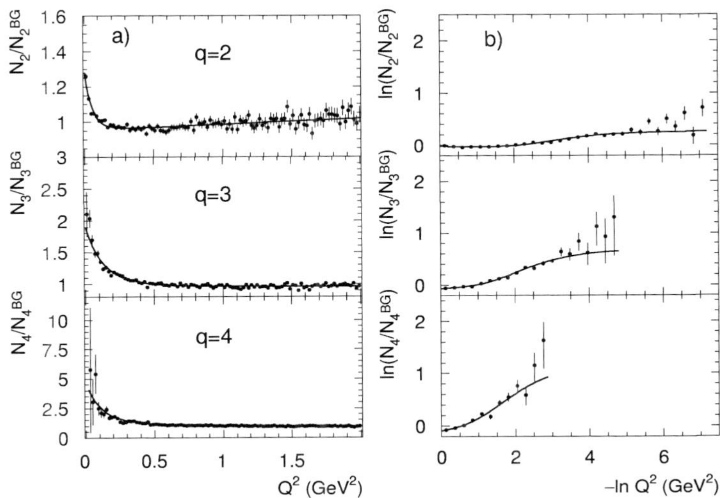

FIGURE 2. The two-, three- and four-particle correlation function as a function of Q^2 (left) and -lnQ^2 (right) [17]. Curves show the multi-Gaussian fits according to [18].

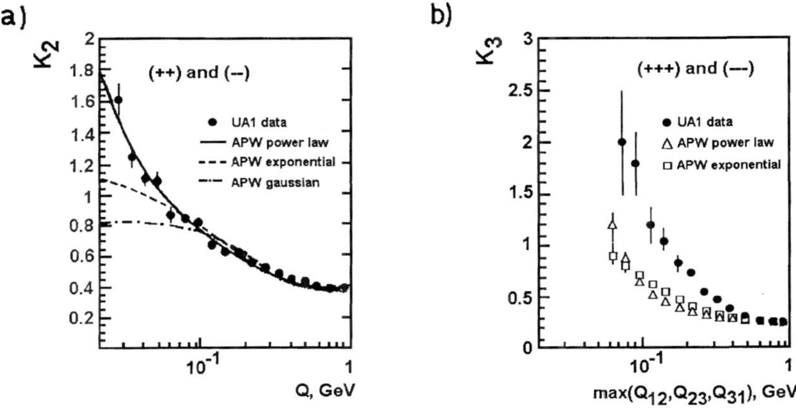

FIGURE 3. a) Second-order cumulant with fits of the forms given. b) Third-order cumulant with APW predictions based on K_2 [19].

$$K_3(Q_{12}, Q_{23}, Q_{31}) = 2\lambda^2(1-\lambda)[d_{12}d_{23} + d_{23}d_{31} + d_{31}d_{12}] + 2\lambda^3 d_{12}d_{23}d_{31} \ . \quad (3)$$

The data are best described by the power-law form of d_{ij} and clearly exclude a Gaussian parametrization. Also K_3, plotted in Fig. 3b, shows a power-law increase which is even

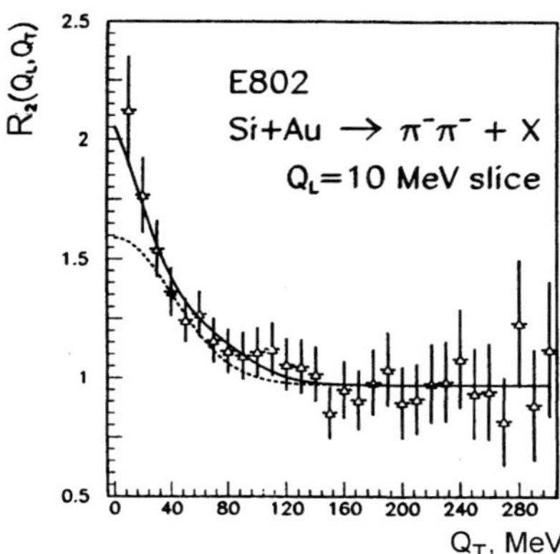

FIGURE 4. Projection onto Q_{T} of two-dimensional Gaussian (dashed) and Edgeworth (solid) fits for a small-Q_{L} slice [23].

stronger than expected in the APW model.

So, there is ample room for improvement of the models and we believe that the recently developed methods of studying the correlations (higher-order cumulants, higher dimensionality, alternative parametrizations of the correlation function) have opened the way for an improvement of our picture of BEC.

A purely mathematical extension of the usual Gaussian approximation of the BE correlation function is the Edgeworth expansion [22] as suggested in [23],

$$R_2(Q) = \gamma(1+\lambda^* \exp(-t^2/2)[1+\frac{\kappa_3}{3!}H_3(t)+\frac{\kappa_4}{4!}H_4(t)+\ldots]) \,, \tag{4}$$

with $t = \sqrt{2}Q \cdot r$, H_n being the n-th order Hermite polynomial, and κ_n the n-th order cumulant moment of the correlation function. The Hermite polynomials of odd order vanish at the origin, so that $\lambda = \lambda^*[1+\kappa_4/8+\ldots]$. Similarly, the exponential approximation can be extended by a Laguerre expansion [23]. A generalization to higher dimensions is straightforward [23], except for possible correlations between the Q_i components.

The influence of the non-Gaussian shapes was studied [23] on AFS [24, 25], E802 [26] and NA44 [27] data. In Fig. 4, the transverse component Q_{T} of a 2D Edgeworth fit is compared to that of a 2D Gaussian fit to the E802 data. The deviation from a Gaussian (dashed) is obvious, and the Edgeworth expansion (full line) is flexible enough to describe it (with $\lambda = 1$!). Furthermore, results more satisfactory than those obtained with either Gaussian or exponential parametrizations were obtained in a 3D analysis of e^+e^- collisions at the Z-mass [28], where the confidence level increased from 3 to 30%.

Fig. 5 shows that even an exponential is not steep enough to reproduce the fast increase of K_2 in hadron-hadron data. An interesting observation of [23] is, that a Laguerre expansion of an exponential can reproduce these UA1 and the NA22 data. However, at low Q^2 data are still systematically above the fit and a power-law fit is reported in [23] to give similarly good χ^2/NDF with a smaller number of fit parameters. This may imply a connection between BE correlations and the power-law scaling (intermittency) observed

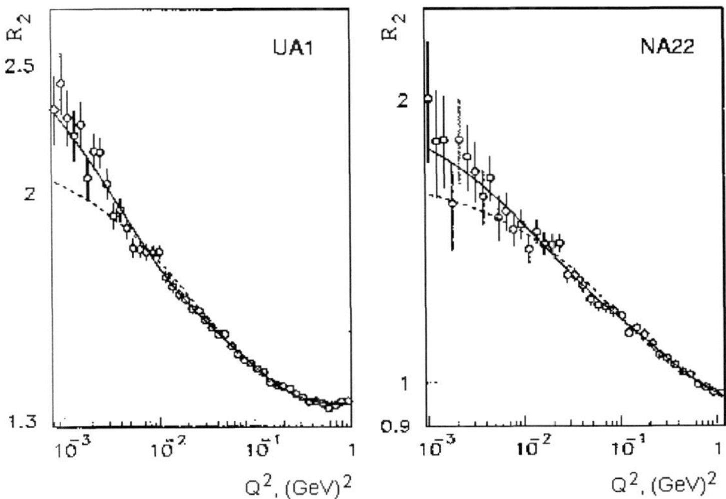

FIGURE 5. The figures show F_2^s which is proportional to the two-particle Bose-Einstein correlation function, as measured by the UA1 and the NA22 Collaborations. The dashed lines stand for the exponential fit, the solid lines for that with the Laguerre expansion [23].

for multiplicity fluctuations [9].

Functions $f(r)$ which behave asymptotically as

$$f(r) \to |r|^{-1-\mu} \quad \text{for} \quad |r| \to \infty \tag{5}$$

with non-finite variance are naturally described by Lévy-stable distributions (see [29] for earlier application to multiplicity fluctuations). A feature particularly useful for the purpose of BEC [30] is that the Fourier transform (characteristic function) of a symmetric ($\beta = 0$) stable distribution is

$$F(Q) = \exp(iQ\delta - |\gamma Q|^{\mu}), \tag{6}$$

from which follows that

$$R_2(Q) = 1 + \exp(-|rQ|^{\mu}) \tag{7}$$

with $r = 2^{1/\mu}\gamma$. Equation (7) is a generalization of the Gaussian ($\mu = 2$) or exponential ($\mu = 1$) distribution to $0 < \mu \leq 2$.

Fitting this form to the data of Fig. 5 gives Lévy indices $\mu = 0.49 \pm 0.01$ for UA1 and $\mu = 0.67 \pm 0.07$ for NA22 [30]. The quality of the fits is similar to that of the Laguerre expansion above. The advantage of the Lévy-stable fits is, however, that they relate the Lévy index μ to a power-law behavior of the configuration-space density fluctuations at large distance. The recent successes of this parametrization are given in the folowing talks [31, 32, 33, 34].

We conclude that Lévy stable distributions have the potential to
1. generalize the conventional simple-minded Gaussian approximation of the BE correlation function and
2. establish the connection to the approximate power-law scaling of multiplicity fluctuations (intermittency).
In our opinion, this allows for a desperately needed new approach to the interpretation and use of BE correlations.

REFERENCES

1. K. Kolehmainen and M. Gyulassy, *Phys. Lett.*, **B180**, 203 (1986).
2. A.N. Makhlin and Yu.M. Sinyukov, *Z. Phys.*, **C39**, 69 (1988); S.V. Akkelin and Yu.M. Sinyukov, *Phys. Lett.*, **B356**, 525 (1995).
3. M.G. Bowler, *Particle World*, **2**, 1 (1991).
4. J.D. Bjorken, *Phys. Rev.*, **D27**, 140 (1983).
5. X. Artru and G. Mennessier, *Nucl. Phys.*, **B70**, 93 (1974).
6. B. Andersson and W. Hofmann, *Phys. Lett.*, **B169**, 364 (1986).
7. A. Białas and R. Peschanski, *Nucl. Phys.*, **B273**, 703 (1986) and **B308**, 857 (1988).
8. W. Kittel and E.A. De Wolf, *Soft Multihadron Dynamics*, World Scientific, Singapore, 2005.
9. A. Białas, *Nucl. Phys.*, **A545**, 285c (1992); *Acta Phys. Pol.*, **B23**, 561 (1992).
10. J. Masarik, A. Nogová, J. Pišút, N. Pišútova, *Z. Phys.*, **C75**, 95 (1997).
11. NA22 Coll., N.M. Agababyan et al., *Z. Phys.*, **C59**, 405 (1993).
12. UA1 Coll., N. Neumeister et al., *Z. Phys.*, **C60**, 633 (1993); H.C. Eggers, B. Buschbeck and P. Lipa in Proc. XXVth Int. Symp. on Multiparticle Dynamics, eds. D. Bruncko et al. (World Scientific, Singapore, 1996), p.650
13. E665 Coll., M.R. Adams et al., *Phys. Lett.*, **B308**, 418 (1993).
14. H1 Coll., C. Adloff et al., *Z. Phys.*, **C75**, 437 (1997).
15. DELPHI Coll., P. Abreu et al., *Phys. Lett.*, **B286**, 201 (1992); *Z. Phys.*, **C63**, 17 (1994).
16. WA98 Coll., M.M. Aggarwal et al., *Eur. Phys. J.*, **C16**, 445 (2000).
17. NA22 Coll., N.M. Agababyan et al., *Z. Phys.*, **C68**, 229 (1995).
18. M. Biyajima et al., *Progr. Theor. Phys.*, **84**, 931 (1990); ibid., **88**, 157A (1992).
19. H.C. Eggers, P. Lipa and B. Buschbeck, *Phys. Rev. Lett.*, **79**, 197 (1997).
20. I.V. Andreev, M. Plümer and R.M. Weiner, *Int. J. Mod. Phys.*, **A8**, 4577 (1993).
21. H. Gyulassy, S. Kauffmann, L.W. Wilson, *Phys. Rev.*, **C20**, 2267 (1979)
22. M.G. Kendall and A. Stuart, The Advanced Theory of Statistics, Vol.1 (Ch. Griffin, London 1958)
23. S. Hegyi and T. Csörgő, Proc. Budapest Workshop on Relativistic Heavy Ion Collisions, eds. T. Csörgő et al. (KFKI-1993-11/A, Budapest, 1991) p.47;
T. Csörgő, Proc. Cracow Workshop on Multiparticle Production, eds. A. Białas et al. (World Scientific, Singapore, 1994) p.175; T. Csörgő, S. Hegyi, *Phys. Lett.*, **B489**, 15 (2000).
24. AFS Coll., T. Åkesson et al., *Z. Phys.*, **C36**, 517 (1987).
25. K. Kulka and B. Lörstad, *Z. Phys.*, **C45**, 581 (1990).
26. E802 Coll., T. Abbott et al., *Phys. Rev. Lett.*, **69**, 1030 (1992).
27. NA44 Coll., B. Lörstad, Proc. Budapest Workshop on Relativistic Heavy Ion Collisons, eds. T. Csörgő et al. (KFKI-1993-11/A, Budapest, 1993) p.36
28. L3 Coll., M. Acciarri et al., *Phys. Lett.*, **B458**, 517 (1999).
29. Ph. Brax and R. Peschanski, *Phys. Lett.*, **B253**, 225 (1991).
30. T. Csörgő, S. Hegyi, W.A. Zajc, *Eur. Phys. J.*, **C36**, 67 (2004)
31. T. Csörgő, these proceedings.
32. T. Novak, these proceedings.
33. M. Bysterský, these proceedings.
34. H. Eggers, these proceedings.

Bose-Einstein or HBT Correlation Signals of a Second Order QCD Phase Transition

T. Csörgő*, S. Hegyi*, T. Novák† and W. A. Zajc**

*MTA KFKI RMKI, H - 1525 Budapest 114, P.O.Box 49, Hungary
†University of Nijmegen, NL - 6525 ED Nijmegen, Toernooiveld 1
**Department of Physics, Columbia University, 538 W 120th Street, New York, NY 10027, USA

Abstract. For particles emerging from a second order QCD phase transition, we show that a recently introduced shape parameter of the Bose-Einstein correlation function, the Lévy index of stability equals to the correlation exponent - one of the critical exponents that characterize the behaviour of the matter in the vicinity of the second order phase transition point. Hence the shape of the Bose-Einstein / HBT correlation functions, when measured as a function of bombarding energy and centrality in various heavy ion reactions, can be utilized to locate experimentally the second order phase transition and the critical end point of the first order phase transition line in QCD.

INTRODUCTION

The study fractal phenomena was initiated in high energy particle and nuclear physics by Bialas and Peschanski in ref. [1], with the motivation of searching for a second order phase transition by studying intermittency or the the power-law behaviour of moments of the multiplicity distribution in narrowing bins of the momentum space, see refs. [2, 3] for excellent reviews on this topic.

The mathematical properties of Bose-Einstein correlation functions for Lévy stable sources were written up by three of us in refs. [4, 5], and are recapitulated in the next section. In ref. [6] we have added a physical interpretation and showed, that in case of jet physics, the fractal properties of QCD cascades can naturally be measured by the Lévy index of stability of the Bose-Einstein correlation functions. Our analytic results were similar in spirit to the numerical investigations of Wilk and collaborators in ref. [7]. Note that these correlations are frequently referred to as Hanbury Brown - Twiss or HBT correlations in the literature of heavy ion physics.

Bialas realized, that Bose-Einstein correlations and intermittency might be deeply connected [8], and considered a distribution of Gaussians where the radius parameter of the Gaussian has a power-law distribution, thus giving a way to the study fractals in coordinate space with the help of Bose-Einstein correlations. Brax and Peschanski were the first to introduce Lévy distributions, in momentum space, to multiparticle production in high energy physics [9]. They have suggested to use the measured value of the Lévy index of stability to signal quark gluon plasma production in heavy ion physics. Here we reconnect these seemingly different topics, and show how the excitation function of the shape parameter of the correlation function can be utilized to locate experimentally the critical end-point of QCD.

CP828, *Multiparticle Dynamics*
edited by V. Šimák, M. Šumbera, Š. Todorova, and B. Tomášik

BOSE-EINSTEIN CORRELATIONS & LÉVY STABLE SOURCES

The two-particle Bose-Einstein correlation function is defined with the help of the two-particle and single-particle invariant momentum distributions as:

$$C_2(\mathbf{k}_1, \mathbf{k}_2) = \frac{N_2(\mathbf{k}_1, \mathbf{k}_2)}{N_1(\mathbf{k}_1) N_1(\mathbf{k}_2)}. \tag{1}$$

If long-range correlations can be neglected or corrected for, and if the short-range correlations are dominated by Bose-Einstein correlations, this two-particle Bose-Einstein correlation function is related to the Fourier-transformed source distribution. For clarity, let us consider the case of a one-dimensional, factorized source,

$$S(x,k) = f(x)\, g(k). \tag{2}$$

In this case [4, 5], the Bose-Einstein correlation function is

$$C_2(k_1, k_2) = 1 + |\tilde{f}(q)|^2, \tag{3}$$

where the Fourier transformed source density (often referred to as the *characteristic function*) and the relative momentum are defined as

$$\tilde{f}(q) = \int \mathrm{d}x \exp(iqx)\, f(x), \qquad q = k_1 - k_2. \tag{4}$$

For the case of the jets decaying to jets to jets and so on, as well as at a second order phase transition, where fluctuations appear on all possible scales with a power-law tailed distribution, the final position of a particle is given by a large number of position shifts, hence the distribution of the final position x is obtained as a convolution,

$$x = \sum_{i=1}^{n} x_i, \qquad f(x) = \int \Pi_{i=1}^{n} \mathrm{d}x_i\, \Pi_{j=1}^{n} f_j(x_j)\, \delta(x - \sum_{k=1}^{n} x_k). \tag{5}$$

In the case of particle emission from QCD jets, that the fractal defining the particle emission is infrared stable: adding one more, very soft gluon does not change the resulting source distributions. A similar property holds for systems at a second order phase transition: the system becomes invariant under a renormalization group transformation. Bose-Einstein correlation functions for such particle emitting sources were evaluated recently by three of us, which we summarize here following refs. [4] and [5].

Various forms of the Central Limit Theorem state, that under certain conditions, the distribution of the sum of large number of random variables converges (for $n \longrightarrow \infty$) to a limit distribution. In case of "normal" elementary processes, that have finite means and variances, the limit distribution of their sum is a Gaussian. In case of random motion in a thermal medium, such position distribution corresponds to normal diffusion. However, near a second order phase transition point, fluctuations appear on all scales and the variance of the elementary process diverges, corresponding to the so-called anomalous diffusion. In this case, the system still may be invariant under convolution, and the shape of the limit distribution becomes independent from the number of elementary steps.

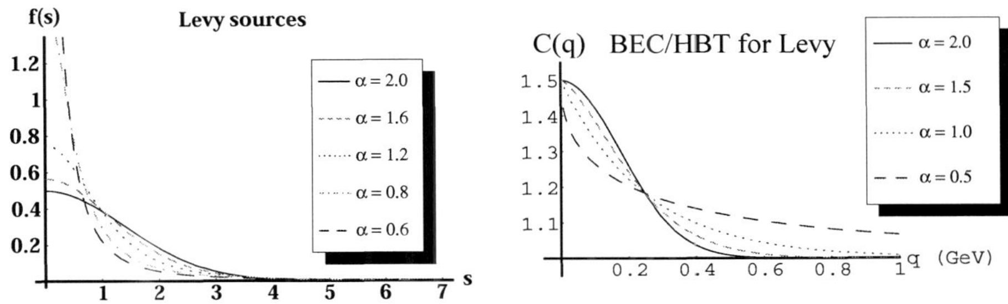

Figure 1. (left) Source functions for univariate symmetric Lévy laws, as a function of the dimensionless variable $s = r/R$, for various values of the Lévy index of stability, α. (right) Bose-Einstein correlation (or HBT) correlation functions for univariate symmetric Lévy laws, for a fixed scale parameter of $R = 0.8$ fm and various values of the Lévy index of stability, α.

Stable distributions are precisely those limit distributions that can occur in Generalized Central Limit theorems. Their study was begun by the mathematician Paul Lévy in the 1920's. The stable distributions can be given in terms of their characteristic functions, as the Fourier transform of a convolution is a product of the Fourier-transforms,

$$\tilde{f}(q) = \prod_{i=1}^{n} \tilde{f}_i(q) \tag{6}$$

and limit distributions appear when the convolution of one more elementary process does not change the shape of the limit distribution, but it results only in a modification of the parameters of the limit distribution. The characteristic function of univariate and symmetric stable distributions is

$$\tilde{f}(q) = \exp\left(iq\delta - |\gamma q|^{\alpha}\right), \tag{7}$$

where the support of the density function $f(x)$ is $(-\infty, \infty)$. Deep mathematical results imply that the index of stability, α, satisfies the inequality $0 < \alpha \leq 2$, so that the source distribution be always positive. These Lévy distributions are indeed stable under convolutions, in the sense of the following relations:

$$\tilde{f}_i(q) = \exp\left(iq\delta_i - |\gamma_i q|^{\alpha}\right), \qquad \prod_{i=1}^{n} \tilde{f}_i(q) = \exp\left(iq\delta - |\gamma q|^{\alpha}\right), \tag{8}$$

$$\gamma^{\alpha} = \sum_{i=1}^{n} \gamma_i^{\alpha}, \qquad \delta = \sum_{i=1}^{n} \delta_i. \tag{9}$$

Thus the Bose-Einstein correlation functions for uni-variate, symmetric stable distributions (after a core-halo correction, and a re-scaling) read as

$$C(q;\alpha) = 1 + \lambda \exp\left(-|qR|^{\alpha}\right). \tag{10}$$

Refs. [4] and [5] discuss further examples and details and generalize these results to three dimensional, hydrodynamically expanding, core-halo type sources, as well as to three-particle correlations.

THE ANOMALOUS DIMENSION OF QCD JETS AND BE/HBT

In QCD, jets emit jets that emit additional jets and so on. The resulting fractal structure of QCD jets was related to intermittency and power-law dependence of multiplicity moments on the bin-size in momentum space with the help of a beautiful geometric interpretation of the color picture in refs. [10, 11, 12], and an infrared stable measure on the parton states, related to the hadronic multiplicity distribution. These ideas were developed further in refs. [13, 14, 15], [16] as well as in refs. [17, 18].

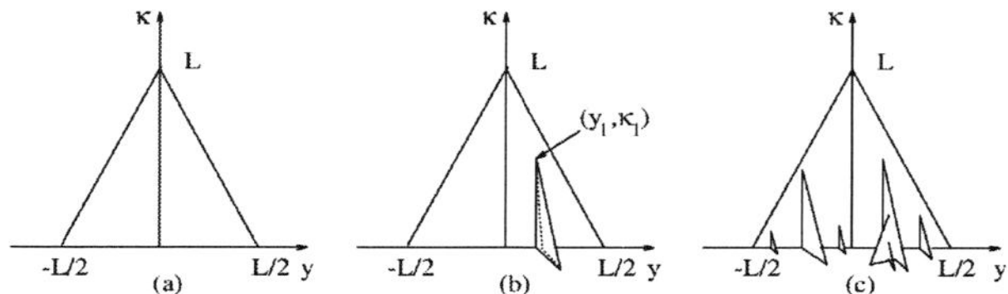

Figure 2. The phase-space of QCD jets in the (y, κ) plane, where $\kappa = \log(k_t^2)$. (a) The phase space available for a gluon emitted by a high energy $q\bar{q}$ system is a triangular region in the (y, κ) plane. (b) If one gluon is emitted at (y_1, κ_1), the phase space for a second (softer) gluon is given by the area of this folded surface. (c) The total gluonic phase space can be described by this multifaceted fractal surface [10, 11, 12].

A high energy $q\bar{q}$ system radiates gluons according to the dipole formula

$$dn = \frac{3\alpha_s}{4\pi^2} \frac{dk_\perp^2}{k_\perp^2} dy d\phi, \qquad (11)$$

hence the phase-space for the emission of a gluon is given by the relation

$$|y| \leq \frac{1}{2} \ln(s/k_\perp^2), \qquad (12)$$

which corresponds to the triangular region in a $(y, \ln k_\perp^2)$ diagram as shown in Fig. 1 (a). If two gluons are emitted, then the distribution of the hardest gluon is described by eq. (1). The distribution of the second, softer, gluon corresponds to two dipoles, the first is stretched between the quark and the first gluon, and the second between the first gluon and the anti-quark. The phase-space available for the second gluon corresponds to the folded surface in Fig. 1 (b), with the constraint $k_{\perp,2}^2 < k_{\perp,1}^2$, as the first gluon is assumed to be the hardest one. This procedure can be generalized so that the emission of a third, still softer gluon corresponds to radiation from three color dipoles, with n gluons emitted already the emission of the $n+1$-th gluon is given by a chain of $n+1$ dipoles. Thus, with many gluons, the gluonic phase space can be represented by a multi-faceted surface as illustrated in Fig. 1 (c). Each gluon adds a fold to the surface, which increases the phase-space for softer gluons. (Note, that in this process the recoils are neglected, as is normal in leading log approximation). Due to its iterative nature, the process generates a Koch-type fractal curve at the base-line. The length of this base-line of the partonic structure on Figure 1.c is proportional to the particle multiplicity. This curve is longer, when studied with higher resolution: it is a fractal curve, embedded into the four-dimensional energy-momentum space, characterized by the fractal dimension

$$d_f = 1 + \sqrt{\frac{3\alpha_s}{2\pi}}, \tag{13}$$

or one plus the anomalous dimension of QCD [10, 11, 12]. With the help of the Lund string fragmentation picture, this fractal in momentum space is mapped into a fractal in coordinate space, and the constant of conversion is the hadronic string tension, $\kappa \approx 1$ GeV/fm. This mapping does not change the fractal properties of the curve.

A walk, where the length of the steps is given by a Lévy distribution, and the direction of the steps is random, corresponds to a fractal curve, in physical terms it can be interpreted as the path of a test particle performing a generalized Brownian motion. This motion is referred to as anomalous diffusion and the probability that the test particle diffuses to distances r greater than a certain value of $|s|$ is given by $P(r > |s|) \propto |s|^{-\alpha}$. This relation is valid for anomalous diffusion in any dimensions. Thus the Lévy index of stability α is the fractal dimension of the trajectory of the corresponding anomalous diffusion [19]. When we apply this result to QCD, there are two key considerations.

First, if gluon radiation is neglected, the $q\overline{q}$ system hadronizes as a 1+1 dimensional hadronic string, which has no fractal structure. If the gluon emission is switched on, the emission of gluon n from one of the n dipoles corresponds to a step of an anomalous diffusion in the plane transverse to the given dipole. Hence the anomalous dimension of QCD equals to the Lévy index of stability of this anomalous diffusion,

$$\sqrt{\frac{3\alpha_s}{2\pi}} = \alpha_{\text{Lévy}}. \tag{14}$$

Second, data on Bose-Einstein correlations are often determined in terms of the invariant momentum difference $Q_{\text{inv}} = \sqrt{-(p_1 - p_2)^2}$. Bose-Einstein correlation functions that depend on this invariant momentum difference can be obtained within the framework of the so-called τ-model. This model assumes a broad proper-time distribution, $H(\tau)$ and very strong correlations between coordinate and momentum in all directions, $x^\mu/\tau \propto p^\mu/m_{(t)}$. Hence $(x_1 - x_2)(p_1 - p_2) \propto \tau Q_{\text{inv}}^2$, see refs. [20, 21] for details. In this case, the Bose-Einstein correlation function measures the Fourier-transformed proper-time distribution $\tilde{H}$ in the following, unusual manner:

$$C_2(Q_{\text{inv}}) \simeq 1 + \lambda Re\tilde{H}^2\left(\frac{Q_{\text{inv}}^2}{2m_{(t)}}\right), \tag{15}$$

where $m_{(t)}$ stands for the (transverse) mass of the pair for (two)- or more jet events. From this relation it follows, that $\alpha_{\text{BEC}} = 2\alpha_{\text{Lévy}}$. Thus we find the following relationship between the running QCD coupling constant α_s and the exponent of an invariant relative momentum dependent Bose-Einstein correlation function α_{BEC}:

$$\alpha_s = \frac{\pi}{6}\alpha_{\text{BEC}}^2. \tag{16}$$

In ref. [6] we have compared this leading log result to NA22 and UA1 correlation data of refs. [22, 23] and found a reasonable agreement with these data.

BOSE-EINSTEIN CORRELATIONS AT A SECOND ORDER QCD PHASE TRANSITION

The main motivation behind the experimental and theoretical programme of high energy physics is to study the phase diagram of hot and dense hadronic matter. According to recent lattice QCD calculations at finite temperature and baryon density, there exist a line of first order phase transitions that separates the hadronic and the quark-gluon plasma (QGP) state. This line of the first order phase transitions ends at the critical end-point (CEP), where the transition from hadron gas to QGP becomes a second order phase transition. Recent lattice QCD calculations located [24] this CEP at $T_E = 162 \pm 2$ MeV and $\mu_E = 360 \pm 40$ MeV. Below these baryochemical values, the transition from a hadron gas to a QGP becomes a cross-over, and at vanishing net baryon density the critical temperature becomes $T_c = 164 \pm 2$ MeV (the errors are statistical only). In this calculation, the quark masses were already at the physical value, but the continuum extrapolation was missing. S. Katz presented improvements at the Quark Matter 2005 conference [25], using physical quark masses and working towards the continuum extrapolation. He reported $T_c = 189 \pm 8$ MeV for the critical temperature at $\mu_B = 0$.

At the CEP, the second order phase transition is characterized by the fixed point of the renormalization group transformations. In a quark-gluon plasma, the vacuum expectation value of the quark condensate $c = \langle \overline{q} q \rangle$ vanishes, while in the hadronic phase, this vacuum expectation value becomes non-zero. The correlation function of the order parameter is defined as $\rho(R) = \langle c(r+R)c(r) \rangle - \langle c \rangle^2$ and measures the spatial correlation between the pions. At the CEP, this correlation function decays as

$$\rho(R) \propto R^{-(d-2+\eta)}, \tag{17}$$

a power-law. The parameter η is called as the exponent of the correlation function.

For Lévy stable sources, corresponding to an anomalous diffusion with large fluctuations in coordinate space, the correlation between the initial and actual positions decays also as a power-law, where the exponent is given by the Lévy index of stability α as

$$\rho(R) \propto R^{-(1+\alpha)}. \tag{18}$$

As we are considering a QCD phase transition in a $d = 3$ three-dimensional coordinate space we find that the correlation exponent equals to the Lévy index of stability, $\alpha = \eta$. Stephanov, Rajagopal and Shuryak pointed out [26], that the universality class of the second order QCD phase transition is that of the 3d Ising model. For this universality class, the correlation exponent has been determined by Rieger [27] as

$$\alpha(\text{Lévy}) = \eta(\text{3d Ising}) = 0.50 \pm 0.05. \tag{19}$$

Fig. 1 indicates that the change in the shape of the correlation function is rather significant, if α decreases from its Gaussian value of 2 to 0.5, its characteristic value at the 2nd order QCD phase transition point. Fig. 3 illustrates how this shape parameter of the Bose-Einstein correlation function may depend on the relative temperature near the critical point. Hence studying the *shape* parameter of the two-particle Bose-Einstein or HBT correlation functions as a function of the bombarding energy or the centrality of the heavy ion collisions, a previously unknown tool is obtained to determine if the pions are emitted from the neighbourhood of the critical end point of the QCD phase diagram.

Furthermore, based on an universality class argument, we have determined that the second order QCD phase transition at the critical end point will be signalled with the value of $\alpha = 0.5$, a very spiky Bose-Einstein correlation function indeed.

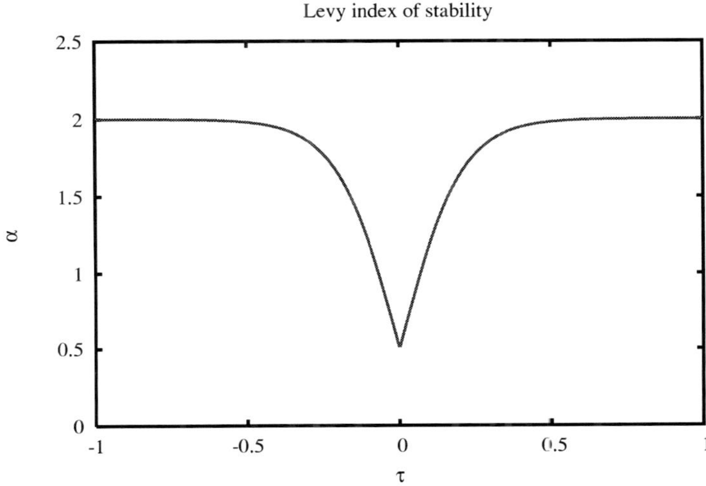

Figure 3. Illustration of the behaviour of the Lévy index of stability of Bose-Einstein correlations as a function of the dimensionless temperature variable $\tau = (T - T_c)/T_c$ in the neighbourhood of th critical endpoint of the 1st order phase transition line in QCD. At the critical endpoint, the phase transition becomes 2nd order and the Lévy index of stability decreases to the correlation exponent of QCD. As this transition has the same universality class as that of the 3d Ising model, one expects a decrease from the $\alpha \approx 2$ values that are characteristic to a Boltzmann gas and normal diffusion to $\alpha = 0.5$, corresponding to the correlation exponent of QCD at the critical endpoint. As shown in Fig. 1, such a change in the shape parameter makes the Bose-Einstein correlation functions much sharper than a simple Gaussian, so the spiky structure of the correlation function could be used to search for this point experimentally.

CONCLUSIONS

We have recapitulated earlier results that indicate, that the general shape of the Bose-Einstein or HBT correlation functions is a streched exponential or Lévy stable form, where the Lévy index of stability becomes a new shape parameter of the correlation function with $0 < \alpha \leq 2$ and the popular Gaussian parameterization corresponds to the $\alpha = 2$ particular, special case. Then we have studied two physically interesting examples.

In case of particle emission from jets, we have recapitulated the connection between the stability index of the Bose-Einstein/HBT correlation functions and the running coupling constant of QCD.

We have also considered a scenario, when the power-law tail of a Lévy distribution of the particle emission *in the coordinate space* appears due to a second-order QCD phase transition. In this case, the Lévy index of stability of the Bose-Einstein or HBT correlation function was shown to be equal to the correlation exponent of QCD. This value is known to be 0.5 ± 0.05 from universality class considerations. Hence by measuring the excitation function of the Lévy index of stability (the shape parameter of the two-particle Bose-Einstein or HBT correlation functions), one can experimentally determine the bombarding energy and centrality range where a heavy ion collision hits the critical end point of QCD. Clearly, more work is necessary to check to what extent this interesting effect can be masked by the decays of various resonances, hydrodynamic expansion, and by the time evolution of the particle emitting source between the second order phase transition point and the freeze-out temperature.

ACKNOWLEDGMENTS

It is our pleasure to acknowledge the inspiring discussions with A. Bialas, R. Glauber, W. Kittel and R. Peschanski. This research was supported by the NATO Collaborative Linkage Grant PST.CLG.980086, by the Hungarian - US MTA OTKA NSF grant INT0089462 and by the OTKA grants T038406 and T049466.

REFERENCES

1. A. Bialas and R. Peschanski, *Nucl. Phys. B* **273**, 703-718 (1986).
2. A. Bialas, *Nucl. Phys. A* **525**, 345-360 (1991).
3. E. A. De Wolf, I. M. Dremin and W. Kittel, *Phys. Rept.* **270**, 1-141 (1996).
4. T. Csörgő, S. Hegyi and W. A. Zajc, *Eur. Phys. J. C* **36**, 67-78 (2004).
5. T. Csörgő, S. Hegyi and W. A. Zajc, `nucl-th/0402035`.
6. T. Csörgő, S. Hegyi, T. Novák and W. A. Zajc, *Acta Phys. Polon. B* **36**, 329-338 (2005).
7. O. V. Utyuzh, G. Wilk and Z. Wlodarczyk, *Phys. Rev. D* **61**, 034007 (2000).
8. A. Bialas, *Acta Phys. Polon. B* **23**, 561-567 (1992).
9. P. Brax and R. Peschanski, *Phys. Lett. B* **253**, 225-230 (1991).
10. P. Dahlqvist, B. Andersson and G. Gustafson, *Nucl. Phys. B* **328**, 76-84 (1989).
11. G. Gustafson and A. Nilsson, *Nucl. Phys. B* **355**, 106-122 (1991).
12. G. Gustafson, *Proc Int. Workshop on Correlations and Multiparticle Production, Marburg, West Germany, 1990* (World Scientific, Singapore, 1991, eds. M. Plümer, S. Raha and R. M. Weiner)
13. G. Gustafson and A. Nilsson, *Z. Phys. C* **52**, 533-542 (1991).
14. G. Gustafson, *Nucl. Phys. B* **392**, 251-278 (1993).
15. B. Andersson, G. Gustafson, J. Samuelsson, *Nucl. Phys. B* **463**, 217-237 (1996).
16. Y. L. Dokshitzer and I. M. Dremin, *Nucl. Phys. B* **402**, 139-165 (1993).
17. W. Ochs and J. Wosiek, *Phys. Lett. B* **305**, 144-150 (1993).
18. W. Ochs and J. Wosiek, *Z. Phys. C* **68**, 269-196 (1995).
19. V. Seshardi, B. J. West, *Proc. Nat. Acad. Sci. USA* **79**, 4501-4505 (1982).
20. T. Csörgő and J. Zimányi, *Nucl. Phys. A* **517**, 588-598 (1990).
21. T. Csörgő, `hep-ph/0301164`.
22. N. M. Agababian *et al.* [EHS/NA22 Coll.], *Z. Phys. C* **59**, 405-426 (1993).
23. N. Neumeister *et al.* [UA1-Min. Bias-Coll.], *Z. Phys. C* **60**, 633-642 (1993).
24. Z. Fodor and S. D. Katz, *JHEP* **0404**, (2004) 050.
25. S. D. Katz, `hep-ph/0511166`. `http://www.kfki.hu/events/hun/qm2005/plenary/Katz/0930_aug5GlobeKatzt.pdf`
26. M. A. Stephanov, K. Rajagopal and E. V. Shuryak, *Phys. Rev. Lett.* **81**, 4816-4819 (1998).
27. H. Rieger *Phys. Rev. B.* **52**, 6659-6667 (1995).

Non-Gaussian Effects in Identical Pion Correlation Function at STAR

M. Bysterský (for the STAR Collaboration)

Nuclear Physics Institute, Academy of Sciences of the Czech Republic, 250 68 Řež near Prague, Czech Republic

Abstract. Preliminary femtoscopic results on identical pions from high statistics data set of Au+Au collisions at $\sqrt{s_{NN}} = 200$ GeV taken during the fourth RHIC run are presented. The measured three-dimensional correlation function is studied at low relative momenta using the Gaussian parametrization and the Lévy stable parametrization. The latter is expected to better describe the data. As the results show, both parametrizations underestimate the peak of the measured correlation function equally.

Keywords: STAR, Femtoscopy, Correlation function, Non-Gaussian, Lévy
PACS: 25.75.-q

INTRODUCTION

Motivation for this study is to check whether a recently proposed parametrization of the correlation function using Lévy stable source distribution [1] brings a significant improvement over the standard Gaussian fit [2]. In addition to this, the non-Gaussian source distribution function is also used in most of the models, but the standard method of fitting experimental correlation function assumes a Gaussian source [2].

Possible methods of studying the non-Gaussian effects in the experimental correlation function include Edgeworth expansion [3] and Lévy stable source distribution parametrization [1].

CORRELATION FUNCTION OF IDENTICAL PIONS

Event and particle selection criteria. We briefly list the values of the event and particle selection criteria used in the present analysis. Detailed description can be found in [4]. Events are binned by centrality in five bins corresponding to 0–5%, 5–10%, 10–20%, 20–30% and 30–80% of the total hadronic Au+Au cross-section. In addition to the track cuts listed in [4], measured specific ionization of pions is required to be farther than $\pm 2\sigma$ from the Bethe-Bloch theoretical value for electrons. This cut removes a contamination due to conversion electrons in the low momentum region. Pairs of identical pions are binned by average transverse pair momentum $k_T = \frac{1}{2}|\vec{p}_{T1} + \vec{p}_{T2}|$ in four bins corresponding to $k_T \in (0.15\text{–}0.25, 0.25\text{–}0.35, 0.35\text{–}0.45, 0.45\text{–}0.60)$ GeV/c and the results are presented as a function of the average k_T in each of these bins.

CP828, *Multiparticle Dynamics*
edited by V. Šimák, M. Šumbera, Š. Todorova, and B. Tomášik

Experimentally, two-particle correlations are studied by constructing the correlation function as a ratio

$$C(\vec{q}) = \frac{A(\vec{q})}{B(\vec{q})}, \tag{1}$$

where $A(\vec{q})$ is the measured distribution of the momentum difference $\vec{q} = \vec{p}_1 - \vec{p}_2$ for pairs of particles from the same event and $B(\vec{q})$ is the corresponding reference distribution for pairs of particles from different events belonging to the same event class as analyzed event [2]. The two-particle correlation function at low relative momentum $\vec{q}$ of identical pion pairs is studied using Bertsch-Pratt parametrization [5, 6, 7] in the longitudinal co-moving system (LCMS) frame, where the relative momentum vector is decomposed into the out, side and long components $\vec{q} = (q_{\text{o}}, q_{\text{s}}, q_{\text{l}})$.

Gaussian parametrization

Standard method of fitting the two-pion correlation function assumes the Gaussian source distribution. The correlation function is usually parametrized by a three-dimensional Gaussian in $\vec{q}$ [2]. Taking into account a repulsive Coulomb interaction between charged identical pions, the measured correlation function (1) is fitted using Bowler-Sinyukov procedure [8, 9],

$$C(\vec{q}) = (1-\lambda) + \lambda K_{\text{c}} \left[1 + \exp\left(-\sum_{\text{i,j}} R_{\text{ij}}^2 q_{\text{i}} q_{\text{j}} \right) \right]. \tag{2}$$

Here, the correlation strength λ equals the fraction of pairs originating in the same spatio-temporal region relevant for Bose-Einstein correlations, K_{c} is the squared Coulomb wave-function integrated over the source with radius 5 fm [4]. R_{ij} are the Gaussian source radius parameters defined as the widths of the source emission function. Let us note, that only the pairs obeying Bose-Einstein statistics are considered to Coulomb interact. For an azimuthally integrated analysis the cross-terms vanish [2], $R_{\text{ij}} = 0$, $\text{i} \neq \text{j}$.

Non-Gaussian parametrization

Detailed analysis of the shape of the correlation function is important because it carries information about the space-time structure of the particle emitting source [1, 2]. Deviations from Gaussian shape can be studied using Edgeworth expansion or Lévy stable source distribution.

Edgeworth expansion [3] provides model-independent approach for an analysis of the shape of the correlation function. In our previous pion interferometry analysis [4] it was shown that the Edgeworth expansion, based on an experimentally preferred Gaussian weight function and a complete orthogonal set of even order Hermite polynomials, up to 6th order is sufficient to describe the data. However, physical interpretation of the higher order (4th, 6th) fit parameters is not clear.

Lévy stable source parametrization

To study possible deviations of the correlation function from a Gaussian shape, we followed the method suggested in [1]. This formalism is relevant for the femtoscopic studies of the expanding systems created in heavy-ion collisions, where the scale of the fluctuations may be characterized by long tails and asymptotic power-law like behavior. The probability distribution of particle emission points then corresponds to a Lévy stable distribution [1]. Using the Bowler-Sinyukov procedure to determine the repulsive Coulomb interaction between identical pions, the two-particle correlation function is then characterized by a stretched exponential parametrization,

$$C(\vec{q}) = (1-\lambda) + \lambda K_{\mathrm{c}} \left[1 + \exp\left(- \left(\sum_{\mathrm{i,j}} R_{\mathrm{ij}}^2 q_{\mathrm{i}} q_{\mathrm{j}} \right)^{\alpha/2} \right) \right]. \tag{3}$$

The additional parameter α, when compared to (2), is the Lévy index of stability, $0 < \alpha \leq 2$. For $\alpha = 2$ the Gaussian form (2) is recovered, while for $\alpha < 2$ the correlation function becomes more peaked than a Gaussian and develops longer tails. For the azimuthally integrated analysis, $R_{\mathrm{ij}} = 0$, $\mathrm{i} \neq \mathrm{j}$.

DISCUSSION OF RESULTS

Here we present results on two-particle correlations of charged identical pions in Au+Au collisions at $\sqrt{s_{\mathrm{NN}}} = 200$ GeV measured in the STAR detector during the fourth RHIC run (2003–2004). Data set of 11×10^6 minimum-bias triggered events is used for this analysis.

The interferometric parameters, correlation strength λ and radii R_{o}, R_{s} and R_{l}, are obtained by fitting the measured correlation function (1) with the Gaussian parametrization (2). Interferometric radius parameters measure the sizes of the homogeneity regions, regions from where the particles are emitted with the same average pair momentum k_{T}, and their k_{T} dependence contains dynamical information of the pion emitting source [2].

Figure 1 shows STAR preliminary results on interferometric parameters as functions of k_{T} for five centrality bins, where the three-dimensional experimental correlation function is fitted using the Gaussian parametrization of the correlation function (2). Results of the analysis of higher statistics data set of Au+Au collisions at $\sqrt{s_{\mathrm{NN}}} =$ 200 GeV are compared with the previously analyzed STAR data [4] for the same system at the same energy taken during the second RHIC run (2001–2002). It can be seen that the extracted interferometric radii R_{o}, R_{s} and R_{l} are consistent within errors with the previous analysis. A small systematic shift in radii can be attributed to the momentum resolution correction which is not included in this analysis. Significant difference in λ is observed in the lowest k_{T} bin. This is explained by an improved purity of the pion sample, where the additional cut on particle specific ionization removes contamination from the conversion electrons in the low momentum region.

The Gaussian parametrized fit (2) and the Lévy stable source parametrized fit (3) to the measured correlation function, each subtracted from the measured three-dimensional correlation function, are projected in the out, side and long coordinates and compared in Figure 2. Correlation function is shown for the k_T bin 0.25–0.35 GeV$/c$, for the most central collisions and the projections are constrained by the unprojected variables $q_o, q_s, q_l < 30$ MeV$/c$. It can be seen that contrary to [1], Lévy stable source distribution parametrization (3) does not fit the three-dimensional experimental correlation function significantly better than the standard Gaussian parametrization (2). Both parametrizations equally underestimate the peak value of the measured correlation function for relative momenta $q_o, q_s, q_l < 20$ MeV$/c$ and underestimate the tail of the correlation function. Both effects are mostly visible in the long projection in Figure 2.

In Figure 3 the interferometric parameters obtained from the Gaussian parametrization (2) are compared to the parameters obtained using Lévy parametrization (3) of the measured three-dimensional correlation function. Interferometric parameters are shown as functions of k_T for the most central and peripheral collisions. The Lévy fit returns significantly larger values of the fit parameters R_o, R_s, R_l and λ. However, these parameters can not be directly compared to the Gaussian ones representing the interferometric radii, because the non-Gaussian parameters do not satisfy the definition of being the widths of the source emission function. The large values of the non-Gaussian fit parameters are strongly anti-correlated with rather low value of the Lévy index of stability α, which stays between $1.2 < \alpha < 1.5$.

It can be seen that the Lévy stable source distribution parametrization (3) does not bring an advantage in describing the detail shape of the measured correlation function, nor in the number of the fitting parameters. Therefore use of the standard Gaussian parametrization (2) is sufficient and preferred.

SUMMARY

The preliminary results on identical pion interferometry using high statistics sample of Au+Au collisions at $\sqrt{s_{NN}} = 200$ GeV from the STAR experiment at RHIC have been presented.

The results of the Gaussian fit to the measured three-dimensional correlation function are consistent within errors with the previously analyzed STAR data.

It has been shown that in the low relative momentum region the Lévy stable source parametrization does not fit the experimental correlation function significantly better when compared to standard Gaussian parametrization.

Edgeworth expansion provides the detailed fit to the measured correlation function [4], but the interpretation of the higher order fit parameters is not clear.

It seems that to represent the experimental correlation function in Au+Au collisions at RHIC, the Gaussian parametrization is sufficient.

ACKNOWLEDGMENTS

This work is supported by the IRP AV0Z10480505 and the Grant Agency of the Czech Republic under contract 202/04/0793.

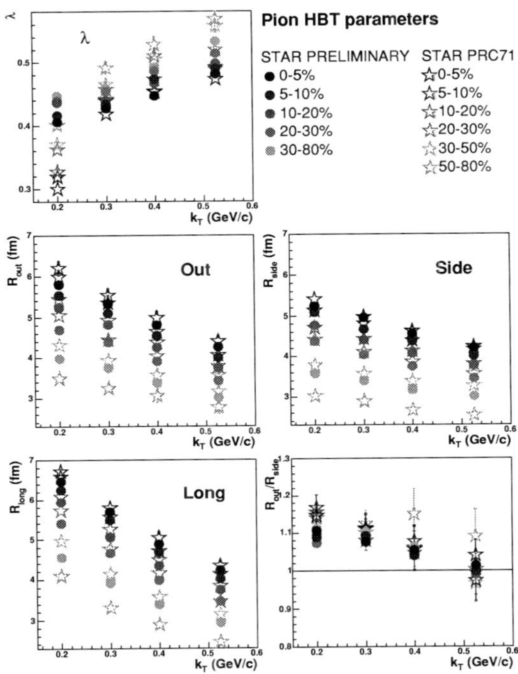

FIGURE 1. Interferometric parameters as functions of k_T and centrality. Results of the Gaussian parametrized fit (2) to the experimental correlation function, Au+Au collisions at $\sqrt{s_{NN}} = 200$ GeV.

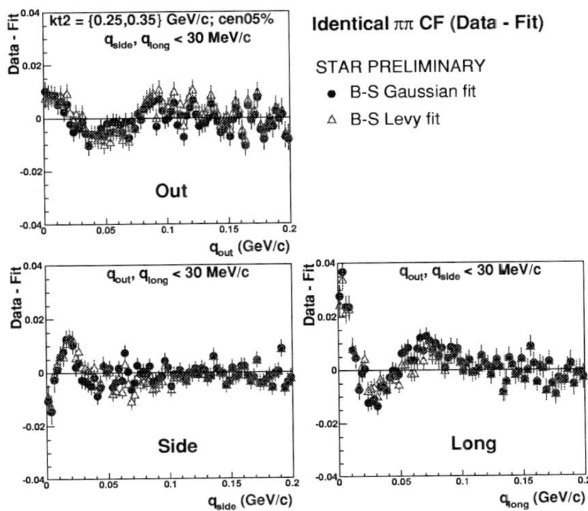

FIGURE 2. Gaussian parametrized fit (2) to data compared to the Lévy stable source parametrized fit (3) to data, each subtracted from the experimental correlation function.

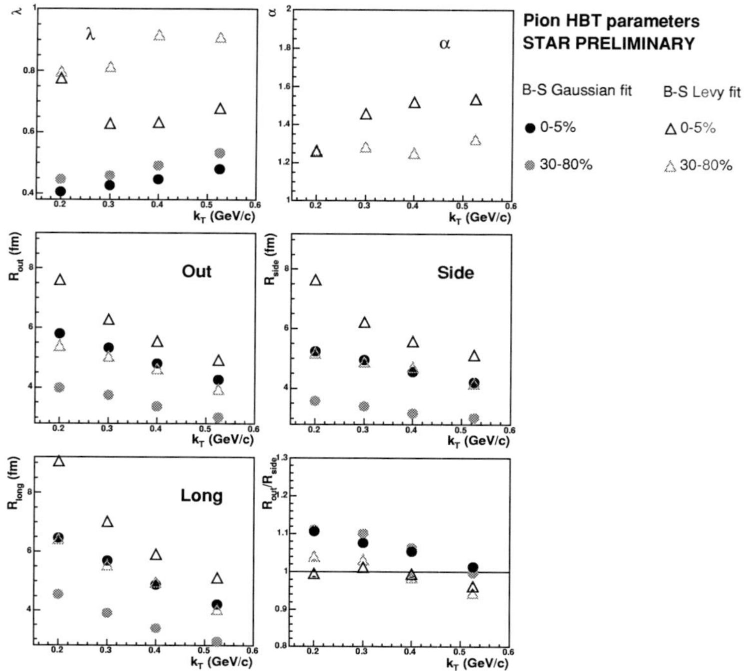

FIGURE 3. Results on interferometric parameters as functions of k_T and centrality obtained using Gaussian parametrized fit (2) compared to the parameters of Lévy stable source parametrized fit (3) of the measured correlation function.

REFERENCES

1. T. Csörgő, S. Hegyi and W. A. Zajc, "Bose-Einstein correlations for Levy stable source distributions", *Eur. Phys. J. C* **36**, 67 (2004) [arXiv:nucl-th/0310042].
2. U. A. Wiedemann and U. W. Heinz, "Particle interferometry for relativistic heavy-ion collisions", *Phys. Rept.* **319**, 145 (1999) [arXiv:nucl-th/9901094].
3. T. Csörgő and S. Hegyi, " Model independent shape analysis of correlations in 1, 2 or 3 dimensions", *Phys. Lett. B* **489**, 15 (2000).
4. J. Adams *et al.* [STAR Collaboration], "Pion interferometry in Au + Au collisions at s(NN)**(1/2) = 200-GeV", *Phys. Rev. C* **71**, 044906 (2005) [arXiv:nucl-ex/0411036].
5. M. I. Podgoretsky, "On The Comparison Of Identical Pion Correlations In Different Reference Frames", *Sov. J. Nucl. Phys.* **37**, 272 (1983) [*Yad. Fiz.* **37**, 455 (1983)].
6. S. Pratt, "Coherence And Coulomb Effects On Pion Interferometry", *Phys. Rev. D* **33**, 72 (1986).
7. G. F. Bertsch, "Pion Interferometry As A Probe Of The Plasma", *Nucl. Phys. A* **498**, 173C (1989).
8. M. G. Bowler, "Coulomb corrections to Bose-Einstein correlations have been greatly exaggerated", *Phys. Lett. B* **270**, 69 (1991).
9. Y. Sinyukov *et al.*, "Coulomb corrections for interferometry analysis of expanding hadron systems", *Phys. Lett. B* **432**, 248 (1998).

Results on Lévy stable parametrizations of Bose-Einstein Correlations

Tamás Novák
(L3 Collaboration)

Radboud University Nijmegen/NIKHEF, Nijmegen, The Netherlands

Abstract. Bose-Einstein correlations of identical charged-pion pairs produced in hadronic Z decays are analyzed in terms of various parametrizations. A good description is achieved using Lévy stable distributions. The source function is reconstructed with the help of the τ-model.

Keywords: Bose-Einstein correlations, stable distributions, source functions
PACS: 13.38.Dg, 13.87Fh, 13.66Bc

INTRODUCTION

In particle and nuclear physics intensity interferometry provides a direct experimental method for the determination of sizes, shapes and lifetimes of particle-emitting sources (for recent reviews see [1, 2]). In particular, boson interferometry provides a powerful tool for the investigation of the space-time structure of particle production processes, since Bose-Einstein correlations (BEC) of two identical bosons reflect both geometrical and dynamical properties of the particle radiating source.

In e^+e^- annihilation BEC are maximal if the invariant momentum difference is small, even when one of the relative momentum components is large, as was seen by TASSO [3] and which we have confirmed. For a hydrodynamical type of source, on the contrary, BEC decrease when any of the relative momentum components is large [2, 4].

Here we investigate various parametrizations and find that a good description of the Bose-Einstein correlation function can be achieved using Lévy stable distributions as the source function. Within the framework of models assuming strongly correlated coordinate and momentum space, we then reconstruct the complete space-time picture of the particle emitting source in hadronic Z decay.

For our analysis we use a sample of about 500 thousand two-jet events, selected by the Durham algorithm [5] with $y_{\text{cut}} = 0.006$, from e^+e^-annihilation data collected by L3 at a center-of-mass energy of 91.2 GeV.

PARAMETRIZATIONS OF BEC

The two-particle correlation function is defined as:

$$R_2(p_1, p_2) = \frac{\rho_2(p_1, p_2)}{\rho_1(p_1)\rho_1(p_2)}, \tag{1}$$

CP828, *Multiparticle Dynamics*
edited by V. Šimák, M. Šumbera, Š. Todorova, and B. Tomášik

where $\rho_2(p_1,p_2)$ is the two-particle invariant momentum distribution, $\rho_1(p_i)$ the single-particle invariant momentum distributions and p_i the four-momentum of particle i. Since we are only interested in BEC, the product of single particles densities is replaced by the so-called reference sample, $\rho_0(p_1,p_2)$, the two-particle density that would occur in the absence of Bose-Einstein interference. Here we use mixed events as a reference sample.

After some assumptions [1, 2], this two-particle correlation function is related to the Fourier transformed source distribution. In this case

$$R_2(p_1,p_2) = 1 + |\tilde{f}(Q)|^2, \tag{2}$$

where $f(x)$ is the density distribution of the source, Q is the invariant four-momentum difference, $Q = -(p_1 - p_2)^2$ and $\tilde{f}(Q)$ is the Fourier transform of $f(x)$.

Gaussian distributed source

The simplest assumption is that the source has a symmetric Gaussian distribution, in which case $\tilde{f}(Q) = \exp\left(i\mu Q - \frac{(RQ)^2}{2}\right)$ and

$$R_2(Q) = \gamma\left[1 + \lambda\exp\left(-(RQ)^2\right)\right](1 + \delta Q), \tag{3}$$

where the parameter γ is a constant of normalization, λ an incoherence factor, which measures the strength of the correlation, and $(1+\delta Q)$ is introduced to parametrize possible long-range correlations not adequately accounted for in the reference sample.

A fit of Eq.(3) to the data results in an unacceptably low confidence level from which we conclude that the shape of the source deviates from a Gaussian. The fit is particularly bad at low Q values.

Lévy distributed source

Adopting Nolan's $S(\alpha, \beta = 0, \gamma, \delta; 1)$ convention [6] for the symmetric Lévy stable distribution with rescaling of the scale parameter γ to R and the location parameter δ to x_0, the Fourier transform (characteristic function) $\tilde{f}(Q)$ has the following general form:

$$\tilde{f}(Q) = \exp(iQx_0 - |RQ|^\alpha). \tag{4}$$

The index of stability, α, satisfies the inequality $0 < \alpha \leq 2$. The case $\alpha = 2$ corresponds to a Gaussian source distribution. For more details see [6].

Then R_2 has the following, relatively simple form [7]:

$$R_2(Q) = \gamma[1 + \lambda\exp(-(RQ)^\alpha)](1 + \delta Q). \tag{5}$$

After fitting Eq.(5) to the data it is clear that the correlation function is far from Gaussian: $\alpha \approx 1.3$. The confidence level, although improved compared to the fit of Eq.(3), is still unacceptably low.

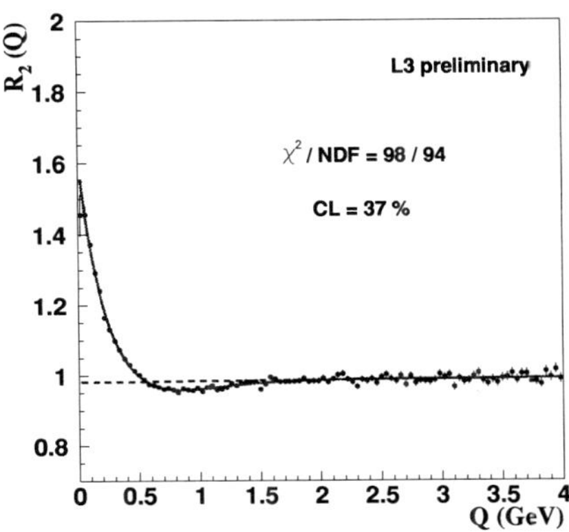

FIGURE 1. The Bose-Einstein correlation function R_2. The curve corresponds to the fit of Eq.(6).

Since there is no particle production before the onset of the collision, a more appropriate form of the source distribution for the time component is the asymmetric stable distribution. In this case, one obtains the following result for the correlation function [8]:

$$R_2(Q) = \gamma[1 + \lambda \cos[(R_a Q)^\alpha] \exp(-(RQ)^\alpha)](1 + \delta Q), \tag{6}$$

where R_a is an additional parameter, a measure of the onset of particle production [6, 7].

The fit of Eq.(6) to the data, shown in Figure 1, is statistically acceptable. The data are well described by the fit. Note that for Q between 0.5 GeV and 1.5 GeV the data points go below the dashed line, which stands for the long-range correlations extrapolated to lower Q values. These data points indicate an anti-correlation in the $Q \approx 1$ GeV region. This property of the data is well reproduced by the fitted curve, which also goes below unity as a result of the cosine term in Eq.(6), which comes from the asymmetric Lévy assumption. The fitted value of α is 0.82 ± 0.03.

THE τ-MODEL

A model of strongly correlated phase-space was developed [9] to explain the experimentally found invariant relative momentum Q dependence of Bose-Einstein correlations in e^+e^- reactions. This model also predicts a specific transverse mass dependence of R_2, that we subject to an experimental test here.

In this model, it is assumed that the average production point $\bar{x}^\mu$ of particles with a given momentum k^μ is given by

$$\bar{x}^\mu(k^\mu) = dk^\mu. \tag{7}$$

In the case of two-jet events, $d = \frac{\tau}{m_T}$, where $\tau = \sqrt{t^2 - k_z^2}$ is the longitudinal proper-time and $m_T = \sqrt{m^2 + p_T^2}$ is the transverse mass. The second assumption is that the distribution of $x^\mu(k^\mu)$ about its average, $\delta_\Delta(x^\mu(k^\mu) - \bar{x}^\mu(k^\mu))$, is narrower than the proper-time distribution. Then the emission function of the τ-model is

$$S(x,k) = \int_0^\infty \mathrm{d}\tau H(\tau)\delta_\Delta(x - dk)N_1(k), \tag{8}$$

where $H(\tau)$ is the longitudinal proper-time distribution, the factor $\delta_\Delta(x-dk)$ describes the strength of the correlations between coordinate space and momentum space variables and $N_1(k)$ is the experimentaly measurable single-particle spectrum.

In the plane-wave approximation the Yano-Koonin formula [10] gives the following two-pion multiplicity distribution:

$$\rho_2(k_1,k_2) = \int \mathrm{d}^4x_1\mathrm{d}^4x_2 S(x_1,k_1)S(x_2,k_2)\left(1+\cos\left[(k_1-k_2)(x_1-x_2)\right]\right). \tag{9}$$

Approximating the δ_Δ function by a Dirac delta function, the argument of the cosine becomes

$$(k_1-k_2)(\bar{x}_1-\bar{x}_2) = -0.5(d_1+d_2)Q^2. \tag{10}$$

Then the two-particle Bose-Einstein correlation function is approximated by

$$R_2(k_1,k_2) = 1+\lambda \mathrm{Re}\widetilde{H}^2\left(\frac{Q^2}{2\overline{m}_\mathrm{T}}\right), \tag{11}$$

where $\widetilde{H}(\omega) = \int \mathrm{d}\tau H(\tau)\exp(i\omega\tau)$ is the Fourier transform of $H(\tau)$. Thus an invariant relative momentum dependent BEC appears.

Guided by the result of the previous section, we use an asymmetric Lévy distribution for the longitudinal proper-time density. Thus the corresponding BEC function has an analytic, although somewhat complicted form [7, 8]:

$$R_2(Q^2,\overline{m}_\mathrm{T}) = \gamma\left[1+\lambda\cos\left(\frac{\tau_0 Q^2}{\overline{m}_\mathrm{T}} + A\left(\frac{\Delta\tau Q^2}{\overline{m}_\mathrm{T}}\right)^{\frac{\alpha}{2}}\right)\exp\left(-\left(\frac{\Delta\tau Q^2}{\overline{m}_\mathrm{T}}\right)^{\frac{\alpha}{2}}\right)\right]B \tag{12}$$

where the parameter τ_0 is the proper-time of the onset of particle production, $\Delta\tau$ is a measure of the width of the proper-time distribution, $A = \tan\left(\frac{\alpha\pi}{4}\right)$ and $B = (1+\delta Q)$

After fitting for various $\overline{m}_\mathrm{T}$ interval we find that the quality of the fits is statistically acceptable and the fitted values of the model parameters are stable and within errors the same in all investigated m_T interval. The τ-model with a one-sided Levy proper-time distribution describes the data with parameters $\tau_0 = 0$ fm, $\alpha \approx 0.80 \pm 0.05$ and $\Delta\tau \approx 2.0 \pm 0.6$ fm.

RECONSTRUCTION OF THE EMISSION FUNCTION

In order to reconstruct the space-time picture of the emitting process we assume that the emission function can be factorized in the following way:

$$S(r,z,t) = I(r)G(\eta)H(\tau), \tag{13}$$

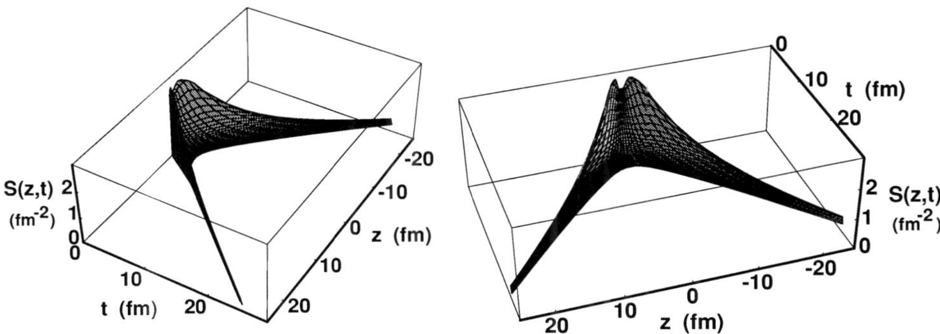

FIGURE 2. Two views of the longitudinal part of the source function normalized to the average number of pions per event.

where $I(r)$ is the single-particle transverse distribution, $G(\eta)$ is the space-time rapidity distribution of particle production, which approximately coincides with the single-particle rapidity distribution, and $H(\tau)$ is the observed proper-time distribution.

With these assumptions one can reconstruct the longitudinal part of the emission function integrated over the transverse distribution. It is plotted as a function of t and z in Figure 2. It exhibits the typical boomerang shape with a maximum at low t and z but with tails reaching out to very large t and z values, a feature already observed for hadron-hadron [11] and heavy ion collisions[12].

The transverse profile, which follows from Eq. (8), has the following form:

$$\frac{\mathrm{d}^4 n}{\mathrm{d}\tau \mathrm{d}^3 r} = \frac{m_\mathrm{T}^3}{\tau^3} H(\tau) N_1\left(k = \frac{m_\mathrm{T} r}{\tau}\right). \tag{14}$$

This equation describes the particle production in coordinate space as a function of the proper-time τ. It describes the expansion of the source as the proper-time increases. The particle production probability is proportional to the proper-time distribution $H(\tau)$. Figure 3 shows the transverse part of the emission function for various proper-times. Particle production starts immediately, increases rapidly and decreases slowly. A ring-like structure, similar to the expanding, ring-like wave created by a pebble in a pond, is reconstructed from L3 data, as shown in Fig. 3. An animated gif file that shows this effect is available from [13].

ACKNOWLEDGMENTS

The author would like to thank T. Csörgő, W. Kittel and W. Metzger for inspiration, support and careful attention, as well as to all members of the L3 collaboration.

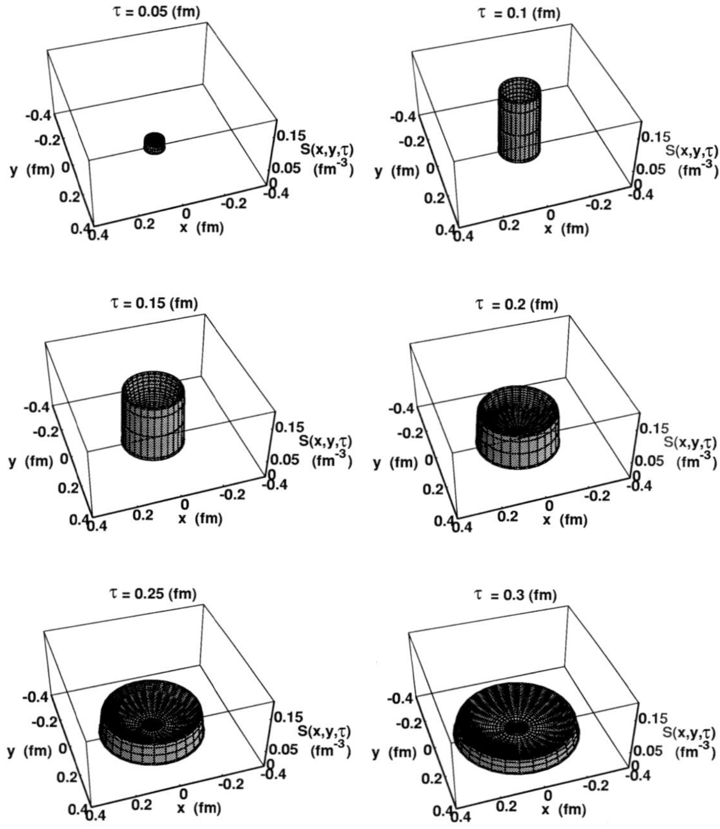

FIGURE 3. The source function normalized to the average number of pions per event for various proper-times.

REFERENCES

1. W. Kittel, *Acta Phys. Pol.* **B32** (2001) 3927.
2. T. Csörgő, *Heavy Ion Physics* **15**, (2002) 1.
3. M. Althoff *et al.* (TASSO Collab.), *Z. Phys* **C30** (1986) 355.
4. T. Csörgő and B. Lörstadt, *Phys. Rev.* **C54** (1996) 1390.
5. S. Cantani *et al.*, *Phys. Lett.* **B279** (1991) 432.
6. J. P. Nolan, Stable distributions: Models for Heavy Tailed Data http://academic2american.edu/ jp-nolan/stable/CHAP1.PDF
7. T. Csörgő, S. Hegyi, W. A. Zajc *Eur. Phys. J.* **C36** (2004) 67. hep-ph/0301164
8. T. Csörgő, private communication
9. T. Csörgő and J. Zimányi, *Nucl. Phys.* **A517** (1990) 588.
10. F. B. Yano amd S. E. Koonin, *Phys. Lett.* textbfB78 (1978) 556.
11. N. M. Agababyan *at al.* (NA22 Collab.), *Phys. Lett.* **B422** (1998) 359.
12. A. Ster, T. Csörgő and B. Lörstad, *Nucl. Phys* **A661** (1999) 419.
13. www.hef.kun.nl/ novakt/movie/movie.gif

COMPARSIONS OF DIFFERENT COLLIDING SYSTEMS INCLUDING COLLISIONS OF ELEMENTARY PARTICLES

Chairperson: Š. Todorova

Bose-Einstein Correlations in e^+e^- Annihilation and $e^+e^- \rightarrow W^+W^-$ [1]

W. J. Metzger

Radboud University, Nijmegen, Netherlands

Abstract. Results on Bose-Einstein correlations in $e^+e^- \rightarrow$ hadrons are reviewed.

Keywords: Bose-Einstein Correlations
PACS: 13.66.Bc, 13.87.Fh

INTRODUCTION

To study Bose-Einstein correlations (BEC) among q identical particles, one usually examines the ratio, R_q, of the q-particle inclusive density, $\rho_q(p_1,...,p_q)$, to that expected when BEC are absent, ρ_{0q}: $R_q(p_1,...,p_q) = \rho_q(p_1,...,p_q)/\rho_{0q}(p_1,...,p_q)$. This ratio is usually regarded as a function of Q, where $Q^2 = M_q^2 - (qm)^2$ with M_q the mass of the q particles and m the mass of each particle. If the particles have identical 4-momenta, $Q = 0$. For 2 particles, Q^2 is simply the 4-momentum difference. Thus, *e.g.*, 2-particle BEC are studied using $R_2(Q) = \rho(Q)/\rho_0(Q)$, where the subscript q has been suppressed.

It can be shown in a variety of ways that R_q is related to the spatial distribution of the particle production [2, 3]. For example, assuming incoherent particle production and a spatial source density of pion emitters, $S(x)$, leads to $R_2(Q) = 1 + |G(Q)|^2$, where $G(Q) = \int \mathrm{d}x\, e^{iQx} S(x) = |G|\, e^{i\phi}$ is the Fourier transform of $S(x)$. Assuming $S(x)$ to be a Gaussian with radius r results in

$$R_2(Q) = 1 + \lambda e^{-Q^2 r^2} , \tag{1}$$

where we have inserted, as is customary, an additional parameter, λ, which is meant to account for several effects such as partial coherence (completely coherent particle production would imply $\lambda = 0$), multiple sources and particle purity.

The lack of time dependence in S is certainly wrong. The assumption of a spherical Gaussian distribution of particle emitters may seem unlikely in e^+e^- annihilation, where there is a definite jet structure. However, we must keep in mind that BEC only occur among particles produced close to each other in phase space. Thus, success of (1) in describing the data does not imply that the hadronization volume is a sphere of radius r.

Other parametrizations have been considered in the literature. Nevertheless, in spite of the above-noted limitations, this Gaussian parametrization (1) is most frequently used by

[1] Given the space limitations of these proceedings, this write-up is very condensed. Almost all figures and tables, as well as much of the discussion, have been eliminated. They can be found in the referenced publications or in the long version of this write-up [1].

CP828, *Multiparticle Dynamics*
edited by V. Šimák, M. Šumbera, Š. Todorova, and B. Tomášik

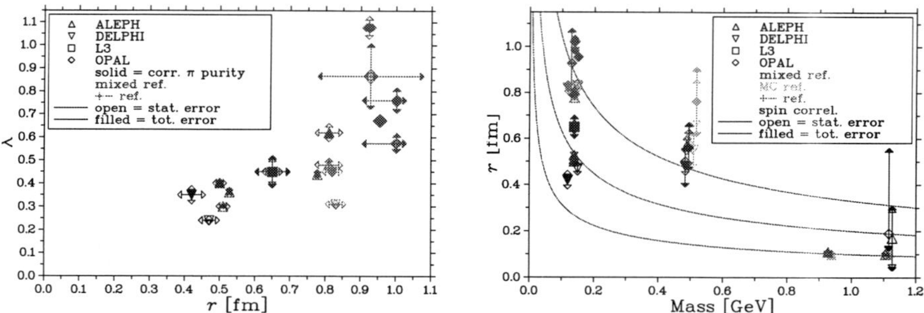

FIGURE 1. (left) λ and r at $\sqrt{s} = M_Z$ found in the LEP experiments [5, 6, 7, 8, 9, 10, 11]. (right) Dependence of r on the mass of the particle as determined at $\sqrt{s} = M_Z$ from 2-particle BEC for charged pions [5, 6, 7, 8, 9, 10, 11], charged kaons [12, 13] and neutral kaons [14, 12, 15] and from Fermi-Dirac correlations for protons [14] and lambdas [16, 17]. The curves illustrate a $1/\sqrt{m}$ dependence.

experimentalists. When it does not fit well, an expansion about the Gaussian (Edgeworth expansion [4]) can be used instead. In the interest of comparison of as much e^+e^- data as possible, I only consider results using the Gaussian or Edgeworth parametrizations.

EXPERIMENTAL DIFFICULTIES

There are several experimental problems affecting BEC results and their interpretation. Particle purity, resonances and weak decays all affect the measured values of λ and r. Other problems are the effect of final-state interactions, both Coulomb and strong, and the choice of the "reference sample," the sample for which ρ_0 is the density. For a discussion of these problems, see [1]. Finally, there is the effect of long-range correlations not adequately taken into account by the reference sample. R_2 is not usually found to be constant at large Q. To account for this the right hand side of (1) is multiplied by an appropriate factor, usually a linear dependence on Q: $\gamma(1+\delta Q)$.

EXPERIMENTAL RESULTS

Dependence on the reference sample. The values of λ and r found for charged-pion pairs from hadronic Z decays by ALEPH [5, 6], DELPHI [7], L3 [8] and OPAL [9, 10, 11] are displayed in Fig. 1. Solid points are corrected for pion purity; open points are not. This correction increases the value of λ but has little effect on the value of r. All of the results with $r > 0.7$ fm were obtained using an unlike-sign reference sample, while those with smaller r were obtained with a mixed reference sample. The choice of reference sample clearly has a large effect on the observed values of λ and r. In comparing results we must therefore be sure that the reference samples used are comparable.

Dependence on the center-of-mass energy. Comparison of values of r obtained using the same reference sample for $\sqrt{s} = 29$–91 GeV shows no evidence of dependence [1].

Dependence on the particle mass. It has been suggested, on several grounds [18], that r should depend on the particle mass as $r \propto 1/\sqrt{m}$. Values of r found at LEP for various types of particle are shown in Fig. 1. Comparing only results using the same type of reference sample (in this case mixed), we see no evidence for a $1/\sqrt{m}$ dependence. Rather, the data suggest one value of r for mesons and a smaller value for baryons. The value for baryons, about 0.1 fm, seems very small; if true it is telling us something unexpected about the mechanism of baryon production.

Dependence on particle multiplicity. The values of λ and r from charged-pion 2-particle BEC depend on the charged particle multiplicity, n_{ch}, of the events: λ decreases with n_{ch} while r increases. However, such a dependence is also seen on the number of jets [10]. When 2-jet events are selected, little, if any, dependence on n_{ch} remains. For 3-jet events, r seems independent of n_{ch}, although λ does decrease with n_{ch}. Thus, the dependence of λ and r on n_{ch} seems largely due to a dependence on the number of jets.

Elongation of the source. The Gaussian parametrization (1) assumes a spherical source. Given the jet structure of e^+e^- events, one might expect a more ellipsoidal shape. To investigate this, the parametrization is generalized to allow different radii along and perpendicular to the jet axis. The analysis uses the longitudinal center-of-mass system (LCMS) with the longitudinal axis along the jet-axis, *e.g.*, the thrust axis. The advantage of the LCMS is that the energy difference, and therefore the difference in emission time of the pions, couples only to the out-component, Q_{out}. Thus Q_L and Q_{side} reflect only spatial dimensions of the source, while Q_{out} reflects a mixture of spatial and temporal dimensions. Three-dimensional analyses parametrizing R_2 as a function of Q_L, Q_{side}, and Q_{out} have been performed by L3 [19] and OPAL [11], and two-dimensional analyses by ALEPH [6] and DELPHI [20], in which the out and side components are replaced by a transverse one, $Q_t^2 = Q_{out}^2 + Q_{side}^2$. However, the interpretation of the corresponding parameter, r_t, as a transverse radius is not unambiguous, since it includes the effect of the difference in emission time. The longitudal radius is found to be about 20% larger than the side (transverse) radius. Further, the amount of the elongation increases when narrower 2-jet events are selected [11].

It should also be mentioned that ZEUS [21] performed a similar 2-dimensional analysis in deep inelastic ep interactions. The ratio r_t/r_L found, similar to that found by DELPHI and ALEPH, is independent of the virtuality of the exchanged photon.

$\pi^0\pi^0$. In hadronization models with local charge conservation, *e.g.*, string models, neutral pions can be produced closer together than identical charged pions. One could expect this to be reflected in a smaller BEC radius for $\pi^0\pi^0$ than for $\pi^\pm\pi^\pm$. Only two e^+e^- experiments have attempted a BEC analysis of $\pi^0\pi^0$, L3 [22] and OPAL [23]. The experimental selections used are quite different. While L3 requires the energy of the pions to be less than 6 GeV, OPAL demands π^0 momenta greater than 1 GeV. Further, OPAL uses only 2-jet events, defined as having thrust larger than 0.9.

In order to make a comparison with charged pions, L3 also analyzes $\pi^\pm\pi^\pm$ with a selection similar to its π^0 selection. Results are listed in Table 1. Comparison of the L3 values for $\pi^0\pi^0$ and $\pi^\pm\pi^\pm$ with the same selection indicates that both r and λ are smaller in the $\pi^0\pi^0$ case, but the significance is only about 1.5 standard deviations.

TABLE 1. Results of fits to R_2 for $\pi^0\pi^0$ [22, 23] and $\pi^\pm\pi^\pm$ [22, 8, 11]. The indicated uncertainties combine statistical and systematic uncertainties.

	Experiment	Selection	Ref. sample	r (fm)	λ
$\pi^0\pi^0$	L3	$E_\pi < 6\,\text{GeV}$	MC	0.31 ± 0.10	0.16 ± 0.09
	OPAL	$p_\pi > 1\,\text{GeV}$, 2-jet	mix	0.59 ± 0.09	0.55 ± 0.14
$\pi^\pm\pi^\pm$	L3	$E_\pi < 6\,\text{GeV}$	MC	0.46 ± 0.01	0.29 ± 0.03
	L3		mix	0.65 ± 0.04	0.45 ± 0.07
	OPAL		+−	$1.00^{+0.03}_{-0.10}$	0.57 ± 0.05

For OPAL, the comparison is more difficult to make, since OPAL's charged-pi results use a different reference sample and different selection. Other experiments have found the ratio of r using a mixed reference sample to that using unlike-sign to be about 0.68 (ALEPH [6]) or 0.56 (DELPHI [7]). Applying such a factor would lower the OPAL value to about 0.62 fm, which agrees well with their result for $\pi^0\pi^0$. However, it is not clear what the effect of the 2-jet and π^0-momentum cuts is. In the L3 case, requiring the pions to have $E < 6\,\text{GeV}$ and using a MC reference sample rather than a mixed one decreased r by about 30%. It is therefore conceivable that the OPAL requirement of $p > 1\,\text{GeV}$ would increase r, in which case r for $\pi^0\pi^0$ would be smaller than for $\pi^\pm\pi^\pm$.

3-particle BEC. Correlations among more than two particles are classified as trivial, a consequence of lower-order correlations, and genuine. We study 3-particle BEC using $R_3(Q_3)$. Note that $Q_3^2 = Q_{12}^2 + Q_{23}^2 + Q_{31}^2$. The same assumptions that lead to (1) for R_2, lead to [24, 25] R_3^{gen}, the R_3 that would occur if there were no 2-particle BEC:

$$R_3^{\text{gen}}(Q_{12}, Q_{23}, Q_{31}) = 1 + 2\lambda^{1.5}\Re\{G(Q_{12})G(Q_{23})G(Q_{31})\}\,, \tag{2}$$

where $\Re$ denotes the real part. Thus R_3^{gen} depends on the phase of G.

Both R_2 and R_3^{gen} are measured by L3 [8] and fit with both a Gaussian and an Edgeworth parametrization, the latter providing a better fit.

By combining these measurements, one can extract $\omega = \cos(\phi_{12} + \phi_{23} + \phi_{13})$. If the particle production is completely incoherent, the phase ϕ_{ij} is expected to be zero, and $\omega = 1$. Incoherence also implies $\lambda = 1$, but λ is affected by many other factors, which should not affect ω. The measurements find ω consistent with unity; we conclude that particle production is completely incoherent.

BEC among three pions have previously been observed at lower energies [26, 27]. At LEP, DELPHI [28] and OPAL [29] have also observed genuine 3-pion BEC. Unfortunately, none of these experiments measured ω.

$\mathrm{W} \to \mathrm{q\bar{q}}$. Having found no evidence for a $\sqrt{s}$ dependence of BEC, and noting that in any case the masses of the Z and W are not much different, there is only one reason to expect BEC to be different in W decays than in Z decays. Whereas about 20% of Z decays are to $\mathrm{b\bar{b}}$, almost no W decays involve a b-quark. The long lifetime of the b results in diminished BEC. Thus, we should expect BEC in W-decay to be like that in the decay of the Z to light (udsc) quarks. This is indeed found to be the case [30, 31].

Inter-string BEC. A more interesting case is $e^+e^- \to W^+W^- \to q\overline{q}q\overline{q}$. The W-bosons are produced not far above threshold and consequently travel only about 0.1 fm before decaying, less than the distance over which hadronization occurs. There can therefore be a significant degree of overlap of the two hadronizing systems, resulting in BEC among particles from different W-bosons. On the other hand, in the string picture no BEC is expected between particles arising from different strings. The existence of inter-string BEC is thus a fundamental question for the string picture.

The LEP experiments have tested the expectation [32] of no inter-string BEC using various test quantities. All four LEP experiments have reported final results [33, 34, 31, 35]. Only DELPHI [34] claims to see significant inter-W BEC. Their distribution of $\delta_I(Q)$, the correlation function of genuine inter-W BEC [36], clearly shows an enhancement at small Q, while no enhancement is seen in the MC where BEC (BE_{32} model [37]) is included only between pions from the same W. When BE_{32} is also applied to pions from different W-bosons, a larger enhancement is seen than that of the data. However, the interpretation of the enhancement in the data as inter-W BEC is somewhat clouded by their observance of an enhancement, albeit smaller, in the δ_I distribution of unlike-sign pions. This enhancement in the data is larger than that for MC with inter-W BEC.

The final results of the four LEP experiments, expressed as the ratio of the effect seen to that expected in the BE_{32} model are combined to give the preliminary LEP result [38] of 0.17 ± 0.13. The data are thus compatible with no inter-W BEC.

BEC IN THE LUND STRING MODEL

In the Lund string model, the longitudinal break-up of the color string is governed by the area law. The matrix element to get a final state depends on the area A spanned by the string. Transverse momentum arises via a tunneling mechanism, which is also related to b. To incorporate BEC, the probability of a final state is given by the square of the sum of the matrix elements corresponding to the areas of all the permutations of identical bosons [39, 40, 41]. This model results in 2- and genuine 3-particle BEC. It predicts elongation along the jet axis and a radius for $\pi^0\pi^0$ smaller than for $\pi^\pm\pi^\pm$. There is no mechanism to create correlations between particles from different strings.

The model has been incorporated in MC for a $q\overline{q}$ string. While the formalism to do so for the more realistic case of a string with multiple gluons has been worked out [42], a successful MC implementation has, unfortunately, so far proved elusive.

CONCLUSIONS

The study of BEC in e^+e^- presents a number of problems, both experimental and theoretical. Values obtained for parameters vary considerably among experiments, even when the same parametrization is used. Nevertheless, certain features are clear: BEC, both 2-particle and genuine 3-particle, exist; they seem independent of cms energy; the source shape is somewhat elongated in the jet direction; and the (Fermi-Dirac) radius for baryons is smaller than the radius for mesons. Experimentally, it is not clear whether the radius for neutral pions is smaller than that for charged pions. BEC in W decay is the same as in light-quark Z decay. The data are compatible with no inter-W BEC.

The implementation of BEC in the Lund string model appears consistent with these experimental findings. However, the experimental evidence that pion production is completely incoherent seems at odds with the coherent addition of amplitudes in the model.

REFERENCES

1. W. Metzger, Bose-Einstein Correlations in e^+e^- Annihilation and $e^+e^- \to W^+W^-$ (2005), hep-ph/0509031.
2. G. Goldhaber, S. Goldhaber, W. Lee, and A. Pais, *Phys. Rev.* **120**, 300–312 (1960).
3. D. H. Boal, C.-K. Gelbke, and B. K. Jennings, *Rev. Mod. Phys.* **62**, 553–602 (1990).
4. F. Edgeworth, *Trans. Cambridge Phil. Soc.* **20**, 36 (1905), see also Harald Cramér, Mathematical Methods of Statistics, Princeton Univ. Press, 1946.
5. ALEPH Collab., D. Decamp, et al., *Z. Phys.* **C54**, 75–85 (1992).
6. ALEPH Collab., A. Heister, et al., *Eur. Phys. J.* **C36**, 147–159 (2004).
7. DELPHI Collab., P. Abreu, et al., *Phys. Lett.* **B286**, 201–210 (1992).
8. L3 Collab., P. Achard, et al., *Phys. Lett.* **B540**, 185–198 (2002).
9. OPAL Collab., P. Acton, et al., *Phys. Lett.* **B267**, 143–153 (1991).
10. OPAL Collab., G. Alexander, et al., *Z. Phys.* **C72**, 389–398 (1996).
11. OPAL Collab., G. Abbiendi, et al., *Eur. Phys. J.* **C16**, 423–433 (2000).
12. DELPHI Collab., P. Abreu, et al., *Phys. Lett.* **B379**, 330–340 (1996).
13. OPAL Collab., G. Abbiendi, et al., *Eur. Phys. J.* **C21**, 23–32 (2001).
14. ALEPH Collab., S. Schael, et al., *Phys. Lett.* **B611**, 66–80 (2005).
15. OPAL Collab., R. Akers, et al., *Z. Phys.* **C67**, 389–401 (1995).
16. ALEPH Collab., R. Barate, et al., *Phys. Lett.* **B475**, 395–406 (2000).
17. OPAL Collab., G. Alexander, et al., *Phys. Lett.* **B384**, 377–387 (1996).
18. G. Alexander, *Rep. Prog. Phys.* **66**, 481–522 (2003).
19. L3 Collab., M. Acciarri, et al., *Phys. Lett.* **B458**, 517–528 (1999).
20. DELPHI Collab., P. Abreu, et al., *Phys. Lett.* **B471**, 460–470 (2000).
21. ZEUS Collab., S. Chekanov, et al., *Phys. Lett.* **B583**, 231–246 (2004).
22. L3 Collab., P. Achard, et al., *Phys. Lett.* **B524**, 55–64 (2002).
23. OPAL Collab., G. Abbiendi, et al., *Phys. Lett.* **B559**, 131–143 (2003).
24. B. Lörstad, *Int. J. Mod. Phys.* **A4**, 2861–2896 (1989).
25. V. Lyuboshitz, *Sov. J. Nucl. Phys.* **53**, 514–2896 (1991).
26. M. A. TASSO Collab., et al., *Z. Phys.* **C30**, 355–369 (1986).
27. I. Juricic, et al., *Phys. Rev.* **D39**, 1–20 (1989).
28. DELPHI Collab., P. Abreu, et al., *Phys. Lett.* **B355**, 415–424 (1995).
29. OPAL Collab., K. Ackerstaff, et al., *Eur. Phys. J.* **C5**, 239–248 (1996).
30. ALEPH Collab., R. Barate, et al., *Phys. Lett.* **B478**, 50–64 (2000).
31. L3 Collab., P. Achard, et al., *Phys. Lett.* **B547**, 139–150 (2002).
32. S. Chekanov, E. De Wolf, and W. Kittel, *Eur. Phys. J.* **C6**, 403–411 (1999).
33. ALEPH Collab., S. Schael, et al., *Phys. Lett.* **B606**, 265–275 (2005).
34. DELPHI Collab., J. Abdallah, et al., *Eur. Phys. J.* **C44**, 161–174 (2005).
35. OPAL Collab., G. Abbiendi, et al., *Eur. Phys. J.* **C36**, 297–308 (2004).
36. E. De Wolf, Correlations in $e^+e^- \to W^+W^-$ hadronic decays (2001).
37. L. Lönnblad, and T. Sjöstrand, *Eur. Phys. J.* **C2**, 165–180 (1998).
38. LEP Electroweak Working Group, A Combination of Preliminary Electroweak Measurements and Constraints on the Standard Model, Tech. Rep. LEPEWWG/2005-01 (2005).
39. B. Andersson, and W. Hofmann, *Phys. Lett.* **B169**, 364–368 (1986).
40. B. Andersson, and M. Ringnér, *Nucl. Phys.* **B513**, 627–644 (1998).
41. B. Andersson, and M. Ringnér, *Phys. Lett.* **B421**, 283–288 (1998).
42. B. Andersson, S. Mohanty, and F. Söderberg, *Eur. Phys. J.* **C21**, 631–647 (2001).

Inter-string Bose-Einstein Correlations in Hadronic Z Decays using the L3 Detector at LEP

Qin Wang
(L3 Collaboration)

Radboud University Nijmegen/NIKHEF, Nijmegen, The Netherlands

Abstract. Inter-string Bose-Eintein correlations are studied in hadronic Z events at $\sqrt{s} = 91.2$ GeV. Preliminary results are presented comparing 1-string and 2-string systems: 2-jet with 3-jet events and quark jets with gluon jets.

Keywords: inter-string Bose-Einstein correlations 2-jet 3-jet quark jet gluon jet
PACS: 13.66.Bc, 13.87.Fh

MOTIVATION OF INTER-STRING STUDY

Phenomenological models are normally used to describe the fragmentation process. Among these models, the string model is quite successful. It uses the break-up of strings spanned between partons to describe the hadronization process, and hadrons are produced coherently within one string. Bose-Einstein correlations (BEC) within one string are predicted by the Lund Area Law[1], but no inter-string BEC are expected.

Inter-string BEC are widely studied in WW events at LEP because of their influence on the measurement of the W mass. Present combined results from LEP give no evidence for inter-string BEC between the two W's[2]. However, the low statistics of WW events limits the possibility of finding it.

Here we investigate inter-string BEC using three-jet Z events, which have higher statistics and which also contain two strings, spanned between the radiated gluon and the primary quark and anti-quark. We compare BEC between 1- and 2-string systems, i.e., between 2- and 3-jet events and between quark and gluon jets.

METHODS AND FORMULAE

BEC between two particles are investigated using the function:

$$R_2 = \frac{\rho_2(Q)}{\rho_{0,2}(Q)} \quad , \tag{1}$$

where Q is the four-momentum difference between the two particles,

$$Q = \sqrt{-(p_1 - p_2)^2} \quad , \tag{2}$$

CP828, *Multiparticle Dynamics*
edited by V. Šimák, M. Šumbera, Š. Todorova, and B. Tomášik

$\rho_2(Q)$ is the two-particle density of the experimental data,

$$\rho_2(Q) = \frac{1}{N_{\text{pairs}}} \frac{\mathrm{d}n_{\text{pairs}}}{\mathrm{d}Q} \quad , \tag{3}$$

and $\rho_{0,2}(Q)$ is the two-particle density of a reference sample which contains all correlations except BEC.

We use two kinds of reference sample: (1) event mixing, where each particle in a mixed event comes from a different experimental event, and where other correlations, which are also removed by the mixing, are compensated for by Monte Carlo; (2) JETSET or PYTHIA Monte Carlo without BEC.

The Edgeworth expansion[3, 4] is used to parametrize R_2:

$$R_2(Q) = \gamma(1+\delta Q+\varepsilon Q^2)\left[1+\lambda e^{-R^2Q^2}\left(1+\frac{\kappa}{3!}H_3(RQ)\right)\right] \quad , \tag{4}$$

where $H_3(RQ) = (\sqrt{2}RQ)^3 - 3\sqrt{2}RQ$ is the third-order Hermite polynomial. The parameter λ measures the strength of correlation and R reflects the radius of source.

The correlation strength λ is expected[5, 6] to decrease when the number of independent overlapping strings increases. The source radius R is expected to increase when the two strings are dependent, since color flux goes from one string over the gluon tip to the other string, so that the distance between two points on the two strings is along this longer curve rather than on a straight line. Thus we expect:

(1) If there are no inter-string BEC but the two strings overlap:

$$\lambda_{2\text{-string}} < \lambda_{1\text{-string}} \qquad R_{2\text{-string}} \approx R_{1\text{-string}}$$

(2) If there are inter-string BEC and overlap between the two strings:

$$\lambda_{2\text{-string}} \approx \lambda_{1\text{-string}} \qquad R_{2\text{-string}} \gtrsim R_{1\text{-string}}$$

(3) If the two strings do not overlap:

$$\lambda_{2\text{-string}} \approx \lambda_{1\text{-string}} \qquad R_{2\text{-string}} \approx R_{1\text{-string}}$$

COMPARISON OF 2-JET AND 3-JET EVENTS

In b-quark events, BEC are suppressed because of the long lifetime of the b quark. So we only use light-quark (udsc-) events to do the analysis. Event anti-b-tagging is used to select light-quark events. For this purpose, the event discriminant δ_{event}[7] is calculated, based on the probability that all tracks originate at the interaction vertex. Light-quark events are chosen by $\delta_{\text{event}} \leq 1.3$. The DURHAM jet clustering algorithm[8] is used to classify events as 2-jet or 3-jet, where the resolution parameter y_{cut} is chosen to be 0.005, 0.006, 0.01, 0.02 or 1. $R_2(Q)$ is calculated using JETSET without BEC, PYTHIA without BEC and mixed events as reference samples, and it is parametrized using Eq. (4).

Fig. 1 shows the parameters λ and R for 2- and 3-jet events as a function of y_{cut} using each of the reference samples. From the plot we see that λ does not change with y_{cut}, while R increases with y_{cut}, especially in the 3-jet events. We also notice that λ and R are reference sample dependent. Using mixed events as reference sample gives the smallest value for λ and the largest value for R. Furthermore, we find λ to be slightly smaller in 3-jet than in 2-jet events, while R is bigger in 3-jet events than in 2-jet events. Similar behavior has been previously observed[9, 10], but the results do not distinguish among the expectations above.

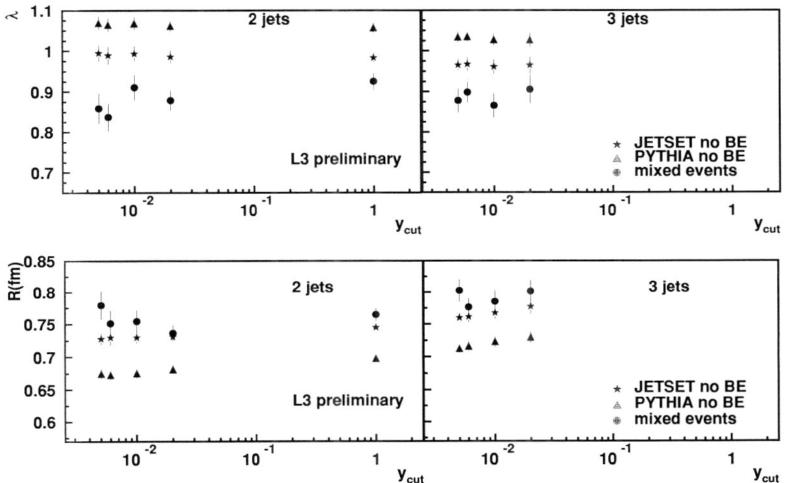

FIGURE 1. λ (above) and R (below) vs. y_{cut} in 2- and 3-jet events. JETSET no BEC (stars), PYTHIA no BEC (triangles) and mixed events (dots) are used as reference sample. Error bars are statistical only.

COMPARISON OF QUARK JETS AND GLUON JETS

The two strings in the 3-jet events do not fully overlap. The overlap is largest in the gluon jet and smallest in the quark jets. So the difference between quark and gluon jets is expected to be more sensitive to inter-string BEC than the difference between 2- and 3-jet events.

In order to select gluon jets, we first use the DURHAM algorithm with $y_{cut} = 0.02$ to select 3-jet events. After that, we apply two methods to select the gluon jet:

1. Energy rank: the least energetic jet in the light-quark 3-jet event is chosen as the gluon jet.
2. Double b-tagging: Firstly, event b-tagging is used to select b-quark events by requiring $\delta_{event} \geq 2.5$, which results in a purity and efficiency of b-quark events of 82% and 45% (see Fig. 2a).
   Secondly, jet b-tagging is used in the 3-jet b-quark events to identify the gluon jet. We have tried two ways for the jet b-tagging:
   (a) Jet b-tagging on 3 jets:
   If $\delta_1 \geq \delta_{jet}$, $\delta_2 \geq \delta_{jet}$ and $\delta_3 \leq \delta_{jet} \Longrightarrow$ jet 3 is the gluon jet

(b) Jet b-tagging on 2 jets: jet 1 is the most energetic jet

If $\delta_2 \geq \delta_{jet}$ and $\delta_3 \leq \delta_{jet} \Longrightarrow$ jet 3 is the gluon jet

Since method (b) results in slightly higher purity and efficiency, we choose to use it. With $\delta_{jet} = 0.8$, we obtain 10,711 gluon jets with 83% purity and an efficiency of 5% (see Fig. 2b).

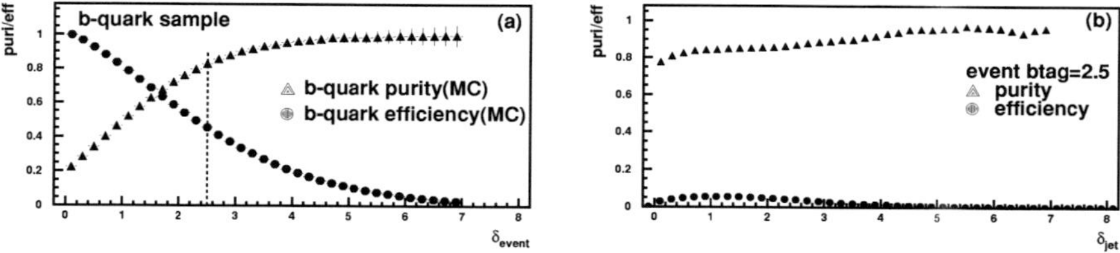

FIGURE 2. (a) Purity and efficiency of b-quark events vs. δ_{event}. (b) Purity and efficiency of gluon jets vs. δ_{jet} for δ_{event}=2.5.

Since BEC in b-quark jets are suppressed, we use quark jets from light-quark 2-jet and 3-jet events, in the latter case assuming the least energetic jet to be the gluon.

We use Monte Carlo without BEC as reference sample and parametrize R_2 using Eq. (4) with parameters $\delta = \varepsilon = 0$.

The gluon jets have, on average, lower energy than the quark jets. Before comparing quark and gluon jets, we check whether the results really reflect the difference of quark and gluon jets, or just the energy difference. Instead of using jet energy, we use jet hardness defined as

$$h = E_{jet} \sin \left(\frac{\theta_{1,2}}{2} \right) \quad , \tag{5}$$

where E_{jet} is the energy of the jet and $\theta_{1,2}$ is the smaller of the two angles between this jet and the other two jets.

Fig. 3 shows the jet hardness distribution for gluon jets, second-most energetic quark jets and most energetic quark jets. This plot is for the light-quark data sample and jets are identified by energy rank.

Fig. 4 shows the values of λ and R for quark and gluon jets in various hardness ranges. We use JETSET as reference sample in the left two plots and PYTHIA in the right two plots. No matter which reference sample we use, λ and R do not have an obvious

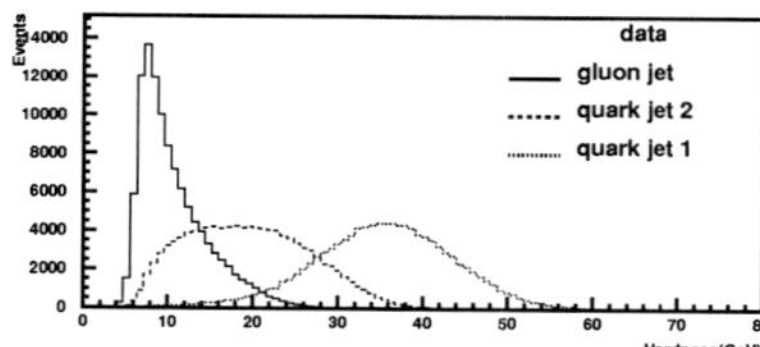

FIGURE 3. Hardness distribution for gluon jets (solid line), second-most energetic quark jets (dashed line) and most energetic quark jets (dotted line) in the data sample.

hardness dependence. λ is slightly higher for the gluon jet, while R is the same for quark and gluon jets.

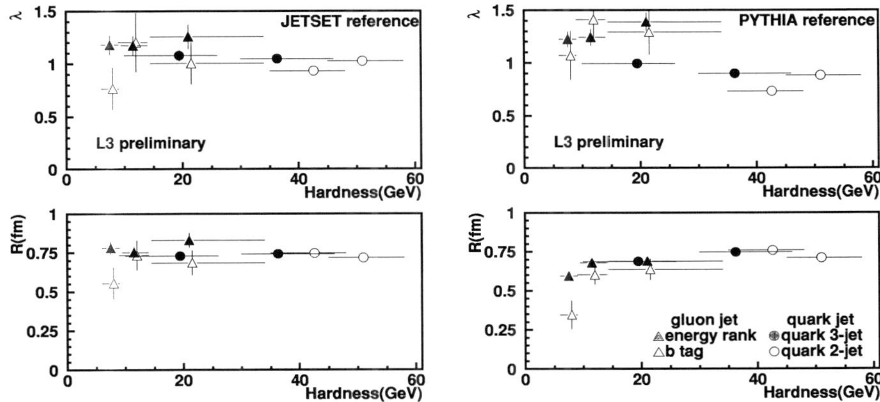

FIGURE 4. λ and R for gluon and quark jets in different hardness ranges. Vertical error bars represent statistical errors and horizontal error bars indicate the range of hardness. JETSET (left two plots) and PYTHIA (right two plots) are used as the reference sample. Triangles are the results for the gluon jet, where full triangles correspond to energy rank and open triangles to jet b-tagging. Solid circles are the results from quark jets in light-quark 3-jet events. Open circles are for the quark jets in light-quark 2-jet events.

Since we do not see any significant dependence on hardness, we continue the comparison of quark and gluon jets ignoring the energy difference. We now study λ and R of quark and gluon jets using only particles in a particular window of x , where x is defined as:

$$x = p_z/E_{\text{jet}} \quad , \tag{6}$$

with p_z the component of the particle's momentum along the jet axis and E_{jet} the energy of the jet.

Since there is more overlap in the tip of the gluon jet, we expect the quark-gluon differences to increase with x. Fig. 5 shows the values of λ and R as a function of x for quark and gluon jets. The quark jets are from light-quark 3-jet events. The gluon jet is identified using energy rank or b-tagging, the latter having a somewhat smaller value of λ. The two quark jets, which cover different hardness ranges, show consistent results. For both quark and gluon jets, λ shows some x dependence, while R does not. Considering the dependence on the gluon jet identification method, neither λ nor R shows significant difference between quark and gluon jets.

CONCLUSIONS

By comparing the correlation strength λ and the source radius R in 2-jet and 3-jet events from hadronic Z decays, we find that λ is slightly smaller in the 3-jet events than in the 2-jet events, and R is slightly larger in 3-jet events. These results do not confirm the

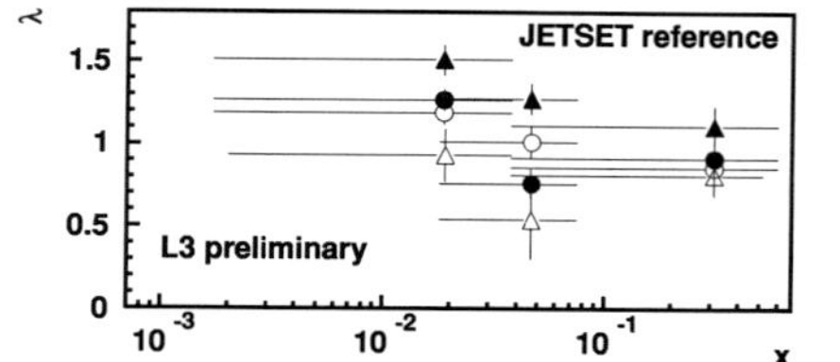

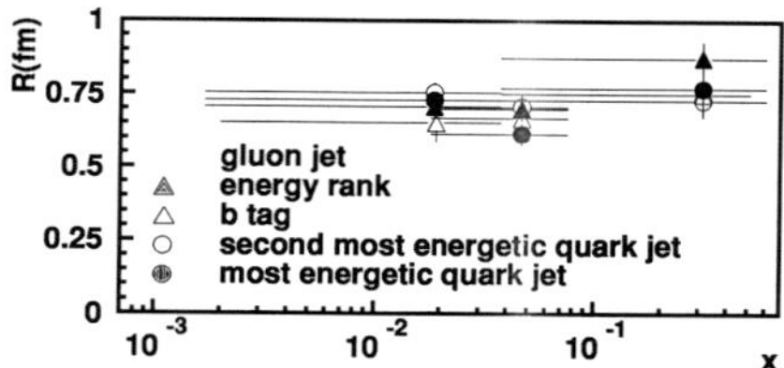

FIGURE 5. λ and R vs. x for the gluon jet (triangles) and quark jet (circles) using JETSET as the reference sample. Vertical error bars represent statistical errors and horizontal error bars indicate the range of x. The quark jets are the second most energetic quark jet (open circles) and most energetic quark jet (solid circles) in the light-quark 3-jet events. The gluon jet is identified by energy rank (solid triangles) and b-tagging (open triangles).

predictions on the behavior of λ and R in 1-string and 2-string systems with or without inter-string BEC.

By comparing quark and gluon jets, we find that λ shows some x dependence, while R does not. Neither λ nor R shows significant difference between quark and gluon jets. From the current results, we can only state that either there may be inter-string BEC or there is no overlap between the two strings in the gluon jet. Further study will continue.

ACKNOWLEDGMENTS

The author would like to thank W. Kittel and W. J. Metzger for useful discussions and suggestions. The author also acknowledges all the efforts of the other members of the L3 collaboration.

REFERENCES

1. B. Andersson and M. Ringner, *Nucl. Phys. B* **513**, 627–644(1998).
2. LEP Electroweak Working Group, *A Combination of Preliminary Electroweak Measurements and Constraints on the Standard Model*, LEPEWWG/2005-01.
3. F. Y. Edgeworth, *Trans. Cambridge Phil. Soc.* **20**, 36–65(1905).
4. T. Csörgő, S. Hegyi, *Phys. Lett. B* **489**, 15–23(2000).
5. P. Lipa and B. Buschbeck, *Phys. Lett. B* **223**, 465–469(1989).
6. B. Buschbeck, H. C. Eggers and P. Lipa, *Phys. Lett. B* **481**, 187–193(2000).
7. L3 Collab., M. Acciarri *et al.*, *Phys. Lett. B*, **411**, 373–386(1997).
8. S. Catani *et al.*, *Phys. Lett. B* **269**, 432–438(1991).
9. OPAL Collab., G. Alexander *et al.*, *Z. Phys. C* **72**, 389–398(1996).
10. N. van Remortel, *Ph.D. thesis*, University of Antwerpen (2003).

Multidimensional HBT correlations in $p\bar{p}$ collisions at $\sqrt{s} = 630$ GeV

H.C. Eggers*, B. Buschbeck† and F.J. October**,*

*Department of Physics, University of Stellenbosch, 7602 Stellenbosch, South Africa
†Institut für Hochenergiephysik, Nikolsdorfergasse 18, A–1050 Vienna, Austria
**Institute for Maritime Technology, 7995 Simonstown, South Africa

Abstract. We analyse second moments R_2 of like-sign pion pairs in the two-dimensional (q_L, q_T) and three-dimensional (q_O, q_S, q_L) decompositions of the three-momentum difference. Conventional fit parametrisations such as gaussian, exponential, power-law and Edgeworth fail miserably, while more elaborate ones such as Lévy do well but fail to yield a unique best-fit solution. A two-component model using a hard cut to separate small- and large-scale parts appears possible but not compelling. In all cases, the data exhibits a strong and hitherto unexplained peak at small momentum differences which exceeds current fits.

Keywords: particle correlations, intensity interferometry
PACS: 13.85.Hd, 13.87.Fh, 13.85.-t, 25.75.Gz

The UA1 experiment, having completed data-taking at the CERN SPS in the late eighties, continues to be relevant and interesting. The current concentrated effort at RHIC to quantify and understand ultrarelativistic nuclear collisions relies extensively on comparisons with baseline scenarios constructed from the corresponding "trivial" hadron-hadron sample. Current experimental energies of 200 AGeV at RHIC are still below those available to UA1 by a factor three, so that UA1 results may also provide a window on possible energy dependencies of current investigations.

In this contribution, we provide preliminary results on HBT analysis mainly in terms of the two-dimensional decomposition of the three-momentum difference, defining in the usual way $q_L = |\mathbf{q}_L| = |(\mathbf{q}\cdot\hat{z})\,\hat{z}|$, with $\hat{z}$ the beam direction, and $q_T = |\mathbf{q} - \mathbf{q}_L|$. Brief reference is also made to the three-dimensional Bertsch-Pratt case, with similar results and issues arising.

Like-sign (LS) pion pairs from approximately 2.45 million minimum-bias events measured by the UA1 central detector were analysed. This represents a twofold increase in statistics over Ref. [1] and a 15-fold increase compared to earlier UA1 HBT analyses [2, 3, 4]. Standard cuts [1] were applied, including single-track cuts $p_\perp \geq 0.15$ GeV/c, $|y| \leq 3$ and, to avoid acceptance problems, $45° \leq |\phi| \leq 135°$. The sample contains an estimated 15% contamination of charged kaons.

The most important among the standard pair cuts is the "ghost cut" which eliminates spurious "split track" LS pairs within a narrow cone but many real LS pairs also. A correction factor compensating for this was determined by passing unlike-sign (US) pairs through the same ghost-cut algorithm. Such correction factors were determined

CP828, *Multiparticle Dynamics*
edited by V. Šimák, M. Šumbera, Š. Todorova, and B. Tomášik

for each (q_L, q_T) bin[1] and charged-multiplicity subsample; for some (q_L, q_T) bins, this correction factor ranges up to 1.7 or even 1.9 for low-multiplicity subsamples.

We corrected for Coulomb repulsion by parametrising the Bowler Coulomb correction in the invariant momentum difference $Q = \sqrt{-(p_1 - p_2)^2}$ [5] with an exponentially damped Gamov factor $G(Q)$ [6], with a best-fit value $Q_{\rm eff} = 0.173 \pm 0.001$ GeV/c,

$$F_{\rm coul}(Q) = 1 + [G(Q) - 1] \exp(-Q/Q_{\rm eff}) \,. \tag{1}$$

The reference sample was formed by randomly combining LS tracks taken from pools of events in the same subsample of event charged multiplicity N as the sibling event currently being analysed. Note that the reference for fixed-multiplicity subsamples is not the poisson distribution but the multinomial, whose second moment is $\rho_2^{\rm mult}(\mathbf{q}\,|\,N) = (1 - N^{-1})\rho_1 \otimes \rho_1(\mathbf{q}\,|\,N)$, so that the appropriate normalised moment is [1, 7]

$$R_2(\mathbf{q}) = \frac{\sum_N P_N\, \rho_2^{\rm LS}(\mathbf{q}\,|\,N)}{\sum_N P_N\, (1 - N^{-1})\, \rho_1 \otimes \rho_1^{\rm LS}(\mathbf{q}\,|\,N)} \,. \tag{2}$$

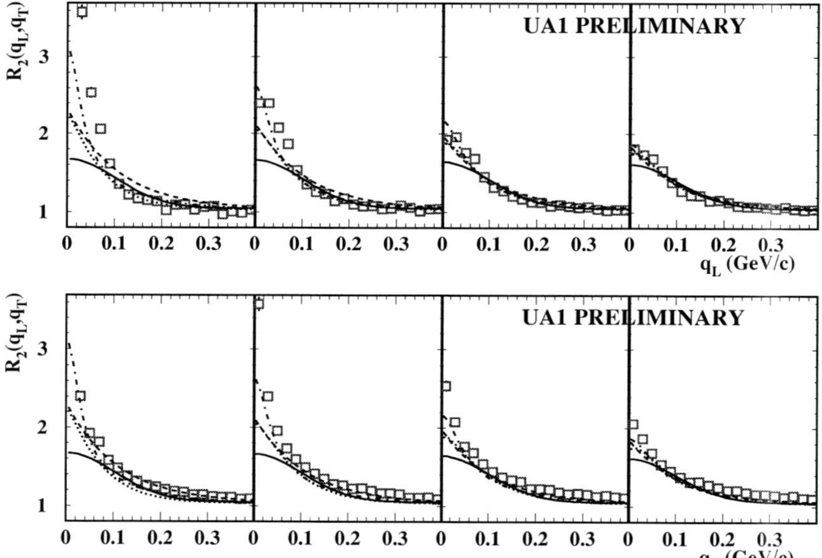

FIGURE 1. Upper panels: $R_2(q_L, q_T)$ data and best fits, shown for the first slices (left to right) q_T = 0.00–0.02, 0.02–0.04, 0.04–0.06 and 0.06–0.08 GeV/c. Lower panels: $R_2(q_L, q_T)$ and the same fits for corresponding fixed-q_L slices. Solid lines: Gauss/Edgeworth fits; dashed: exponential; dotted: exponential with cross term; dash-dotted: Lévy. The bin $(q_L, q_T) < (0.02, 0.02)$ GeV/c is omitted from all fits and plots.

---

[1] The (q_L, q_T) bins for ghost corrections use q_L measured in the detector rest system ($p\bar{p}$ CMS), while all HBT quantities are measured in the LCMS.

The ghost- and coulomb-corrected normalised moment $R_2(q_L, q_T)$ is shown in Fig. 1, together with fits to parametrisations $R_2 = \gamma[1 + \lambda\, |S_{12}|^2]$, with $|S_{12}|^2$ parametrised respectively as $\exp(-R_L^2 q_L^2 - R_T^2 q_T^2 - 2R_{LT} q_L q_T)$ (gauss with cross term), $\exp(-R_L q_L - R_T q_T)$ (exponential), $\exp(-R_L q_L - R_T q_T - 2R_{LT}\sqrt{q_L q_T})$ (exponential with cross term) and $R_2 = \gamma\left[1 + (R_L q_L)^{-\alpha_L} (R_T q_T)^{-\alpha_T}\right]$ (product power law). Note that all results shown are preliminary.

It is immediately apparent that none of these fits reproduces the data, with χ^2/NDF ranging from 3 to 9. A parametrisation based on an Edgeworth expansion [8, 9],

$$|S_{12}|^2 = \exp(-R_L^2 q_L^2 - R_T^2 q_T^2) \prod_{d=L,T} \left[1 + \kappa_{4,d}\, H_4(\sqrt{2} R_d\, q_d)/24\right], \qquad (3)$$

(with $\kappa_{4,d}$ the fourth-order cumulant in q_d and H_4 the corresponding hermite polynomial) fares no better as best-fit values for $\kappa_{4,d}$ in both directions turn out to be negligible. We omit the third-order terms in $\kappa_{3,L}$ and $\kappa_{3,T}$ as they are antisymmetric in q_d.

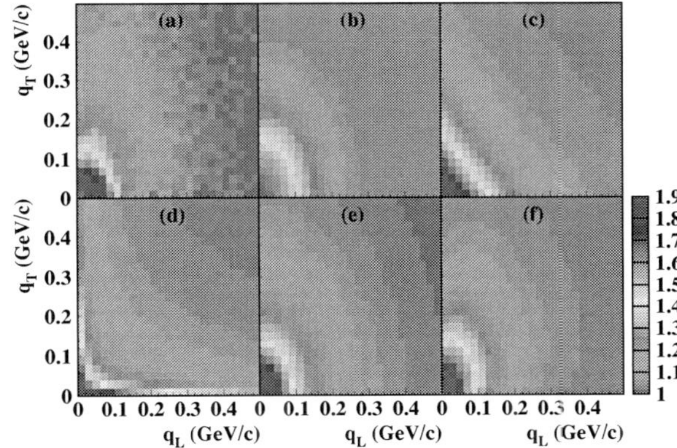

FIGURE 2: Comparing the shapes of $R_2(q_L, q_T)$ at intermediate scales. Panels show (a)) UA1 data, (b) gauss, (c) exponential, (d) power-law, (e) Lévy, (f) exponential with cross term fits. The gauss-type fits have the right shape, but end up far below the data peak at small (q_L, q_T). While the simple exponential and product power-law may approximate the peak reasonably, both fail miserably when shapes at intermediate scales are considered. Lévy does well but no unique best fit can be found. Shape-wise, the exponential with cross term (f) appears to come out on top. All plots are truncated at $R_2 \leq 1.9$ in order to bring out structure at intermediate scales.

A Lévy-based parametrisation [10],

$$|S_{12}|^2 = \exp(-R_L^2 q_L^2 - R_T^2 q_T^2)^{\alpha/2}, \qquad (4)$$

yields better results in reproducing the strong peak observed in the data; however, four of the five fit parameters, viz. λ, R_L, R_T and α, are strongly correlated so that no unique best fit can be achieved. One example of many equivalent "best" fits is shown in Fig. 1. Omitting a second small-(q_L, q_T) bin from the Lévy fit renders the fit even more unstable (the innermost bin $(q_L, q_T) < (0.02, 0.02)$ GeV/c is excluded from all analysis as a matter of course). This is hardly surprising, as the four abovementioned parameters collectively depend strongly on the exact shape of the peak in the very small (q_L, q_T) region — the very region that experimental measurement struggles to resolve.

Interesting, nonetheless, is the observation that fits fail for different reasons: As shown in Fig. 2, the gauss and Edgeworth fits reproduce the shape of R_2 at intermediate (q_L, q_T) rather well but lacks the strong peak exhibited in the data,[2] while the exponential and product power-law parametrisations are more peaked but fail to describe the shape at intermediate (q_L, q_T). Judging by shape alone, the exponential with a $2R_{LT}\sqrt{q_L q_T}$ cross term does best (Fig. 2(f)).

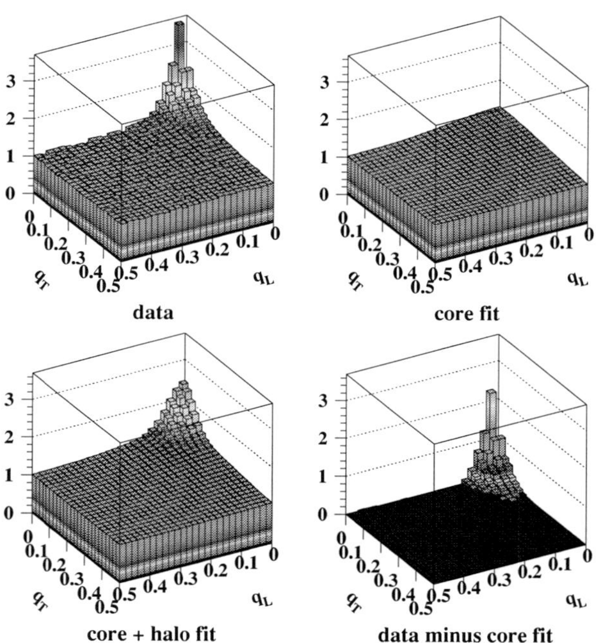

FIGURE 3: Two-scale butcher's model. Of the $R_2(q_L, q_T)$ data (left upper panel) only bins with $(q_L, q_T) > (0.16, 0.16)$ GeV/c are fit to the "core" gaussian of Eq. (5) as shown in the right upper panel. The result is subtracted from the data to reveal the "halo" (right lower) which is then fit to the halo part of Eq. (5). The combined core-halo fit is shown in the left lower panel.

Either way, it is clear that the superiority of the power-law fit to the one-dimensional invariant four-momentum moment $R_2(Q)$ over Gauss and exponential fits [4] is not repeated in the $R_2(q_L, q_T)$ case due to the hyperbolic shape of our "product power-law" parametrisation[3] compared to the elliptic form of the $R_2(q_L, q_T)$ data shown in

[2] Edgeworth fits are indistinguishable from the normal gaussian fits throughout this analysis.

[3] Other forms, such as a "sum power law" $R_2 = \gamma[1 + (R_L q_L)^{-\alpha_L} + (R_T q_T)^{-\alpha_T}]$ remain to be tested.

Fig. 2(a). The analysis of shapes at intermediate scales appears to provide valuable information. Shape analysis in the form of Refs. [11, 12] may help to quantify these qualitative observations.

The failure of conventional parametrisations to reproduce the elusive peak at small (q_L, q_T) suggests that there may be two scales in the system. Borrowing the (strictly speaking inappropriate) terms "core" and "halo" from the literature [13], we try a two-scale "butcher's model", similar to simpler precursors in Refs. [14, 15],

$$R_2(q_L, q_T) = \gamma \left[1 + \lambda_C \exp(-R_{LC}^2 q_L^2 - R_{TC}^2 q_T^2) + \lambda_H \exp(-R_{LH}^2 q_L^2 - R_{TH}^2 q_T^2)\right], \quad (5)$$

and proceed as follows: First, we fit only bins with momentum differences larger than a hard cutoff, $(q_L, q_T) > (q_{\text{cut}}, q_{\text{cut}})$, to the core gaussian. The resulting best-fit is subtracted from all data. The remaining halo "data" with $(q_L, q_T) \leq (q_{\text{cut}}, q_{\text{cut}})$ is then fit with the halo gaussian, and the resulting "core" and "halo" best fits combined. Data and fit histograms at various steps of this procedure are shown in Fig. 3.

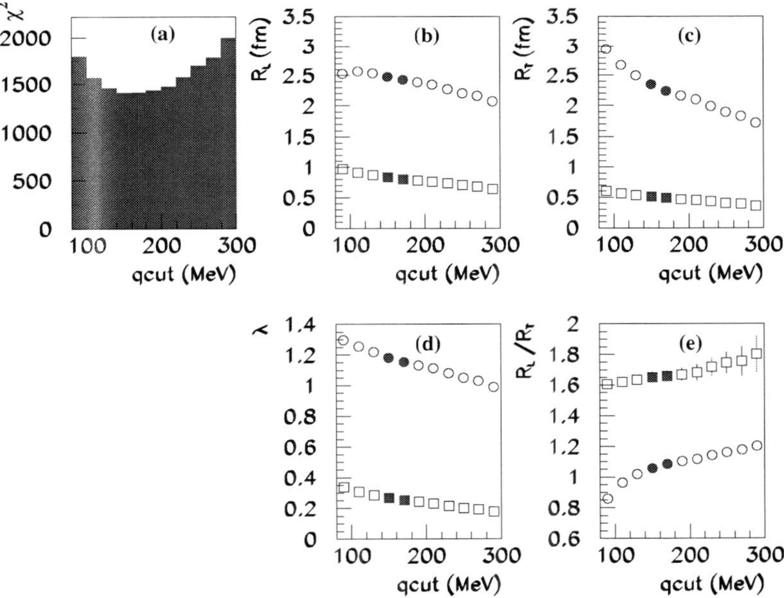

FIGURE 4: Dependence on q_{cut}. (a) Total χ^2 for both core and halo fits as a function of q_{cut}. The two lowest χ^2 values correspond to $q_{\text{cut}} = 0.16$ and 0.18 GeV/c, giving $\chi^2/\text{NDF} = 2.28$. Panels (b)–(e): Dependence of best-fit parameter values on q_{cut}. Squares and circles represent core and halo fit parameters respectively. A clear separation of scales is observed as assumed in the model. The halo continues to have a chaoticity parameter λ_H that exceeds its theoretical limit of 1.

In Fig. 4, the dependence of the two-scale model on q_{cut} is tested. As shown in Fig. 4(a), the joint χ^2 for both fits as a function of q_{cut} is found to have a well-defined minimum for $q_{\rm cut} = 160$–180 MeV/c. The best estimates for R_L, R_T, λ and R_L/R_T are the two filled points in Fig. 4(b)–(e) corresponding to the two best $q_{\rm cut}$ values. Averaging these two numbers, we estimate $R_{LC} = 0.82 \pm 0.02$ fm, $R_{TC} = 0.49 \pm 0.02$ fm, $R_{LH} = 2.45 \pm 0.03$ fm, $R_{TH} = 2.29 \pm 0.05$ fm, $\lambda_C = 0.26 \pm 0.01$ and $\lambda_H = 1.17 \pm 0.02$, signalling a prolate "core" and a roughly spherical "halo". The clear separation between the sizes of the core and halo radii *a posteriori* support the assumption of the presence of two scales, i.e. signal that the two-scale model is consistent. We note that λ_H continues to exceed the theoretical limit of 1.00, albeit not as strongly as the huge intercept $R_2(0,0) > 2.7$ seen in the data itself. *The joint best* $\chi^2/\mathrm{NDF} = 2.28$ *for the two-scale model is still rather large, however, so that all numbers and conclusions should be treated with caution.*

Turning briefly to the more common three-dimensional Bertsch-Pratt representation [16], we find that the data once again has a strong peak at small (q_O, q_S, q_L), and that the simple gaussian parametrisation fails completely. In Fig. 5, we show the result of fitting the Lévy and Edgeworth parametrisations,

$$|S_{12}|^2 = \exp(-R_O^2 q_O^2 - R_S^2 q_S^2 - R_L^2 q_L^2)^{\alpha/2}, \tag{6}$$

$$|S_{12}|^2 = \exp(-R_O^2 q_O^2 - R_S^2 q_S^2 - R_L^2 q_L^2) \prod_{d=O,S,L} \left[1 + \kappa_{4,d} H_4(\sqrt{2} R_d q_d)/24\right], \tag{7}$$

to $R_2(q_O, q_S, q_L)$, with slices plotted along the three axes of the three-dimensional space. Even these parametrisations appear not to describe the data well for small $|\mathbf{q}|$, and in this threedimensional case the discrepancies are spread more widely than in the (q_L, q_T) case. Also, while the Lévy fit does appear to do reproduce the data to a reasonable degree, it again suffers from strong inherent correlations between the parameters λ, R_O, R_S, R_L and α. Such correlation again implies that no unique minimum for the χ^2 exists and thereby no unique best-fit values for its parameters either.

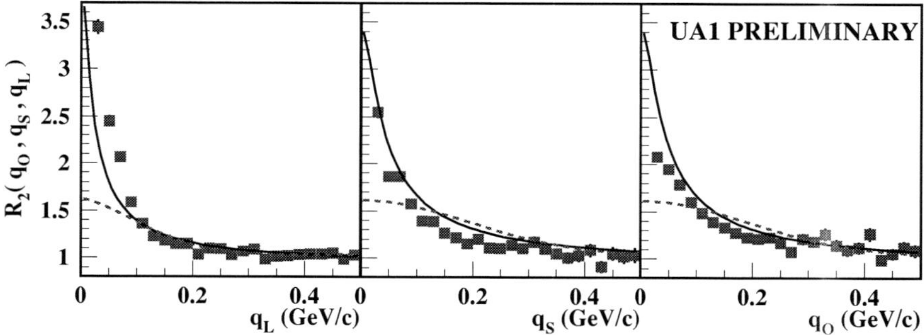

FIGURE 5: Second moment $R_2(q_O, q_S, q_L)$ of three-dimensional momentum difference decomposition, plotted along the three axes q_d, with the other two variables $(q_e, q_f) < (0.02, 0.02)$ GeV/c. Solid line: Lévy fit; dashed line: Edgeworth.

It is, of course, always possible to try even more elaborate parametrisations, for example using separate α-exponents for each of the (q_O, q_S, q_L) directions. While a better fit might theoretically be achieved, this will invariably come at the price of even more highly correlated parameters. The data cannot distinguish between many combinations of "best" values for such parameters.

We note that, even within the arguably artificial method of the two-scale model, the strong peak in the data is not reproduced, resulting in the quoted low confidence level. Nevertheless, it may provide a useful hint that "something else is going on", be it the influence of jets, clustering effects or some other unknown factor.

We stress that is is unlikely that the strong peak seen in our data at small momentum differences, over and above simple gaussian or exponential parametrisations, is due to bias. First, the peak persists even without the Coulomb or split-track corrections. Second, we believe that the same peak may well be responsible for the power-law form and higher-order effects seen in our earlier work with one-dimensional distributions [4]. Third, the presence of unidentified kaons and protons in the sample imply that the real peak should exceed the one shown here, so that the present results may be lower limits. It should also be noted that a number of other hadronic experiments have previously seen significant deviation from gaussian behaviour, especially at small $|\mathbf{q}|$ [17, 18]. The challenge is clearly now to find a convincing physical cause and explanation.

ACKNOWLEDGMENTS

This work was supported in part by the National Research Foundation of South Africa. BB thanks the University of Stellenbosch for kind hospitality.

REFERENCES

1. B. Buschbeck, H. C. Eggers and P. Lipa, *Phys. Lett.* B**481**, 187–193 (2000), hep-ex/0003029.
2. UA1 Collaboration; C. Albajar et al., *Phys. Lett.* B**226**, 410–416 (1989).
3. UA1 Collaboration, N. Neumeister et al., Phys. Lett. B**275**, 186 (1992).
4. H. C. Eggers, P. Lipa and B. Buschbeck, *Phys. Rev. Lett.* **79**, 197–200 (1997), hep-ph/9702235.
5. M.G. Bowler, *Phys. Lett.* B**270**, 69–74 (1991).
6. D. Brinkmann, PhD Thesis, University of Frankfurt (1995).
7. P. Lipa, H.C. Eggers and B. Buschbeck, *Phys. Rev.* D**53**, 4711–4714 (1996), hep-ph/9604373.
8. T. Csörgő, in: *Proc. Cracow Workshop on Multiparticle Production*, 1993, edited by A. Białas, K. Fiałkowski, K. Zalewski and R.C. Hwa, World Scientific (1994); pp. 175–186.
9. STAR Collaboration, J. Adams et al., nucl-ex/0411036.
10. T. Csörgő, S. Hegyi and W. A. Zajc, *Eur. Phys. J.* C**36**, 67–78 (2004), nucl-th/0310042.
11. P. Danielewicz and S. Pratt, *Phys. Lett.* B**618**, 60–67 (2005), nucl-th/0501003.
12. Z. Chajecki, T.D. Gutierrez, M.A. Lisa and M. Lopez-Noriega (STAR Collaboration), in: *21st Winter Workshop on Nuclear Dynamics*, Breckenridge CO, February 2005, nucl-ex/0505009.
13. T. Csörgő, B. Lörstad and J. Zimányi, *Z. Phys.* C**71**, 491 (1994), hep-ph/9411307.
14. R. Lednitsky and M.I. Podgoretzkii, *Sov. J. Nucl. Phys.* **30**, 432 (1979);
R. Lednitsky et al., *Sov. J. Nucl. Phys.* **38**, 147 (1983).
15. EHS/NA22 Collaboration, N.M. Agababyan et al., *Z. Phys.* C**59**, 195–210 (1993).
16. G.F. Bertsch, M. Gong and M. Tohyama, Phys. Rev. C**37**, 1896–1900 (1988).
17. AFS Collaboration, T. Åkesson et al., *Phys. Lett.* B**129**, 269–272 (1983).
18. EHS/NA22 Collaboration, N. M. Agababyan et al., *Z. Phys.* C**68**, 229 (1995);
Z. Phys. C**71**, 405 (1996).

Pion Interferometry from p+p to Au+Au in STAR

Z. Chajęcki (for the STAR Collaboration)

The Ohio State University, 191 W. Woodruff Avenue, Columbus, Ohio 43210, USA

Abstract. The geometric substructure of the particle-emitting source has been characterized via two-particle interferometry by the STAR collaboration for all energies and colliding systems at RHIC. We present systematic studies of charged pion interferometry. The collective nature of the source is revealed through the m_T dependence of HBT radii for all particle types. Preliminary results suggest a scaling in the pion HBT radii with overall system size, as central Au+Au collisions are compared to peripheral collisions as well as with Cu+Cu and even with d+Au and p+p collisions, naively suggesting comparable flow strength in all systems. To probe this issue in greater detail, multidimensional correlation functions are studied using a spherical decomposition method. This allows clear identification of source anisotropy and, for the light systems, the presence of significant long-range non-femtoscopic correlations.

Keywords: HBT, femtoscopy, heavy ion collisions, intensity interferometry
PACS: `25.75Gz`

INTRODUCTION

Particle interferometry is an useful technique that provides information on the space-time properties of the particle emitting source and may be helpful in understanding the dynamics of the system created in high energy collisions by studying the transverse mass dependence ($m_T = \sqrt{k_T^2 + m_\pi^2}, k_T = \frac{1}{2}(p_{T_1} + p_{T_2})$) of HBT radii (for the latest review article see [1]). In this article femtoscopic results from a small system (p+p collisions) and a large system (Au+Au collisions) measured by the same experiment, at the same collision energy and detector acceptance are presented for the first time in high energy physics. The particular focus is on the m_T dependence of HBT radii and an attempt to understand its origins for different initial sizes of the emitting source is made.

ANALYSIS DETAILS

The STAR Time Projection Chamber (TPC) [2] was used to reconstruct particles of interest. Particle identification was achieved by measuring momentum and specific ionization losses of charged particles in the gas of TPC (dE/dx technique). A large data statistics in Au+Au, Cu+Cu and d+Au collisions allows to do an analysis for different centralities. Additionally, d+Au data set allows to extract p+Au collisions. It is performed by selecting events with a single neutron tagged in the Zero Degree Calorimeter (ZDC) in the deuteron beam direction.

CP828, *Multiparticle Dynamics*
edited by V. Šimák, M. Šumbera, Š. Todorova, and B. Tomášik

In this study pions with transverse momentum 0.10 GeV/c $< p_T <$ 1.00 GeV/c were used and the analysis was done for four bins of k_T within a range of $[0.15, 0.60]$ GeV/c. Two-track effects due to splitting (one particle reconstructed as two tracks) and merging (two particles reconstructed as one track) were removed from the data.

The dependence of the correlation function on the transverse momentum is studied as a function of three components of pair relative momentum in a Pratt-Bertsch coordinate system [3, 4] in the longitudinally co-moving frame. The fit was performed using a method suggested by Bowler [5] and Sinyukov [6] assuming Gaussian parametrization of the source. For more details on analysis technique see [11].

SYSTEM EXPANSION AND MULTIPLICITY SCALING

One of the differences that may be expected between such a large system as Au+Au and a small system as p+p is the expansion. This can be studied in a model-dependent approach when the final RMS of the system is compared to the initial one. The first value is equal, with good accuracy, to R_{side} calculated for the lowest k_T range ([0.15, 0.25] GeV/c). The second one is calculated with Glauber model for nuclei and a proton initial size was taken from an e^- scattering [12] as a reference. The result of this comparison for p+p, d+Au, Cu+Cu and Au+Au collisions at the same energy of the collision ($\sqrt{s_{NN}}$ = 200 GeV) are combined on the left panel of Figure 1a.

As seen, the most central Au+Au collisions undergo an expansion by a factor of two while p+p collisions show no or a little expansion. Data points from peripheral d+Au collisions show similarity to p+p while central d+Au points exhibit an expansion like in peripheral Cu+Cu. Finally, central Cu+Cu expands similarly like peripheral Au+Au.

Figure 1b presents the HBT radii dependence on $(dN_{ch}/d\eta)^{1/3}$ (dN_{ch} - number of charged particles at midrapidity) for different colliding systems and at different energies

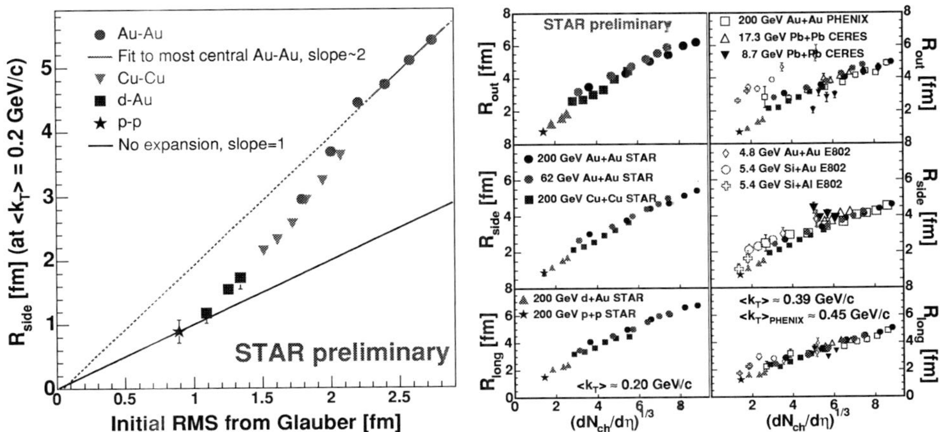

FIGURE 1. a) Final size of the source vs. initial radii. STAR Au+Au data from [11]; b) Femtoscopic radii dependence on the number of charged particles.

of the collisions. The motivation for studying such a relation is its connection to the final state geometry through the particle density at freeze-out. All STAR results, from p+p, d+Au, Cu+Cu and Au+Au collisions, are combined on the left panel of this figure and, as seen, all radii exhibit a scaling with $(dN_{ch}/d\eta)^{1/3}$. On the right panel of this figure STAR radii, this time for different k_T range, are plotted together with AGS/SPS/RHIC systematics [1]. It is impressive that the geometric radii (R_{side} and R_{long}) follow the same curve for different collisions over a wide range of energies and, as it was checked, this observation is valid for all k_T bins studied by STAR. It is a clear signature that the multiplicity is a scaling variable that drives geometric HBT radii at midrapidity. R_{out} mixes space and time information. Therefore it is unclear whether to expect the simple scaling with the final state geometry. Although, because of the finite intercepts of the linear scaling [1, 7], results do not confirm predictions that freeze-out takes place at the constant density [8]. Additionally, this scaling was verified at midrapidity only, and some dependence on rapidity outside this region may be expected [9, 8] so it is not obvious if the scaling holds then.

As a result of this study, one can venture to predict the size of the source at midrapidity without knowing anything about the collision (like energy, N_{part}, impact parameter, etc.) except for the multiplicity [9, 8, 1]. This scaling is expected to persist for all systems that are meson dominated but is violated for low energy collisions that are dominated by baryons [9, 1, 10].

TRANSVERSE MASS DEPENDENCE OF HBT RADII

In heavy ion collisions a decrease of HBT radii with an increase of m_T is commonly associated with the transverse flow of a bulk matter [11]. Natural question would be whether this dependence looks different in smaller systems like p+p or d+Au and what is the origin of this dependence. On Figure 2a the three dimensional radii from p+p and d(p)+Au collisions are plotted vs m_T. For these systems femtoscopic sizes decrease with the increase of the transverse mass and d+Au results show also the dependence on the centrality like it is observed in Au+Au collisions [11]. Additionally, the value of R_{side} and R_{long} for p+Au collisions is similar to p+p collisions while R_{long} is more like in d+Au collisions. Although it has to be emphasized that due to the way of extracting p+Au events from d+Au sample p+Au results correspond rather to peripheral p+Au collisions so the size of the source is expected to be larger for central collisions. Therefore, results suggest that the size of the source in p+Au collisions is not the same as in p+p. Comparison of the peripheral d+Au collisions, that include about 15% of p+Au collisions, with and without extracted p+Au events show no significant difference but that may be due to a fact that the d+Au sample still includes n+Au events that cannot be excluded from the data.

In elementary particle collisions, resonance production contributes significantly to the m_T dependence of the HBT radii [13], while in heavy ion collisions, flow effects dominate this dependence [14]. The other scenarios that can give the similar dependence are the Heisenberg uncertainty and the string fragmentation [15].

On Figure 2b the ratio of the three dimensional radii in Au+Au, Cu+Cu and d+Au collisions to p+p radii is plotted vs m_T. Surprisingly, these ratios look flat although it is

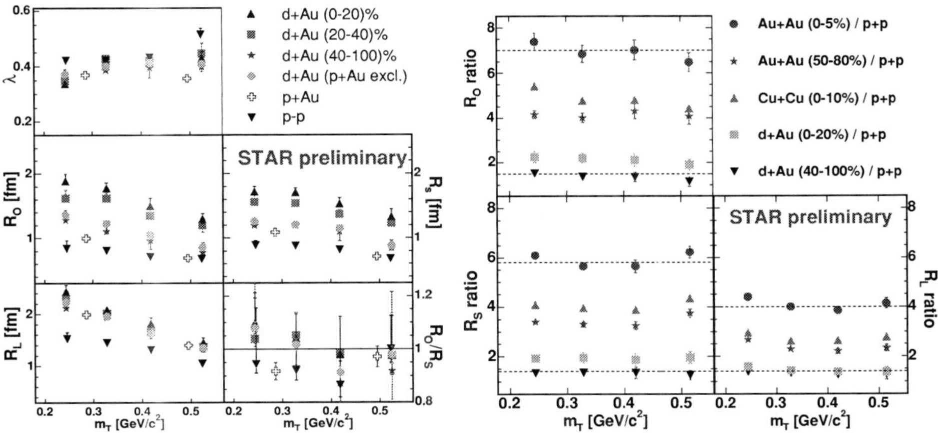

FIGURE 2. a) m_T dependence of HBT radii and λ in p+p and d(p)+Au collisions at $\sqrt{s_{NN}} = 200$ GeV; b) Ratio of HBT radii from Au+Au, Cu+Cu and d+Au by p+p collisions at $\sqrt{s_{NN}} = 200$ GeV.

expected that different origins drive the transverse mass dependences of the HBT radii in Au+Au and p+p collisions. If these expectations are correct the data show that one may not distinguish different physics between p+p and Au+Au collisions studying pion interferometry.

An alternative explanation of this phenomena came from a work done by Csörgő *et al.* [16]. Authors using a Buda-Lund hydrodynamic model, that successfully describes the momentum correlations in Au+Au collisions, were able to fit STAR p+p spectra and HBT radii. But in this case they claim that the transverse mass dependence of the femtoscopic sizes is not generated by the transverse flow, but by the temperature inhomogeneities of hadron-hadron collisions due to the freezing scale. Then the conclusion from this study would be that in p+p collisions the system has similar bulk properties as in Au+Au collisions.

Non-identical particle correlations like π-K or π-p in Au+Au collisions show a difference in then average emission points of two particles that is due to flow [17]. Therefore, femtoscopic study of particles with different masses in p+p collisions could be used to verify a flow hypothesis in small systems like p+p and d(p)+Au.

EVIDENCE OF NON-FEMTOSCOPIC CORRELATIONS

When doing femtoscopic analysis in p+p and d+Au collisions a problem with non-femtoscopic correlations has been observed. It is manifested in a non-vanished tail of the correlation function to unity, for large $\vec{Q}$. In elementary particle collisions [15, 18] these non-femtoscopic correlations were also observed and taken into account by adding an *ad-hoc* component to the parametrization of the correlation function that assumes that the correlation function for large $\vec{Q}$ depends linearly on the three components of the two-

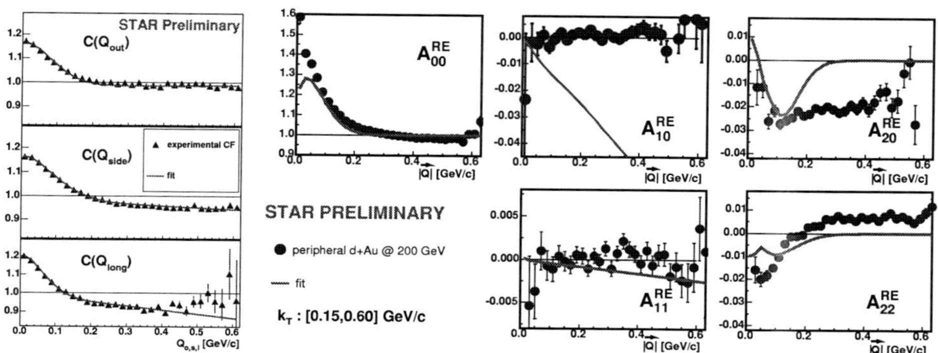

FIGURE 3. a) 1D projections of 3D correlation function. b) First five non-vanished components of the spherical harmonics decomposition of the correlation function. In a) and b) 3D correlations function for d+Au peripheral collisions was fitted with parametrization given by Eq.(1)

particle relative momentum (see equation (1)).

$$C_2(Q_{out}, Q_{side}, Q_{long}) = (1 + \lambda e^{-(Q_{out}^2 R_{out}^2 + Q_{side}^2 R_{side}^2 + Q_{long}^2 R_{long}^2)})(1 + \alpha Q_{out} + \eta Q_{side} + \zeta Q_{long}) \tag{1}$$

Using this parametrization the fit to the STAR p+p and d+Au collisions was performed and the femtoscopic radii turned out to be larger up to 30% in comparison to the standard parametrization. Figure 3a shows the projections of the 3D correlation function for the most peripheral d+Au collisions (that is STAR worst case) and the projections of the fit described above. It looks like the fit matches experimental data with good accuracy but more careful study is required to judge on the correctness of the new parametrization and will be performed with method described below.

A common approach to present 3D correlation function is to project it onto the three components of $\vec{Q}$ separately, as shown on Figure 3a. The disadvantage of this approach is that when doing such projections one has to constrain non-projected components to keep a signal but then the full information on the correlation function in the 3D space is lost. To eliminate this inconvenience a new approach of studying correlations was applied that is based on a decomposition of the correlation function into spherical harmonics (for detailed description of this method see [19]). In this method no cuts are performed on $\vec{Q}$'s components what allows to recognize symmetries in $\vec{Q}$-space to see, looking at 1D plots, relevant aspects of 3D correlation functions.

Figure3b shows the first few components of the decomposition of the correlation function onto the spherical harmonics for peripheral d+Au collisions. The fitted correlation function, that includes a new term to account for non-femtoscopic effect, was decomposed using this method. As shown on Figure 3a the new parametrization fits the correlation function with good accuracy but with the spherical harmonic method it is seen that the fit is not correct. Distributions for l=1 are non-zero and $A_{1,0}$ shows a strong dependence on $|\vec{Q}|$. In a system like $\pi - \pi$ at midrapidity all odd components should vanish by symmetry. Additionally, the new parametrization does not fit the baseline of the correlation function that has an evidence in non-vanished $A_{2,0}$ and $A_{2,2}$ distributions for large $|\vec{Q}|$.

This study shows that it is not sufficient to look at the Cartesian projections of the correlation function to judge about the quality of the fit and the correctness of the used parametrization. It is required to use spherical harmonic method to see the experimental data and the fit in the 3D space.

The analytic formulas of spherical harmonics are well-known so the $A_{2,0}$ and $A_{2,2}$ distributions may be parametrized and included in the fit. Such study was presented in [20] and it showed a good agreement with experimental data. Due to the lack of the space the results are not presented here, but the radii in p+p and d+Au collisions are changed up to 10% the most, although the m_T scaling described in the previous section persists.

CONCLUSIONS

The results of pion interferometry for several energies and colliding systems at RHIC have been presented. In agreement with data at SPS and AGS, STAR indicates that the multiplicity is the scaling variable that determines the size of the source at freeze-out at midrapidity. The m_T dependence of HBT radii seems to be independent of collision species or multiplicity. Finally, a problem with the baseline of the correlation function for low multiplicity collisions has been reported, and a promising tool based on the spherical harmonic decomposition of the correlation function has been used in order to address it. The physics of this structure remains under investigation. The advantage of this method in studying the correctness of the parametrization of the correlation function used in a fit has been shown.

REFERENCES

1. M. Lisa, S. Pratt, R. Solz, U. Wiedemann, *Annu. Rev. Nucl. Part. Sci.* (2005) 55:357-402.
2. M. Anderson *et al.*, *Nucl. Instrum. Meth.* **A 499** (2003) 659 .
3. S. Pratt, T. Csörgö, and J. Zimanyi, *Phys. Rev.* **C 42** (1990) 2646.
4. G. Bertsch, *Nucl. Phys.* **A 498** (1989) 173c.
5. M. G. Bowler, *Phys. Lett.* **B 270** (1991) 69.
6. Yu. M. Sinyukov *et al.*, *Phys. Lett.* **B 432** (1998) 249.
7. S.S. Adler *et al.* (PHENIX Collaboration), *Phys. Rev. Lett.* **93** (2004) 152302.
8. T. Csörgő and L. P. Csernai, *Phys. Lett.* **B 333** (1994) 494.
9. R. Stock, *Annalen der Physik,* **48** (1991) 195.
10. D. Adamová *et al.* (CERES Collaboration), *Phys. Rev. Lett.* **90** (2003) 022301.
11. J. Adams *et al.* (STAR Collaboration), *Phys. Rev.* **C 71** (2005) 044906.
12. I. Sick, *Eur. Phys. J.* **A 24**, s1, (2005) 65-67.
13. R.M. Weiner, *Phys. Rep.* **237** (2000) 249-346.
14. U. A. Wiedemann, U. Heinz, *Phys. Rev.* **C 56** (1997) 3265.
15. G. Alexander, ArXiv:hep-ph/0302130.
16. T. Csörgő, M. Csanád, B. Lörstad, A. Ster, ArXiv:hep-ph/0406042.
17. J. Adams *et al.* (STAR Collaboration), *Phys. Rev. Lett.* **91** (2003) 262302.
18. N.M. Agababyan *et al.* (NA22 Collaboration), *Z. Phys.* **C 71** (1996) 405.
19. Z. Chajęcki *et al.* (for the STAR Collaboration), ArXiv:nucl-ex/0505009.
20. Z. Chajęcki (for the STAR Collaboration), *Proceedings of the Quark Matter 2005 Conference*, to appear in *Nucl. Phys. A*, ArXiv:nucl-ex/0510014.

Comparison of Emission Functions in h+p, p+p, A+A Reactions

A. Ster* and T. Csörgő†

*MTA KFKI RMKI, MFA, H - 1525 Budapest 114, P.O.Box 49, Hungary
†MTA KFKI RMKI, H - 1525 Budapest 114, P.O.Box 49, Hungary

Abstract. The space-time evolution of colliding systems is compared at energies of SPS and RHIC experiments. Previously, the Buda-Lund hydrodynamical model was able to reconstruct the final states of h+p, A+A reactions. Now, we use these hydro results, together with preliminary ones obtained from RHIC p+p collisions, to calculate the particle emission functions. Their shapes and the parameters characterizing the source of emission are shown and discussed in detail. The comparison gives that at RHIC energies we are above the critical temperature ($T_c = 172 \pm 3\ MeV$) of deconfinement. Moreover, we can see fairly different types of emission dynamics in case of hadron-hadron and heavy ion reactions.

INTRODUCTION

The Buda-Lund hydro model [1] is successful in describing the BRAHMS, PHENIX, PHOBOS and STAR data on identified single particle spectra and the transverse mass dependent Bose-Einstein or HBT radii as well as the pseudorapidity distribution of charged particles in Au + Au collisions both at $\sqrt{s_{\rm NN}} = 130$ GeV [2] and at $\sqrt{s_{\rm NN}} = 200$ GeV [3]. The result of the simultaneous fit to all these datasets indicate the existence of a very hot region, with a temperature significantly greater than 170 MeV. Fodor and Katz calculated the phase diagram of lattice QCD at finite net barion density [4]. These lattice results, obtained with light quark masses four times heavier than the physical value, indicated that in the $0 \leq \mu_B \leq 300$ MeV region the transition from confined to deconfined matter is a cross-over, with $T_c \simeq 172 \pm 3$ MeV. This value is, within one standard deviation, independent of the bariochemical potential in the $0 \leq \mu_B \leq 300$ MeV region. The Buda-Lund fits, combined with these lattice results, provide an indication for quark deconfinement in Au + Au collisions with $\sqrt{s_{NN}} = 130$ GeV, as well as, with $\sqrt{s_{NN}} = 200$ GeV colliding energies at RHIC.

In this paper we compare the the different reactions via the particle emission probabilities for which we use the Buda-Lund hydrodynamical model parameters extracted from the final data of the above experiments. In this analyses we have included our earlier hydro results we obtained fitting data of CERN SPS experiments of Pb+Pb and h+p collisions, as well [5, 6]. Our preliminary hydro results on p + p collisions at $\sqrt{s_{\rm NN}} = 200$ GeV of PHENIX and STAR were also used, as an addition. However, we have made the first attempt to fit d+Au data, as well, this time those results are not part of the study because such sort of collisions represent an asymmetric case. Currently, the model describes axially or ellipsoidally symmetric cases.

CP828, *Multiparticle Dynamics*
edited by V. Šimák, M. Šumbera, Š. Todorova, and B. Tomášik

EMISSION FUNCTION IN THE BUDA-LUND HYDRO MODEL

The Buda-Lund hydro model was introduced in refs. [1, 7]. This model was defined in terms of its emission function $S(x,k)$, for axial symmetry, corresponding to central collisions of symmetric nuclei. The observables are calculated analytically, see refs. [2, 8] for details and key features. Here we summarize the Buda-Lund emission function in terms of its fit source parameters.

The single particle invariant momentum distribution, $N_1(k_1)$, is obtained as

$$N_1(k_1) = \int d^4x S(x,k_1). \tag{1}$$

For chaotic (thermalized) sources, in case of the validity of the plane-wave approximation, the two-particle invariant momentum distribution $N_2(k_1,k_2)$ is also determined by $S(x,k)$, the single particle emission function, if non-Bose-Einstein correlations play negligible role or can be corrected for, see ref. [8] for a more detailed discussion. Then the two-particle Bose-Einstein correlation function, $C_2(k_1,k_2) = N_2(k_1,k_2)/[N_1(k_1)N_1(k_2)]$ can be evaluated in a core-halo picture [9], where the emission function is a sum of emission functions characterizing a hydrodynamically evolving core and a surrounding halo of decay products of long-lived resonances, $S(x,k) = S_c(x,k) + S_h(x,k)$. Consequently, the single particle spectra can also be given as a sum, $N_1(k) = N_{1,c}(k) + N_{1,h}(k)$. In the correlation function, an effective intercept parameter $\lambda \equiv \lambda_*(K)$ appears,

$$C_2(k_1,k_2) = 1 + \frac{|\tilde{S}(q,K)|^2}{|\tilde{S}(0,K)|^2} \simeq 1 + \lambda_*(K)\frac{|\tilde{S}_c(q,K)|^2}{|\tilde{S}_c(0,K)|^2}, \tag{2}$$

where the relative and the momenta are $q = k_1 - k_2$, $K = 0.5(k_1 + k_2)$, and the Fourier-transformed emission function is defined as $\tilde{S}(q,K) = \int d^4xS(x,K)\exp(iqx)$.

The measured λ_* parameter of the correlation function is utilized to correct the core spectrum for long-lived resonance decays [9]: $N_1(k) = N_c(k)/\sqrt{\lambda_*(k)}$. The emission function of the core is assumed to have a hydrodynamical form,

$$S_c(x,k)d^4x = \frac{g}{(2\pi)^3}\frac{k^\nu d^4\Sigma_\nu(x)}{B(x,k)+s_q}, \tag{3}$$

where g is the degeneracy factor ($g = 1$ for pseudoscalar mesons, $g = 2$ for spin=1/2 barions). The particle flux over the freeze-out layers is given by a generalized Cooper–Frye factor: the freeze-out hypersurface depends parametrically on the freeze-out time τ and the probability to freeze-out at a certain value is proportional to $H(\tau)$, $k^\nu d^4\Sigma_\nu(x) = m_t\cosh(\eta - y)H(\tau)d\tau\tau_0 d\eta\, dr_x dr_y$. Here $\eta = 0.5\log[(t+r_z)/(t-r_z)]$, $\tau = \sqrt{t^2 - r_z^2}$, $y = 0.5\log[(E+k_z)/(E-k_z)]$ and $m_t = \sqrt{E^2 - k_z^2}$. The freeze-out time distribution $H(\tau)$ is approximated by a Gaussian, $H(\tau) = \frac{1}{(2\pi\Delta\tau^2)^{3/2}}\exp\left[-\frac{(\tau-\tau_0)^2}{2\Delta\tau^2}\right]$, where τ_0 is the mean freeze-out time, and the $\Delta\tau$ is the duration of particle emission, satisfying $\Delta\tau \ll \tau_0$. The (inverse) Boltzmann phase-space distribution, $B(x,k)$ is given by

$$B(x,k) = \exp\left(\frac{k^\nu u_\nu(x)}{T(x)} - \frac{\mu(x)}{T(x)}\right), \tag{4}$$

and the term s_q is 0, -1, and $+1$ for Boltzmann, Bose-Einstein and Fermi-Dirac statistics, respectively. The flow four-velocity, $u^\nu(x)$, the chemical potential, $\mu(x)$, and the temperature, $T(x)$ distributions for axially symmetric collisions were determined as follows:

The expanding matter is assumed to follow a three-dimensional, relativistic flow, characterized by transverse and longitudinal Hubble constants, $u^\nu(x) = (\gamma, H_t r_x, H_t r_y, H_z r_z)$, where γ is given by the normalization condition $u^\nu(x) u_\nu(x) = 1$.

The Buda-Lund hydro model characterizes the inverse temperature $1/T(x)$, and fugacity, $\exp[\mu(x)/T(x)]$ distributions of an axially symmetric, finite hydrodynamically expanding system with the mean and the variance of these distributions, in particular

$$\frac{\mu(x)}{T(x)} = \frac{\mu_0}{T_0} - \frac{r_x^2 + r_y^2}{2R_G^2} - \frac{(\eta - y_0)^2}{2\Delta\eta^2}, \tag{5}$$

$$\frac{1}{T(x)} = \frac{1}{T_0}\left(1 + \frac{r_t^2}{2R_s^2}\right)\left(1 + \frac{(\tau - \tau_0)^2}{2\Delta\tau_s^2}\right). \tag{6}$$

Here R_G and $\Delta\eta$ characterize the spatial scales of variation of the fugacity distribution, $\exp[\mu(x)/T(x)]$, that control particle densities. Hence these scales are referred to as geometrical lengths. These are distinguished from the scales on which the inverse temperature distribution changes, the temperature drops to half if $r_x = r_y = R_s$ or if $\tau = \tau_0 + \sqrt{2}\Delta\tau_s$. Different combinations may also be used to measure the flow, temperature and fugacity profiles [1, 8]: $H_t \equiv b/\tau_0 = \langle u_t \rangle / R_G = \langle u_t' \rangle / R_s$, $H_l \equiv \gamma_t/\tau_0$, where $\gamma_t = \sqrt{1 + H_t^2 r_t^2}$ is evaluated at the point of maximal emissivity, and

$$\frac{1}{R_s^2} = \frac{a^2}{\tau_0^2} = \langle \frac{\Delta T}{T} \rangle_r \frac{1}{R_G^2} = \frac{T_0 - T_s}{T_s} \frac{1}{R_G^2}, \tag{7}$$

$$\frac{1}{\Delta\tau_s^2} = \frac{d^2}{\tau_0^2} = \langle \frac{\Delta T}{T} \rangle_s \frac{1}{\Delta\tau^2} = \frac{T_0 - T_e}{T_e} \frac{1}{\Delta\tau^2}. \tag{8}$$

EMISSION FUNCTIONS OF SPS AND RHIC COLLISIONS

In these analyses these emission functions are calculated for CERN SPS Pb+Pb and h+p collisions, along with RHIC experiments BRAHMS, PHENIX, PHOBOS and STAR. All SPS and RHIC datapoints were fitted simultaneously, using the analytic expressions and the CERN Minuit fitting package. The fitting package used in this analysis is version 1.5, made public at [14]. Table 1 shows the collected source parameters obtained from Buda-Lund hydrodynamical analysis of the data quoted. Where it was applicable, some old source parameters of SPS data have been transformed to new ones, like R_s where the temperature drops to half and the evaporation temperature (T_e). The emission probabilities $S(x,y)$ and $S(z,t)$ shown in Figs. 1, 2, 3, 4, 5 were calculated by equations defined in the previuos section using the the source parameters collected in Table 1.

The figures give comprehensive quantitative and qualitative picture about the nature of the nuclear processes in question. In general, small size systems can be characterized by ring of fire type particle emission, while in heavy ion collisions fireball like evolution

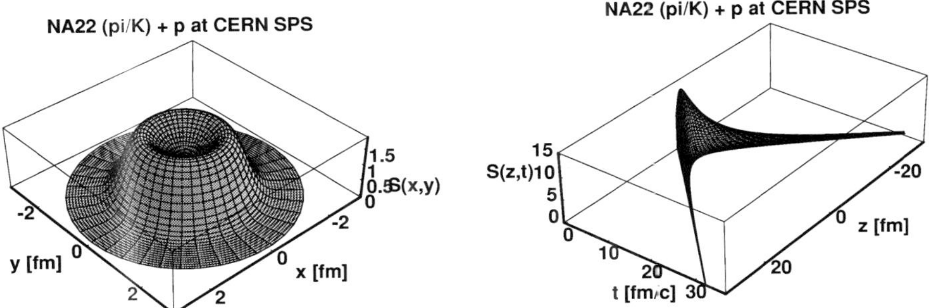

Figure 1. The emission function of h+p reactions at SPS. The left panel shows the value of S in plane (x,y), while the right one shows it in plane (t,z). The strong temperature gradient drives out particles in waves from the center of the emission.

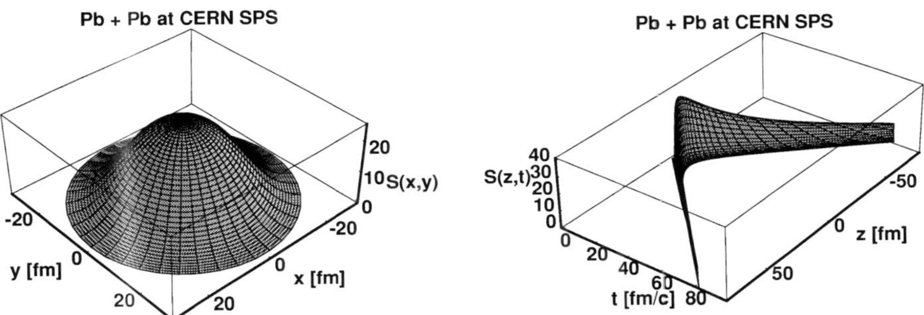

Figure 2. The emission function of Pb+Pb reactions at SPS. The left panel shows the value of S in plane (x,y), while the right one shows it in plane (t,z). The temperature gradient competes with the the strengh of the particle flow which results in fireball type of emission.

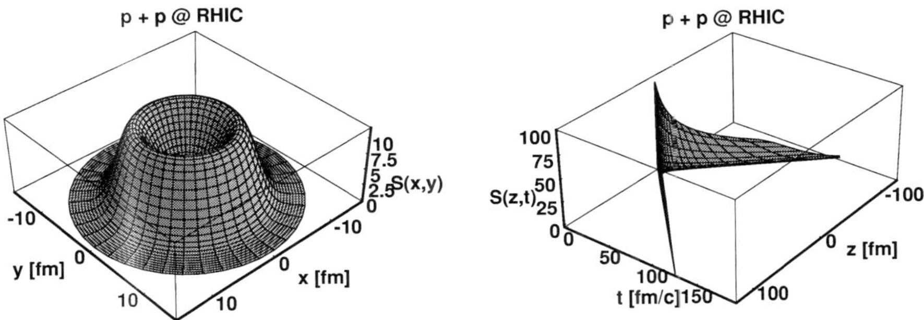

Figure 3. The emission function of p+p reactions at RHIC. The left panel shows the value of S in plane (x,y), while the right one shows it in plane (t,z). The very strong temperature gradient drives out particles from the emission center.

Table 1. Buda-Lund hydro model v1.5 source parameters, corresponding to fits to BRAHMS, PHENIX, PHOBOS and STAR data for Au+Au collisions at $\sqrt{s_{NN}} = 130$ GeV and $\sqrt{s_{NN}} = 200$ GeV shown in ref. [2], as well as, to preliminary PHENIX and STAR single particle spectra and HBT radii data for p+p collisions at $\sqrt{s_{NN}} = 200$ GeV. Pb+Pb data were fitted in ref. [6], while h+p data at CERN SPS in ref. [5]. In the latter case the parameters that were not calculated at the time of publication are left out or transformed to the new ones.

Buda-Lund parameter	SPS h+p	SPS Pb+Pb	RHIC p+p 200 GeV	RHIC Au+Au 130 GeV	RHIC Au+Au 200 GeV
T_0 [MeV]	140 ± 3	139 ± 6	289 ± 8	214 ± 7	196 ± 13
T_e [MeV]	-	87 ± 24	90 ± 42	102 ± 11	117 ± 12
μ_B [MeV]	0 fixed	0 fixed	8 ± 76	77 ± 38	61 ± 52
R_G [fm]	0.88 ± 0.13	7.1 ± 0.2	1.2 ± 0.3	28.0 ± 5.5	13.5 ± 1.7
R_s [fm]	1.4 ± 0.3	28 ± 21	1.13 ± 0.16	8.6 ± 0.4	12.4 ± 1.6
$\langle u_t' \rangle$	0.2 ± 0.07	0.55 ± 0.06	0.04 ± 0.26	1.0 ± 0.1	1.6 ± 0.2
τ_0 [fm/c]	1.4 ± 0.1	5.9 ± 0.6	1.1 ± 0.2	6.0 ± 0.2	5.8 ± 0.3
$\Delta\tau$ [fm/c]	1.3 ± 0.3	1.6 ± 1.5	0.1 ± 0.5	0.3 ± 1.2	0.9 ± 1.2
$\Delta\eta$	1.36 ± 0.02	2.1 ± 0.4	3.0 fixed	2.4 ± 0.1	3.1 ± 0.1
χ^2/NDF	642 / 683	342 / 277	89 / 71	158 / 180	114 / 208

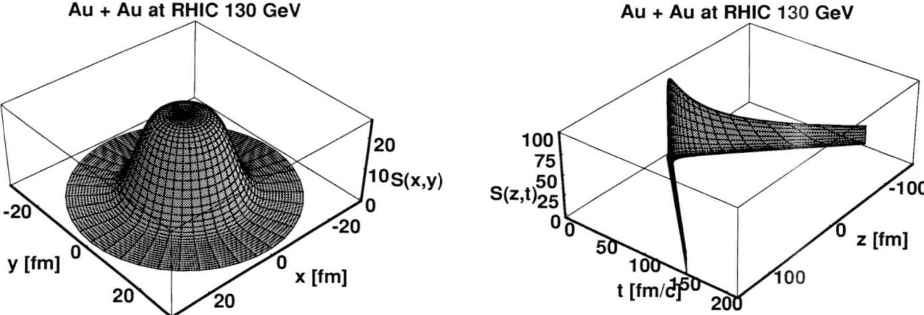

Figure 4. The emission function of Au+Au 130 GeV reactions at SPS. The left panel shows the value of S in plane (x,y), while the right one shows it in plane (t,z). The temperature gradient competes with the strengh of the particle flow which results in fireball type of emission.

takes place. The two types of behaviour are result of two competing processes: whether the collective motion (flow) of particles is fast enough to overcome the pressure caused by the temperature gradient. Consult the corresponding radii and temperature components in the table of source parameters and the figures to recognize such relations.

CONCLUSIONS

It was shown that the Buda-Lund hydrodynamical model describes single particle distributions, rapidity distributions, HBT correlation function radii without puzzle in experiments of h+p and Pb+Pb at SPS, p+p and Au+Au at RHIC.

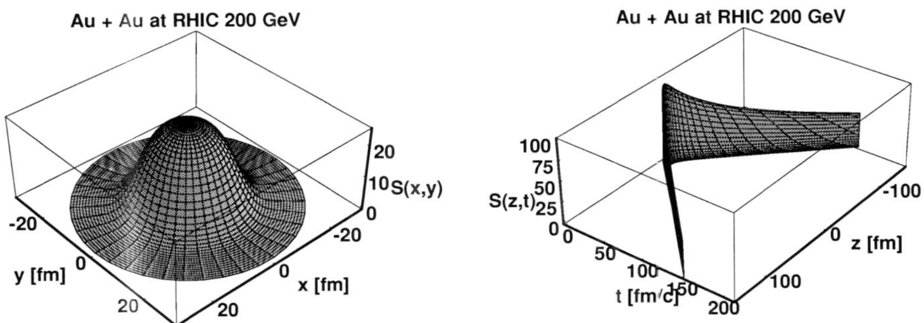

Figure 5. The emission function of Au+Au 200 GeV reactions at SPS. The left panel shows the value of S in plane (x,y), while the right one shows it in plane (t,z). The temperature gradient competes with the strengh of the particle flow which results in fireball type of emission.

We have calculated the emission functions of the reactions that show rings of fire in case of h+p at SPS and p+p at RHIC, whereas, for Pb+Pb at SPS and Au+Au at RHIC their space-time dependence show fireballs.

The quantitative comparison of the parameters of the emission sources extracted by Buda-Lund hydrodynamical model calculations has revealed that the freeze-out temperature in the center of the reaction zone is lower then the critical temperature of deconfinement of quarks ($T < T_c = 172 \pm 3$ MeV) both in h+p and in Pb+Pb reactions at SPS. But, it is always higher then that ($T > T_c$) by more or about 3 standard deviations at RHIC both in p+p and in Au+Au collisions.

ACKNOWLEDGMENTS

This research was supported by the Hungarian - US MTA OTKA NSF grant INT0089462 and by the OTKA grant T038406.

REFERENCES

1. T. Csörgő and B. Lörstad, *Phys. Rev. C* **54**, 1390-1403 (1996).
2. M. Csanád, T. Csörgő, B. Lörstad, A. Ster, *Act. Phys. Pol. B* **35**, 191-196 (2004).
3. M. Csanád, T. Csörgő, B. Lörstad and A. Ster, *J. Phys. G* **30**, S1079-S1082 (2004).
4. Z. Fodor and S. D. Katz, *JHEP* **0203**, 014 (2002).
5. N. M. Agababian *et al.* [EHS/NA22 Coll.], *Phys. Lett. B* **422**, 359-368 (1998).
6. A. Ster, T. Csörgő and B. Lörstad, *Nucl. Phys. A* **661**, 419-422 (1999).
7. T. Csörgő and B. Lörstad, *Nucl. Phys. A* **590**, 465c-468c (1995).
8. T. Csörgő, *Heavy Ion Phys.* **15**, 1-80 (2002).
9. T. Csörgő, B. Lörstad and J. Zimányi, *Z. Phys. C* **71**, 491-497 (1996).
10. S. V. Akkelin, T. Csörgő, B. Lukács, Y. M. Sinyukov, M. Weiner, *Phys. Lett. B* **505**, 64-70 (2001).
11. T. Csörgő, `hep-ph/0111139`.
12. T. Csörgő and J. Zimányi, *Heavy Ion Phys.* **17**, 281-293 (2003).
13. M. Csanád, T. Csörgő and B. Lörstad, *Nucl. Phys. A* **742**, 80-94 (2004).
14. `http://www.kfki.hu/~csorgo/budalund/budalund1.5.qm04.tar.gz`

FEMTOSCOPY WITH PENETRATING PROBES

Chairperson: T. Csörgő

Direct Photon Interferometry

Dmitri Peressounko

RRC "Kurchatov Institute", Kurchatov sq. 1, 123182, Moscow, Russia

Abstract. We consider recent developments in the theory of the two-photon interferometry in ultrarelativistic heavy ion collisions with emphasis on the difference between photon and hadron interferometry. We review the available experimental results and discuss possibilities of measurement of the photon Bose-Einstein correlations in ongoing and future experiments.

Keywords: photon,interferometry, heavy ion collisions
PACS: 25.75.-q, 25.75.Gz

Direct photon interferometry is one of the most interesting and informative tools for exploring properties of the hot matter created in heavy ion collisions. Photons have extremely large free path length in the hot matter and deliver direct information about space-time dimensions of the inner hottest part of the collision. Moreover, the direct photons, emitted at different stages of the collision, dominate in the direct photon spectrum in different ranges of transverse momentum, therefore, measuring correlation radii at different K_T one can extract space-time dimensions of the system at the different stages of the collision and thus access the equation of state of the hot matter.

Direct photons contribute only a small fraction of the total photon yield while the dominant part of inclusive photons comes from decays of final hadrons, mainly π^0 and η mesons. Fortunately, the lifetime of these hadrons is extremely large, and the width of Bose-Einstein correlations between decay photons is of the order of a few eV, so that it can not be observed and it does not obscure the direct photon correlations. So, when one talks about the photon interferometry, one means the correlations of direct photons.

Technically the interferometry of direct photons in most respects is similar to the hadron interferometry, but still it has several specific features, which make it special. These features, which will be discussed in details below, are related to the following properties of a photon:

- The penetrating nature of the direct photons. Since direct photons are emitted from the central zone of the collision and photons with different K_T are emitted at the different stages of the collision, photon interferometric correlation radii do not follow the M_T scaling, usual for hadrons, but have more complicated shape.
- Zero mass of a photon results in specific interpretation of the invariant correlation radius and correlations strength parameter, and requires some special one-dimensional parameterization of the photon correlations.
- Small proportion of direct photons in the total photon yield makes photon correlation strength parameter very small which in turn leads to importance of background photon correlations.

First predictions of the direct photon intensity correlations in heavy ion collisions have

CP828, *Multiparticle Dynamics*
edited by V. Šimák, M. Šumbera, Š. Todorova, and B. Tomášik

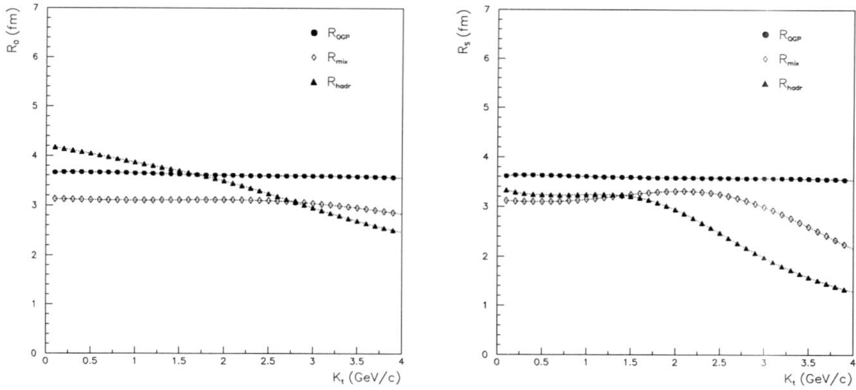

FIGURE 1. Two-photon correlation radii in Au+Au collisions at RHIC energy [4] for "side" (left plot) and "out" (right plot) directions. Contributions of QGP, mixed and hadronic phase are shown separately.

been published long ago [1, 2, 3, 4], while the next bunch of calculations [5, 6, 7] appeared after publishing of the first experimental results on direct photon interferometry by WA98 collaboration [8]. Despite the large number of publications, up to now there is no agreement between predictions not only in the absolute values of the correlation radii of direct photons, but even on the shape of K_T dependence of the correlation radii. For example, in the Fig. 1 we present predictions of the "out" and "side" correlation radii of photons, emitted in Au+Au collisions at RHIC energy [4]. These predictions are made within 2+1 Bjorken hydrodynamics with the first order phase transition. Contributions from the different phases: QGP, mixed and hadronic phase are shown separately. We find no K_T dependence for photons, emitted from the QGP phase (including pQCD photons), radii from the mixed phase are constant up to $K_T \sim 3$ GeV where Doppler-shifted contributions from a few accelerated pieces became important, and photons emitted from the hadronic phase exhibit some K_T dependence in agreement with the large collective flow developed in this phase. We compare two predictions of direct photon correlation radii for central Au+Au collisions at RHIC energy ($\sqrt{s_{NN}} = 200$ GeV), in Fig. 2. In the left plot we present result of the calculations, made within 2+1 Bjorken hydrodynamics [4]. In the right plot we show results obtained using parameterization of the evolution with a constant acceleration [7]. We find the bump and the region with $R_s > R_o$ in the first case and the M_T scaling in the second case. Whether this discrepancy can be attributed to the difference in evolution or it is related to the details of extraction of correlation radii is not clear yet.

Strength of direct photon correlations is usually extremely small, being on the level of tenth of percent, therefore it is difficult to gather sufficient statistics to construct the full three-dimensional correlation function. So one has to deal with averaged one-dimensional distributions. However, for massless particles the averaging from the full three-dimensional correlation function to those, depending on $Q_{inv} = \sqrt{-(k_1 - k_2)^2}$

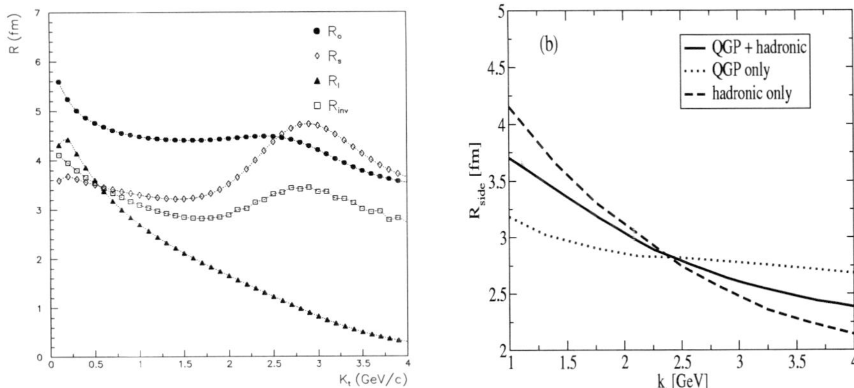

FIGURE 2. K_T dependence of different components of photon correlation radii in Au+Au collisions at RHIC energy, obtained with 2+1 Bjorken hydrodynamics [4] (left plot) and R_{side} dependence within model with parameterization with constant acceleration [7] (right plot).

has some unexpected features. To illustrate this, let us consider a simple toy model: assume, that photons are emitted from the symmetric Gaussian source with the radii $R_x = R_y = R_z = R$ and the emission duration τ so that the full correlation function of photons is

$$C_2(q,K) = 1 + \lambda \exp\left(-q_x^2 R^2 - q_y^2 R^2 - q_z^2 R^2 - q_e^2 \tau^2\right),$$

where q and K are relative and average momenta of the pair and λ is the correlations strength parameter. To go from the full three-dimensional to the one-dimensional parameterization, we have to integrate over the components of the relative momentum under the condition $\delta(Q_{inv}^2 + q^2)$. The result can be expressed as follows:

$$\begin{aligned} C_2(Q_{inv},K) &= \frac{1}{4\pi}\int C_2(\hat{q},\hat{K})\, d\Omega = \frac{1}{4\pi}\int \left[1 + \lambda \exp\left(-Q_{inv}^2 R^2 - 4K_T^2 \cos^2\theta (R^2 + \tau^2)\right)\right] d\Omega \\ &= 1 + \lambda_{inv} \exp\left(-Q_{inv}^2 R^2\right), \end{aligned}$$

where $\hat{q}$ and $\hat{K}$ are the relative and average pair momentum in pair CM frame, the integration $d\Omega$ is done over directions of the relative momentum and K_T is average transverse momentum of the pair. We find that the correlation strength of the one-dimensional correlation function is considerably reduced:

$$\lambda_{inv} = \frac{\lambda}{2}\int d\cos\theta \exp\left(-4K_T^2 \cos^2\theta (R^2 + \tau^2)\right) = \frac{\sqrt{\pi}\, \mathrm{erf}(2K_T \sqrt{R^2 + \tau^2})}{2K_T \sqrt{R^2 + \tau^2}}.$$

Calculations with the more realistic source demonstrate that the invariant correlation radius of massless particles is (using out-side-long three-dimensional radii) an average of the R_s and R_l correlation radii and almost independent on the R_o component, while the

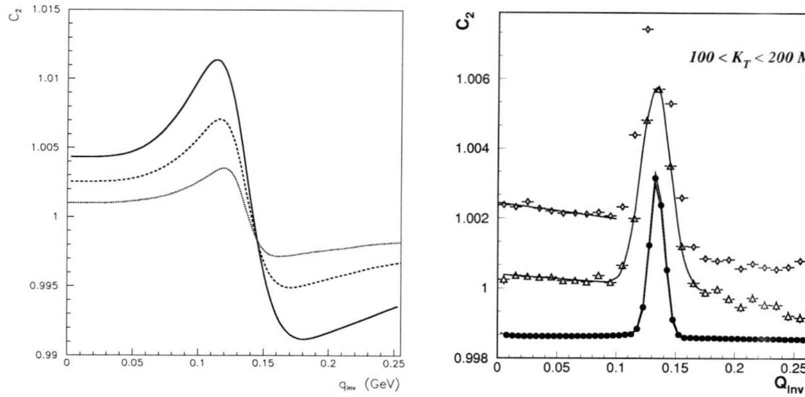

FIGURE 3. Decay photon background correlations. Left plot is residual correlations between products of Bose-Einstein correlated π^0, calculated for 3 different radii of pion correlations: 4, 5 and 6 fm (solid, dashed and dotted lines correspondingly) [4]. Right plot presents result of Monte-Carlo simulations of different residual correlations within WA98 acceptance and experimental cuts [8] due to π^0 BE correlations (diamonds), elliptic flow (triangles) and kinematic correlations (boxes).

invariant correlation strength parameter depends on the product $(K_T R_o)$ and is considerably smaller than the three-dimensional correlation strength parameter. This takes place because Q_{inv} for massless particles depends mainly on the opening angle and is almost independent on the q_o so we average over this component in the range $|q_o| \leq 2K_T$.

The extremely small strength of the direct photon correlations leads to the importance of the photon background correlations: even small correlations between decay photons may completely hide the direct photon correlations. Since the decay photons originate in decays of final hadrons, they may carry some residual correlations. Keeping in mind that the main part of the decay photons comes from π^0 decays, one can classify the background correlations as following: (1) residual correlations between the decay photons originated from Bose-Einstein correlated neutral pions; (2) residual correlations between products of kinematically correlated pions or photons, e.g. photon correlations in the processes $K_S^0 \to 2\pi^0 \to 4\gamma$ or $\omega \to \gamma\pi^0 \to 3\gamma$; (3) residual correlations due to collective (elliptic) flow of parent pions. The first point (1) is the most dangerous since if the shape of these residual correlations will repeat the shape of Bose-Einstein correlations of the parent pions, this background will completely hide the direct photon correlations. Fortunately, this is not the case. One can analytically demonstrate, that the residual correlations due to pion Bose-Eeinstein correlations have a characteristic wave-like shape with the plateau at small Q_{inv} (see Fig. 3) and can be disentangled from the direct photon correlations. Monte-Carlo simulations made by WA98 collaboration [8] (fig. 3, right plot) and by Utyuzh et al. [9] support this conclusion. The shape of the background correlations due to the kinematic correlations (case 2) and the elliptic flow (case 3) strongly depends on the apparatus acceptance and the experimental cuts used in analysis, but usually it appears as a long range correlations, as presented in Fig. 3,

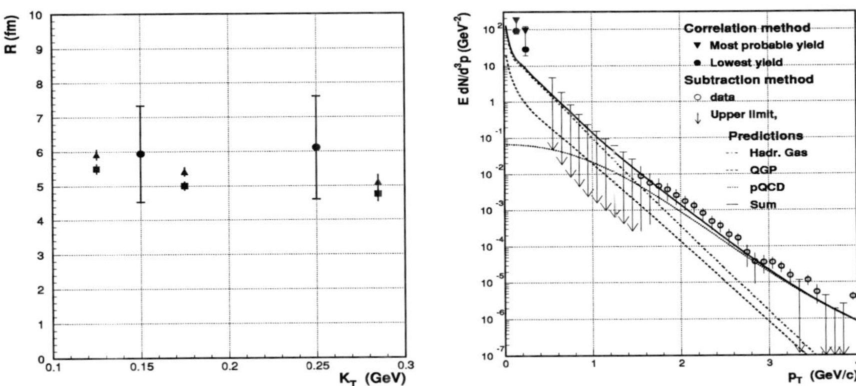

FIGURE 4. Left plot: direct photon invariant correlation radius in Pb+Pb collisions at 158 AGeV (circles), compared to side (triangles) and long (boxes) charged pions correlation radii. Right plot: direct photon yield, extracted from correlation strength parameter, compared to theoretical predictions and yield, obtained withe statistical method [8].

right plot. So, the background photon correlations in nucleus-nucleus collisions can be disentangled from the direct photon correlations but should be accounted in calculation of the correlation parameters.

The first measurement of the direct photon correlations in ultrarelativistic heavy ion collisions was performed by WA98 collaboration [8]. Its unique electromagnetic calorimeter, consisting of $4 \cdot 4 \cdot 40$ cm^3 lead glass blocks, was situated at a distance of 21 m from the interaction point. This provided an excellent opportunity to measure photon pairs with very small relative momenta. The main difficulty in this analysis was the separation between the apparatus effects like cluster merging and splitting and the real physical correlations, since the former strongly distort photon correlation function at small q. This separation was done by introducing cuts on a minimal distance between clusters and exploring the dependence of the final result on these cuts. However, at $K_T \geq 0.5$ GeV photons with relative momenta $q < 50$ MeV have so small relative angle that it was not possible to perform such an analysis any more and the invariant correlation radius and the strength parameter were measured only at small K_T, see Fig. 4. The photon invariant correlation radius was close to the pion "side" and "long" radii and was considerably above the theoretical predictions [4]. In addition to the photon correlation radius, using the correlation strength parameter, a proportion of the direct photons was extracted and the direct photon yield was measured at a very small p_T, where other methods can not be applied. Since the relation between the full three-dimensional and the one-dimensional invariant correlation strength parameters involves R_o – radius which can not be estimated using R_{inv} – a *lower* limit, corresponding to $R_o = 0$ and the most probable yield ($R_o = 6$ fm) of direct photons was extracted.

Presently there is a possibility to extract the direct photon correlations in A+A collisions at RHIC energy with the ongoing PHENIX and STAR experiments and at

LHC with building ALICE experiment. Experiment PHENIX has an electromagnetic calorimeter, consisting of two parts, one of them is the same calorimeter used in WA98 installed at a distance of 540 cm from the interaction point while the rest of calorimeter has coarser granularity $5.5 \cdot 5.5$ cm^2 and situated at a distance of 510 cm from the interaction point. Although in PHENIX the calorimeter is 4 times closer to the interaction point than one was in the WA98 experiment, this is compensated by smaller energies of photons since PHENIX is a collider experiment so that it is able to access photon pairs at small $q \sim 30$ MeV up to $K_T \sim 1$ GeV. Experiment STAR is going to get advantage of their tracking chamber and extract the direct photon correlations between photons, one of which has converted into electron-positron pair on a material of the detector and the second is detected with calorimeter. The ALICE experiment at LHC will be even more suitable for measuring of two-photon correlations. Its highly granulated PHOS calorimeter made of $2 \cdot 2 \cdot 20$ cm^3 $PbWO_4$ crystals will be installed at a distance of 460 cm from the interaction point and will have 4 times more channels per solid angle and much better energy and position resolutions than existing calorimeter in PHENIX. Simulations show that PHOS will be able to measure direct photon correlations up to $K_T \sim 2$ GeV.

To summarize, the direct photon correlations are very important tool for exploring space-time evolution of the hot matter in ultrarelativistic heavy ion collisions. Although there are plenty of predictions, there is no agreement neither in absolute value of the photon correlation radii nor even in the shape of their K_T dependence. We considered several remarkable differences between photon and hadron interferometry, related to the penetrating nature and zero mass of a photon and the small yield of direct photons. First results from the WA98 experiment demonstrated possibility of measurement of the direct photon correlations in ultrarelativistic heavy ion collisions. Ongoing experiments PHENIX and STAR at RHIC as well as building experiment ALICE at LHC have real possibility to measure the direct photon correlations so one can expect more results soon.

ACKNOWLEDGMENTS

This work was supported by MPN of Russian Federation under the grant NS-1885.2003.2 and by the INTAS grant No 04-83-4050.

REFERENCES

1. A.N. Makhlin, *JETP Lett.* **46**, 55-58 (1987); A.N. Makhlin, *Sov. J. Nucl. Phys.* **49**, 151-156 (1989).
2. D.K. Srivastava and J. Kapusta, *Phys. Rev. C* **48**, 1335-1345 (1993); D.K. Srivastava, C. Gale, *Phys. Lett. B* **319**, 407-411 (1993); D.K. Srivastava, *Phys. Rev. D* **49**, 4523-4531 (1994); D.K. Srivastava and J. Kapusta, *Phys. Rev. C* **50**, 505-508 (1994).
3. L.V. Razumov, R.M. Weiner, *Phys. Lett. B* **319**, 431-437 (1993); U. Ornik et al., hep-ph/9509367; L.V. Razumov, H. Feldmeier, hep-ph/9508318; A. Timmermann, *Phys. Rev. C* **50**, 3060-3063 (1994).
4. D. Peressounko, *Phys. Rev. C* **67**, 014905 (2003).
5. J. Alam et al., *Phys. Rev. C* **67**, 054902 (2003); J. Alam et al., *Phys. Rev. C* **70**, 054901 (2004).
6. D.K. Srivastava, *Phys. Rev. C* **71**, 034905 (2005); S. Bass, B. Muller, D.K. Srivastava, *Phys. Rev. Lett.* **93**, 162301 (2004).
7. T. Renk, hep-ph/0408218.
8. M.M. Aggarwal et al., *Phys. Rev. Lett.* **93**, 022301 (2004).
9. O.V. Utyuzh and G. Wilk, *Nukleonika* **49**, S15-S17 (2004); hep-ph/0312364.

MULTIPLE FSI INTERACTIONS

Chairperson: T. Csörgő

Analyses of Third Order Bose-Einstein Correlation by Means of Coulomb Wave Function

Minoru Biyajima*, Takuya Mizoguchi† and Naomichi Suzuki**

*Department of Physics, Shinshu University, Matsumoto, 390-8621, Japan
†Toba National College of Maritime Technology, Toba 517-8501, Japan
**Department of Comprehensive Management, Matsumoto University, Matsumoto 390-1295, Japan

Abstract. In order to include a correction by the Coulomb interaction in Bose-Einstein correlation (BEC), the wave function for the Coulomb scattering were introduced in the quantum optical approach to BEC in the previous work. If we formulate the amplitude written by Coulomb wave functions according to the diagram for BEC in the plane wave formulation, the formula for $3\pi^-$BEC becomes simpler than that of our previous work. We re-analyze the raw data of $3\pi^-$BEC by NA44 and STAR Collaborations by this formula. Results are compared with the previous ones.

Keywords: Bose-Einstein correlation, Coulomb wave function, three-body
PACS: 25.75.-q, 25.75.Gz

INTRODUCTION

Recently, in addition to the data on the charged two-body Bose-Einstein correlation (BEC), data on the three-body charged BEC have been reported [1, 2, 3, 4]. In some papers [2, 3], Coulomb correction is done with fixed interaction region, 5 fm. On the other hand, raw data with acceptance correction are reported [1, 4]. In Ref. [5], the $3\pi^-$BEC is calculated by the use of the Coulomb wave function with fixed source radius R. We have analyzed the BEC without assuming the size of source radius using the Coulomb wave function [6, 7]. In this paper, we would like to refine the formula for the $3\pi^-$BEC using the diagrammatic representation for BEC in the quantum optical (QO) approach [8, 9, 10].

In order to describe the two-body charged BEC (for example, $2\pi^-$system), we should solve the Schrödinger equation of Coulomb scattering. The regular solution at the origin of the Coulomb potential is given by,

$$\psi^C_{\mathbf{k}_i\mathbf{k}_j}(\mathbf{x}_i,\mathbf{x}_j)=\Gamma(1+i\eta_{ij})e^{-\pi\eta_{ij}/2}e^{i\mathbf{k}_{ij}\cdot\mathbf{r}_{ij}}F[-i\eta_{ij},1;(k_{ij}r_{ij}-\mathbf{k}_{ij}\cdot\mathbf{r}_{ij})], \tag{1}$$

where, $\mathbf{r}_{ij}=\mathbf{x}_i-\mathbf{x}_j$, $\mathbf{k}_{ij}=(m_j\mathbf{k}_i-m_i\mathbf{k}_j)/(m_i+m_j)$, $r_{ij}=|\mathbf{r}_{ij}|$, $k_{ij}=|\mathbf{k}_{ij}|$ and $\eta_{ij}=e_ie_j\mu_{ij}/k_{ij}$. μ_{ij} is the reduced mass of m_i and m_j, $i,j=1,2,3$ and $i\neq j$. $F[a,b;x]$ is the confluent hypergeometric function, and $\Gamma(x)$ is the Gamma function.

CP828, *Multiparticle Dynamics*
edited by V. Šimák, M. Šumbera, Š. Todorova, and B. Tomášik

The wave function of identical Bose particles should be symmetrized. For two particle momentum density we have,

$$
\begin{aligned}
N^{(2\pi^-)} &= \frac{1}{2}\prod_{i=1}^{2}\int \rho(\mathbf{x}_i)d^3\mathbf{x}_i\left|\psi^C_{\mathbf{k}_1\mathbf{k}_2}(\mathbf{x}_1,\,\mathbf{x}_2)+\psi^C_{\mathbf{k}_1\mathbf{k}_2}(\mathbf{x}_2,\,\mathbf{x}_1)\right|^2 \\
&= \prod_{i=1}^{2}\int \rho(\mathbf{x}_i)d^3\mathbf{x}_i(G_1+G_2), \\
G_1 &= \frac{1}{2}\Big(\left|\psi^C_{\mathbf{k}_1\mathbf{k}_2}(\mathbf{x}_1,\,\mathbf{x}_2)\right|^2+\left|\psi^C_{\mathbf{k}_1\mathbf{k}_2}(\mathbf{x}_2,\,\mathbf{x}_1)\right|^2\Big), \\
G_2 &= \mathrm{Re}\Big(\psi^C_{\mathbf{k}_1\mathbf{k}_2}(\mathbf{x}_1,\,\mathbf{x}_2)\psi^{C*}_{\mathbf{k}_1\mathbf{k}_2}(\mathbf{x}_2,\,\mathbf{x}_1)\Big),
\end{aligned}
$$

where,

$$
\rho(\mathbf{x}) = \frac{1}{(2\pi R^2)^{3/2}}\exp[-\frac{\mathbf{x}^2}{2R^2}]. \tag{2}
$$

The three particle momentum density for $3\pi^-$ BEC is written as [5],

$$
N^{(3\pi^-)} = \frac{1}{6}\prod_{i=1}^{3}\int \rho(\mathbf{x}_i)d^3\mathbf{x}_i\left|\sum_{j=1}^{6}A(j)\right|^2, \tag{3}
$$

$$
\begin{aligned}
A(1) &= A_1 = \psi^C_{\mathbf{k}_1\mathbf{k}_2}(\mathbf{x}_1,\,\mathbf{x}_2)\psi^C_{\mathbf{k}_2\mathbf{k}_3}(\mathbf{x}_2,\,\mathbf{x}_3)\psi^C_{\mathbf{k}_3\mathbf{k}_1}(\mathbf{x}_3,\,\mathbf{x}_1), \\
A(2) &= A_{23} = \psi^C_{\mathbf{k}_1\mathbf{k}_2}(\mathbf{x}_1,\,\mathbf{x}_3)\psi^C_{\mathbf{k}_2\mathbf{k}_3}(\mathbf{x}_3,\,\mathbf{x}_2)\psi^C_{\mathbf{k}_3\mathbf{k}_1}(\mathbf{x}_2,\,\mathbf{x}_1), \\
A(3) &= A_{12} = \psi^C_{\mathbf{k}_1\mathbf{k}_2}(\mathbf{x}_2,\,\mathbf{x}_1)\psi^C_{\mathbf{k}_2\mathbf{k}_3}(\mathbf{x}_1,\,\mathbf{x}_3)\psi^C_{\mathbf{k}_3\mathbf{k}_1}(\mathbf{x}_3,\,\mathbf{x}_2), \\
A(4) &= A_{123} = \psi^C_{\mathbf{k}_1\mathbf{k}_2}(\mathbf{x}_2,\,\mathbf{x}_3)\psi^C_{\mathbf{k}_2\mathbf{k}_3}(\mathbf{x}_3,\,\mathbf{x}_1)\psi^C_{\mathbf{k}_3\mathbf{k}_1}(\mathbf{x}_1,\,\mathbf{x}_2), \\
A(5) &= A_{132} = \psi^C_{\mathbf{k}_1\mathbf{k}_2}(\mathbf{x}_3,\,\mathbf{x}_1)\psi^C_{\mathbf{k}_2\mathbf{k}_3}(\mathbf{x}_1,\,\mathbf{x}_2)\psi^C_{\mathbf{k}_3\mathbf{k}_1}(\mathbf{x}_2,\,\mathbf{x}_3), \\
A(6) &= A_{13} = \psi^C_{\mathbf{k}_1\mathbf{k}_2}(\mathbf{x}_3,\,\mathbf{x}_2)\psi^C_{\mathbf{k}_2\mathbf{k}_3}(\mathbf{x}_2,\,\mathbf{x}_1)\psi^C_{\mathbf{k}_3\mathbf{k}_1}(\mathbf{x}_1,\,\mathbf{x}_3).
\end{aligned} \tag{4}
$$

In the plane wave approximation, each amplitude $A(i)$ approaches to the following form;

$$
\begin{aligned}
A(1) &= A_1 \xrightarrow{\mathrm{PW}} e^{i\mathbf{k}_{12}\cdot\mathbf{r}_{12}}e^{i\mathbf{k}_{23}\cdot\mathbf{r}_{23}}e^{i\mathbf{k}_{31}\cdot\mathbf{r}_{31}} = e^{(3/2)i(\mathbf{k}_1\cdot\mathbf{x}_1+\mathbf{k}_2\cdot\mathbf{x}_2+\mathbf{k}_3\cdot\mathbf{x}_3)}, \\
A(2) &= A_{23} \xrightarrow{\mathrm{PW}} e^{i\mathbf{k}_{12}\cdot\mathbf{r}_{13}}e^{i\mathbf{k}_{23}\cdot\mathbf{r}_{32}}e^{i\mathbf{k}_{31}\cdot\mathbf{r}_{21}} = e^{(3/2)i(\mathbf{k}_1\cdot\mathbf{x}_1+\mathbf{k}_2\cdot\mathbf{x}_3+\mathbf{k}_3\cdot\mathbf{x}_2)}, \\
A(3) &= A_{12} \xrightarrow{\mathrm{PW}} e^{i\mathbf{k}_{12}\cdot\mathbf{r}_{21}}e^{i\mathbf{k}_{23}\cdot\mathbf{r}_{13}}e^{i\mathbf{k}_{31}\cdot\mathbf{r}_{32}} = e^{(3/2)i(\mathbf{k}_1\cdot\mathbf{x}_2+\mathbf{k}_2\cdot\mathbf{x}_1+\mathbf{k}_3\cdot\mathbf{x}_3)}, \\
A(4) &= A_{123} \xrightarrow{\mathrm{PW}} e^{i\mathbf{k}_{12}\cdot\mathbf{r}_{23}}e^{i\mathbf{k}_{23}\cdot\mathbf{r}_{31}}e^{i\mathbf{k}_{31}\cdot\mathbf{r}_{12}} = e^{(3/2)i(\mathbf{k}_1\cdot\mathbf{x}_2+\mathbf{k}_2\cdot\mathbf{x}_3+\mathbf{k}_3\cdot\mathbf{x}_1)}, \\
A(5) &= A_{132} \xrightarrow{\mathrm{PW}} e^{i\mathbf{k}_{12}\cdot\mathbf{r}_{31}}e^{i\mathbf{k}_{23}\cdot\mathbf{r}_{12}}e^{i\mathbf{k}_{31}\cdot\mathbf{r}_{23}} = e^{(3/2)i(\mathbf{k}_1\cdot\mathbf{x}_3+\mathbf{k}_2\cdot\mathbf{x}_1+\mathbf{k}_3\cdot\mathbf{x}_2)}, \\
A(6) &= A_{13} \xrightarrow{\mathrm{PW}} e^{i\mathbf{k}_{12}\cdot\mathbf{r}_{32}}e^{i\mathbf{k}_{23}\cdot\mathbf{r}_{21}}e^{i\mathbf{k}_{31}\cdot\mathbf{r}_{13}} = e^{(3/2)i(\mathbf{k}_1\cdot\mathbf{x}_3+\mathbf{k}_2\cdot\mathbf{x}_2+\mathbf{k}_3\cdot\mathbf{x}_1)},
\end{aligned} \tag{5}
$$

where PW means the plane wave approximation of the amplitude. It should be noted that factor 3/2 does not appear in the formulation of $3\pi^-$BEC using the plane wave.

QUANTUM OPTICAL APPROACH

The amplitude squared in Eq. (3) can be classified into the following groups,

$$
\begin{aligned}
F_1 &= (1/6)[A_1A_1^* + A_{12}A_{12}^* + A_{23}A_{23}^* + A_{13}A_{13}^* + A_{123}A_{123}^* + A_{132}A_{132}^*], \\
F_{12} &= (1/6)[A_1A_{12}^* + A_{23}A_{123}^* + A_{13}A_{132}^* + c.c.], \\
F_{23} &= (1/6)[A_1A_{23}^* + A_{12}A_{132}^* + A_{13}A_{123}^* + c.c.], \\
F_{31} &= (1/6)[A_1A_{13}^* + A_{23}A_{132}^* + A_{12}A_{123}^* + c.c.], \\
F_{123} &= (1/6)[A_1A_{132}^* + A_{132}A_{123}^* + A_{13}A_{12}^* + A_{12}A_{23}^* + A_{23}A_{13}^* + A_{123}A_1^*] \\
F_{132} &= (1/6)[A_1A_{123}^* + A_{23}A_{12}^* + A_{12}A_{13}^* + A_{123}A_{132}^* + A_{132}A_1^* + A_{13}A_{23}^*], \qquad (6)
\end{aligned}
$$

where, c.c. denotes the complex conjugate. In the plane wave approximation, F_1 reduces to 1, F_{ij} corresponds to exchange between i and j charged particles, and F_{123} correspond to exchange among three charged particles.

In the previous work [6], we introduced the Coulomb wave function to the core-halo model [11], where following parameters are used; the fraction of multiplicity from the core part, $f_c = \langle n_{core}\rangle / \langle n_{tot}\rangle$, and the fraction of coherently produced particles from the core part, $p_c = \langle n_{co}\rangle / \langle n_{core}\rangle$. In the quantum optical approach, chaoticity parameter is defined as $p = \langle n_{chao}\rangle / \langle n_{core}\rangle$, where $p = 1 - p_c$. In the following, we use f_c and p. The radius of halo part is assumed to be infinitely large. Therefore, particles emitted from the halo part do not contribute to BEC, namely exchange of particles. In Ref. [10], the higher order BEC is formulated, where a contamination effect is included. If we can identify the contribution of the halo part to the contamination, both formulations coincide.

In the QO approach, the two particle momentum density $\rho(\mathbf{p}_1, \mathbf{p}_2)$ and the three particle momentum density $\rho(\mathbf{p}_1, \mathbf{p}_2, \mathbf{p}_3)$ are given respectively as [9, 10],

$$
\begin{aligned}
\rho(\mathbf{p}_1,\mathbf{p}_2) &= \rho(\mathbf{p}_1)\rho(\mathbf{p}_2) + g(\mathbf{p}_1,\mathbf{p}_2), \\
\rho(\mathbf{p}_1,\mathbf{p}_2,\mathbf{p}_3) &= \rho(\mathbf{p}_1)\rho(\mathbf{p}_2)\rho(\mathbf{p}_3) + g(\mathbf{p}_1,\mathbf{p}_2)\rho(\mathbf{p}_3) \\
&\quad + g(\mathbf{p}_2,\mathbf{p}_3)\rho(\mathbf{p}_1) + g(\mathbf{p}_3,\mathbf{p}_1)\rho(\mathbf{p}_2) + g(\mathbf{p}_1,\mathbf{p}_2,\mathbf{p}_3).
\end{aligned}
$$

The second order cumulant $g(\mathbf{p}_1, \mathbf{p}_2)$ and the third order cumulant $g(\mathbf{p}_1, \mathbf{p}_2, \mathbf{p}_3)$ are given in Fig. 1 [9, 10]. In Fig. 1, the chaotic component $r(\mathbf{p}_1, \mathbf{p}_2)$ is shown by the solid line with arrow from 1 to 2, and the coherent component $c(\mathbf{p}_1, \mathbf{p}_2)$ is shown by dotted line. In the formulation of BEC with the Coulomb wave function, momentum and coordinate cannot be decoupled. Therefore, interpretation of diagram is somewhat modified; The source function $\rho(\mathbf{x})$ is attached to the starting point of solid line, and the delta function $\delta(\mathbf{x})$ is to that of dashed line. In addition, p is replaced by $f_c p$, and $1-p$ is by $f_c(1-p)$ to include the contribution from the halo part.

In the $2\pi^-$BEC, $N^{2\pi^-}$ corresponds to $\rho(\mathbf{p}_1, \mathbf{p}_2)$, G_1 to $\rho(\mathbf{p}_1)\rho(\mathbf{p}_2)$, and G_2 to $g(\mathbf{p}_1, \mathbf{p}_2)$ beside the normalization factor. Then, the formula for $2\pi^-$BEC is written as,

$$
\begin{aligned}
\frac{N^{2\pi^-}}{N^{BG}} &= C\prod_{i=1}^{2}\int \rho(\mathbf{x}_i)d^3\mathbf{x}_i(G_1 + f_c^2p^2G_2) \\
&\quad + C\int d^3\mathbf{x}_1 \int d^3\mathbf{x}_2 \rho(\mathbf{x}_1)\delta(\mathbf{x}_2)2f_c^2p(1-p)G_2
\end{aligned}
$$

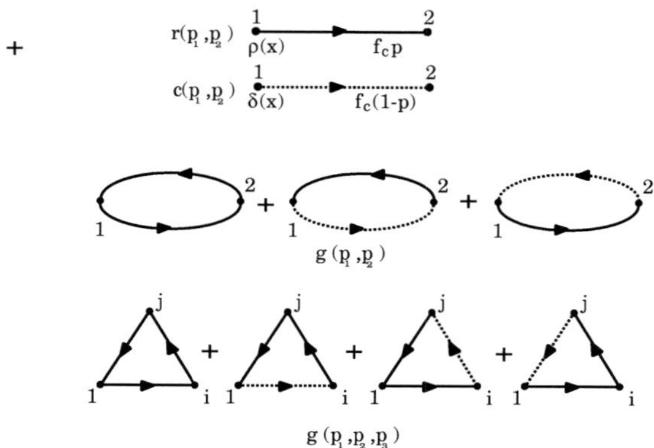

FIGURE 1. Cumulant up to the third order, $(i,j) = (2,3)$ or $(3,2)$.

In the third order BEC, $N^{3\pi^-}$ corresponds to $\rho(\mathbf{p}_1,\mathbf{p}_2,\mathbf{p}_3)$, F_1 to $\rho(\mathbf{p}_1)\rho(\mathbf{p}_2)\rho(\mathbf{p}_3)$, F_{ij} to $g(\mathbf{p}_i,\mathbf{p}_j)$, and $(F_{123}+F_{123})$ to $g(\mathbf{p}_1,\mathbf{p}_2,\mathbf{p}_3)$, beside the normalization factor. Then, the formula for $3\pi^-$BEC is given by

$$\begin{aligned}\frac{N^{3\pi^-}}{N^{BG}} &= C\prod_{i=1}^{3}\int \rho(\mathbf{x}_i)d^3\mathbf{x}_i\left[F_1+3f_c^2p^2F_{12}+2f_c^3p^3\cdot Re[F_{123}]\right] \\ &+ C\prod_{i=1}^{3}\int d^3\mathbf{x}_i\delta(\mathbf{x}_1)\rho(\mathbf{x}_2)\rho(\mathbf{x}_3)6f_c^2p(1-p)\left(F_{12}+f_cpRe[F_{123}]\right). \qquad (7)\end{aligned}$$

Compare Eq.(7) with Eq.(12) in Ref. [7], where $F_2 = F_{12}+F_{23}+F_{31}$ and $F_3 = F_{123}+F_{132}$.

ANALYSIS OF $3\pi^-$BEC

The formula (7) is applied to the analysis of raw data on $3\pi^-$BEC by STAR Collaboration [4] and NA44 Collaboration [1]. The results for $3\pi^-$BEC by STAR Collaboration [4] are shown in Table 1 and Fig. 2. For comparison, the results in the previous work [6] are also shown in the lower part of Table 1.

TABLE 1. Analyses of raw data on $3\pi^-$ BEC by STAR Collaboration [4].

	p	1.0	0.8	0.6	0.40±0.02
present work	f_c	0.75±0.02	0.78±0.02	0.85±0.02	1.0(fixed)
	R (fm)	5.34±0.24	6.21±0.30	6.88±0.35	7.43±0.34
	χ^2/N_{dof}	2.68/34	0.93/34	0.65/34	0.58/34
Ref. [6]	f_c	0.75±0.02	0.78±0.02	0.85±0.02	
	R (fm)	5.34±0.24	5.99±0.28	6.48±0.32	
	χ^2/N_{dof}	2.80/34	1.39/34	0.98/34	

Parameters f_c and p are almost the same with the previous calculations. The radius R is almost the same at $p = 1.0$, and the present result gradually becomes larger than the previous one. At $p = 0.6$, the new result becomes about 6% larger than the previous one. Fitting to the STAR data is slightly improved. In Fig.2, the result at $p = 1$ is shown, and the possible region of p and f_c estimated from the analysis of $2\pi^-$BEC and $3\pi^-$BEC is shown. Each hatched region is defined by the fitted value $\pm$ 2 standard deviation.

The results for $3\pi^-$BEC by NA44 Collaboration [1] are shown in Table 2 and Fig. 3. Previous results [7] are also shown in Table 2. Parameters f_c and p are almost the same with the previous calculations. As for the radius R, the present result gradually becomes larger than the previous one, as the chaoticity parameter decreases from 1 to 0.6.

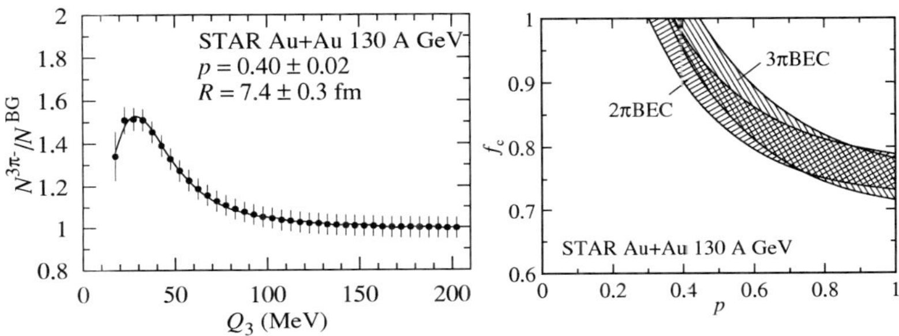

FIGURE 2. (a) Analysis of raw data on $3\pi^-$BEC by STAR Collaboration; f_c is fixed at 1. (b) Possible region of f_c and p estimated from the analyses of $2\pi^-$ BEC and $3\pi^-$BEC.

TABLE 2. Analyses of raw data on $3\pi^-$ BEC by NA44 Collaboration [1].

	p	1.0	0.8	0.6	0.32±0.05
present work	f_c	0.67±0.04	0.70±0.04	0.76±0.05	1.0(fixed)
	R (fm)	2.88±0.39	3.32±0.49	3.66±0.54	4.05±0.57
	χ^2/N_{dof}	6.64/15	6.78/15	6.81/15	6.80/15
Ref. [7]	f_c	0.67±0.04	0.70±0.04	0.76±0.05	
	R (fm)	2.89±0.39	3.22±0.46	3.47±0.51	
	χ^2/N_{dof}	6.7/15	6.8/15	6.8/15	

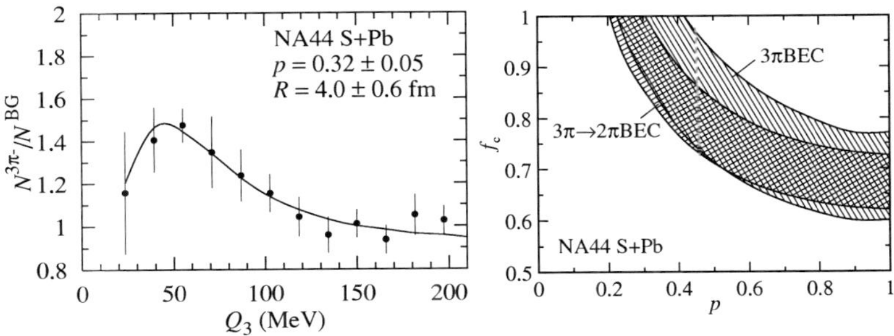

FIGURE 3. (a) Analysis of $3\pi^-$BEC by NA44 Collaboration; f_c is fixed at 1. (b) Possible region of f_c and p estimated from the analyses of $2\pi^-$ BEC and $3\pi^-$BEC.

At $p = 0.6$, the new result becomes about 5% larger than the previous one. Fitting of the present calculations to the NA44 data is almost the same with that of the previous results.

CONCLUDING REMARKS

We refine the formula for $3\pi^-$BEC according to the diagram in the QO approach. In our formulation three parameters, fraction of core part f_c, chaoticity parameter p, and radius of interaction range R are included. We apply the formula to the analysis of data on $3\pi^-$BEC by STAR and NA44 Collaborations, and compare with the previous calculations. Parameters f_c and p are almost the same with the previous calculations. The radius R is almost the same at $p = 1.0$, and the present result gradually becomes larger than the previous one, as p decreases. At $p = 0.6$, the new result becomes about 5% larger than the previous one. Fitting to the STAR data is slightly improved. That to the NA44 data is almost the same.

From the comparison of the results on $2\pi^-$BEC [6, 7] with those on $3\pi^-$BEC, we have a relation,

$$R_{2nd} \simeq 1.5 R_{3rd}. \tag{8}$$

Factor 1.5 in Eq. (8) corresponds to $3/2$ in Eq. (5). This result suggests that our approach should be equivalent to the plane wave formulation applied to the data with Coulomb correction. Detail calculation is reported in Ref. [12]

ACKNOWLEDGMENTS

Authors would like to thank J.R. Glauber for variable comments. They also would like to thank RCNP at Osaka University, Faculty of Science, Shinshu University, Toba National College of Maritime Technology and Matsumoto University for financial support.

REFERENCES

1. H. Boggild et al., NA44 Collaboration,*Phys. Lett. B* **455**, 77-83 (1999).
2. K. Adcox et al., PHENIX Collaboration, *Phys. Rev. Lett.* **88**, 192302 (2002).
3. M.M. Aggarwal et al., WA98 Collaboration, *Phys. Rev. C* **67**, 014906 (2003).
4. J. Adams et al., STAR Collaboration, *Phys. Rev. Lett.* **91**, 262301 (2003).
5. E.O. Alt, T. Csörgő, B. Lörstad and J. Schmidt-Sorensen, *Phys. Lett. B* **458**, 407-414 (1999).
6. T. Mizoguchi and M. Biyajima, *Phys. Lett. B* **499**, 245-252 (2001); [addendum] `hep-ph/0005326`.
7. M. Biyajima, M. Kaneyama and T. Mizoguchi, *Phys. Lett. B* **601**, 41-50 (2004).
8. M. Biyajima, A. Bartl, et al., *Prog. Theor. Phys.* **84**, 931-940 (1990); [addenda] *Prog. Theor. Phys.* **88**, 157-158 (1992).
9. N. Suzuki and M. Biyajima, *Prog. Theor. Phys.* **88**, 609-614 (1992); *Phys. Rev. C* **60**, 034903 (1999).
10. N. Suzuki, M. Biyajima and I.V. Andreev, *Phys Rev. C*, **56**, 2736-2746 (1997).
11. T. Csörgő, B. Lörstad, J. Schmidt-Sorensen and A. Ster, *Eur. Phys. J. C* **9**, 275-281 (1999).
12. N. Suzuki, K. Ide, M. Biyajima and T. Mizoguchi, *Soryushiron Kenkyu* (Kyoto), in Japanese, (2005).

Coulomb Final State Interactions and Modelling B-E Correlations

O.V. Utyuzh

The Andrzej Sołtan Institute for Nuclear Studies, Hoża 69, 00-681 Warsaw, Poland

Abstract. The different procedures of accounting for Coulomb corrections to two-boson correlation functions for systems formed in ultra-relativistic heavy ion collisions are presented. Both the description of Coulomb interactions of detected particles with each other in terms of Bowler-Sinyukov procedure and influence of the "central" Coulomb potential on the pair correlation function are discussed.

Keywords: Bose-Einstein correlations, Statistical models, Fluctuations
PACS: 25.75.Gz, 12.40.Ee, 03.65.-w, 05.30.Jp

Introduction. Originally boson interferometry analysis was introduced in particle physics by Goldhaber, Goldhaber, Lee and Pais [1] to provide us with information on the space-time characteristics of hadronizing source produced in high-energy collisions (see [2] for details). However, correlations in momentum space are caused not only by quantum statistics but also by final state interactions (FSI) between the two particles (Coulomb interaction for charged particles and strong interaction for hadrons). In heavy ion collisions the effect of short-range strong interactions is usually small and can be neglected. On the contrary, the effect of the Coulomb FSI dominates behavior of the spectra of two charged particles at very small relative momenta. Therefore the quantitative interpretation of experimental results depends critically on the proper understanding of the role of Coulomb interactions of detected pairs of particles with each other, as well as Coulomb interactions of the pair with the system of remaining particles.

Coulomb corrections. Traditionally, as compared to the idealized case of pure quantum-statistical correlations, Coulomb correlations modify the momentum distribution of hadron pairs in the final state by some weight factor $K_{Coul}(Q)$ which is the square of the relative Coulomb wave function, $\psi_C(r)$:

$$C_2(Q) = K_{Coul}(Q)C_2^{QS}(Q) = (1+G)K_{Coul}(Q), \tag{1}$$

where

$$G = \exp\left[-\sum_i^{dim} Q_i^2 R_i^2\right]; \qquad K_{Coul}(Q) = |\psi_C(r)|^2, \tag{2}$$

and

$$\psi_C(r) = e^{i\delta_c}\sqrt{A_C(\eta)}e^{ipr}F[-i\eta,1;i(pr-\mathbf{pr})]. \tag{3}$$

Therefore, for a long time the simplest form of Coulomb correction was the Gamow factor [3]—the square of the relative Coulomb wave function $\psi_C(0)$ of the produced

CP828, *Multiparticle Dynamics*
edited by V. Šimák, M. Šumbera, Š. Todorova, and B. Tomášik

pair at zero separation—a procedure, which is still well founded if the effective size of the emitting source is much smaller than the Bohr radius $|a|$

$$K_{Gamow} \equiv |\psi_C(r=0)|^2 = \frac{2\pi\eta(Q)}{e^{2\pi\eta(Q)}-1}. \tag{4}$$

However, in the case of heavy ion collisions where expected average distance between any two emitted particles in their c.m.s. is larger, the Gamow factor is not physically correct. In this case Coulomb interaction should be weaker and the simple Gamow correction substantially deviates from the true Coulomb effect and leads to a big overestimation of the value of two-pion correlation function. Therefore the corrections due to Coulomb interactions was accomplished by applying correction weights (determined by calculating the Coulomb correlation function $K_{Coul}(Q)$ for a spherical Gaussian source $S(\vec{r})$ [4, 5]) to all pairs:

$$K_{Coul}(Q) = \int_\Omega d\Omega |\psi_C(r)|^2 S(\vec{r}). \tag{5}$$

This method is applied iteratively, successively fitting distributions of the correlation function $C(Q)$ and iteratively applying the fit value R to a new $S(\vec{r})$.

Recently, CERES Collaboration [6] noted that this approach overcorrects the Coulomb effect and advocated an improved procedure [7] which applies the Coulomb weight only to the fraction of pairs that participate in the Bose-Einstein correlation (BEC),

$$C_2(Q) = \underbrace{(1-\lambda)\times 1}_{halo} + \underbrace{\lambda(1+G)K_{Coul}}_{core}, \tag{6}$$

where the $1-\lambda$ and λ terms account for the nonparticipating and participating fractions of pairs, respectively. In essence, it was argued that only those pairs which contribute to the measured BE correlation strength also contribute to Coulomb repulsion. It was found that the parameters of Pratt-Bertsch parametrization [9] R_{long} and R_{side} depend very little on the assumed strength of the Coulomb repulsion, cf. Fig. 1(left panel), however, the results for R_{out} (and correspondingly R_{out}/R_{side}) are very sensitive to the procedure employed.

The above procedure was extended by PHENIX collaboration [8] to allow for intermediate range decay pions, such as those emerging from decays of ω, which may contribute to the Coulomb strength without being resolved in the measured Bose-Einstein correlation,

$$C_2(Q) = \underbrace{(1-\lambda_{+-})\times 1}_{halo} + \underbrace{(\lambda_{+-}-\lambda)K_{Coul}}_{Coulomb\ only} + \underbrace{\lambda(1+G)K_{Coul}}_{core} \tag{7}$$

where a new parameter, λ_{+-}, used to decouple the Coulomb and Bose-Einstein fractions is determined by fitting Coulomb correlation functions calculated with several values of strength, λ_{+-}[1]. As was argued in [8] the differences arising from the use of Eq. (7) are

[1] It was shown that the best value of Coulomb strength, λ_{+-}, minimizing χ^2 was $\lambda_{+-} = 0.50$.

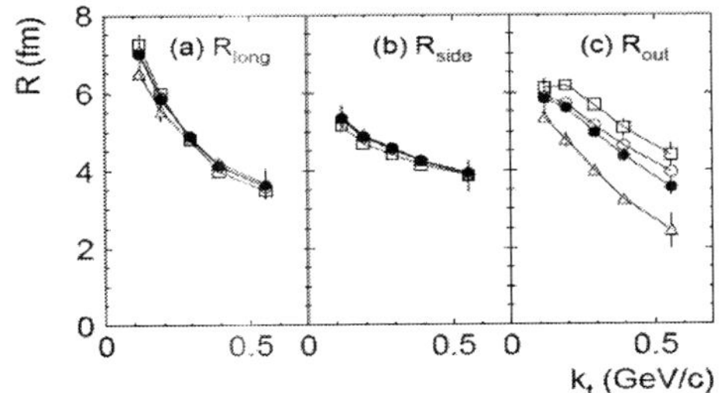

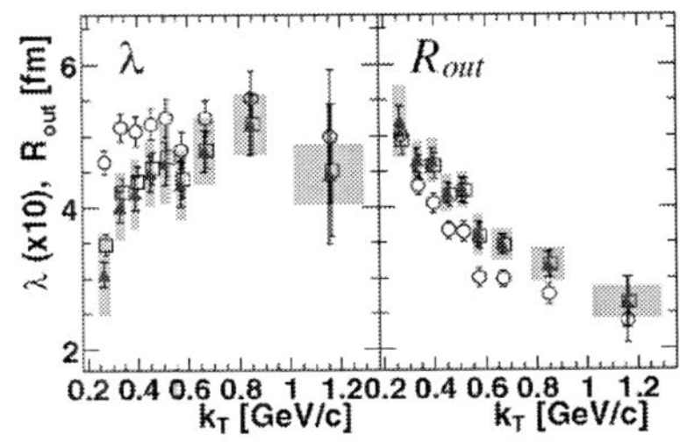

FIGURE 1. Left panel (reprinted from [6] with permission from ©2003 Elsevier) - CERES data [6] - full Coulomb strength Eq. (1) (open triangles), no Coulomb repulsion (open squares), partial Coulomb repulsion Eq. (6) (full circles). Right panel (reprinted figure with permission from [8]. Copyright ©2004 by the American Physical Society) - PHENIX data [8] - full Coulomb strength Eq. (1) (open circles), partial Coulomb repulsion Eq. (6) (open squares), partial Coulomb repulsion Eq. (7) (filled triangles).

small and can be incorporated into the total systematic errors, cf. Fig. 1(right panel).

Nearly equivalent correction methods was recently applied by other experiments: STAR [10] and PHOBOS [11]. The results showed no significant change[2] when using either of the correction methods, cf. Fig. 2 (left panel).

At the end of this section it is worth to mention that complexity of the problem of Coulomb corrections increases enormously with the order of correlation function. In the literature there is so far no consistent and systematic treatment of the FSI of charged multi-boson systems. Therefore, the experimental removal of the Coulomb effects from the n-particle Bose-Einstein correlation functions is very difficult and is based only on some *ad hoc* generalization of Gamow formula to the case of multi-bosonic systems. First attempts to do it in a proper way were made by T. Csörgő [12] by developing a new method to correct for many-body Coulomb effects based on explicit, analytically given form of the n-body Coulomb wave function.

The "central" Coulomb potential. In this section we turn to the question of the effects of the Coulomb interactions of the pair with the remaining particles of the source.

According to ideas presented in [4] and later in [13] this extremely difficult many-body problem can be simplified by assuming that the remaining particles can be described by some central Coulomb potential, $Z_{eff}e^2/r$, where the effective charge $Z_{eff} \simeq (Z_A + Z_B)$ (with Z_A and Z_B being charges of colliding nucleolus). This central potential accelerates positive mesons away and slows down the negatives ones. Therefore, the final momenta difference of two particles moving inside Coulomb potential is changed in the following way [13]:

[2] Notice that for $\lambda = 1$, all procedures used for Coulomb corrections are equivalent.

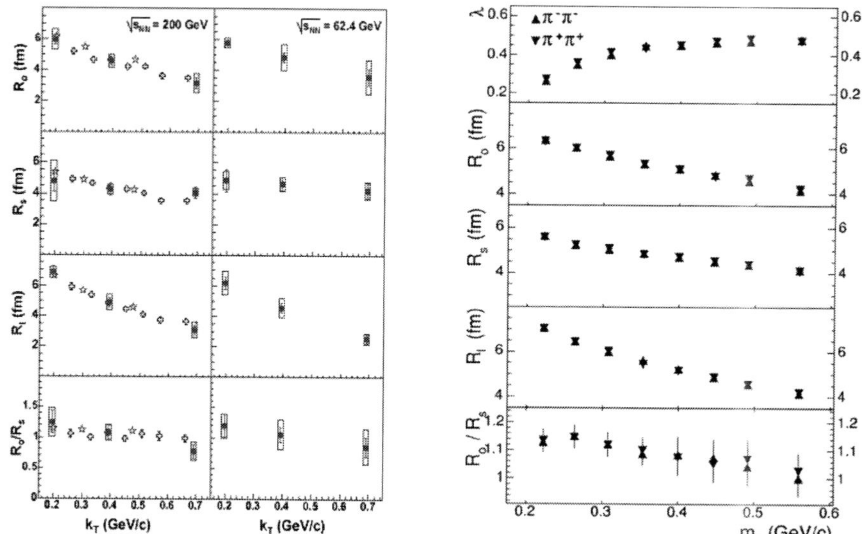

FIGURE 2. Left panel - STAR, PHENIX and PHOBOS data [11] - Pratt-Bertsch parameters R_o, R_s, R_l and the ratio R_o/R_s for $\pi^-\pi^-$ pairs as a function of $\langle k_T \rangle$. Right panel (reprinted figure with permission from [10]. Copyright ©2005 by the American Physical Society)- STAR data [10] - HBT parameters for $\pi^+\pi^+$ and $\pi^-\pi^-$ correlation functions.

$$Q_i' \sim Q_i \left(1 + \frac{Z_{eff} e^2/r}{p_i} \right). \tag{8}$$

However, recently STAR Collaboration [10] has reported excellent agreement between parameters of Pratt-Bertsch parametrization extracted from the positively and negatively charged pion analyzes. Therefore Coulomb interaction between outgoing charged pions and residual positive charge in the source is negligible.

Numerical modelling of Bose-Einstein correlations. Now we would like to focus attention on problem of modeling of Bose-Einstein correlations. Referring to [14, 15] for details and further references, let us recapitulate here the main aspects of the problem.

Despite long history of BEC study the question of numerical modeling of the BEC by means of Monte-Carlo (MC) event generators still remains open. The main reason why it is so is the fact that BEC are quantum-mechanical phenomenon and therefore it is extremely difficult to model them by using MC event generators which are probabilistic by the construction. In most approaches used at present the effect of BEC is obtained by suitable changing the original output of MC generators used and introducing this way (more or less artificially) desired bunching in the phase-space of the finally produced identical particles. This is achieved by one of two approaches:

(a) shifting (in each event) momenta of adjacent like-charged particles in such a way as to get desired $C_2(Q)$ (one has to correct afterwards for the energy-momentum imbalance introduced this way) [17],

(*b*) scrutinize all events obtained from a particular MC generator against the possible amount of bunching they are already showing and counting them as many times as necessary to get desirable $C_2(Q)$ [18].

However, such approaches lead (depending on the method used) to such unwanted features as violation of the energy-momentum conservation or changes in the original (i.e., obtained directly form MC generators) multiparticle spectra. Moreover, as was shown in [15] each afterburner *inevitably* changes physics of the original MC code to which it is attached. This observation is one of our motivations (presented in detail in [14, 15, 16]) to look for such MC scheme, which would be based on accounting first for effects of quantum statistics (especially, the Bose-Einstein statistics) and only later for other aspects of particle production (hadronization). The cornerstone of such scheme is the observation that Bose-Einstein statistics forces identical particles to be produced in bunches [19]. It was argued in [14] that such bunches are represented by producing identical particles in cells in phase-space and particles in the cells are distributed according to geometrical (Bose-Einstein) distribution. In [15] extension to this algorithm is presented, which—in addition to bunching—accounts also for the proper symmetrization of the corresponding two-particle wave function (not used before) and allows for 3-dimensional extension of our algorithm. This extension is based on the observation that symmetrization correlates in a specific way the energy-momenta of particles with their space-time locations (see [15, 16] for details). It is important to stress here that such extension allows, in a simple way, to properly account for effects of Coulomb and strong final state interactions discussed here [15, 16].

Summary. To summarize—we have presented review of the Coulomb correction methods intensively used to analyze the recent experimental data. The results obtained by different experiments are consistent and show that parameter R_{out} (and therefore also the ratio R_{out}/R_{side}) crucially depends on the strength of Coulomb corrections. In what concerns Coulomb interactions between outgoing charged pions and residual positive charge remaining in the source the STAR Collaboration provided arguments that it is negligible.

In what concerns numerical modelling of Bose-Einstein correlations we notice that, according to [15], any afterburner procedure changes in some uncontrollable way the physics of the Monte-Carlo code used. It seems that proper incorporation of BEC should start at the very beginning of construction of any Monte-Carlo code (see [14, 15, 16] for first attempts in this direction).

ACKNOWLEDGMENTS

OU is grateful for support and for the warm hospitality extended to him by organizers of the WPCF2005. Author is indebted to G. Wilk for reading the manuscript and helpful suggestions.

REFERENCES

1. G. Goldhaber et al., *Phys. Rev.* **120**, 300-312 (1960).
2. T. Csörgő, *Heavy Ion Phys.* **15**, 1-80, (2002); W. Kittel, *Acta Phys. Polon. B* **32**, 3927-3972 (2001); G. Alexander, *Rep. Prog. Phys.* **66**, 481-522 (2003) (and references therein).
3. M. Gyulassy, S. K. Kauffmann, and L. W. Wilson, *Phys. Rev. C* **20**, 2267-2292 (1979).
4. G. Baym and P. Braun-Munzinger, *Nucl. Phys. A* **610**, 286c-296c (1996).
5. C. Adler et al., *Phys. Rev. Lett.* **87**, 082301 (2001).
6. D. Adamová *et al.*, (CERES Collab.) *Nucl. Phys. A* **714**, 124-144 (2003).
7. M. G. Bowler, *Phys. Lett. B* **270**, 69-74 (1991); Yu. M. Sinyukov, R. Lednicky, S. V. Akkelin, J. Pluta, and B. Erazmus, *Phys. Lett. B* **432**, 248-257 (1998).
8. S. S. Adler *et al.*, (PENIX Collab.) *Phys. Rev. Lett.* **93**, 152302 (2004).
9. G. F. Bertsch, M. Gong, and M. Tohyama, *Phys. Rev. C* **37**, 1896-1900 (1988); S. Pratt, *Phys. Rev. D* **33**, 1314-1327 (1986); S. Chapman, P. Scotto, and U. W. Heinz, *Phys. Rev. Lett.* **74**, 4400-4403 (1995).
10. J. Adams *et al.*, (STAR Collab.) *Phys. Rev.* **C71**, 044906 (2004).
11. B. B. Back *et al.*, (PHOBOS Collab.) nucl-ex/0409001.
12. E. O. Alt, T. Csörgő, B. Lörstad and J. Schmidt-Sørensen, *Phys. Lett. B* **458**, 407-414 (1999); E.O.Alt, T. Csörgő, B. Lörstad and J. Schmidt-Sørensen, *Eur. Phys. J. C* **13**, 663-670 (2000).
13. D. Hardtke and T. J. Humanic, *Phys. Rev. C* **57**, 3314-3318 (1998).
14. O. V. Utyuzh, G. Wilk, Z. Wlodarczyk, *Quantum Clan Model description of Bose Einstein Correlations*, hep-ph/0503046, to be published in *Acta Phys. Hung. A - Heavy Ion Phys.* (2005); cf. also "Numerical modelling of quantum statistics in high-energy physics", hep-ph/0410398.
15. O. V. Utyuzh, G. Wilk and Z. Włodarczyk, "Bose-Einstein correlations from "within"" *(this volume)*, hep-ph/0509320.
16. O. V. Utyuzh, G. Wilk, Z. Włodarczyk, "Proposition of numerical modelling of BEC" (poster presented at Quark Matter 2005, Budapest, Hungary, 4-9 August 2005), hep-ph/0509342.
17. L. Lönnblad and T. Sjöstrand, *Eur. Phys. J. C* **2**, 165-180 (1998).
18. A. Białas and A. Krzywicki, *Phys. Lett. B* **354**, 134-137 (1995); K. Fiałkowski and R. Wit, *Eur. Phys. J. C* **2**, 691-695 (1998); K. Fiałkowski, R. Wit and J. Wosiek, *Phys. Rev. D* **58**, 094013 (1998); T. Wibig, *Phys. Rev. D* **53**, 3586-3590 (1996).
19. E. E. Purcell, *Nature* **178**, 1449 (1956).

NON-IDENTICAL PARTICLE CORRELATIONS

Chairperson: R. Ledincký

Non-Identical Particle Femtoscopy in Heavy-Ion Collisions

Adam Kisiel

Wydział Fizyki, Politechnika Warszawska, ul. Koszykowa 75, 00-662 Warsaw, Poland

Abstract. Non-identical particle correlations have been used to study the properties of the source emitting particles in heavy-ion collisions. In this work the status of non-identical particle correlations technique and latest results from ultra-relativistic heavy-ion collision experiments are discussed. Some specific features of the Coulomb interaction, important for the asymmetry analysis, are discussed.

Keywords: non-identical particle correlations, pair wave function, Coulomb wave function
PACS: 13.75.Lb, 13.75.Gx, 24.10.Nz, 24.10.Pa, 25.75.-q, 25.75.Gz, 25.75.Ld

INTRODUCTION

Non-identical particle correlations technique [1, 2, 3, 4] has been used successfully in intermediate energy collisions to study the details of the time ordering of the emission of particles from these systems [5, 6, 7]. The measurements were usually done with baryons, for which both Coulomb and strong interaction were important. The measured time differences were of the order of tens or hundreds of fm/c.

In ultra-relativistic heavy-ion collisions the matter produced is meson dominated. For meson interactions Coulomb becomes the dominant contribution in two-particle interaction, therefore its properties will be studied in more detail. Also in such collisions very short lifetimes of the source are observed [8, 9, 10], therefore large emission time differences between particle species are not expected. This opens up a possibility to study space asymmetries as well [11], which turn out to be closely connected to the dynamics of the emitting source [4, 11]. In the last section, the experimental results in non-identical particle correlations and their interpretation will be discussed.

The topic of measuring pair interaction potential through non-identical particle correlations is covered in a separate talk at this conference [12, 4] and will not be discussed here.

TWO PARTICLE WAVE-FUNCTION AND THE CORRELATION FUNCTION

In case of two non-identical particles their pair wave function may include contributions from two types of interaction[4, 11, 2, 13]. With no interaction it is a plane wave. The strong interaction leads to a scattered spherical wave. The Coulomb interaction modifies both the plane and the scattered wave. The movement of the center-of-mass of the pair can be separated out and does not enter into our consideration. We then have the relative

CP828, *Multiparticle Dynamics*
edited by V. Šimák, M. Šumbera, Š. Todorova, and B. Tomášik

wave function:

$$\Psi_{-\mathbf{k}^*}(\mathbf{r}^*) = e^{i\delta_c}\sqrt{A_c(\eta)}[e^{i\mathbf{k}^*\mathbf{r}^*}F(-i\eta,1,i\xi) + f_c(\mathbf{k}^*)\tilde{G}(\rho,\eta)/\mathbf{r}^*] \tag{1}$$

where $\xi = \mathbf{k}^*\mathbf{r}^* + k^*r^* \equiv \rho(1+\cos(\theta^*))$, $\rho = k^*r^*$, $\eta = (k^*a)^{-1}$, $a = (\mu z_1 z_2 e^2)^{-1}$. The $A_c(\eta)$ is the Coulomb Gamov factor, F is the confluent hyper-geometric function, G is the combination of of the regular and singular s-wave Coulomb functions, f is the s-wave scattering amplitude. For illustration of the emission asymmetry measurement we will analyze only the Coulomb part of the interaction:

$$\Psi_{-\mathbf{k}^*}(\mathbf{r}^*) = e^{i\delta_c}\sqrt{A_c(\eta)}[e^{i\mathbf{k}^*\mathbf{r}^*}F(-i\eta,1,i\xi)] \tag{2}$$

Apart from the asymmetry analysis, Coulomb and Strong interaction may also be used to study the shape of the source [14].

In the theoretical formulation the correlation function is defined as the modulus squared of the two-particle wave function averaged over the source emission function:

$$C(\vec{q},\vec{K}) = \frac{\int d^4x_1 S(x_1,p_1) d^4x_2 S(x_2,p_2)\,|\Psi_{-\mathbf{k}^*}(\mathbf{r}^*)|^2}{\int d^4x_1 S(x_1,p_1)\int d^4x_2 S(x_2,p_2)} \tag{3}$$

where $\mathbf{q} = \mathbf{p_1} - \mathbf{p_2}$, $\vec{K} = (\vec{p}_1 + \vec{p}_2)/2$. Therefore by analysing the behaviour of the pair wave function we can also infer the properties of the correlation function itself.

THE COULOMB INTERACTION AND THE ASYMMETRY MEASUREMENT

The F function from (2) is of particular interest in the asymmetry analysis, because it depends on the source spatial distribution. It is an infinite series of the form:

$$F(\alpha,1,z) = 1 + \alpha z/1!^2 + \alpha(\alpha+1)z^2/2!^2 + \ldots \tag{4}$$

which in our case corresponds to:

$$F = 1 + r^*(1+\cos\theta^*)/a + (r^*(1+\cos\theta^*)/a)^2/4 + ik^*r^{*2}(1+\cos\theta^*)^2/4a + \ldots \tag{5}$$

where a is the Bohr radius of the system and θ^* is the angle between the relative momentum and the relative separation in the pair rest frame. The dependence of the wave function on this angle is crucial for the asymmetry measurement. An example of the contribution of the F function to the pair wave function modulus for the same value of r^* and different values of θ^* is seen on Fig. 1.

Let's restrict ourselves to the transverse plane. For a pair of particles we can define three directions: the direction of the pair relative momentum $\mathbf{k}^*$, the direction of the pair relative position $\mathbf{r}^*$ and some third direction that we measure for the pair, for example the pair average momentum $\vec{K}$. We can then define the angles between these three directions: angle Ψ between $\mathbf{k}^*$ and $\vec{K}$, angle ϕ between $\vec{K}$ and $\mathbf{r}^*$ and θ^* as the third angle. We can trivially write: $\Psi = \phi + \theta^*$, which leads to:

$$\cos(\Psi) = \cos(\phi)\cos(\theta^*) - \sin(\phi)\sin(\theta^*) \tag{6}$$

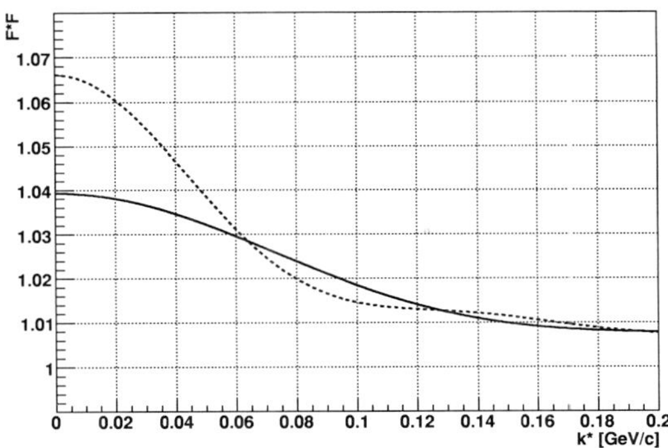

FIGURE 1. The behaviour of the modulus squared of the confluent hyper-geometric function F for pion-pion pair, at a fixed relative separation $r^* = 10fm$ and two values of θ^* - 60deg (solid line) and 120deg (dashed line)

Later we consider the sign of the average cosines of these angles, for which we have:

$$\text{sign}\langle\cos(\Psi)\rangle = \text{sign}\langle\cos(\phi)\rangle\,\text{sign}\langle\cos(\theta^*)\rangle \tag{7}$$

We also notice, that in the usual Bertsch-Pratt decomposition we have $k^*_{out} \equiv k^*\cos(\Psi)$ and $k^*_{side} \equiv k^*\sin(\Psi)$. To study emission asymmetries, it was proposed [1] to study not one, but two separate correlation functions. The first one, C_+, is constructed only with pairs for which $k^*_{out} > 0$. The second, C_-, only with pairs with $k^*_{out} < 0$. The first case corresponds in the classical picture to the scenario where the first particle is faster, the second—to the second particle being faster (in the transverse direction). Then, the two functions are compared, e.g. by studying the "double ratio": C_+/C_-. The "double ratio" can be different from unity. This means that in one of the samples the correlation is stronger. This can only happen if the $\langle\cos(\theta^*)\rangle$ is different in both samples. In other words e.g. by selecting $\cos(\Psi) > 0$ we obtained a sample of pairs for which $\langle\cos(\theta^*)\rangle > 0$ as well, while in the other sample for $\cos(\Psi) < 0$ we have $\langle\cos(\theta^*)\rangle < 0$. From (7) one sees this can only happen if $\langle\cos(\phi)\rangle > 0$. This is the essence of the asymmetry measurement. By selecting pairs based on direction of it's relative momentum k^* with respect to the pair velocity, and knowing that the interaction strength depends on the angle θ^* we can infer what is the direction of the pair relative separation $\mathbf{r}^*$ with respect to the pair velocity. Moreover we can replace the pair velocity in the consideration with any other direction defined for the pair (for example - the "Side" direction) and measure $\mathbf{r}^*$ with respect to that direction as well. This is an observable which is not accessible in any other way. Quantitative considerations show that one can measure not only wheather r^* is (on the average) parallel or anti-parallel to v, but also the the value of $r^*_{out} \equiv r^*\cos(\phi)$. As an example, the sensitivity of the pion-kaon correlation function to the size of the system as well as to the average shift between particle species is shown on Fig. 2. It is interesting to note that the asymmetry between C_+ and C_- does not vanish at $k* = 0$ [3]. It comes directly from (5), where the first term giving the asymmetry does not depend on $k*$. However one must also remember that the

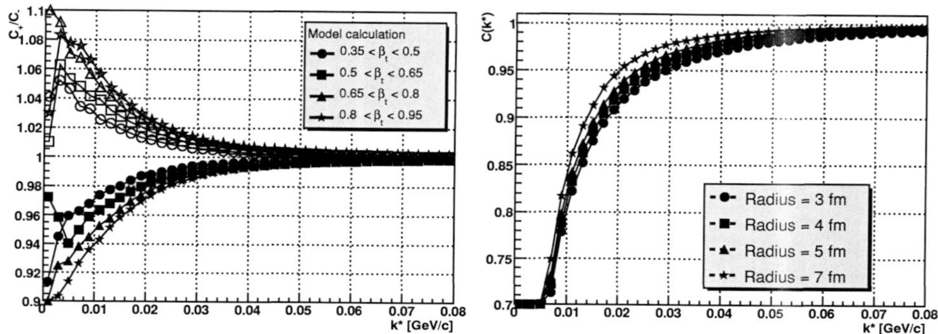

FIGURE 2. The dependence of the model pion-kaon correlation function on the radius of the source (right panel) and the shift between particle species (left panel). On the right panel "double ratios" are shown for like sign pion-kaon pairs (filled) and unlike-sign pairs (open)

A_c factor will mask this asymmetry by forcing the whole pair wave-function to go to zero for same-charge and infinity for opposite charge particles at $k^* = 0$.

The analysis of experimental data of non-identical particle correlations is significantly complicated by the fact that there is no analytical form of the correlation function that can be fit to the data, as is the case for identical particle interferometry. One is forced to perform numerical integration of the source function taking into account the two-particle momentum phase-space and experimental acceptance [2, 13, 15] a method similar to the one used before to fit e.g. proton-proton correlation functions [16, 17, 18, 19]. The "best-fit" values are found by finding the source parameters which produce the correlation function which most closely matches the experimental one.

THE ORIGINS OF ASYMMETRY

The asymmetry analysis was developed as a tool to study the time sequence of the emission of particles [1] and it has been used in this manner in analysis of heavy-ion collisions at intermediate energy [5, 6, 7]. The time differences measured were on the order of tens or hundreds of fm/c. However in relativistic collisions such long-lived source is not observed [8, 9, 10]. Parameterizations which assume a sudden break-up of the rapidly expanding fireball seem to describe the data self-consistently. Calculations based on hadron rescattering, like the RQMD model [20], also predict time shifts of the order of only a few fm/c [21]. It is also well known that particles coming from short-lived resonance decay will appear as coming from a delayed source. A simple calculation in THERMINATOR model [22] shows that this is indeed the case (see Fig. 3), however the overall difference between the average emission times of e.g. pions and kaons is again only a few fm/c.

Later it was realized that another mechanism could produce shifts in average positions of particles of different masses [24, 4]. This mechanism is a well-known radial flow. In a simple model of the source, a particle's velocity is composed of the radial velocity

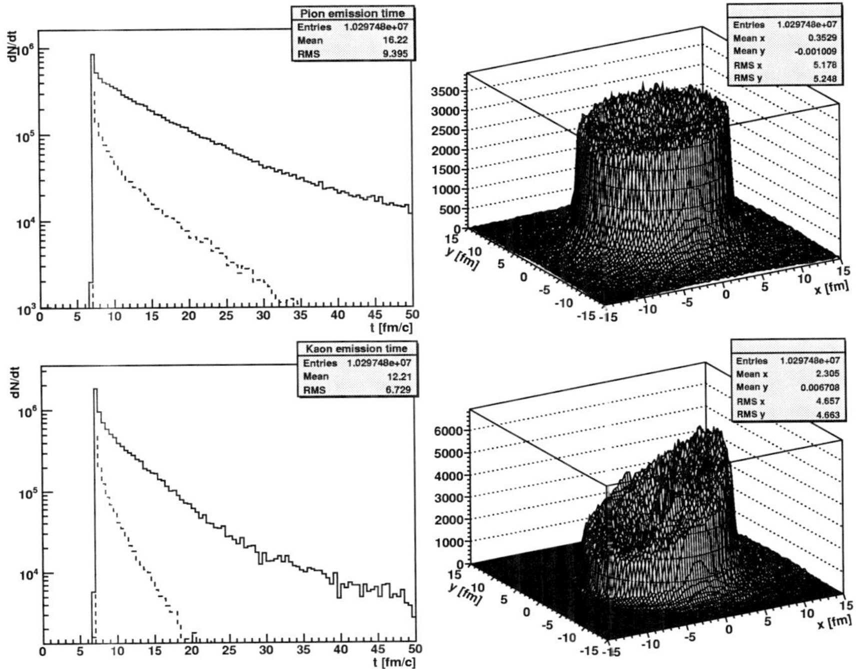

FIGURE 3. The origins of asymmetry for pion-kaon pairs. On the left the emission times modelled in THERMINATOR are shown, pions at the top, kaons at the the bottom. The solid lines show all particles, the dashed lines - only the primordial ones. On the right hand side the average emission points of particles for pairs with average momentum along the x axis are shown. Top plot is for pions, bottom for kaons.

v_r, coming from radial flow and thermal velocity v_T. v_r always points in the outwards direction, and is the same for all particles, or in other words we assume the matter is behaving collectively. This means that particle's velocity direction will be uniquely determined by the space-time point of its emission. The thermal velocity, in contrast, is directed randomly. Therefore it will randomize the observed velocity, and the correlation between particle's emission point and velocity direction will be somewhat reduced. Just how big this reduction will be depends on the relative strength of v_r and v_T. Since temperature matters less for particles with higher mass, we expect the lighter particles to have much more randomized velocity than heavier particles. This is seen on Fig. 3. It shows the distribution of emission points of particles, obtained from the THERMINATOR calculation, whose velocity is in the "out" direction - that is along the x axis. We see that for pions this distribution fills the source almost evenly, which means that the average emission point of pions is close to the center. In contrast for kaons the distribution is sharply peaked in the "out" direction, leading to the average emission point closer to the edge. For protons this effect is even more pronounced. We also note that there is no such shift in the "side" direction.

It is known that radial flow produces the "size"-p correlations, which are observed in the data [8, 9] as the k_T dependence of the interferometry radii. However alternative explanations, like e.g. temperature gradients [23] were proposed. What is unique about radial flow though is that it also produces x-p correlations, which manifest themselves as shifts in average position between particles of different masses, as we discussed above. This phenomenon can be measured by non-identical particle correlations, which will provide an additional constraint on the dynamical models of the expanding source.

THE EXPERIMENTAL RESULTS

The first experimental results on non-identical particle correlations in relativistic heavy-ion collisions that concerned not only time ordering of the emission of baryons were obtained at the SPS by the the NA49 experiment [24, 25] and at AGS by the E877 experiment [26]. The first included $\pi^+ - \pi^-$, $\pi^+ - p$ and $\pi^- - p$ correlations and were compared to the RQMD predictions. They concentrated on the analysis of the size of the emitting system and concluded that RQMD underpredicts the size of the emitting system at theses energies. They have also observed the asymmetry between pions and protons, in agreement with predictions from the RQMD model. The second concluded the existence of the shift between pion and proton sources. A study similar to the second one has been also recently performed by CERES collaboration [27].

Next results came from RHIC [28]. Pion-kaon correlations have been analyzed in detail, taking into account both the size of the system and the shifts between particles of different masses. The results showed significant shift in average emission points between pions and kaons and were found to be consistent with the flow scenario. More preliminary analysis followed, resulting in the presentation by STAR of the complete set of $\pi - K$, $\pi - p$ and $K - p$ correlation functions at both 130 and 200 AGeV Au-Au collisions [21]. Again significant shifts between particle species of different masses were observed. The flow hypothesis was found to be consistent with all data. Time shifts were also found to be important in the analysis. Recent analysis of correlations between non-identical particles of same masses [29]: $\pi^+ - \pi^-$ and $p - \bar{p}$ are consistent with zero asymmetry which is consistent with the flow scenario.

Recently STAR has also presented preliminary results from $\pi - \Xi$ correlations [30]. Ξ particle is heavy and has a small cross-section for the interaction with the hadronic matter. Blast-wave fits suggest that the particle undergoes an early freeze-out, decoupling from the system earlier than e.g. protons or pions. This hypothesis can be directly tested by the asymmetry measurement through non-identical particle correlations.

The analysis was also extended into a more exotic systems of $p - \Lambda$ and $p - \bar{\Lambda}$ systems [21]. The size of the system emitting these particles was compared to source sizes obtained from regular interferometry analysis and found to be consisted with the observed m_T systematics.

SUMMARY

The history and theoretical basis of non-identical particle femtoscopy has been discussed. Coulomb interaction has been discussed in detail, showing how it can be used to measure emission asymmetries. The possible origins of asymmetries have been discussed and the link to the hydrodynamical flows has been established. Recent non-identical particle correlations data from relativistic heavy-ion experiments have been discussed and showed to support the creation of strong radial flows in such collisions.

ACKNOWLEDGMENTS

The author wishes to thank Richard Lednický for many discussions concerning all aspects of non-identical particle correlations. Research carried out within the scope of the ERG (GDRE): Heavy ions at ultrarelativistic energies- a European Research Group comprising IN2P3/CNRS, Ecole des Mines de Nantes, Universite de Nantes, Warsaw University of Technology, JINR Dubna, ITEP Moscow and Bogolyubov Institute for Theoretical Physics NAS of Ukraine.

REFERENCES

1. R. Lednický, V. I. Lyuboshitz., B. Erazmus, D. Nouais, *Phys. Lett. B* **373**, 30-34 (1996).
2. R.Lednický, V.L.Lyuboshitz, *Sov. J. Nucl. Phys.* **35**, 770 (1982); *Proc. CORINNE 90, Nantes, France*, ed. D.Ardouin, World Scientific, 1990, p 42; *Heavy Ion Physisc* **3** 93 (1996).
3. S. Voloshin, R. Lednický, S. Panitkin, N. Xu; *Phys. Rev. Lett.* **79**, 4766-4769 (1997).
4. R.Lednický, nucl-th/0305027.
5. S. Gaff et al. *Phys. Rev. C* **58**, 2161-2166 (1998).
6. R. Ghetti et al., *Phys. Rev. C* **62**, 037603 (2000).
7. P. Chung et al., *Phys. Rev. Lett.* **91**, 162301 (2003).
8. C. Adler et al. (STAR Collaboration), *Phys. Rev. Lett.* **87**, 082301 (2001).
9. J. Adams et al. (STAR Collaboration), *Phys. Rev. C* **71**, 044906 (2005).
10. F. Retière, M. A. Lisa, *Phys. Rev. C* **70**, 044907 (2004).
11. R. Lednicky, S. Panitkin and N. Xu, nucl-th/0304062.
12. F. Retière, *these proceedings*.
13. R. Lednický, nucl-th/0501065.
14. S. Pratt, S. Petriconi; *Phys. Rev. C* **68**, 054901 (2003).
15. A. Kisiel, *Nukleonika* **49 (Suplement 2)**, S81-S82 (2004).
16. S. Padula, M. Gyulassy, S. Gavin *Nucl. Phys. B* **329**, 357-375 (1990).
17. M. A. Lisa, *Phys. Rev. C* **49**, 2788-2791 (1994).
18. S. Padula, M. Gyulassy, *Phys. Lett. B* **348**, 303-308 (1995).
19. S. Panitkin, D. Brown, *Phys. Rev. C* **61**, 021901 (2000).
20. H. Sorge, *Phys. Rev. C* **52**, 3291-3314 (1995).
21. A. Kisiel (for the STAR Collaboratio), *J. Phys. G, Nucl. Phys.* **30**, S1059 - S1063 (2004).
22. A. Kisiel, T. Tałuć, W. Broniowski, W. Florkowski, nucl-th/0504047.
23. A. Ster, T. Csorgo, B. Lorstad; *Proc. CF'98 (WSCI 1999)*, ed. T. Csorgo et al, p. 137 (1999); hep-ph/9809571.
24. R. Lednický; *NA49 Note number 210*.
25. Ch. Blume et al. (NA49 Collaboration), *Nucl. Phys. A* **715**, 55-64 (2003).
26. D. Miskowiec, nucl-ex/9809003, *in proc. CRIS'98, June 8-12, 1998, Acicastello, Italy, "Acicastello 1998, Measuring the size of things in the universe"*, pp. 279-288.
27. D. Antonczyk, *Poster at Quark Matter 2005 conference*.
28. J. Adams et al. (STAR Collaboration), *Phys. Rev. Lett.* **91**, 262302 (2003).
29. H. Gos, *QM 2005 proceedings*; H. Gos, *these proceedings*.
30. P. Chaloupka, *QM 2005 presentation*, P. Chaloupka, *these proceedings*.

$\pi-\Xi$ Correlations at RHIC

Petr Chaloupka (for the STAR collaboration)

Nuclear Physics Institute, Academy of Sciences of the Czech Republic, 250 68 Rez near Prague, Czech Republic

Abstract. We report on $\pi-\Xi$ correlation analysis in Au+Au collisions at $\sqrt{s_{NN}}$=200 GeV and $\sqrt{s_{NN}}$=62 GeV from year 2004 high statistics run of the STAR experiment at RHIC. Correlation analysis of $\pi-\Xi$ system is presented addressing independently the issue of flow of multi-strange baryons in heavy-ion collisions. We observe effects of both Coulomb and strong interactions in the $\pi-\Xi$ final state, the latter proceeding via $\Xi^*(1530)$ resonant interaction. The technique of spherical harmonics decomposition of the correlation function is used to study space-time separation of the pion and Ξ sources.

Keywords: STAR, femtoscopy, non-identical correlations, $\pi-\Xi$
PACS: 25.75.-q

INTRODUCTION

Ultra-relativistic heavy-ion collisions study strongly interacting QCD matter in extreme conditions. At energy densities achieved in Au+Au collisions at RHIC the produced matter is governed by partonic degrees of freedom. Its nearly ideal fluid flow, marked by constituent interactions of very short mean free path, is established most probably at a stage preceding hadron formation [1].

Current data [1] on spectra and elliptic flow from Au+Au collisions at RHIC energies suggest that the hot and dense system created in the collision builds up substantial collectivity leading to rapid transverse expansion. Moreover, properties of the induced flow strongly depend on whether the collectivity is achieved at partonic or hadronic level.

Because of their presumably small hadronic cross-section Ξs are expected to undergo few interactions in the hadronic phase [2], and thus providing more direct probe into the early partonic stage.

Measurements of momentum correlations between non-identical particles at small relative velocities are used to study space-time characteristics of the heavy-ion collisions [3]. Compared to identical particle correlations, these measurements are not only sensitive to the space-time extent of the particle-emitting source, but also, as suggested in [4], to the relative emission asymmetry among the two particles. Spatial asymmetry can arise as a consequence of collective expansion since flow induces a strong correlation between particles' velocities and emission points [5], such that heavier particles are on average emitted closer to the surface of the source. This in turn leads to an effective decrease of measured HBT radii and different average emission points for particle species with non-equal masses [5]. Therefore non-identical particle correlations can be used as an independent cross-check of flow measurements in heavy-ion collisions. This method is especially suitable for systems like $\pi-\Xi$, where the mass-difference is large.

CP828, *Multiparticle Dynamics*
edited by V. Šimák, M. Šumbera, Š. Todorova, and B. Tomášik

EXPERIMENTAL DATA, ANALYSES TECHNIQUE

We present results from three different datasets: d+Au and Au+Au collisions at $\sqrt{s_{NN}}$=200 GeV, and Au+Au collisions at $\sqrt{s_{NN}}$=62 GeV. The data were measured by the STAR experiment at RHIC using its main detector - the Time Projection Chamber (TPC) [6] which records charged primary and secondary particles at mid-rapidity. The particles are identified via specific energy loss (dE/dx) in the TPC gas. We are able to topologically reconstruct charged $\Xi(\bar{\Xi})$-hyperons using decay chain $\Xi \rightarrow \Lambda + \pi$, and subsequently $\Lambda \rightarrow \pi + p$.

Only events with longitudinal primary vertex position within 25 cm of the TPC center are accepted for this analyses. We select primary pions and Ξs in rapidity region $|y| < 0.5$. Our selection method and detector acceptance then limits the p_t range of our pion-sample to $0.08 < p_t < 0.6$ GeV/c and Ξ-sample to $0.7 < p_t < 3.0$ GeV/c.

The $\pi-\Xi$ correlation functions $C(\vec{k^*})$ were analyzed in the pair rest frame, where $\vec{k^*} = \vec{p_\pi} = -\vec{p_\Xi}$. Small magnitude of k^* then means that both particles move with a small relative velocity in their pair rest frame.

The correlation function is constructed as a ratio $C(\vec{k^*}) = A(\vec{k^*})/B(\vec{k^*})$ of two distributions of $\pi-\Xi$ pairs: $A(\vec{k^*})$ - pairs obtained from a single event, and $B(\vec{k^*})$ - the corresponding "mixed-events" distribution, where each particle is taken from a different event. In order to ensure a well-defined baseline, events for mixing are divided into classes with similar multiplicity and primary vertex position. Since both π and Ξ-samples contain non-primary or misidentified particles, which reduces the strength of the measured correlation, all presented correlation functions were corrected for pair purities. The measured correlation functions were corrected according to a relation

$$C_{true}(\vec{k^*}) = \frac{C_{measured}(\vec{k^*}) - 1}{\text{Pair Purity}} + 1, \quad (1)$$

where Pair Purity is a product of purities of individual particles' sample. The $C_{measured}(\vec{k^*})$ is the uncorrected correlation function as it was measured, and $C_{true}(\vec{k^*})$ is the corrected one after the purities were taken into the account. The purity of the Ξ-sample, was estimated from the signal to combinatoric background ratio in the invariant mass distribution as a function of p_t. The pion-sample purity was estimated from λ as measured in π-π HBT analyses [7].

STAR RESULTS ON $\pi-\Xi$ CORRELATIONS

In Fig. 1 we present STAR preliminary results on $\pi-\Xi$ correlations as a function of $|\vec{k^*}|$ for all four charge combinations of the $\pi-\Xi$ pairs from the 10% most central Au+Au collisions at $\sqrt{s_{NN}}$=200 GeV. While in the low k^*-region ($k^* < 0.05$ GeV/c) of $C(k^*)$ all charge combinations are dominated by the Coulomb interaction, the strong interaction manifests itself in unlike-sign pairs as the $C(k^*)$ peaks at $k^* = 0.15$ GeV/c, corresponding to $\Xi^*(1530)$ resonance. Since results, within statistical errors, are the same for $\pi^+ - \Xi^-, \pi^- - \Xi^+$ and $\pi^+ - \Xi^+, \pi^- - \Xi^-$ pairs, correspondingly, the data with like and unlike-sign pairs can be combined in order to improve the statistics.

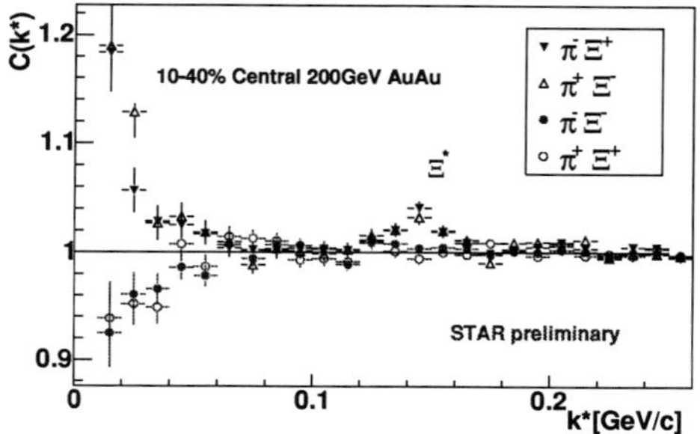

FIGURE 1. $\pi - \Xi$ correlation function for 10% most central Au+Au collisions at $\sqrt{s_{NN}}$=200 GeV. Peak at $k^* = 0.15$ GeV/c in unlike-sign pairs corresponds to $\Xi^*(1530)$ resonance.

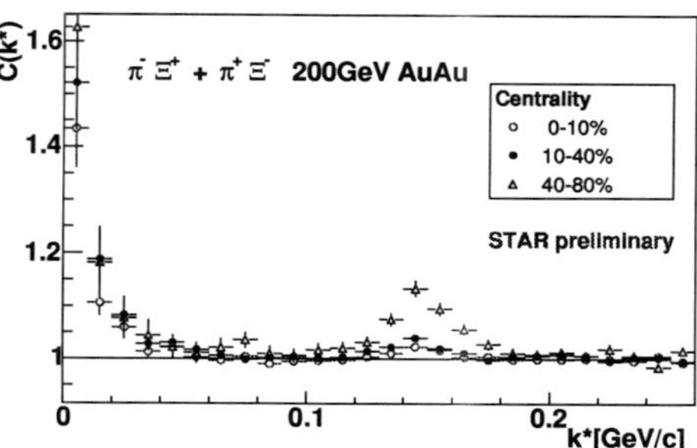

FIGURE 2. The centrality dependence of the correlation function in $\sqrt{s_{NN}}$=200 GeV Au+Au collisions. The shown data are combined unlike-sign $\pi^+ - \Xi^-$, and $\pi^- - \Xi^+$ pairs.

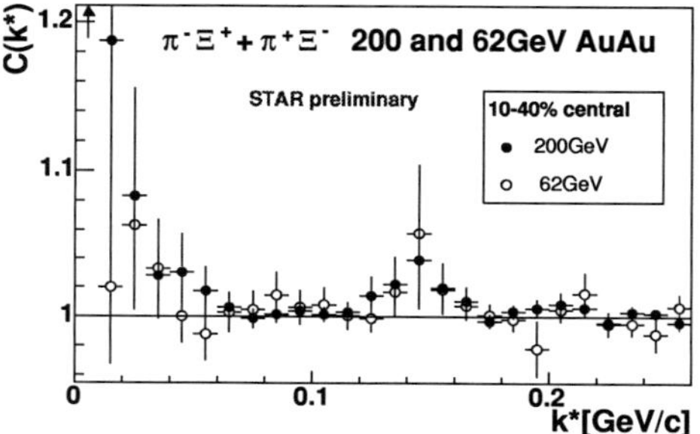

FIGURE 3. The comparison of combined unlike-sign $\pi - \Xi$ $C(k^*)$ for two different energies in Au+Au collisions: open circles - $\sqrt{s_{NN}}$=62 GeV; full circles - $\sqrt{s_{NN}}$=200 GeV.

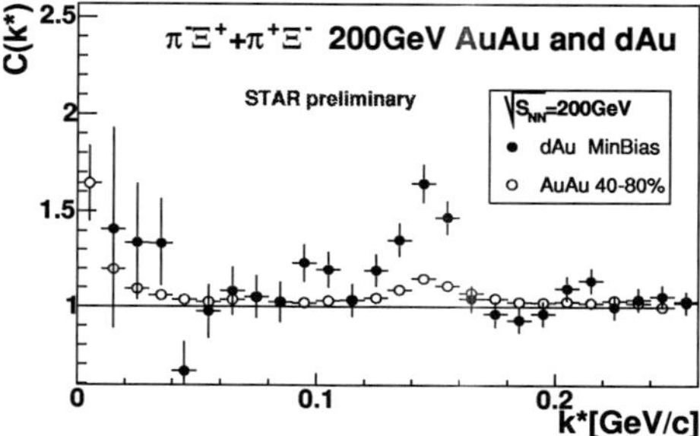

FIGURE 4. The comparison of combined unlike-sign $\pi - \Xi$ $C(k^*)$ for Au+Au and d+Au collisions at $\sqrt{s_{NN}}$=200 GeV: open circles - peripheral Au+Au; full circles - minimum bias d+Au collisions.

Figure 2 shows $C(k^*)$ of combined unlike-sign pairs at $\sqrt{s_{NN}}$=200 GeV for three different centrality bins. Comparison of $C(k^*)$ for combined unlike-sign pairs from centrality 10-40% in Au+Au collisions at $\sqrt{s_{NN}}$=200 GeV and $\sqrt{s_{NN}}$=62 GeV is presented in Fig. 3. In Fig. 4 we compare two different systems: Au+Au (centrality 40-80%) and d+Au minimum bias collisions at the same energy $\sqrt{s_{NN}}$=200 GeV.

From Figs. 2, 3, and 4 we observe a clear dependence of the $\pi - \Xi$ correlation on the type of the system and centrality. Moreover, in the region of Ξ^* resonance, the $C(k^*)$ shows much stronger sensitivity to the source size than in the Coulomb region, which is also affected by a low available statistics. On the other hand, the strength of the correlation signal does not seem to be significantly dependent on the collision energy.

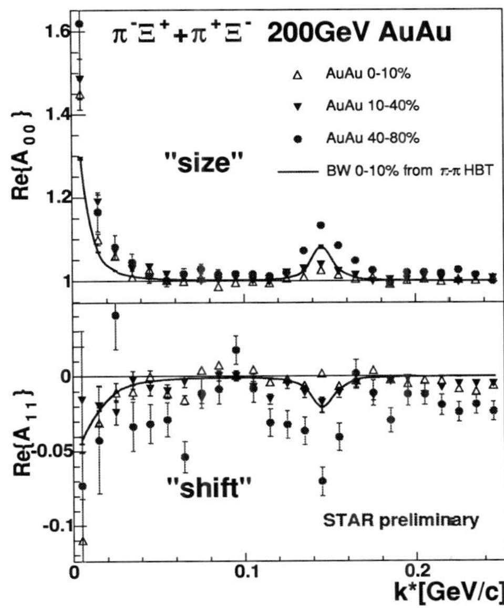

FIGURE 5. Combined unlike-sign $\pi - \Xi$ pairs: centrality dependence of $A_{0,0}(k^*)$ and $A_{1,1}(k^*)$ coefficient of the spherical projections of $C(\vec{k^*})$ in $\sqrt{s_{NN}}$=200 GeV Au+Au. Solid line is a theoretical prediction for the most central data assuming Ξ flow.

Spherical harmonics decomposition

As mentioned before, current data suggest that Ξ develops substantial flow, thus creating an asymmetry between average emission points of pions and Ξs. Independent test of this hypothesis can be pursued via decomposition of $C(\vec{k^*})$ into spherical harmonics, as suggested in [8]. This method takes an advantage of symmetries of the correlation function by constructing k^* dependent projection coefficients $A_{\lambda,\mu}(k^*)$ in a following way:

$$A_{\lambda,\mu}(|\vec{k^*}|) = \sum_{i}^{\text{all bins}} C(|\vec{k^*}|, \cos\theta_i, \varphi_i) Y_{\lambda,\mu}(\theta_i, \varphi_i) F_{\lambda,\mu}(\theta_i, \Delta_{\cos\theta}, \Delta_\varphi) \tag{2}$$

where $(|\vec{k^*}|, \theta, \varphi)$ are spherical coordinates of $\vec{k^*}$, and $F_{\lambda,\mu}(\theta_i, \Delta_{\cos\theta}, \Delta_\varphi)$ represents a numerical factor correcting for finite bin sizes $\Delta_{\cos\theta}$ and Δ_φ. Most of the information carried by the 3-dimensional correlation function then reflects in a small number of 1-dimensional plots as $A_{\lambda,\mu}(k^*)$ coefficients with increasing λ are statistically diminishing. It has been pointed out [9] that $A_{\lambda,\mu}(k^*)$ components correspond one-to-one with anisotropies of the same order λ, μ of the spatial homogeneity region. While the coefficient $A_{0,0}(k^*)$ of (2) can be interpreted as an angle-averaged $C(k^*)$, the coefficient $A_{1,1}(k^*)$ is sensitive to the pair emission asymmetry [9]. It vanishes when both particles are on average emitted from the same space-time point.

In Fig. 5 we present a centrality dependence of $A_{0,0}(k^*)$ and $A_{1,1}(k^*)$ coefficients for $\sqrt{s_{NN}}$=200 GeV Au+Au collisions data. The solid line shows a theoretical prediction for the most central bin. The final state interaction was calculated using the S. Pratt's approach [10]. Emission coordinates were generated by the hydro-inspired Blastwave Model [5]. Blastwave parameters for the pion source were taken from a previous STAR $\pi-\pi$ correlation analyses [7]. The same parameters were used also for the Ξ source, thus assuming Ξ flow comparable to that of pion. As can be seen, the coefficient $A_{0,0}(k^*)$ follows the same centrality dependence and sensitivity as the correlation function in Fig. 2. Most importantly, the coefficient $A_{1,1}(k^*)$ differs significantly from zero in both, Coulomb and Ξ^*, regions implying that pions and Ξs are not emitted from the same average space-time point. The measured $A_{1,1}(k^*)$ qualitatively follows the theoretical prediction which is based on the assumption of a significant Ξ flow.

CONCLUSIONS

We have presented preliminary measurements of $\pi-\Xi$ correlations in heavy-ion collisions from the STAR experiment at RHIC. All presented results from Au+Au collisions at two different collision energies and d+Au collisions show the effects of final state Coulomb and strong interactions between the produced π and Ξ. The presented correlation functions clearly show sensitivity of the correlation function to the size of the system, the most statistically significant part coming from the region of $\Xi^*(1530)$ resonance. We have used a novel technique of spherical harmonics decomposition of the correlation function, which shows that pions and Ξs are not emitted from the same average space-time point. Such a space-time shift can be attributed to the Ξ flow.

ACKNOWLEDGMENTS

The author wishes to thank S. Pratt for making his code of model calculations available to us.

The work is supported by the IRP AV0Z10480505 and by the Grant Agency of the Czech Republic grant 202/04/0793.

REFERENCES

1. J. Adams, et al., *Nucl. Phys. A* **757**, 102–183 (2005), `nucl-ex/0501009`.
2. H. van Hecke, H. Sorge, and N. Xu, *Phys. Rev. Lett.* **81**, 5764 (1998).
3. M. A. Lisa, S. Pratt, R. Soltz, and U. Wiedemann (2005), `nucl-ex/0505014`.
4. R. Lednicky, V. L. Lyuboshits, B. Erazmus, and D. Nouais, *Phys. Lett. B* **373**, 30–34 (1996).
5. F. Retiere, and M. A. Lisa, *Phys. Rev. C* **70**, 044907 (2004).
6. K. H. Ackermann, et al., *Nucl. Instrum. Meth. A* **499**, 624–632 (2003).
7. J. Adams, et al. (2004), `nucl-ex/0411036`.
8. Z. Chajecki, T. D. Gutierrez, M. A. Lisa, and M. Lopez-Noriega (2005), `nucl-ex/0505009`.
9. D. A. Brown, et al. (2005), `nucl-th/0507015`.
10. S. Pratt, and S. Petriconi, *Phys. Rev. C* **68**, 054901 (2003), `nucl-th/0305018`.

OTHER NEW RESULTS

Chairperson: S. Pratt

Quantum Treatment of the Multiple Scattering and Collective Flow in Intensity Interferometry

Cheuk-Yin Wong

Physics Division, Oak Ridge National Laboratory, Oak Ridge, Tennessee 37830, USA
Physics Department, University of Tennessee, Knoxville, Tennessee 37996, USA

Abstract. We apply the path-integral method to study the multiple scattering and collective flow in intensity interferometry in high-energy heavy-ion collisions. We show that the Glauber model and eikonal approximation in an earlier quantum treatment are special examples of the more general path-integral method. The multiple scattering and collective flow lead essentially to an initial source at a shifted momentum, with a multiple collision absorption factor that depends on the pion absorption cross section and a phase factor that depends on the deviations of the in-medium particle momenta from their asymptotic values.

Keywords: Relativistic heavy-ion collisions, Particle correlations
PACS: 25.75.Gz

INTRODUCTION

The intensity interferometry, as first proposed by Hanbury-Brown-Twiss (HBT) to measure the angular diameter of a star [1], has been applied to optical coherence [2], subatomic physics, and nuclear collisions [3]. In HBT measurements in high-energy heavy-ion collisions, the initial source particles undergo multiple scattering and collective flow. As a result, it is conventionally assumed that the initial chaotic source evolves into a different chaotic source distribution at thermal freeze-out which becomes the distribution measured in intensity interferometry. This conventional assumption is recently subject to question. Because the intensity interferometry is purely a quantum-mechanical phenomenon, the multiple scattering and collective flow must be investigated within a quantum-mechanical framework [4, 5, 6]. Applying the Glauber multiple scattering theory at high energies and the optical model at lower energies, we find that multiple scatterings lead essentially to an initial source distribution with absorption [4, 5, 6]. Using the Feynman path integral method, we further find that the collective flow leads to a phase factor that depends on the deviations of the in-medium particle momenta from their asymptotic values. Subsequent work by Kapusta and Li [7] supports qualitatively some of the earlier results of Ref. [4]. Following this suggestion of a quantum treatment, Cramer and his collaborators [8] later considered the effects of produced pions in a static optical model potential. The work of Cramer *et al*. [8] is however incomplete, as the effects of particle collisions and the collective flow have not been taken into account.

In this manuscript, we apply the path-integral method for the quantum treatment of the multiple scattering and collective flow in intensity interferometry. We show that the Glauber model and eikonal approximation in an earlier quantum treatment of the multiple scattering [4, 5] are special examples of the more general path-integral method.

CP828, *Multiparticle Dynamics*
edited by V. Šimák, M. Šumbera, Š. Todorova, and B. Tomášik

PION ENVIRONMENT AND ELASTIC SCATTERING

Before we present a quantum treatment of intensity interferometry, it is illuminating to investigate the pion environment after a chaotic pion source is produced in the phase transition of a quark-gluon plasma.

At the phase transition temperature of $T \sim 180$ MeV, the average pion density is ~ 0.3 pions/fm^3, the pion energy is ~ 0.5 GeV, and the average C.M. energy, $\langle \sqrt{s_{\pi\pi}} \rangle$, in a π-π collision is ~ 0.7 GeV. Thus the pion source is dense and pions are energetic after their production. As $\langle \sqrt{s_{\pi\pi}} \rangle$ is only slightly lower than the ρ mass, a substantial fraction of π-π collisions will go through the I=1, ρ resonance. The width of the ρ resonance is 150 MeV in free space. In a thermalized medium at $T = 150$ MeV, the width increases substantially to ~ 300 MeV [9], and the ρ resonance mean lifetime in the medium becomes $\sim \hbar/(300\text{MeV})$ or 0.67 fm/c. The orbiting time for the ρ meson is of order $2\pi r$ and $2r \sim 0.77$ fm [10]. As the ρ meson mean lifetime in the medium (~ 0.67 fm/c) is much shorter than its orbiting time (~ 2.4 fm/c), it is unlikely for an $I = 1$, π-π pair to complete an orbital revolution before the ρ meson breaks apart into two pions. The scattering of the two pions through the intermediary ρ meson is essentially an elastic scattering with an energy-dependent amplitude. Furthermore, the $\pi\pi \to K\bar{K}$ threshold is ~ 1 GeV, substantially greater than $\langle \sqrt{s_{\pi\pi}} \rangle$. Hence, chemical reactions of pions in the medium are essentially completed very shortly after the phase transition. From the state of chemical freeze-out to thermal freeze-out, the scatterings suffered by the pions are predominately elastic.

PATH INTEGRAL METHOD FOR INTENSITY INTERFEROMETRY

As a pion propagates from the state of chemical freeze-out to thermal freeze-out, the elastic $\pi\pi$ scattering can be described by short-ranged scalar and vector interactions, $v_{\rm col}^{(s,v)}(q - q_i)$, where q is the coordinate of the propagating pion and q_i is the coordinate of a pion in the medium. The propagating pion is also subject to a collective flow which can be described by a long-range density-dependent mean-field scalar and vector interactions, $V_{\rm mf}^{(s,v)}(q)$, as in similar cases in the dynamics of the nuclear fluid [11]. The Lagrangian for the propagating pion is given by

$$L(q,\dot{q}) = L_{\rm mf}(q,\dot{q}) + L_{\rm col}(q,\dot{q}), \tag{1}$$

where
$$L_{\rm mf}(q,\dot{q}) = -[m_\pi + V_{\rm mf}^{(s)}(q)]\sqrt{1-\dot{\mathbf{q}}^2} + \dot{\mathbf{q}} \cdot \mathbf{V}_{\rm mf}^{(v)}(q) - V_{\rm mf}^{0(v)}(q), \tag{2}$$

$$L_{\rm col}(q,\dot{q}) = -V_{\rm col}^{(s)}(q)\sqrt{1-\dot{\mathbf{q}}^2} + \dot{\mathbf{q}} \cdot \mathbf{V}_{\rm col}^{(v)}(q) - V_{\rm col}^{0(v)}(q), \tag{3}$$

$$V_{\rm col}^{(s,v)}(q) = \sum_i^{N_\pi - 1} v_{\rm col}^{(s,v)}(q - q_i). \tag{4}$$

From the Lagrangian of the propagating pion, one obtains the pion three-momentum,

$$\mathbf{p} = \partial L/\partial \dot{\mathbf{q}} = \gamma[m_\pi + V^{(s)}(q)]\dot{\mathbf{q}} + \mathbf{V}^{(v)}(q), \tag{5}$$

the pion Hamiltonian,

$$H = p^0 = \mathbf{p}\cdot\dot{\mathbf{q}} - L = \gamma[m_\pi + V^{(s)}(q)] + V^{0(v)}(q), \tag{6}$$

and the pion mass-shell condition,

$$[p^0 - V^{0(v)}(q)]^2 - [\mathbf{p}^2 - \mathbf{V}^{(v)}(q)]^2 - [m_\pi + V^{(s)}(q)]^2 = 0, \tag{7}$$

where $\gamma = 1/\sqrt{1-\dot{\mathbf{q}}^2}$ and $V^{(s,v)}(q) = V_{\rm mf}^{(s,v)}(q) + V_{\rm col}^{(s,v)}(q)$. For a pion produced at x with momentum κ to propagate in the pion medium to the thermal freeze-out point x_f and be detected at the detecting point x_d with momentum k, the probability amplitude is [12]

$$K(\kappa x \to k x_d) = \int \mathscr{D}q\, e^{iS(\kappa x \to k x_d; q)}, \tag{8}$$

where $\int\mathscr{D}q...$ is the sum over all paths q from x to x_d, and the action $S(\kappa x \to k x_d; q)$ is

$$S(\kappa x \to k x_d; q) = \int L(q,\dot{\mathbf{q}})\, dt = -\int_x^{x_d,(\text{path } q)} p(q')\cdot dq'. \tag{9}$$

We can separate $S(\kappa x \to k x_d; q)$ into different contributions,

$$S(\kappa x \to k x_d; q) = -k\cdot(x_d - x) + \delta_{\rm mf}(\kappa x, k x_d; q) + \delta_{\rm col}(\kappa x, k x_d; q), \tag{10}$$

$$\delta_{\rm mf}(\kappa x, k x_d; q) = -\int_x^{x_d,(\text{path } q)} [p_{\rm mf}(q') - k]\cdot dq' = -\int_x^{x_f,(\text{path } q)} [p_{\rm mf}(q') - k]\cdot dq', \tag{11}$$

$$\delta_{\rm col}(\kappa x, k x_d; q) = -\int_x^{x_d,(\text{path } q)} p_{\rm col}(q')\cdot dq' = -\int_x^{x_f,(\text{path } q)} p_{\rm col}(q')\cdot dq', \tag{12}$$

$$p_{\rm mf}(q) = \left(\gamma[m_\pi + V_{\rm mf}^{(s)}(q)] + V_{\rm mf}^{0(v)}(q),\ \gamma[m_\pi + V_{\rm mf}(q)]\dot{\mathbf{q}} + \mathbf{V}_{\rm mf}^{(v)}(q)\right), \tag{13}$$

$$p_{\rm col}(q) = \left(\gamma V_{\rm col}^{(s)}(q) + V_{\rm col}^{0(v)}(q),\ \gamma V_{\rm col}^{(s)}(q)\dot{\mathbf{q}} + \mathbf{V}_{\rm col}^{(v)}(q)\right). \tag{14}$$

Because of the additivity of the collision potentials in Eq. (4), the phase shift for multiple collision $\delta_{\rm col}$ is a sum of the phase shifts for individual collisions, similar to the case of the Glauber wave function in multiple scattering [4, 13],

$$\delta_{\rm col}(\kappa x, k x_d; q) = \sum_i^{N_\pi - 1} \delta_{{\rm col},i}(\kappa x, k x_d; q), \tag{15}$$

where $\delta_{{\rm col},i}(\kappa x, k x_d; q)$ is obtained from Eqs. (12) and (14) with the potential $v_{\rm col}(q - q_i)$ in place of the total collision potential $V_{\rm col}(q)$. The propagation amplitude is therefore

$$K(\kappa x \to k x_d) = \int \mathscr{D}q \exp\{-ik\cdot(x_d - x) + i\delta_{\rm mf}(\kappa x, k x_d; q) - i\delta_{\rm col}(\kappa x, k x_d; q)\}. \tag{16}$$

If one makes the approximation that the dominant contribution to the path integral comes from the trajectory along the classical path q_c for mean-field motion (which need not be a straight line), then the amplitude is approximately

$$K(\kappa x \to k x_d) \approx \exp\{-ik\cdot(x_d - x) + i\delta_{\rm mf}(\kappa x, k x_d; q_c) + i\delta_{\rm col}(\kappa x, k x_d; q_c)\}. \tag{17}$$

For the propagation of an energetic pion, the phase shifts along a straight-line trajectory are just those considered in Ref. [4].

For a pion with momentum κ_i produced at x_i to propagate to momenta k_i at the detecting point x_{di}, the amplitude is

$$\Psi(\kappa_i x_i \to k_i x_{di}) = A(\kappa_i x_i) e^{\phi_0(x_i)} K(\kappa_i x_i \to k_i x_{di}), \tag{18}$$

where $A(\kappa_i x_i)$ is the production amplitude, and $\phi_0(x_i)$ is a random and fluctuating production phase for the chaotic source. The probability amplitude for the production of two identical pions (κ_1, κ_2) at (x_1, x_2) to be detected subsequently as k_1 at x_{d1} and k_2 at x_{d2} is

$$\frac{1}{\sqrt{2}}\left\{\Psi_1(\kappa_1 x_1 \to k_1 x_{d1})\Psi_1(\kappa_2 x_2 \to k_2 x_{d2}) + (x_1 \leftrightarrow x_2)\right\}. \tag{19}$$

The probability $P(k_1 k_2)$ for the detection of two pions with momenta (k_1, k_2) is the absolute square of the sum of the above amplitudes from all x_1 and x_2 source points. Because of the random and fluctuating phase $\phi_0(x_i)$ for a chaotic source, the absolute square of the sum of the amplitudes becomes the sum of the absolute squares of the amplitudes [3]. One obtains $P(k_1 k_2) = P(k_1)P(k_2)[1 + R(k_1 k_2)]$, where $P(k_i)$ is the probability of detecting a pion of momentum k_i, and

$$\begin{aligned} P(k_1)P(k_2)R(k_1 k_2) = &\sum_{x_1 x_2} A(\kappa_1 x_1)A(\kappa_2 x_2)A(\kappa_1 x_2)A(\kappa_2 x_1) \\ &\times K(\kappa_1 x_1 \to k_1 x_{d1})K(\kappa_2 x_2 \to k_2 x_{d2})K^*(\kappa_1 x_2 \to k_1 x_{d1})K^*(\kappa_2 x_1 \to k_2 x_{d2}). \end{aligned} \tag{20}$$

We can evaluate the product $K(\kappa_1 x_1 \to k_1 x_{d1})K^*(\kappa_2 x_1 \to k_2 x_{d2})$. It is equal to

$$\begin{aligned} K(\kappa_1 x_1 \to k_1 x_{d1})K^*(\kappa_2 x_1 \to k_2 x_{d2}) &= e^{-ik_1\cdot(x_{d1}-x_1)+ik_2\cdot(x_{d2}-x_2)} \\ \times \int \mathscr{D}q \int \mathscr{D}q' \exp\{&i\delta_{\rm mf}(\kappa_1 x_1, k_1 x_{d1}; q) - i\delta^*_{\rm mf}(\kappa_2 x_1, k_2 x_{d2}; q') \\ &+ i\delta_{\rm col}(\kappa_1 x_1, k_1 x_{d1}; q) - i\delta^*_{\rm col}(\kappa_2 x_1, k_2 x_{d2}; q')\}. \end{aligned} \tag{21}$$

The real part of the phase difference $\delta_{\rm col}(\kappa_1 x_1, k_1 x_{d1}; q) - \delta^*_{\rm col}(\kappa_2 x_1, k_2 x_{d2}; q')$ is stationary when $q - q' = 0$ and is random and fluctuating when $q - q' \neq 0$. The sum over $\exp\{i\delta_{\rm col}(\kappa_1 x_1, k_1 x_{d1}; q) - \delta^*_{\rm col}(\kappa_2 x_1, k_2 x_{d2}; q')\}$ is approximately zero when $q - q' \neq 0$. Thus, the phase factor $\exp\{i\delta_{\rm col}(\kappa_1 x_1, k_1 x_{d1}; q) - \delta^*_{\rm col}(\kappa_2 x_1, k_2 x_{d2}; q')\}$ operationally behaves approximately as $\delta(q - q')\exp\{-2\mathscr{I}m\, \delta_{\rm col}(\kappa_1 x_1, k_1 x_{d1}; q')\}$. Upon integrating over $\mathscr{D}q'$, we obtain

$$\begin{aligned} &K(\kappa_1 x_1 \to k_1 x_{d1})K^*(\kappa_2 x_1 \to k_2 x_{d2}) \approx e^{-ik_1\cdot(x_{d1}-x_1)+ik_2\cdot(x_{d2}-x_2)} \\ &\times \int \mathscr{D}q \exp\{i\delta_{\rm mf}(\kappa_1 x_1, k_1 x_{d1}; q) - i\delta^*_{\rm mf}(\kappa_2 x_1, k_2 x_{d2}; q) - 2\mathscr{I}m\, \delta_{\rm col}(\kappa_1 x_1, k_1 x_{d1}; q)\}. \end{aligned}$$

In the sum over paths in the above integral, the dominant contribution comes from the trajectory q_c that minimizes the action difference of the mean field, $\{i\delta_{\rm mf}(\kappa_1 x_1, k_1 x_{d1}; q) - i\delta^*_{\rm mf}(\kappa_2 x_1, k_2 x_{d2}; q)\}$. We therefore have

$$\begin{aligned} K(\kappa_1 x_1 \to k_1 x_{d1})K^*(\kappa_2 x_1 \to k_2 x_{d2}) \approx \exp\{&-ik_1\cdot(x_{d1}-x_1) + ik_2\cdot(x_{d2}-x_2) \\ &+ i\delta_{\rm mf}(\kappa_1 x_1, k_1 x_{d1}; q_c) - i\delta^*_{\rm mf}(\kappa_2 x_1, k_2 x_{d2}; q_c) - 2\mathscr{I}m\, \delta_{\rm col}(\kappa_1 x_1, k_1 x_{d1}; q_c)\}. \end{aligned} \tag{22}$$

The other factor in Eq. (20) can be approximated in a similar way. When all the factors are collected, we obtain

$$R(k_1,k_2) = \left| \int d^4x e^{i(k_1-k_2)\cdot x + i\phi_{\rm mf}(k_1k_2,x;q_c) + i\phi_{\rm col}(k_1k_2,x;q_c)} \rho_{\rm eff}(k_1k_2,x) \right|^2, \qquad (23)$$

where

$$\rho_{\rm eff}(k_1k_1,x) = \frac{\sqrt{f(\kappa_1 x)f(\kappa_2 x)}}{P(k_1)P(k_2)}, \qquad (24)$$

$$\phi_{\rm mf}(k_1k_2,x;q_c) = \int_x^{x_f,(\text{path } q_c)} \{[p_{1\,\rm mf}(q) - k_1] - [p^*_{2\,\rm mf}(q) - k_2]\} \cdot dq, \qquad (25)$$

$$\phi_{\rm col}(k_1k_2,x;q_c) = \delta_{\rm col}(\kappa_1 x, k_1 x_{d1};q_c) - \delta^*_{\rm col}(\kappa_2 x, k_2 x_{d2};q_c) \approx 2i\mathscr{I}m\delta_{\rm col}(\kappa_1 x, k_1 x_{d1};q_c),$$

and $f(\kappa x)$ is the momentum distribution of the initial chaotic source at x with momentum κ that evolves asymptotically to k [4]. The collective flow (mean-field interaction) 'distorts' the initial momentum κ into the final detected momentum k. The above results from the path-integral method contain those of Ref. [4] as special cases.

The multiple scattering phase $\phi_{\rm col}(k_1k_2,x;q_c)$ can be simplified to be [13]

$$-2\mathscr{I}m\; \delta_{\rm col}(\kappa_1 x, k_1 x_{d1};q_c) = -\int_x^{x_f,(\text{path } q_c)} n_\pi(q')\; \sigma_{\rm abs}\; dq', \qquad (26)$$

where $n_\pi(q')$ is the pion density at (q') and $\sigma_{\rm abs}$ is the pion absorption cross section.

CONCLUSIONS AND DISCUSSIONS

In the environment after the phase transition, chemical reactions are essentially completed, and the collisions between pions are predominately elastic. The elastic propagation of a pion can be studied by the Feynman path-integral method. We describe the collective flow of the pion by its dynamical motion in a density-dependent long-range mean-field potential, and its scatterings with other pions by short-range π-π interactions. We find that HBT correlation measurements lead to an effective source distribution that depends on the initial source distribution at a shifted momentum, with a multiple scattering absorption factor $e^{i\phi_{\rm col}}$ and a collective flow phase factor $e^{i\phi_{\rm mf}}$.

The multiple scattering absorption factor depends on the pion absorption cross section. A substantial fraction of the π-π collisions will go through the I=1, ρ resonance. The width of the ρ meson increases substantially in the medium [9] and the ρ meson mean lifetime in the medium is much shorter than its orbiting time. The scattering of two pions through the intermediary ρ meson is essentially an elastic scattering and does not represent an absorption process in intensity interferometry. Therefore, if one considers the π^+-π^+ correlation, a detected π^+ can be absorbed in its propagation through a pion medium only by interacting with a π^- in the $\pi^+\pi^- \to \pi^0\pi^0$ reaction. The cross section $\sigma(\pi^+\pi^- \to \pi^0\pi^0)$ is equal to $(8\pi/9k^2)\sin^2[\delta(I=0) - \delta(I=2)]$, where k is the magnitude of the pion momentum in the π-π C.M. system and $\delta(I)$ is the π-π phase shift for the state of total isospin I [14]. One finds an absorption cross section $\sigma(\pi^+\pi^- \to \pi^0\pi^0)$ of $\sim$8 mb and a mean absorption path length $\sim$12 fm in a thermalized pion medium at T=180 MeV. The mean absorption path length increases as the temperature decreases. The degree of π^+ absorption by the $\pi^+\pi^- \to \pi^0\pi^0$ reaction is small.

The collective flow leads to a net phase shift $\phi_{\rm mf}(k_1k_2,x;q_c)$ in Eq. (25) that depends on the deviations of the in-medium particle momenta $p_{i{\rm mf}}(q)$ from their asymptotic values k_i. The contributions of the two terms from $p_{1{\rm mf}}$ and $p_{2{\rm mf}}$ in Eq. (25) tend to cancel and give rise only to a small effect, as indicated by similar hydrodynamical calculations where one evaluates $\phi_{\rm mf}$ by following pion trajectories [6]. Because of the small absorption due to multiple scattering and the small net phase shift due to the collective flow, the effective density measured in HBT measurements is expected to depend on a source distribution close to the initial (chemical freeze-out) source distribution, at a shifted momentum. We expect that as the initial pion source transverse dimension is approximately the spatial dimension of the colliding nuclei, the effective distribution measured by HBT should be nearly independent of the collision energy. We also expect that in the initial source prior to the collective flow, the transverse size in the "out" direction should be approximately the same as that in the "side" direction. Hence, $R_{\rm out}/R_{\rm side}$ should be approximately close to 1. These expectations are consistent with the gross features of HBT transverse radii in high-energy heavy-ion collisions. It will therefore be of great interest to carry out numerical calculations of the evolution of the produced hadron matter to study the above effects.

ACKNOWLEDGMENTS

The author thanks Profs. H. Crater, R. J. Glauber, Chi-Sing Lam, and Weining Zhang for helpful discussions. This research was supported in part by the Division of Nuclear Physics, U.S. Department of Energy, under Contract No. DE-AC05-00OR22725, managed by UT-Battelle, LLC and by the National Science Foundation under contract NSF-Phy-0244786 at the University of Tennessee.

REFERENCES

1. R. Hanbury-Brown and R. Q. Twiss, *Phil. Mag.* **45**, 663-682 (1954).
2. R. J. Glauber, *Phys. Rev. Lett.* **10**, 84-86 (1963); R. J. Glauber, *Phys. Rev.* **130**, 2529-2539 (1963).
3. For a general review of the Hanbury-Brown-Twiss intensity interferometry, see Chapter 17 of C. Y. Wong, *Introduction to High-Energy Heavy-Ion Collisions*, World Scientific Pub. Company, 1994.
4. C. Y. Wong, *J. Phys. G* **29**, 2151-2168 (2003).
5. C. Y. Wong, *J. Phys. G* **30**, S1053-S1058 (2004).
6. W. N. Zhang, M. J. Efaaf, C. Y. Wong, and M. Khaliliasr, *Chin. Phys. Lett.* **21**, 1918-1921 (2004).
7. J. Kapusta and Y. Li, *J. Phys.* **G30**, S1069-S1072 (2004).
8. J. G. Cramer, G. A. Miller, J. M. S. Wu, and Jin-Hee Yoon, *Phys. Rev. Lett.* **94**, 102302 (2005); J. G. Miller and G. A. Cramer, `nucl-th/0507004`.
9. M. Urban, M. Buballa, and J. Wambach, *Phys. Rev. Lett.* **88**, 042002 (2002).
10. C. Y. Wong, E. S. Swanson, and T. Barnes, *Phys. Rev. C* **65**, 014903 (2002).
11. C. Y. Wong, J. Maruhn, and T. Welton, *Nucl. Phys. A* **253**, 469-489 (1975); C. Y. Wong and H. H. K. Tang, *Phys. Rev. Lett.* **40**, 1070-1073 (1978); C. Y. Wong and H. H. K. Tang, *Phys. Rev.* **20**, 1419-1452 (1979); C. Y. Wong, *Phys. Rev.* **25**, 1460-1475 (1982).
12. For a recent review on the path-integral method, see H. Kleinert, *Path Integrals in Quantum Mechanics, Statistics, Polymer Physics, and Financial Markets*, World Scientific Pub. Company, Singapore 2004, Fourth Extended Edition.
13. C. Y. Wong and R. J. Glauber, to be published.
14. B. R. Martin, D. Morgan, and G. Shaw, *Pion-Pion Interaction in Particle Physics*, Academic Press, 1976, p.101.

CORRELATIONS FROM EVENT GENERATORS

Chairperson: J. Pluta

HBT Results from a Rescattering Model

Thomas J. Humanic

Department of Physics, The Ohio State University, Columbus, Ohio, 43210 USA

Abstract. Preliminary azimuthal pion HBT results obtained from a hadronic rescattering model are compared with experimental RHIC results. The model is also used to obtain preliminary predictions for radial and elliptic flow and pion HBT parameters for LHC Pb+Pb collisions. A test of the calculational method used for the rescattering code is shown using the subdivision method.

Keywords: hadronic rescattering model
PACS: 25.75.-q

INTRODUCTION

It has been found that the predictions from a simple hadronic rescattering model agree reasonably well with flow and HBT measurements for Pb+Pb collisions at the SPS [1] and Au+Au collisions at RHIC [2]. In the present talk rescattering calculations for azimuthal HBT will be compared with recent RHIC STAR experimental results [3]. The rescattering model will also be used to make flow and HBT predictions for Pb+Pb collisions at the LHC. Before showing these predictions, results from a test of the rescattering model calculational method using the subdivision method [4] will be presented.

SUBDIVISION TEST OF THE RESCATTERING MODEL CALCULATIONAL METHOD

Two criticisms which have been made against using the present hadronic rescattering model to make RHIC-energy predictions are 1) the initial state for the calculation is too hot and too dense to be considering hadrons, thus the results are meaningless, and 2) the calculational results may have reasonable agreement with data but that is only accidental because the calculation is dominated by computational artifacts which strongly influence the results. A response to 1) is that the rescattering calculation should be viewed as a limiting case study of how far one can get with an extreme and simple model such that maybe we can learn something about the true initial state from this unexpected agreement with data, e.g. maybe hadron-like objects exist in the QGP, and/or the QGP has a short lifetime and then quickly hadronizes. The response to 2) is that the rescattering model has been tested for Boltzmann-transport-equation-like behavior and the influence of superluminal artifacts using the subdivision method [5]. Although in that test the rescattering model results were shown to not be significantly affected by using a subdivision of $l = 5$, it was not studied whether $l = 5$ was sufficiently large to significantly reduce the superluminal artifacts to make the test meaningful.

CP828, *Multiparticle Dynamics*
edited by V. Šimák, M. Šumbera, Š. Todorova, and B. Tomášik

In the present test, subdivisions of $l = 1,5$, and 8 are used in rescattering calculations of RHIC-energy Au+Au collisions with $b = 8$ fm centrality. Plots of the transverse signal velocity distributions for these subdivision are shown in order to determine how effective these subdivisions are in reducing the superluminal effects. The pion HBT radius parameters are also shown for these subdivisions in order to determine how sensitively they depend on them. The left plot in Fig. 1 shows the transverse signal velocity distributions for all particles in the calculation. As shown, there are indeed superluminal effects present for $l = 1$, but the higher subdivisions significantly reduce these effects. From this, it is seen that using $l = 5$ and $l = 8$ should each provide a valid test of the effects of these artifacts on the results of the calculation. Results from calculating pion HBT parameters vs. p_T for $b = 8$ fm is shown in the right plot of Fig. 1. As is seen, the higher subdivisions do not significantly affect the HBT results. Radial and elliptic flow results from the rescattering calculations can also be shown to not be affected significantly by using these higher subdivisions [5]. Thus one can conclude from this test that the previously published results and present results from the rescattering model are not affected by superluminal artifacts, and criticism 2) above is answered.

AZIMUTHAL HBT RESULTS FROM THE RESCATTERING MODEL

The STAR collaboration has recently published experimental results on the azimuthal dependence of pion HBT parameters in central and non-central RHIC collisions [3]. Preliminary calculations have been made with the rescattering model to extract azimuthal HBT parameters to compare with the STAR results. These comparisons are shown in Figs. 2 and 3 below. In Fig. 2 the azimuthal dependence of HBT parameters is shown for $b = 0$ fm rescattering model calculations compared with results from STAR central collisions ($0-5\%$ centrality). As seen for both the calculation and data, no oscillations occur with respect to ϕ for any of the parameters except the cross-term parameter, $R^2_{outside}$, as would be expected for an azimuthally symmetric system. Figure 3 shows a similar com-

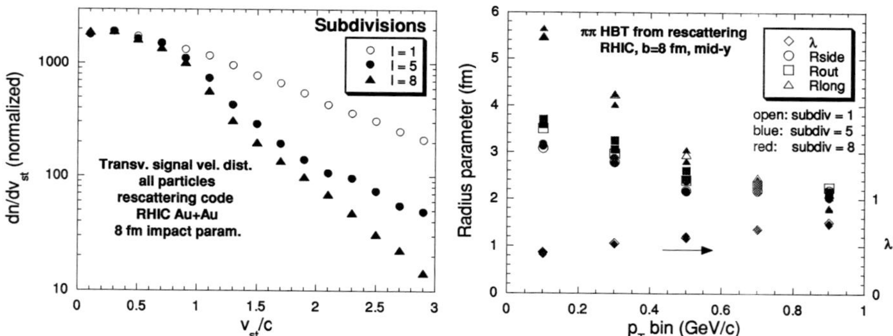

FIGURE 1. Subdivision test of the rescattering code, showing transverse signal velocity distribution and pion HBT parameters vs. p_T for $b = 8$ fm centrality and subdivisions 1, 5, and 8.

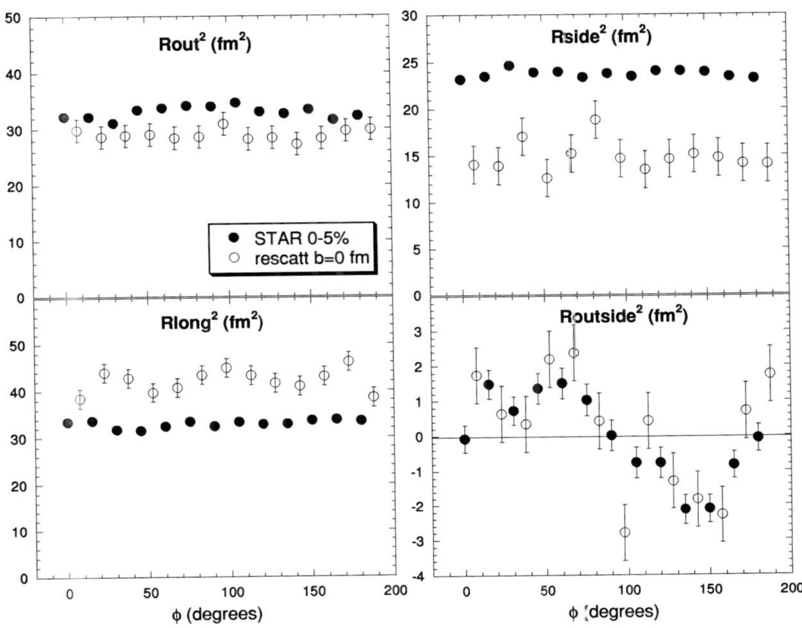

FIGURE 2. Azimuthal pion HBT parameters from the rescattering model with $b = 0$ fm centrality compared with central STAR results

parison for $b = 4$ fm calculations and STAR medium-central collisions ($10 - 20\%$ centrality). As would be expected, the non-central collisions break the azimuthal symmetry of the pion source and oscillations are now seen in all parameters. The calculations are seen to be in reasonable agreement with the data.

HBT AND FLOW PREDICTIONS FOR THE LHC FROM THE RESCATTERING MODEL

Preliminary calculations for the LHC have been carried out with the rescattering model, the results from which are shown below. In performing LHC calculations, the following parameters were used in the code: 1) a collision impact parameter of $b = 8$ fm, 2) an initial temperature parameter of 500 MeV, 3) a hadronization proper time for the initial system of 1 fm/c, 4) a dn/dy at mid-rapidity for central collisions for all particles of 4000, and 5) an initial rapidity width of 4.2. These parameters were judged to be reasonable guesses to simulate LHC Pb+Pb collisions. They at least satisfy the self-consistency check that summing over the energy of all particles in an event at the end of the calculation agrees with the input total energy of a LHC Pb+Pb collision with an impact parameter of $b = 8$ fm. An impact parameter of $b = 8$ fm was chosen for the present preliminary study both to obtain non-negligible elliptic flow values and

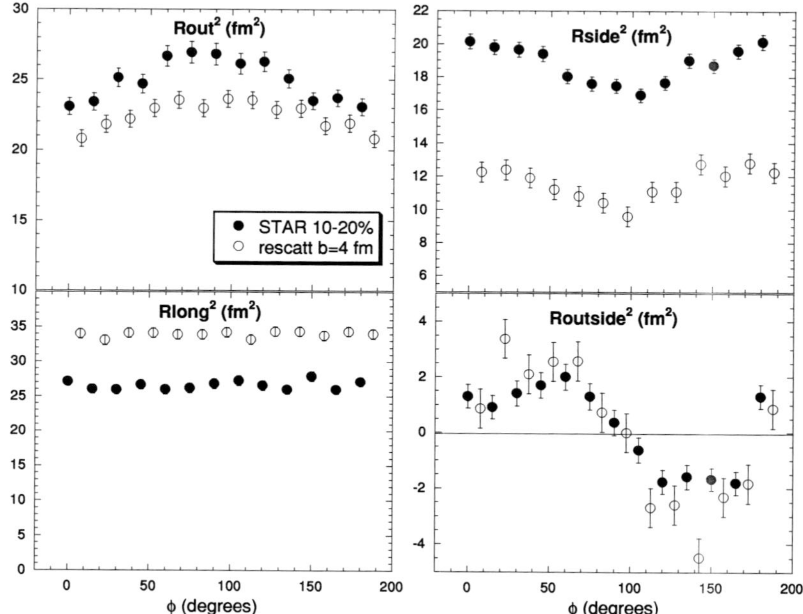

FIGURE 3. Azimuthal pion HBT parameters from the rescattering model with $b = 4$ fm centrality compared with medium-central STAR results

for calculational convenience (even for this impact parameter the cpu time used by the code for each LHC Pb+Pb event was about 60 hours). For item 3) above, the hadronization proper time was taken to be the same as was used in the SPS and RHIC calculations. Results of these calculations are compared with similar calculations at $b = 8$ fm centrality for RHIC Au+Au collisions and are shown in Figs. 4, 5, and 6. All of these results are obtained at mid-rapidity, i.e $-2 < y < 2$.

Figure 4 shows the pion freezeout time and z-position distributions for LHC Pb+Pb and RHIC Au+Au from the rescattering model. As seen, the average pion freezeout time and z-position for LHC are about twice as large as those for RHIC. Although the tails of the freezeout time distributions extend beyond 100 fm/c, the peaks for the LHC and RHIC occur at fairly short times in the collision, at about 10 fm/c and 5 fm/c, respectively. Thus, effects from earlier times in general have the greatest influence on the results from these calculations.

Figure 5 shows the radial and elliptic flow predictions for LHC Pb+Pb compared with RHIC Au+Au from the rescattering model. As seen in the m_T-distribution plot on the left, although all species of particles start from a common temperature in the calculation, after rescattering the exponential slope parameters follow the usual radial flow pattern of $slope(\pi) < slope(K) < slope(N)$ for both LHC and RHIC. The slopes are seen to be consistently larger at LHC than RHIC, as well as for the degree of radial flow which is built up. Looking at the plot of pion elliptic flow vs. p_T on the right, it is seen that LHC

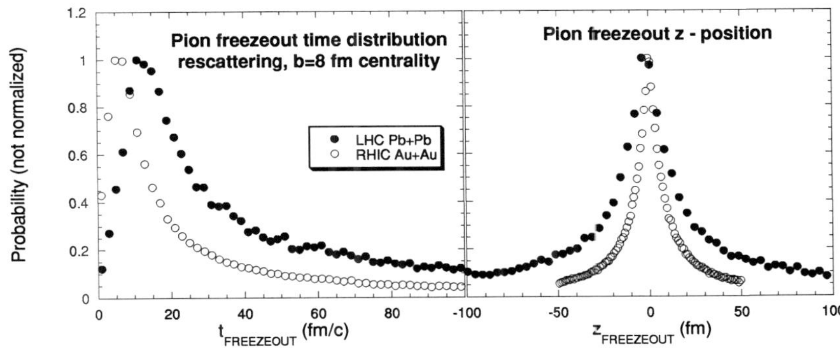

FIGURE 4. Pion freezout time and z-position distributions for LHC Pb+Pb and RHIC Au+Au for $b = 8$ fm centrality collisions at mid-rapidity from the rescattering model.

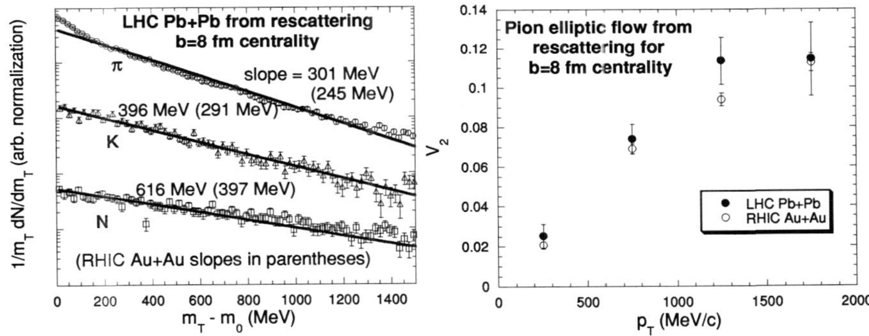

FIGURE 5. m_T distribution and pion elliptic flow predictions for LHC Pb+Pb compared with RHIC Au+Au for $b = 8$ fm centrality collisions at mid-rapidity from the rescattering model. Note that in the plot on the left, corresponding RHIC slope parameters from the rescattering model are show in parentheses for comparison with the LHC values.

and RHIC give about the same values. This is due to the elliptic flow stabilizing at a very early stage in the rescattering calculation.

Figure 6 shows the pion HBT parameters vs. p_T for LHC Pb+Pb compared with RHIC Au+Au from the rescattering model. HBT parameters are extracted by fitting the usual gaussian 3D parameterization of the pion source to the 3D correlation function as calculated from the rescattering model. The transverse radius parameters, R_{Tside} and R_{Tout}, are seen to be somewhat larger and show a stronger p_T dependence for LHC as compared with RHIC. The longitudinal radius parameter, R_{Long} is seen to be significantly larger for LHC as compared with RHIC, clearly reflecting that the pion freezeout times at LHC are twice as long as at RHIC according to the rescattering model. The lambda parameter is seen to increase with increasing p_T in the same way for both LHC and RHIC, reflected the reduced influence of long-lived resonances at the higher p_T values.

Summarizing the results of this preliminary study, it is predicted from the rescattering

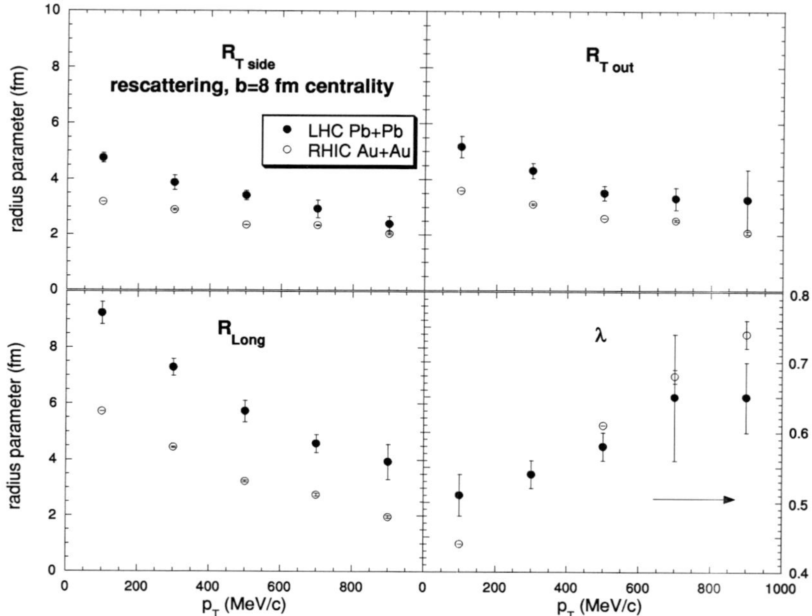

FIGURE 6. Pion HBT predictions for LHC Pb+Pb compared with RHIC Au+Au for $b = 8$ fm centrality collisions at mid-rapidity from the rescattering model.

model that medium-peripheral ($b = 8$ fm) LHC Pb+Pb collisions will produce more radial flow and larger HBT radii than the analogous RHIC Au+Au collisions, although elliptic flow and the lambda parameter values will look the same. It would be interesting to carry out a similar study for central collisions, i.e. $b = 0$ fm, with this model.

ACKNOWLEDGMENTS

The author would like to acknowledge the National Science Foundation for supporting this work under grant number PHY-0355007.

REFERENCES

1. T. J. Humanic, *Phys. Rev. C* **57**, 866-876 (1998).
2. T. J. Humanic, *Nucl. Phys. A* **715**, 641-647 (2003).
3. J. Adams et al, STAR Collaboration, *Phys. Rev. Lett.* **93**, 012301 (2004).
4. D. Molnar and M. Gyulassy, *Phys. Rev. C* **62**, 054907 (2000).
5. T. J. Humanic, nucl-th/0301055.

CONDENSATION AND SQUEEZED STATES

Chairperson: R. Glauber

Variational Approach in Quantum Field Theories: To Dynamical Chiral Phase Transition

Yasuhiko Tsue

Physics Division, Faculty of Science, Kochi University, Kochi 780-8520, JAPAN

Abstract. The time-dependent variational method with a squeezed state is presented to investigate the dynamics of quantum scalar-field theories. It is shown that this method is identical to the time-dependent variational method with a Gaussian trial functional in the functional Schrödinger picture. This variational method is applied to the O(4) linear sigma model in order to investigate the dynamical process of chiral phase transition.

Keywords: chiral phase transition, squeezed state
PACS: 11.30.Rd, 05.70.Fh, 12.39.Fe, 12.38.Mh, 25.75.-q

INTRODUCTION AND SUMMARY

One of the recent interests in the experimental and theoretical studies about the physics of relativistic heavy ion collisions is to clarify the nature of matter at high temperature and/or baryon number density and very high energy density. Also, it is interesting to investigate the dynamical processes of the various phase transitions of a system governed by the quantum chromodynamics (QCD). Especially, in connection with the problem of the formation of a disoriented chiral condensate (DCC) [1-10], the dynamics of chiral phase transition has an attractive interest. Recently, Ikezi, Asakawa and the present author have shown that there is a possibility of the formation of DCC by taking account of both the quantum fluctuations around the chiral condensate and the mode-mode coupling of quantum meson modes [11]. In that study, to describe the time evolution of the order parameter of the chiral phase transition, namely the chiral condensate, the time-dependent variational approach has been adopted. It has been shown that the use of the squeezed state has been quite essential in order to include the higher order quantum effects around the chiral condensate self-consistently.

In this paper, we present the time-dependent variational approach with a squeezed state in order to introduce the appropriate quantum effects around the mean field in both quantum mechanical systems [12-14] and quantum scalar field theoretical systems [13,15-17]. Especially, a collective rotation of the chiral condensate in the isospin space is treated in this method in the context of the formation and the decay of DCC [15,16]. The amplification of quantum meson modes can also be investigated in the late time of the chiral phase transition, in which the resonance by forced oscillation as well as the parametric resonance caused by the oscillation of chiral condensate is realized [18,19], although this subject is not mentioned in this paper.

CP828, *Multiparticle Dynamics*
edited by V. Šimák, M. Šumbera, Š. Todorova, and B. Tomášik

TIME-DEPENDENT VARIATIONAL APPROACH WITH A SQUEEZED STATE IN QUANTUM MECHANICS

Let us start with a quantum mechanical system in which the Hamiltonian is written as

$$\hat{H}=\frac{1}{2}\hat{P}^2+V(\hat{Q}). \tag{1}$$

We introduce the creation and annihilation operators, $\hat{a}^{+}$ and $\hat{a}$:

$$\hat{Q}=\sqrt{\frac{\hbar}{2\omega}}(\hat{a}+\hat{a}^{+}),\quad \hat{P}=-i\sqrt{\frac{\hbar\omega}{2}}(\hat{a}-\hat{a}^{+}). \tag{2}$$

The squeezed state is constructed as

$$|\Phi(t)\rangle=\exp\left(\alpha\hat{a}^{+}-\alpha^{¿}\hat{a}\right)\exp\left(\frac{1}{2}\left(B\hat{a}^{+2}-B^{¿}\hat{a}^{2}\right)\right)|0\rangle$$

$$=\left(2\mathrm{G}\right)^{-1/4}\exp\left(\frac{i}{\hbar}\left(p(t)\hat{Q}-q(t)\hat{P}\right)\right)\exp\left(\frac{1}{2\hbar}\left(\omega-\frac{1}{2\mathrm{G}(t)}+i2\Sigma(t)\right)\hat{Q}^2\right)|0\rangle, \tag{3}$$

which is a vacuum state with respect to the shifted and the Bogoliubov-transformed operator $\hat{c}$ as

$$\hat{c}|\Phi\rangle\equiv\left[(\hat{a}-\alpha)\cosh|B|-(\hat{a}^{+}-\alpha^{¿})\frac{B}{|B|}\sinh|B|\right]|\Phi\rangle=0. \tag{4}$$

Here, in (3), q and p are related to α and α and G and Σ are related to B and B , respectively. The expectation values are calculated easily and are given as

$$\langle\Phi(t)|\hat{Q}|\Phi(t)\rangle=q(t),\ \langle\Phi(t)|\hat{P}|\Phi(t)\rangle=p(t),$$

$$\langle\Phi(t)|\hat{Q}^2|\Phi(t)\rangle=q(t)^2+\hbar G(t),\langle\Phi(t)|\hat{P}^2|\Phi(t)\rangle=p(t)^2+\hbar\left(\frac{1}{4\mathrm{G}}+4\mathrm{G}(t)\Sigma(t)^2\right). \tag{5}$$

As is seen in (5), q and p correspond to the expectation values of $\hat{Q}$ and $\hat{P}$. Thus, these variables present the classical images or classical counterparts of the quantum mechanical systems. On the other hand, the variables G and Σ always appear with $\hbar$. Thus, we can say that these variables represent the quantum fluctuations around the classical counterpart q and p. The uncertainties are given by

$$(\Delta q)^2=\langle\Phi(t)|\left(\hat{Q}-q(t)\right)^2|\Phi(t)\rangle=\hbar G(t),$$

$$(\Delta p)^2=\langle\Phi(t)|\left(\hat{P}-p(t)\right)^2|\Phi(t)\rangle=\hbar\left(\frac{1}{4\mathrm{G}(t)}+4\mathrm{G}(t)\Sigma(t)^2\right). \tag{6}$$

Thus, the uncertainties are not fixed due to the existence of dynamical variables G and Σ , while the uncertainties are fixed in the case of the coherent state because $G=1/(2\omega)$ and $\Sigma=0$ in the coherent state.

The squeezed state (3) is identical to a Gaussian wave function in the wave function representation:

$$\langle Q|\Phi_{sq}\rangle\equiv\Phi(Q)=\left(2\pi\hbar G\right)^{-1/4}e^{ipq/2\hbar}\exp\left\{\frac{i}{\hbar}p(t)(Q-q(t))-\frac{1}{\hbar}\left(\frac{1}{4\mathrm{G}(t)}-i\Sigma(t)\right)(Q-q(t))^2\right\}. \tag{7}$$

The time evolution of the squeezed state is governed by the time-dependent variational principle:

$$\delta S=\delta\int\langle\Phi(t)|i\hbar\partial_t-\hat{H}|\Phi(t)\rangle dt=0. \tag{8}$$

As a result, the derived equations of motion have the canonical equations of motion as

$$\dot{q}=\frac{\partial H}{\partial p}\ ,\ \dot{p}=-\frac{\partial H}{\partial q}\ ,\ \hbar\dot{G}=\frac{\partial H}{\partial\Sigma}\ ,\ \hbar\dot{\Sigma}=-\frac{\partial H}{\partial G}. \tag{9}$$

Here, $H=\langle\Phi(t)|\hat{H}|\Phi(t)\rangle$. Again, G and Σ appear with $\hbar$, and these dynamical variables describe the dynamics of quantum fluctuations.

TIME-DEPENDENT VARIATIONAL APPROACH WITH A SQUEEZED STATE IN QUANTUM SCALAR FIELD THEORIES

Next, let us formulate the time-dependent variational method with a squeezed state in the quantum scalar field theories. We can extend the squeezed state formulated in quantum mechanical system to one in the quantum field theory. The squeezed state is constructed as

$$|\Phi(t)\rangle=\exp[S(t)]\exp[T(t)]|0\rangle, \tag{10}$$

$$S(t)=i\int d^3x\sum_a\left[\bar{\pi}_a(\vec{x},t)\hat{\varphi}_a(\vec{x})-\bar{\phi}_a(\vec{x},t)\hat{\pi}_a(\vec{x})\right],$$

$$T(t)=\int\int d^3x d^3y\sum_{ab}\hat{\varphi}_a(\vec{x})\left[\frac{1}{4}\left[G_{ab}^{(0)-1}(\vec{x},\vec{y})-G_{ab}^{-1}(\vec{x},\vec{y},t)\right]+i\Sigma_{ab}(\vec{x},\vec{y},t)\right]\hat{\varphi}_b(\vec{y}).$$

Here, $\hat{\varphi}_a$ is a field operator and $\hat{\pi}_a$ is its conjugate operator, where suffix a denotes the internal degrees of freedom. The functions $\bar{\phi}_a(\vec{x},t)$, $\bar{\pi}_a(\vec{x},t)$, $G_{ab}(\vec{x},\vec{y},t)$ and $\Sigma_{ab}(\vec{x},\vec{y},t)$ correspond to variational dynamical variables. Here, $G_{ab}^{(0)}=\langle 0|\hat{\varphi}_a\hat{\varphi}_b|0\rangle$. This squeezed state in the field theory is equivalent to the Gaussian wavefunctional in the functional Schrödinger picture which was originally introduced by Jackiw and Kerman [20], and has been developed by several authors [21-23]. In the functional Schrödinger picture, the action of field operator $\hat{\varphi}_a$ on the quantum state corresponds to multiplying the field variable $\phi_a(\vec{x})$ on the wavefunctional and its conjugate operator $\hat{\pi}_a$ has to be replaced into the functional derivative $-i\delta/\delta\phi_a(\vec{x})$. The Gaussian wavefunctional, which is equivalent to the squeezed state (10), is written as

$$\Psi[\phi(\vec{x})]=\langle\phi|\Phi(t)\rangle=N\exp\left[i\int d\vec{x}\,\bar{\pi}_a(\vec{x},t)\tilde{\phi}_a(\vec{x},t)\right]$$

$$*\exp\left\{\int\int d\vec{x}\,d\vec{y}\,\tilde{\phi}_a(\vec{x},t)\left(\frac{1}{4}G_{ab}^{-1}(\vec{x},\vec{y},t)-i\Sigma_{ab}(\vec{x},\vec{y},t)\right)\tilde{\phi}_b(\vec{y},t)\right\}, \tag{11}$$

where we define $\tilde{\phi}_a(\vec{x},t)\equiv\phi_a(\vec{x})-\bar{\phi}_a(\vec{x},t)$. Here, N is a normalization factor. The expectation values with respect to the squeezed state or the Gaussian wavefunctional are easily calculated:

$$\langle \Phi(t)|\phi_a(\vec{x})|\Phi(t)\rangle=\bar{\phi}_a(\vec{x},t),\ \langle \Phi(t)|\pi_a(\vec{x})|\Phi(t)\rangle=\bar{\pi}_a(\vec{x},t),$$
$$\langle \Phi(t)|\hat{\phi}_a(\vec{x},t)\hat{\phi}_b(y,t)|\Phi(t)\rangle=G_{ab}(\vec{x},\vec{y},t)\left(\dot{\iota}\langle\hat{\phi}_a(\vec{x},t)\hat{\phi}_b(y,t)\rangle\right),$$
$$\langle \Phi(t)|\hat{\pi}_a(\vec{x},t)\hat{\pi}_b(y,t)|\Phi(t)\rangle=\frac{1}{4}G^{-1}_{ab}(\vec{x},\vec{y},t)+4\Sigma G\Sigma_{ab}(\vec{x},\vec{y},t),$$
$$\hat{\phi}_a(\vec{x},t)\equiv\phi_a(\vec{x})-\bar{\phi}_a(\vec{x},t)\ ,\ \hat{\pi}_a(\vec{x},t)\equiv\pi_a(\vec{x})-\bar{\pi}_a(\vec{x},t). \tag{12}$$

Thus, $\bar{\phi}_a(\vec{x},t)$ is identical with the mean field and the two-point function $G_{ab}(\vec{x},\vec{y},t)$ represents the quantum fluctuations around the mean field, whose time evolution is determined by the time-dependent variational principle (8).

APPLICATION TO O(4) LINEAR SIGMA MODEL

In this section, let us describe the dynamical process of the chiral phase transition in the O(4) linear sigma model in the time-dependent variational method with a squeezed state developed in the previous section [15-19,24-26]. Especially, we treat a collective rotation of the chiral condensate in the isospin space [15,16].

The Hamiltonian density of O(4) linear sigma model is as follows:

$$H=\frac{1}{2}\pi_a^2(\vec{x})+\frac{1}{2}\left(\nabla\phi_a(\vec{x})\right)^2+\frac{m_0^2}{2}\phi_a{}^2(\vec{x})+\frac{\lambda}{24}\left(\phi_a(\vec{x})^2\right)^2-c\phi_a(\vec{x})\delta_{a0}, \tag{13}$$

where a runs from 0 to 3 in which suffix 0 means the sigma field and 1-3 mean the pi fields. The Gaussian trial wavefunctional (11), or equivalently, the squeezed state (10) is introduced and the time evolution of this quantum state is governed by the time-dependent variational principle (8). In the case that the chiral condensate has a value in the sigma direction only, the equation of motion for chiral condensate is derived as

$$\left(\partial_\mu\partial^\mu+m_0^2+\frac{\lambda}{6}\bar{\phi}^2(\vec{x},t)+\left(\frac{\lambda}{6}\mathrm{Tr}\ G(\vec{x},\vec{x})+\frac{\lambda}{3}G(\vec{x},\vec{x})\right)\right)\bar{\phi}(\vec{x},t)=cI, \tag{14}$$

where I represents $(I)_a=\delta_{a0}$ and $G_{ab}(\vec{x},\vec{x})$ here represents the thermal average of $\hat{\phi}_a(\vec{x},t)\ \hat{\phi}_b(\vec{x},t)$. As for the fluctuations, the equations of motion for G and Σ are reformulated [15] by introducing the reduced density matrix, M, as is similar to the time-dependent Hartree-Bogoliubov theory developed in the nuclear theory:

$$M_{ab}(\vec{x},\vec{y};t)=\begin{pmatrix}-i\langle\hat{\phi}_a(\vec{x},t)\hat{\pi}_b(\vec{y},t)\rangle & \langle\hat{\phi}_a(\vec{x},t)\hat{\phi}_b(\vec{y},t)\rangle\\ \langle\hat{\pi}_a(\vec{x},t)\hat{\pi}_b(\vec{y},t)\rangle & i\langle\hat{\pi}_a(\vec{x},t)\hat{\phi}_b(\vec{y},t)\rangle\end{pmatrix}-\frac{1}{2}\delta(\vec{x}-\vec{y})$$
$$=\begin{pmatrix}2iG\Sigma & G\\ G^{-1}/4+4\Sigma G\Sigma & -2i\Sigma G\end{pmatrix}. \tag{15}$$

Then, the equation of motion for the fluctuations is recast into the Liouville-von-Neumann type equation of motion:

$$i\dot{M}=[H\ ,\ M] \tag{16}$$
$$H=\begin{pmatrix}0 & 1\\ \Gamma & 0\end{pmatrix},\ \Gamma_{ab}=\left(-\nabla^2+m_0^2+\frac{\lambda}{6}\bar{\phi}^2+\frac{\lambda}{6}\langle\hat{\phi}_c\hat{\phi}_c\rangle\right)\delta_{ab}+\frac{\lambda}{3}\left(\bar{\phi}_a\bar{\phi}_b+\langle\hat{\phi}_a\hat{\phi}_b\rangle\right).$$

Here, let us consider the isospin rotation of chiral condensate. Then, the use of the isospin rotating frame is available to describe the collective isospin rotation. The chiral condensate ϕ and the reduced density matrix M together with the Hamiltonian matrix H are gauge-transformed with respect to the isospin rotation, which is parameterized by q^{μ} :

$$\phi(x)=U(x)\begin{pmatrix}\phi_0\\0\\0\\0\end{pmatrix},\quad \tau_y=\begin{pmatrix}0&-i&0&0\\i&0&0&0\\0&0&0&0\\0&0&0&0\end{pmatrix},\quad U(x)=\exp(iqx\tau_y),\quad q^{\mu}=(\omega,\vec{q}), \tag{17}$$

$$M_r(\vec{x},\vec{y},t)=U^{+}(\vec{x},t)M(\vec{x},\vec{y},t)U(\vec{y},t), H(q)=U^{+}(x)\left(-i\frac{\partial}{\partial t}+H\right)U(x)=\begin{pmatrix}\omega\tau_y & 1\\ \Gamma(\vec{q}) & \omega\tau_y\end{pmatrix}.$$

As a result, the dynamics of chiral condensate is governed by the following equation of motion:

$$\left(-q^2+m_0^2+\frac{\lambda}{6}\phi_0^2+\frac{\lambda}{6}\sum_{c=0}^{3}\langle\phi_c(\vec{x})\phi_c(\vec{x})\rangle+\frac{\lambda}{3}\langle\phi_0(\vec{x})\phi_0(\vec{x})\rangle\right)\phi_0=c. \tag{18}$$

The phase diagrams with c=0 case are given in Figs.1 and 2 in Ref.[15]. In the pure time-like isospin rotation ($q^2=\omega^2>0$), the chiral symmetry breaking is enhanced. On the other hand, in the pure space-like isospin rotation ($q^2=-|\vec{q}|^2<0$), the chiral symmetry breaking is recovered. Then, the critical momentum exists. In classical calculation [1], the critical momentum is evaluated as $|\vec{q}_c^2|=M_\sigma^2/2\approx(354\ \mathrm{MeV})^2$ at T=0. However, in our treatment where quantum effects are introduced, the critical momentum is about 50 MeV (See, Table 1). It is thus to say that the quantum fluctuations smear out the effective potential and make symmetry breaking more difficult to reach. In conclusion, the quantum fluctuations play an important role for the dynamical chiral phase transition.

The lifetime of collective isospin rotation of chiral condensate can be evaluated as follows [16]: The energy density of collective rotating condensate is

$$E_0=\frac{1}{2}\bar{\pi}_a^2+\frac{1}{2}\left(\nabla\bar{\phi}_a\right)^2=\frac{1}{2}\phi_0^2\left(\omega^2+\vec{q}^2\right), \tag{19}$$

while the energy density of two meson excitation is calculated by the reduced density matrix as

$$\Delta E=\frac{1}{V}\left[\mathrm{Tr}(HM)-\mathrm{Tr}(H_0M_0)\right]\approx\gamma(q)\ t, \tag{20}$$

where subscript 0 means the variables with c=0 and V denotes the volume that the classical configuration occupies in chiral symmetry restoration region. Thus, the damping time $\tau(q)$ of collective isospin rotation and the number of emitted pions N_π is evaluated as

TABLE . Critical momentum

T (MeV)	0	20	40	60	80
$\lvert q_c\rvert$ (MeV)	50.0	49.9	49.7	47.2	37.7

$$\tau(q)=E_0/\left(\frac{\Delta E}{t}\right), \qquad N_\pi=\left(\frac{\Delta E_{\pi\pi}}{t}\times V\right)/M_\pi, \tag{21}$$

where M_π is the pion mass. If $E_0 \approx (160 \text{ MeV})^4$ and $V \approx (10 \text{ fm})^3$, then

$$\tau=\tau_{\pi\pi}\approx 40 \text{ fm}/c, \qquad N_\pi=15 \text{ mesons per fm}/c \tag{22}$$

are obtained [16].

ACKNOWLEDGMENTS

The author would like to express his sincere thanks to the organizers of WPCF2005 and conveners of condensation and squeezed states for giving an opportunity of his talk. He also thanks to Professors D. Vautherin, T. Matsui, M. Asakawa for collaborations about dynamical chiral phase transition. He acknowledges Professors M. Yamamura and J. da Providência for collaborations about the time-dependent variational approach with squeezed states in quantum many-particle systems. This work was partially supported by Grants-in-Aid from the Japanese Ministry of Education, Culture, Sports, Science and Technology [No.15740156].

REFERENCES

1. A. A. Anselm, *Phys. Lett. B* **217**, 169-172 (1989); A. A. Anselm, M. G. Ryskin, *Phys. Lett. B* **266**, 482-484 (1991); A. A. Anselm, M. G. Ryskin and A. G. Shuvaev, *Z. Phys. A* **354**, 333-341 (1996).
2. J. P. Blaizot and A. Krzywicki, *Phys. Rev. D* **46**, 246-251 (1992).
3. J. D. Bjorken, *Int. J. Mod. Phys. A* **7**, 4189-4257 (1987).
4. K. Rajagopal and F. Wilczek, *Nucl. Phys. B* **404**, 577-589 (1993).
5. M. Asakawa, Z. Huang and X.-N. Wang, *Phys. Rev. Lett.* **74** , 3126-3129 (1995).
6. D. Boyanovsky, H. J. de Vega and R. Holman, *Phys. Rev. D* **49**,2769-2785 (1994); D. Boyanovsky, H. J. de Vega, R. Holman, D. S. Lee and A. Singh, *Phys. Rev. D* **51**, 4419-4444 (1995) ; D. Boyanovsky, H. J. de Vega, R. Holman and J. Salgado, *Phys. Rev. D* **59**, 125009 (1999).
7. F. Cooper, S. Habib, Y. Kluger, E. Mottola, J. P. Paz and P. R. Anderson, *Phys. Rev. D* **50**, 2848-2869 (1994); F.Cooper, Y. Kluger, E. Mottola and J. P. Paz, *Phys. Rev. D* **51**, 2377-2397 (1995); J. Berges, *Nucl. Phys. A* **699**, 847-886 (2002); F. Cooper, J. F. Dawson and B. Mihaila, *Phys. Rev. D* **67**, 056003 (2003).
8. A. Krzywicki and J. Serreau, Phys. Lett. B **448**, 257-264 (1999); G. Aarts, D. Ahrensmeier, R. Baier, J. Berges and J. Serreau, *Phys. Rev. D* **66**, 045008 (2002).
9. M. Salle, J. Smit and J. C. Vink, *Phys. Rev. D* **64**, 025016 (2001); M. Salle, J. Smit and J. C. Vink, *Nucl. Phys. B* **625**, 495-511 (2002); G. Aarts and J. Smit, *Phys. Rev. D* **61**, 025002 (2001).
10. L. M. A. Bettencourt, K. Pao and J. G. Sanderson, *Phys. Rev. D* **65**, 025015 (2002).
11. N. Ikezi, M. Asakawa and Y. Tsue, *Phys. Rev. C* **69**, 032202(R) (2004).
12. Y.Tsue, Y.Fujiwara, A.Kuriyama and M.Yamamura, *Prog. Theor. Phys.* **85**, 693-698 (1991).
13. Y.Tsue and Y.Fujiwara, *Prog. Theor. Phys.* **86**, 443-467; 469-489 (1991).
14. Y.Tsue, *Prog. Theor. Phys.* **88**, 911-932 (1992).
15. Y.Tsue, D.Vautherin and T.Matsui, *Prog. Theor. Phys.* **102**, 313-332 (1999).
16. Y.Tsue, D.Vautherin and T.Matsui, *Phys. Rev. D* **61**, 076006-1-11 (2000) .
17. Y.Tsue, A.Koike and N.Ikezi, *Prog. Theor. Phys.* **106**, 807-822 (2001).
18. Y.Tsue, *Prog. Theor. Phys.* **107**, 1285-1290 (2002).
19. K.Watanabe, Y.Tsue and S.Nishiyama, *Prog. Theor. Phys.* **113**, 369-384 (2005).
20. R.Jackiw and A.Kerman, *Phys. Lett. A* **71**, 158-162 (1979).
21. R.Balian and M.Vènéroni, *Phys. Rev. Lett.* **47**, 1353-1356, 1765 (1981).
22. O.Eboli, R.Jackiw and S.-Y.Pi, *Phys. Rev. D* **37**, 3557-3581 (1988).
23. R.Jackiw, *Physica A* **158**, 269-290 (1989).
24. D.Vautherin and T.Matsui, *Phys. Rev. D* **55**, 4492-4495 (1997).
25. D.Vautherin and T.Matsui, *Phys. Lett. B* **437**, 137-145 (1998).
26. T. Matsui, "Variational approach to the dynamics of quantum fields," in *QCD Perspectives on Hot and Dense Matter,* edited by J.-P. Blaizot and E. Iancu, Dordrecht: Kluwer Academic Publishers, 2002, pp. 419-431.

Nonequilibrium Chiral Dynamics and Two-Particle Correlations in the Time-Dependent Variational Approach with Squeezed States

N. Ikezi*, M. Asakawa† and Y. Tsue**

*Research Center for Nuclear Physics (RCNP), Osaka University, Ibaraki 567-0047, Japan.
†Department of Physics, Osaka University, Toyonaka 560-0043, Japan
**Physics Division, Faculty of Science, Kochi University, Kochi 780-8520, Japan

Abstract. We study the dynamics of chiral phase transition in the O(4) linear sigma model by using the time-dependent variational approach with squeezed states. Our numerical simulations show that large domains of the disoriented chiral condensate (DCC) are formed through the mode-mode correlation. We also present a result of an analysis of the two-particle correlation function for the pion fields, which reflects unique nature of the squeezed states. In particular, we will show that the chaoticity parameter is not close to zero even if DCC domains are produced.

Keywords: Chiral phase transition, Disoriented chiral condensate, Squeezed states
PACS: 25.75.-q, 11.30.Rd, 11.10.Lm

INTRODUCTION

The possibility of the formation of the disoriented chiral condensate (DCC) in relativistic heavy ion collisions has been investigated in a number of theoretical studies. Within the classical approximation it has been recognized that large domains of DCC are produced during the nonequilibrium process in the course of the time evolution from the quench initial condition [1]. On the other hand, in calculations including quantum mechanical effects, mainly homogeneous systems have been studied to avoid difficulty in calculating two-point Green's function which represents quantum fluctuation and correlation. In fact, there have been several attempts to include spatial inhomogeneity [2] in the quantum mechanical time evolution of the fields. However, the correlation between modes with different momenta (mode-mode correlation) was not taken into account in these works and the effect of interactions has been included only through the mean fields. It has not been settled whether large domain structure is formed when quantum effects are taken into account. We show that the direct coupling of modes or the mode-mode correlation is important for the time evolution of the system when the system does not possess translational invariance.

We also study the two-particle correlation function for the pion fields which reflects unique nature of the squeezed states. The two particle correlation function in relativistic heavy ion collision experiments is used for the spatial and temporal size measurements of a source emitting particles. Usually it is defined as the ratio of the two particle distribution function and the single particle distribution function. If the source is chaotic,

CP828, *Multiparticle Dynamics*
edited by V. Šimák, M. Šumbera, Š. Todorova, and B. Tomášik

namely, if particles are emitted independently from the source, two particle correlation function reduces to a product of the single particle distribution functions. The Hanbury Brown - Twiss or the Goldhaber - Goldhaber - Lee - Pais effect (HBT) uses this fact to find the size of the particle emitting source from the two particle correlation function. However, if the source is not chaotic, then the two particle correlation function can not be used for the measure of the source any longer. For example, the particle emission from a source described by a coherent state is highly correlated due to phase coherence and it is easily shown that the HBT effect does not exist for particle emitted with phase coherence. Thus the coherent states are considered as the opposite limit of the emission from a chaotic source. In practice, deviation from the fully chaotic behavior of the two particle correlation function is considered as the measure of coherence in a source, or merely owing to contamination from particles from far outside the source volume. In this paper, without assuming the chaotic source, we analyze the two particle correlation function in the time-dependent variational approach (TDVA) with squeezed states, which includes the coherent states as a special case.

EQUATION OF MOTION AND INITIAL CONDITION

We take the O(4) linear sigma model as a low energy effective theory of QCD and apply the method of the time dependent variational approach (TDVA) with squeezed states as trial states as in the previous paper [3]. This method makes it possible to solve the time-evolution of the order parameters (mean fields), and the quantum fluctuations and correlations in a self-consistent manner.

We denote the sigma ($\sigma(\vec{x})$) and pion ($\vec{\pi}(\vec{x})$) field operators as a four dimensional vector, $\phi_a(\vec{x}) = (\sigma(\vec{x}), \vec{\pi}(\vec{x}))$, where a runs from 0 to 3. Then the Hamiltonian H of the O(4) linear sigma model is given as

$$H = \int \sum_{a=0}^{3} \left\{ \frac{1}{2}\pi_a(\vec{x})^2 + \frac{1}{2}\vec{\nabla}\phi_a(\vec{x}) \cdot \vec{\nabla}\phi_a(\vec{x}) + \lambda \left[\phi(\vec{x})^2 - v^2\right]^2 - h\phi_0(\vec{x}) \right\} d\vec{x}, \quad (1)$$

where $\phi(\vec{x})^2 = \sum_{a=0}^{3} \phi_a(\vec{x})^2$, and λ, v, and h are constants. Note that $\pi_a(\vec{x})$ is the conjugate field operator of $\phi_a(\vec{x})$. operator. We determine the three model parameters, λ, v, and h, so that they give the pion mass $M_\pi = 138$ MeV, the sigma meson mass $M_\sigma = 500$ MeV, and the pion decay constant $f_\pi = 93$ MeV in the one loop level in the "broken symmetry" vacuum state as in the previous study, i.e., $\lambda = 3.44$, $v = 110$ MeV, and $h = (103\text{ MeV})^3$ [3]. The squeezed states used as trial states have the following form:

$$\begin{aligned} |\Phi(t)\rangle = \mathcal{N}(t) \exp\left\{ i \sum_{a=0}^{3} \int (D_a(\vec{x},t)\phi_a(\vec{x}) - C_a(\vec{x},t)\pi_a(\vec{x}))\, d\vec{x} \right\} \exp\left\{ \sum_{a=0}^{3} \int \phi_a(\vec{x}) \right. \\ \left. \left(\frac{1}{4}G_a^{(0)-1}(\vec{x},\vec{x}') - \frac{1}{4}G_a^{-1}(\vec{x},\vec{x}',t) + i\Pi_a(\vec{x},\vec{x}',t)\right)\phi_a(\vec{x}) d\vec{x} d\vec{x}' \right\} |0\rangle. \end{aligned} \quad (2)$$

Here $|0\rangle$ is the vacuum state, and $G_a^{(0)}(\vec{x},\vec{y}) = \langle 0|\phi_a(\vec{x})\phi_a(\vec{y})|0\rangle$. $\mathcal{N}(t)$ is a normalization constant. $C_a(\vec{x},t)$ and $D_a(\vec{x},t)$ are the c-number mean fields of $\phi_a(\vec{x})$ and $\pi_a(\vec{x})$, respectively. $G_a(\vec{x},\vec{y},t)$ and $\Pi_a(\vec{x},\vec{y},t)$ are the quantum correlation functions and the canonical conjugate variable for $G_a(\vec{x},\vec{y},t)$. Thus, the squeezed state as the trial state is specified by mean field variable $C_a(\vec{x},t)$ (chiral order parameter), its conjugate variable $D_a(\vec{x},t)$, quantum fluctuation and correlation $G_a(\vec{x},\vec{y},t)$ and its conjugate variable $\Pi_a(\vec{x},\vec{y},t)$. All of these variables are real. Their time evolutions are determined through the time dependent variational principle. The equations of motion in momentum space read

$$
\begin{aligned}
\ddot{C}_a(\vec{k},t) &= -\vec{k}^2 C_a(\vec{k},t) - \mathcal{M}_a^{(1)}(\vec{k},t), \\
\dot{G}_a(\vec{k},\vec{k}',t) &= 2\langle\vec{k}|\Big[G_a(t)\Pi_a(t) + \Pi_a(t)G_a(t)\Big]|\vec{k}'\rangle, \\
\dot{\Pi}_a(\vec{k},\vec{k}',t) &= \frac{1}{8}\langle\vec{k}|G_a^{-1}(t)G_a^{-1}(t)|\vec{k}'\rangle - 2\langle\vec{k}|\Pi_a(t)\Pi_a(t)\,\vec{k}'\rangle \\
&\quad - \frac{1}{2}(2\pi)^3\vec{k}^2\delta^3(\vec{k}-\vec{k}') - \frac{1}{2}\mathcal{M}_a^{(2)}(\vec{k}-\vec{k}',t), \\
\mathcal{M}_a^{(1)}(\vec{k},t) &= -m^2 C_a(\vec{k},t) + \int \frac{d\vec{l}d\vec{l}'}{(2\pi)^6} C_a(\vec{k}-\vec{l}-\vec{l}',t)\Big(4\lambda\sum_{b=0}^{3} C_b(\vec{l},t)C_b(\vec{l}',t) \\
&\quad + 12\lambda G_a(\vec{l},\vec{l}',t) + 4\lambda\sum_{b(\neq a)} G_b(\vec{l},\vec{l}',t)\Big) - h\delta_{a0}V, \\
\mathcal{M}_a^{(2)}(\vec{k},t) &= -m^2 V + \int \frac{d\vec{l}}{(2\pi)^3}\Big(12\lambda\left(C_a(\vec{k}-\vec{l},t)C_a(\vec{l},t) + G_a(\vec{k}-\vec{l},\vec{l},t)\right) \\
&\quad + 4\lambda\sum_{b(\neq a)}\left(C_b(\vec{k}-\vec{l},t)C_b(\vec{l},t) + G_b(\vec{k}-\vec{l},\vec{l},t)\right)\Big),
\end{aligned} \tag{3}
$$

where $m^2 = 4\lambda v^2$ and $V = \mathrm{Tr}\mathbf{1} = \int d\vec{x}$. The one-point functions $C_a(\vec{k},t)$ and $D_a(\vec{k},t)$ and the two-point functions $G_a(\vec{k},\vec{k}',t)$ and $\Pi_a(\vec{k},\vec{k}',t)$ are the momentum space representations of $C_a(\vec{x},t)$, $D_a(\vec{x},t)$, $G_a(\vec{x},\vec{x}',t)$, and $\Pi_a(\vec{x},\vec{x}',t)$, respectively. In Eq. (3), we have used the following notation, $\langle\vec{k}|H(t)I(t)|\vec{k}'\rangle = \frac{1}{(2\pi)^3}\int H(\vec{k},\vec{k}'',t)I(\vec{k}'',\vec{k}',t)d\vec{k}''$. $\mathcal{M}_{ab}^{(2)}(\vec{k},t)$ in Eq. (3) which originates from the nonlinear coupling term in the Hamiltonian, tells us that mode-mode correlations in momentum space arise through $\mathcal{M}_{ab}^{(2)}(\vec{k}-\vec{k}',t)$ with $\vec{k}\neq\vec{k}'$ even if there is initially no such correlation.

As the initial condition, we adopt the quench scenario: System undergoes a rapid change of the effective potential from the chirally symmetric phase to the chirally broken phase, and the chiral order parameters are stranded around the top of the Mexican hat potential. In order to actualize this situation in the numerical simulation, $C_a(\vec{x},0)$ and $D_a(\vec{x},0)$ are randomly distributed according to the Gaussian forms with the following parameters,

$$
\begin{aligned}
\langle C_a(\vec{x},0)\rangle &= 0, \qquad \langle C_a(\vec{x},0)^2\rangle - \langle C_a(\vec{x},0)\rangle^2 = \delta^2, \\
\langle D_a(\vec{x},0)\rangle &= 0, \qquad \langle D_a(\vec{x},0)^2\rangle - \langle D_a(\vec{x},0)\rangle^2 = \frac{D}{d^2}\delta^2,
\end{aligned} \tag{4}
$$

where δ is the Gaussian width and D is the spatial dimension. We use the Gaussian width $\delta = 0.19v$ below. In relating the Gaussian widths of $C_a(\vec{x},0)$ and $D_a(\vec{x},0)$, we

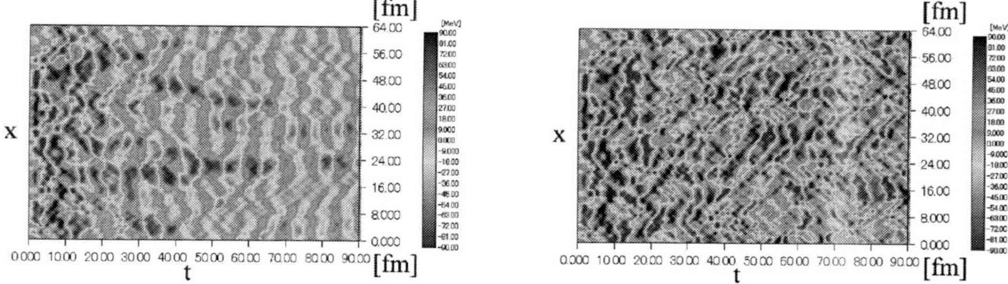

FIGURE 1. Time evolution of the mean fields of the third component of the pion field. The left figure corresponds to the case with mode-mode correlation (case (i)), while the right one corresponds to the case without mode-mode correlation (case (ii)).

have taken advantage of the virial theorem [4]. As for the quantum correlation and fluctuation, we have assumed that each of the sigma and pion states in momentum space is independently in a coherent state with a degenerate mass m_0,

$$G_a(\vec{x},\vec{y},0) = \frac{1}{(2\pi)^3}\int_0^{\Lambda} \frac{e^{i\vec{k}\cdot(\vec{x}-\vec{y})}}{2\omega_k} d\vec{k}, \qquad \Pi_a(\vec{x},\vec{y},0) = 0, \tag{5}$$

with $\omega_k = \sqrt{m_0^2 + \vec{k}^2}$. We adopt $m_0 = 200$ MeV. The two-point functions $G_a(\vec{x},\vec{y},0)$ and $\Pi_a(\vec{x},\vec{y},0)$ are thus diagonal in momentum space in the initial state.

NUMERICAL RESULT

We have calculated the mean fields, which represent chiral order parameter, and the two-point functions, which implies quantum fluctuation and correlation, on a discrete lattice with the total length in the z direction $L = 64$ fm and the lattice spacing $d = 1.0$ fm and assumed translational invariance in the x and y directions. To understand the role of the mode-mode correlation, we have made a comparison of two cases : (i) With the (momentum) mode-mode correlations. (ii) Neglecting the mode-mode correlations.

Figure 1 illustrates the time evolution of the third component of pion field. The value of the mean field for the third component of the pion field in the isospin space is represented by different colors at each position and time. In the case (i), we can see the domain formation of disoriented chiral condensate (DCC), which is denoted by region with deep color along the vertical axis, and it continues up to $t = 60$ fm beyond the time scale of the rolling-down of the order parameter. On the contrary, in the case (ii), we see that short range fluctuations are dominant throughout the time evolution and no qualitative change is observed after a few fm. These results imply that the mode-mode correlation is a key ingredient for the DCC domain formation. As for the behavior of quantum fluctuation, we have found that the quantum fluctuation is substantially amplified in the cases with the mode-mode correlation, i.e., case (i). On the other hand, such substantial amplification of the quantum fluctuation is absent in the cases without mode-mode correlation, i.e., case (ii). In the space of the trial states, the amplification of the quantum fluctuation implies squeezing of the states. Thus, the main part of the DCC domain formation occurs in concurrence with squeezing of the states. Also we confirmed that, in the case (i), the off-diagonal components of the two points function in momentum space which represent the mode-mode correlations indeed appear as time elapses.

Next, we show the result of analysis of two particle correlation function. In order to analyze the two particle correlation function and the particle numbers associated with the pion fields, we use a set of free field operator and its conjugate field operator $\{\tilde{\phi}_a(x), \tilde{\pi}_a(x)\}$. Then the annihilation $\tilde{a}_a(\vec{k})$ and creation operators $\tilde{a}_a^\dagger(\vec{k})$ are given as the expansion coefficients of Fourier transformation of the field operator and its conjugate field operator. We define the two particle correlation function with these annihilation and creation operators as follows,

$$\begin{aligned} C_a(\vec{k},\vec{k}',t) &= \frac{\langle\Phi(t)|\tilde{a}_a^\dagger(\vec{k})\tilde{a}_a^\dagger(\vec{k}')\tilde{a}_a(\vec{k})\tilde{a}_a(\vec{k}')|\Phi(t)\rangle}{N_a(\vec{k},t)N_a(\vec{k}',t)} = 1+\lambda_a(\vec{k},\vec{k}',t), \\ N_a(\vec{k},t) &= \langle\Phi(t)|\tilde{a}_a^\dagger(\vec{k})\tilde{a}_a(\vec{k})|\Phi(t)\rangle. \end{aligned} \tag{6}$$

Here $\lambda_a(\vec{k},\vec{k}',t)$ is called "chaoticity parameter". This parameter $\lambda_a(\vec{k},\vec{k}',t)$ takes typical value in some states. For example, in infinite size system, $\lambda = 0$ for a coherent state, and $\lambda = \delta_{\vec{k}\vec{k}'}$ for a thermal state. For squeezed states, λ varies from 0 to infinity and its off-diagonal components ($\vec{k} \neq \vec{k}'$) are also finite. This reflects the fact that the squeezed states include the coherent state as a special case and span a wide subspace in the physical Hilbert space. Therefore, the squeezed states are expected to be able to describe a variety of quantum features of the system.

Figure 2 shows the time evolution of the chaoticity parameter $\lambda_a(\vec{k}',\vec{k}',t)$ of two particle correlation functions for the third component of the pion field in the case (i) and (ii). At the initial time, two particle correlation has a small value. This is because the initial states are chosen to be in a coherent state. However, as we have already mentioned above, the chaoticity parameter $\lambda_a(\vec{k}',\vec{k}',t)$ is 0 in coherent states. This small value of our the chaoticity parameter at the initial states comes from the mass difference between the mass in the initial states M_0 and the mass which is used in analyzing the two particle correlation function and the single particle spectrum in Eq. (6). In the left parts of Fig. 2, we see that the diagonal components as well as off-diagonal components evolve in time. We also find that the chaoticity parameter approaches to "thermal" (fully chaotic) value around $t = 40$ fm. This result does not mean that the system is in a thermal state. In fact, as we have already seen in Fig. 1, around $t = 40$ fm, DCC domains are still formed and the growth of mode-mode correlation is also not negligible. Therefore this merely shows that the correlation between the two particles approaches the thermal value. On the contrary, we see that the chaoticity parameter in the right parts of Fig. 2 is rather small throughout the time evolution. Our results indicate that the existence of coherent domain does not necessarily imply that the chaoticity parameter is close to zero as in the case of coherent states if the states are in a squeezed states. From these results, we can conclude that inclusion of the quantum fluctuation and correlation is important in both calculating the time evolution of the system and analyzing the chaoticity parameter, or two particle correlation function.

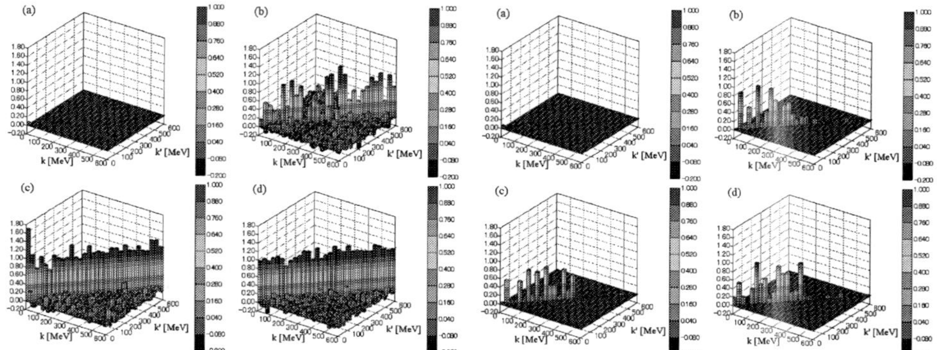

FIGURE 2. Time evolution of chaoticity parameter $\lambda_a(\vec{k}',\vec{k}',t)$ of two particle correlation function. The four figures on the left side correspond to the case (i), and the four figures on the right side correspond to the case (ii). In the each parts, (a), (b), (c), and (d) correspond to $t = 0$ fm, $t = 10$ fm, $t = 40$ fm and $t = 60$ fm, respectively.

SUMMARY

In summary, we have calculated the two particle correlation function of the O(4) linear sigma model in the TDVA with squeezed states. We have not assumed any chaotic source in the analysis and shown the time evolution of chaoticity parameter for the first time. In the case including the mode-mode correlation, we have observed that DCC domain formation, which implies the existence of coherent domains of the pion field, and the growth of the mode-mode correlation as well as the quantum fluctuation. However, in spite of these continuing DCC domain formation and the mode-mode correlation, we have found that the chaoticity parameter approaches thermal value. Therefore, we conclude that coherent domain does not always mean coherence in the phase space if the quantum fluctuation and correlation, which is represented as the squeezing degree of freedom in our approach, is included. On the other hand, no significant DCC domain formation or squeezing of the states was observed in the case without the mode-mode correlation.

In order to implement more realistic discussion, we plan to include finite size effect and carry out calculations in 2+1 or 3+1 dimension in future work.

REFERENCES

1. K. Rajagopal and F. Wilczek, *Nucl. Phys. B* **404**, 577-589 (1993); M. Asakawa, Z. Huang, and X.-N. Wang, *Phys. Rev. Lett.* **74**, 3126-3129 (1995).
2. F. Cooper, J. F. Dawson, and B. Mihaila, *Phys. Rev. D* **67**, 056003 (2003); M. Salle, J. Smit, and J. C. Vink, *Phys. Rev. D* **64**, 025016 (2001); M. Salle, J. Smit, and J. C. Vink *Nucl. Phys. B* **625**, 495-511 (2002); G. Aarts and J. Smit, *Phys. Rev. D* **61**, 025002 (2001); L. M. A. Bettencourt, K. Pao, and J. G. Sanderson, *Phys. Rev. D* **65**, 025015 (2002).
3. N. Ikezi, M. Asakawa and Y. Tsue, *Phys. Rev. C* **69**, 032202(R) (2004).
4. M. Asakawa, H. Minakata, and B. Müller, *Phys. Rev. D* **58**, 094011 (1998).

$\phi\phi$ Back-to-Back Correlations in Finite Expanding Systems

S. S. Padula*, Y. Hama†, G. Krein*, P. K. Panda** and T. Csörgő‡

*Inst. de Física Teórica, UNESP - Rua Pamplona 145, 01405-900 São Paulo, SP - Brazil
†Instituto de Física, USP, Caixa Postal 66318, 05389-970 São Paulo, SP - Brazil
**Depto. de Física-CFM, UFSC - C. P. 476, 88040-900 Florianópolis, SC - Brazil
‡MTA KFKI RMKI, H - 1525 Budapest 114, P.O. Box 49, Hungary

Abstract. Back-to-Back Correlations (BBC) of particle-antiparticle pairs are predicted to appear if hot and dense hadronic matter is formed in high energy nucleus-nucleus collisions. The BBC are related to in-medium mass-modification and squeezing of the quanta involved. Although the suppression of finite emission times were already known, the effects of finite system sizes and of collective phenomena had not been studied yet. Thus, for testing the survival and magnitude of the effect in more realistic situations, we study the BBC when mass-modification occurs in a finite sized, thermalized medium, considering a non-relativistically expanding fireball with finite emission time, and evaluating the width of the back-to-back correlation function. We show that the BBC signal indeed survives the expansion and flow effects, with sufficient magnitude to be observed at RHIC.

Keywords: Modified mass in hot-dense medium, squeezed states, particle-antiparticle correlation
PACS: 25.75.-q, 25.75.Gz, 25.75.Ld

In the late 1990's the *Back-to-Back Correlations* (*BBC*) between boson-antiboson pairs were shown to exist if the particles masses were modified in a hot and dense medium [1]. Not much longer after that, it was also shown that a similar BBC existed between fermion-antifermion pairs with medium-modified masses [2]. A similar formalism is applicable to both BBC cases, related to the Bogoliubov-Valatin transformations of in-medium and asymptotic operators. Both the bosonic (bBBC) and the fermionic (fBBC) Back-to-Back Correlations are positive and have unlimited magnitude, thus differing from the identical-particle correlations, also known as HBT (Hanbury Brown & Twiss) correlations, which are limited for both cases, being negative in the fermionic sector. BBC were expected to be significant for $p_T < 2$ GeV/c. Nevertheless, already in the Ref. [1], it has been shown that the duration of the emission process significantly suppresses its magnitude. The effects of finite system sizes and of collective phenomena were not known, which motivated us to investigate their consequences. Some preliminary results will be discussed here and illustrated in some particular cases.

1. INFINITE HOMOGENEOUS MEDIUM

In our analysis, we assume the validity of local thermalization and hydrodynamics, from the beginning up to the system freeze-out, as well as a short duration of the particle emission. We also consider that the effective in-medium Hamiltonian can be written as

$$H = H_0 - \int d\mathbf{x} d\mathbf{y} \phi(\mathbf{x}) \delta M^2(\mathbf{x}-\mathbf{y}) \phi(\mathbf{y}), \qquad (1)$$

CP828, *Multiparticle Dynamics*
edited by V. Šimák, M. Šumbera, Š. Todorova, and B. Tomášik

where

$$H_0 = \int d\mathbf{x}(\dot{\phi}^2 + |\nabla\phi|^2 + m^2\phi^2) \tag{2}$$

is the asymptotic (free) Hamiltonian in the matter rest frame, and the second term in Eq. (1) describes the medium modifications. The scalar field ϕ represents quasi-particles propagating with momentum-dependent medium-modified mass $m_\star$, related to the vacuum mass, m, by

$$m_\star^2(|\mathbf{k}|) = m^2 - \delta M^2(|\mathbf{k}|).$$

This implies that the dispersion relations in the vacuum and in-medium are given, respectively, by $\omega_k^2 = m^2 + \mathbf{k}^2$ and $\Omega_k^2 = m_\star^2 + \mathbf{k}^2 = \omega_k^2 - \delta M^2(|\mathbf{k}|)$, where Ω is the frequency of the in-medium mode with momentum $\mathbf{k}$.

The annihilation (creation) operator, b ($b^\dagger$), for the in-medium, thermalized quasi-particles is related to the annihilation (creation) operator of the asymptotic, observed quanta with momentum $k^\mu = (\omega_k, \mathbf{k})$, a ($a^\dagger$), by the Bogoliubov-Valatin transformation

$$a_k^\dagger = c_k^* b_k^\dagger + s_{-k} b_{-k} \; ; \; a_k = c_k b_k + s_{-k}^* b_{-k}^\dagger, \tag{3}$$

where $c_k = \cosh(f_k)$ and $s_k = \sinh(f_k)$; $f_k = \frac{1}{2}\log(\frac{\omega_k}{\Omega_k})$ is the so-called *squeezing parameter*, since the Bogoliubov transformation is equivalent to a squeezing operation.

On the other hand, the two-particle probability distribution is given by

$$\langle a_{\mathbf{k}_1}^\dagger a_{\mathbf{k}_2}^\dagger a_{\mathbf{k}_2} a_{\mathbf{k}_1} \rangle = \left[\langle a_{\mathbf{k}_1}^\dagger a_{\mathbf{k}_1} \rangle \langle a_{\mathbf{k}_2}^\dagger a_{\mathbf{k}_2} \rangle + \langle a_{\mathbf{k}_1}^\dagger a_{\mathbf{k}_2} \rangle \langle a_{\mathbf{k}_2}^\dagger a_{\mathbf{k}_1} \rangle + \langle a_{\mathbf{k}_1}^\dagger a_{\mathbf{k}_2}^\dagger \rangle \langle a_{\mathbf{k}_2} a_{\mathbf{k}_1} \rangle \right]. \tag{4}$$

The first term on the r.h.s. is proportional to the product of the two single-inclusive distributions, with momenta k_i, i.e., $N_1(\mathbf{k}_i) = \omega_{\mathbf{k}_i} \frac{d^3N}{d\mathbf{k}_i} = \omega_{\mathbf{k}_i} \langle a_{\mathbf{k}_i}^\dagger a_{\mathbf{k}_i} \rangle$, while the second term is proportional to the square modulus of $G_c(\mathbf{k}_1, \mathbf{k}_2) = G_c(1,2) = \sqrt{\omega_{\mathbf{k}_1} \omega_{\mathbf{k}_2}} \langle a_{\mathbf{k}_1}^\dagger a_{\mathbf{k}_2} \rangle$, the so-called **chaotic amplitude**. The last term is related to this new contribution, $G_s(\mathbf{k}_1, \mathbf{k}_2) = G_s(1,2) = \sqrt{\omega_{\mathbf{k}_1} \omega_{\mathbf{k}_2}} \langle a_{\mathbf{k}_1} a_{\mathbf{k}_2} \rangle$, which is called **squeezed amplitude.**

In cases where the particle is its own anti-particle (for $\pi^0\pi^0$ or $\phi\phi$ boson pairs, for instance), both terms contribute, and the full correlation function is written as

$$C_2(\mathbf{k}_1, \mathbf{k}_2) = 1 + \frac{|G_c(\mathbf{k}_1, \mathbf{k}_2)|^2}{G_c(\mathbf{k}_1, \mathbf{k}_1) G_c(\mathbf{k}_2, \mathbf{k}_2)} + \frac{|G_s(\mathbf{k}_1, \mathbf{k}_2)|^2}{G_c(\mathbf{k}_1, \mathbf{k}_1) G_c(\mathbf{k}_2, \mathbf{k}_2)}, \tag{5}$$

where the first two terms correspond to the HBT correlation, and last term, represents this additional contribution to the correlation function, i.e., the squeezing part.

In the above equations, the amplitudes were written in terms of the asymptotic operators. However, the averages in Eq. (4) are estimated by means of the density operator $\hat{\rho}$, as the thermal average for globally thermalized gas of b quanta that is homogeneous in the system with a certain volume V, with $\hat{\rho} = (1/Z)\exp[-(V/T)(1/(2\pi)^3) \int \Omega b^\dagger b]$. Being so, the above equations should be expressed in terms of the in-medium operators by means of Eq. (3), prior to performing the thermal averages.

We know that the maximum value of the HBT correlation function is attained when $k_1 = k_2 = k$, resulting in $C_c(k,k) = 2$. Accordingly, the maximum value of the BBC correlation function is attained for $k_1 = -k_2 = k$, resulting, after performing the thermal averages, into

$$C_s(\mathbf{k}, -\mathbf{k}) = 1 + \frac{|c^*_{\mathbf{k}} s_{\mathbf{k}} n_{\mathbf{k}} + c^*_{-\mathbf{k}} s_{-\mathbf{k}} (n_{-\mathbf{k}} + 1)|^2}{n_1(\mathbf{k})\, n_1(-\mathbf{k})}, \tag{6}$$

where $n_1(\mathbf{k}) = \left[|c_{\mathbf{k}}|^2 n_{\mathbf{k}} + |s_{-\mathbf{k}}|^2 (n_{-\mathbf{k}} + 1)\right]$ is related to the spectral function by $N_1(\mathbf{k}) = \frac{V}{(2\pi)^3}\, \omega_{\mathbf{k}}\, n_1(\mathbf{k})$; $n_{\mathbf{k}}$ is the Bose-Einstein distribution function of the in-medium quanta with energy $\Omega_{\mathbf{k}}$ at temperature T. Strictly speaking Eq. (6) is valid only in the rest frame of the medium.

2. FINITE SIZE MEDIUM MOVING WITH COLLECTIVE VELOCITY

For a hydrodynamical ensemble, both the chaotic and the squeezed amplitudes, G_c and G_s, respectively, can be written in the special form derived by Makhlin and Sinyukov [3] (see Eqs. (22) and (23) of Ref. [1]), namely

$$G_c(\mathbf{k}_1, \mathbf{k}_2) = \int \frac{d^4\sigma_\mu(x)}{(2\pi)^3} K^\mu_{1,2}\, e^{i q_{1,2}\cdot x} \left\{ |c_{1,2}|^2\, n_{1,2}(x) + |s_{-1,-2}|^2 \left[n_{-1,-2}(x) + 1\right] \right\}, \tag{7}$$

$$G_s(\mathbf{k}_1, \mathbf{k}_2) = \int \frac{d^4\sigma_\mu(x)}{(2\pi)^3} K^\mu_{1,2}\, e^{2iK_{1,2}\cdot x} \left\{ s^*_{-1,2}\, c_{2,-1}\, n_{-1,2}(x) + c_{1,-2}\, s^*_{-2,1} \left[n_{1,-2}(x) + 1\right] \right\}. \tag{8}$$

In Eq. (7) and (8), $d\sigma^4_\mu(x) = d^3\Sigma_\mu(x,\tau) F(\tau) d\tau$ is the product of the normal-oriented volume element depending parametrically on τ (freeze-out hyper-surface parameter) and on its invariant distribution, $F(\tau)$; σ^μ is the hydrodynamical freeze-out surface.

For studying the expansion of the system we adopt the non-relativistic hydrodynamical model of Ref. [4]. In this model the fireball expands in a spherically symmetric manner with a local flow vector given by the four-velocity $u^\mu = \gamma(1, \mathbf{v})$, assumed to be non-relativistic, with $\gamma = (1 - \mathbf{v}^2)^{-1/2} \approx 1 + \mathbf{v}^2/2$, where

$$\mathbf{v} = \langle u \rangle \mathbf{r}/R,$$

being $\langle u \rangle$ and R the mean expansion velocity and the radius of the fireball, respectively. We then divide the inhomogeneous medium into independent cells and assume that the expressions for G_c and G_s can be evaluated locally within each cell using Eq. (3). The squeezing coefficient can be written, in more detail, as

$$f_{i,j}(x) = \frac{1}{2} \log\left[\frac{K^\mu_{i,j}(x)\, u_\mu(x)}{K^{*\nu}_{i,j}(x)\, u_\nu(x)}\right] = \frac{1}{2} \log\left[\frac{\omega_{\mathbf{k}_i}(x) + \omega_{\mathbf{k}_j}(x)}{\Omega_{\mathbf{k}_i}(x) + \Omega_{\mathbf{k}_j}(x)}\right] \equiv f_{\pm i, \pm j}(x), \tag{9}$$

where, as in HBT, the pair momentum difference and the pair average momentum are given, respectively, by $q^\mu_{i,j}(x) = k^\mu_i(x) - k^\mu_j(x)$, and $K^\mu_{i,j}(x) = \frac{1}{2}\left[k^\mu_i(x) + k^\mu_j(x)\right]$. In addition, we consider the Boltzmann limit of the Bose-Einstein distribution for n_k, i.e.,

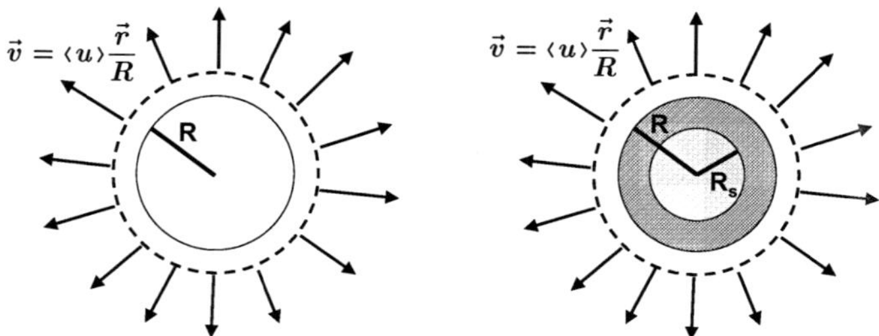

FIGURE 1. Schematic representation of the region where the mass-shift occurs: on the l.h.s., the modification is extended to the whole system, whereas on the r.h.s. it happens only in a smaller portion of the thermalized medium. The figures represent cross-sectional areas of the Gaussian profiles.

$n_{i,j} \sim \exp[-(K^{\mu}_{i,j}u_{\mu} - \mu(x))/T(x)]$, and assume a time-dependent parametric solution of the hydrodynamical equations, i.e., $\mu(x)/T(x) = \mu_0/T_0 - r^2/(2R^2)$, as in Ref. [4]. Furthermore, we consider two possible scenarios for the freeze-out: the first, corresponds to a sudden freeze-out, in which $F(\tau) \propto \delta(\tau - \tau_0)$. The second scenario corresponds to a smeared freeze-out, for which $\frac{\theta(\tau-\tau_0)}{\Delta\tau}e^{-(\tau-\tau_0)/\Delta\tau}$. This last more realistic scenario has a dramatic effect on the Back-to-Back Correlation function, as already showed in Ref.[1], by reducing severely the signal's magnitude, even for a smearing of about $\Delta\tau \sim 2$ fm/c.

In discussing finite-size effects, we distinguish between the volume of the entire thermalized medium, denoted by V (with radius R), and the volume filled with mass-shifted quanta, denoted by V_s (with radius R_s). Naturally, $V_s \leq V$ in the general case. In the derivation of the expressions for $G_c(1,2)$ and $G_s(1,2)$, for simplicity, we introduce a Gaussian profile function in the integrands, i.e., $\sim \exp[-\mathbf{r}/(2R)^2]$. In Fig. 1 we illustrate this by showing cross-sectional areas corresponding to Gaussian profiles, for the cases with $V = V_s$ and $V > V_s$.

In the non-relativistic limit, the accounting for the squeezing effects can be simplified for small mass shifts $(m_\star - m)/m \ll 1$, such that the squeezing parameter in Eq. (9) can be written simply as $f(i,j,\mathbf{r}) \approx \frac{1}{2}\log\left(\frac{m}{m_\star}\right)$. This limit is important, because in this case the coordinate dependence enters the squeezing parameter f only through the possible position dependence of the mass-shift which, in principle, could be calculated from thermal field models in the local density approximation. Therefore, in an approximation such that the position dependence of the in-medium mass can be neglected, the $c(i,j) = c_0$ and $s(i,j) = s_0$ factors can be removed from the integrands in Eq. (7) and (8) and all what remains to be done are Fourier transforms of Gaussian functions.

For completeness, we write below the expression of the Back-to-Back Correlation function for the case where the mass shift occurs in entire volume of the system, V. A detailed discussion and more complete formulation of the problem, including the case of mass-shift in a smaller portion of the system, can be found in Ref.[5]. In what follows, we will concentrate on the value of momenta of the participant pair that maximizes the BBC signal, i.e., the case in which $\mathbf{k}_1 = -\mathbf{k}_2 = \mathbf{k}$, using the fact that the single-inclusive distribution depends only on the absolute value of the momentum,

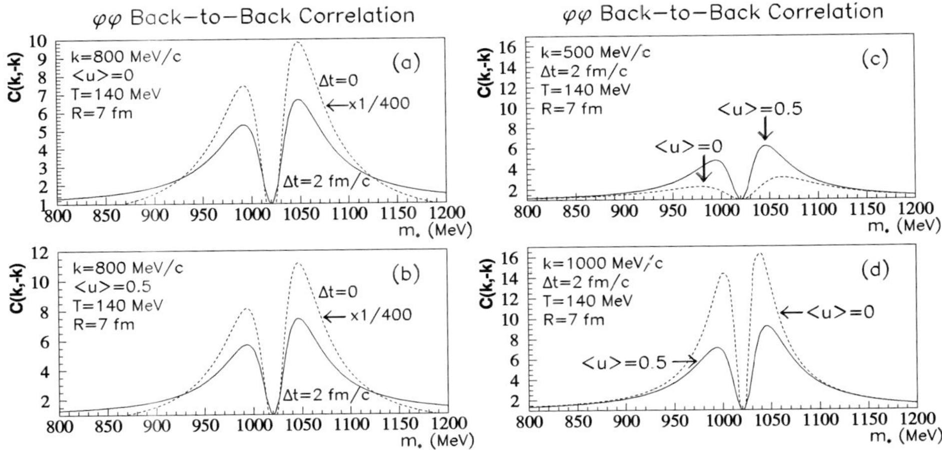

FIGURE 2. In the left panel, parts (a) and (b) illustrate the effect of finite emission times. The dashed curves, corresponding to a sudden emission ($\Delta t = 0$), have been decreased by a factor of 400, and the solid curves show the suppression caused by a finite emission duration, of about $\Delta t \simeq 2$ fm/c, which drastically reduces the BBC magnitude. In the right panel we illustrate the cases with and without flow, for two values of the momentum of the back-to-back pair (in (c), $k = 500$ MeV/c and in (d), $k = 1000$ MeV/c). The mass-shifting is supposed to occur in the entire system volume. In the plots, $m_\star$ is the in-medium modified mass and T stands for the freeze-out temperature.

i.e., $G_c(\mathbf{k},\mathbf{k}) = G_c(-\mathbf{k},-\mathbf{k})$. Nevertheless, since the strict condition of back-to-back pair holds only in the rest frame of the medium, we implicitly are considering the case corresponding to a weak flow coupling in the expanding system, which is expected to be fulfilled for pair emitted near the center of the system. In this case, the BBC correlation function can then be written as

$$C^V_{BBC}(\mathbf{k},-\mathbf{k}) = 1 + \left|c_0\, s_0\right|^2 \left[2R^3\left(1+\frac{m^2 <u>^2}{m_\star T}\right)^{-3/2} \exp\left(-\frac{m_\star}{T} - \frac{\mathbf{k}^2}{2m_\star T}\right) + R^3\right]^2$$
$$\times \left[\frac{R^3\left(|c_0|^2+|s_0|^2\right)}{\left(1+\frac{m^2<u>^2}{m_\star T}\right)^{3/2}} \exp\left(-\frac{m_\star}{T} - \frac{\mathbf{k}^2}{2m_\star T} + \frac{m^2 <u>^2 \mathbf{k}^2}{2m_\star^2 T^2 (1+\frac{m^2<u^2>}{m_\star T})}\right) + R^3|s_0|^2\right]^{-2}. \quad (10)$$

In Fig. 2 we illustrate some of the results found in this non-relativistic approximation, in the particular case of weak flow coupling. We see that the cases corresponding to the absence of flow and to its inclusion produce similar results within the limits of our illustration. However, depending on the value of $k_1 = -k_2 = k$, there are noticeable differences. Being so, we see that, for smaller values of k, the presence of flow seems to slightly enhance the signal, whereas at large values of k, the non-flow case wins. Nevertheless, the non-flow case grows faster with increasing k. In this particular example discussed here, the effect is mainly due to the denominator in Eq. (10), since it contains the flow parameter in the exponential's coefficient.

3. SUMMARY AND CONCLUSIONS

Our main goal on presenting these new results here was to revive the discussion on the search of the squeezed BBC. For fulfilling this purpose, we estimated the strength of the BBC signal in a more realistic situation, considering the mass-shifting in a finite region, and the emission occurring during a finite time interval. We also considered that the system expands non-relativistically and analyzed the simplified situation of weak flow dependence (i.e., back-to-back pairs emitted close to central system region) of the BBC. Finally, we employed $\phi\phi$ pair correlation for illustration. We showed in Fig. 2 the back-to-back correlation function versus the in-medium shifted mass, $m_\star$. We also saw that both the non-flow and the flow cases produced similar results, with pronounced maxima around $m \approx m_\star$. Although we did not show here results corresponding to the case of mass-shift occurring only in a smaller portion of the system, it is shown elsewhere [5, 6] to decrease proportionally to the size of the mass-shift region. However, the effect of decreasing the system size is far less significant than the finite emission time for reducing the BBC magnitude. We also saw that, at least in its weak-coupling limit, the flow may work for slightly enhance the BBC signal for small values of the momentum k. Our main conclusion, however, is that, in the particular framework discussed above, a sizeable strength of the squeezed BBC signal could be seen, making it a promising effect to be searched for experimentally at RHIC.

Naturally, for a more refined calculation, it is mandatory to introduce a model-based mass-shift. After that, it is also essential to perform more realistic calculations with flow, in a less particular kinematic region, while simultaneously searching for those which could optimize the observation of the BBC signal. Also an estimate of the shape and width of the BBC around the direction $k_1 = -k_2 = k$ should be implemented. Finally, for being able to make predictions closer to the experimental conditions, it will be extremely important to obtain some feed-back on the experimental acceptance, conditions, and restrictions that could finally lead to the BBC discovery.

ACKNOWLEDGMENTS

One of us (SSP) would like to express her gratitude to the Fundação de Amparo à Pesquisa do Estado de São Paulo (FAPESP) for the financial support (Proc. N. 05/52190-1 and 2004/10619-9 - *Projeto Temático*), which permitted her to participate in the WPCF, and in the ISMD 2005. T. Csörgő was supported by the grant OTKA T038406.

REFERENCES

1. M. Asakawa, T. Csörgő and M. Gyulassy, *Phys. Rev. Lett.* **83**, 4013-4016 (1999).
2. P. K. Panda, T. Csörgő, Y. Hama, G. Krein and S. S. Padula, *Phys. Lett. B* **512**, 49-56 (2001).
3. A. Makhlin and Yu. Sinyukov, *Yad. Phys.* **46**, 637-646 (1987); Yu. Sinyukov, *Nucl. Phys. A* **566**, 589c-592c (1994).
4. T. Csörgő, B. Lörstad and J. Zimányi, *Phys. Lett. B* **338**, 134-140 (1994); P. Csizmadia, T. Csörgő and J. Zimányi, *Phys. Lett. B* **443**, 21-25 (1998).
5. S. S. Padula, Y. Hama, G. Krein, P. K. Panda, and T. Csörgő, in preparation.
6. S. S. Padula, Y. Hama, G. Krein, P. K. Panda, and T. Csörgő, nucl-th/0510064.

List of participants

Alam Jan-e	VECC Kolkata, India	jane@veccal.ernet.in
Ajitanand Nuggehalli	SUNY Stony Brook, U.S.A.	ajit@mail.chem.sunysb.edu
Aneva Boyka	INRNE, Bulgaria	blan@inrne.bas.bg
Bartels Joachim	U.Hamburg, Germany	bartels@desy.de
Bialas Andrzej	Jagellonian U., Poland	bialas@th.if.uj.edu.pl
Bleicher Marcus	ITP Frankfurt, Germany	bleicher@th.physik.uni-frank
Bopp Fritz Wilhelm	U.Siegen, Germany	bopp@physik.uni-siegen.de
Brown David L.	LLNL, U.S.A.	dlb@llnl.gov
Buschbeck Brigitte	IHEP Vienna, Austria	brigitte@hephy.oeaw.ac.at
Bysterský Michal	NPI ASCR, Czech Republic	bystersky@ujf.cas.cz
Cammin Jochen	U.of Rochester, U.S.A.	cammin@fnal.gov
Campanelli Mario	U.of Geneva, Switzerland	mario.campanelli@cern.ch
Černý Karel	Charles U., Czech Republic	kcerny@ipnp.troja.mff.cuni.cz
Chajecki Zbigniew	Ohio State U., U.S.A.	chajecki@mps.ohio-state.edu
Chaloupka Petr	NPI ASCR, Czech Republic	petrchal@rcf.rhic.bnl.gov
Chen Gang	U.of Geosciences Wuhan, China	cheng@iopp.ccnu.edu.cn
Christakoglou Panos	U.of Athens, Greece	Panos.Christakoglou@cern.ch
Chung Paul	SUNY, U.S.A.	pchung@mail.chem.sunysb.edu
Chýla Jiří	IP ASCR, Czech Republic	chyla@fzu.cz
Cramer John G.	U.of Washington, U.S.A.	cramer@phys.washington.edu
Csanád Máté	Eotvos U., Hungary	csanad@elte.hu
Csörgő Tamás	MTA KFKI RMKI, Hungary	csorgo@sunserv.kfki.hu
Danielewicz Pawel	NSCL/Michigan State U., U.S.A.	danielewicz@nscl.msu.edu
De Wolf Eddi	U.of Antwerpen, Belgium	eddi.dewolf@ua.ac.be
Delcourt Benoit	L.A.L. Orsay, France	delcourt@lal.in2p3.fr
Dong Xin	U.of Science&Techn., China	dongx@mail.ustc.edu.cn
Dremin Igor	Lebedev Phys.Inst., Russia	Dremin@td.lpi.ru
Eggers Hans	U.of Stellenbosch, South Africa	eggers@physics.sun.ac.za
Fabbri Fabrizio	INFN Bologna, Italia	fabbri@bo.infn.it
Fialkowski Krzysztof	Jagellonian U., Poland	fialkowski@th.if.uj.edu.pl
Fiala Lukáš	IP ASCR, Czech Republic	fialal@fzu.cz
Field Rick	U.of Florida, U.S.A.	rfield@phys.ufl.edu
Floris Michele	INFN&U.Cagliari, Italy	michele.floris@ca.infn.it
Fu Jinghua	IPP, Huazhong Normal U., China	fjh@mail.tsinghua.edu.cn
Fujii Hirotsugu	U.of Tokyo, Japan	hfujii@phys.c.u-tokyo.ac.jp
Gagliardi Carl	Texas A&M U., U.S.A.	cggroup@comp.tamu.edu
Gazda Daniel	CTU , Czech Republic	gazda@linux.fjfi.cvut.cz
Glauber Roy	Harvard U., U.S.A.	glauber@physics.harvard.edu
Gos Hanna	SUBATECH/U.of Techn., Poland	gos@if.pw.edu.pl
Gustafson Gösta	Lund U., Sweden	gosta.gustafson@thep.lu.se
Hallman Timothy	BNL, U.S.A.	hallman@bnl.gov
Hama Yogiro	U.of Sao Paulo, Brazil	hama@fma.if.usp.br
Holzmann Wolf	SUNY Stony Brook, U.S.A.	wholz@mail.chem.sunysb.edu

Homma Kensuke	Hiroshima U., Japan	homma@hepl.hiroshima-u.ac.jp
Huang Huan Zhong	UCLA, U.S.A.	huang@physics.ucla.edu
Huang Yanping	IPP, Huazhong Normal U., China	huangyp@iopp.ccnu.edu.cn
Hubáček Zdeněk	CTU, Czech Republic	zdenek.hubacek@cern.ch
Humanic Thomas	Ohio State U., U.S.A	humanic@mps.ohio-state.edu
Ikezi Naoko	RCNP Osaka U., Japan	ikezi@rcnp.osaka-u.ac.jp
Jia Jiangyong	Columbia U., U.S.A.	jjia@nevis.columbia.edu
Jönsson Leif	Lund U., Sweden	leif.jonsson@hep.lu.se
Jung Hannes	DESY, Germany	hannes.jung@desy.de
Juran Josef	Silesian U., Czech Republic	josef.juran@seznam.cz
Khoze Valeri	IPPP, Durham U., U.K.	v.a.khoze@durham.ac.uk
Kisiel Adam	Warsaw U.of Techn., Poland	kisiel@if.pw.edu.pl
Kittel Wolfram	Radboud U., Netherlands	wolfram@hef.ru.nl
Kniege Stefan	U.Frankfurt a.M., Germany	kniege@ikf.uni-frankfurt.de
Ko Che-Ming	Texas A&M U., U.S.A.	ko@comp.tamu.edu
Kokoulina Elena	GSTU, Belarus/JINR, Russia	elena.kokoulina@sunse.jinr.ru
Kozlov Gennady	JINR, Russia	kozlov@jinr.ru
Kuchin Igor	IPT, Kazakhstan	kuchin@satsun.sci.kz
Kundrát Vojtěch	IP ASCR, Czech Republic	kundrat@fzu.cz
Kupčo Alexander	IP ASCR, Czech Republic	kupco@fnal.gov
Lacey Roy	SUNY Stony Brook, U.S.A.	Roy.Lacey@Stonybrook.edu
Lednický Richard	IP ASCR /JINR Russia	lednicky@fzu.cz
Leroy Claude	U.of Montreal, Canada	leroy@lps.umontreal.ca
Levonian Sergey	DESY, Germany	sergey.levonian@desy.de
Li Na	IPP, Huazhong Normal U., China	lin@iopp.ccnu.edu.cn
Li Zhiming	IPP, Huazhong Normal U., China	lizm@hef.ru.nl
Lietava Roman	U.of Birmingham, U.K.	rl@hep.ph.bham.ac.uk
Lisa Michael	Ohio State U., U.S.A.	lisa@mps.ohio-state.edu
Liu Lianshou	IPP, Huazhong Normal U., China	liuls@iopp.ccnu.edu.cn
Lokajíček Miloš	IP ASCR, Czech Republic	lokajick@fzu.cz
Lörstad Bengt	Lund U., Sweden	bengt@quark.lu.se
Lungwitz Benjamin	U.of Frankfurt, Germany	lungwitz@ikf.uni-frankfurt.de
Maj Radoslaw	Swietokrzyska Academy, Poland	radmaj@pu.kielce.pl
Manly Steven	U.of Rochester, U.S.A.	manly@pas.rochester.edu
Marquet Cyrille	CEA Saclay, France	marquet@spht.saclay.cea.fr
McLerran Larry	BNL, U.S.A.	mclerran@bnl.gov
Metzger Wesley	Radboud U., Netherlands	W.Metzger@hef.ru.nl
Mrázová Kristina	Silesian U., Czech Republic	kristina.mrazova@fpf.slu.cz
Nikolaev Nikolai	NRW, Germany	N.Nikolaev@fz-juelich.de
Nouicer Rachid	BNL, U.S.A.	rachid.nouicer@bnl.gov
Novak Tamás	HEFIN Nijmegen, Netherlands	novakt@hef.kun.nl
Ohnishi Hiroaki	RIKEN, Switzerland	Hiroaki.Onishi@cern.ch
Okorokov Vitaly	MEPI, Russia	okorokov@bnl.gov
Orava Risto	U.of Helsinki, Finland	risto.orava@cern.ch
Padula Sandra	UNESP, Brasil	padula@ift.unesp.br
Pachr Miloš	CTU, Czech Republic	pachrmilos@hotmail.com
Panitkin Sergey	BNL, U.S.A.	panitkin@bnl.gov

Pejchal Ondřej	Charles U., Czech Republic	pejchal@matfyz.cz
Peressounko Dmitri	RRC Kurchatov Institute, Russia	D.Y.Peressounko@cern.ch
Peschanski Robi	CEA Saclay, France	pesch@spht.saclay.cea.fr
Peters Krisztian	U.of Manchester, U.S.A.	petersk@fnal.gov
Pinfold James	U.of Alberta, Canada	pinfold@phys.ualberta.ca
Pluta Jan	Warsaw U.of Techn., Poland	pluta@if.pw.edu.pl
Praszalowicz Michal	Jagellonian U., Poland	michal@if.uj.edu.pl
Pratt Scott	Michigan State U., U.S.A.	prattsc@msu.edu
Rafelski Johann	U.of Arizona, U.S.A.	rafelski@physics.arizona.edu
Rak Jan	UNM, U.S.A.	janrak@bnl.gov
Renk Thorsten	Duke U./U.of Jyväskylä, Finland	Thorsten.Renk@phys.jyu.fi
Retiere Fabrice	TRIUMF, Canada	fretiere@triumf.ca
Ripp-Baudot Isabelle	IReS Strasbourg, France	ripp@in2p3.fr
Ristea Catalin	NBI, Denmark	ristea@nbi.dk
Robbins Simon	Bergisches U., Germany	robbins@physik.uni-wuppertal.de
Rogachevsky Oleg	JINR/PNPI, Russia	rogach@sunhe.jinr.ru
Roland Gunther	Cambridge U., U.S.A.	rolandg@mit.edu
Royon Christophe	CEA Saclay, France	royon@hep.saclay.cea.fr
Rybczynski Maciej	Swietokrzyska Academy, Poland	Maciej.Rybczynski@pu.kielce.pl
Šafařík Karel	CERN, Switzerland	Karel.Safarik@cern.ch
Šálek David	Charles U., Czech Republic	salekd@yahoo.com
Šándor Ladislav	Slovak AS, Košice, Slovakia	sandor@saske.sk
Sarkisyan Edward	CERN/U.of Manchester, U.K.	E.Sarkisyan-Grinbaum@cern.ch
Schmitz Norbert	Max-Planck-Institute, Germany	nschmitz@mppmu.mpg.de
Schweda Kai	Berkeley Lab., U.S.A.	koschweda@lbl.gov
Sedlák Kamil	Oxford U., U.K.	k.sedlak1@physics.ox.ac.uk
Shao Ming	U.of Science&Technology, China	swing@ustc.edu.cn
Shephard William	U.of Notre Dame, U.S.A.	Shephard.1@nd.edu
Šimák Vladislav	IP ASCR/CTU, Czech Republic	simak@fzu.cz
Sinyukov Yuriy	Bogolyubov ITP, Ukraine	sinyukov@bitp.kiev.ua
Soluk Richard	U.of Alberta, Canada	soluk@phys.ualberta.ca
Srivastava Dinesh	W. Bengal, India	dinesh@veccal.ernet.in
Stavinsky Alexey	ITEP Moscow, Russia	stavinsk@jlab.org
Ster Andras	KFKI-MFA, Hungary	ster@mfa.kfki.hu
Straub Bruce	Oxford U., U.K.	straub@desy.de
Šumbera Michal	NPI ASCR, Czech Republic	sumbera@ujf.cas.cz
Surrow Bernd	MIT, U.S.A.	surrow@mit.edu
Sutton Mark	U.College London, U.K.	sutt@mail.desy.de
Suzuki Naomichi	Matsumoto U., Japan	suzuki@matsu.ac.jp
Švec Jan	IP ASCR , Czech Republic	svecj@fzu.cz
Tai An	UCLA, U.S.A.	atai@physics.ucla.edu
Taranenko Arkady	Dep.of Chemistry, U.S.A.	arkadij@ram0.i2net.sunysb.edu
Taševský Marek	IP ASCR, Czech Republic	Marek.Tasevsky@cern.ch
Thews Robert	U.of Arizona, U.S.A.	thews@physics.arizona.edu
Todorova Šárka	Tufts U., U.S.A.	sarka.todorova@cern.ch
Tokarev Mikhail	JINR, Russia	tokarev@sunhe.jinr.ru
Tomášik Boris	NBI, Denmark	boris.tomasik@cern.ch

Torrieri Giorgio	McGill U., Canada	torrieri@physics.mcgill.ca
Trainor Tom	U.of Washington, U.S.A.	trainor@hausdorf.npl.washington.edu
Tsue Yasuhiko	Kochi U., Japan	tsue@cc.kochi-u.ac.jp
Utyuzh Oleg	Soltan INS, Poland	utyuzh@fuw.edu.pl
Valkárová Alice	Charles U., Czech Republic	alice@ipnp.troja.mff.cuni.czăă
Vojik Martin	Silesian U., Czech Republic	fx999034@axpsu.fpf.slu.cz
Wang Meng	Bonn U., Germany	wangm@physik.uni-bonn.de
Wang Qin	HEFIN Nijmegen, Netherlands	wangq@hef.ru.nl
Wobisch Markus	Fermilab, U.S.A.	wobisch@fnal.gov
Wong Cheuk-Yin	Oak Ridge NL, U.S.A.	wongc@ornl.gov
Wu Yuanfang	IPP, Huazhong Normal U., China	wuyf@iopp.ccnu.edu.cn
Yan Wenbiao	DESY, Germany	yanwb@mail.desy.de
Zawiejski Leszek	INP PAS, Poland	Leszek.Zawiejski@ifj.edu.pl
Zalewski Kacper	Jagellonian U., Poland	zalewski@th.if.uj.edu.pl
Zborovský Imrich	NPI ASCR, Czech Republic	zborovsky@ujf.cas.cz
Zotov Nikolai	Skobeltsyn INP, MSU, Russia	zotov@theory.sinp.msu.ru

WPCF 2005

Author Index

A

B

C

D

E

L

M

N

O

P

Q

R

S

T

U